Handbook on Characterization of Biomass, Biowaste and Related By-products

Ange Nzihou
Editor

Handbook on Characterization of Biomass, Biowaste and Related By-products

Volume 2

Springer

Editor
Ange Nzihou
RAPSODEE CNRS, UMR-5302
Université de Toulouse, IMT Mines Albi
Albi Cedex, 09, France

ISBN 978-3-030-35019-2 ISBN 978-3-030-35020-8 (eBook)
https://doi.org/10.1007/978-3-030-35020-8

This Springer imprint is published by the registered company Springer Nature Switzerland AG
The registered company address is: Gewerbestrasse 11, 6330 Cham, Switzerland

Dedicated to
Emilie, Arthur and Laurence
Emilienne
Daniel, who would love to hear how the idea of this book blossoms

Preface

In the challenging transition toward sustainable development, one of the key pillars is the valorization of non-food biomass and biowaste into energy and valuable molecules and materials with emphasis on practices that prevent harmful emissions. The range of biomass and biowaste feedstocks available is much larger than ever before. This opens new opportunities for innovation and development of new technologies, processes, approaches and market in the context of sustainable development and the circular economy. In this context, understanding the behavior and properties of these streams is essential. It requires a combination of techniques to get a good insight of the complex materials that biomass and biowaste usually are. The handbook provides insights on standard characterization techniques, together with the advanced and emerging ones.

With this contribution to the state of the art, we planned this work to fill an important gap in the literature. We expect this reference work to be the mortar that holds the building blocks on understanding the properties and characteristics of biomass and biowaste for their better selection and use for dedicated purposes.

It is divided into 15 chapters written by 87 authors from 16 countries.

The creation of this handbook, gathering various fields of engineering sciences (chemical, environmental and mechanical), has been a complex and demanding venture. It would not have been possible without the dedication, countless effort and time from *Dr. Augustina Ephraim and Dr. Nathalie Lyczko* who helped to develop the concept of the book and addressed a number of issues raised by the contributors. *I would like to express my deepest gratitude to them both.*

Special thanks to chapter chairs, *Patricia Arlabosse, Hélène Carrere, Marco J. Castaldi, Jun Dong, Capucine Dupont, Catherine Dupré, Augustina Ephraim, Thierry Ghislain, Deepu Gopakumar, Carol Sze Ki Lin, Shengyong Lu, Hélène Métivier, Doan Pham Minh, Elsa Weiss-Hortala*, for guiding the work of the respective chapter that they handled to their outstanding completion.

I'm grateful to the *87 authors and to the reviewers* who dedicated their valuable time and expertise to create this extraordinary source of information for a wide audience.

The enthusiastic support from *Dolorès Liret and Kattialyn Gossiaux* is gratefully acknowledged.

My warm thanks to *Fritz Schmuhl* for his unswerving support and advices throughout this project. My thanks go to *the Springer Nature editorial and production teams* helping to complete this project.

The high rate of acceptance on my solicitation from the biomass and biowaste community is a strong indicator of the importance of the need that the *handbook on characterization of biomass, biowaste and related by-products* fills.

Albi, France

Ange Nzihou
Distinguished Professor of Chemical
and Environmental Engineering

Contents

Contributors

Philippe Accart RAPSODEE CNRS, UMR-5302, Université de Toulouse, IMT Mines Albi, Albi Cedex, 09, France

Andrés Anca-Couce Institut für Wärmetechnik/Institute of Thermal Engineering, TU Graz, Graz, Austria

Georgia Antonopoulou Institute of Chemical Engineering Sciences (FORTH/ICE-HT), Patras, Greece

Patricia Arlabosse RAPSODEE CNRS, UMR-5302, Université de Toulouse, IMT Mines Albi, Albi Cedex, 09, France

Muthupandian Ashokkumar Faculty of Science, School of Chemistry, The University of Melbourne, Melbourne, VIC, Australia

Hassen Benbelkacem Laboratoire DEEP—EA 7429 (Déchets Eaux Environnement Pollutions—Waste Water Environment Pollutions), Université de Lyon—INSA Lyon, Villeurbanne, France

Céline Boachon RAPSODEE CNRS, UMR-5302, Université de Toulouse, IMT Mines Albi, Albi Cedex, 09, France

Alfons Buekens State Key Laboratory of Clean Energy Utilisation, Zhejiang University, Hangzhou, China;
Department of Chemical Engineering, Vrije Universiteit Brussel, Brussels, Belgium

Hugh D. Burrows CQC, Department of Chemistry, University of Coimbra, Coimbra, Portugal

Rachel Calvet RAPSODEE CNRS, UMR-5302, Université de Toulouse, IMT Mines Albi, Albi Cedex, 09, France

Maria G. Campos CQC, Department of Chemistry, University of Coimbra, Coimbra, Portugal

Hélène Carrère Laboratoire de Biotechnologie de l'Environnement, INRA, Narbonne, France

Marco J. Castaldi Department of Chemical Engineering, The City College of New York, New York, NY, USA

Vincent Chatain Laboratoire DEEP—EA 7429 (Déchets Eaux Environnement Pollutions—Waste Water Environment Pollutions), Université de Lyon—INSA Lyon, Villeurbanne, France

Tong Chen State Key Laboratory of Clean Energy Utilisation, Zhejiang University, Hangzhou, China

Anthony Chesnaud MINES ParisTech, PSL Research University, MAT—Centre des matériaux, CNRS UMR 7633, Evry, France

Yong Chi Department of Energy Engineering, Zhejiang University, Yuquan Campus, Hangzhou, China

Simona Ciuta RRT Design and Construction, New York, USA

Jean-Michel Commandré Biomass Energy Unit, CIRAD, Montpellier Cedex 5, France

Lucy Culleton National Physical Laboratory, Teddington, UK

Sylvie Del Confetto RAPSODEE CNRS, UMR-5302, Université de Toulouse, IMT Mines Albi, Albi Cedex, 09, France

Cédric Delattre Université Clermont Auvergne, CNRS, Institut Pascal, Aubière Cedex, France

Alba Dieguez-Alonso Institute of Energy Engineering, Technische Universität Berlin, Berlin, Germany

Papa Niokhor Diouf SEREX, Rimouski College Technology Transfer Center, Amqui, Canada

Jean-Louis Dirion RAPSODEE CNRS, UMR-5302, Université de Toulouse, IMT Mines Albi, Albi Cedex, 09, France

Jun Dong RAPSODEE CNRS, UMR-5302, Université de Toulouse, IMT Mines Albi, Albi Cedex, 09, France;
Department of Energy Engineering, Zhejiang University, Yuquan Campus, Hangzhou, China

Céline Druilhe UR OPAALE, Irstea, Rennes, France

Nathalie Dumont Laboratoire DEEP—EA 7429 (Déchets Eaux Environnement Pollutions—Waste Water Environment Pollutions), Université de Lyon—INSA Lyon, Villeurbanne, France

Capucine Dupont Department of Environmental Engineering and Water Technology, IHE Delft Institute for Water Education, Delft, The Netherlands

Catherine Dupré GEPEA UMR 6144 CNRS, Saint-Nazaire Cedex, France

Xavier Duret Department of Chemical Engineering and Biotechnological Engineering, Université de Sherbrooke, Sherbrooke, Canada

Telma Encarnação CQC, Department of Chemistry, University of Coimbra, Coimbra, Portugal

Augustina Ephraim RAPSODEE CNRS, UMR-5302, Université de Toulouse, IMT Mines Albi, Albi Cedex, 09, France

Marilyne Fauchon Biodimar, LEMAR UMR 6539, Institut Européen de la Mer, Université de Bretagne Occidentale, Brest, France

Clément Gaignard Université Clermont Auvergne, CNRS, Institut Pascal, Aubière Cedex, France

Thierry Ghislain Department of Chemical Engineering and Biotechnological Engineering, Université de Sherbrooke, Sherbrooke, Canada

Maria Gonzalez Martinez Department of Environmental Engineering and Water Technology, IHE Delft Institute for Water Education, Delft, The Netherlands

Deepu Gopakumar School of Industrial Technology, Universiti Sains Malaysia, Penang, Malaysia

Laurène Haurie RAPSODEE CNRS, UMR-5302, Université de Toulouse, IMT Mines Albi, Albi Cedex, 09, France

Pinjing He College of Environmental Science and Engineering, Tongji University, Shanghai, China

Claire Hellio Biodimar, LEMAR UMR 6539, Institut Européen de la Mer, Université de Bretagne Occidentale, Brest, France

Junko Ito Algae Biomass and Energy System R&D Center, University of Tsukuba, Tsukuba, Japan

Julie Jimenez Laboratoire de Biotechnologie de l'Environnement, INRA, Narbonne, France

Guneet Kaur School of Energy and Environment, City University of Hong Kong, Kowloon, Hong Kong;
Sino-Forest Applied Research Centre for Pearl River Delta Environment, Department of Biology, Hong Kong Baptist University, Kowloon Tong, Hong Kong

Céline Laroche Université Clermont Auvergne, CNRS, Institut Pascal, Aubière Cedex, France

Eric Latrille Laboratoire de Biotechnologie de l'Environnement, INRA, Narbonne, France

Jean-Michel Lavoie Department of Chemical Engineering and Biotechnological Engineering, Université de Sherbrooke, Sherbrooke, Canada

Jack Legrand GEPEA UMR 6144 CNRS, Saint-Nazaire Cedex, France

Chong Li School of Energy and Environment, City University of Hong Kong, Kowloon, Hong Kong;
Agricultural Genomics Institute at Shenzhen, Chinese Academy of Agricultural Sciences, Shenzhen, Guangdong, People's Republic of China

Xiaotong Li School of Energy and Environment, City University of Hong Kong, Kowloon, Hong Kong

Carol Sze Ki Lin School of Energy and Environment, City University of Hong Kong, Kowloon, Hong Kong

Xiaoqing Lin State Key Laboratory of Clean Energy Utilisation, Zhejiang University, Hangzhou, China

Shengyong Lu State Key Laboratory of Clean Energy Utilisation, Zhejiang University, Hangzhou, China

Gerasimos Lyberatos Institute of Chemical Engineering Sciences (FORTH/ICE-HT), Patras, Greece;
School of Chemical Engineering, National Technical University of Athens, Athens, Greece

Nathalie Lyczko RAPSODEE CNRS, UMR-5302, Université de Toulouse, IMT Mines Albi, Albi Cedex, 09, France

Hanna J. Maria International and Inter University Centre for Nanoscience and Nanotechnology, Mahatma Gandhi University, Kottayam, Kerala, India;
School of Chemical Sciences, Mahatma Gandhi University, Kottayam, Kerala, India

Hélène Métivier Laboratoire DEEP—EA 7429 (Déchets Eaux Environnement Pollutions—Waste Water Environment Pollutions), Université de Lyon—INSA Lyon, Villeurbanne, France

Philippe Michaud Université Clermont Auvergne, CNRS, Institut Pascal, Aubière Cedex, France

Rajesh Munirathinam RAPSODEE CNRS, UMR-5302, Université de Toulouse, IMT Mines Albi, Albi Cedex, 09, France

Ioanna Ntaikou Institute of Chemical Engineering Sciences (FORTH/ICE-HT), Patras, Greece

Ange Nzihou RAPSODEE CNRS, UMR-5302, Université de Toulouse, IMT Mines Albi, Albi Cedex, 09, France

Stewart Oakley Department of Civil Engineering, California State University, Chico, CA, USA

Khai Lun Ong School of Energy and Environment, City University of Hong Kong, Kowloon, Hong Kong

Alberto A. C. C. Pais CQC, Department of Chemistry, University of Coimbra, Coimbra, Portugal

Konstantina Papadopoulou School of Chemical Engineering, National Technical University of Athens, Athens, Greece

Séverine Patry RAPSODEE CNRS, UMR-5302, Université de Toulouse, IMT Mines Albi, Albi Cedex, 09, France

Doan Pham Minh RAPSODEE CNRS, UMR-5302, Université de Toulouse, IMT Mines Albi, Albi Cedex, 09, France

Guillaume Pierre Université Clermont Auvergne, CNRS, Institut Pascal, Aubière Cedex, France

Daniel Pleissner Faculty of Sustainability, Sustainable Chemistry (Resource Efficiency), Institute for Sustainable and Environmental Chemistry, Leuphana University Lüneburg, Lüneburg, Germany

Yohan Richardson Biomass Energy and Biofuels Laboratory 2iE-CIRAD (LBEB), International Institute for Water and Environmental Engineering (2iE Foundation), Ouagadougou, Burkina Faso

Christine Rolland RAPSODEE CNRS, UMR-5302, Université de Toulouse, IMT Mines Albi, Albi Cedex, 09, France

Lina María Romero Millán RAPSODEE CNRS, UMR-5302, Université de Toulouse, IMT Mines Albi, Albi Cedex, 09, France

Louise Roques RAPSODEE CNRS, UMR-5302, Université de Toulouse, IMT Mines Albi, Albi Cedex, 09, France

Abdoul Razac Sane RAPSODEE CNRS, UMR-5302, Université de Toulouse, IMT Mines Albi, Albi Cedex, 09, France

Rababe Sani RAPSODEE CNRS, UMR-5302, Université de Toulouse, IMT Mines Albi, Albi Cedex, 09, France

Martial Sauceau RAPSODEE CNRS, UMR-5302, Université de Toulouse, IMT Mines Albi, Albi Cedex, 09, France

Benoît Serive USR 3151, CNRS Sorbonne Université, Station Biologique de Roscoff, Roscoff Cedex, France

Patrick Sharrock RAPSODEE CNRS, UMR-5302, Université de Toulouse, IMT Mines Albi, Albi Cedex, 09, France

Sabu Thomas International and Inter University Centre for Nanoscience and Nanotechnology, Mahatma Gandhi University, Kottayam, Kerala, India;
School of Chemical Sciences, Mahatma Gandhi University, Kottayam, Kerala, India

Eric Trably Laboratoire de Biotechnologie de l'Environnement, INRA, Narbonne, France

Anne Trémier UR OPAALE, Irstea, Rennes, France

Christelle Tribout UPS, INSA, LMDC, Université de Toulouse, Toulouse Cedex 4, France

Fei Wang College of Energy Engineering, Zhejiang University, Yuquan Campus, Hangzhou, China

Makoto M. Watanabe Algae Biomass and Energy System R&D Center, University of Tsukuba, Tsukuba, Japan

Elsa Weiss-Hortala RAPSODEE CNRS, UMR-5302, Université de Toulouse, IMT Mines Albi, Albi Cedex, 09, France

Claire E. White Department of Civil and Environmental Engineering and the Andlinger Center for Energy and the Environment, Princeton University, Princeton, NJ, USA

Mingxiu Zhan College of Metrology and Measurement Engineering, China Jiliang University, Hangzhou, China

Mengmei Zhang State Key Laboratory of Clean Energy Utilisation, Zhejiang University, Hangzhou, China

Yaqian Zhao School of Civil Engineering, UCD Dooge Centre for Water Resource Research, University College Dublin, Dublin 4, Ireland;
State Key Laboratory of Eco-Hydraulic Engineering in Arid Area, Xi'an University of Technology, Xi'an, People's Republic of China

Chapter 6
Municipal Solid Waste

Marco J. Castaldi, Yong Chi, Simona Ciuta, Jun Dong, Pinjing He, Nathalie Lyczko, Ange Nzihou and Fei Wang

Abstract In this chapter, the sampling, preconditioning, as well as physico-chemical, biochemical and thermal characterization methods for municipal solid waste are introduced. The typical equipment used in the sample preparation and characterization are illustrated. Furthermore, an overview of the characteristics of municipal solid waste is presented which provides a useful starting database.

6.1 Sampling and Preconditioning

6.1.1 Sampling

Municipal solid waste (MSW), commonly known as trash or garbage, is defined by the U.S. EPA (Environmental Protection Agency) as a waste type consisting of everyday

M. J. Castaldi (✉)
Department of Chemical Engineering, The City College of New York, 160 Convent Avenue, New York, NY 10031, USA
e-mail: mcastaldi@ccny.cuny.edu

S. Ciuta
RRT Design and Construction, 1 Huntington Quadrangle, Melville, New York, USA

Y. Chi · J. Dong
Department of Energy Engineering, Zhejiang University, Yuquan Campus, 38 Zheda Road, Hangzhou 310027, China

P. He
College of Environmental Science and Engineering, Tongji University, 1239 Siping Road, Shanghai 200092, China

J. Dong · N. Lyczko · A. Nzihou
RAPSODEE CNRS, UMR-5302, Université de Toulouse, IMT Mines Albi, Campus Jarlard, 81013 Albi Cedex, 09, France

F. Wang
College of Energy Engineering, Zhejiang University, Yuquan Campus, 38 Zheda Road, Hangzhou 310027, China

A. Nzihou (ed.), *Handbook on Characterization of Biomass, Biowaste and Related By-products*, https://doi.org/10.1007/978-3-030-35020-8_6

items discarded by the public, such as product packaging, grass clippings, furniture, clothing, bottles and cans, food scraps, newspapers, appliances, consumer electronics, and batteries [1]. These wastes come from households and similar mixed waste generated in institutions such as schools and hospitals, and commercial sources such as restaurants and small businesses. Once generated, MSW must be collected by municipalities or by the order of them [2], followed by proper management including material recycling, biochemical treatment (composting, anaerobic digestion), thermal treatment (pyrolysis, gasification, incineration with the resulting energy used to generate electricity and/or steam in some cases), and landfill disposal. Due to the high heterogeneity of MSW, accurate and reliable data on MSW generation and composition are crucial for decision-making concerning estimating physical, mechanical, chemical, biological and thermal characteristics of wastes, the development of waste technologies, as well as planning sound waste management systems [3, 4]. In this way, waste sampling activities are essential for obtaining appropriate waste composition and characteristics data.

Internationally, sampling and analyzis of MSW is required by different waste related directives, standards and methods. On European level the existing sampling standards include the European Commission "Methodology for the analyzis of solid waste (SWA-Tool)", EN 14899 "Characterization of waste—Sampling of waste materials—Framework for the preparation and application of a Sampling Plan", CEN/TR 15310-1-5 "Characterization of waste—Sampling of waste materials" and EN 14735 "Characterization of waste—Preparation of waste samples for ecotoxicity tests". The U. S. has introduced "MSW Characterization Methodology", in China CJ/T 313-2009 "Sampling and analyzis methods for domestic waste", in France NF X30-413 "Constitution of a sample of household waste contained in a waste collection vehicle" and NF X30-408 "Characterization of a sample of household related waste". The variety of methods used is due to different source separation systems, sorting guidelines, waste collection systems and information needed about the composition of mixed MSW [5].

As a general rule, guidance for carrying out waste sampling procedures depends on the objectives of the client. For instance, waste producers want to know the type of recovery/disposal available, waste treatment plants managers need to understand if they can accept and will be able to treat the waste, and authorities are interested in the environmental impacts associated with a specified waste [6]. The different needs of the concerned actors require a carefully planned sampling procedure to get a representative sample suitable for further analyzis.

The sampling method introduced here is mainly based on the European Commission "Methodology for the analyzis of solid waste (SWA-Tool)". However, the differences related to various standards will be mentioned to enable the comparability of different waste analyzes, and be capable of adaptation to differing local circumstances. It should be considered that the provided list of information sources is not intended to be exhaustive. Besides, the indicated documents and standards may be updated to their new versions in the near future.

The "Methodology for the analyzis of solid waste (SWA-Tool)" aims to provide an approach for the representative sampling of the waste. It also includes an approach for the manual sorting and analyzis of such waste to determine: waste characteristics (composition) and waste quantification (amount produced) based on waste composition.

It should be highlighted before introducing the sampling procedures that the actual constitution of municipal waste may vary between countries. The "Methodology for the analyzis of solid waste (SWA-Tool)" covers the waste fractions of:

- Residual household waste which may be described as mixed solid waste from households, which is collected, transported, and disposed of, either by the household, the municipality or by any other third party in any kind of containers and/ or plastic bags; and
- Residual co-collected commercial waste which may be described as mixed solid waste from commerce, which is co-collected, transported, and co-disposed of, either by the household, the municipality or by any other third party in any kind of containers and/ or plastic bags. The composition of daily residual commercial waste is similar to the composition of residual waste from households. But the amount and composition arises in spatial clusters and depends on the business sector.

Not included in this kind of definition are:

- Separately collected household and commercial material streams such as glass, paper, plastics;
- Separately collected municipal waste streams which may include small scale hazardous waste, electrical/electronic waste, street cleanings, garden/park waste;
- Any other waste stream, which is not produced from routine activity such as bulky waste.

By comparison, the U.S. "MSW Characterization Methodology" [7] does not include municipal sludges, non-hazardous industrial wastes, residue from automobile salvage operations, and construction and demolition debris along with MSW. The Finnish NT ENVIR 001 "solid waste, municipal: sampling and characterization" method [8] concerns consumer waste (means normal waste, including larger objects such as furnishings etc., from households, small shops etc., and offices. This is also the case for waste of similar nature and quantity from other activity) and parts of production waste (means waste from industrial activity and service activities which in type or quantity is significantly different from consumer waste).

According to "Methodology for the analyzis of solid waste (SWA-Tool)", generally, there are four recommended stages for waste characterization as follows: (1) Pre-investigation; (2) Analyzis design and planning; (3) Execution of waste analyzis; and (4) Evaluation of waste analyzis. Regarding sampling steps, the procedures could be summarized as follows.

6.1.1.1 Before Sampling: Pre-investigation

The pre-investigation aims at providing the basis of necessary information for the municipality, county or country intending to undertake a waste analyzis. The following background information preceding the waste analyzis is suggested.

General Description of the Area Under Investigation

A general overview of the area under investigation and the portion of the area to be involved in the waste analyzis is recommended to provide a useful background context to the proposed study and assist in the planning stage. The following minimum information is suggested:

(i) Identification of the area or portion of the area to be assessed, its location and surface area;
(ii) Identification of the various relevant geopolitical districts and the levels at which relevant waste management data may be available.

General Population Information and Waste Management Information

General population information and waste management information is necessary to create a selection basis from which the representative sample can be generated. The following data could be important to collect:

General population information

(i) Number of inhabitants
(ii) Number of households
(iii) Income (such as GNP per capita)
(iv) Types and proportions of residential structures

Waste management information

(i) General description of the organization of the waste management system (actors, responsibilities etc.)
(ii) Type of waste streams produced and collected especially mixed residual household, and co-collected household and commercial waste
(iii) Description of waste container systems in use such as household bins, communal bins and bin storage capacities
(iv) Average numbers of households and/or persons using bins
(v) Total bin volume; spatial distribution of bins; collection intervals
(vi) Method of waste collection such as open truck or refuse collection vehicles compactor and types of waste collected

(vii) Description of collection rounds
(viii) Weighing data of collection vehicles
(ix) Disposal methods such as landfill, energy from waste, reuse/recycling and the types of waste and quantities involved.

6.1.1.2 Sampling Process Design and Planning

Sampling Strategy

Sampling is a process used in statistical analyzis in which a predetermined part of information is taken from an entire population. It is necessary to work with waste samples because it is not possible to analyze the whole population of waste for the area under investigation. A number of sampling methods have been developed, which can be classified into two categories shown below and one should refer to [9] or follow the NT ENVIR 001:

- Probability sampling: Sample has a known probability of being selected
- Non-probability sampling: Sample does not have known probability of being selected as in convenience or voluntary response surveys.

The most common methods are described in the following [8, 10]:

(i) **Random sampling:** belongs to "probability sampling". Each individual of a population has an equal probability of being sampled. Planning a random sampling operation begins with a complete listing and numbering of possible participants or units of analyzis. A table of random numbers or a set of computer-generated random numbers is then applied to choose sample targets until reaching the intended sample size.

(ii) **Systematic random sampling:** belongs to "probability sampling". In systematic random sampling, cases are selected at a specific preset interval after a random starting point. That means, a systematic random sample begins just as with simple random sampling, by making a list of all the possible cases to be included in the sample and assigning each case a number. Next, select the size of interval, by dividing the total number of cases by the preset number of sample.

(iii) **Stratified random sampling:** belongs to "probability sampling". This form of sampling is used if the population is non-randomly heterogeneous. Drawing a stratified random sample is done in two stages: first the population is divided into groups or strata based on a specific characteristic that is central to the study purpose. In such a way each part (stratum) is homogeneous, and random sampling is then performed in each stratum created. Stratified random sampling can be particularly useful when the aim is to accurately reflect the proportional representation of certain groups within a population.

(iv) **Cluster random sampling:** belongs to "probability sampling". This sampling draws random samples in stages from a series of naturally occurring groups (clusters), which is most often used when the sizes or spread of the population being studied is too large that it is not feasible to draw a simple or systematic random sample without first dividing the population into smaller more manageable units.

In practice, the "Methodology for the analyzis of solid waste (SWA-Tool)" recommends the use of stratified random sampling, where possible, as the basis for sample selection for a local or regional waste analyzis program.

Number and Type of Strata

Stratification refers to the process of statistical subdivision of the in-homogenous parent population (e.g. waste arising of an area) into (more) homogenous sub populations (non overlapping groups, e.g. waste from a certain residential structure), called strata. A statistical stratification allows reducing the sample size, since the waste within each stratum varies less than in the parent population. Therefore, stratification allows increasing the accuracy of results.

Stratification is not compulsory for a waste analyzis program according to the European Commission "Methodology for the analyzis of solid waste (SWA-Tool)", but may have advantages for both, accuracy of results and additional waste management information. Potential stratification criteria may include several factors that may influence the composition or the amount of waste to be analyzed and these may in turn vary in effect between municipalities; examples include:

(i) Residential structure
(ii) Heating systems
(iii) Seasonal variations
(iv) Bin size
(v) Availability of civic amenity sites
(vi) Holiday periods
(vii) Type of collection system (separate collection)
(viii) Levels of public education and awareness on waste issues
(ix) Etc.

The number of strata improves the quality of data, however will increase the costs and an excessive number of necessary samples (larger sample size) in order to achieve the required accuracy of results for each stratum [11]. In practice, the "Methodology for the analyzis of solid waste (SWA-Tool)" recommends not more than 5 relevant strata be used; in addition it will be useful to set up a stratification matrix at the initial planning stage. To develop a stronger understanding of the methodology one is referred to Annex II Part B where an example is presented.

Level of Sampling

There are three principal levels at which sampling may take place, namely:

(i) Sampling inside the household/business such as from an internal waste bin
(ii) Sampling outside the household/business such as from an external waste bin/ container such as used in kerbside collection
(iii) Sampling from refuse collection vehicles.

The "Methodology for the analyzis of solid waste (SWA-Tool)" recommends sampling at the external waste bin/container outside the households or business properties. Sampling from the refuse collection vehicles is mentioned as the lowest level of sampling. A main drawback of sampling from refuse collection vehicles may be that the sampled waste cannot be accurately attributed to the geographical areas and/or household types generating the waste [4]. The waste is likely to be heterogeneous, which may make the visual identification of waste composition more difficult and time consuming, also increase the costs of the sorting and analyzis stage. Additionally, as most refuse collection vehicles may be equipped with mechanical stress and blending processes (compaction), individual waste fractions may suffer from cross-contamination, leading to increased inaccuracies during further analyzing procedures [4, 12].

Determination of Sampling Unit, Number and Size

Sampling units are the smallest sub groups of the parent population which are separately selected, collected, sorted and analyzed, and for which separate analyzis results are produced.

There are three main sampling units that could be used to obtain the necessary waste samples for analyzis:

(i) A specific waste bin volume
(ii) A specific weight of household/commercial waste
(iii) A specific number of persons who generate relevant waste.

The "Methodology for the analyzis of solid waste (SWA-Tool)" recommends the type of sampling unit based on the volume of the waste bin (Please note this does not mean the volume of the waste contained within the bin). The choice of the sampling unit will usually depend on the bin type which is most commonly used in the waste analyzis study area. Note that the smaller the volume of the sampling unit the greater the statistical accuracy of results; however, the greater the time required for sorting and analyzis for the equivalent sample size. "Methodology for the analyzis of solid waste (SWA-Tool)" recommends the sampling units be of similar size. The recommended sampling unit used for a waste analyzis should be the lowest common denominator of bin size from the following: 120; 240; 360; 660; 1100; 2400; and 3600 L. Bins sampled for analyzis, which are less than 120 L volume, should be aggregated to one of these sample sizes.

The total number of sampling units required depends on 2 main criteria:

(i) The variation (heterogeneity) of the waste, expressed by the natural variation coefficient. This variation coefficient is usually unknown and has to be estimated on the basis of results from past waste analyzes.
(ii) The desired accuracy of the results.

The required total number of sampling units for a waste analyzis campaign can be easily estimated by using Table 6.1.

According to the recommendation of "Methodology for the analyzis of solid waste (SWA-Tool)":

- If the variation coefficient of the analyzed waste is known (e.g. from former waste analyzes in the same area) it should be used to calculate the required number of sampling units by using Table 6.1.
- If the variation coefficient of the analyzed waste is unknown the following sample sizes for the different types of waste are recommended (regardless of the size of the chosen sampling unit):

Table 6.1 Calculation of necessary number of sampling units (95% confidence level)

Coefficient of variation gauge for variation in parent population (%)	Necessary number of sampling units n (95% confidence level) with maximumnecessary to adjust the results to account for the allowance for random sampling error					
	2.5%	5%	10%	15%	20%	30%
15	138	35	9	4	2	1
20	246	61	15	7	4	2
25	384	96	24	11	6	3
30	553	138	35	15	9	4
35	753	188	47	21	12	5
40	983	246	61	27	15	7
45	1245	311	78	35	19	9
50	1537	384	96	43	24	11
55	1859	465	116	52	29	13
60	2213	553	138	61	35	15
70	3012	753	188	84	47	21
80	3934	983	246	109	61	27
90	4979	1245	311	138	78	35
100	6147	1537	384	171	96	43
120	8851	2213	553	246	138	61
140	12,047	3012	753	335	188	84
160	15,735	3934	983	437	246	109
200	24,586	6147	1537	683	384	171

household waste: 45 m^3
mixture household/commercial waste: 80 m^3
commercial waste: 100 m^3

In case of stratified sampling, strata are usually analyzed as a part of a total waste analyzis and as such, with smaller sample sizes. If the sample size for strata goes beyond a certain level, results will probably be biased and are not representative. The "Methodology for the analyzis of solid waste (SWA-Tool)" thus suggests a sample size not less than 6 m^3 (bin volume) for household waste divided into at least 6 number of sample units as the smallest possible sample for one stratum; for commercial waste, the number of sampling units of one stratum shall be 15 and the sample size not less than 15 m^3 (bin volume). As comparison, the Finnish NT ENVIR 001 "solid waste, municipal: sampling and characterization" method [8] recommends a sample size of 100–200 households for household waste in a defined community. If the investigation also involves variations between types of houses, types of municipality, sizes of households, social and economic groups etc., the size of the sample from each area should be of 40–100 households. The Irish "Municipal waste characterization" method [13] recommends a sampling size in a range of roughly 50–250 households, resulting in a sample of approximately 1000–5000 kg. In larger areas where the sample size will be greater than 250 households, it is recommended that a survey be split into several sub-surveys.

Sampling Period

According to the recommendation of "Methodology for the analyzis of solid waste (SWA-Tool)":

- Where the normal municipal waste collection for the relevant parent population is repeated on a daily or weekly basis it is recommended that the duration for waste sampling and sample collection covers a minimum of one week waste. This will allow the sampling of waste to be spread over each working day (Monday to Friday) covering the full collection cycle and any potential variation due to non-collection of waste at a weekend.
- Where the normal municipal waste collection for the relevant parent population is repeated on a bi-weekly (fortnightly) basis it is recommended that the duration for waste sampling and sample collection cover a minimum of two weeks waste i.e. the full collection cycle.

Seasonal variations in waste generation should also be considered in sampling. Where a seasonal analyzis involving less than four seasonal campaigns has been undertaken it may also be necessary to adjust the results to account for the missing seasonal investigation(s).

6.1.1.3 After Sampling: Execution and Evaluation of Waste Analyzis

The predetermined analyzis planning may be used for samples collection. Each sample unit collected then has to be sorted separately for composition analyzis.

As quality assurance, "Methodology for the analyzis of solid waste (SWA-Tool)" provides a comprehensive list of statistical values to be calculated for each waste category, each campaign and for the total result:

(i) Mean
(ii) Median
(iii) Standard deviation
(iv) Variation coefficient
(v) Confidence coefficient
(vi) Relative confidence interval (%)
(vii) Confidence interval (kg)
(viii) Composition (%).

If stratification is applied, the results of the single strata have to be aggregated to obtain the total sample result. The total sample result has to be calculated as the weighted mean of the results from a single stratum. A key objective is to enable waste composition studies to achieve results at or above a minimum statistical accuracy in a cost effective way [3].

Extrapolation comprises the conclusion from the obtained sample results to the parent population. It is recommended to report and present the final results including the following data: raw data, statistical calculations, evaluation of single results of strata, extrapolation of the overall results and of the waste quantification, and graphical presentation of results.

6.1.2 Size Reduction

Size reduction, belongs to a method of MSW preconditioning, may benefit several advantages from several advantages as described in [14]:

(i) **Recovery materials for recycling**: reducing the size of bulk waste stream breaks the raw MSW into its basic components. This helps a more effective separation and materials recycling step including paper, cardboard, plastic, glass, ferrous metal, aluminum and other residual metals.
(ii) **Facilitating waste processing system**: shredding MSW reduces the particle size of raw waste to a more workable size, which could be better handled by any subsequent processing system. For example, separation of combustible organic materials for biochemical or thermal treatment.
(iii) **Improving efficiency of material recovery facilities**: shredding produces different size distributions for the different material components of MSW in material recovery facilities (MRFs), allowing for automated material separation such as air classifier, screens and optical sorters.

Table 6.2 Equipment used for size reduction

Type	Mode of action	Application
Small grinders	Grinding, mashing	Organic residential solid wastes
Chippers	Cutting, slicing	Paper, cardboard, tree trimmings, yard waste, wood, plastics
Large grinders	Grinding, mashing	Brittle and friable materials, used mostly in industrial operation
Jaw crushers	Grinding, breaking	Large solids
Rasp mills	Shredding, tearing	Moistened solid wastes
Shredders	Shredding, tearing	All types of municipal wastes
Cutters, Clippers	Shredding, tearing	All types of municipal wastes
Hammer mills	Shredding, tearing, cutting, crushing	All types of municipal wastes, most commonly used equipment for reducing size and homogenizing composition of wastes
Hydropulper	Shredding, tearing	Ideally suited for use with pulpable wastes, including paper, wood chips. Used primarily in the papermaking industry. Also used to destroy paper records

The machines used for MSW size reduction are variously called shredders, crushers, pulverizers, mills, and hoggers; where shredder is the most frequently used term [15]. Generally, various equipment capable of MSW size reduction are available on the market, as listed in Table 6.2, together with their modes and applications [16].

The shredders could be classified into two categories: high-speed, low-torque hammer mills; and low-speed, high-torque shear shredders [14]. The differences in their principles may lead to distinctions in the acceptable MSW feed as well as the size distribution of products and overall process capacity.

6.1.2.1 High-Speed, Low-Torque Grinders (The Hammer Mills)

Hammer mills are the most often used in large commercial operations. The principle of hammer mills relies almost entirely on impact and abrasive forces to smash the incoming feedstock into small particles. The MSW feed, as it enters the mill, is hit by the impact forces of the hammer which rotates at a very high speed. Being struck and cut to a required size, the waste then falls out of the bottom of the mill. The product size could be controlled by in-stalling screens with large or small openings.

Hammer mills are generally more effective in processing brittle materials due to their reliance on impact force. Because of the variety of materials contained in MSW, this will cause inconsistent size reduction among different components of the waste, for example, a greater percentage of size reduction received by ‘brittle’ glass and less size reduction of 'ductile' paper and petroleum based plastics. Problems may occur for rags and stringy materials, since they may wrap around the shaft and cause overloading and disruption of the operation.

6.1.2.2 Low-speed, High-torque Grinders (Shear Shredders)

Low-speed, high-torque grinders, such as rotary shear shredders, rely on shear cutting and tearing forces with little to no impact force involved. Shear shredders used for MSW preconditioning usually consist of two counter rotating shafts. Each shaft contains cutters that intermesh and create large shear forces to tear and shear the materials. The advantages of the use of shear shredders are a more uniform product particle size, and lower power consumption.

The tolerances between cutting knives and the number of shafts are main factors determining the feature of a shear shredder, i.e., more shafts and smaller tolerances would lead to smaller product particle sizes. Compared with the hammer mills, advantages by the use of shear shredders attain at a more uniform particle size of the outcome products, as well as a lower power consumption. On the other hand, shear shredders have a limited throughput compared to hammer mills, due to their low shaft speed. Another potential problem with the shear shredders is the difficulty in processing tough metals [14].

6.1.2.3 MSW Size Reduction Equipment Operation and Selection

Several crucial parameters may affect the operation of different shredders [14]:

(i) **Capacity**: Generally, high-speed hammer mills may have capacities ranging from less than 1 ton/h up to 300 tons/h. This figure is higher than that of low-speed shear shredders, which are commonly designed at 20–70 tons/h. The capacity of the subsequent MSW treatment faculties could be one of the important aspects determining the capacity of a shredder. For example, MSW incineration plants could reach capacities up to 3000 tons/day. This total plant capacity may be met by several separate lines, for each shredding unit, the maximum capacity should be designed at around 130% the nominal demand to allow for downtime and the processing of non-ideal materials of high durability.

(ii) **Size**: The size of the shredders should be designed compact to be energy efficient, and smoothly integrated into the existing waste management systems.

(iii) **Energy consumption**: The energy consumption of high-speed hammer mills is in a range of around 6–22 kWh/ton, while the low-speed shear shredders tend to have a lower value at 3–11 kWh/ton. However, the data are dependent on rotor speed and feed composition.

(iv) **Safety concerns**: High-speed hammer mills may have the potential of violent projectiles of shrapnel from the shredding chamber due to the high rotor speed. Low-speed shear shredders are generally safer owing to the absence of high speed components; however, both types of shredders can lead to explosions. Special concerns should be paid for safety issues. For example, the addition of a solid metal plate or a chain curtain outside of the hammer mills shredder chamber.

(v) **Operation and maintenance**: Both high-speed hammer mills and low-speed shear shredders undergo severe wear and tear when processing MSW; thus replacement and maintenance of components is essential. Generally, hammer mills may lead to more maintenance than shear shredders as a result of operating at high speeds.

Based on the aforementioned operation parameters, the factors that decide the selection of MSW shredders could include the following [16]:

(i) The properties of incoming MSW components before and after shredding
(ii) Size requirements for shredded MSW by component
(iii) Method of feeding shredders, provision of adequate shredder hood capacity (to avoid bridging) and clearance requirement between feed and transfer conveyors and shredders
(iv) Type of operation (continuous or intermittent)
(v) Operational characteristics including energy requirements, routine and specialized maintenance requirement, simplicity of operation, reliability, noise output, and air and water pollution control requirements
(vi) Site consideration, including space and height, access, noise and environmental limitations
(vii) Metal storage after size reduction for the next operation.

6.1.3 Separation

MSW represents a rich source of useful materials such as metals, paper, plastic, and glass. Therefore, waste separation is a key step for the recycling of materials. The waste separation activity could be approached in two ways:

- Source segregation (at individual households) at the waste collection stage;
- Mechanical processing and sorting using automated techniques at central facilities.

6.1.3.1 MSW Source Separation

MSW source separation is often performed as a preliminary and effective way for the sorting of recyclables. The municipality generally provides different easily identifiable containers for householders to deposit segregated MSW fractions. Some European countries have already long-term experience with separate collection since late 1970s. It is also the case in the U.S., where recycling had gained increased attention in the late 1980s and 1990s, and by 1991, 39 states and the District of Columbia had passed some recycling legislations [17]. In developing countries such as China and India, source separation programs have been established in the last decade.

The source separation systems could be classified into several categories based on the location of collection [18]:

(i) **Household/Kerbside collection**: consists of bins allocated to individual families. Usually, each family receives a single bin where all types of recyclable waste is placed, or, multiple bins for different kinds of recyclables. This separation and collection approach is very near to the source of waste generation, and could guarantee a high level of citizen participation.
(ii) **Neighborhood collection**: Usually, each family receives a single bin where all types of recyclable wastes are placed and is responsible for delivering their waste to prescribed destination. This system is typically applied for apartment complexes. This separation and collection approach is also very close to the source of waste generation.
(iii) **Zone collection**: consists of large bins for different waste types located in central areas that serve one or multiple neighborhoods. This system often applies to cases where there is kerbside collection for general waste, but no kerbside collection of recyclables.
(iv) **Green point**: refers to a segregated waste collection and storage point, which is specifically designed to collect not only separated waste, but also to selectively collect materials not covered by the other systems, such as hazardous waste, different classes of household electrical appliances, computer equipment and clothes, etc.

Practically, most European countries apply kerbside collection to separate recyclables at household level. The U.S. also applies recyclable drop-off sites (recyclable collection centers) on a regular basis. These are different from the zone collection approach as these centers are open to the public at set times, and only certain specific, predefined types of waste can be disposed of there.

The source separation systems could also be classified according to different waste streams:

(i) **Single-stream commingled**: all recyclables, including paper (old newspapers, advertisement and magazines, office paper), cardboard, plastic bottles and containers, aluminum and steel cans, glass and liquid carton containers, are collected in a single separate box.

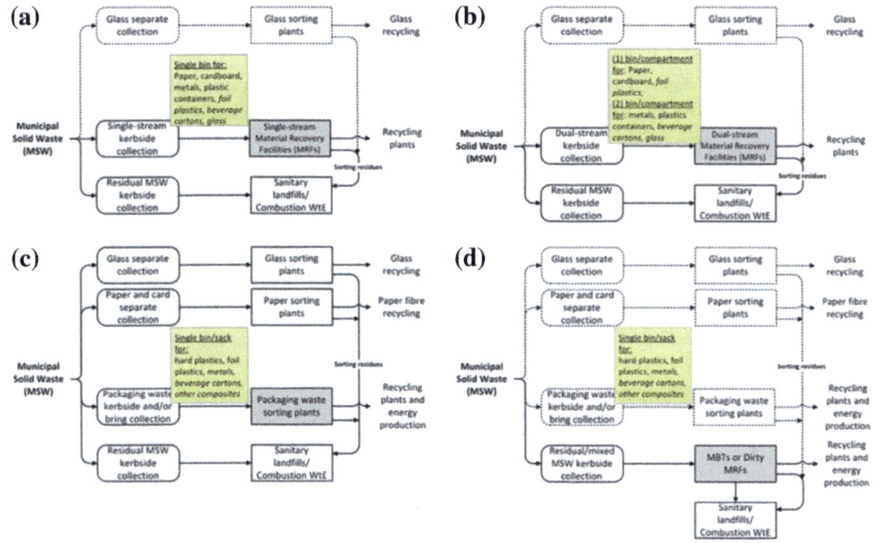

Fig. 6.1 Different MSW source separation systems: **a** single-stream; **b** dual-stream; **c** mixed packaging waste; and **d** recovery from residual/mixed MSW. Dotted lines and processes denote other collection and sorting systems that may or may not be present. Waste fractions in italic may or may not be included in the commingled stream. The collection of biowaste (i.e. yard waste and food waste), bulky waste and hazardous households waste are not included. Reprinted with permission from Elsevier [17]

(ii) **Dual-stream commingled**: all recyclables are separated into two boxes, i.e., (1) paper and cardboard, so called "fibres" and (2) plastic, metal and glass, so called "containers".

(iii) **Mixed packaging collection**: entails the establishment of a parallel system for the collection of lightweight packaging waste (hard plastics, foil plastics, metals, beverage cartons, etc.), alongside older existing municipal systems (i.e. paper and glass).

(iv) **Wet-Dry collection**: replaces the existing residual waste into a two-bins i.e., (1) a Dry bin, containing the mixed packaging materials plus the rest of dry potentially recyclable materials found in the residual bin; and (2) a Wet bin, containing the biowaste, plus all other wet fractions from the residual bin.

Figure 6.1 provides an overview of the different collection systems. A brief summary of each system is given in Table 6.3.

6.1.3.2 MSW Automated Sorting Techniques

Either source separated waste or non-separated waste can be then further separated by automated techniques. Various comminution and sorting techniques are

Table 6.3 Household MSW collection systems and separate collection and materials in italic denote that these are not always present or included

Collection systems	Materials in highlighted (bold) collection systems	Form of collection	Predominant in
Glass collection *Biowaste collection* **Single-stream commingled** Residual MSW	One bin: paper, cardboard, metals, rigid plastics, *foil plastics, beverage cartons, glass*	Kerbside	US, UK, Ireland, Canada, Australia, France
Glass collection *Biowaste collection* **Dual-stream commingled** Residual MSW	Bin/compartment 1: paper, cardboard, *foil plastics* Bin/compartment 2: metals, rigid plastics, *foil plastics, beverage cartons, glass*	Kerbside	US, UK, Canada, Australia, France
Mixed paper and *cardboard* Glass collection *Biowaste collection* **Lightweight packaging waste** Residual MSW	One bin/sack: rigid plastics, foil plastics, *metals, beverage cartons, other composite packaging*	Mostly Kerbside (except glass) Mostly Bring/public systems Mixture of Kerbside and Bring/public systems	Austria, Germany Spain, Portugal Belgium, Italy, Netherlands, Sweden, and other European countries
Paper collection Glass collection *Biowaste collection and other* **mono-material streams** Residual MSW	Separate bins/bags or multi-compartment bins for: metals, cardboard, glass (sometime by colour), plastics, beverage cartons etc.	Mostly Kerbside	Sweden, Finland, Japan, the UK (as kerbside-sort systems)
Paper collection Glass collection **Wet-Dry system**	(broad definition) Bin 1: dry MSW items Bin 2: wet and soiled MSW items	Kerbside	Germany (trial), some communities in the US (but without the separate collection of paper and glass)

Reprinted with permission from Elsevier [17]

available, such as magnetic separation, gravity separation, etc. The aim is to segregate recyclable materials from the waste. Some of these techniques will be introduced in brief here.

Generally, automated waste sorting techniques can be divided into two categories: direct sorting and indirect sorting. Direct sorting techniques use properties of the waste components to respond to the application of magnetism, eddy currents or gravity to separate them into fractions, while the mechanism of indirect sorting is based on sensors to detect the presence and often the location of specified waste fractions. Table 6.4 has summarized some of these techniques, along with their applicability to different waste fractions [19].

- **Screening**: screening is commonly used as MSW pre-treatment to separate small and heavy organic waste from the dry fractions. For example, Fig. 6.2 illustrates a disc screen where the organic fraction falls between the discs and the large, light and dry fractions are conveyed to the edge of the disc.
- **Magnetic separation**: magnetic technique is commonly used to separate ferrous scrap from the waste stream by taking advantage of the magnetic susceptibility of ferrous metals. Figure 6.3 shows a typical suspended magnet. The mixed waste is transported through a conveyor belt; while the ferrous metal pieces are attracted and removed to a separate container.
- **Eddy current**: this technique is used to separate non-ferrous metals (assuming ferrous metals have been separated beforehand) from non-metallic waste streams. An example is given in Fig. 6.4, where the waste mixture is transported by a conveyor and the non-magnetic electrically conductive metals are separated by the external magnetic flux.
- **Plastic separation**: Several techniques exist for further separating of various plastic fractions from the non-metallic waste. For example, triboelectrostatic separation uses the mechanism of 'contact electrification' or 'frictional electrification'. Hydrocyclone utilizes centrifugal force for density separation. Froth flotation uses the properties of hydrophobicity of plastic for its separation.
- **Indirect sorting**: various indirect sorting techniques are also available, which use sensors for detecting specified waste fractions followed by segregation using actuators. The mostly used techniques are: (i) laser induced breakdown spectroscopy (LIBS), which applies laser to ablate waste materials and separate fractions such as metals, plastics and treated wood waste; (ii) X-ray transmission (XRT), which uses the properties of X-ray to detect the atomic density of different waste fractions; (iii) optical sorting, which uses camera based sensors for the identification of waste visual/tactile cues like colour, shapes, texture and size; and (iv) spectral imaging, which combines both spectral reflectance measurement and image processing technologies.

The various automated sorting techniques have been reviewed by the latest researchers, as illustrated in Fig. 6.5.

6.1.4 Drying

In the field of MSW management, drying preconditioning is used primarily for thermal treatment (pyrolysis, gasification, incineration). The aim is to decrease the moisture content of the waste, in order to increase its calorific value. Drying and

Table 6.4 A summary of various direct and indirect automated sorting techniques and applicability

	Sorting technique												
	Direct sorting												
	Screw press	Disc screen	Shredder + magnet	Magnetic drum	Magnetic head pulley	Magnetic overhead belt	Eddy current	Magnetic density separation	Triboelectrostatic	Hydrocyclone	Jigging	Froth	Air separator
Organic	√	√	√										
Ferrous metal			√	√	√	√					√		
Non-ferrous metal							√	√	√		√		
Plastic								√	√	√	√	√	√
Paper													√
Glass													
Wood													

	Sorting technique				
	Indirect sorting				
	Eddy current	LIBS	X-ray sort	Optical sort	Spectral sort
Organic					
Ferrous metal					
Non-ferrous metal	√	√	√	√	√
Plastic		√	√		√
Paper				√	
Glass				√	√
Wood		√	√		

Reprinted with permission from Elsevier [19]

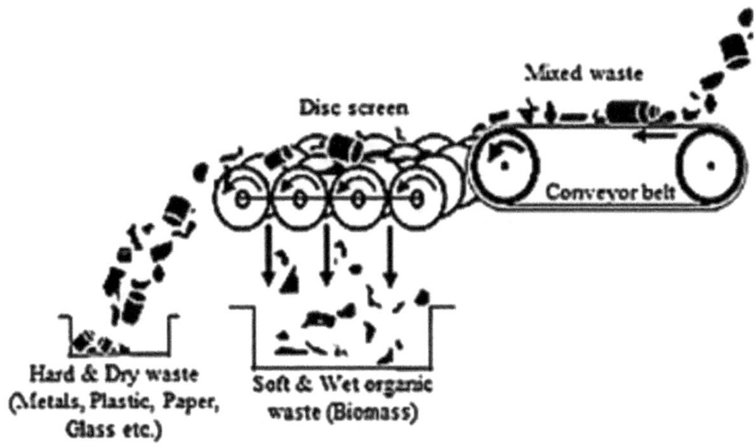

Fig. 6.2 Typical screening equipment: disc screen, an illustration. Reprinted with permission from Elsevier [19]

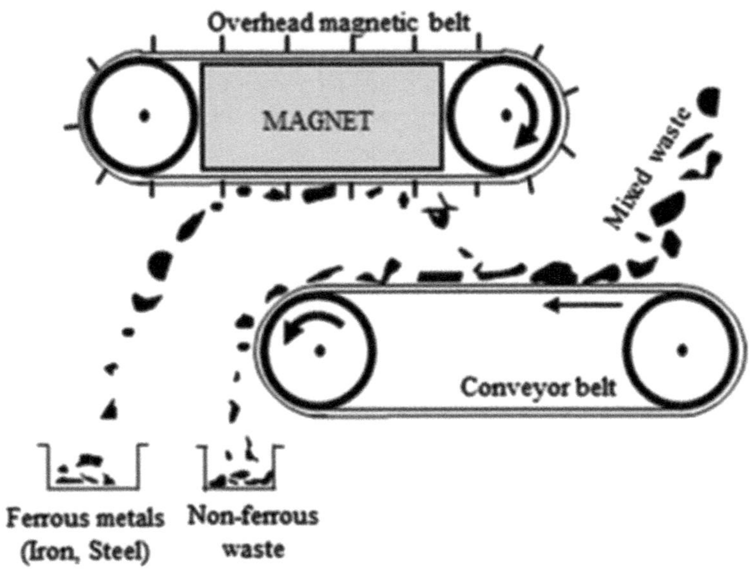

Fig. 6.3 Typical magnetic equipment: suspended magnet, an illustration. Reprinted with permission from Elsevier [19]

dewatering operations are widely used in sludge treatment (see details in Chap. 10). Drying applied in MSW is mainly due to the fact that MSW in many countries is characterized by high water content derived from a relatively high proportion of food waste. Table 6.5 lists the moisture content of typical waste composition. A high moisture content can make it difficult to sustain auto-thermal combustion.

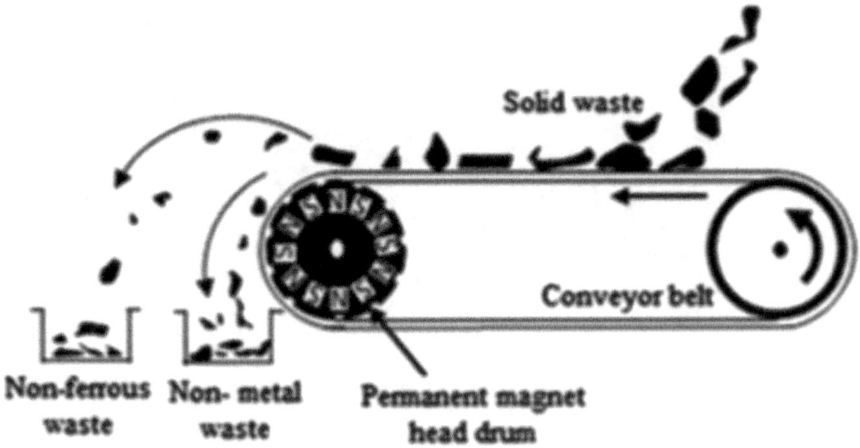

Fig. 6.4 Typical eddy current equipment: an illustration. Reprinted with permission from Elsevier [19]

It can also prevent steady-state operating conditions being achieved in pyrolysis and gasification systems, even Solid Recovered Fuel (SRF) or Refuse Derived Fuel (RDF). Other advantages of drying include increasing the sorting efficiency of the waste, adding value to the dried fractions of waste, and favoring the application of MSW recycling and reuse [20].

In general, solid waste drying technologies could be traditionally divided by two main groups: dewatering and thermal drying [22]. Dewatering refers to waste dehydration by means of individual physical steps, which is widely used in wastes that contains a high quantity of liquid in their composition, for example, sewage sludge. Thermal drying refers to traditional drying using a hot gas as a heat source, normally air. Currently, new drying methods have appeared and expanded the technologies applicable to waste treatment: fry-drying and bio-drying. The former refers to the evaporation of moisture from solid waste through the contact of boiling oil at a temperature above the water boiling temperature; while the latter refers to moisture removal by means of aerobic biological processes through the degradation of organic matter. The main solid waste drying technologies are illustrated in Fig. 6.6, while the following will provide a brief introduction to the thermal drying and bio-drying of MSW.

6.1.4.1 MSW Thermal Drying

MSW thermal drying relies on an external supply of heat (generally obtained from auxiliary fuel). This could be achieved based on heat convection, conduction, radiation or combination of heat transfer modes. In most drying case, convection is the usual choice, with hot air directed onto the wet waste stream in order to apply

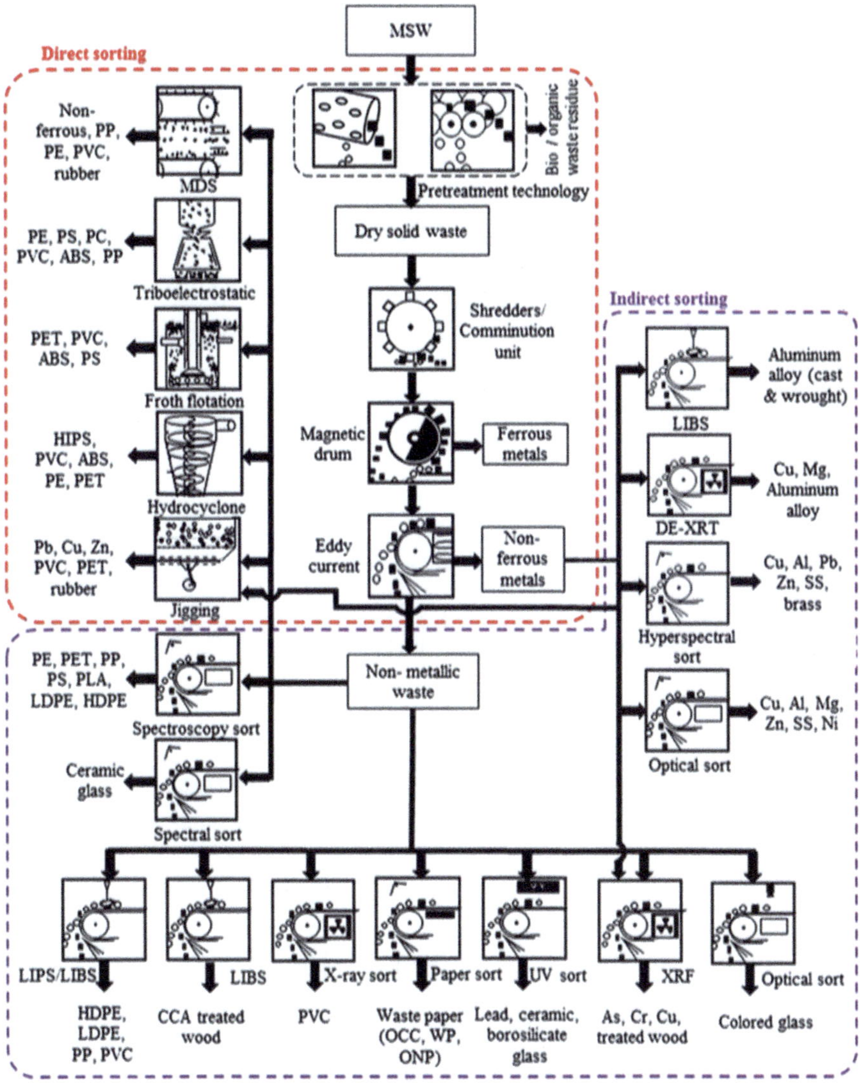

Fig. 6.5 A summary of various direct and indirect automated sorting techniques. Reprinted with permission from Elsevier [19]

heat, and to remove the evaporated moisture. The process can be operated under vacuum, where heat can be applied by microwave and/or direct radiation. Freeze drying and lyophilisation are possible.

Thermal dryers can be classified by the gas/solid contact configuration, such as rotary, tray, fluidised bed, tunnel etc. The heat can be introduced by direct contact or via an externally-heated surface, and the drying medium can be air, flue gas or superheated steam, see e.g. [23, 24].

Table 6.5 Moisture content of typical waste composition

Waste composition	Moisture content (%)
Asphalt	6–12
Cardboard, corrugated, paper box	4–10
Brick/Concrete/Tile/dirt	6–12
Electronic equipment	~0
Food waste	50–80
Garden trimmings	30–80
Glass	1–4
Leather	8–12
Metal-Ferrous	2–6
Metal-Non Ferrous	2–4
Municipal solid waste/biomedical waste	15–40
Paper	4–10
Plastic	1–4
Rubber	1–4
Sawdust	10–15
Textile	6–15
Wood	15–40

Reprinted with permission from Springer Nature [21]

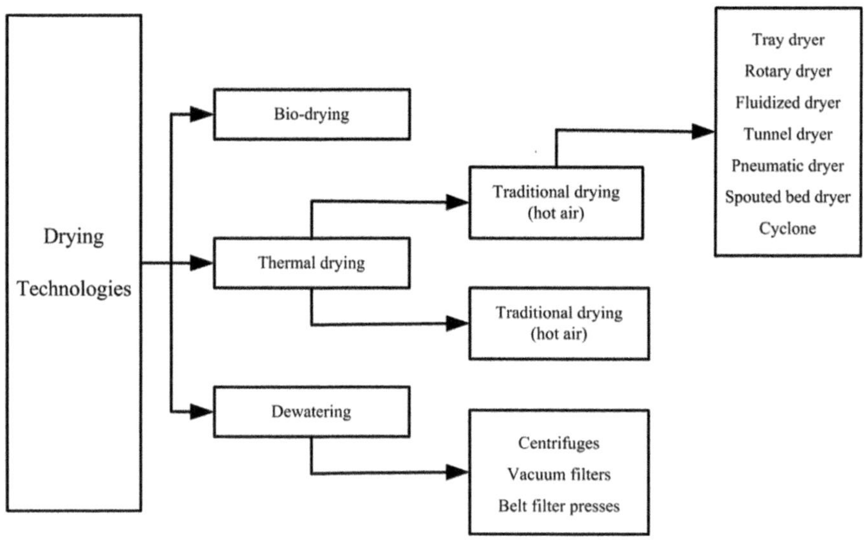

Fig. 6.6 Common solid wastes drying technologies. Reprinted with permission from Taylor and Francis [22]

Direct dryers are by far the most common, accounting for about 85% of industrial dryers, despite their relatively low thermal efficiency caused by the difficulty in recovery of latent heat of vaporization at the dryer exhaust. A typical direct rotary dryer is composed of a rotating cylindrical tube in which a hot stream of air flows. The slope inclines slightly to conveniently convey the material through the dryer using gravity. The wet material in the rotary shell contacts the hot air directly; the evaporated moisture is carried away by the same flowing air. Usually the drying gas temperatures may range from 50 to 400 °C depending on the feed materials, retention time in the rotary drum is around 30–45 min. The required energy input will depend on the moisture content of the feedstock [16].

An indirect rotary dryer is also made of rotating cylinder with concentric inner shell. The incoming feedstock is rotated in the inner coil without direct contact with the outer shell which the hot air flows, i.e., heat is transferred by conduction. Since no gas flow is presented on the wet feedstock, it is necessary to apply either vacuum or use gentle gas flow to remove the evaporated moisture inside the dryer chamber. Normally, the temperature of heat transfer surfaces may range from −40 °C (as in freeze drying) to about 300 °C. Advantages of indirect dryer include rapid evaporation of waste at a relatively fast heat exchange rate; and drying of heat-sensitive materials without thermally degrading them.

6.1.4.2 MSW Bio-Drying

Bio-drying uses bio-conversion technology to degrade organic matter contained in the waste and reduces the moisture content. Its main advantage is the use of the thermal energy generated by the aerobic biodegradation of waste organic matter to evaporate water, thus achieving self-drying [22]. The MSW bio-drying still remains a relatively new technology. Bio-drying is typically employed in mechanical-biological treatment plants (MBT) for the processing of residual MSW, and the final product can be used as a biofuel. Currently, around 20 commercial plants are operational in Europe, with overall capacity of around 2,000,000 Mg/year [25].

Figure 6.7 illustrates a typical bio-drying system [25]. The core of a bio-drying reactor is a container coupled with an aeration system. Aerobic biodegradation of readily decomposable organic matter occurs inside the container, convective moisture is removed through controlled, excessive aeration. The bio-drying container could be either enclosed, open tunnel-hall, or rotating drum. The process duration is up to two weeks depending on specified technique, waste humidity and putrescibility, while the air temperature is normally maintained at 40 °C at the outlet of the biological reactor [26].

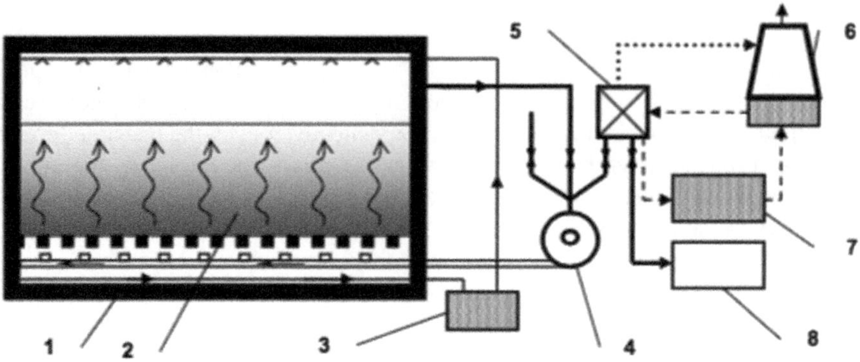

Fig. 6.7 MSW bio-drying system: an illustration based on a Herhof system: 1. enclosed box; 2. air forced through the waste matrix, heated by the exothermic aerobic biodegradation of readily decomposable waste fragments; 3. leachate collection and circulation system; 4. forced aeration system with partial air recirculation, mixing ambient air and conditioned process air; 5. heat exchanger; 6. cooling tower; 7. water (vapour condensate); 8. exhaust air treatment through biofilter or regenerative thermal oxidation (RTO). Reprinted with permission from Elsevier [25]

6.1.4.3 Selection of Drying Technologies and Equipment

Both thermal drying and bio-drying have their own advantages and disadvantages, which are listed in Table 6.6. Those factors should be considered when selecting proper drying technologies [26].

The following aspects should be taken into consideration when choosing a suitable drying system:

(i) Physical, chemical and biochemical properties of the wet feedstock to be dried
(ii) Specifications of the desired product, including moisture content, expected variability in feedstock characteristics
(iii) Drying characteristics of the feedstock, including maximum material temperature, drying kinetics, moist solid sorption isotherms, toxicological properties of the product
(iv) Dryer throughput, mode of feedstock production (batch/continuous), upstream and downstream processing operations
(v) Safety aspects, e.g., fire hazard and explosion hazards, toxicity
(vi) Operational characteristics, e.g., type and cost of fuel, maintenance requirements, need for automatic control
(vii) Site consideration, e.g., space in plant, noise and environmental regulations.

Table 6.6 Comparison between thermal drying and bio-drying technologies

	Thermal drying	Bio-drying
Advantages	Easy regulation of the final humidity of the waste exiting the plants Possibility of reaching humidity values lower than 10% Short duration of the process and rapidity of starting Possibility of process air recycling Independence from the selective collection strategies	No auxiliary fuel is needed The volume of the reactor is potentially available for temporary waste storing A reduction of mass to be transported in case of decentralized treatment plants can decrease the number of tracks on the territory The treated waste is suitable for inert, metals and glass separation at high efficiency The increase of waste calorific value enables Refuse Derived Fuel production and waste modification, when not suitable for a direct combustion
Disadvantages	Auxiliary fuel use makes thermal drying not viable for decentralized treatment The necessity of a high flow of process air maintained at <105 °C to avoid risks of softening of the plastics and phenomena of ignition of the material; this causes a high energy cost for the operation of the blowers The thermal energy consumption for unit of mass of evaporated water is higher than the one for sewage sludge thermal drying, as the granulometry of MSW is less favourable and the technology is less optimized No volume of storage for pre-treated waste is available because of the low retention time in the dryer	A fly's reproductive cycle must be taken into account; a specified removal stage must be implemented, but the obtainable efficiency is not 100% The long duration of the process causes the need of high reactor volumes There is limited possibility of regulation of the humidity and of the waste caloric value A final humidity lower than 10% is not obtainable Process air recycling is not viable because of the role of oxygen The process air flowrate to be treated is significant The consumed volatile solids are no longer available for energy generation Selective collection of organic fractions can significantly decrease its content in the residual MSW

6.2 Physical and Mechanical Characterization of Municipal Solid Waste

As a society grows in number and wealth, the amount of waste it generates also grows, thus threatening its environment, its public health and its quality of life. Therefore, the disposal of MSW must be taken seriously through a scientific management system. A successful MSW management system not only depends on the adequate funds but also on proper strategies and design of the system using accurate data of MSW characterizations. The physical and mechanical characteristics of municipal solid waste are the key factors for the planning, designing, and implementation of municipal solid waste management systems [27]. This chapter describes the physical and mechanical characterizations of municipal solid waste and test methods of determining them. The details are as follows:

- Physical composition
- Particle size
- Bulk density
- Moisture content
- Compressibility
- Permeability.

6.2.1 *Physical Composition*

MSW is a heterogeneous mixture and a large percentage of MSW is the result of the products after use. Its composition varies greatly from municipality to municipality, and it changes significantly with time. The physical composition is one of the vital formulations of MSW management.

6.2.1.1 Test Methods for Physical Composition

Regarding the measurement of the physical composition of MSW, there are three commonly-used methods: material flow, manual sampling and photogrammetry.

Test method by material flow method

The material flow method is based on the data from published industry production statistics to estimate waste composition. The EPA (Environmental Protection Agency) uses material flow method to estimate the national waste composition.

However, collecting the data is expensive and time-consuming and the institutions that report the data must be authoritative. Besides, the material flow method is only suitable for a national level. At local and regional level, the material flow is complex with incomplete statistics, and the applicable measure methods for physical composition are always manual sorting and photogrammetry.

Test method by manual sampling

Manual sampling is the most common method for the measure of waste composition. The sample is manually separated into the specified categories; the sorted components are then individually weighted and the weight percentage can be calculated. However, its accuracy is dependent on the sample and the sample must represent the average compositions of waste in the target area.

The American Society for Testing and Materials (ASTM D5231-92:2016) has established a Standard Test for Determination of the Compositions of Unprocessed Municipal Solid Waste. The test method can be used to define and report the composition of municipal solid waste through the selection and manual sorting of waste sample. This method describes the procedures for measuring the composition of unprocessed municipal solid waste. The procedures include the collection of a representative sorting sample of unprocessed waste, manual sorting of the waste into individual waste components, data reduction, and reporting of the results; weights of 200–300 lb (91–136 kg) for sorting samples are recommended; the sample from vehicles is selected at random.

In additional to the Standard Test for Determination of the Compositions of Unprocessed Municipal Solid Waste by ASTM, there are many other test methods developed in different area with different aims [11]. For example, the University of Central Florida has developed a standard method that was economical and practical in implementation and adapted to the waste in the State of Florida [28]; the Environment Agency of England and Wales focused on the total municipal solid waste flow [29]; International Energy Agency aimed to promote harmonization of sampling and analytical protocols in areas related to municipal solid waste conversion to energy [30].

The sorting categories are important factors for measuring the waste composition. For most test methods, a limited number of primary categories (also called main components) are listed and each primary category has its description and subcategories. However, the primary categories in different methods differ widely: the number of primary categories varies a considerably; the same material is given different category names in different methods; the same primary category may have different subcategories. This makes it difficult to compare between different studies.

Test method by photogrammetry

Photogrammetry takes photographs of representative samples of the waste and analyzes the resulting photograph, which is projected onto a screen and divided into a 10×10 grid. The material at each grid intersection is identified and recorded, so that the overall mass fraction of each component can be calculated as a bulk density [31].

Photogrammetry technique is a safer way for measuring the waste compositions than manual sampling technique. This method can avoid practitioners to touch or smell waste and protect them from disease transmission. However, the bulk densities are important factors for photogrammetry and bulk densities must be accurate. Thus, analyzing the photograph and obtain the bulk densities are also time-consuming work.

6.2.1.2 Studies on Physical Composition

The physical composition of MSW is closely related to the economy and population and varies with time and region. Table 6.7 provides the physical compositions of

Table 6.7 Overview of physical compositions of MSW in various countries (%)

Country	Organic waste	Paper/ cardboard	Plastics	Glass	Metal	Textile	Wood	Ash	Others
China (2003)	52.6	6.9	7.3	1.6	0.5	4.7	6.9	19.2	–
Singapore (2008)	9.5	21.2	11.5	1	14.6	1.6	4.5	9.5	26.7
USA (2005)	25	34	12	5	8	–	–	–	16
Japan (2000)	34	33	13	5	3	–	–	–	12
Korea (2005)	28	24	8	5	7	–	–	–	28
Canada (2005)	24	47	3	6	13	–	–	–	8
France (2005)	32	20	9	10	3	–	–	–	26
Netherlands (2005)	35	26	19	4	4	–	–	–	12
Germany (2005)	14	34	22	12	5	–	–	–	12
Switzerland (2005)	29	20	15	4	3	–	–	–	29
Australia (2005)	47	23	4	7	5	–	–	–	13
Mexico (2005)	51	15	6	6	3	–	–	–	18
Slovak R. (2005)	38	13	7	8	3	–	–	–	31
Portugal (2005)	34	21	11	7	4	–	–	–	23
Hungary (2005)	29	15	17	2	2	–	–	–	35
Europe (2003)	30	32	7	10	8	4	–	9	–

Reprinted with permission from Elsevier [32]

MSW in various countries. As shown in Table 6.7, the distribution of MSW components varies widely in different countries. Organic waste accounts for more than 50% in China and Mexico, while only 9.5% in Singapore. Similarly, cardboard varies widely from 6.9 to 47%. The test methods for composition are different and the category 'other' includes different components. Table 6.8 also illustrates the overview of physical compositions of MSW in various continents. Based on the economy, the physical compositions of MSW in various countries are summarized in Table 6.9. The proportion of organic waste goes down with the improvement of economy, and low-income countries have the highest proportion (between 40 and 85%); With the decline of proportion of organic waste, the proportions of other categories increase, especially the paper/cardboard.

Even within one country, different cities have different levels of development, resulting in different MSW compositions. The distributions of MSW composition in different cities are similar to the one of China's average, but each has subtle differences.

In additional to region, the evolution of physical composition with time is also obvious. Refer to [32, 34] for the MSW composition in different cities of China,

Table 6.8 Overview of physical compositions of MSW in various continents, 2000 (%)

Region	Food waste	Paper/ cardboard	Wood	Textiles	Rubber/ leather	Plastics	Metal	Glass	Other
Asia									
Eastern Asia	26.2	18.8	3.5	3.5	1	14.3	2.7	3.1	7.4
South-Central Asia	40.3	11.3	7.9	2.5	0.8	6.4	3.8	3.5	21.9
South-Eastern Asia	43.5	12.9	9.9	2.7	0.9	7.2	3.3	4	16.3
Western Asia and Middle East	41.1	18	9.8	2.9	0.6	6.3	1.3	2.2	5.4
Africa									
Eastern Africa	53.9	7.7	7	1.7	1.1	5.5	1.8	2.3	11.6
Middle Africa	43.4	16.8	6.5	2.5	–	4.5	3.5	2	1.5
Northern Africa	51.1	16.5	2	2.5	–	4.5	3.5	2	1.5
Southern Africa	23	25	15	–	–	–	–	–	–
Western Africa	40.4	9.8	4.4	1	–	3	1	–	–
Europe									
Eastern Europe	30.1	21.8	7.5	4.7	1.4	6.2	3.6	10	14.6
Northern Europe	23.8	30.6	10	2	–	13	7	8	–
Southern Europe	36.9	17	10.6	–	–	–	–	–	–
Western Europe	24.2	27.5	11	–	–	–	–	–	–
Oceania									
Australia and New Zealand	36	30	24	–	–	–	–	–	–
Rest of Oceania	67.5	6	2.5	–	–	–	–	–	–
America									
North America	33.9	23.2	6.2	3.9	1.4	8.5	4.6	6.5	9.8
Central America	43.8	13.7	13.5	2.6	1.8	6.7	2.6	3.7	12.3
South America	44.9	17.1	4.7	2.6	0.7	10.8	2.9	3.3	13
Caribbean	46.9	17	2.4	5.1	1.9	9.9	5	5.7	3.5

Reprinted with permission from 2006 IPCC [33]

Table 6.9 Comparison of physical compositions of MSW with economy (%)

Country	Organic waste	Paper/ cardboard	Plastics	Glass	Metal	Textile
Low-income countries	40–85	1–10	1–5	1–10	1–5	1–5
Middle-income countries	20–65	8–30	2–6	1–10	1–5	2–10
Upper-income countries	6–30	25–66	2–8	4–12	3–13	2–6

Reprinted with permission from Elsevier [34]

including Beijing. Some composition data in one year have two versions and it is because the data come from different researchers. As we can see, the proportion of food waste rises and the proportion of non-combustibles declines obviously with time. The main reason is that people's living standard improves with economy and there is a tradition for Chinese residents to accumulate and sell the high-value recyclables to established and regularly-appearing collectors. Thus, the evolution of physical composition in one region with time can reflect its urbanization and MSW management strategies.

Within one year and one city, people from different communities have different habits in different seasons, resulting in different distributions of MSW composition. Variation of physical compositions of MSW in Zielona Góra with season and place for sample collection, respectively are found in [32]. Thus, a scientific test of physical composition should contain the different seasons and places of sample collection.

6.2.2 Particle Size

Particle size is a basic physical characteristic, and can influence the design and operation of material storage, transportation and processing. Urban waste is a complex mixture of particles with a wide range of sizes, which makes their measurement extremely difficult [31]. The size is reported as a distribution based on mass fractions.

ASTM provides a test method for particle size of refuse derived fuel (RDF). RDF is the solid fuel from MSW and its form is always pellets, cubettes, or briquettes [35]. The pellets, cubettes, or briquettes all have recognizable lengths. Because of their geometric regularity, a particle size distribution (psd) can be obtained by measuring the dimensions of a suitable number of individual particles (~ 1.0 kg) and allocating them to bins according to volume (and hence mass) [35]. Manual measurement is superior to sieving techniques because the latter may break the particles.

The technique used for obtaining a psd of MSW is generally by a set of manually-manipulated coarse screens, progressing from larger to smaller openings. In water engineering, a uniformity coefficient is employed as a simple description of psd. It is defined as [36]:

$$UC = \frac{D_{60}}{D_{10}} \tag{6.1}$$

where

UC is the uniformity coefficient;
D_{60} is the particle size where 60% of the particles are smaller than that size;
D_{10} is the particle size where 10% of the particles are smaller than that size

The use of a set of manually manipulated screens is also common for determination of particle-size distribution of MSW. The MSW is placed on the screen and shaken until no particles pass through the opening; the particle remaining on the screen are collected and weighed; the process is repeated from the screen with largest opening to smallest opening. The UC does not give absolute dimensions, but provides a crude method of comparing the spread of size, with a value of 1.0 indicating uniform size and higher values for a greater size range. Despite the paper materials being generally much larger, the UCs of all three components lie between 2.3 and 2.8.

6.2.3 Bulk Density

Bulk density is the weight per unit volume that particulate matter occupies. The volume includes particle volume, interparticle void volume, and internal pore volume. Bulk density is a basic parameter for MSW and important for the design of MSW management system, such as storage, types of collection and transport vehicles, etc. The test method is described as follows [37] or ASTM E1109-18:

a. Measure the internal dimensions of the box and calculate its volume.
b. Determine the weight of the box empty.
c. Prepare the sample by cone and quartering into four piles. Select a test sample at random from one pile. Fill the box to overflowing and avoid segregation, compaction or spillage, especially loss of the fine particulate material. Next, tamp the box three times by lifting it 6 cm above the ground and dropping squarely. Then, draw a rigid straight-edge across the top of the box to level the contents.
d. Weigh the filled box and calculate the bulk density by Eq. (6.2).
e. Repeat the procedures of c and d with another test samples from the remaining three.
f. Calculate the average of the two bulk densities. If there are more than 10% different between the two values, it is necessary to measure a third bulk density.

$$\text{Bulk density} = \frac{\mathrm{W} - \mathrm{W_T}}{\mathrm{V}} \tag{6.2}$$

where

W is the weight of the box full;
W_T is the weight of the box empty;
V is the volume of the box

For measuring box, internal dimensions approximately 60 by 60 by 60 cm (24 by 24 by 24 in) is suggested and its interior surfaces shall be nonabsorbent to

Table 6.10 Bulk density of different MSW categories

MSW category	Bulk density (kg/m^3)
MSW—Commercial	
Commercial—dry waste	56–73
Commercial—all waste, uncompacted	138
Mixed MSW—Residential, Institutional, Commercial	
Uncompacted	250–300
Compacted	400–700
Mixed MSW—Multifamily uncompacted	95
MSW—Landfill	
Compacted—MSW small landfill with best management practices	1200–1700
Compacted—MSW large landfill with best management practices	1700–2000
Compacted—MSW very large landfill with best management and cover practices	>2000

moisture [37]. However, for the sample with large particles, larger measuring box is needed and no particle larger than two-thirds the length, height, or width of the box.

Unlike the density of material, the bulk density is not an absolute material property and changes with pressure, as shown in Table 6.10. When the MSW is placed into a garbage can, it is loose and its density might be between 90 and 150 kg/m^3; if the MSW is pushed into can, its density can be at 180 kg/m^3; the bulk density is between 350 and 600 kg/m^3 in a collection truck and can reach 700 kg/m^3 after compacted with machinery in a landfill [32, 38]. Besides, owing to its own weight, MSW placed loosely will settle with time and its bulk density will increase.

EPA provides a list on the bulk densities of waste composition and different MSW categories [39, 40].

6.2.4 Moisture Content

The moisture content is an important factor for estimating heat value, landfill size, and transport requirements [41]. Moisture content is always defined as the percentage of the weight of water to the total MSW. Moisture content can influence waste degradation and stabilization; high moisture content leads to more leachate and makes it difficult when MSW is fired directly; leachate usually contains both dissolved and suspended material and causes environmental impact. The common test method for moisture content is described as follows:

a. Choose a clean and dry aluminum dish and weigh it.
b. Put the measured sample into the dish and re-weigh it.
c. Move the dish into the oven and dried at 105 °C for 24 h.

d. Remove the dish from the oven, weigh it after it cool down in a desiccator.
e. Calculate the moisture content of the sample using Eqs. (6.3) and (6.4) which give the wet and dry weight respectively.

$$M_{wet} = \frac{w - d}{w - m} \times 100\% \tag{6.3}$$

$$M_{dry} = \frac{w - d}{d - m} \times 100\% \tag{6.4}$$

where

M_{wet} is the moisture content on a wet basis, %;
M_{dry} is the moisture content on a dry basis, %;
w is the weight of dish and sample before drying, kg;
d is the weight of the dish and sample after drying, kg;
m is the weight of dish, kg

The moisture content is always described on a wet basis unless otherwise indicated. The moisture content on a dry basis can exceed 100%, but is useful in some cases, e.g. geotechnical engineering.

Drying in 105 °C for 24 h can make sure the moisture evaporates completely and avoid the organic decomposition and vaporization of volatile material. The process of cooling down must be in the desiccator and thus avoid errors by water absorption.

The moisture content of MSW can be estimated if its composition is known by using a pro rata calculation based on the moisture of the components. The values for common materials are listed in Table 6.11.

Table 6.11 Moisture content of MSW

MSW composition	Moisture content (%)
Paper/cardboard	10
Textiles	20
Food waste	60
Wood	15
Garden and Park waste	60
Nappies	60
Rubber and Leather	16
Plastics	0
Metal	0
Glass	0
Other, inert waste	10

Reprinted with permission from 2006 IPCC [33]

As shown in Table 6.11, food waste, nappies, garden and park waste are the main contribution to the total moisture content. Thus, the MSW with high proportion of organic waste has more moisture. The moisture content of MSW varies from 15 to 40% in the United States and from 38 to 62% in China [34–41].

The weather condition can also influence the moisture content. A pile of loose refuse has about 20% of moisture content in the sunny day, while the moisture content can reach 40% when there is a rain [42].

Of course, the moisture content of one component can be affected by other components. Newsprint has about 7% of moisture content on wet basis when it is sent to receptacle, but the average moisture content of newsprint often exceeds 20% after transportation by refuse truck [31].

6.2.5 Compressibility

The compressibility is an important engineering parameter regarding mechanical treatment, such as RDF production, and landfill. Landfill is closely related to the compressibility on the short-term and long-term performance, as well as its expansion, closure, and post-closure [43]. Compressibility is a measure of the deformation of the material as a response to a pressure. When a stress is applied on MSW, there will be a relative volume change, associating with a reduction of pore volume between particles. However, MSW is a complex and highly compressible mixture under load, when it will deform, compact and increase in density. Not only will the components rearrange to occupy interstices (primary consolidation), but individual components will inherently lose porosity (secondary compression).

The test method for compressibility is based on the theory of one-dimensional consolidation [44]. The stress-strain relationship involves a constrained modulus D, which is given by [45]:

$$\Delta\varepsilon_{vi} = \frac{\Delta\sigma_v}{D} \tag{6.5}$$

where

$\Delta\sigma_v$ is the increment of vertical stress;
D is the constrained modulus and has units of stress;
$\Delta\varepsilon_{vi}$ is the immediate vertical strain increment

ASTM provides two standard test methods for one-dimensional consolidation properties of soils using incremental loading, which are applicable to the test for compressibility of MSW. The two methods are introduced as follow [46]:

- Test method A: This method is performed with constant load increment duration 24 h, or multiples thereof; time-deformation readings are required on a minimum of two load increments; this test method provides only the compression curve of the specimen and the results combine both primary consolidation and secondary compression deformations [46].

- Test method B: Time-deformation readings are required on all load increments; successive load increments are applied after 100% primary consolidation is reached, or at constant time increments as described in Test method A; this test method provides the compression curve with explicit data to account for secondary compression, the coefficient of consolidation for saturated materials, and the rate of secondary compression [46].

To study the compressibility of MSW, the MSW are divided into three parts: incompressible solid materials, reinforced material that is difficult to be biodegraded and the material that is easy to be biodegraded [47, 48]. The detail of the three parts are listed as follow:

- Incompressibility solid materials: stone, sand, soil, brick, glass, ceramics, metals, etc.
- Reinforced material that is difficult to be biodegraded: textile, rubber, plastics, animal bones, etc.
- The material that is easy to be biodegraded: food, vegetable, fruit, paper, garden wastes, trees, wood, etc.

The process of MSW settlement by physical and biochemical mechanisms is always divided into three processes: immediate compression, mechanical creep, and biodegradation-induced compression [49, 50]. The mechanical compression of MSW is based on physical mechanisms, while the biodegradation is mainly based on biochemical.

The scale effects have few effects on MSW compression; the immediate compression ratio, mechanical creep ratio, and bio-compression ratio were nearly identical in a laboratory-scale (305-mm cell) experiment and field-scale experiment [51].

Shredding of MSW can affect the compressibility and strength properties. The immediate compression settlement is mainly effected by shredding, while creep and biological strain rates are not significantly affected by R-value, which is the ratio of waste particle size to apparatus size [52].

The compressibility of MSW in Hangzhou are analyzed [47]: the compression process was also composed of stress compression, consolidation compression and creep compression; the former two were about 80–85% of the total compression, while the creep compression 15–20%; in addition, the consolidation compression could be divided into two part, the process that discharging of water from the void of the MSW, and the process that discharging of water and air which were generated from the degradation of organic substance in MSW.

The compressive parameters, such as $C_{C\varepsilon}$ the compression ratio for primary settlement and $C_{\alpha\varepsilon}$ the secondary compression ratio, are impacted by the waste structure, waste composition, and unit weight [47]. Based on the mathematical formulas, the compressibility parameters can be estimated reliably by the percentage of <20-mm material and the unit weight of the material.

6.2.6 Permeability

Permeability (or hydraulic conductivity) is an ability of a material to transmit fluids/gases and depends on the pore size distribution, surface area and porosity of the material. The permeability of compacted MSW is an important physical property because it reflects the movement of liquids and gases, especially in a landfill.

ASTM published a standard test method for permeability of granular soils, under constant-head conditions [53]:

- Continuity of flow with no sample volume change during a test.
- Flow with the sample voids saturated with water and no air bubbles in the sample voids.
- Flow is steady state with no change in hydraulic gradients.
- Direct proportionality of velocity of flow with hydraulic gradients below certain values, at which turbulent flow starts.

This method adopts a constant-head method for the laminar flow of water through the measured sample, which is also available for MSW; the sample should contain less than 10% of the material passing the No. 200 sieve (74 microns The specific procedure is described as follows [53]:

a. Open the inlet value from the water tank slightly for the first test; allow the water to flow through the measured sample until a stable head is established and there is no change in the manometer levels.
b. Measure and record the time, t, for a predetermined quantity of water, Q, to flow through the sample.
c. Measure and record the head, h (difference between the manometer readings).
d. Repeat the test with a different value of the head, h, at least five times.
e. Calculate the coefficient of permeability by Eq. (6.6).

$$k = \frac{QL}{A\,th} \tag{6.6}$$

where

k is the coefficient of permeability;
Q is quantity of water discharged, cm^3/s;
L is distance between manometers openings, cm;
A is cross-sectional area of specimen, cm^2;
t is total time to collect the quantity Q, s;
h is difference in head on manometers, cm

However, this test method is withdrawn by ASTM in 2015.

The permeability of MSW is related to the physical properties, such as density, particle size, porosity, material type, degree of saturation, stage of decomposition [45].

Many researchers studied the relationship between permeability and density; the results found that increase of density results in the decrease of permeability [45, 54, 55]. At same density, the permeability results from different studies have a wide variation, owing to other factors, e.g. degree of decompositions [56].

Regarding the permeability of MSW in landfill, placement of waste in layers and the use of daily cover soil resulted in higher permeability in the horizontal direction because of a structure of sub-horizontal layers; the permeability reduced from $1e^{-3}$ to $1e^{-6}$ cm/s if depth varied from 0 to 60 m [55, 57].

6.3 Chemical Characterization of Municipal Solid Waste

6.3.1 Elemental Composition

6.3.1.1 Definition

Elemental composition of municipal solid waste (MSW) is usually described in three ways: (1) organic elements, generally including C, H, N, O and S, a few of which contain Cl; (2) macro inorganic elements, mainly Na, K, Cl, S, Mg, Al, Fe, Ca and other elements; and (3) trace heavy metal elements including As, Ba, Cd, Cr, Cu, Hg, Ni, Pb and Zn. The trace heavy metal elements are the main pollutants in MSW. Among the macro-elements, N, P and K are required for plant growth and are also called plant nutrients. The elemental composition of MSW can be used to estimate a number of engineering design parameters such as calorific value and flue gas composition for incineration, mass ratio of C/N for composting, and biochemical methane potential for anaerobic digestion.

6.3.1.2 Analytical Methods

Organic elements: The content of C, H, N, and S in MSW can be directly determined using an elemental analyzer. The analytical principle is shown in Fig. 6.8. The sample should be first crushed into a powder after which 5–20 mg of the sample is weighed into a tin/silver boat for testing in the elemental analyzer.

The standard test method for Cl in MSW is BS EN 15408:2011 (oxygen bomb calorimeter combined with ion chromatographic method). A 2–5 g sample is

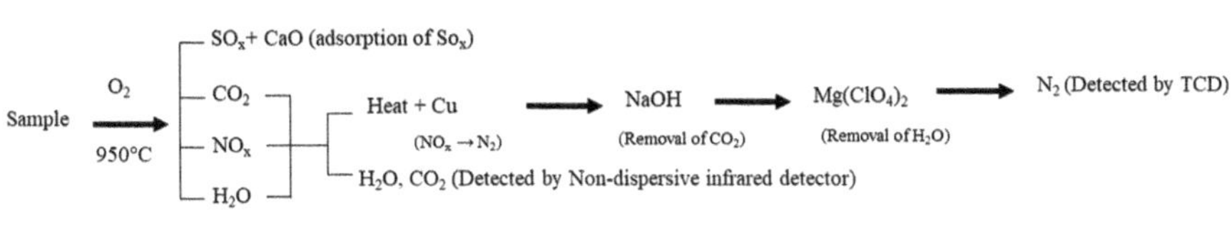

Fig. 6.8 The analytical principle for determining organic elements

burned out in an oxygen bomb calorimeter containing 10 mL of absorption liquid (2.52 g/L Na_2CO_3, 2.54 g/L $NaHCO_3$, 25 mL 30% (w/w) H_2O_2 plus deionized water to 1 L). The total chlorine content in the MSW sample is the Cl content in the absorption liquid as determined by ion chromatography. Currently some elemental analyzers can directly measure the concentration of flammable chlorine in solid samples. Ion chromatography is used to determine the concentration of chlorine in the water rinsing solution after the elemental analyzis, which is regarded as non-flammable chlorine. The sum of flammable chlorine and non-flammable chlorine is the total chlorine content. Cl also can be detected by X-ray fluorescence (XRF).

Plant nutrients

The nutrient elements N, P and K in solid waste are generally represented by total nitrogen (TN), total phosphorus (TP), and total potassium (TK).

According to the standard DIN EN 13654-2, TN in MSW can be determined using the Kjeldahl method. The nitrogenous substances in MSW can be transferred to NH_4^+–N by dissolving a sample in concentrated sulfuric acid, catalyzed by a mixture of potassium sulfate, anhydrous copper sulfate and selenium powder. Free ammonia is absorbed by a boric acid solution and then measured by a titration test, the result of which is TN.

TP in MSW can be determined by digesting a sample with nitric acid using microwaves, during which the phosphorous salts and organic phosphorus are converted to orthophosphate, and the digested liquid can by analyzed by inductively coupled plasma optical emission spectrometry (ICP-OES) or inductively coupled plasma-mass spectrometry (ICP-MS) after reaching a constant volume [58].

Two test methods (No. 3 and No. 4, below) can be used to determine TK in MSW.

Macro inorganic elements

The macro inorganic elements of solid waste can be determined using X-ray fluorescence (XRF). XRF is a method for elemental analyzis (and chemical analyzis) that measures the wavelength of X-ray radiation and the associated energy emitted from elements. Quantitative methods generally can be divided into two major categories, namely the single element method and multi-element method. XRF can be used to qualitatively and quantitatively determine Na, P, Cl, K, Mg, Al, Fe, Ca and other macro metal elements in MSW.

The sample analyzed using XRF should not contain water, oil, volatile components or any corrosive solvent. The dried MSW sample is ground to < 50 μm. Approximately 1.0 g of the powdered sample is compressed for analyzis in the XRF instrument.

Macro metal and trace heavy metals

Macro metals and trace heavy metals (including As) in MSW can be dissolved by digestion in strong acid or alkali (Cr^{6+}). Following digestion Na, K, Mg, Al, Fe, Ca, As, Ba, Cd, Cr, Cu, Hg, Ni, Pb, Zn and other elements can be detected by inductively coupled plasma-atomic emission spectrometry (ICP-AES) or inductively coupled plasma-mass spectrometry (ICP-MS). Na, K, Mg, Al, Fe, Ca, Ba, Cd, Cr, Cu, Hg, Ni, Pb, Zn and other elements also can be detected by flame atomic absorption spectrometry (FLAA). Alternatively, Ar, Cr, Cd, Ca, Co, Fe, Pb, Mo, Se and other elements can be detected by graphite furnace atomic absorption spectroscopy (GFAAS), and As and Hg can be detected by atomic fluorescence spectrometry (AFS).

Digestion is the most important pretreatment method for the determination of metal elements in MSW. Digestion is a method of dissolving a MSW sample to a liquid form by oxidation and decomposition enhanced by the addition of acids and oxidants (hydrochloric acid, nitric acid, hydrofluoric acid and perchloric acid, hydrogen peroxide, etc.). The digestion process is enhanced by heating, using either electric heating devices or microwave heating devices.

Acid digestion uses the strong oxidizing properties of acids and oxidants to destroy the structure of minerals or organic matter and thereby release heavy metals into solutions for measurement. It is currently the most commonly used MSW digestion method (EPA method 3050B, 3051A and 3052). Acids for digestion of different solid samples should be chosen according to their oxidizing abilities. Hydrochloric acid (HCl) and nitric acid (HNO_3) are used to decompose substance that are easily oxidized. Hydrofluoric acid (HF) is used to dissolve silicates. Perchloric acid ($HClO_4$) is used to remove hydroxyl-containing organic substances. Hydrogen peroxide (H_2O_2) is used to oxidize organic substances in solids that remain insoluble in acids. If none of the acids are suitable when used alone, aqua regia can be added to achieve complete dissolution of a sample. However, it should be noted that the addition of acid may interfere with subsequent tests of the elements. For specific samples, the specific acid digestion method can be found in EPA method 3051A or EPA method 3052.

The EPA method 3051A is designed to mimic extraction using conventional heating with HNO3, or alternatively, with a mixture of $HNO_3$and HCl. This method is applicable to the microwave-assisted acid extraction/dissolution of sediments, sludges, soils, and oils for the following elements: Na, K, Mg, Al, Fe, Ca, As, Ba, Cd and Cr. In this method, a dried sample (0.25–0.5 g) is crushed to <150 μm and then digested by adding 10 mL of concentrated HNO_3 or 9 mL of concentrated HNO_3 and 3 mL of concentrated HCl. The procedure involves heating a sample to 175 ± 5 °C in 5.5 ± 0.25 min and maintaining the target temperature for 4.5 min. After cooling for 5 min, the digested liquid is collected and stored at 4 °C until analyzis.

The EPA method 3052 is applicable to the microwave-assisted acid digestion of siliceous matrices, as well as organic matrices and other complex matrices by using concentrated HNO_3 and HF. This method is suitable for determination of Na, K,

Mg, Al, Fe, Ca, As, Ba, Cd, Cr, Cu, Hg, Ni, Pb, Zn and other elements contained in ashes, biological tissues, oils, oil contaminated soils, sediments, sludges, or soils. A dried sample (0.25–0.5 g) is crushed to <150 μm and then digested by adding 9 mL of concentrated HNO_3 and 3 mL of concentrated HF. The amount of HF can be adjusted according to the silicate content of the actual sample. If the solution contains high concentrations of Fe, Ag, Al, Ba and Sb, 2 ± 2 mL of concentrated HCl can be added to achieve complete digestion. The procedure involves heating a sample to 180 ± 5 °C in 5.5 ± 0.25 min and maintaining the target temperature for 9.5 min. After cooling for 5 min, the digested liquid is collected and stored at 4 °C until analyzis.

6.3.1.3 Elemental Composition of MSW

Typical values for the elemental composition of MSW are listed in Tables 6.12 and 6.13. Organic elements in MSW from different locations are shown in Table 6.12. Organic elements in the components that comprise MSW are shown in Table 6.13. The tables are presented as typical of various countries, but some values, such as nitrogen, can vary widely. In many areas of countries like India, the MSW is mostly ash.

6.3.2 Biochemical Composition

6.3.2.1 Definition

The biochemical composition of solid waste, also known as biomass composition, is the main characteristic describing the biodegradability of MSW. The biodegradability of solid waste can be evaluated by analyzing the total organic matter of the waste and its composition.

Table 6.12 Organic elements in MSW from different locations

States	Organic elements content (% of dw[a])						Ash (% of dw[a])
	C	H	O	N	S	Cl	
China [59]	41.8	6.3	22.5	0.85	0.85	0.90	–
Korea [60]	51.0	6.0	39.2	3.5	0.2	–[a]	5.3
Netherlands [61]	51.9	6.7	38.7	2.1	0.5	0.3	–
United Kingdom [62]	47.6	7.4	33.3	3.4	0.15	–	–
Denmark [59]	49.5	6.9	35.1	2.95	0.22	0.78	–
America [59]	48.0	6.4	37.6	2.6	0.4	–	–

Note
[a]The content is based on dry waste

Table 6.13 Organic elements in MSW components

MSW component		Organic element content (% of dw[a])						Moisture content (% of ww[b])
		C	H	O	N	S	Cl	
China	Paper	41.3 (3.2[c])	5.9 (0.4)	38.6 (2.9)	2.6 (0.5)	0.2 (0.1)	0.46	43.2 (14.1)
	Plastics	60.4 (4.0)	7.9 (0.5)	22.3 (6.2)	0.5 (0.7)	0.1 (0.05)	1.9	43.5 (15.2)
	Wood	42.9 (4.7)	6.1 (0.1)	42.0 (1.0)	1.6 (1.1)	0.1 (0.03)	0.36	48.0 (14.3)
	Textiles	46.3 (2.5)	6.6 (0.3)	40.8 (2.5)	4.4 (3.2)	0.5 (0.5)	0.46	43.5 (13.6)
	Food waste	36.8 (2.3)	5.5 (0.5)	38.6 (2.9)	2.6 (0.5)	0.3 (0.1)	0.82	68.0(5.8)
Denmark	Paper	45.5	6.3	36.2	0.44	0.08	0.13	24.9
	Plastics	67.0	9.4	9.0	0.90	0.12	3.0	9.9
	Wood	43.9	5.5	25.4	1.8	0.22	0.24	47.6
	Textiles	53.1	6.4	31.4	2.6	0.38	0.58	14.2
	Food waste	49.9	6.9	34.4	3.1	0.23	0.82	72.2
	Non-combustibles	4.0	0.58	3.0	0.25	0.11	0.43	14.8
	Glass	0	0	0	0	0.04	0	7.2
	Metal	24.5	3.8	3.6	0.18	0.02	0.04	11.0

Note
[a]The content is based on dry waste
[b]The content is based on wet waste
[c]The number in the bracket is the standard deviation
Reprinted with permission from Elsevier [63]

Generally, the organic matter in solid waste can be divided into three categories:

(1) Easily biodegradable organic components, or putrescible organics, such as starch, protein, fat, etc., that are usually abundant in the food waste component of MSW.
(2) Recalcitrantly biodegradable organic components, i.e., organic components that are relatively difficult to biodegrade, such as cellulose, hemicellulose, lignin, pectin, wax, humic acid, etc., usually abundant in the paper, textile, and wood (as well as bamboo and other plant material) that comprises MSW, as well as in the compost or digestate from MSW.
(3) Non-biodegradable organic components, such as plastics and rubber, which are excluded from the measure of biomass.

6.3.2.2 Analytical Methods

Total Organic Matter

Total organic matter (OM) in MSW can be determined by loss on ignition (Chinese standard CJ/T 96-2013). Organic matter in solid waste can be regarded as the weight loss that occurs during burning at 600 °C. The analytical procedure is as follows.

Accurately weigh the sample and place it in a porcelain crucible that has been heated for 2 h and cooled to a constant weight. Heat the crucible and sample in a muffle furnace to 600 °C and maintain the target temperature for 6–8 h. Then, transfer the crucible to a desiccator for cooling. After cooling, weigh the crucible. Then, place the crucible in the muffle furnace again at 600 °C for 10 min, and weigh the crucible after cooling. Repeat the above operation until the weight is constant.

The content of organic matter is calculated according to Eq. (6.7):

$$C_{om} = \frac{m_1 - m_2}{m_3(1 + c_i)} \times 100\% \tag{6.7}$$

where

C_{om} The content of organic matter in the sample, %;
m_1 Total weight of crucible and dry sample, g;
m_2 Total weight of crucible and burnt sample, g;
c_i Proportion of plastic in dry waste, wt%;
m_3 The weight of dry sample, g

Note that some salts will be oxidized or decomposed at 600 °C e.g. alkali metal compounds can be volatilized at temperatures of 300–400 °C. BS EN 13137:2001 measures TOM from the amount of carbon dioxide released when the sample is combusted at higher temperatures (>900 °C).

Fat

The fat content of MSW can be measured using the Soxhlet extraction method (Chinese standard GB/T 5009.6-2016). Fat is very soluble in organic solvents. After fat in a sample is directly extracted using a solvent such as anhydrous ether or petroleum ether, the solvent is removed by evaporation and the residual is dried to obtain the free fat content. The test procedure is as follows.

Weigh a certain amount of sample and place it into a filter paper tube. Put the filter paper tube into the extraction tube of the Soxhlet extractor, then connect the distillation flask that has been previously dried to constant weight. Add enough anhydrous ether or petroleum ether from the upper end of the condensation tube of the extractor to fill two-thirds of the distillation flask. Heat the flask on a water bath and continuously reflux the extraction with anhydrous ether or petroleum ether (6–8 times/h). The flask containing the leachate is then evaporated to dryness to recover the fat. ASTM G120–15 gives general guidance for Soxhlet operation. Generally, the extraction lasts for 6–10 h. At the end of the extraction, one drop of extract is taken using a frosted glass rod. When no spots appear on the frosted glass rod, the extraction is complete.

Remove the distillation flask and recover the anhydrous ether or petroleum ether, which now contains the extracted fat. When the remaining solvent in the distillation flask is about 1–2 mL, evaporate it to dryness in a water bath. Then, dry the bottle at 100 ± 5 °C for 1 h, and cool it in a desiccator for 0.5 h. Repeat the above operation until constant weight (±2 mg between successive measurements).

The content of fat is calculated according to Eq. (6.8):

$$X = \frac{m_1 - m_0}{m_2} \times 100\% \tag{6.8}$$

where

X The concentration of fat in the sample, g/100 g;
m_1 The weight of the distillation flask and the fat it contains after constant weight, g;
m_0 The weight of the clean and dry distillation flask, g;
m_2 The weight of the sample, g;

Protein

Protein can be measured using the Kjeldahl method (Chinese standard GB/T 5009.5-2010) as follows. Weigh a certain amount of sample and use an automatic Kjeldahl analyzer to perform the test. The protein in the sample is decomposed by catalytic heating and the resulting ammonia is combined with sulfuric acid to produce ammonium sulfate. Alkaline distillation transforms the ammonium sulfate into free ammonia, which is then absorbed by boric acid. Then, titrate the boric acid containing the absorbed ammonia using a standard titration solution of sulfuric acid or hydrochloric acid. Multiply the consumption of acid by the conversion factor (the general value is 6.25) to determine the protein content.

The European Committee for Standardization recommends different test methods for determining proteins in different classes of substances. For example, the protein in livestock feed can be measured by digestion and distillation (European standard EN ISO 5983-2-2009) and the protein in textiles can be measured by the hypochlorite method (European standard EN ISO 1833-4:2017). The American Materials and Testing Association uses the modified Lowry method to analyze water-soluble proteins in natural rubber and elastic products (American standard ASTM D5712-2015). Therefore, a suitable method for determining the content of protein in MSW can be chosen according the different components of solid waste.

Starch

Starch in MSW can be determined using the acid hydrolysis method (Chinese standard GB/T 5009.9-2016). After removing fat and soluble sugars from the sample, the starch is hydrolyzed with acid into a reducing monosaccharide, which is then measured as reducing sugar and converted into starch. The test procedure is as follows.

Weigh a certain amount of sample, accurately add an appropriate amount of water and pound the mixture into a homogenate using a tissue masher. Weigh some homogenate (equivalent to 2.5–5.0 g of the original sample) and place it in an Erlenmeyer flask, first washing out the fat in the sample with ether five times, then washing out the soluble sugars in the residue with ethanol several times. Add hydrochloric acid to the resulting sample, connect the condenser tube, and place it in a boiling water bath. Then, cool the tube after refluxing. After the hydrolyzed sample solution is cooled, add 2 drops of methyl red indicator solution. First, adjust the sample hydrolyzate to yellow with sodium hydroxide solution, and then correct it with hydrochloric acid (1 + 1) to red. Then add lead acetate solution and shake the sample for 10 min. Sodium sulfate solution is added to remove excess lead. After shaking, transfer the solution and residue to a volumetric flask. Filter the solution and residue, discard the initial filtrate, and use the remaining filtrate for analyzis.

Then use a previously calibrated alkaline copper tartrate solution to determine the concentration of starch in the sample solution. First add alkali copper tartrate solution and alkaline copper tartrate solution to an Erlenmeyer flask, then add some water and two glass beads. Use a burette to add some filtered sample solution, which is 1 mL less than the predicted volume, into the Erlenmeyer flask and heat it to boiling within 2 min. Keep boiling and continue titrating with sample solution at a rate of one drop every two seconds until the blue color just fades to the end. Record the sample fluid consumption volume.

The content of starch is calculated according to Eq. (6.9):

$$X = \frac{(A_1 - A_2) \times 0.9}{m \times \frac{V}{500} \times 1000} \times 100 \tag{6.9}$$

where

X The concentration of starch in the sample, g/100 g;
A_1 The weight of glucose in the hydrolyzate of the test sample, mg;
A_1 The weight of glucose in the blank reagent, mg;
0.9 Conversion factor of glucose to starch;
m The weight of the sample, g;
V The volume of hydrolyzate of the test sample, mL

Total Cellulose

Total cellulose refers to the total amount of hemicellulose and cellulose retained after lignin is removed from the plant fiber in a raw sample. The test method (Chinese standard GB/T 2677.10-1995) involves measuring the amount of the residue (which is total cellulose) quantitatively by removing the lignin contained in the sample, which has been extracted using sodium chlorite at a pH of 4–5. The test procedure is as follows.

Accurately weigh a sample, wrap it in qualitative filter paper and tie it with cotton thread for phenyl alcohol extraction. After this extraction, open the filter paper package and transfer the entire sample into the Erlenmeyer flask of the total cellulose analyzer. Add distilled water, glacial acetic acid, and sodium chlorite. Shake well. Heat the solution in a 75 °C constant temperature water bath for 1 h. Repeat the addition of glacial acetic acid and sodium chlorite as well as heating in the water bath until the sample turns white, indicating that the content of lignin is 2–4%. Remove the Erlenmeyer flask and cool it in an ice-water bath. Thereafter, filter the flask contents by suction, and wash the filtrate repeatedly with distilled water until the filtrate does not show an acidic reaction. Finally, wash the filtrate with acetone three times, and vacuum filter the resulting mixture. Wash the filter externally with distilled water, then dry the filter in an oven at 105 ± 2 °C to a constant weight.

The content of total cellulose is calculated according to Eq. (6.10):

$$X_1 = \frac{m_1}{m_0} \times 100 \tag{6.10}$$

where

X_1 The content of total cellulose in the sample, %;
m_1 The weight of total cellulose after drying, g;
m_0 The weight of the sample after drying, g.

The lignocellulose content of MSW also can be measured using the method proposed by Van Soest [64, 65], in short, by fractional distillation using a FiberCap™ instrument (FOSS, Denmark). In the first step, the soluble fractions in the sample of waste are removed using a neutral detergent at 100 °C for 60 min. In the second step, the hemicellulose analogs in the sample are extracted and removed using acid detergent at 100 °C for 60 min. In the third step, the cellulose analogs in the material are extracted using 72% H_2SO_4 for 3 h. After each step, the sample is washed with deionized water and acetone. After the third step, the sample is dried at 105 °C for 5 h and weighed to measure the VS (volatile solid) of the third-step residue. The difference of VS in the third step is the lignin content in the material.

In addition, fluorescence and infrared spectroscopy are widely used to study the contents of major biochemical components such as carbohydrates, proteins, fats, lignin, and cellulose in soil organic matter [66]. However, fluorescence and infrared spectroscopy are qualitative methods and do not provide quantitative measurements.

6.3.2.3 Biochemical Composition of the Solid Waste

Due to the differences in socioeconomic conditions and living habits, the composition of MSW in different countries, and even in different regions of the same country, is very different. Table 6.14 shows the composition of MSW in some countries.

Yang et al. [63] compared the physical composition of MSW from China and other developing and developed countries, and found that the main differences in compositions were the contents of waste that degrade slowly (including paper, wood, and textiles) and those that degrade rapidly (including food waste). In developed countries MSW components that degrade slowly contribute more than 30% to the total wet weight; the contribution is less than 30% in developing countries. Correspondingly, the content of rapidly degradable waste is also very different, ranging from 20 to 45% in developed countries and from 35 to 75% in developing countries.

Cellulose is the major biodegradable waste component. In 2011, the high content of lignocellulose waste from MSW in the USA was contributed by paper (28.0%), yard waste (13.5%), and wood (6.4%) [76]. Corrugated boxes with high lignin content accounted for 42% of paper waste. In another study, food waste accounted for only 14.5% of MSW [77]. In 2010, MSW in Turin, Italy, was shown to contain 32% food waste, 14% garden waste and 34% paper [78]. MSW in the United Kingdom contained 18% food waste and 14% garden waste in 2006 [79]. In the USA, the lignin content of garden waste (21.6–43.8%) is much higher than the lignin content in kitchen waste (11.4%) [76]. Therefore, yard wastes with high

Table 6.14 The physical composition of MSW in some countries

Country	%ww						Year	References
	Organic matters	Paper	Plastic	Glass	Metal	Others		
Brazil	36.1	17.1	23.3	3.5	2.4	17.6	2007	[67]
China	59.0	8.0	10.0	3.0	1.0	19.0	2002	[68]
Egypt	60.0	10.0	12.0	3.0	2.0	13.0	1998	[69]
India	40.0	10.0	2.0	0.2	–	47.8	2008	[70]
United States	34.4	28.0	12.7	4.6	8.8	11.5	2011	[71]
Australia	47.0	23.0	4.0	7.0	5.0	14.0	2002	[72]
Austria	35.0	22.0	11.0	8.0	5.0	19.0	2004	[73]
United Kingdom	35.7	22.7	10.0	6.6	4.3	20.7	2006	[74]
Germany	30.0	24.0	13.0	10.0	1.0	22.0	2005	[75]
Italy	29.0	28.0	5.0	13.0	2.0	23.0	2005	[73]

Note
ww = wet weight

lignin content comprise a large proportion of MSW in developed countries such as those in Europe and the USA; conversely the relative proportion of food waste with low lignin content is smaller.

However, the content of lignin in MSW in China is quite different from that in Europe and the USA. Taking Shanghai, China as an example, from 2008 to 2009 organic matter in MSW included 64% food waste, 11% paper, and 1% courtyard waste; furthermore, toilet paper with low lignin content accounted for 82% of all paper. In general, MSW in China can be classified as MSW having low lignin content, or alternatively, MSW having high food waste content.

Table 6.15 shows 29 kinds of typical biodegradable materials in MSW, including discarded paper, vegetables, meat, fish, agricultural products, fruits, plants, etc., which have been studied and analyzed for their biochemical composition. The experimental results show that the fat content of pork fat is the highest of the materials analyzed, reaching 99.6%. The fat content of pork bone (42.2%) and soybean (23.2%) are the next highest, and the fat content of other materials is less than 10%. Except for pork fat, the protein content of animal-derived food wastes

Table 6.15 Biochemical composition of 29 typical types of biomass

Materials	VS volatile solids (g g TS)	Protein (g g TS)	Lipid (g g TS)	Hemicellulose (g g TS)	Cellulose (g g TS)	Lignin (g g TS)
Camphor tree branch	0.968(0.004)	0.018 (0.004)	0.028 (0.011)	0.187 (0.001)	0.426 (0.016)	0.205 (0.007)
Camphor tree leaf	0.917 (0.002)	0.093 (0.005)	0.060 (0.023)	0.175 (0.003)	0.188 (0.002)	0.201 (0.002)
Metasequoia branch	0.963 (0.005)	0.033 (0.004)	0.032 (0.012)	0.126 (0.001)	0.349 (0.008)	0.337 (0.005)
Metasequoia leaf	0.917 (0.007)	0.113 (0.001)	0.090 (0.039)	0.121 (0.004)	0.113 (0.006)	0.108 (0.002)
Bamboo branch	0.973 (0.002)	0.010 (0.001)	0.006 (0.001)	0.262 (0.004)	0.426 (0.003)	0.168 (0.005)
Bamboo leaf	0.869 (0.004)	0.140 (0.001)	0.044 (0.001)	0.386 (0.004)	0.258 (0.001)	0.050 (0.005)
Reed	0.934 (0.006)	0.061 (0.003)	0.011 (0.001)	0.330 (0.008)	0.381 (0.002)	0.090 (0.001)
Cynodon dactyton	0.912 (0.002)	0.090 (0.001)	0.030 (0.001)	0.465 (0.001)	0.286 (0.004)	0.061 (0.002)
Celery	0.738 (0.002)	0.197 (0.002)	0.063 (0.022)	0.110 (0.010)	0.237 (0.003)	0.042 (0.005)
Lettuce	0.816 (0.005)	0.299 (0.002)	0.070 (0.008)	0.129 (0.005)	0.149 (0.004)	0.041 (0.009)
Peanut shell	0.970 (0.002)	0.047 (0.002)	0.006 (0.001)	0.129 (0.001)	0.400 (0.003)	0.338 (0.003)
Grapefruit peel	0.958 (0.014)	0.073 (0.002)	0.010 (0.001)	0.048 (0.001)	0.161 (0.007)	0.008 (0.004)

(continued)

Table 6.15 (continued)

Materials	VS volatile solids (g g TS)	Protein (g g TS)	Lipid (g g TS)	Hemicellulose (g g TS)	Cellulose (g g TS)	Lignin (g g TS)
Watermelon peel	0.823 (0.010)	0.276 (0.005)	0.055 (0.027)	0.118 (0.011)	0.265 (0.012)	0.033 (0.013)
Apple core and peel	0.985 (0.005)	0.035 (0.001)	0.030 (0.002)	0.029 (0.002)	0.075 (0.002)	0.017 (0.004)
Orange peel	0.999 (0.001)	0.048 (0.004)	0.013 (0.001)	0.034 (0.002)	0.119 (0.009)	0.002 (0.001)
Banana peel	0.838 (0.024)	0.091 (0.003)	0.086 (0.029)	0.174 (0.007)	0.190 (0.001)	0.116 (0.012)
Tea residue	0.965 (0.001)	0.239 (0.001)	0.038 (0.001)	0.336 (0.001)	0.152 (0.001)	0.113 (0.001)
Sugarcane residue	0.966 (0.004)	0.034 (0.001)	0.035 (0.018)	0.211 (0.002)	0.279 (0.002)	0.045 (0.001)
Potato	0.956 (0.001)	0.087 (0.001)	0.002 (0.001)	0.452 (0.009)	0.025 (0.002)	0.005 (0.001)
Soybean	0.945 (0.001)	0.359 (0.006)	0.232 (0.032)	0.040 (0.002)	0.064 (0.002)	0.000 (0.000)
Cotton	0.983 (0.003)	0.015 (0.001)	0.004 (0.001)	0.013 (0.007)	0.965 (0.001)	0.000 (0.000)
Fabric	0.999 (0.001)	na	na	na	na	na
Fat pork	0.998 (0.001)	0.028 (0.003)	0.996 (0.001)	na	na	na
Lean pork	0.995 (0.001)	0.801 (0.033)	0.074 (0.001)	na	na	na
Pig bone	0.534 (0.006)	0.478 (0.007)	0.422 (0.014)	na	na	na
Fish bone	0.447 (0.009)	0.619 (0.008)	0.196 (0.003)	na	na	na
Toilet	0.993 (0.004)	0.003 (0.001)	0.256 (0.001)	0.101 (0.005)	0.848 (0.007)	0.042 (0.012)
Office paper	0.807 (0.006)	0.006 (0.001)	0.004 (0.001)	0.108 (0.003)	0.796 (0.006)	0.000 (0.000)
Bleached newspaper	0.915 (0.002)	0.007 (0.001)	0.008 (0.001)	0.112 (0.001)	0.662 (0.009)	0.166 (0.016)

Reprinted with permission from Elsevier [80]

(45%) is higher than that of other biodegradable components. The highest protein content in plants is that of soybeans (35.9%); the protein content of other materials is lower than 30%. Potatoes, apple core and peel, orange peel, grapefruit peel and sugarcane residue have the highest polysaccharide content of the materials analyzed, all exceeding 35%. The highest cellulose content is 96.5% for cotton, followed by that of paper (>65%). The highest lignin content is that of the

Metasequoia branch (a species of softwood, 33.7%) and the peanut shell (a type of leguminous seed waste, 33.8%). The biochemical composition of these 29 biodegradable materials represents the biochemical composition of degradable materials in MSW to a certain extent. Therefore, the mean content of fat (protein, cellulose and lignin) in materials from the same category of biodegradable material can be weighed and accumulated according to the composition of the MSW to approximate the fat content (protein content, etc.) of the MSW.

6.3.3 Mineral Composition

6.3.3.1 Definition

The composition and content of minerals in MSW vary considerably, and are influenced by characteristics of different regions such as differences in lifestyle, consumption level, fuel structure, and climatic environment.

The physical composition of MSW in various countries varies with different levels of economic development, but in general, the mineral composition of MSW mainly includes calcareous soil (including ash), brick and ceramics, and glass. Thus, the mineral composition of MSW can be classified on the basis of:

(1) Ash type: ash, ash dust, etc.;
(2) Brick and tile ceramics: blocky products such as discarded bricks, tiles, porcelain, stones and cement blocks;
(3) Glass: various types of glass and glass products.

6.3.3.2 Analytical Methods

X-Ray Diffraction

X-ray diffraction (XRD) is a commonly used technique for determining the composition of materials, including the structure or morphology of internal atoms or molecules. The procedure involves analyzing the diffraction pattern when a material is subjected to X-rays in a diffractometer.

Mineral matter can be identified using appropriate software (such as Diffract Plus software employing the ICDP PDF-2 database; Bruker Corp., Billerica, MA, USA) to provide a reference for the diffraction patterns of samples. The grain size distributions of the materials can be determined according to the sieve analyzis method for fine aggregates (C132, ASTM) [81].

Data analyzis

Qualitative analyzis

A diffraction pattern can indicate the chemical bonding state of the elements in a sample. By photographing the diffraction patterns of many crystals, a standard reference of the diffraction patterns can be produced. In the qualitative analyzis of XRD patterns, the diffraction pattern of the sample is compared with standard diffraction patterns to identify the unknown substances in the sample.

Quantitative analyzis

According to diffracted ray intensity theory, the diffraction intensity of a phase in a multiphase mixture increases as the relative content of the phase increases.

Scanning Electron Microscopy (SEM)

Scanning electron microscopy is an analytical technique that uses a finely focused electron beam to bombard the surface of a sample, from which secondary electrons and backscattered electrons generated by the interaction between the electron and the sample are used to observe and analyze the surface or fracture morphology of the sample.

When a powder sample is prepared for SEM, the powder can be directly sprinkled on the double-sided carbon conductive adhesive of the sample holder and pressed with a flat object. Any weakly adhering particles can then be blown away. When a bulk sample is prepared, it can be first ground and then polished with an epoxy resin [82].

Transmission Electron Microscopy (TEM)

Transmission electron microscopy is an analytical technique that uses an electron source instead of a light source to observe a sample. An electromagnetic field serves as a lens, and the electron beam passes through the sample. Due to the low penetrating ability of the electron beam, the sample analyzed by TEM must be very thin. According to the sample's atomic number, the sample thickness is generally between 5 and 500 nm [82].

Electron MicroProbe Analyzis (EMPA)

Electron microprobe analyzis is a high-efficiency analytical technique based on electron optics and X-ray spectroscopy, i.e. the principles governing the interaction between high-energy electrons and a solid substance. EMPA bombards the surface of the sample with a beam of highly focused electrons having high enough energy to excite characteristic X-ray signals within a limited depth and laterally expanding microdomain volume. The wavelength (or energy) and intensity of these signals provide important information for characterizing the elements and their concentrations contained in the micro-domains. The system for detecting X-ray signals in an electron probe is usually a spectrometer or an energy spectrometer [83].

Other Methods

Secondary Ion Mass Spectrometry (SIMS)

Secondary ion mass spectrometry is a technique for achieving high-sensitivity microanalyzis by analyzing the mass spectrum of secondary ions emitted from the surface of a sample as it is bombarded by an ion beam.

The preparation of a sample for SIMS analyzis is basically the same as that for EMPA, and the preparation of the microparticles is the same as the preparation of such samples used for SEM and TEM. In fact, in most cases, the samples used for ion probe analyzis can be used interchangeably with electron probes, scanning electron microscopes, transmission electron microscopes, and samples used in a series of surface analyzes [84].

Atomic Force Microscopy (AFM)

Atomic force microscopy transcends the resolutions of other microscopes that are limited by light and electron wavelengths. AFM uses microscopic probes to observe the morphology of matter in three dimensions and obtains information on the interaction between the probe and the sample. The typical AFM has a lateral resolution of 2 nm and a vertical resolution of <0.1 nm.

6.3.3.3 Typical Mineral Composition of Municipal Solid Waste and Its Residues After Treatment

The mineral composition of MSW includes NaCl in organic waste (food waste), $CaCO_3$ in inorganic waste (masonry, ash), and SiO_2 in various waste products. Due to the high moisture content of MSW, there are few data on its mineral composition and content. Solid residues such as fly ash (FA) and bottom ash (BA), which are generated from the incineration of MSW (MSWI), are mainly the residues of inorganic components in the MSW after combustion. Although some of these residues contain relatively high concentrations of heavy metals and toxic organics, all contain a relatively high content of minerals.

MSW fly ash contains 24–27% lime, as well as some silicates and alumino-silicates [85]. The main oxides in FA are SiO_2, CaO, and Al_2O_3. XRD analyzis of MSWI fly ash has shown that the main crystalline compounds, in order of size, are quartz (SiO_2), sylvite (KCl) and halite (NaCl), anhydrite ($CaSO_4$), calcite ($CaCO_3$) and lime (CaO) [86].

XRD used in conjunction with microscopic analyzis has shown that MSWI bottom ash components can be divided into seven groups of components: (i) crystalline silicates in a glass matrix, such as hedenbergite; (ii) ferrohedenbergite, melilite, wollastonite, quartz and feldspar (which can appear as rounded grains with many fissures and cracks); (iii) oxides, sometimes in a siliceous glass matrix, such as magnetite, hematite and periclase; (iv) bottle glass, mostly composed of Na_2O, CaO and SiO_2 sometimes fused with quartz and feldspar; (v) metals such as aluminium, brass, steel, copper and zinc; (vi) ceramic fragments including china, brick, plaster and mortar; and (vii) components of organic origin such as apatite (bone fragments), charcoal, plant fibers and polymers [87]. The major crystalline phases identified in milled ash are quartz (SiO_2) and calcite ($CaCO_3$), ghelenite ($Ca_2Al_2SiO_7$) and hematite (Fe_2O_3). Analyzis of milled ash sintered at the maximum density temperature has revealed that diopside ($CaMgSi_2O_6$) is the principal crystalline phase, and that significant amounts of clinoenstatite ($MgSiO_3$) and wollastonite ($CaSiO_3$) also are present, as well as minor amounts of albite ($NaAlSi_3O_8$) and hematite [88].

6.3.4 Heavy Metal Speciation

6.3.4.1 Definition

The bioavailability and toxicity of heavy metals and their migration and transformation in the environment are not only related to how much is present in the environment, but also (and to a greater extent) on their chemical and physical forms. Therefore, the heavy metal speciation has attracted great attention.

Heavy metal speciation which means the actual form (ion or molecule) of a certain heavy metal existing in the environment refers to four aspects including valence state, chemical form, combined form and structure. Heavy metals exhibit different toxicities and environmental behavior due to speciation differences. Speciation analyzis is the determination and characterization of physical and chemical forms of heavy metals in environmental samples. There are three main methods for heavy metal speciation: non-destructive instrument analyzis, sequential selective extraction (SSE), and chemical equilibrium analyzis. Only SSE will be discussed in detail.

Non-destructive instrument analyzis provides information about the distances, coordination number and species of atoms surrounding the selected element, as well as the formal oxidation state and coordination chemistry. These methods include XRD, X-ray photoelectron spectroscopy (XPS) and X-ray absorption fine structure spectroscopy (XAFS), which includes X-ray absorption near-edge structure (XANES) spectroscopy and extended X-ray absorption fine structure (EXAFS) spectroscopy. The sequential selective extraction procedure (SSE) uses selected chemical reagents to extract samples under certain environmental conditions sequentially. Chemical equilibrium model analyzis, such as the models MINTEQ and NICA-Donnan, use thermodynamic analyzis to provide information about the geophysical behavior of heavy metals under equilibrium conditions.

The SSE procedure has been widely employed to investigate heavy metals speciation in soils, sediments, and other solid wastes. Each specific extraction in the sequence is related to a specific chemical entity (e.g., carbonates, iron/manganese (hydro-oxides) or form (e.g., soluble, exchangeable) associated with the metal. SSE has two main advantages in evaluating metal status in soils and sediments. One is the ability to relate a specific extraction in the sequence to a specific chemical entity (e.g., carbonates, iron oxides) or form (e.g., soluble, exchangeable, sorbed). The other is the ability to use the solid-phase speciation to make inferences about the behavior of the metal in the environment (e.g., mobility and bioavailability), or its response to changes in environmental conditions (e.g., solubility changes due to pH or Eh changes) [89]. However, the SSE procedure has received rigorous criticism because it seldom achieves any of the following: (a) complete dissolution of a specific target component, (b) selective extraction or dissolution of the target component, or (c) full recovery of dissolved metals without their resorption or precipitation. Therefore, this method can only provide operationally defined phases/components, and cannot demonstrate the actual heavy metal species or structures.

The specific chemical entity determined in the SSE procedure can be divided into various components (soluble, exchangeable, carbonate, oxyhydroxides, organic and residual).

(1) The soluble component. Determining the soluble component is the first step in the SSE procedure. However, because samples often have a relatively low content (sometimes below the detection limit) of soluble metals, this step is often omitted, and the soluble component is included with the exchangeable component (determined in the next step).
(2) The exchangeable component. The exchangeable component is attributed to metals that are associated with solid particles in the sample by electrostatic exchange mechanisms. The extraction is accomplished by shaking the sample in the presence of a concentrated solution of an inert electrolyte. The electrolytes commonly used are nitrate or chloride salts of ammonium, calcium, potassium, or magnesium.
(3) The carbonate component. The dissolution of carbonates is usually the first acid-extraction step in the SSE procedure. Acid dissolution in 1 M ammonium acetate (NH_4Ac) at pH 5, or by ligand exchange in 0.1 M EDTA at pH 6.5, are the two often-used techniques to characterize metal carbonates.
(4) Iron and manganese oxyhydroxides. Oxyhydroxides are separated into two or three main components using the SSE procedure: manganese oxyhydroxides, amorphous iron oxyhydroxides, and well-crystallized iron oxyhydroxides. Reducing agents, ligands, and acids are used to dissolve oxyhydroxides.

(5) The organic component. Metals associated with organic matter are extracted by destroying the organic material with hydrogen peroxide (H_2O_2) or sodium perchlorate ($NaClO_4$), or by dispersing the organic matter with complexing agents (e.g., DTPA, $Na_4P_2O_7$, EDTA).

(6) The residual component. The residual component is usually attributed to metals within the mineral lattice of primary and secondary minerals that are resistant to all previous extraction steps. Extraction of this component is performed using strong mineral acids as the last step in the SSE procedure.

6.3.4.2 Analytical Methods

In 1979, Tessier et al. developed a five-step SSE procedure designed to characterize metals by their pollution potential [90]. This protocol was widely used as developed or with adaptations. The European Commission's Bureau Community of Reference (BCR) [91], Kersten et al. [92] and Krishnamurti et al. [93] modified the SSE protocol developed by Tessier et al.

Leaching procedures and reagents of the Tessier five-step SSE are as follows.

(1) The exchangeable component (F1): The sample (2 g) is extracted at room temperature for 1 h using 8 mL of either magnesium chloride solution (1 M MgC12, pH 7.0) or sodium acetate solution (1 M NaOAc, pH 8.2) and continuous agitation (3000 g). The supernatant liquid is determined by ICP-OES. The residue is washed by adding 20 mL of deionized water and then shaking for 10 min. The washed residue is used in the next step.

(2) The carbonate component (F2): The residue from step (1) is leached at room temperature for 8 h using 8 mL of 1 M NaAc adjusted to pH 5.0 with centrifugation. The determination is performed as described for the previous step.

(3) Iron and manganese oxyhydroxides (F3): The residue from step (2) is extracted using either 20 mL of 0.3 M $Na_2S_2O_4$ + 0.175 M Na-citrate + 0.025 M H-citrate, or 0.04 M NH_2OH–HCl in 25% (v/v) HAc. The extraction is performed at 96 ± 3 °C with occasional agitation and the time needed for complete dissolution of the free iron oxides must be evaluated. The determination is performed as described for the previous steps.

(4) The organic component (F4): To the residue from step (3) is added 3 mL of 0.02 M HNO_3 and 5 mL of 30% H_2O_2 adjusted to pH 2 with HNO_3, and the mixture is heated to 85 ± 2 °C for 2 h with occasional agitation. A second 3-mL aliquot of 30% H_2O_2 (adjusted to pH 2 with HNO_3) is then added and the sample is heated again to 85 ± 2 °C for 3 h with intermittent agitation. After cooling, 5 mL of 3.2 M NH_4OAc in 20% (v/v) HNO_3 is added and the sample is diluted to 20 mL and agitated continuously for 30 min. The addition of NH_4OAc is designed to prevent adsorption of extracted metals onto the oxidized sediment. The determination is performed as described for the previous steps.

(5) The residual component (F5): The residue from step (4) is digested using a HF-$HClO_4$ mixture

The BCR modification to the Tessier SSE procedure involves four steps, as follows.

(1) Exchangeable and acid soluble fraction (F1): Add 40 mL of CH3COOH (0.11 M) to 1 g of sample in a 100 mL centrifuge tube and extract by shaking for 16 h at ambient temperature. No delay should occur between the addition of the extractant solution and beginning of the shaking. The supernatant liquid is analyzed by ICP-OES. Wash the residue by adding 20 mL of deionized water and shaking for 15 min. The washed residue is used in step (2).
(2) Reducible fraction (F2): Add 40 mL of $NH_2OH{\cdot}HCl$ (0.5 M) to the residue from step (1) in the centrifuge tube and extract by shaking for 16 h at ambient temperature. The determination is performed as described for step (1).
(3) Oxidizable fraction (F3): To the residue from Step 2 (in the centrifuge tube) add 10 mL of H_2O_2 (8.8 M) carefully in small aliquots to avoid losses due to violent reaction. Cover the tube with a watch glass and digest the contents at room temperature for 1 h with occasional manual shaking. Continue the digestion by heating the covered tube for 1 h at 85 °C in a water bath, then remove the cover and reduce the volume to a few mL. Add another aliquot (10 ml) of H_2O_2 (8.8 M). Cover the tube with the watch glass and heat the tube again (85 °C for 1 h). Then, remove the watch glass and reduce the volume to a few mL. Add 50 mL of CH_3COONH_4 (1 M, pH 2) to the cool residue. The determination is performed as described for the preceding steps.
(4) Residual fraction (F4): Digest the solid residue from step (3) in aqua regia.

6.3.4.3 Heavy Metal Speciation of the Byproducts from MSW Treatment

SSE procedures have been widely employed to investigate heavy metals speciation in soils and sediments. Given that the characterizations between fly ash (FA), bottom ash (BA), soil and sediments are similar, SSE procedures (especially the Tessier five-step extraction and BCR modification) have been used in determining the heavy metal speciation of FA and BA.

Table 6.16 contains data about heavy metal speciation of the byproducts generated from MSW treatment.

Table 6.16 Heavy metal speciation of the byproducts produced from MSW treatment

City	Object	Method	Speciation	Mass distribution%					
				Cu	Zn	Cd	Ni	Pb	As
China [94]	FA	Four-step SSE	Sol	12.2	10.7	8.9	12	13.90	–
			Red	25.4	33.5	7.8	18.5	39.70	–
			Oxi	10.2	8	6.2	13.5	8.30	–
			Res	60.9	47.8	83.1	56.1	39.00	–
Shanghai city in China [95]	FA	BCR	Sol	10.77	8.79	15.35	0.96	12.75	27.83
			Red	61.61	42.96	74.89	10.66	11.10	39.42
			Oxi	17.04	28.37	6.80	22.73	27.79	0
			Res	10.58	19.87	2.98	65.66	48.36	32.75
Wuhan city in China [96]	FA	BCR	Sol	3.21	21.53	42.82	–	0.00	–
			Red	31.74	33.87	30.88	–	29.10	–
			Oxi	23.65	9.13	24.74	–	17.22	–
			Res	41.60	35.17	1.88	–	51.76	–
Japan [97]	FA	Tessier	Sol	0.00	0.28	0.62	–	7.12	–
			Exc	0.00	2.23	0.93	–	13.66	–
			Car	2.24	3.36	14.64	–	1.54	–
			Fe–Mn	36.97	46.31	39.06	–	18.40	–
			Org	31.67	1.96	1.53	–	5.87	–
			Res	29.02	45.89	43.26	–	53.16	–
Sewden [98]	FA	Four-step SSE	Exc	58.37	30.37	90.21	–	0.00	19.92
			Sol	19.10	26.51	1.32	–	58.37	2.99
			Red	5.01	21.13	2.42	–	19.10	13.75
			Oxi	1.87	3.84	0	–	5.01	8.37
			Res	15.57	16.71	5.06	–	15.57	54.01
Wuhan city in China [96]	BA	BCR	Sol	13.05	39.53	16.49	–	2.60	–
			Red	28.24	0.00	60.82	–	27.33	–
			Oxi	21.23	24.74	2.60	–	16.94	–
			Res	37.03	34.83	18.34	–	53.52	–
Hangzhou city in China [99]	BA	Tessier	Exc	0	0	0	0	0	2.26
			Car	24.74	23.42	16.37	7.60	10.63	2.69
			Fe–Mn	2.24	24.65	11.92	14.04	12.78	3.50
			Org	45.74	5.73	1.72	6.65	1.78	1.78
			Res	26.88	45.80	69.13	71.71	75.26	90.23

6.3.5 pH

6.3.5.1 Definition

pH refers to the hydrogen ion activity when a solid and water (or solution) reach equilibrium, generally expressed as $-10\log[H^+]$. pH is an important indicator for MSW treatment and beneficial utilization. The $[H^+]$ is ionized hydrogen found in aqueous solutions of all acids.

6.3.5.2 Analytical Methods

The electrode method (including composite electrodes) [100] is the commonly used method for determining pH. Both fresh samples and dried samples can be tested, but the same procedure must be used on samples that are to be compared. A certain amount of MSW is weighed and then enough distilled water or 0.1 M KCl or 0.01 M $CaCl_2$ is added to form a liquid-to-solid ratio of 5:1 or 8:1. After shaking, the suspension should be allowed to stand for approximately 1 h. Then the pH is measured using a calibrated pH meter. During the measurement, the container and sample should be gently agitated to promote a homogenous solution and electro-chemical equilibrium.

Temperature affects the accuracy of pH measurements by the electrode method. The output of the electrode changes with temperature, but this variation can be controlled with the instrument's temperature compensation or by calibrating the electrode and the instrument at the sample temperature. The pH of the sample itself also is affected by temperature. This error cannot be controlled; thus, when reporting a pH value, both the pH and the temperature at the time of analyzis must be indicated.

In general, a glass electrode is not easily affected by the color, turbidity, colloidal matter, oxidants or reducing substances, or salt content in the solution. However, coating of the electrode by oily substances and the presence of particulate matter may affect the electrode response.

pH indicator paper also can be used to determine the pH of samples, either as a supplement to the electrode method, or when the electrode method cannot be used. However, pH indicator paper cannot determine pH accurately; it can merely determine whether the sample is corrosive. Moreover, pH indicator paper is not suitable for the determining the pH of substances that can mask or affect the color change of the paper.

6.3.5.3 The pH of Municipal Solid Waste

The pH of raw MSW is generally near-neutral, fluctuating between 5 and 8 [101–105]. The indicative pH of MSW can be affected by many factors, such as the location of the sampling point in the waste collection and transportation process, the differences in the sources of MSW generation, and operational practices used at MSW collection and transportation facilities.

6.3.5.4 The pH of by-Products of Municipal Solid Waste Treatment

Composting is an important treatment technique for managing MSW, and pH is an important indicator for evaluating compost maturity. The composting rate reaches its maximum value when the pH is 7.5–8.5 [106]. The pH of fresh waste, and of compost in the initial stage of fermentation, ranges from weakly acidic to neutral, generally 5–7. During the composting process, pH gradually increases, such that at the end of composting pH can reach approximately 8.0–9.0 [107].

Anaerobic digestion also is used to treat MSW. The pH of MSW during anaerobic digestion shows a trend of decline before increasing toward the end of the process [108]. With the progress of hydrolysis and acidification in the initial stage of anaerobic digestion, the pH value decreases rapidly and reaches its minimum value within 2–6 days; thereafter, with the progress of methanogenesis, the pH value gradually increases.

Residues from MSW incineration are strongly alkaline. However, the pH of these residues is usually not determined in isolation; rather, it is determined jointly with leaching toxicity. The preparation of leachates for pH measurement is the same as that for the toxicity measurement.

6.3.6 Leaching Toxicity

6.3.6.1 Definition

During landfilling or during temporary storage of MSW, infiltrating rainwater, as well as liquids inherent in the waste, can dissolve some of the soluble harmful substances and escape from the waste as leachate. Uncontrolled, the leachate may enter the surrounding environment (soil and water resources), and when the concentration of the dissolved substances exceeds a certain threshold, the leachate will cause harm to the environment. When the amount of harmful sub-stances in the leachate exceeds the threshold set by relevant regulations, the waste is said to have "leaching toxicity". The identification of leaching toxicity is one of the criteria for judging hazardous wastes, and is also an important technical procedure for developing protocols for solid waste management, treatment and disposal, as well as for resource utilization.

The main toxicity indicators for MSW and its by-products include pH, organic mercury, total mercury, lead, cadmium, total chromium, hexavalent chromium, copper, zinc, antimony, bismuth, nickel, arsenic, inorganic fluoride, and cyanide (in terms of CN), as well as other inorganic pollutants and some organic pollutants. Aside from these chemical metrics, leachate can serve as a biomonitoring sample for assessing the bioavailability of some components.

The leaching toxicity of wastes under specific environmental conditions is difficult to obtain by on-site sampling. Usually, specific leaching experiments are used to assess the leaching risk of wastes under certain conditions. Leaching tests attempt to replicate specific natural environmental conditions. The waste is placed in contact with specific leaching agents, and the subsequent release of pollutants is evaluated. The basic objective of leaching tests is to provide a risk assessment.

When landfill treatment involves relatively stable materials (such as bottom ash and fly ash), the effect of infiltrating rain is similar to that of pure water; thus, in leaching tests natural rainfall is often simulated using distilled water. In the case of landfilling raw MSW or mixed landfilling of both inert waste and raw MSW, the leachate typically contains a large amount of organic acids (inevitable products of organic degradation, mostly acetic acid). Thus, in leaching tests of mixed MSW, the leaching liquid is often modeled using organic acids. However, in some tests a mixture of sulfuric acid and nitric acid may be used, on the assumption that when waste is stored, it is exposed to acid rain.

The leaching of heavy metals in MSW can be affected by a number of factors: the speciation and solubility of heavy metals, the physical properties of the solid waste (such as particle size, specific surface area, mineral composition) and the type and concentration of leaching agents. Other factors include the liquid-to-solid ratio and renewal rate, properties of leachate (pH, conductivity, ionic strength, temperature), contact between solids and liquid (with or without mixing, presence or absence of leaching agent renewal), leaching time, complexation, adsorption, and oxidation/reduction in the leaching process. Therefore, a large number of leaching test methods [109–115] have been developed.

6.3.6.2 Analytical Methods

This section lists some commonly used leaching test methods (Tables 6.17 and 6.18). These leaching test methods identify waste leaching toxicity for different protection targets, and include batch extraction experiments and continuous extraction experiments. Batch extraction experiments, also called static extraction experiments, mix a quantitative leaching agent and waste within a certain period of time. The mixture is shaken to promote dissolution of contaminants, and then the liquid and solids are separated for analyzis.

Batch extraction methods include the Toxicity Characteristic Leaching Procedure (TCLP) (EPA method 1313), followed by analyzing the pH and concentrations of pollutants in leachate using an inductively coupled plasma optical emission spectrometer (ICP-OES) or atomic absorption spectrophotometer (AAS) and other equipment. During batch extraction, the system is assumed to be in a stable state (equilibrium state) throughout the experiment, and there is no renewal of the extractant.

In contrast, continuous extraction experiments are conducted by continuously or intermittently renewing the leaching agent to promote pollutant leaching. Leachate samples are collected at different times for analyzis to provide information on the dynamics of pollutant migration (change over time, liquid-solid ratio). The representative methods are EPA method 1314 and 1315.

The TCLP procedure is designed to determine the mobility of both organic and inorganic analytes present in liquid, solid and multiphasic wastes. The particle size of the sample is reduced to less than 9.5 mm, if necessary. Then, no less than 100 g of waste is extracted with an amount of extraction fluid (CH_3COOH solution with a pH of 4.93 ± 0.05 or 2.88 ± 0.05) equal to 20 times the mass of the solid phase.

Table 6.17 Leaching methods for toxicity testing of solid waste

Method	JLT14	JLT13	ALT	TVA	AF NOR X31-210	DIN38414 S4	EPA1313	EPA1316
Nation	Japan		Netherlands	Switzerland	France	Germany	USA	
Particle size	<2 mm	<5 mm	<125 μm	According to the final disposal of waste	<4 mm	<4 mm	<5 mm	<5 mm
Mass	>50 g	>50 g	16 g	100–200 g	150 g	100 g	>20 g	20–400 g
Leaching agent	Distilled water/HCl	Distilled water/HCl	Deionized water/HNO_3	CO_2 saturated distilled water	Deionized water	Deionized water	KOH/HNO_3/ Deionized water	Deionized water
pH	5.8–6.3	5.8–6.3	7.4	5.6	–	–	2–13	–
L/S (L/kg)	10:1	10:1	100:1	10:1	10:1	10:1	10:1	0.5–10:1
Temperature	Ambient temperature	Ambient temperature	–	–	Hot water bath with constant temperature	–	20 ± 2 °C	20 ± 2 °C
container	–	–	Bottle	Bottle	2L Container	2L Container	Jars	Jars
Stirring	Horizontal oscillation	Horizontal oscillation	Magnetically stirring	100 mlCO_2/min bubble	Stirring or oscillation	Horizontal oscillation	End-over-end 28 ± 2 rpm	End-over-end 28 ± 2 rpm
Time	6 h	6 h	3 h × 2 times	35 h	16 h	24 h	24–72 h	24–72 h
steps	1	1	1–2	1–2	1–3	1–2	1	1
Separation	1000 rpm centrifuge for 45 min or 0.45 μm filter	1 μm filter	0.45 μm filter	0.45 μm filter	0.45 μm filter or centigrade	0.45 μm filter	0.45 μm filter or centigrade	0.45 μm filter or centigrade

Method	EP	TCLP	HJ 557–2010	HJ/T 299–2007	HJ/T 300–2007
Nation	USA/Canada		China		

(continued)

Table 6.17 (continued)

Method	EP	TCLP	HJ 557–2010	HJ/T 299–2007	HJ/T 300–2007
Nation	USA/Canada		China		
Particle size	<9.5 mm	<9.5 mm	<3 mm	<9.5 mm	<9.5 mm
Mass	100 g	100 g	100 g	150–200 g	75–100 g
Leaching agent	Deionized water/0.5 N acetic acid	Acetic acid buffer solution	Distilled water	H_2SO_4 and HNO_3/ deionized water	Acetic acid buffer solution
pH	5	2.88/4.93	–	3.20	2.64/4.93
L/S (L/kg)	20:1	20:1	10:1	10:1	20:1
Temperature	20–40 °C	22.3 ± 3 °C		23 ± 2 °C	23 ± 2 °C
container	–	Plastic bottle	2LJars	2LJars	2LJars
Stirring	–	Up-to-end 30 ± 2 rpm	Horizontal oscillation	Up-to-end 30 ± 2 rpm	Up-to-end 30 ± 2 rpm
Time	24 h	18 h	8 h + 16 h standing	18 ± 2 h	18 ± 2 h
steps	1	1	1	1	1
Separation	0.45 μm filter	0.6–0.8 μm filter	0.45 μm filter or centigrade	0.6–0.8 μm filter	0.45 μm filter

Table 6.18 Leaching methods of solid waste

Method	EPA1314	EPA1315 (monolithic samples)	EPA 1315 (compacted granular samples)	Column leaching test
Country	USA	USA	USA	Netherlands
Particle size	<2.5 mm	–	<2 cm	<3 mm
Mass	–	>50 g	16 g	40 g
Leaching agent	Reagent water/$CaCl_2$	Reagent water	Reagent water	Deionized water
pH	–	–	–	4
L/S (L/kg)	0–10	L/A = 9±1 ml/cm^2	L/A = 9±1 ml/cm^2	20:1
Temperature	20 ± 2 °C	20 ± 2 °C	20 ± 2 °C	–
Container	Column	Support resistant to acid and alkane	Support resistant to acid and alkane	1 L plastic bottle
Stirring	–	–	–	–
Time	15 d	63 d	63 d	24 h × 5 times
Step	–	9	9	1–5
Separation	0.45 μm filter	0.45 μm filter	0.45 μm filter	0.45 μm filter

The mixture is rotated end-over-end at 30 ± 2 rpm in a container for 18 ± 2 h. Following extraction, the liquid for analyzis is separated from the solid phase by filtration through a 0.6–0.8 μm glass fiber filter.

The TCLP (EPA method 1313) is intended to provide a pH of the leachate using a parallel batch extraction procedure as part of an environmental leaching assessment for use in the evaluation of disposal, beneficial use, treatment effectiveness, and site remediation options. Dilute solutions of nitric acid and potassium hydroxide in reagent water is used at a liquid-to-solid ratio of 10:1. A total of 12 bottles (i.e., nine test positions and three method blanks) are tumbled in an end-over-end fashion for a specified contact time, which depends on the particle size of the sample.

In EPA method 1314, the eluent (the default eluent is reagent water) is introduced into a column of moderately packed granular material in an up-flow mode, with eluate collection per-formed as a function of the cumulative liquid-to-solid ratio (L/S). The flow rate is maintained be-tween 0.5 and 1.0 L/S per day to increase the likelihood of local equilibrium between the solid and liquid phases, due to residence times longer than one day. Eluate volumes are chemically analyzed for a combination of inorganic and non-volatile organic analytes depending on the constituents of potential concern.

EPA method 1315 comprises leaching of continuously water-saturated monolithic or compacted granular material in an eluent-filled tank with periodic renewal of the leaching solution. Samples and reagent water remain in contact at a specified liquid-surface-area ratio (L/A). The leaching solution is exchanged with fresh reagent water at nine pre-determined intervals. The mass of sample, eluate pH and specific conductance should be recorded at the end of each time interval; at these intervals analytical samples also are collected and preserved. Eluate concentrations are plotted as a function of time, as a mean interval flux, and as cumulative release as a function of time. These data are used to estimate mass transfer parameters for each constituent of potential concern.

6.3.6.3 Leaching Toxicity of Municipal Solid Waste

Because MSW generally has a low content of heavy metals, batch experiments are generally not necessary for determining leaching toxicity. In the landfilling process, both the leaching environment formed by infiltration of rainfall and surface water and the liquid-soaked environment produced by self-degradation are observed for MSW. The dissolution of heavy metals over a long period can be simulated by long-term dynamic leaching experiments of waste using simulated rainwater of different pH values.

When a simulated rainwater with initial pH of 4.3, 6.5, and 8.8 was used for long-term soaking of municipal solid waste [116], the pH of the leachate gradually increased with the passage of time. However, the influence of the initial pH was not obvious, and the final pH of all samples was 7.5–8.0. The concentration of heavy metals in the leachate was very low [116], and far lower than the limits for landfills.

6.3.6.4 Leaching Toxicity of Residues from Municipal Solid Waste Incineration

The residual by-products of MSW incineration (such as fly ash and bottom ash) are enriched with heavy metals from the MSW, which also change during the high temperature thermal treatment process. Therefore, it is necessary to analyze the leaching toxicity of incineration residues in order to reuse or dispose of them rationally. Refer to [116–120] for some of the experimental results from studies of MSW incineration fly ash and bottom ash using specific leaching experiments. The results of the leaching experiments vary sharply due to influences by the composition of MSW, parameters in the incineration process, differences in the air pollution control systems, and variations in the leaching tests themselves.

6.4 Thermal Characterization of Municipal Solid Waste

This section is designed to enable one to characterize MSW samples using well-known thermal techniques. The results will provide a complete data set that will enable a research engineer to understand their MSW samples and predict how the material will perform in various thermal processing systems such as combustion, gasification and pyrolysis.

6.4.1 Proximate and Ultimate Analyzis

Proximate analyzis results in the determination of moisture, volatile matter, ash and fixed carbon of a sample. Ultimate analyzis yields the weight percent of carbon, hydrogen, oxygen, sulfur and ash.

6.4.1.1 Proximate Analyzis

Standards

The following standards should be followed to determine the proximate analyzis of MSW samples:

- ASTM D3173, ISO 11722, AS1038.3 determines the sample moisture which directly impacts the heating value and combustion efficiency of MSW as a fuel [121–123].
- ASTM D4239, ISO 351, AS1038.6 determines the sulfur content of MSW sample, important for the evaluation of sulfur emissions from MSW thermal treatment [124–126].
- ASTM D3175, ISO 562 determines the volatile matter evolved after the moisture loss. This provides a good estimate of the total gas that will be produced during conversion [127, 128].
- ASTM D5142, ISO 17246, AS1038.3 determines the fixed carbon which either results in a solid char residual or in a final product that can be used for numerous applications [129–131].
- ASTM D5373-08 determines the total carbon that includes the carbon in volatile matter [132].
- ASTM D3174 determines the ash content and is important for slag formation [133].

All the procedures described below require a laboratory balance with a sensitivity of minimum 0.1 mg. The preparation of the MSW samples for each test is performed in accordance to Sect. 6.1 of this Chapter.

Moisture Content

The moisture content (MC) is determined by weighing the waste before and after drying. Drying often takes place at 77 or 105 °C until constant weight. At 105 °C, in addition to water, also volatile fatty acids, specific organic components and mercury may volatilize. The moisture content usually is expressed as a percentage of the wet weight. The wet weight of a waste sample is made up of the dry weight plus the moisture content. There are typically two ways to determine the moisture content, the "green basis" and "oven-dry basis". The green basis is considered a wet method of MC measurement where the percent moisture is expressed as a percentage of the total weight of the wet material, including both the dry material and any water or moisture in the sample. This method is most commonly used to determine the MC of woody biomass but is applicable to fresh MSW. The equation to determine the wet-basis MC is the following:

$$MC_{wet-basis} = \left(\frac{wt.\,of\,water}{wt.of\,water + wt.of\,dry\,MSW}\right) \times 100\% \tag{6.12}$$

In the oven-dry basis the percent moisture in the MSW is expressed as a percentage of the dry weight of material. This is a more common method and normally used as the standard. The equation to determine the dry basis MC is the following:

$$MC_{oven-dry} = \left(\frac{wt.\,of\,water}{wt.\,of\,dry\,MSW}\right) \times 100\% \tag{6.13}$$

Apparatus

The moisture content of MSW samples is determined following ASTM D3173-11 method [121]. The experiment is performed in a uniformly heated furnace with temperature regulation capabilities between 104 and 110 °C. Preheated air is constantly supplied to the oven at a rate of two to four oven volume exchanges per minute. The oven must have openings for good air circulation and ventilation. The air should be preheated and dried to 1.9 mg/L (8% relative humidity at 25 °C) moisture content before entering the furnace, by passing it through desiccants such as Drierite® (www.drierite.com) or using other drying methods.

Procedure

Grind or shred MSW samples to a 250 μm particle size and place in crucible (either ceramic or preferably platinum) with covers, 22 mm (7⁄8 in.) in depth and 44 mm (13⁄4 in.) in diameter. Alternatively similar shaped fused silica crucibles can be used. The empty crucibles must be conditioned, prior to any tests with MSW, in the oven under the same conditions that the MSW sample will undergo. Place the cover on top of the crucibles after removal from the oven and weigh after cooling in a desiccator or over a desiccant for 15–30 min.

Now that the crucible is weighed, using a spatula place approximately 1 g of MSW sample in the crucible and re-weigh it. Move the crucible and ~1 g sample to the oven without the cover on for at least 24 h at 105 °C. Remove the crucible and sample from the oven and immediately put the cover on it. Allow for cooling in a desiccator or over desiccant and weigh after 15–30 min. Calculate the percentage of the moisture in the samples as the percentage of mass loss over the initial mass as follows:

$$Moisture\% = \left(\frac{wt.\,of\ sample\ used - wt.\,of\ sample\ after\ heating}{wt.\,of\ sample\ used}\right) \times 100\% \tag{6.14}$$

This is the standard moisture content reported in proximate analyzes of carbonaceous materials.

Volatile Matter

Apparatus

The volatile matter (VM) method described herein determines the mass loss occurring under controlled conditions. The furnace used should be a vertical electrically heated tube furnace regulated to maintain 950 ± 20 °C in the crucible. Temperature measurements must be using a suitable thermocouple that is in close proximity to the MSW sample.

Procedure

Crucibles used in this method shall be platinum, to avoid reactions, with a fitted cover sized 10–20 mL. A similar size nickel-chrome alloy may be used after 4 h of heat pre-treatment at 500 °C. Again, weigh approximately 1 g of MSW sample ground to 250 μm in a previously weighed crucible; same as the previous step. Close the crucible cover and insert into a preheated oven at 950 ± 20 °C. The oven should have similar flow provisions as the moisture test. After 2 or 3 min quickly inspect if the crucible cover is still in the proper position (i.e. it did not come loose or fall off) and adjust if necessary. In some cases the liberation of highly volatile samples causes the cover to move or material to spill out of the crucible. In such cases the analyzis should be considered invalid, the test should be aborted, and the procedure must be re-started from the beginning.

After 7 min remove the covered crucible from the furnace and allow it to cool down, preferably in a desiccator. After cooling weigh the crucible. The percentage loss of weight minus the percentage moisture equals the volatile matter. When testing material that may have certain metals or contaminants it is important to minimize sparking or flaming. If it is suspected that the sample contains sparking materials, they must be subjected to a preliminary gradual heating of the samples for 6 min at temperature of 600 ± 50 °C after which the sample is heated for additional 6 min at 950 ± 20 °C. If sparking is observed, the determination shall be rejected and the test shall be repeated with new samples until no sparking occurs during the preliminary stage or the heating stage to 950 °C.

Table 6.19 TGA steps for proximate analyzis [133]

Step	1	2	3	4	5	6	7	8	9	10	11	12	13
T_{start} (K)	303	343	363	378	378	418	418	773	773	873	873	873	973
T_{end}(K)	343	363	378	378	418	418	773	773	873	873	873	973	973
SR[a] (K/min)	30	15	2	0	10	0	10	0	20	0	0	20	0
Time (s)	80	80	450	1800	240	600	2130	3600	300	600	2400	300	600
Atmosphere	N_2	N_2	N_2	N_2	N_2	N_2	N_2	N_2	N_2	N_2	N_2	N_2	N_2

[a]Scan Rate

Calculate the weight loss percent as follows:

$$weight\,loss\% = \left(\frac{wt.\,of\,sample\,used - wt.\,of\,sample\,after\,heating}{wt.\,of\,sample\,used}\right) \times 100\% \tag{6.15}$$

Calculate the volatile matter percent in the analyzis samples as follows:

$$VM\% = weight\,loss\% - moisture\% \tag{6.16}$$

Another technique to obtain the volatile matter is to use a thermogravimetric analyzer (TGA). This unit provides a dynamic weight change as a function of temperature, heating rate or time.

Using the TGA is not the best method to determine the proximate analyzis of a material, but if carried out carefully in 13 sequential steps as detailed in Table 6.19, will give results comparable to those obtained by standard methods. The various inflexion temperatures, hold times and temperature ramp (scan) rates are programmed into the machine, so that the total expired time per test is 3.7 h.

Steps 1–4 determine the moisture content, steps 5–10 yield the volatile matter, and steps 11–13 provide the ash content. The target samples must be weighed (or tared) while in the TGA crucible and evenly distributed to ensure no heat transport effects. The sweep gas should be a pure inert for steps 1–10, such as N_2, at adequate flow rates to efficiently remove the vapors; typically the flow rate ranges between 10 and 100 ml/min. For the ash determination (steps 11–13) air must be used at the same flow rate that was used for the inert. The moisture content can be obtained by subtracting the initial mass from the steady mass obtained during step 4. The volatile matter will then be the difference between the mass immediately at the start of step 5 (i.e. at the end of step 4) and the mass at the end of step 10 (i.e. at 600 s). The ash is then determined from the steady mass at step 11 to the residual inorganic matter remaining after step 13. Finally the fixed carbon can be determined via the formula FC = 100 − MC − VM − Ash, where VM is calculated on a dry weight basis and MC is calculated on a wet basis.

Ash

Apparatus

Ash is the remaining residue after burning the MSW sample under controlled conditions. The determination of ash of a sample is performed in an electrically heated furnace with adequate air circulation controlled at temperatures of 700–750 °C, similar to the MC and VM furnace configurations. The air flow through the furnace should be maintained at two to four changes per minute. The configuration of the furnace must allow for ventilation of the exhaust gases and uniform air distribution inside the furnace. The air flow rate should be measured using a wet test meter or equivalent equipment to provide accurate readings. The sample should be put into porcelain or platinum crucibles 22 mm (7/8 in.) in depth, and 44 mm (1¾ in.) in diameter with covers.

Procedure

Perform a temperature calibration test prior to the sample test to ensure there is minimal overshoot during the heating phase. Set the furnace temperature to 750 ± 10 °C. Insert the crucible with approximately 1.0 g of sand in the furnace while the temperature at the center of the sample is measured using a thermocouple. Wait until the sand sample temperature is in equilibrium with the furnace temperature. This will provide the time necessary to estimate the heat transfer from the furnace to the center of the MSW test sample and will be used as a guideline for keeping time during the test.

Once the calibration is complete, cool the oven back to room temperature. Place approximately 1.0 g of MSW sample, ground to a particle size of 250 mm, into the crucible then into the furnace with the cover. Gradually heat the furnace to a target temperature of 450–500 °C during a 1.0 h period. Regulate the furnace temperature so that the sample reaches 700–750 °C by the end of second hour. If the sample was previously subjected to the volatile matter test, discussed above, heat the sample up to 950 °C. Keep the sample at the final temperature for additional 2 h. Remove the sample from the furnace and allow it to cool in a desiccator before weighing. The ash percent in the sample is expressed as:

$$Ash\% = \left(\frac{wt.\,of\,crucible, cover, ash - wt.\,of\,empty\,crucible\,\&\,cover}{wt.\,of\,sample\,used}\right) \times 100\% \tag{6.17}$$

Fixed Carbon

The fixed carbon is a calculated value and represents the remaining fraction after subtracting the percentages of moisture, volatile matter and ash from the original mass of MSW sample. All percentages shall be on the same moisture reference base.

$$Fixed\ Carbon\% = 100\% - (Moisture\% + Ash\% + Volatile\ Matter\ \%) \tag{6.18}$$

6.4.1.2 Ultimate Analyzis

The ultimate analyzis is used to determine the elemental composition of MSW sample. Analyzers, such as elemental, ultimate or carbon-sulfur analyzer perform the analyzis by combusting the sample before measuring the percent of carbon, hydrogen, nitrogen, sulfur or ash. Oxygen is usually measured by difference. The standard followed to determine the ultimate analyzis of MSW sample is ASTM D3176. The procedure to determine carbon, hydrogen, nitrogen and sulfur is referred to as CHNS and measures the carbon percent as total carbon. The combustion gas includes carbon as CO_2 from solid carbon, organic hydrocarbons and in some cases from mineral origin. Total carbon includes inorganic and organic carbon released as volatile matter during combustion and represents the weight percent of carbon in the entire MSW sample.

The total organic carbon (TOC) is determined as the difference between the total carbon and inorganic carbon measured in a coulometer.

Total hydrogen includes the weight percentage of hydrogen released during combustion as water vapor or hydrocarbons. These values need to be corrected for the hydrogen derived from moisture, by using the following equation (ASTM D3176-09):

$$Hydrogen_{fuel-based} = Hydrogen_{measured\ as\ water} - (0.1119 \times MC) \tag{6.19}$$

where MC is the moisture content as determined above.

The total oxygen cannot be determined analytically since oxygen is introduced in the analyzer combustion chamber for the complete combustion of the MSW sample. The total oxygen is determined by difference, following the equation below (ASTM D3176-09):

$$\begin{aligned} Total\ oxygen\ weight\ percent &= 100 \\ &-(total\ carbon\ weight\ \% + total\ hydrogen\ weight\ \% \\ &+ total\ nitrogen\ weight\ \% + total\ sulfur\ weight\ \% \\ &+ total\ ash\ weight\%) \end{aligned} \tag{6.20}$$

To correct for fuel based oxygen and remove the oxygen derived from moisture in the sample, the equation below is applied:

$$\begin{aligned} Total\ oxygen\ (fuel-based, weight\ \%) &= Total\ oxygen\ (analyzer\ measured, weight\ \%) \\ &-((0.8881 * moisture\ (analyzer\ measured,\ as\ determined, weight\ \%) \end{aligned} \tag{6.21}$$

Total nitrogen is expressed as the measured weight percent of nitrogen in the MSW sample (ASTM method D5373-08). A simple method that uses common acids and easily found equipment in the laboratory is the Kjeldahl method. It is used to determine the nitrogen content in organic and inorganic samples and has been use for nearly 100 years for the determination of nitrogen in a wide range of samples. It consists of three steps, digestion of the sample in acid, distillation of the sample and titration. The three steps provide the nitrogen content via organic nitrogen converted into NH_4^+ during digestion, then NH_3 is distilled and retained in a receiver vessel and finally the titration yields the amount of nitrogen quantitatively.

Procedure

Weigh approximately 10 g of an MSW sample that is as homogeneous as possible. This is done preferably by grinding the MSW sample to a powder if possible.

- Place the sample into a digestion flask and add 20 ml of 98% sulfuric acid and mix gently.
- Heat the mixture for about 180 min to between 350 and 380 °C until white fumes can be seen.
- The digestion is finished when the sample will be totally transparent
- Cool the sample to room temperature and slowly add about 100 ml of water.
- Once cooled, add 50 ml of a 50% sodium hydroxide solution to the solution. This neutralizes the sample and converts the ammonium ion (NH_4^+) to ammonia (NH_3)
- Transfer the contents to a standard laboratory distillation unit.
- The distillate is then captured in 50 ml of a 4% boric acid solution including an indicator such as Tashiro's indicator.
- The boric acid solution should change from reddish-blue to green over the course of about 5 min.
- Titrate the boric acid solution, now with dissolved NH_3 with 0.25 mol/l of HCl has a slight violet color.
- Titrate another 50 ml of a 4% boric acid solution including an indicator such as Tashiro's indicator that does not have any contact with the distillation vapors—this is a "blank". It will be used to determine the amount of nitrogen atoms in the boric acid solution.

Calculation

The volume of HCl and concentration enables the calculation of the moles of nitrogen atoms in the boric acid, titrated solution. The calculations for % nitrogen are specific to the boric acid solution only, if other acids are used a different calculation and procedure is needed that is not presented here.

$$\%N = \frac{\{(ml\,of\,standard\,acid) - (ml\,blank) \times N\} \times 1.4007}{weight\,of\,sample\,in\,grams} \tag{6.22}$$

where

N is the normality of the acid solution, "ml blank" is the milliliters of standard acid needed to titrate the boric acid is the receiving solution without the distillation vapors.

Total sulfur is determined by ASTM methods D5373-08, or D4239-02. The analyzis is performed with a sulfur analyzer which combusts the MSW sample at a temperature of 1,370 °C in an oxygen atmosphere. The oxygen that reacts with sulfur and forms sulfur dioxide passes through an infrared absorption detector which analyzes the sulfur concentration. Typical sulfur content in MSW ranges between 0.1 and 0.4 w%, therefore those values can be used without much impact on the heating value or mass balance.

Elemental Analyzers

To analyze for the individual elements of an MSW sample it is best to use a dedicated apparatus that is specifically designed for the elemental analyzis. However it is possible to configure a standard gas chromatograph (GC) that properly separates H_2, O_2, N_2, CO_2, CO, SO_2, SO_3 (e.g. SO_x) and H_2O and is equipped with proper detectors, such as a TCD. This section will briefly describe the dedicated elemental analyzer and then briefly describe a GC configuration that permits a good estimate of the elemental analyzis. If such instruments exist in the laboratory it is best to consult the manufacturer's manual and technical support for setting up these systems.

Apparatus

The section below discusses the operating principle of a EuroEA 3000 elemental analyzer, however the method to determine CHNS of a MSW sample is performed following similar steps on other analyzers, such as Thermo Fisher or Perkin Elmer. The analyzer combusts the samples in a reactor chamber in the presence of oxygen at temperatures above 1350 °C. To reach complete combustion some reactors use catalysts such as tungsten trioxide but for MSW that is not necessary and due to the ash content is not recommended. The gas mixture released after combustion is a mixture of CO_2, H_2O, NOx, SO_x and some excess O_2 that is passed through a silica tube with copper granules. The copper granules reduce the nitrogen oxides and sulfur trioxide to elemental nitrogen and sulfur dioxide and any excess oxygen is retained. The gas is carried by high purity helium through a gas chromatographic column where separation of the gaseous species occurs. A thermal conductivity detector (TCD) at the end of the GC column registers a staircase type of signal proportional to the amount of each gas component. The analyzer software provides the percentages of nitrogen, carbon, hydrogen, and sulfur contained in the sample

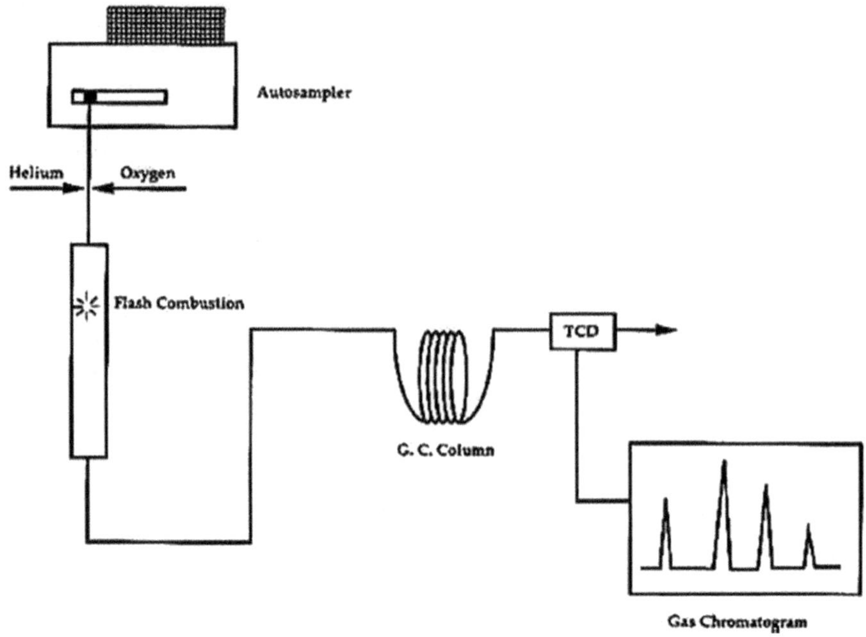

Fig. 6.9 The general schematic of an elemental analyzer

based on the calibration performed according to the instrument manual. The calibration is performed using a standard to establish precise calibrations and is routinely verified before running test samples. Figure 6.9 is a general layout of elemental analyzers.

Procedure

Most of modern analyzers are provided with an auto-sampler which facilitates the analyzis of a sequence of samples automatically without assistance from the operator. A sequence will include a group of MSW samples as well as blank metallic capsules and calibration capsules. The blank capsules do not contain samples, however they are folded to the same size as the MSW and calibration samples and loaded into the auto-sampler. The purpose of the blank samples is to verify any contamination present within the capsules. Sometimes the blank capsules will contain V_2O_5 catalyst to assess the effect the capsule and catalyst have on the results. The operator should expect a stable baseline without unexpected peaks.

The calibration capsules contain standard samples provided with the instrument, not exceeding 10 mg. These samples are aimed to verify the instrument calibration. The MSW samples are prepared following the methods presented in Sect. 6.1 making sure that are representative given the heterogeneous character of the MSW. The samples are dried and ground into a powder weighing 5–8 mg and placed into capsules provided with the instrument. The samples should not exceed 10 mg.

Standard calibration sample shall be kept within the same weight range of 5–8 mg for consistent results. The capsules are then folded carefully and placed in the analyzer's auto-sampler. Best practices require verifying and adjusting when necessary the calibration with each sequence run of MSW samples for accurate and reliable results. The standard samples are tested in duplicate or triplicate and checked against the existing calibration points. Adjustments to the integration parameters may be needed.

Gas Chromatograph (GC) Configuration

To determine the elemental composition of an MSW sample using a standard GC system one must have a furnace connected to a water-condensing system and containment system large enough to contain the combustion gases from the furnace. The recommended containment is a Tedlar® sample bag that can be easily connected to the GC injection port. The sample preparation is the same as discussed above for the elemental analyzer, but in this case the MSW sample will be placed in a furnace capable of achieving 1000 °C with adequate ventilation for air to completely combust the sample. It is important to use a small amount of sample so the total volume of product gases is not too large. For example, using a standard formula for MSW as $C_6H_{10}O_5$ and taking a 0.5 g sample will result in approximately 3 L of product gas if all the water is condensed.

Procedure

Place approximately a 0.5 g sample into a vertical, electrically heated furnace where the outlet of the furnace is connected to a water-condensing train. Figure 6.10 shows a general setup for a condensing train.

The gases exiting the train are then completely captured in the containment bag. Periodically, every 5 min, inspect the crucible with the sample. Once the sample

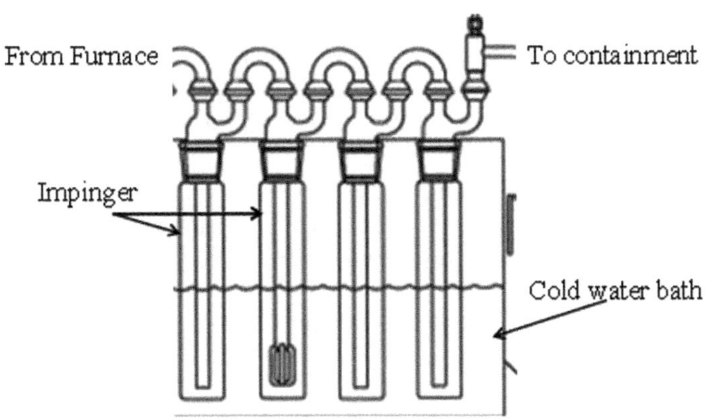

Fig. 6.10 Example of water condensing train

takes on a gray or white color, continue for another 5 min. The test should be completed in approximately 20 min with only ash remaining. Remove the sample bag and connect to the GC for analyzis.

Remove the ash sample and allow it to cool in a desiccator and weigh. Return the ash sample to the furnace and allow it to remain at 1000 °C for another 40 min (or two times the duration of the test). During this re-heating of the ash sample it is not necessary to capture any gases. Remove the ash sample and allow it to cool again in a desiccator. Re-weigh the sample. If there is no change in weight the procedure is complete. If there is a change in weight, the test must be redone with a longer time for the initial MSW sample to remain in the furnace with gas capture.

6.4.2 Lower and Higher Calorific Value

6.4.2.1 Standards

The following standards shall be followed to determine the calorific value of MSW samples: ASTM D5865 [134], ISO 1928 [135], AS1038.5 [136] to determine the calorific value.

6.4.2.2 Heating Value Background

Heating value determination is broken into two broad categories, (i) higher heating value and (ii) lower heating value. There are other terms typically used such as gross heating value, which is the same as higher heating value and net heat value which is the same as the lower heating value. There are correlations and formulas to estimate the heating value using the ultimate and proximate analyzis but the most accurate is the use of a bomb (constant volume) calorimeter.

The higher heating value (HHV), gross heating value or heat of combustion represents the amount of heat released during the complete combustion of a unit of fuel. The byproducts of the MSW complete combustion are CO_2 and water vapor. The water vapor stores a fraction of the energy released during combustion as latent heat of vaporization. The HHV includes the latent heat of vaporization which is recovered in some applications by condensing the water vapor. The lower or net heating value (LHV) represents the HHV with the subtraction of the latent heat of vaporization of the water vapor formed. In most applications the temperature of the exhaust gas is above the condensation temperature of water vapor hence the energy stored in the water vapor is lost.

The HHV and LHV are usually expressed in Joules per kilogram, calories per gram or British Thermal Units (BTU) per pound. The following conversions are applied:

$$1\,cal/g = 1.8\,BTU/lb;\, 1\,cal/g = 4186.8\,J/kg \tag{6.23}$$

In the oxygen bomb calorimeter, the combustion occurs in an oxygen environment under high pressure (25–30 atm) at an initial ambient temperature close to 25 °C so that one can ignore temperature corrections when comparing calculations at ambient temperature, $T_{ambient}$ with standard state energies or enthalpies at 25 °C. For highly accurate energy and enthalpy changes one would need to use a pressure correction term since the internal energy (U) and enthalpy (H) of combustion is presented in the tables at standard state conditions. It is acceptable to ignore the difference between ΔU and ΔUo (standard state) due to a number of published corrections indicate errors of the order of 0.1–0.4%. An estimate of the heat given off during combustion of the sample and absorbed by the water reservoir of the bomb in terms of an effective heat capacity of the entire bomb calorimeter system can be determined

$$Q_{bomb} = C_{bomb,effective}(\Delta T) \tag{6.24}$$

and

$$\Delta H = \Delta U + RT(\Delta n_{gas}) \tag{6.25}$$

where

$$\Delta H \sim \Delta Ho \tag{6.26}$$

and one can estimate the enthalpy of combustion of the fuel sample. Here the Δ(pV) from the liquid and solid components of the fuel introduces a negligible error in the ΔH calculation (thus can be ignored), and one only has the gaseous components term.

6.4.2.3 Calorimetry

Apparatus

The typical calorimeter used is the constant volume (bomb) calorimeter that provides a direct measurement of HHV of any material that can sustain combustion with oxygen. The main components of calorimeters are: containment vessel (bomb), jacketed vessel, temperature measuring device, ignition leads and stirrer as shown in the figure below. One can find sufficient detail by downloading the Parr 1341 Oxygen Bomb Calorimeter instrument manual from the internet which shows good detail on the apparatus and sample data and calculations.

Calorimeters can be automated using a microprocessor to record temperature readings that are used to immediately calculate and report the HHV. All calorimeters work on the same principle by combusting a sample of known weight in a high-pressure rated vessel charged with approximately 20–30 atmospheres of oxygen. The bomb is immersed in a vessel filled with a known amount of water, insulated from the exterior environment. The amount of heat released is determined from the measured rise of water temperature after combustion. The calorimeter is provided with a stirrer that uniformly mixes the water in the vessel. The ambient temperature changes in the test area must be minimal and the calorimeter must not be located in direct sunlight or near hot or cold sources.

The result will be reported as energy released per unit mass of sample. Therefore an accurate balance to weigh the samples is required with a resolution to the nearest 0.0001 g. However, less sensitive balances can be used but must have resolution to 0.1 g or better. The bomb is designed to be gas tight to pressures of 50 MPa at room temperature and should be constructed of stainless steel. The most accurate calorimeters have platinum resistance or linear thermistor thermometers capable of measuring temperatures to the nearest 0.0001 °C. However, reasonable analyzes can be made using liquid in glass thermometers (i.e. Hg or EtOH) or Beckman differential thermometers to measure temperature. The crucible that contains the sample should be platinum, quartz or metal alloy. If using a metal based crucibles it must be treated in an oven with good air circulation for 4 h at 500 °C before the test to remove any possible combustible material.

The sample must be ignited remotely and is done through an ignition fuse of nickel-chrome alloy (typical trade name is Chromel C) or iron wire. The ignition wire serves as a resistor that becomes hot when a current is passed through it, thus igniting the sample in the crucible. The heating value of the ignition fuse used for the test must be known and is normally obtained from a reliable source such as the manufacturer or a reference table. The fuse's heating value will be subtracted from the heating value calculated after measuring the temperature rise of the water bath.

Procedure

For calibration purposes a material with known heating value must be used. Typically benzoic acid is used as the calibration "fuel" to determine the calorimeter constant. The calorimeter constant is the heat loss associated with the apparatus. The combustion equation for benzoic acid (the calibration standard), is

$$2\,C_7H_6O_2(s) + 15\,O2(g) \rightarrow 14\,CO_2(g) + 6\,H_2O(l) \tag{6.27}$$

In general an empirical equation for the approximate composition of any hydrocarbon fuel could be written that would enable one to calculate Δn_{gas}.

$$C_xH_yO_z(s) + (2x + y/2 - z)/2O_2(g) \rightarrow xCO_2(g) + y/2H_2O(l) \tag{6.28}$$

After the calorimeter is calibrated with the standard compound benzoic acid that has been compressed into a pellet, it is recommended to then conduct a test with another material with a well-known heating value such as wood or coal.

If the calibration standard chosen is benzoic acid, the heat of combustion for the reference substance benzoic acid is 6318 cal/gm of benzoic acid burned. The heat of combustion of the nickel-chrome fuse wire must be precisely measured and is typically 2.3 cal/cm where the 10 cm length of ignition wire needs to be measured both before combustion and after to subtract the residual unburned wire length and account for the wire's contribution to the total heat of combustion in the bomb. A length of 10 cm for the apparatus shown above is sufficient to wrap the wire around the electrical posts so that it reaches down to press against the pellet yet does not touch the cup.

A constant characteristic effective heat capacity of the calorimeter must be determined accurately to ensure subsequent tests yield high quality data. The effective heat capacity that is calculated from the calibration test lumps all the pieces of the apparatus that effect the temperature change measured. They are: the bomb, leads, sample cup, bucket, thermometer, stirrer, water reservoir. Normally the water reservoir is most susceptible to changes during the testing. It is imperative that a precise quantity of water for a given set of tests using a calibrated calorimeter is used. The initial water temperature should be nearly constant between subsequent test varying less than 1 °C in temperature and the volume of water measured for the calorimeter bucket should be 2000 gm varying less than 0.5 gm of water for each run.

The charging of the bomb is performed according to the operating instructions of the unit. To ensure complete combustion (all enthalpy release), nearly pure O_2 (99%+) environment must be used in the bomb. Initially purge the bomb with O_2 for about 30 s and then refill the bomb to 25–30 atm. Do not overload the bomb with too much sample or oxygen since the bomb could then rupture. Fill the bomb with only a sufficient amount of O_2 necessary for complete combustion as evidenced by no carbon residue remaining after combustion of the sample. Lower O_2 gas pressures enable higher combustion temperatures and enhanced turbulence that provide for a better overall combustion process in the bomb.

The bomb calorimeter only approximates an adiabatic instrument since a very small, although finite, amount of heat is lost through the insulating surface. Also, the water bath temperature rises as a result of the mechanical work and resulting frictional forces introduced into the fluid by the stirrer. This "radiative correction" for non-adiabatic operation is accomplished by calculating a "corrected firing time" with an instantaneous jump in temperature that attempts to equate the two bounded areas. For the calorimeter being use, the characteristic temperature change ΔT is typically calculated at a time between 1 and 3 min after the ignition is engaged.

Method to determine the calorimeter heat capacity (i.e. $C_{bomb,effective}$)

- Weigh the benzoic acid pellet and place into the sample holder
- Add 1 ml of water to the bomb before assembly
- Connect the ignition fuse in such way that it will be in contact with the benzoic acid pellet.
- After the sample holder assembly is put into place, close the bomb tightly and inject oxygen until the pressure inside reaches 20–30 atmospheres (2–3 MPa).

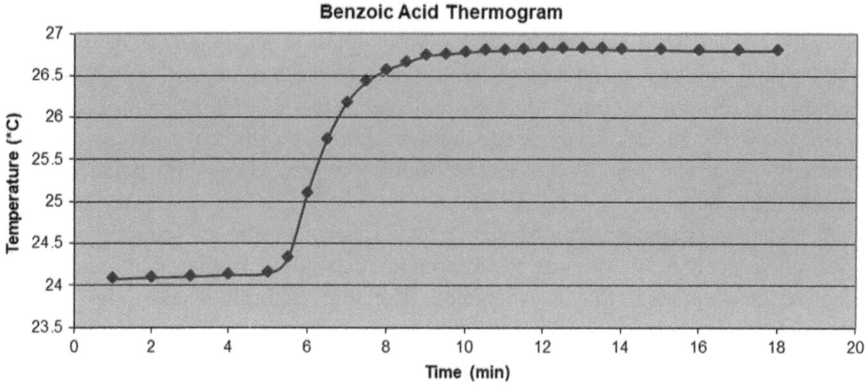

Fig. 6.11 General example of real calorimeter data using a standard of benzoic acid

- Fill the calorimeter vessel with cold water, approximately 2 °C below room temperature and place the bomb in the vessel. Inspect for any bubbles created in the water, which are sign of the bomb leaking oxygen.
- Start the calorimeter stirrer and allow for stabilization of temperature. Read the thermometer and ensure that the temperature remains stable for 30 s before engaging the ignition. For calorimeters without a stirrer allow for 5 min for temperature stabilization before ignition. After ignition, the temperature increase should be recorded to the nearest 0.002 °C (or as accurate as possible) until stabilization of ±0.001 °C for a period of 30 s or until three successive readings do not change by more than ± 0.001 °C.
- Remove the bomb and release the pressure at uniform rate. Inspect the sample for unburned parts and discard if not completely burned (Fig. 6.11).

$$C_{bomb,effective} = \frac{Q_{ba}M_{ba} + C_{wire}L_{wire}}{\Delta T_{corrected}} \quad (6.29)$$

Using the ASTM method to determine the radiation heat correction, the heat gained by the calorimeter from the surroundings during the first 63% of the temperature rise at a linear rate determined by the 5 min pre-period is compensated for by the 37% cooling of the temperature at a linear rate determined by the 5 min post-period. This 63% transition point is the time t63 that needs to be determined since the ΔT value for the heat of combustion of the sample is determined at this transition time. Often this transition time is calculated as the time on the thermogram where the temperature has reached approximately 60% of the total rise from the firing temperature (ta) to the leveling off temperature (tc) at which the temperature starts to decline at a constant rate.

The ΔT is determined after identifying the time t_i at which the bomb was actually fired and the time t_f at which the thermogram first started to level off (or plateau) linearly and calculating r_1 the rate of linear decrease in temperature and r_2 the rate of linear increase in temperature. The temperature at which the ΔT is calculated is Td and the time at which the ΔT is calculated is t_d

$$T_d = (T_f - T_i) \times 0.6 + T_i \tag{6.30}$$

$$T_f = T_{f_{corrected}} + r_2(t_f - t_d) \tag{6.31}$$

$$T_i = T_{i_{corrected}} + r_1(t_i - t_d) \tag{6.32}$$

$$\Delta T = T_{f_{corrected}} - T_{i_{corrected}} = (T_f - T_i) - r_2(t_f - t_d) + r_1(t_i - t_d) \tag{6.33}$$

Corrections to the heating value

There are corrections that can be applied to the heating value determination just described from the formation of acids and the combustion of sulfur. However typically, the corrections amount to less than 1.5%.

It is also possible to determine the heating value from the ultimate analyzis through various correlations. There are a number of possible formulas and correlations to use, see Table 6.20. It is recommended that more than one appropriate formula is used and an average is taken to obtain a final value with quantitative uncertainty that has statistical significance.

6.4.3 Loss on Ignition

Loss on ignition is widely used to approximate the organic carbon content and the carbonate content of an MSW sample. The method has been adapted from the geology field [138] but is useful for analyzis of MSW due to the simplicity of the procedure. This method combined with the formulae from Wilson in the table above can allow a very rough approximation of heating value of a given MSW sample with only a muffle furnace and scale as equipment. The primary assumption necessary is the oxygen and hydrogen both come from water only (i.e. the moisture content as discussed above) and the nitrogen content (N) can be considered to be 0.75 weight %.

Apparatus

The main piece of equipment is a simple muffle furnace or oven capable of achieving up to 1000 °C at a controlled rate without exceeding the target temperature. Sufficient air flow and exhaust is required to ensure adequate oxygen levels to achieve complete conversion of the organic material to carbon dioxide.

Table 6.20 Models and equations to approximate higher heating value

Name	Equation	Units	Remarks	Application
1. Models based on ultimate analyzis				
Dulong	HHV = 81C + 342.5(H − O/8) + 22.5S − 6(9H − *W*)	kcal/kg	Modified (wt%)	MSW/Coal
Steuer Scheurer–Kestner Chang	HHV = 81(C − 3 × O/8) + 57 × 3 × O/8 +345(H − O/10) + 25S − 6(9H + *W*) HHV = 81(C − 3 × O/4) + 342.5H + 22.5S + 57 × 3 × O/4−6(9H + *W*) HHV = 8561.11 + 179.72H − 63.89S − 111.17O − 91.11Cl − 66.94 N	kcal/kg kcal/kg kcal/kg	(wt%) (wt%) (wt%)	MSW MSW MSW
Boie	HHV = 83.22C + 274.3H − 25.8O + 15 N + 9.4Cl + 65P	kcal/kg	(wt%)	Refuse
Vondracek	HHV = C (89.17 − 0.0622C1) + 270(H_2O/10) + 25S (C1—carbon content on moisture and ash free basis)	kcal/kg	(wt%)	Refuse
Wilson Mott and Spooner	HHV = 7831C_{org} + 35,932Substrates and co-digestions main characterization (H − O/8) + 2212S − 3545C_{inorg} + 1187O + 578 N HHV = 0.336C + 1.418H − 0.0145O − 0.0941S	kcal/kg MJ/kg	(wt fraction) (wt%)	MSW Coal/refuse
Inst. for Gas Tech., USA	HHV = 0.3417C + 1.3221H + 0.1232S −0.1198 (O + N) − 0.0153*A*	MJ/kg	(wt%)	Coal/refuse
2. Models based on proximate analyzis				
Goutal	HHV = 147.6 × FC + *K* × VM (*K* is a constant that varies with the value of VM)	Btu/lb	(wt%)	Coal/refuse
Bento	HHV = 44.75 × VM − 5.85 × *W* + 21.2	kcal/kg	(wt%)	Refuse
Traditional	HHV = 45 × VM − 6 × *W*	kcal/kg	(wt%)	Refuse
3. *Models based on physical composition*				
Conventional	HHV = 88.2Pl + 40.5(Ga + Pa) − 6 *W*	kcal/kg	(wt%)	Refuse
Tokyo Ali Khan	HHV = [(100 W)/100]{38.8(Pa + Ga + *T* + Oc) + 50.9 (Te + Ru) + 73.7Pl} − 6 *W* HHV = [23(Ga + 3.6 × Pa)] + [160(Pl + Ru)]	kcal/kg Btu/lb	(wt%) (wt%)	MSW MSW

HHV net calorific value; *W* wt% of water, dry basis; *A* wt% of ash, dry basis; *VM* %volatile matter; *FC* %fixed carbon; *W* total moisture; *Pa* paper; *Ga* garbage/food; *Te* textile; *Ru* rubber and leather; Pl ¼ plastics; *Oc* other combustibles; *T* wood and grass.Loss on ignition
Reproduced with permission from Elsevier [137]

Procedure

A powdered sample is dried in an oven at 90–100 °C in a pre-weighed ceramic crucible for one hour. After cooling to room temperature in a desiccator, the sample and crucible are weighed. This gives the dry weight of the sample which is the basis for all weight loss calculations.

- The sample and crucible are then placed in a muffle furnace and heated to 550 °C for one hour. After cooling to room temperature, the sample is again weighed. The difference between this weight and the dry weight is the amount of organic carbon ignited.
- The sample is returned to the muffle furnace and heated to 1000 °C for one hour. The weight loss between 550 and 1000 °C is the amount of CO_2 evolved from carbonate minerals

6.4.4 Thermogravimetry/Thermogravimetric Analyzis (TGA)

Thermogravimetric (TG) analyzis (TGA) is a technique that combines the use of a temperature programmed furnace that encloses a sample holder that is connected to a micro-balance. The main information obtained is the mass change as a function of temperature and time. That information can then be used to determine a wide range of information provided numerous assumptions are made. Since the data obtained is weight change per degree or weight change per time, it becomes very easy to calculate rates of weight change per mass or time, thus the slope, differential thermogravimetric (DTG), of the process. TG and DTG have several advantages but one of the major advantages is that, by conducting one experiment, it is possible to characterize, over a very wide temperature range, the performance of a sample resulting an overview of thermal behavior at all temperatures of interest. It is also possible to determine kinetic parameters of the various MSW samples with several reacting gas environments over a wide temperature range and varying the heating rates.

Dynamic thermogravimetry is the most common form of thermogravimetric analyzis. A continuous recording of weight change of the sample in a flowing or static gas atmosphere is made as a function of time or temperature at a fixed heating rate. Its use has expanded to determination of decomposition temperatures of solids to methods for determining proximate analyzes and development of kinetic expressions for the decomposition of solid materials. This section will focus only on a general description of TGA, the type of information that can be obtained from MSW samples and simple methods to obtain useful kinetic expressions.

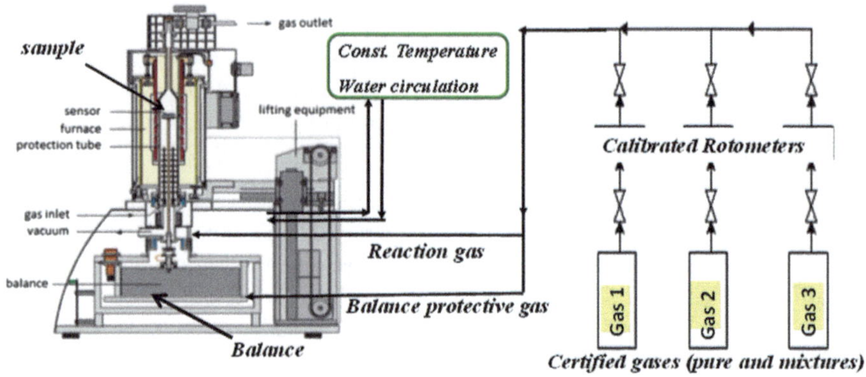

Fig. 6.12 TGA configuration with a typical flow configuration

Apparatus

There are many TGA configurations by numerous companies. They divide into three basic categories, (i) cantilever (DuPont, TA Instruments), (ii) sample hang down (Cahn) and (iii) sample carrier where the balance is below the furnace (Netszch). The figure below shows the Netszch configuration with the position of reactive and protective gases and the relation of the sample and the furnace. Nearly all systems have a similar set-up with regard to relative position of the furnace and sample (Fig. 6.12).

Procedure

The procedure for operating a TGA should follow the manufacturer's instructions. However, a general rule to follow for all systems is to have a sample that is representative of the MSW composition in the region of interest. It has been shown that testing the individual components and calculating a weighted average does not always yield accurate results. Therefore the MSW should be prepared by grinding individual components of the MSW stream to particles less than 1.0 mm. The total sample weight and size should allow for a uniform flow of heat into the sample and adequate flow of the evolved gases from the sample. This should be discussed in the manufacturer's documentation.

Application to kinetic extraction

A typical TG or DTG temperature/mass curve under a flow of air is given in Fig. 6.13. Although there is only one point of inflection at the conclusion of water loss, the curve can be divided into three sections in which the sequential processes of drying, devolatilization and residual carbon combustion take place, resulting in mass lost as vapor or gas. The remaining mass is taken as ash. The stages are indistinguishable because multiple reactions are occurring simultaneously.

However, when attempting to extract the kinetics from the curves it is best to use the DTG (slope) curves as shown below. In Fig. 6.14 plotted below, the TG curve

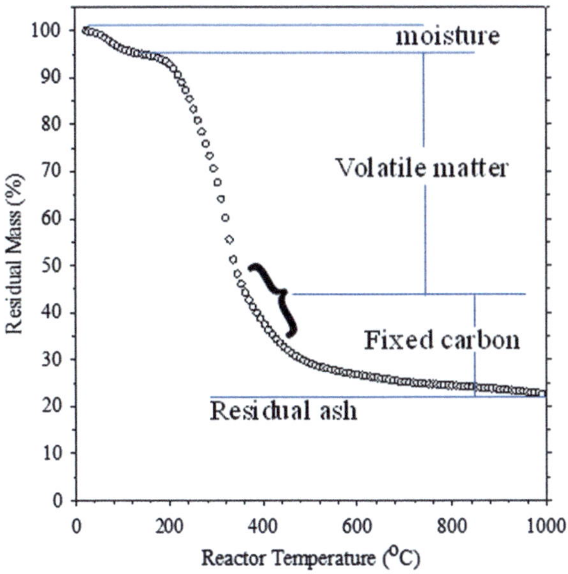

Fig. 6.13 Typical TGA decomposition curve of biomass

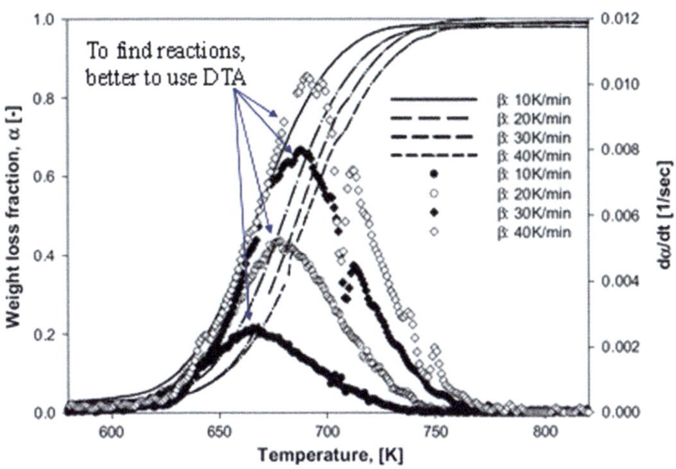

Fig. 6.14 Combined TG and DTG curves a various heating rates

is shown as a function of conversion of initial mass at four different heating rates (β). In that figure one can see there is one major decomposition region, done for simplicity, which is very clearly distinguished by the peak in the DTG curve.

Here we will show the classic method of Coats and Redfern to demonstrate how kinetics can be obtained using TG data. Other methods, show in the Figure below, are modifications of the Coats-Redfern method to better approximate the reaction sequence occurring and to avoid guessing of the reaction order. It should be noted that multiple heating rates will provide more reliable kinetics. The rate is a function of the conversion of material, temperature and pressure of test conducted. Usually most TGA experiments are conducted close enough to atmospheric pressure that the pressure can be considered constant.

Coats and Redfern method

The assumed reaction is $A(s) \rightarrow B(s) + C(g)$ and the conversion of solid material (MSW) is written as

$$\frac{dX_A}{dt} = k(1 - X_A)^n \tag{6.34}$$

where X_A is the fractional conversion and k is the rate coefficient that can be expressed in an Arrhenius form. The equation calculates the rate of conversion dX_A/dt is proportional to the amount of remaining unreacted component A, raised to the power n which is the order of reaction, as in Eq. (6.34). However it should be clear that there is no physical meaning that can be ascribed to that form of the equation, it is a good data fit of the temperature dependence. The n in Eqs. (6.34) and (6.35) is the reaction order. There is no theoretical justification provided for frequency factor, ϑ, the E in the exponent and reaction order, n, in the case of solid decomposition therefore, it is recommended that terms "pre-exponential factor' and "exponent term" respectively are used.

Typically the heating rate for each test should be constant, between 1 and 10 K/min, thus for a linear heating rate, α = dT/dt, the substitution for K in Eq. 6.34 is done in the following steps.

$$\frac{dX_A}{dT} = \frac{\vartheta}{\alpha} e^{\frac{-E}{RT}} (1 - X_A)^n \tag{6.35}$$

Rearranging gives

$$\frac{dX_A}{(1 - X_A)^n} = \frac{\vartheta}{\alpha} e^{\frac{-E}{RT}} dT \tag{6.36}$$

To integrate Eq. (6.36) within limits of X = 0 to X and T = 0 to T, first substitute for E/RT with u. The approximate solution for the right hand side of this integral is then given as

$$\int e^{-u} u^{-b} du = u^{(1-b)} e^{-u} \frac{\sum(-1)^n (b)^n}{u^{(n+1)}} \tag{6.37}$$

Substituting this solution into Eq. (6.36) for n not equal to 1 gives

$$\frac{\left(1-(1-X_A)^{(1-n)}\right)}{(1-n)} = \frac{\vartheta R T^2}{\alpha E} e^{\frac{(1-2RT)}{E}} e^{\frac{-E}{RT}} \tag{6.38}$$

and for n = 1 Eq. (6.38) becomes

$$-log\left\{\frac{(1-X_A)}{T^2}\right\} = \frac{\vartheta R}{\alpha E} e^{\frac{(1-2RT)}{E}} e^{\frac{-E}{RT}} \tag{6.39}$$

To determine the kinetic parameters for any reaction by using this method one must guess at the reaction order. There is a large body of evidence that suggests the reaction order should be an integral number between 0 and 3. Subsequently, by taking log of both sides, Eqs. (6.38) and (6.39) transform into the linear function y = mx + c. The reaction order that most closely matches the liner function can then be used to determine the kinetic parameters by using graphical techniques. This method has been used successfully for a variety of materials, however the reaction order must be known (i.e. guessed) beforehand. Another limitation is that for cases where there is step wise conversion which is commonly observed in MSW decomposition requires the treatment of the data to be done for separate steps individually. An example is provided in Fig. 6.15 that shows the individual models for a decomposing MSW sample.

Another way to process TGA data without guessing the reaction one can use the differential method originally develop by Freeman and Carroll [139]. This method helps in determining the kinetics of a reaction over the entire range of temperature. Again, using the same reaction expression A (s) → B (s) + C (g), the kinetic parameters are determined by using equation

$$-\frac{dX_A}{dt} = kX_A^n \tag{6.40}$$

Fig. 6.15 TG curve of typical MSW with sectioned fitted kinetic parameters

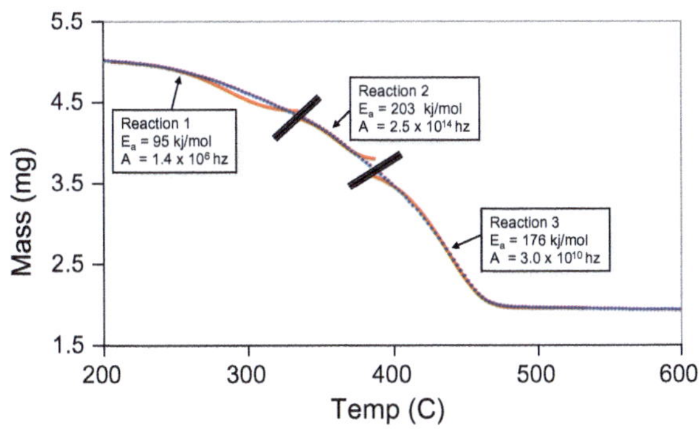

where n is the reaction order and k is the specific rate coefficient that again can follow the Arrhenius form. Integrating and substituting in the exponential kinetic expression yields the following:

$$\frac{-E}{R}\frac{\Delta\left(\frac{1}{T}\right)}{\Delta lnX_A}=\frac{\Delta\ln\left(-\frac{dX_A}{dt}\right)}{\Delta lnX_A}-n \tag{6.41}$$

Therefore a plot of the following equation will result in a straight line where the slope value is equal to –E/R and the intercept will be –n, the reaction order.

$$\frac{\Delta\left(\frac{1}{T}\right)}{\Delta lnX_A}\ vs\ \frac{\Delta\ln\left(-\frac{dX_A}{dt}\right)}{\Delta lnX_A} \tag{6.42}$$

Importantly this method must be used during discernable decomposition regimes of the TG profile. Typically the DTG can assist in identifying the regimes where the point of maximum decomposition rates occur (i.e. normally the center or inflection point). Then one can better identify the bounds of the individual reaction zones.

Finally there are numerous forms that can be used that may better approximate the thermal decomposition of the MSW depending on the composition. The equations in Table 6.21, taken from Vyazovkin et al. [140] provides several forms based on the following expressions, again assuming the test conducted is done at constant pressure.

Table 6.21 Example of kinetic models used for TGA analyzes

Reaction model	Code	f (α)	g (α)
Power law	P4	$4\alpha^{3/4}$	$\alpha^{1/4}$
Power law	P3	$3\alpha^{2/3}$	$\alpha^{1/3}$
Power law	P2	$2\alpha^{1/2}$	$\alpha^{1/2}$
Power law	P2/3	$2/3\alpha^{-1/2}$	$\alpha^{3/2}$
One-dimensional diffusion	D1	$1/2\alpha^{-1}$	α^2
Mampel (first order)	F1	$1-\alpha$	$-\ln(1-\alpha)$
Avrami-Erofeev	A4	$4(1-\alpha)[-\ln(1-\alpha)]^{3/4}$	$[-\ln(1-\alpha)]^{1/4}$
Avrami-Erofeev	A3	$3(1-\alpha)[-\ln(1-\alpha)]^{2/3}$	$[-\ln(1-\alpha)]^{1/3}$
Avrami-Erofeev	A2	$2(1-\alpha)[-\ln(1-\alpha)]^{1/2}$	$[-\ln(1-\alpha)]^{1/2}$
Three-dimensional diffusion	D3	$3/2(1-\alpha)^{2/3}[1-(1-\alpha)^{1/3}]^{-1}$	$[1-(1-\alpha)^{1/3}]^2$
Contracting sphere	R3	$3(1-\alpha)^{2/3}$	$1-(1-\alpha)^{1/3}$
Contracting sphere	R2	$2(1-\alpha)^{1/2}$	$1-(1-\alpha)^{1/2}$
Two-dimensional diffusion	D2	$[-\ln(1-\alpha)]^{-1}$	$(1-\alpha)\ln(1-\alpha)+\alpha$

Reproduced with permission from Elsevier [141]

$$\frac{dX_A}{dt} = k(T)f(X_A) \qquad g(X_A) = \int_0^{\alpha} \frac{dX_A}{f(X_A)} \tag{6.43}$$

6.4.5 Differential Scanning Calorimetry

Apparatus

The differential scanning calorimetry (DSC) is a thermal analyzis technique in which the physical sample properties are measured as a function of temperature. Similar to the TGA described in the section above the sample is subjected to controlled, predefined heating or cooling programs, under specified purge gas at controlled flow rate.

The procedure described in this section relates to a heat flux DSC instrument, where the sample and reference material are enclosed in the same furnace with a metallic block with high thermal conductivity, ensuring a good heat flow path between sample and reference. Other DSC systems consist of two identical ovens, one for the sample and the other for the reference material.

The DSC measures the difference in energy required to maintain a MSW sample and a reference at an identical temperature program.

The changes in temperature between the sample and the reference are related to the enthalpy or heat capacity changes in the sample, which are quantified using calibration experiments.

The instrument consists of a furnace capable of providing controlled heating or cooling at constant rate, temperature sensor, and differential sensor capable of detecting the heat flow difference between the sample and reference equivalent to 1 μW. The furnace is purged with minimum 99% inert gas, such as nitrogen, argon or helium at flow rates varying from 10 to 50 ml/min. The instrument is provided with temperature controller, recording device and software for integrating areas under endothermic and exothermic peaks.

Calibration

The instrument shall be calibrated first time after installation and then periodically before experiments. The calibration procedures depend on the type of instrument used, however most of calibration modes will include the procedures listed below. Determine deviations in the instrument baseline by running the predefined temperature program without samples or crucibles.

Another calibration mode is performed using large sapphire disks, provided with the instrument on the sample and reference positions. Both modes include preheat followed by equilibration to an initial temperature, holding isothermal for five minutes, heating at contact rate to final temperature and holding isothermal for 5 min. The range of temperatures shall be as broad as the range of experimental tests.

Calculate the slope and offset values needed to flatten the baseline and zero the heat flow signal by heating the empty furnace and holding isothermal at the temperature limits. Calibrate the instrument using standard metals, such as indium by heating thought the metal melting transition and calculating the heat of fusion. The instrument constant is the ratio between the calculated heat of fusion and the theoretical value. Perform the experiments using the same purge gas used for the calibration modes.

Sample preparation

Prepare samples according to the procedures descried in 6.1 ensuring the sample is homogeneous.

- Use inert crucibles provided with the instrument.
- Use a balance with capacity of 100 mg or grater and accuracy of ± 10 μg.
- Weigh samples of 5–10 mg.
- Purge the cell/furnace with inert gas at 10–50 ml/min ± 5 mL/min.
- Set up the temperature program, heating rate and flow rate of purge gas.

Subject the sample to the temperature program that will usually include heating, isotherms, i.e. holding the sample at specified temperatures for a defined amount of time or cooling the sample.

The results will look similar to the ones shown in Fig 6.16.

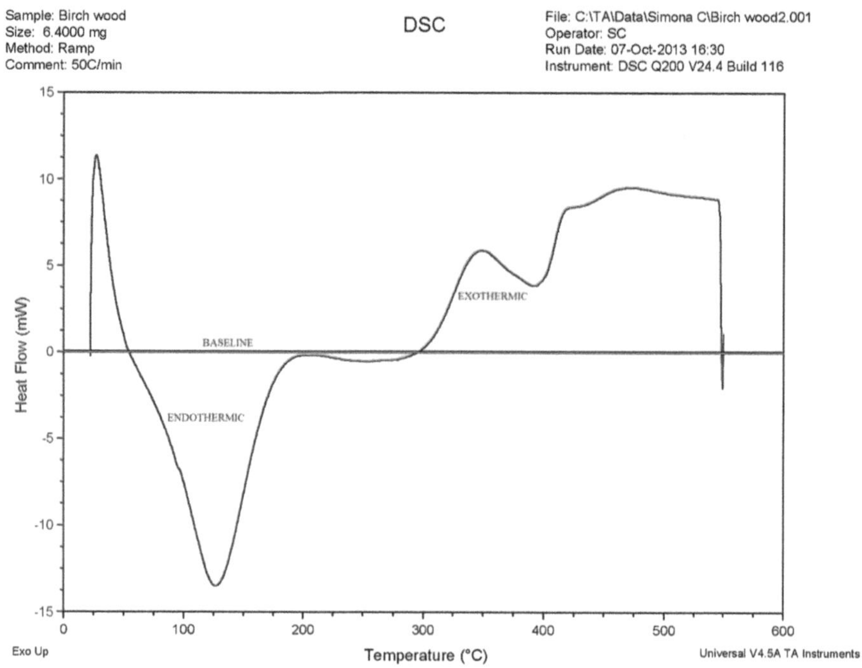

Fig. 6.16 Example of TA Instruments DSC results of Birch wood sample

Figure 6.16 presents the results of DSC analyzis performed on a 6.4 mg birch wood sample in a TA Instruments equipment, heated at constant rate of 50 °C/min from room temperature to 550 °C, followed by 30 min isothermal holding at 550 °C. The peaks above the baseline are exothermic peaks, while the peaks under the baseline are endothermic peaks. The first peak is caused by the difference in temperatures between the sample and the empty reference crucibles measured by the instrument thermocouples during initial heating and shall be ignored. Integrate the areas under the exothermic and endothermic peaks as a function of time to determine the enthalpy or heat of transition in mJ. Calculate the mass normalized enthalpy by diving the enthalpy obtained during the test to the mass of the sample.

6.5 Characteristics Database of Municipal Solid Waste

Characteristics database of municipal solid waste include 4 sections: physical and mechanical characterization, chemical characterization, biochemical characterization, thermal characterization. The structure of classification characteristics data of municipal solid waste are as Fig. 6.17.

The characteristics database can provide details of MSW from any source, including continent, country and city. The structure is shown in Fig. 6.18.

Characteristics database of municipal solid waste can also search the data for a given period as shown in Fig. 6.19 for the characteristics data of municipal solid waste from 1990 to 2017.

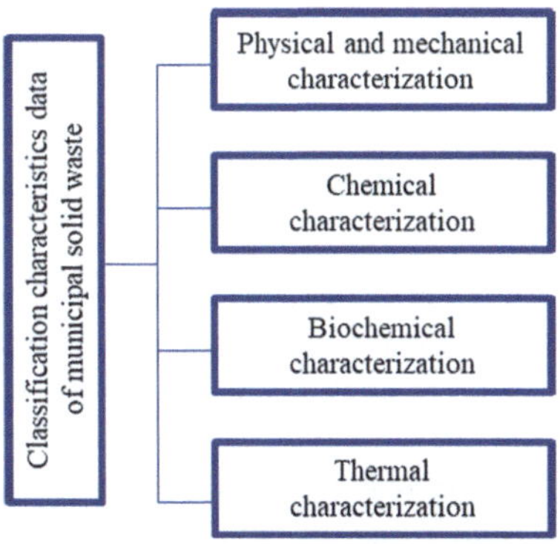

Fig. 6.17 The structure of characteristics database of municipal solid waste

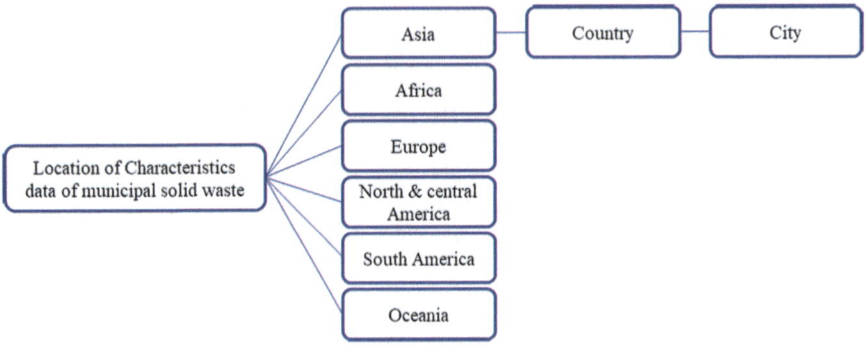

Fig. 6.18 The structure of location characteristics database of municipal solid waste

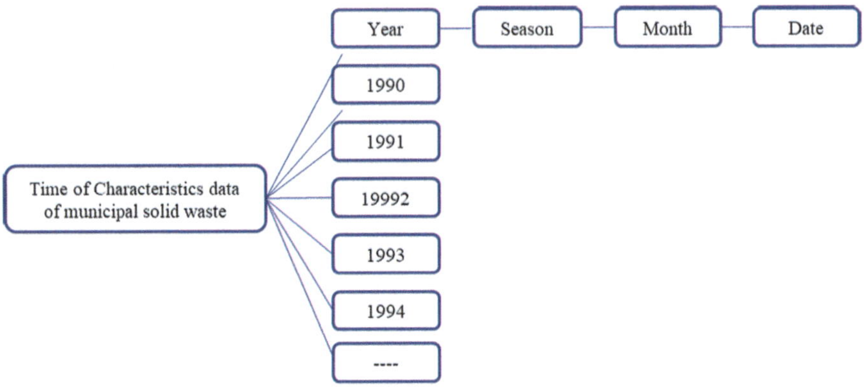

Fig. 6.19 The structure of time characteristics data of municipal solid waste

6.5.1 Physical and Mechanical Characteristics

The data of physical characteristics of municipal solid waste in database include organic garbage, paper, plastics, glass, metal, textile fiber, wood timber, ash.

For example, data of physical compositions of MSW in various cities in China from 1997 to 2003 are presented in Table 6.22, in some US city in Table 6.23 and in some worldwide countries in Table 6.24.

Other data of physical and mechanical characteristics of municipal solid waste include particle size (mmPS), bulk density (t/m^{3}BD), moisture content (%MC), ash (%A), compressibility (CP), permeability (cm/sPM). The structure of other data of physical and mechanical characteristics of municipal solid waste are shown in Table 6.26.

For example, data of physical and mechanical characteristics of MSW (%) are as Table 6.25.

Table 6.22 Data of physical compositions of MSW in various cities, China

No.	City	Time	Organic garbage	Paper	Plastics	Glass	Metal	Textile fiber	Wood timber	Ash	Moisture
1	Beijing	2000	51.83	5.43	10.37	5.43	1.00	2.97	5.75	17.22	/
2	Jinan	1999	48.91	3.65	7.26	0.48	0.16	1.48	0.00	36.56	39.70
3	Shanghai	2003	56.09	4.58	8.59	2.88	0.89	2.29	11.6	6.78	/
4	Nantong	2005	58.90	8.77	11.28	0.50	0.37	4.51	0.63	15.04	50.76
5	Hangzhou	2005	29.49	10.56	17.61	1.58	0.41	7.69	4.19	19.54	52.16
6	Jinhua	2002	35.60	12.15	15.70	2.03	3.80	5.06	6.33	10.00	38.63
7	Shenzhen	2000	59.40	11.00	14.00	5.00	/	3.90	/	6.70	/
8	Guangzhou	2001	63.76	5.48	20.84	1.97	0.30	3.53	2.24	1.88	57.15
9	Dongguan	2002	22.23	6.44	19.28	3.37	6.10	16.06	7.83	20.00	30.33
10	Kunming	2000	61.67	11.43	9.43	2.02	0.87	1.06	/	13.52	/
11	Hongkong	1997	25.50	25.70	17.10	3.20	3.60	5.00	5.00	15.70	33.56
12	Xingtai	1999	33.10	3.90	4.50	1.70	0.40	1.40	1.20	53.80	31.20
13	Wuhan	1999	57.44	5.06	9.51	3.02	3.18	1.15	0.90	19.72	51.37
14	Shenyang	2000	67.50	7.60	11.00	2.80	0.50	1.70	1.50	2.30	58.07
15	Panjin	1999	64.50	4.10	5.80	2.30	0.50	1.50	1.30	20.00	55.54
16	Yanan	2003	38.08	8.61	11.22	5.24	0.54	1.12	0.00	35.21	/
17	Xian	1997	15.74	3.35	7.93	1.20	2.48	2.48	3.94	1.84	/

Table 6.23 Data of physical compositions of MSW in some US city [142, 143]

No.	City	Time	Organic garbage	Paper	Plastics	Glass	Metal	Textile fiber	Wood timber
1	California	2003	37.00	27.40	12.40	3.00	4.80	3.10	0.40
2	Delaware	2006	27.50	33.40	14.10	2.90	6.50	3.30	0.30
3	Florida	2000	27.30	34.60	9.10	4.20	10.90	4.60	
4	Iowa	2005	22.50	39.50	18.20	2.10	5.60	5.90	
5	Kansas	2001	36.60	39.50	6.30	3.60	4.50	11.00	6.80
6	Minnesota	2000	27.90	37.50	12.00	2.80	5.30	2.80	7.70
7	Missouri	1998	30.10	36.00	14.00	5.50	6.80		
8	Oregon	2002	38.50	27.90	14.00	2.90	6.80	3.80	5.00
9	Pennsylvania	2001	29.60	40.00	13.20	3.40	6.20	4.30	2.80
10	Wisconsin	2001	25.80	31.70	15.30	3.30	8.90	3.50	2.50
11	Tennessee	2005	21.66	32.09	17.08	5.06	5.73	5.35	1.34
12	Georgia	2005	14.70	38.70	15.80	3.70	7.40	5.60	4.40
13	The united state	2003	24.00	26.30	15.40	6.20	7.30	9.00	7.50

Table 6.24 Data of physical compositions of MSW in some urban city

No.	Country	Time	Organic garbage	Paper	Plastics	Glass	Metal	Textile fiber	Wood timber
1.	Japan [144]	2017	40.50	32.14	10.42	0.86	2.04	0.63	0.99
2.	Germany [145]	2003	15.00	20.00	23.00	7.00	2.00	1.00	13.00
3	Singapore [145]	2008	9.51	21.15	11.46		14.56	1.56	4.52
4	Poland [146]	2014	37.27	11.27	12.13	10.07	2.17	2.80	0.40
5	India [147]	2007	45.30	4.69	2.86	0.73	0.73	2.22	
6	Malaysia [148]	2013	41.06	20.93	22.23	3.56	1.94	7.74	2.45
7	Pakistan [149]	2016	42.18	1.32	1.89	0.69	2.49	0.88	
8	Romania [150]	2010	61.8	12.4	12.55	3.55	1.35	2.55	0.6
9	Bulgaria [150]	2010	59.95	6.75	9.9	3.60	0.75	2.90	1.55
10	Spain [151]	2011	56.30	13.80	10.70	3.30	3.00	3.60	1.30
11	Turkey [152]	2008	45.00	20.00	10.00	10.00	8.00	2.00	
12	Wales [153]	2002	36.00	24.00	10.00	7.00	5.00	2.00	
13	Gaza Strip [154]	2012	52.00	11.00	13.00	3.00	3.00		
14	Mexico	2012	49.50	9.92	13.16	2.65		3.64	

Table 6.25 Data of physical and mechanical characteristics of MSW (%)

No.	City	Bulk density (t/m^3BD)	Moisture content (%MC)	Ash (%A)
1	Chongqing [155]	0.312	54.15	
2	Tangshan [156]	4.46	48.4	14.9
3	Lasa [157]	0.3	24	38
4	Chengdu [158]	0.219	55.18	18.29
5	Shenzhen [159]		48.17	14.03

6.5.2 Chemical Characteristics

Chemical characteristics of municipal solid waste are separated into six tables, the first table is the organic element of municipal solid waste, the second table is macro inorganic elements of municipal solid waste incineration fly ash, the third table is mineral composition of municipal solid waste incineration residue, the fourth table is heavy metal speciation in flue gas of municipal solid waste, the fifth table is heavy metal speciation of municipal solid waste incineration, the sixth table is pH value and leaching toxicity of municipal solid waste incineration residue.

The parameters of organic element of municipal solid waste are shown in Table 6.26.

Mineral composition of municipal solid waste incineration residue is presented in Table 6.27.

Table 6.26 Organic element of municipal solid waste [160–163]

	Organic element(%)						Plant nutrients		
	C	H	O	N	S	Cl	Total nitrogen (TN)	Total phosphorus (TP)	Total potassium (TK)
1	44.9	5.80	28.2				1.20	0.09	0.36
2	31.33	4.62	19.18	1.33	0.15		1.33		
3	29.31	3.88	15.19				0.75	0.06	0.24
4	43.26	7.87	45.21	2.45	0.08	1.13			
5	48.84	8.25	31.82	1.31	1.55	0.76			

Table 6.27 Mineral composition of municipal solid waste incineration residue (%) [164, 165]

No.	SiO_2	CaO	Al_2O_3	Fe_2O_3	MgO	Na_2O	K_2O	TiO_2	SO_3	P_2O_5
1	28.91	24.85	13.34	6.26	2.38	10.68	9.42	/	/	1.52
2	48.87	10.32	17.84	5.86	1.44	1.88	1.77	0.88	0.70	1.54
3	16.90	34.60	8.06	3.94	2.09	0.40	0.46	0.70	14.16	0.76
4	61.51	1.22	16.29	5.23	1.16	1.53	2.18	0.73	0.37	1.57
5	12.6	43.44	11.3	2.72	2.54	2.34	0.65	0.72	/	17.57
6	36.84	9.74	19.49	7.45	2.67	2.31	1.96	1.54	/	/

Table 6.28 Total metal and leaching concentrations of cyclone ash, scrubber ash and bottom ash

Heavy metal	Cyclone ash	Scrubber ash	Bottom ash	Taiwan regulatory limits
Total metal (mg/kg)				
Pd	650 ± 15	1229 ± 84	1230 ± 33	–
Cd	53 ± 2	130 ± 6	93 ± 1	–
Cr	274 ± 25	26 ± 2	248 ± 12	–
Cu	274 ± 25	740 ± 14	1130 ± 48	–
Zn	5540 ± 163	7780 ± 256	8210 ± 184	–
Leaching concentration (mg/L)				
Pd	0.03 ± 0.01	0.06 ± 0.02	ND[a]	5.00
Cd	ND[b]	1.04 ± 0.02	ND	1.00
Cr	0.07 ± 0.03	ND[c]	0.02 ± 0.01	5.00
Cu	ND[d]	0.08 ± 0.01	0.53 ± 0.03	15.00
Zn	0.89 ± 0.30	4.08 ± 0.10	0.18 ± 0.08	–

Mean ± standard deviation (n = 5)
[a]Pb: detection limits < 0.016 mg/L/ [b] Cd: detection limits < 0.014 mg/L
[b]Cd: detection limits < 0.014 mg/L
[c]Cr: detection limits < 0.014 mg/L
[d]Cu: detection limits < 0.013 mg/L
[d]Cu: detection limits < 0.013 mg/L
Reprinted with permission from Elsevier [166]

Total metal and leaching concentrations of cyclone ash, scrubber ash and bottom ash are in Table 6.28.

It should be noted that Cr, Pb and Zn can vary considerably because these elements are found in many consumer products that exist. Although there is a phasing out of these in developed regions, there remains a legacy stock that periodically enters the waste stream.

6.5.3 Biochemical Characteristics

Biochemical characteristics of municipal solid waste include aerobic biodegradability, anaerobic degradation, content of pathogens and bio-toxicity. The characteristics and different parameters of substrates are shown in Table 6.29.

6.5.4 Thermal Characteristics

Thermal characteristics of municipal solid waste are separated to two table, one is proximate and ultimate analyzis of municipal solid waste, another is other data of thermal characteristics of municipal solid waste. Proximate and ultimate analyzis of

Table 6.29 Substrates and co-digestions main characterization (dw: dry weight)

	TS g/kg	VS g/kg	CODt g/kg	C dw %	H dw %	O dw %	N dw %	C/N
Biological sludge	69.8	56.9	77.1	5.4	9.1	36.4	0.6	8.9
OFMSW	467.9	393.5	542.2	20.3	7.8	31.0	0.4	51.8
Co-digestion 1	388.3	326.2	449.2	17.3	8.0	32.1	0.4	39.8
Co-digestion 2	229.0	191.5	263.2	11.4	8.6	34.3	0.5	29.9
Co-digestion 3	149.4	124.2	170.1	8.4	8.8	35.3	0.6	14.8
Inoculum	31.78	19.72	31	–	–	–	–	–

OFMSW Organic fraction of MSW, *TS* Total solids, *VS* Volatile solids, *COD* Chemical oxygen demand

Reprinted with permission from Elsevier [167]

Table 6.30 Proximate and ultimate analyzis of municipal solid waste of Hangzhou

	Proximate analyzis(%)				Ultimate analyzis (%)					Qnet,ar (KJ/kg)
	Mt	Aar	Var	FCar	Car	Har	Nar	St, ar	Oar	
1	61.19	16.61	27.89	3.35	17.55	2.18	1.37	0.12	7.68	6745
2	52.16	13.31	27.52	3.31	18.32	2.17	0.77	0.10	9.48	7623
3	60.11	11.14	25.60	2.96	16.62	1.86	0.52	0.10	9.47	5227
4	60.03	10.79	25.21	3.77	17.15	1.83	0.66	0.12	9.23	5962
5	43.35	7.31	42.16	4.33	28.37	3.81	0.40	0.08	13.82	10,970
6	43.89	11.95	39.83	4.13	26.78	3.83	0.50	0.10	12.69	8409
7	53.96	16.05	25.74	2.25	16.94	1.65	0.50	0.08	8.84	5431
8	61.91	6.57	27.36	1.89	16.88	1.98	0.53	0.06	8.80	5385
9	60.41	10.85	23.64	2.54	16.89	1.98	0.20	0.07	7.05	6345
10	58.36	12.55	23.54	2.11	17.22	1.98	0.31	0.10	6.03	5546
11	48.58	5.66	21.64	3.43	17.00	1.77	1.41	0.06	4.83	5784
12	58.36	9.80	23.00	2.01	15.93	1.85	0.27	0.04	6.92	4780
13	40.03	10.66	21.91	2.51	15.12	1.80	0.43	0.10	6.98	5051
14	27.93	4.30	16.73	1.49	11.37	1.36	0.20	0.05	5.25	4074
Average value	52.16	10.54	26.56	2.86	18.01	2.15	0.58	0.08	8.36	6238

municipal solid waste include the proximate analyzis: moisture, ash, volatile, fixed carbon, ultimate analyzis: Carbon, hydrogen, nitrogen,sulfur,oxygen,chlorine, and low heating value. Parameters are Proximate and ultimate analyzis of municipal solid waste as Tables 6.30 and 6.31.

Other data of thermal characteristics of municipal solid waste include loss on ignition, fusion point of ash, thermogravimetry and Differential scanning calorimetry (DSC). Fusion point of ash includes 4 data of De-formation temperature

Table 6.31 Ultimate analyzis and HHV of various sorted municipal solid wastes

	Material	Ultimate analyzis (%)					Experimental HHV (MJ/kg)
		Car	Har	Nar	St, ar	Oar	
Paper waste							
1	Coated paper	30.5	4.6	2.9	1.5	37.7	12.22
2	Newspaper	51.58	5.84	0.11	0.03	41.44	19.3
3	Cardboard	44.52	5.68	0.1	0.12	41.18	16.9
4	Glossy paper	32.83	3.46	0.1	0.04	35.58	10.4
5	Paper	41.44	8.19	0.11	0.34	43.2	17.68
6	High-grade paper	38.1	5.6	0.15	0.07	46.9	15.235
7	Paper mixture	43	6	0.36	0.17	43.8	17.647
8	Newspaper	43.8	5.9	0.29	0.24	44.4	17.987
9	Corrugated and kraft	46	6.4	0.28	0.22	44.8	18.999
10	Other paper waste	42.7	6.1	0.5	0.14	43.3	17.58
11	Papers	44.49	6.58	0.57	0	41.29	17.018
12	Magazines	35	5	0.05	0.08	39.4	13.551
13	Corrugated paper	45.99	6.35	0.14	0.29	44.26	18.119
14	Glossy paper	43.4	5.3	0.62	0.25	27.5	14.663
15	Magazine stock	32.93	4.64	0.11	0.21	32.85	15.133
16	Brown Paper	44.9	6.08	0	0.11	47.84	17.914
17	Cardboard	49.6	6.4	0.72	0.24	35.7	18.463
18	Corrugated boxes	43.73	5.7	0.09	0.21	44.93	17.278
19	Junk mail	37.87	5.41	0.17	0.09	42.74	14.837
20	Mixed paper	43.41	5.82	0.25	0.2	44.32	17.602
21	Newsprint I	49.14	6.1	0.05	0.16	43.03	19.717
22	Newsprint II	48.82	6.21	0.15	0.25	42.35	19.296
23	Non-glossy paper	47.3	6.1	1.58	0.25	32	19.661
24	Other paper	42.09	5.86	0.4	0.25	39.88	16.555
25	Paper food cartons	44.74	6.1	0.15	0.16	41.92	17.978
26	Plastic-coated paper	45.3	6.17	0.18	0.08	45.5	17.914
27	Trade magazine	32.91	4.95	0.07	0.09	38.55	12.744
28	Waxed milk cartons	59.18	9.25	0.12	0.1	30.13	27.281
29	Paper	43.5	6	0.3	0.2	44	16.809
30	Non-recyclable paper	41.64	5.62	0.34	0.14	44.91	15.58
31	Non-recyclable paper	42.85	5.79	0.1	0.12	45.48	17.37
Wood waste							
32	Block wood	46.9	6.07	0.95	0	43.99	18.261
33	Wood chips	48.1	5.99	0.08	0	45.74	19.916
34	Red wood	50.64	5.98	0.05	0.03	42.88	20.72
35	Soft wood	52.1	6.1	0.2	0	41	20
36	Spruce wood	51.9	6.1	0.3	0	40.9	20.1

(continued)

Table 6.31 (continued)

	Material	Ultimate analyzis (%)					Experimental HHV (MJ/kg)
		Car	Har	Nar	St, ar	Oar	
37	Subabul wood	48.15	5.87	0.03	0	44.75	19.777
38	Eucalyptus saw dust	49.37	6.4	2.02	0	42.01	18.502
39	Fibreboard	48.59	6.3	3.6	0.02	41.05	19.81
40	Spruce wood	47.4	6.3	0.07	0	46.2	19.3
41	Wood	44.76	8.81	0.28	0.45	43.82	23.01
42	Yard waste	40.47	7.47	1.53	0.44	41.85	28.57
43	Yard waste mixture	45	5.6	1.5	0.17	37.7	17.982
44	Wood	46.7	6	0.71	0.16	43.4	19.608
45	Yard waste—grass clippings	43.3	5.9	2.6	0.3	37.6	17.917
46	Yard waste—leaves	50	5.7	0.82	0.1	36	18.678
47	Yard waste—other yard waste	40.7	5	1.3	0.1	40	17.182
48	Oak wood (small branch)	46.78	6.09	2.7	0	40.38	19.2
49	Oak wood (medium branch)	47.16	6.32	2.5	0	41.01	19.24
50	Oak wood (large branch)	47.56	6.67	2.34	0	41.36	19.17
51	Bamboo wood	48.76	6.32	0.2	0	42.77	20.55
52	Forest residue	53.16	6.25	0.3	0.09	40	19.5
53	Pine chips	49.66	5.67	0.51	0.08	38.07	19.79
54	Wood	47.8	5.8	0.3	0.01	45.1	19
55	Wood	50.3	5.6	0.2	0	43.6	20.3
56	Wood	47.2	5.5	0.2	0.01	46.8	18.7
57	Wood	47.7	6.1	0.2	0.04	45.6	19.2
58	Wood and bark	50.46	5.97	0.15	0.05	42.37	20.063
59	Balsam spruce	53.3	6.66	1.49	0.2	35.17	22.187
60	Furniture wood	49.7	6.1	0.1	0.1	42.6	19.33
61	Brush	42.52	5.9	2	0.05	41.2	18.4
62	Demolition softwood	51	6.2	0.1	0.1	41.8	19.61
63	Evergreen shrubs	48.51	6.54	1.71	0.19	40.44	20.316
64	Flowering plants	46.65	6.61	1.21	0.26	40.18	18.667
65	Grass, dirt, leaves	36.2	4.75	2.1	0.26	26.61	14.61
66	Green logs	50.12	6.4	0.14	0.08	42.26	19.56
67	Lawn grass I	46.18	5.96	4.46	0.42	36.43	19.33
68	Lawn grass II	43.33	6.04	2.15	0.05	41.68	17.874
69	Mixed greens	40.31	5.64	2	0.05	39	16.463
70	Ripe leaves I	52.15	6.11	6.99	0.16	30.34	22.915

(continued)

Table 6.31 (continued)

	Material	Ultimate analyzis (%)					Experimental HHV (MJ/kg)
		Car	Har	Nar	St, ar	Oar	
71	Ripe leaves II	40.5	5.95	0.2	0.05	45.1	16.442
72	Rotten timbers	52.3	5.5	0.2	1.2	39	20.232
73	Waste hardwood	49.4	6.1	0.1	0.1	43.7	19.295
74	Wood	49.05	5.99	0.29	0.08	41.13	19.194
75	Wood and textiles	53.8	6	1.07	0.28	31.5	21.002
76	Yard waste	42.35	5.33	1.62	0.24	31.89	16.935
77	Wood	49.5	6	0.2	0.1	42.7	23.26
78	Yard	47.8	6	3.4	0.3	38	16.283
79	Wood	48.4	6	0.1	0	45.3	20.1
80	Wood	47.6	5.9	0.2	0.01	45.8	19.7
81	Wood	47.1	4.9	0.2	0.02	47.7	18.7
82	Wood	46.2	5.8	0.2	0.02	47.2	17.6
83	Wood	46.7	5.9	0.2	0.02	46.8	19.1
84	Wood	48.6	5.8	0.2	0.05	45.3	18.3
85	Wood	47.2	5.6	0.4	0	44.4	18.2
86	Wood	47.8	6	0.2	0.03	45.4	18.8
87	Wood	46.8	5.8	0.2	0.06	46.1	18.6
88	Wood	50.7	6	0.2	0	40.4	20.3
89	Wood	50.4	6.2	0.5	0.01	42.5	20.4
90	Wood	52.3	6.1	0.2	0.03	41.3	20.7
91	Wood	48.3	5.7	0.2	0.04	45.1	19.2
92	Wood waste	41.73	5.39	1.53	0.14	47.48	19.247
93	Wood waste	48.75	5.93	0.6	0.12	39.32	18.426
Plastic waste							
94	Plastic-PA6/PE	79.7	13.3	2.6	0.01	4.2	39.5
95	Plastic-PVC	38	5	0	0	0	19.2
96	Plastic-polyethylene	86	14	0	0	0	40.5
97	Plastic-polystyrene	92	8	0	0	0	44.5
98	Plastics	62.9	4.9	0	0.2	31.18	22.24
99	Plastic-polypropylene	86	14	0	0	0	43.4
100	Plastic-PMMA	60	8	0	0	32	25.8
101	Plastic-LDPE	85.7	14.2	0.05	0	0.05	46.6
102	Plastic-HDPE	86.1	13	0	0	0.9	46.4
103	Plastic-PP	86.1	13.7	0	0	0.2	46.4
104	HDPE bottles	81.6	13.6	0.1	0.2	1.9	43.794
105	Plastic mixture	76.3	11.5	0.26	0.2	4.4	38.377
106	Plastic—PET bottles	68.5	8	0.16	0.08	21.9	32.008
107	Polystyrene	86.3	7.9	0.28	0.3	3.4	39.479

(continued)

Table 6.31 (continued)

	Material	Ultimate analyzis (%)					Experimental HHV (MJ/kg)
		Car	Har	Nar	St, ar	Oar	
108	PVC bottles	44.2	5.9	0.26	0.89	7.6	23.632
109	Polyethylene bags and films	77.4	12.9	0.1	0.12	1.8	39.779
110	Other plastic waste	72.9	11.4	0.45	0.24	5.5	36.662
111	Plastics	79.53	10.64	0.46	0	6.98	39.798
112	Plastic	83.93	12.84	0	0	0.8	37
113	Plastics	66.39	9.17	1	0.34	9.47	31.699
114	Plastic film	67.21	9.72	0.46	0.07	15.82	32.2
115	Polyethylene I	84.54	14.18	0.06	0.03	0	45.972
116	Polyethyline II	84.57	14.13	0.07	0.03	0	45.876
117	Polystyrene I	87.1	8.45	0.21	0.2	3.96	38.337
118	Polystyrene II	86.87	8.42	0.2	0.2	4	38.261
119	Polyurethane I	63.27	6.26	5.99	0.2	17.65	26.162
120	Polyurethane II	63.23	8.32	6.01	0.2	17.64	26.106
121	Polyvinyl chloride I	45.14	5.61	0.08	0.14	1.56	22.776
122	Polyvinyl chloride II	45.09	5.61	0.1	0.1	1.6	22.729
123	Mixed plastics	60	7.2	0	0	22.6	34.092
124	Plastics	60	7.2	0	0	22.8	33.229
125	Plastic-rigid	71.23	7.62	0.08	0.39	19.58	35.721
126	Plastic-rigid	85.75	7.18	0.22	0.31	6.48	43.341
127	Plastic-film and styrofoam	90.38	7.53	0.69	0.4	1	42.461
128	Plastic-film and styrofoam	85.4	9.67	3.02	0.78	0.58	47.06
Rubber and leather waste							
129	Scrap tyre	80.3	5.18	0	0	10.33	33.3
130	Waste tire	84.22	7.58	0.4	1.41	2.67	35.588
131	Tire	60.9	5.3	0.28	2.46	7.1	29
132	Rubber	52.77	9.44	0.78	0.41	13.52	25.04
133	Leather shoe	42.01	5.32	5.98	1	22.83	19.667
134	Shoe heel and sole	53.22	7.09	0.5	1.34	7.76	25.938
135	Tires	79.1	6.8	0.1	1.5	5.9	32.673
136	Leather	60	8	10	0.4	11.5	22.867
137	Rubber and leather	47.88	5.97	1.49	1.3	12.86	21.791
138	Rubber	43.94	5.13	0.18	0.01	49.64	20.834
139	Rubber	53.44	7.2	3.69	0.38	23.19	22.672
Textile waste							
140	Linoleum	48.06	5.34	0.1	0.4	18.7	19.745
141	Oils, paints	66.85	9.65	2	0	5.2	31.16

(continued)

Table 6.31 (continued)

	Material	Ultimate analyzis (%)					Experimental HHV (MJ/kg)
		Car	Har	Nar	St, ar	Oar	
142	Rags	55	6.6	4.62	0.13	31.2	19.778
143	Textile	55.57	6.02	0.08	0.25	36.27	21.22
144	Textiles I	46.19	6.41	2.18	0.2	41.85	18.69
145	Upholstery	47.1	6.1	0.3	0.1	43.6	18.679
146	Textiles II	49.64	6.69	4.15	0.37	36.15	20.449
147	Textiles	55	6.6	4.6	0.2	31.2	19.161
148	Carpet	61.68	8.71	5.46	0.55	21.1	28.698
149	Plastic-textile	56.7	6.36	0.37	0.21	34.15	26.381
150	Plastic-textile	59.37	4.26	0.07	0	35.25	20.743
151	Carpet	62.49	9.7	0.36	0.72	1.97	25.793
MSW mixture							
152	MSW Thai	37.14	5.41	0.22	0.09	24.93	15.59
153	Leather, wood, textile, rubber	44.99	6.01	0.55	0	39.59	17.98
154	MSW total	53	7.32	1.32	0	31.06	25.504
155	RDF (refuse derived fuel)	59.2	8.22	0.52	0	26.75	27.633
156	MSW in Taiwan	40.84	6.27	0.83	0.17	27.68	18.188
157	Raw refuse	46.68	6.61	1.03	0.1	35.94	20.252
158	MSW in Taiwan	38.96	5.86	1.13	0.23	25.9	16.311
159	Textiles/rubber/ leather	50.3	6.4	3.3	0.33	31.3	23.202
160	MSW UK	35.81	4.82	0.78	0.41	24.43	15.17
161	Raw municipal refuse	47.6	6	1.2	0.3	32.9	19.878
162	Processed municipal refuse	47.3	5	1.4	0.2	32.6	21.353
163	Heil mill industrial refuse	33.9	4.6	0.7	0.4	22.4	13.13
164	Gondard mill infustrial refuse	31.2	3.8	0.7	0.1	21.8	11.991
165	MSW	51.81	5.76	0.26	0.36	35.88	22.527
166	MSW	41.22	9.07	1.23	0.79	24.18	22.44
167	Raw refuse	45.93	6.28	0.54	0.24	35.34	19.597
168	Raw refuse	47.18	6.23	0.44	0.22	39.82	19.39
169	Raw refuse	47.16	6.38	0.6	0.14	37.92	19.636
170	Raw refuse	46.75	6.35	0.67	0.1	38.05	19.485
171	Raw refuse	44.27	6.25	0.75	0.2	33.87	19.055
172	Raw refuse	46.33	6.02	0.75	0.2	35.54	19.071
173	MSW in Taiwan	35.89	5.51	1.03	0.4	23	18.122

(continued)

Table 6.31 (continued)

	Material	Ultimate analyzis (%)					Experimental HHV (MJ/kg)
		Car	Har	Nar	St, ar	Oar	
174	MSW	41.21	6.62	1.11	0.09	32.13	18.581
175	MSW	41.07	6.55	1.09	0.09	32.07	18.489
176	MSW	40.87	6.45	1.05	0.09	31.99	18.368
177	MSW	40.6	6.32	1.01	0.09	31.88	18.198
178	MSW	40.2	6.12	0.95	0.1	31.77	17.943
179	MSW	39.54	5.79	0.84	0.11	31.45	17.521
180	Rubber, leather, and hard plastics	53.8	8.9	0.83	0.57	23.3	25.412
Other wastes							
181	BSG (Brewer's Spent Grains) waste	49	6.71	3.94	0.22	35.1	20.83
182	Coffee waste	47.95	6.34	2.82	0.2	36.11	19.82
183	Organics/ Combustibles mixture	48.6	6.8	0.94	0.22	35	21.292
184	Other organics mixture	46.2	6.1	1.9	0.36	33.3	20.232
185	Fines	37.3	5.3	1.6	0.45	29.5	16.231
186	Disposable diapers	48.4	7.6	0.51	0.35	38.8	22.611
187	Other organics	44.2	5.3	1.8	0.81	14.4	17.301
188	Putrescribles	45.56	6.24	2.29	0	36.19	18.63
189	Cotton residue	47.03	5.96	1.79	0.19	38.42	16.9
190	Steet sweepings	34.7	4.76	0.14	0.2	35.2	17.438
191	Vacuum cleaner dirt	35.69	4.73	6.26	1.15	20.08	16.619
192	Household dirt	20.62	2.87	0.5	0.01	4	9.101
193	Other organic	34.6	4.3	1.07	0.38	41.1	16.521

A ash; *V* volatile; *FC* fixed carbon; *HHV* high heating value; *d* dry ba-sis; *daf* dry ash free basis
[a]It is Cl for PVC
Reprinted with permission from Elsevier [168]

(DT), softening temperature (ST), hemispherical temperature (HT), flow temperature (FT). Thermogravimetry include activation energy (E) and Arrhenius pre-exponential factor (A). Differential scanning calorimetry (DSC) include specific heat (Cp) and thermal conductivity.

The values presented in the above two tables further demonstrate the wide variation that can be measured for similar or the same material. Therefore care must be taken to fully understand the conditions of the testing or experiment and the exact nature and source of the material.

Table 6.32 Kinetic parameters of the cellulosic fraction and plastics

	First peak		Second peak		Third peak	
	E (kJ/mol)	log A (log s^{-1})	E (kJ/mol)	log A (log s^{-1})	E (kJ/mol)	log A (log s^{-1})
Cellulosic fraction						
Newspaper	96.7	6.12	246.1	18.03	47.4	1.06
Cardboard	112.3	7.82	230.6	17.25	46	0.99
Recycled paper	136.2	10.11	214	15.67	36.2	0.14
Glossy paper	99.4	6.33	274.7	20.9	44.4	0.65
Spruce	109.8	7.35	255.2	18.85	41.3	0.62
Juice carton[a]	107.5	7.15	248.3	18.85	55.1	1.52
Average	110.3	7.5	244.8	18.3	45.1	0.8
Standard deviation	14	1.4	20.8	1.8	6.3	0.5
Plastics						
HDPE	445.1	29.1				
LDPE	340.8	21.98				
PP	336.7	22.18				
PS	311.5	21.84				
PVC	388.2	34.13	110.4	7.59	149.8	8.5
Juice carton[b]	339.7	21.94				

[a]Only paperboard
[b]Only polyethylene
Reprinted with permission from Elsevier [169]

Table 6.32 provides some results of the pyrolysis characteristics of different components, representing the dry cellulosic fraction and plastics of MSW. Here it can be observed that the different PE fractions provide different activation energies and pre-exponential factors within the same study.

References

1. U.S. EPA (U.S. Environmental Protection Agency): Quantity of municipal solid waste generated and managed. EPA's Report on the Environment (ROE). https://cfpub.epa.gov/roe/indicator.cfm?i=53. Accessed Oct 2016
2. Fischer, C., Crowe, M.: Household and Municipal Waste: Comparability of Data in Eea Member Countries. European Environment Agency, Copenhagen, Denmark (2000)
3. Sahimaa, O., Hupponen, M., Horttanainen, M., Sorvari, J.: Method for residual household waste composition studies. Waste Manage. **46**, 3–14 (2015)
4. Edjabou, M.E., Jensen, M.B., Götze, R., Pivnenko, K., Petersen, C., Scheutz, C., Astrup, T. F.: Municipal solid waste composition: Sampling methodology, statistical analyzes, and case study evaluation. Waste Manage. **36**, 12–23 (2015)

5. Liikanen, M., Sahimaa, O., Hupponen, M., Havukainen, J., Sorvari, J., Horttanainen, M.: Updating and testing of a Finnish method for mixed municipal solid waste composition studies. Waste Manage. **52**, 25–33 (2016)
6. Zettl, E., Milunov, M., Potrykus, A., Kostadinova, T.: Waste Classification, Sampling and Analyzis. Federal Environment Agency, Germany (2015)
7. U.S. EPA (U.S. Environmental Protection Agency). MSW Characterization Methodology. https://www.epa.gov/sites/production/files/2015-09/documents/06numbers.pdf. Accessed Sep 2015
8. Nordtest Method. Solid waste, municipal: sampling and characterisation. Nordtest method NT ENVIR, Finland. http://www.nordtest.info/images/documents/nt-methods/environment/NT%20envir%20001_Solid%20waste,%20municipal_Sampling%20and%20characterisation_Nordtest%20Method.pdf. Accessed May 1995
9. Haque, M.: Sampling methods in social research. Ph.D. Scholar, Visva Bharati, Santiniketan, West Bangal (2010)
10. Dodd, S.-J., Epstein, I.: Practice-based Research in Social Work: A Guide for Reluctant Researchers. Routledge, New York (2012)
11. Dahlén, L., Lagerkvist, A.: Methods for household waste composition studies. Waste Manage. **28**, 1100–1112 (2008)
12. Nilsson, P.: Waste collection: equipment and vehicles. Solid Waste Technol. Manage. **1** and **2**, 251–76 (2010)
13. Environmental Protection Agency: Municipal Waste Characterisation. Wexford, Ireland (1996)
14. Fitzgerald, G.C., Themelis, N.: Technical and economic analysis of pre-shredding municipal solid wastes prior to disposal. Unpublished MS thesis Columbia University, New York (2009)
15. Cheremisinoff, N.P.: Handbook of Solid Waste Management and Waste Minimization Technologies. Butterworth, Heinemann (2003)
16. Ramachandra, T.: Management of Municipal Solid Waste. Environmental Engineering Series, The Energy and Resources Institute (TERI) (2006)
17. Cimpan, C., Maul, A., Jansen, M., Pretz, T., Wenzel, H.: Central sorting and recovery of MSW recyclable materials: a review of technological state-of-the-art, cases, practice and implications for materials recycling. J. Environ. Manage. **156**, 181–199 (2015)
18. Gonzalez-Torre, P.L., Adenso-Dıaz, B., Ruiz-Torres, A.: Some comparative factors regarding recycling collection systems in regions of the USA and Europe. J. Environ. Manage. **69**, 129–138 (2003)
19. Gundupalli, S.P., Hait, S., Thakur, A.: A review on automated sorting of source-separated municipal solid waste for recycling. Waste Manage. **60**, 56–74 (2017)
20. Shao, L.-M., Ma, Z.-H., Zhang, H., Zhang, D.-Q., He, P.-J.: Bio-drying and size sorting of municipal solid waste with high water content for improving energy recovery. Waste Manage. **30**, 1165–1170 (2010)
21. Chandrappa, R., Das, D.B.: Waste quantities and characteristics. Solid Waste Management, pp. 47–63. Springer, Berlin, Heidelberg (2012)
22. Perazzini, H., Freire, F.B., Freire, F.B., Freire, J.T.: Thermal treatment of solid wastes using drying technologies: a review. Dry Technol. **34**, 39–52 (2016)
23. McCormick, P., Porter, H., Lucas, R., Wells, D.: Solids drying fundamentals. In: Chemical Engineers Handbook, 5th edn. pp. 20–4. Mc Graw-Hill Co, New York (1983)
24. Mujumdar, A.S.: Principles, classification, and selection of dryers. In Handbook of Industrial Drying 4th edn, pp. 3–29. CRC Press, Boca Raton, FL (2015)
25. Velis, C., Longhurst, P.J., Drew, G.H., Smith, R., Pollard, S.J.: Biodrying for mechanical–biological treatment of wastes: a review of process science and engineering. Bioresource Technol. **100**, 2747–2761 (2009)
26. Ragazzi, M., Rada, E., Panaitescu, V., Apostol, T.: Municipal solid waste pre-treatment: a comparison between two dewatering options. WIT Trans. Ecol. Environ. **102** (2007)

27. Phuntsho, S., Dulal, I., Yangden, D., Tenzin, U.M., Herat, S., Shon, H.: Studying municipal solid waste generation and composition in the urban areas of Bhutan. Waste Manage. Res. **28**, 545–551 (2010)
28. Sfeir, H., Reinhart, D.R., McCauley-Bell, P.R.: An evaluation of municipal solid waste composition bias sources. J. Air Waste Manag. Assoc. **49**, 1096–1102 (1999)
29. Burnley, S., Ellis, J., Flowerdew, R., Poll, A., Prosser, H.: Assessing the composition of municipal solid waste in Wales. Resour. Conserv. Recycl. **49**, 264–283 (2007)
30. Scott, P.: The international energy agency's (IEA) work in harmonising sampling and analytical protocols related to municipal solid waste (MSW) conversion to energy. Biomass Bioenerg. **9**, 415–439 (1995)
31. Worrell, W.A., Vesilind, P.A.: Solid Waste Engineering, Cengage Learning, SI Version (2011)
32. den Boer, E., Jędrczak, A., Kowalski, Z., Kulczycka, J., Szpadt, R.: A review of municipal solid waste composition and quantities in Poland. Waste Manag **30**, 369–377 (2010)
33. Eggleston H.S., Buendia L., Miwa K., Ngara T., Tanabe K. (eds.): IPCC Guidelines for National Greenhouse Gas Inventories. Prepared by the National Greenhouse Gas Inventories Programme of the Intergovernmental Panel on Climate Change. IGES, Japan (2006)
34. Zhang, D.Q., Tan, S.K., Gersberg, R.M.: Municipal solid waste management in China: status, problems and challenges. J. Environ. Manage. **9**, 1623–1633 (2010)
35. ASTM. Standard Test Method for Measuring Particle Size Distribution of RDF-5. United States: American Society for Testing and Materials, E1037–15 (2017)
36. Zhou, H., Meng, A., Long, Y., Li, Q., Zhang, Y.: An overview of characteristics of municipal solid waste fuel in China: physical, chemical composition and heating value. Renew. Sustain. Energy Rev. **36**, 107–122 (2014)
37. ASTM. Standard Test Method for Determining the Bulk Density of Solid Waste Fractions. United States: American Society for Testing and Materials, E 1109–86 (2004)
38. Ruf, J.A.: Particle size spectrum and compressibility of raw and shredded municipal solid waste, University of Florida (1974)
39. EPA: Volume-to-Weight Conversion Factors. US: Office of Resource Conservation and Recovery (2016)
40. EPA: Waste Materials—Density Data. http://www.epa.vic.gov.au/business-and-industry/lower-your-impact/~/media/Files/bus/EREP/docs/wastematerials-densities-data.pdf
41. Pichtel, J.: Waste Management Practices: Municipal, Hazardous, and Industrial. CRC press (2005)
42. Vesilind, P.A., Worrell, W.A., Reinhart, R.: Solid Waste Engineering, India Library (2002)
43. Zekkos, D., Fei, X., Grizi, A., Athanasopoulos, G.: Response of municipal solid waste to mechanical compression. J. Geotech. Geoenviron. Eng. **143**, 04016101 (2016)
44. Hossain, M., Gabr, M., Barlaz, M.: Relationship of compressibility parameters to municipal solid waste decomposition. J. Geotech. Geoenviron. Eng. **129**, 1151–1158 (2003)
45. Oweis, I.S., Khera, R.P.: Geotechnology of waste management (1990)
46. ASTM: standard test methods for one-dimensional consolidation properties of soils using incremental loading. united states: american society for testing and materials, D2435/D2435M-1 (2011)
47. Zhang, Z., Dazhi, W.: Study on the compressibility of municipal solid waste in Hangzhou, China. In: International Conference on Mechanic Automation and Control Engineering (MACE): IEEE, pp. 1566–8 (2010)
48. Zhang, Z., Dazhi, W.: Study on the compressibility for municipal solid waste with higher content of organic matter. In: 2nd International Conference on Information Science and Engineering (ICISE): IEEE, pp. 4078–81 (2010)
49. Chen, Y., Ke, H., Fredlund, D.G., Zhan, L., Xie, Y.: Secondary compression of municipal solid wastes and a compression model for predicting settlement of municipal solid waste landfills. J. Geotech. Geoenviron. Eng. **136**, 706–717 (2010)

50. Gourc, J.-P., Staub, M., Conte, M.: Decoupling MSW settlement into mechanical and biochemical processes–modelling and validation on large-scale setups. Waste Manage. **30**, 1556–1568 (2010)
51. Bareither, C.A., Benson, C.H., Edil, T.B.: Recent findings on compressibility of municipal solid waste. GeoCongress State Art Pract. Geotech. Eng. **2012**, 4212–4221 (2012)
52. Hossain, M.S., Gabr, M., Asce, F.: The effect of shredding and test apparatus size on compressibility and strength parameters of degraded municipal solid waste. Waste Manage. **29**, 2417–2424 (2009)
53. ASTM: Standard test method for permeability of granular soils (Constant Head). American Society for Testing and Materials, United States, D2434-68 (2006)
54. Taufiq, T.: Characteristics of fresh municipal solid waste: The University of Texas at Arlington (2010)
55. Chen, T.-H., Chynoweth, D.P.: Hydraulic conductivity of compacted municipal solid waste. Bioresour. Technol. **51**, 205–212 (1995)
56. Hossain, M.S., Penmethsa, K.K., Hoyos, L.: Permeability of municipal solid waste (MSW) in bioreactor landfill with degradation. GeoCongress 2008: Geotechnics of Waste Management and Remediation. pp. 120–7 (2008)
57. Dixon, N., Jones, D.R.V.: Engineering properties of municipal solid waste. Geotext. Geomembr. **23**, 205–233 (2005)
58. Ivanov, K.; Zaprjanova, P.; Angelova, V.; Bekjarov, G.; Dospatliev, L.; Gilkes, R. J.; Prakongkep, N.: ICP determination of phosphorous in soils and plants. In: Proceedings of the 19th World Congress of Soil Science: Soil solutions for a changing world, Brisbane, Australia, 1–6 August 2010. Working Group 1.5 Soil sense: rapid soil measurements (2010)
59. Yang, N.: Environmental analysis of municipal solid waste landfilling in China based on life cycle perspective. Ph.D. Thesis, Shanghai: Tongji university (2014)
60. Kim, M.H., Kim, J.W.: Comparison through a LCA evaluation analysis of food waste disposal options from the perspective of global warming and resource recovery. Sci. Total Environ. **408**(19), 3998–4006 (2010)
61. Faaij, A., van Ree, R., Waldheim, L., Olsson, E., Oudhuis, A., van Wijk, A., DaeyOuwens, C., Turkenburg, W.: Gasification of biomass wastes and residues for electricity production. Biomass Bioenerg. **12**(6), 387–407 (1997)
62. Zhang, Y., Banks, C.J., Heaven, S.: Anaerobic digestion of two biodegradable municipal waste streams. J. Environ. Manage. **104**, 166–174 (2012)
63. Yang, N., Damgaard, A., Scheutz, C., Shao, L.M., He, P.J.: A comparison of chemical MSW compositional data between China and Denmark. J. Environ. Sci. **74**, 1–10 (2018)
64. Soest, P.J.V.: Development of a comprehensive system of feed analyzes and its application to forages. J. Anim. **26**(1), 119–128 (1967)
65. Van Soest, P.J.: Use of detergents in the analyzis of fibrous feeds. 2. A rap-id method for the determination of fiber and lignin. J. Assoc. Off. Anal. Chem. **49**, 546–551 (1963)
66. Provenzano, M.R., Malerba, A.D., Pezzolla, D., Gigliotti, G.: Chemical and spectroscopic characterization of organic matter during the anaerobic digestion and successive composting of pig slurry. Waste Manage. **34**(3), 653–660 (2014)
67. Machado, S.L., Carvalho, M.F., Gourc, J.P., Vilar, O.M., Nascimento, J.C.F.D.: Methane generation in tropical landfills: simplified methods and field results. Waste Manage. **29**(1), 153–161 (2009)
68. Huang, Q., Wang, Q., Dong, L., Xi, B., Zhou, B.: The current situation of solid waste management in China. J. Mater. Cycles Waste Manage. **8**(1), 63–69 (2006)
69. M, B.: Regional study on policies and institutional assessment of solid waste management in Egypt, France: Blue Plan Regional Activity Centre (2000)
70. Unnikrishnan, S., Singh, A.: Energy recovery in solid waste management through CDM in India and other countries. Resour. Conserv. Recycl. **54**(10), 630–640 (2010)
71. EPA: Municipal solid waste in The United States: 2011 facts and figures. Environmental Protection Agency, U.S.A. (2013)

72. ABS: Australia's environment: issues and trends. Australian Bureau of Statistic, Australia (2003)
73. OECD: OECD environmental data compendium. Environment Directorate, France, 2006/ 2007
74. DEFRA: Municipal waste composition: review of municipal waste component analyzes—final report, London: Department for Environment Food and Rural Affairs (2008)
75. Mühle, S., Balsam, I., Cheeseman, C.R.: Comparison of carbon emissions associated with municipal solid waste management in Germany and the UK. Resour. Conserv. Recycl. **54** (11), 793–801 (2010)
76. Eleazer, W.E., Odle, W.S., Wang, Y.S., Barlaz, M.A.: Biodegradability of municipal solid waste components in laborato-ry-scale landfills. Environ. Sci. Technol. **31**(3), 911–917 (1997)
77. Chan, C., Jia, C.Q., Graydon, J.W., Kirk, D.W.: The behaviour of selected heavy metals in MSW incineration electrostatic precipitator ash during roasting with chlorination agents. J. Hazard. Mater. **50**(1), 1–13 (1996)
78. Krogmann, U.: Composting–basics of collection and treatment of biogenic wastes of different compositions. Economica-Verlag, Germany (1994)
79. Badami, M., Mittica, A., Poggio, A.: MSW incineration capacity evaluations for the province of Turin (Northen Italy). Ann. N. Y. Acad. Sci. **456**(12), 436–437 (2008)
80. Zheng, W., Phoungthong, K., Lü, F., Shao, L.M., He, P.J.: Evaluation of a classification method for biodegradable solid wastes using anaerobic degradation parameters. Waste Manage. **33**(12), 2632–2640 (2013)
81. Saikia, N., Cornelis, G., Mertens, G., Elsen, J., Van Balen, K., Van Gerven, T., Vandecasteele, C.: Assessment of Pb-slag, MSWI bottom ash and boiler and fly ash for using as a fine aggregate in cement mortar. J. Hazard. Mater. **154**(1–3), 766–777 (2008)
82. Qiang, L.S., Zhao, J.P., Yang, Y.L.: Preparation and characterization of novel functional materials. Harbin Institute of Technology Press, Harbin (2017)
83. Qiu, P.S., Wang, G.F., Guo, L.W.: Material modern analysis test method experiment guide. Harbin Institute of Technology Press, Harbin (2001)
84. Ji, T.D., Lin, Z.R., Guo, L.W.: Secondary Ion Mass Spectrometry and Ion Probes. Science Press, Beijing (1989)
85. Mangialardi, T., Paolini, A.E., Polettini, A., Sirini, P.: Optimization of the solidification/ stabilization process of MSW fly ash in cementitious matrices. J. Hazard. Mater. **70**(1–2), 53–70 (1999)
86. Rémond, S., Pimienta, P., Bentz, D.P.: Effects of the incorporation of municipal solid waste incineration fly ash in cement pastes and mortars: I. Exp. Study. Cem. Concr. Res. **32**(2), 303–311 (2002)
87. Müller, U., Rübner, K.: The microstructure of concrete made with municipal waste incinerator bottom ash as an aggregate component. Cem. Concr. Res. **36**(8), 1434–1443 (2006)
88. Bethanis, S., Cheeseman, C.R., Sollars, C.J.: Properties and micro-structure of sintered incinerator bottom ash. Ceram. Int. **28**(8), 881–886 (2002)
89. Hass, A., Fine, P.: Sequential selective extraction procedures for the study of heavy metals in soils, sediments, and waste materials-a Critical Review. Crit.Al Rev. Environ. Sci. Technol. **40**(5), 365–399 (2010)
90. Tessier, A., Campbell, P.G.C., Bisson, M.: Sequential extraction procedure for the speciation of particulate trace metals. Anal. Chem. **51**(7), 844–851 (1979)
91. Ure, A.M., Quevauviler, P.H., Muntau, H., Griepink, B.: Speciation of heavy metals in soils and sediments. An account of the improvement and harmonization of extraction techniques undertaken under the auspices of the BCR of the Commission of the European Communities. Int. J. Environ. Anal. Chem. **51**, 135–151 (1994)
92. Kersten, M., Schulz, B.: Speciation of Cr in leachates of a MSWI bottom ash landfill. Environ. Sci. Technol. **32**, 1398–1403 (1998)
93. Krishnamurti, G.S., Naidu, R.: Solid-solution speciation and phytoavailability of copper and zinc in soils. Environ. Sci. Technol. **36**(12), 2645–2651 (2002)

94. Pan, Y., Wu, Z., Zhou, J., Zhao, J., Ruan, X., Liu, J., Qian, G.: Chemical characteristics and risk assessment of typical municipal solid waste incineration (MSWI) fly ash in China. J. Hazard. Mater. **261**, 269–276 (2013)
95. Luan, J., Chai, M., Liu, Y., Ke, X.: Heavy-metal speciation redistribution in solid phase and potential environmental risk assessment during the conversion of MSW incineration fly ash into molten slag. Environ. Sci. Pollut. Res. **25**(4), 3793–3801 (2018)
96. Yu, J., Sun, L., Xiang, J., Jin, L., Hu, S., Su, S., Qiu, J.: Physical and chemical characterization of ashes from a municipal solid waste incinerator in China. Waste Manage. Res. **31**(7), 663–673 (2013)
97. Jiao, F., Zhang, L., Dong, Z., Namioka, T., Yamada, N., Ninomiya, Y.: Study on the species of heavy metals in MSW incineration fly ash and their leaching behavior. Fuel Process. Technol. **152**, 108–115 (2016)
98. Saqib, N., Bäckström, M.: Chemical association and mobility of trace elements in 13 different fuel incineration fly ashes. Fuel **165**, 193–204 (2016)
99. Yao, J., Li, W.B., Kong, Q.N., Wu, Y.Y., He, R., Shen, D.S.: Content, mobility and transfer behavior of heavy metals in MSWI bottom ash in Zhejiang province, China. Fuel **89**(3), 616–622 (2010)
100. Soil, sludge and treated biowaste—Determination of pH. https://horizontal.ecn.nl
101. Sun, C.Y., Wen, C.X., Liu, X.L., Liu, B.: Effects of initial pH on anaerobic digestion of biogas from domestic refuse and sludge mixtures. Chin. J. Agric. Technol. **9**, 194–196 (2012)
102. Liu, Q., Sun, Y.H., Kou, W., Cao, Y.X.: Compositions analyzis and an-aerobic digestion characteristics on seasonal municipal solid waste. Chin. J. Environ. Eng. **11**, 4507–4512 (2013)
103. Yue, B., Zhang, Z.B., Huang, Q.F., Li, H.: Study on the physico-chemical properties of living solid waste in several typical villages and towns in China. Environ. Eng. **7**, 105–110 (2014)
104. Medina-Salasa, L.D., Castillo-Gonzáleza, E., Romero-Lópezb, R.: Physical and chemical characteristics of municipal solid waste in a rural locality-study case: cosautlán de carvajal, Veracruz, Mexico. Int. J. Bus. Hum. Ities Technol. **31**(8), 117–122 (2013)
105. Zhang, A.P., Li, M., Chen, Y.M., Hu, L.Z.: Study on characteristics, peasants' awareness and disposal mode of rural household solid waste around Chengdu. Environ. Pollut. Control. **3**, 307–313 (2017)
106. Chen, S.H., Zhang, S.M.: Municipal Waste Compost Management and Technology. Fudan University Press, Shanghai (1990)
107. Wei, Z.M., Wang, S.P., Xi, B.D., Zhao, Y.: Changes of humic substances and organic nitrogen forms during municipal solid waste composting. Acta Sci. Circum. **2**, 235–240 (2007)
108. Lv, B., Pu, G.B.: Study on transformation behavior of nitrogen in an-aerobic digestion of municipal solid waste. Chem. Biol. Eng. **9**, 77–81 (2010)
109. EPA: Liquid-solid partitioning as a function of liquid-solid ratio for constituents in solid materials using an up-flow percolation column procedure. Environmental Protection Agency, U.S.A. (2009)
110. EPA: Mass transfer rates of constituents in monolithic or compacted granular materials using a semi-dynamic tank leaching procedure. U.S.A.: Environmental Protection Agency (2017)
111. EPA: Liquid-solid partitioning as a function of liquid-to-solid ratio in solid materials using a parallel batch procedure. U.S.A.: Environmental Protection Agency (2012)
112. Gawlik, B., van der Sloot, H., Ulberth, F., Nortcliff, S., Simonart, T., Cooper, B., Leschber, R., Andersen, K., Wichman, K.: On the development of horizontal CEN standards supporting the implementation of EU directives on sludge, soil and biowaste—project horizontal. TracTrends Anal. Chem. **23**(10–11), 686–703 (2004)
113. Chinese Standard: Solid waste-extraction procedure for leaching toxicity-sulphuric acid and nitric acid method HJT299–2007 (2007)

114. Chinese Standard: Solid waste-extraction procedure for leaching toxicity-acetic acid buffer solution method HJ/T300-2007 (2007)
115. Chinese Standard: Solid waste-Extraction procedure for leaching toxicity-Horizontal vibration method HJ 557-2010 (2010)
116. Liu, H.H., Sang, S.X., Zhou, X.Z., Lian, C.X.: Leaching characteristics of heavy metals in municipal solid waste under simulated rain. Geochimica **6**, 587–594 (2008)
117. He, P.J., H., Zhang, H., Wang, Z.D., Zhang, C.G.: Pollution characteristics of air pollution control residues from municipal solid waste incineration plant. J. Tongji Univ. (Nat. Sci.) **31** (8), 972–976 (2003)
118. Shen, D.S., Zhen, Y.G., Yao, J., Wang, M.Z., Zhang, Y.: Analysis of pollution characteristics of solid waste incinerator fly ash in Zhejiang province. Chin. J. Environ. Sci. **9**, 2610–2616 (2011)
119. Zhang, H., Yu, S.Y., Shao, L.M., He, P.J.: Influence of air pollution control (APC) systems and furnace type on the characteristics of APC residues from municipal solid waste incinerators. Chin. J. Environ. Sci. **1**, 467–476 (2018)
120. Yu, L., Xie, Y.D., Ji, H.B., Pang, X.L., Liu, J.S.: Leaching tests of domestic garbage slag heavy metals under different pH. Environ. Sci. Manag. **1**, 111–114 (2016)
121. ASTM: Standard Test Method for Moisture in the Analysis Sample of Coal and Coke. ASTM D3173/D3173M-17a (2011)
122. ISO: Solid mineral fuels—Hard coal—Determination of moisture in the general analysis test sample by drying in nitrogen. ISO 11722:2013 (2013)
123. AS: Coal and coke—Analysis and testing Proximate analysis of higher rank coal. AS 1038.3 (2000)
124. ASTM: Standard Test Method for Sulfur in the Analysis Sample of Coal and Coke Using High-Temperature Tube Furnace Combustion. ASTM D4239-14 (2014)
125. ISO: Solid mineral fuels—Determination of total sulfur—High temperature combustion method. ISO 351:1996 (1996)
126. AS: Coal and coke—Analysis and testing Higher rank coal and coke—Ultimate analysis—Total sulfur—High-temperature combustion method. AS 1038.6.3.2-2003 (R2013)
127. ASTM: Standard Test Method for Volatile Matter in the Analysis Sample of Coal and Coke. ASTM D3175-18 (2018)
128. ISO: Hard coal and coke—Determination of volatile matter. ISO 562:2010 (2010)
129. ASTM: Standard Test Methods for Proximate Analysis of the Analysis Sample of Coal and Coke by Instrumental Procedures. ASTM D5142-09 (2009)
130. ISO: Coal—Proximate analysis. ISO 17246:2010 (2010)
131. AS: Coal and coke—Analysis and testing Proximate analysis of higher rank coal. AS 1038.3-2000 (2000)
132. ASTM: Standard Test Methods for Instrumental Determination of Carbon, Hydrogen, and Nitrogen in Laboratory Samples of Coal. ASTM D5373-08 (2008)
133. ASTM: Standard Test Method for Ash in the Analysis Sample of Coal and Coke from Coal. ASTM D3174-12(2018)
134. ASTM: Standard Test Method for Gross Calorific Value of Coal and Coke. D5865-13 (2013)
135. ISO: Solid mineral fuels—determination of gross calorific value by the bomb calorimetric method and calculation of net calorific value. ISO 1928 (2009)
136. AS: Coal and coke-Analysis and testing Gross calorific value. AS1038.5 (1998)
137. Kathiravale, S., Yunus, M.N.M., Sopian, K., Samsuddin, A.H., Rah-man, R.A.: Modeling the heating value of municipal solid waste. Fuel **82**(9), 1119–1125 (2003)
138. Dean, W.: Determination of carbonate and organic matter in calcareous, sediments and sedimentary rock by loss on ignition: comparison with other methods. J. Sediment. Petrol. **44**, 242–248 (1974)
139. Eli, SF., Benjamin, C.: The application of thermoanalytical techniques to reaction kinetics: the thermogravimetric evaluation of the kinetics of the decomposition of calcium oxalate monohydrate. J. Phys. Chem. **62**, 394–397 (1958)

140. Stanier, D.: An introduction to differential scanning calorimetry (2013)
141. Vyazovkin, S., Burnham, A.K., Criado, J.M., Pérez-Maqueda, L.A., Popescu, C., Sbirrazzuoli, N.: ICTAC kinetics committee recommendations for performing kinetic computations on thermal analyzis data. Thermochim. Acta **520**(1–2), 1–19 (2011)
142. Anderson, M.B.: Composition of municipal solid waste in the united states and implications for carbon sequestration and methane yield. J. Environ. Eng. **135**(10), 901–909 (2009)
143. Painter, R., Watson, W.: Tennessee waste characterization study, pp. 16–20. Tennessee State University Department of Civil and Environmental Engineering, Tennessee (2008)
144. Yamada, T., Asari, M., Miura, T.: Municipal solid waste composition and food loss reduction in Kyoto City. J. Mater. Cycles Waste Manage. **3**, 1–10 (2017)
145. Zhang, D., Keat, T.S., Gersberg, R.M.: A comparison of municipal solid waste management in Berlin and Singapore. Waste Manage. **30**(5), 921 (2010)
146. Alwaeli, M.: An overview of municipal solid waste management in Poland. The current situation, problems and challenges. Environ. Prot. Eng. **4**, 181–193 (2015)
147. Sharholy, M., Ahmad, K., Vaishya, R.C.: Municipal solid waste characteristics and management in Allahabad. India. Waste Manage. **27**(4), 490 (2007)
148. Tan, S.T., Lee, C.T., Hashim, H.: Optimal process network for municipal solid waste management in Iskandar Malaysia. J. Clean. Prod. **71**(4), 48–58 (2014)
149. Hussain, M., Haider, S., Abbas, Y.: A study of source specific quantification, composition and disposal methods of municipal solid waste at Konodas Gilgit City, Pakistan. J. Biodivers. Environ. Sci. **8**(5), 20162222–20166663 (2016)
150. Inglezakis, V., Dvorsak, S., Varga, J.: Municipal solid waste experimental studies in Romania and Bulgaria. Cogenttech **3**(3), 64–73 (2012)
151. Montejo, C., Costa, C., Ramos, P.: Analysis and comparison of municipal solid waste and reject fraction as fuels for incineration plants. Appl. Therm. Eng. **31**(13), 2135–2140 (2011)
152. Ersoy, H., Bulut, F., Ersoy, A.F.: Municipal solid waste management and practices in coastal cities of the Eastern Black Sea: a case study of Trabzon City, NE Turkey. Bull. Eng. Geol. Env. **67**(3), 321–333 (2008)
153. Burnley, S.J., Ellis, J.C., Flowerdew, R.: Assessing the composition of municipal solid waste in Wales. Resour. Conserv. Recycl. **49**(3), 264–283 (2007)
154. Abdalqader, A., Hamad, J.: Municipal solid waste composition determination supporting the integrated solid waste management in Gaza strip. Int. J. Environ. Sci. Dev. **3**(2), 172–176 (2012)
155. Peng, Z., Peng, L., Zhang, X.: Study on the composition and physical characteristics of municipal solid waste in Chongqing. Environ. Sci. Manag. **39**(2), 14–17 (2014). (in Chinese)
156. Wang, Y., Dong, X., Zhang, Y.: Analyzis on the characteristics of municipal solid waste in tangshan. J. Tangshan Coll.E **20**(6), 62–63 (in Chinese) (2007)
157. Dan, Z., Han, Z.: Study on the characteristics of municipal solid waste in tibetan plateau area at the dry season. China Biogas **30**(6), 33–36 (2012). (in Chinese)
158. Tao, X., Huang, T., Yang, H.: Investigation and analysis of municipal solid waste in downtown area of Chengdu. Guangdong Agric. Sci. (1), 94–96 (2009). (in Chinese)
159. Huang C.: Statistical study on physical and chemical components of municipal solid waste in Shenzhen. Huazhong University of Science and Technology (2012). (in Chinese)
160. Moradian, F., Pettersson, A., Richards, T.: A thermodynamic equilibrium model applied to predict the fouling tendency in a commercial fluidized-bed boiler, combusting solid waste. Energy Fuels 150422082504003 (2015)
161. Moradian, F., Pettersson, A., Richards, T.: Bed agglomeration characteristics during cocombustion of animal waste with municipal solid waste in a bubbling fluidized-bed boiler-A thermodynamic modeling approach. Energy Fuels **28**(3), 2236–2247 (2014)
162. Ma, W.: Thermal behavior of chlorine during municipal solid waste combustion. Tianjin University (2010). (in Chinese)
163. Tang, L.: Study on the properties of pyrolysis products and elements distribution during Municipal Solid Waste pyrolysis. Chongqing University (2015). (in Chinese)

164. Wang, X., Jiao, Y., Jin, B., Xu, B., Wu, J.: The effect of SiO_2 addictive on melting characteristics of MSW incineration fly ash. Power Syst. Eng. **5**, 17–20 (2007). (in Chinese)
165. Xu, M.: The experimental study on ash sintering characteristics of heated surface on MSW incineration. Zhejiang University (2007)
166. Chen, B.-Y., Lin, K.-L.: Biotoxicity assessment on reusability of municipal solid waste incinerator (MSWI) ash. J. Hazard. Mater. **B136**, 741–746 (2006)
167. Nielfa, A., Cano, R., Fdz-Polanco, M.: Theoretical methane production generated by the co-digestion of organic fraction municipal solid waste and biological sludge. Biotechnol. Rep. **5**, 14–21 (2015)
168. Shi, H., Mahinpey, N., Aqsha, A.: Characterization, thermochemical conversion studies, and heating value modeling of municipal solid waste. Waste Manag **48**, 34–47 (2015)
169. Sørum, L., Grønli, M.G., Hustad, J.E.: Pyrolysis characteristics and kinetics of municipal solid wastes. Fuel **80**(9), 1217–1227 (2000)

Chapter 7
Extraction and Characterization of Nanomaterials from Agrowaste

Deepu Gopakumar, Nathalie Lyczko, Hanna J. Maria, Ange Nzihou and Sabu Thomas

Abstract Green chemistry started for the search of benign methods for the development of nanoparticles from nature and their use in the field of antibacterial, antioxidant, and antitumor applications. Biowastes are eco-friendly starting materials to produce typical nanoparticles with well-defined chemical composition, size, and morphology. Cellulose, starch, chitin and chitosan are the most abundant biopolymers around the world. All are under the polysaccharides family in which cellulose is one of the important structural components of the primary cell wall of green plants. Cellulose nanoparticles can be extracted from agrowaste resources such as jute, coir, bamboo, and pineapple leaves, coir etc. Chitin is the second most abundant biopolymer after cellulose, it is a characteristic component of the cell walls of fungi, the exoskeletons of arthropods and nanoparticles of chitin can be extracted from shrimp and crab shells. Chitosan is the derivative of chitin, prepared by the removal of acetyl group from chitin. Starch nanoparticles can be extracted from tapioca and potato wastes. The preparation of these nanoparticles includes both series of chemical as well as mechanical treatments; crushing, grinding, alkali, bleaching and acid treatments. Transmission electron microscopy, scanning electron microscopy and atomic force microscopy are used to investigate the morphology of nanoscale biopolymers. Fourier transform infrared spectroscopy and X-ray diffraction are being used to study the functional group changes, crystallographic texture of nanoscale biopolymers respectively. Since large quantities of bio wastes are produced annually, further utilization of cellulose, starch and chitins as functionalized materials is very much desired. These

D. Gopakumar (✉)
School of Industrial Technology, Universiti Sains Malaysia, Penang, Malaysia
e-mail: deepu1789@gmail.com

N. Lyczko · A. Nzihou
RAPSODEE CNRS, UMR-5302, Université de Toulouse, IMT Mines Albi, Campus Jarlard, 81013 Albi Cedex, 09, France

H. J. Maria · S. Thomas
International and Inter University Centre for Nanoscience and Nanotechnology, Mahatma Gandhi University, Priyadarshini Hills P.O., Kottayam, Kerala 686560, India

School of Chemical Sciences, Mahatma Gandhi University, Priyadarshini Hills P.O., Kottayam, Kerala 686560, India

A. Nzihou (ed.), *Handbook on Characterization of Biomass, Biowaste and Related By-products*, https://doi.org/10.1007/978-3-030-35020-8_7

nanoparticles can be converted into smart and functional materials by functionalization through chemical modifications due to the presence of large amount of hydroxyl group on the surface. The cellulose, starch and chitin nanoparticles are currently obtained as aqueous suspensions which are used as reinforcing additives for high performance environment-friendly bio-degradable polymer materials. These nanocomposites are being used as biomedical composites for drug/gene delivery, nano-scaffolds in tissue engineering and cosmetic orthodontics. The reinforcing effect of these nanoparticles results from the formation of a percolating network based on hydrogen bonding forces.

7.1 Introduction

Science and technology now focus on producing more sustainable and environmentally friendly materials due to the problems involved in plastic recycling and scarcity of the petroleum-based materials. Nanoscale bio-fillers have attracted much attention in the field of academic research and practical applications due to its excellent reinforcing capability to the polymer matrix. Bio-nanofillers show considerable applications in different fields because of larger surface area and greater aspect ratio, with excellent properties. Being environmentally friendly, applications of nanocomposites using bio-nanofillers, offer new technology and business opportunities for several sectors, such as automotive, electronics and bio-technology industries. Generally, bio-mass based nanofillers can be classified into: (1) Nano cellulose, which includes both cellulose nanocrystals (CNCs), cellulose nanofibers (CNFs) and bacterial nano cellulose (BNC), (2) Nano chitin, which includes both chitin nanocrystals and chitin nanofibers and (3) Starch nanocrystals. The development in the production of bio-fillers like cellulose, chitin, and starch at the nanoscale, holds great promise to mitigate many of the sustainability problems, offering the potential of renewability, biodegradation and a path away from harmful additives. In this context, it has been essential to increase the current knowledge regarding the potential use of these bio-based fillers in various industrial applications. There is a growing trend to use bio-fibers as fillers and or reinforces in polymer composites due to their flexibility during processing, highly specific stiffness, and low cost make attractive to manufacturers.

Our society are increasingly demanding products made from renewable and sustainable resources that are biodegradable, non-petroleum based and have low environmental, animal/human health and safety risks. There is a growing trend to use bio-fibers as fillers and/or reinforces in plastics composites. Their flexibility during processing, high specific stiffness, and low cost make them attractive to manufacturers. As a renewable material, cellulose and its derivatives have been widely studied, focusing on their biological, chemical, as well as mechanical properties. The materials based on cellulose and its derivatives have been used for more than 150 years in a wide variety of applications, such as food, paper

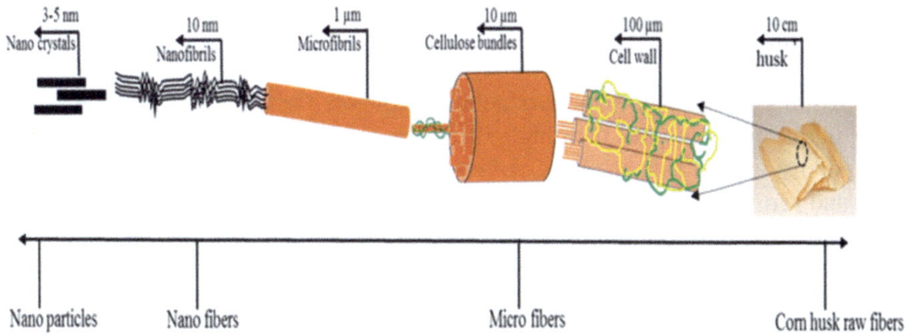

Fig. 7.1 Hierarchical structure of cellulose and its derivatives in nanoscale. Reproduced with permission from American Chemical Society [1]

production, biomaterials and pharmaceuticals. Cellulose is one of the most abundant biopolymers on earth, occurring in wood, cotton, hemp and other plant-based materials and serving as the dominant reinforcing phase in plant structures. Natural cellulose-based materials (wood, hemp, cotton, linen, etc.) have been used by our society as engineering materials for thousands of years and their use continues today as verified by the enormity of the worldwide industries in forest products, paper, textiles, etc. In this context, this review will be focused on the extraction and the characterization of nanoparticles extracted from the agrowaste. Nanocellulose can be classified into (1) cellulose nanofibers (CNFs), cellulose nanocrystals (CNCs) and bacterial nanocellulose (BNC). The major difference between the cellulose nanocrystals and cellulose nanofibers lies in their amorphous portion and dimensions. The Cellulose nanocrystals have nano dimensions in both length and diameter wise, whereas the nanofibers have length in micro dimension and diameter is in the nano dimension. The most important technique involved for the extraction of the cellulose nanocrystals is acid hydrolysis. In this process strong acid like H_2SO_4 is employed for the reaction and it destroys all the amorphous portions (disordered region) and leads to the nanocrystal structure, shown in Fig. 7.1.

7.2 Extraction of Cellulose Nanofibers (CNFs)

Many investigations have been done on the isolation and characterization of cellulose nanofibers from numerous sources. Cellulose nanofibers can be extracted from the cell walls via simple mechanical shearing or by chemo-mechanical routes. Renewable resources are considered as a potential and advantageous source of CNF. The use of such natural materials is becoming more completely understood and appreciated due to their high strength, biocompatibility etc. The extraction of the cellulose nanofibers from the plant cell wall can be accomplished by a variety of methods.

7.2.1 Cryocrushing Treatment

Cryocrushing is a method for producing nanofibers, where the fibers are frozen using liquid nitrogen, and then high shear forces are applied. This process exerts high energy on the frozen fibers, sufficient to break the microfibrils off the fiber walls and form individual fibrils of very low diameter. Cryocrushing combined with a high pressure fibrillation process was used by Wang and Sain for the isolation of nanofibers with diameters in the range 50–100 nm from soybean stock [2]. Alemdar and Sain reported the extraction of cellulose nanofibers from wheat straw and soy hulls by mechanical treatment involving cryocrushing [3]. They investigated the Chemical composition, morphology and physical and thermal properties of the nanofibers to study their usability in bio composite applications. They showed that the produced wheat straw nanofibers diameters were in the range of 10–80 nm with lengths of a few thousand nanometers. Chakraborty et al. also reported a novel technique to produce cellulose microfibrils through mechanical methods. The technique involved a combination of severe shearing in a refiner, followed by high-impact crushing under liquid nitrogen [4].

7.2.2 Grinding Procedure

Grinding process comprises of passing the cellulose fibers through a gap between two specially modified grooved discs, one is static, and the other is revolving at about 1500 rpm as shown in Fig. 7.2. During this process, it has been shown that the cell wall structure is broken down by the high shearing forces and it results in nanosized individualized fibers. Distance between the discs, morphology of the disc channels and the number of passages through the grinder plays an important role in the extent of the fibrillation of cellulose fibers.

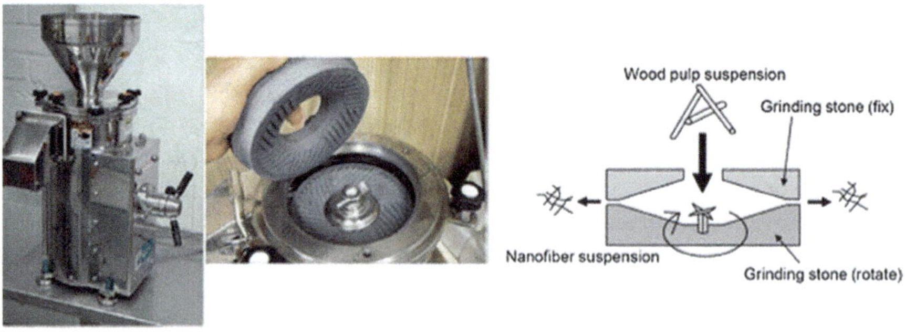

Fig. 7.2 Diagram for working principle of grinder. Reproduced with permission from Springer Nature [5]

Cellulose nanofibers with a uniform width of 15 nm from wood via grinding in an undried state were reported by Abe et al. They obtained the nanocellulose fiber with diameter of 15 nm. This report clearly demonstrated that grinding is an efficient simple method for the production of cellulose nanofibers from biomass [6]. When homogenized cellulosic pulp was subjected to a grinder treatment by Iwamoto et al. (2005) the fibril bundles were further fibrillated and 10 repetitions of the grinder treatment resulted in uniform nanofibers 50–100 nm wide [7].

7.2.3 Electrospinning Technique

Electrospinning is a quite simple and cost-effective method for the fabrication of nanofibers. Here the solution is extruded and electrospun under the action of a high electric field. Once the voltage is high enough, a charged stream of matter is ejected following a rather complicated loop. During this process, the solvent evaporates leaving behind randomly oriented nanofibers accumulating on the collector. Cellulose nanofibers have been produced by using electrospinning technique by dissolving cellulose fibers in solvents, such as ethylene diamine, with a salt selected from the group consisting of potassium thiocynate, potassium iodide. Only few fundamental studies are currently dedicated to this method. Ramakrishna et al. prepared cellulose acetate (CA) nanofibers with fiber diameter ranging from 200 nm to 1 μm by electrospinning technique and solvent employed was acetone/DMF/trifluoroethylene (3:1:1) mixture [8].

7.2.4 Enzymatic Pretreatments

In nature, cellulose is not degraded by a single enzyme but a set of cellulases are involved. These can be classified as A- and B-type cellulases, which are able to attack highly crystalline cellulose and C- and D-type cellulases, which generally require some disorder structure in order to degrade cellulose. Henriksson et al. and Pääkko et al. reported that endoglucanase pre-treatment facilitates disintegration of cellulosic wood fiber pulp into cellulose nanofibers [9, 10]. Another work was done by Tibolla et al. and they isolated cellulose nanofibers from banana peel bran using chemical treatment (involves alkaline treatment, bleaching and acid hydrolysis) and enzymatic treatment (alkaline treatment and hydrolysis with xylanase) [11]. Nanofibers produced from chemical treatment and enzymatic treatment had an average diameter of 10.9 and 7.6 nm and a length of 454.9 and 2889.7 nm respectively. They showed that the nanoparticles isolated from both of the treatments have potential application as reinforcing elements in composites.

7.2.5 Ultrasonic Technique

This process comprises of combination of chemical pretreatment and high intensity ultrasonication. The ultrasonic impact can gradually disintegrate the micron-sized cellulose fibers into nanofibers. Before ultra-sonication the plant fibers are being purified to prepare the cellulose fibers by mild acid hydrolysis followed by alkali treatment and bleaching treatment. This process comprises of the removal of non-cellulosic materials like lignin, hemicelluloses etc. After chemical pretreatment, the purified cellulose fibers are soaked in distilled water. About 110 ml of solution containing chemical-purified cellulose fibers are then placed in an ultrasonic generator of 20–25 kHz in frequency. The ultrasonication is conducted for 30 min to isolate the nanofibers. Li et al. prepared nanocellulose fibers by pretreating cellulose in NaOH/urea/Thiourea solution and then defibrillating the fibers by ultrasonication [12]. They achieve the high yield of 85.4%. The obtained nanocellulose fibers were about 30 nm in diameter with cellulose II crystal structure. They possess high thermal stability with an onset of thermal degradation at 270 °C and a maximum degradation temperature of 370 °C. Chen et al. reported that the Cellulose nanofibers were individualized from poplar wood by explosive chemical pretreatment and high-intensity ultrasonication [13]. When the output power of ultrasonic treatment used for the chemical-purified cellulose fibers was greater than 1000 W, cellulose nanofibers were obtained, 5–20 nm wide and several microns long.

7.2.6 Steam Explosion Coupled with Mild Acid Hydrolysis

This process comprises of short time vapor phase cooking at temperatures in the range of 180–210 °C, followed by explosive decompression and sudden release of pressure. In this process the cellulosic biomass is pressurized for a short period of time in an autoclave with water steam, and then explosively discharged to atmospheric pressure which results in a sudden disintegration of the starting material into a fibrous dispersed solid. This process results in a substantial breakdown of lignocellulosic structure, hydrolysis of the hemicellulose fraction, depolymerization of the lignin components and defibrillation. Deepa et al. reported the extraction of the cellulose nanofibers from the banana plant by the steam explosion process in an autoclave [14]. Chirayil et al. employed the steam explosion process for extracting cellulose nanofibers from isora fiber in an autoclave [15]. This technique involves alkaline treatment, bleaching, acidic steam treatment and homogenization. The results showed that the prepared cellulose nanofibers have nanofibrillar network-like structure with high crystallinity and good thermal stability. Cherian et al. also reported that the Steam explosion process was employed for the successful extraction of cellulose nanofibrils from pineapple leaf fibers for the first time [16]. Steam coupled acid treatment on the pineapple leaf fibers was found to be

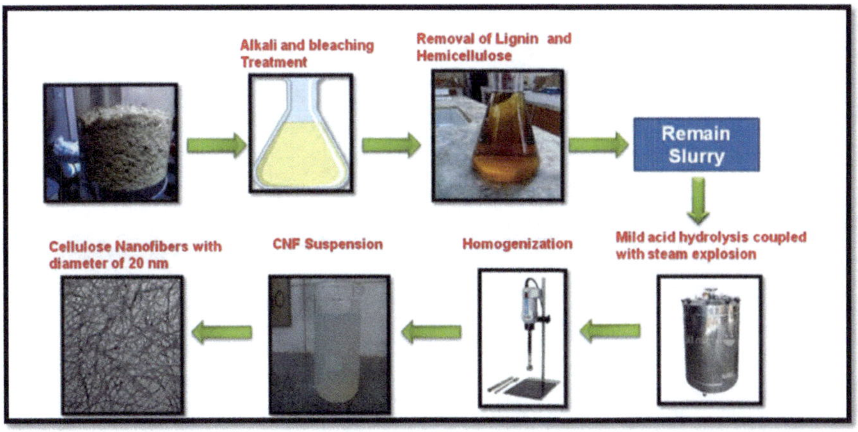

Fig. 7.3 Scheme of the extraction procedure of cellulose nanofibrils from sunflower stalks via steam explosion coupled with mild acid hydrolysis. Reproduced with permission from Elsevier [17]

effective in the depolymerization and defibrillation of the fiber to produce nanofibrils of these fibers. Recently Fortunati et al. successfully extracted cellulose nanofibers from sunflower stalks via steam explosion coupled with mild oxalic acid hydrolysis and the scheme of procedure is shown in Fig. 7.3 [16].

7.2.7 *Homogenization via High Pressure*

This is the extensively used method for the enormous scale production of cellulose nanofibers. This method has been widely used by the dairy and food industry mainly to stabilize food emulsion. This process comprises of the pushing of the cellulose fibers at high force and fed through a spring-loaded valve assembly as shown in Fig. 7.4. As this valve opens and closes at a very fast rate, the fibers are exposed to a large pressure drop with shearing and impacting forces. This combination of forces will lead to a high degree of micro fibrillation of cellulose fibers. Number of homogenization cycles and the applied pressure plays a vital role in the cellulose fibrillation. Thus, the high pressure, leads to the micro fibrillation of cellulose fibers.

Herrick et al. and Turbak et al. first reported the procedure for microfibrillation of cellulose [18]. They achieved the micro fibrillation of cellulose via passing the dilute cellulosic wood pulp-water suspensions through a mechanical homogenizer, in which a large pressure drop facilitated micro fibrillation. Another work was reported by Li et al. [12]. They employed the combination of chemical treatments

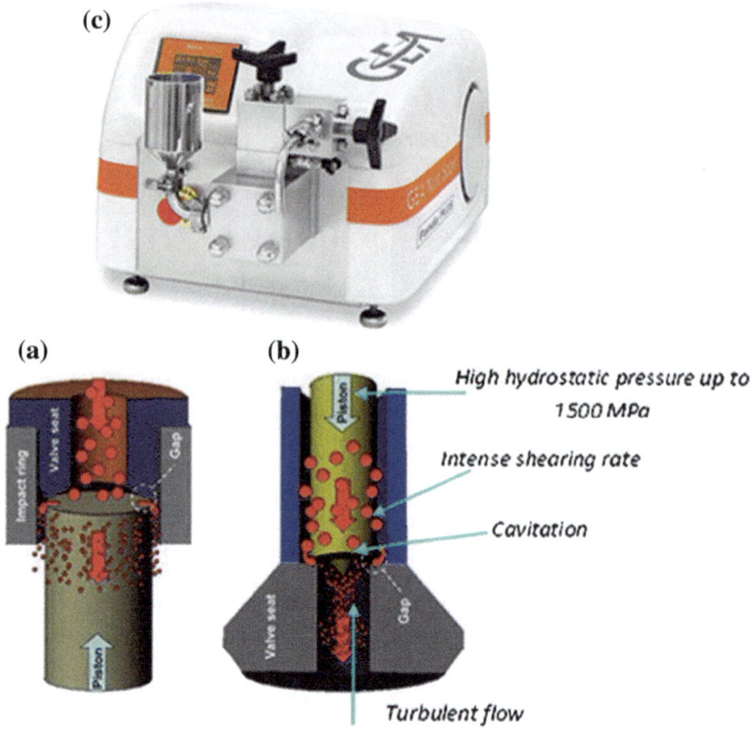

Fig. 7.4 **a**, **b** Scheme illustration of the principle of homogenization and **c** example of lab scale homogenizer (GEA Niro Soavi). Reproduced with permission from Springer Nature [5]

and high pressure homogenization to extract the cellulose nanofibers from de-pectinated sugar beet pulp. The diameter of the obtained cellulose nanofibers was from several nm to 70 nm. They showed that the crystallinity of nanofibers increased significantly after treatments. Similar work was reported by Chen et al. [13]. They fibrillated raw dried cotton fibers into individual cellulose nanofibers by chemical purification and pretreatment by a high speed blender combined with high pressure homogenization (nanofibrillation). They found that resultant nanofibers were in the range of 10–30 nm in diameter with high aspect ratio. Cellulose nanofibers were isolated from banana peels using a combination of chemical treatments by Pelissari et al. using alkaline treatment, bleaching and acid hydrolysis. It was done by passing the suspensions of chemically treated fibers through a high pressure homogenizer [19]. The morphological images of four different NC from corn husk (Fig. 7.5a), tunicate (Fig. 7.5b), bacterial cellulose (Fig. 7.5c), and electrospun cellulose nanofibers (Fig. 7.5d) are shown in Fig. 7.5.

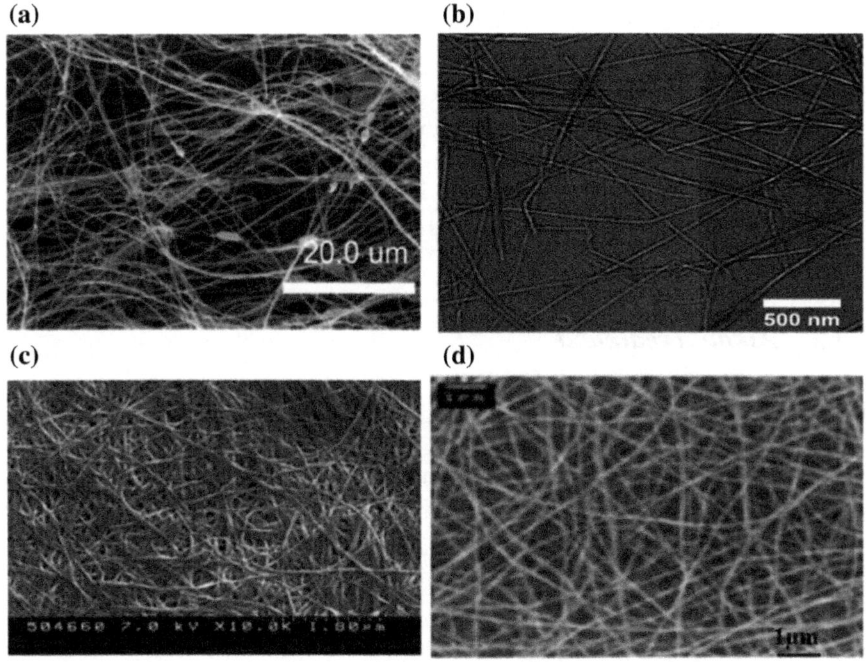

Fig. 7.5 Micrographs of nanocellulose **a** cellulose nanofibers from corn husk, **b** cellulose nanocrystals from tunicate, **c** bacterial cellulose, **d** electrospun cellulose nanofibers. Reproduced with permission from American Chemical Society [1]

7.3 Isolation of Cellulose Nanocrystals (CNCs)

CNC can be obtained from various starting materials, including algae cellulose, cotton linters, wood pulp and microcrystalline cellulose via acid hydrolysis of the macro cellulose. It has been shown that the nanocrystal could be extracted from purified cellulose fibers after a whole dissolution of the non-crystalline fractions by chemical hydrolysis. The amorphous region in the cellulose microfibrils could be hydrolyzed by the strong acid treatment, leaving behind a highly crystalline residue. The resulting rod-like crystalline cellulose are known as cellulose nanocrystals and usually their diameter ranges from 5 to 20 nm depending on the nature of the cellulosic plant source, purification pre-treatment and controlled conditions of acid hydrolysis [20, 21].

7.3.1 Mechanical Size Reduction

The first step prior to extract the CNCs from plant source material is size reduction and washing of the raw lignocellulosic biomass [22]. Normally, the cleaned raw fibers are broken down by milling or grinding. The resultant milled powder has uniform size of millimeter level which is more effective material for chemical treatment.

7.3.2 Alkali Treatment

This is the process of treating the raw lignocellulosic mass into the concentrated strong base solution (aqueous KOH or NaOH solution) in order to remove the alkali soluble materials like hemicelluloses and other impurities covering the external surface of the fiber cell wall [23, 24]. The main mechanism behind the alkaline treatment is the interruption of OH bonding in the fiber network structure leading to the separation of the interfibrillar regions from the cellulose fibers [25].

The mechanism is as follows:

$$[\text{Fiber-OH} + \text{NaOH} \rightarrow \text{Fiber-O-Na} + (\text{alkoxide}) + H_2O]$$

Generally, alkali treatments can be divided into two types: (1) alkali solution heating and (2) alkali treatment in an autoclave. The former is carried out using a combination of high temperature (70–90 °C) and mechanical stirring. The latter is a cooling process of the fibers is done using a combination of high temperature (160 °C) with high pressure [26, 27].

7.3.3 Bleaching Treatment

This process is called delignification and it is an additional important step for the further removal of residual cementing material, mainly lignin, in the alkali treated fibers. Various bleaching treatments have been reported from various cellulosic sources [26, 28]. In most cases, the alkali treated fibers are bleached by boiling the fibers with sodium chlorite ($NaClO_2$) solution under acidic condition. The acidic condition is created by acetate buffer solution which is made up of NaOH and glacial acetic acid, diluted with distilled water [29]. When put in an acidic buffer solution, sodium chlorite would break down into chlorine dioxide (ClO_2) under the presence of buffer salts which help to liberate the chlorine dioxide from sodium chlorite during the bleaching process. The common function of these acids is their ability to release hydronium ions (H^+) for the hydrolytic cleavage of glycosidic bonds in cellular molecular chains within amorphous regions along with the cellulose fibrils, therefore breaking down the hierarchical structure of the nanofibrils bundles into crystalline nanocrystals [30].

7.3.4 Acid Hydrolysis

Among the acid hydrolysis treatments, Sulphuric acid (H_2SO_4) treatment of cellulosic materials has been widely used to extract the CNCs. Acid hydrolysis of cellulose microfibrils is a well-known effective process used to obtain the well-defined cellulose nanocrystals. In fact, the most of the crystalline regions are insoluble in acids [31]. A lot of acids have been used for the acid hydrolysis of cellulose microfibrils that includes H_2SO_4, HCl, oxalic acid, hydrobromic acid (HBr) and HNO_3 [32–34]. Danial et al. extracted the CNCs from waste paper by acid hydrolysis using 60% (V/V) H_2SO_4 solution at 45 °C with constant stirring [35]. They obtained the CNCs with length ranged in 100–300 nm and they observed that newspaper can serve as a precursor for the effective production of CNCs.

Similar results were reported by Kargarzadeh et al. using H_2SO_4 acid hydrolysis of kenaf fibers [36]. They extracted the CNCs from kenaf bast fibers by mechanical stirring using 65 wt% of H_2SO_4 at various reaction times. They concluded that H_2SO_4 (65 wt%), 45 °C and 40 min represent the optimal preparation condition for the effective hydrolysis of kenaf bast fibers. Cellulose nanocrystals have been successfully extracted by Jiang et al. from tomato peels [37]. They used acidified sodium chlorite or chlorine-free alkaline peroxide routes. They isolated CNCs at a 15.7% yield by H_2SO_4 acid hydrolysis (64% H_2SO_4, 8.75 ml/g, 45 °C and 30 min). They obtained spindle shaped CNCs averagely 3.3 nm thick and 7.2 nm wide. They demonstrated that pure cellulose was effectively isolated from under-utilized food processing residues. Lamaming et al. also reported the extraction of CNCs from oil palm trunk by sulphuric acid hydrolysis [38]. They obtained CNCs with an average diameter of 7.67 nm and length of 397.03 nm. Bettaieb et al. extracted CNCs from the *Posidonia oceanica* leaves by sulphuric acid hydrolysis [39]. They obtained CNCs with an average diameter of 6.8 and 520 nm length and a similar result was observed by Fortunati et al. [40]. Mohamed et al. have done the extraction of CNCs from recycled newspaper by sulphuric acid hydrolysis and they investigated the thermal stability of the extracted CNCs [41]. They observed that the thermal stability of the extracted CNCs was affected due to the dehydration reaction by the sulphate groups from the sulphuric acid hydrolysis process. One of the major drawbacks of the sulphuric acid hydrolysis is that the sulphate groups introduced to the CNC through acid hydrolysis accelerate the degradation of cellulose which results in the limited thermal stability of cellulose, thus exerting a negative effect on the various applications of CNCs as reinforcement in Nanocomposites. Therefore, in recent years there was an increasing effort to use the mineral acids rather than sulphuric acid for the hydrolysis of cellulose. Camarero Espinosa et al. reported the comparative study of the CNCs extracted via both sulphuric acid and phosphoric acid hydrolysis [42]. They observed that the CNCs via phosphoric acid hydrolysis have high thermal stability than CNCs from sulphuric acid hydrolysis.

Tang et al. extracted CNCs from old corrugated container fiber by a combined process involving phosphoric acid hydrolysis, enzymatic hydrolysis and sonication [43]. They showed that the enzymatic hydrolysis was effective in enhancing CNC yield after phosphoric acid hydrolysis. Moreover, they showed that the presence of enzymatic hydrolysis imparted the obtained CNC with improved dispersion, increased crystallinity and thermal stability. Another work was reported by Liu et al. and they demonstrated a green and sustainable method for the preparation of CNC from bleached hardwood pulp by concentrated phosphotungstic acid ($H_3PW_{12}O_{40}$) hydrolysis [44]. They obtained rod-like CNCs with the size of 15–40 nm in width and hundreds of nanometers in length. They concluded that the resulting CNCs exhibited much higher thermal stability than the partially sulphated CNC prepared from sulphuric acid hydrolysis. Facile extraction of thermally stable CNCs via hydrochloric acid hydrolysis of cellulose raw material under hydrothermal conditions was reported by Yu et al. [45]. They obtained thermally stable CNCs with a high yield of 93.7% via combining hydrothermal hydrolysis and neutralization of the acid with ammonia. They achieved good stability in aqueous CNC suspensions due to the existence of ammonium groups. Table 7.1 illustrates the nomenclature and size distribution of nanocellulose from various sources. Figure 7.6 shows the flowchart which gives a brief idea about the extraction to characterization stages of the nanocellulose from biomass.

Table 7.1 Nomenclature and size distribution of nanocellulose from various sources

Mode of preparation (sources)	Type of nanocellulose	Size distribution (nm)	References
Sulphuric acid hydrolysis (waste paper, kenaf fiber, tomato peels, oil palm trunk, recycled newspaper)	Cellulose nanocrystals, nanocrystalline cellulose, cellulose nanowhiskers	Length: 100–300 nm Width: 5–20 nm	[35–38, 41]
Cryocrushing Treatment (soyabean stock, wheat straw)	Cellulose nanofibers, microfibrillated cellulose	Diameter: 50–100 nm	[2, 3]
Grinding (cellulose pulp, chemical treated cellulose)	Cellulose nanofibers, microfibrillated cellulose	Diameter: 50–100 nm	[6, 7]
Enzymatic pre-treatment (cellulose wood fiber pulp, banana peel bran)	Cellulose nanofibers, microfibrillated cellulose	Diameter 5–10 nm	[9, 10, 46]
Ultrasonic technique (cellulose pulp, poplar wood)	Cellulose nanofibers, microfibrillated cellulose	Diameter 5–30 nm	[12, 13]
Steam explosion (banana leaf, isora fiber, pineapple leaf, sunflower stalks)	Cellulose nanofibers, microfibrillated cellulose	Diameter: 10–40 nm	[14–17]
High pressure homogenization (cellulose pulp, de-pectinated sugar beet pulp, raw dried cotton fibers, banana peels)	Cellulose nanofibers, microfibrillated cellulose	Diameter: 10–70 nm	[12, 13, 18, 19]

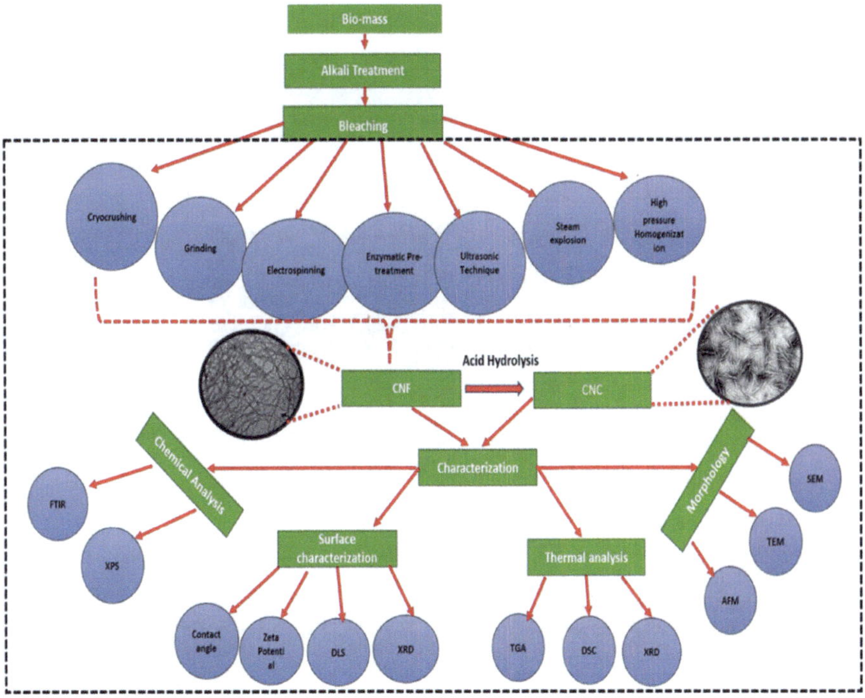

Fig. 7.6 Flowchart illustrating the steps followed from the nanocellulose extraction to characterization

7.4 Extraction of Chitin Nanofibers and Crystals

Chitin is the second most abundant natural polymer after the cellulose. Chitin nanofibers and nanocrystals can be obtained at a relatively low price from the organic resources like shells of selfish (mainly shrimps, lobsters, crabs and krill) and waste of sea food processing industries (as shown in Fig. 7.7). Chitin and cellulose are almost similar polysaccharide compounds, cellulose contains hydroxyl group, whereas chitin contains acetamide group. Chitin is a high molecular weight polymer, specifically β (1-4) (N-acetyl-D-glucoamine) as shown in Fig. 7.8. The high crystallinity and mechanical strength of nanochitin make it as a suitable candidate for reinforcement in polymer nanocomposites. Chitin is natural, non-toxic, non-allergic, antimicrobial, and bio-degradable and it is insoluble in water.

Chitin and its derivatives have great economic value because of their biological activities and their industrial and biomedical applications. It can be extracted from three sources, namely crustaceans, insects and microorganisms. Figure 7.9 shows schematic representations of hierarchical structure of cuticles and existence of chitin in them. However, the main commercial sources of chitin are shells of crustaceans such as shrimps, crabs, lobsters and krill that are supplied in large quantities by the

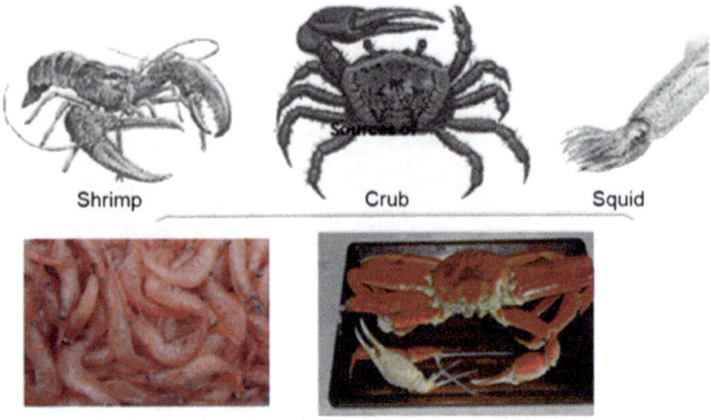

Fig. 7.7 Some sources of chitin. Reproduced with permission from Elsevier [47]

Fig. 7.8 Structure of the chitin molecule, showing two of the N-acetyl glucosamine units that repeat to form long chains in β-1, 4 linkage

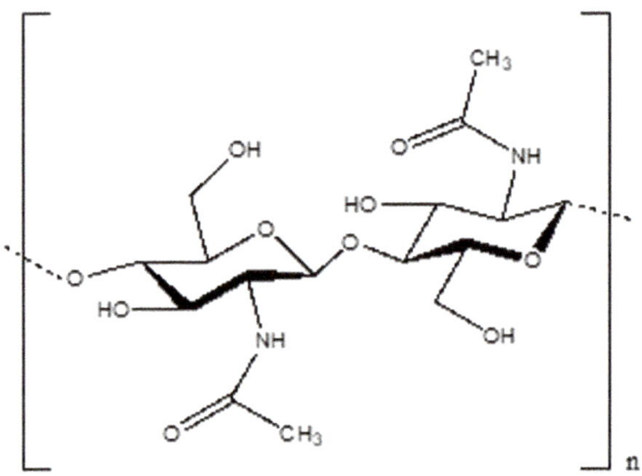

shellfish processing industries. Different cell types are capable in chitin synthesis. Thus, via different crustacean species and preparation methods, various chitin types and derivatives can be obtained. This is due to different modes of living, post-transcriptional regulation and processing of the enzymes, although they might have common principles [47].

Figure 7.10 shows the overall process for preparation of chitin from these shell wastes. It usually includes four main steps: (1) removal of proteins with dilute NaOH or by proteases digest (deproteinization), (2) removal of Ca_2CO_3 in the shell flakes by dissolving in dilute HCl (demineralization), (3) extraction of astaxanthin pigments and lipids with organic solvents such as acetone and ethanol (decolorization), and (4) treatment of residues with NaClO to obtain white chitin products (bleaching) [48].

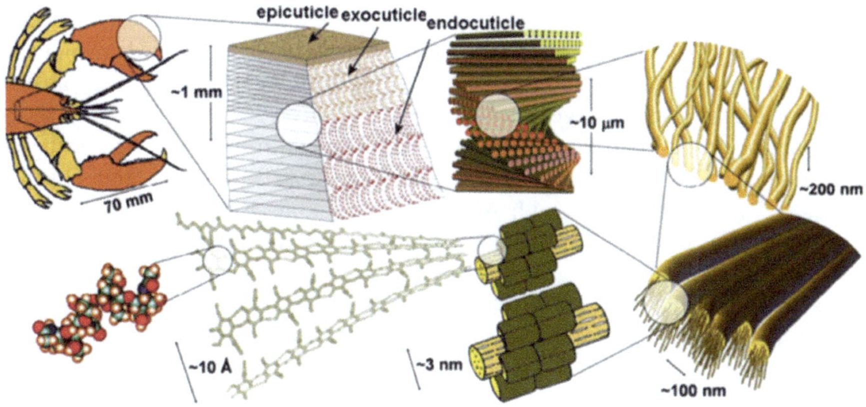

Fig. 7.9 Schematic representations of hierarchical structure of cuticles and existence of chitin in them. Reproduced with permission from Elsevier [47]

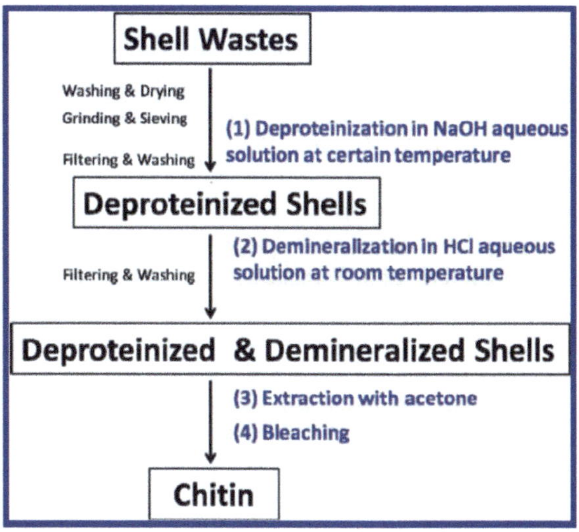

Fig. 7.10 Overall process for preparation of chitin from waste shells. Reprinted with permission from American Chemical Society [48]

There are several works in literature that have been reported about the extraction of the nanochitin from the different sources. Acid hydrolysis is the well-known procedure to extract nanochitin from sea food wastes. Paillet and Dufresne have successfully extracted the chitin nanowhiskers from the crab shells and squid pens by hydrochloric acid hydrolysis [49]. Another interesting work was reported by Ifuku et al. and they extracted the chitin nanofibers from dried crab shells via simple grinding treatment in a never-dried state under an acidic condition after the removal of proteins and minerals [50]. They obtained the chitin nanofibers in the range of 10–20 nm. They demonstrated that the grinding treatment in a never-dried state after the removal of the matrix was applicable to not only wood but also crab shell to isolate nanofibers. Moreover, they concluded that this demonstrated technique is an effective method to obtain homogeneous chitin nanofibers in their original state from waste crab shell in large amounts.

Kadokawa et al. reported the preparation of the chitin nanowhiskers using an ionic liquid 1-allyl-3-methylimidazolium bromide (AMIMBr) [51]. Initially they prepared 9.1 wt% chitin gel with AMIMBr by soaking at room temperature and heating at 100 °C. Then, the gel was soaked in methanol to regenerate chitin, followed by sonication, to give a chitin dispersion. They obtained chitin nanowhiskers with in the range of 20–60 nm in width and several hundred nanometers in length.

7.5 Extraction of Starch Nanoparticles

Starch is a renewable and biodegradable polymer produced by many plants as a source of storage energy. Starch has received considerable attention during the past two decades due to its high strength and biodegradability. Adding to its renewable nature, relatively low cost and worldwide availability, starch is also perceived as an attractive filler to replace fossil-based ones such as carbon black in nanocomposites applications [52]. Starch is a mixture of two main components: amylose, a linear or slightly branched (1 → 4)-α-D-glucan, and amylopectin, a highly branched macromolecule consisting of (1 → 4)-α-D glucan short chains linked through α-(1 → 6) linkages. Figure 7.11 shows a schematic illustration of the starch multi scale structure and it consists of (a) granule (2–100 μm), into which we find (b) growth rings (120–500 nm) composed of (d) blocklets (20–50 nm) made of (c) amorphous and crystalline lamellae (9 nm) containing (g) amylopectin and (h) amylose chains (0.1–1 nm) [52].

There are various techniques have been adopted to extract starch nanocrystals from various sources like wheat, potato, onion etc. Chen et al. extracted starch nanocrystals having the width of 10–20 nm and the length of 40–70 nm from potato starch granules [53]. 36.725 g of potato starch granules was dispersed into 250 mL of 3.16 M sulphuric acid aqueous solution and then stirred for 5 days at 40 °C with a stirring speed of 100 rpm. Another work was reported by Hebeish et al. and they synthesized starch nanoparticles using native maize starch [54]. The synthesis was done through modification of the Solvent displacement method by using alkaline aqueous solution as solvent; strong mechanical stirring broke the bonding interaction between starch macromolecules including hydrogen bond in swollen state. Moreover, they obtained starch nanoparticles in the range of 135–155 nm.

Gonçalves et al. extracted starch nanoparticles from *Araucaria angustifolia* seeds via combined acid hydrolysis with ultrasound technique [55]. They obtained the starch nanoparticles in the range of 22 nm. They concluded that the combined acid hydrolysis with ultrasound technique is an effective method to extract the

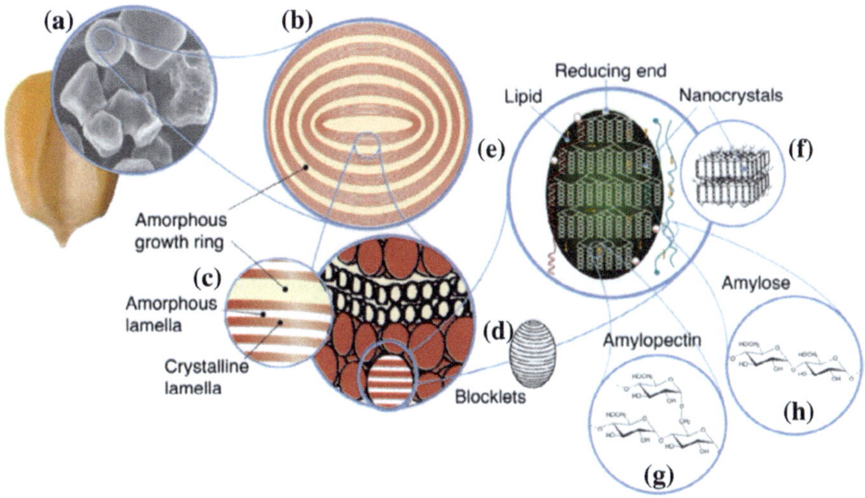

Fig. 7.11 Starch multi-scale structure. **a** Starch granules from normal maize (30 μm), **b** amorphous and semi crystalline growth rings (120–500 nm), **c** amorphous and crystalline lamellae (9 nm): magnified details of the semi crystalline growth ring, **d** blocklets (20–50 nm): constituting unit of the growth rings, **e** amylopectin double helices forming the crystalline lamellae of the blocklets, **f** nanocrystals: other representation of the crystalline lamellae called SNC when separated by acid hydrolysis, **g** amylopectin's molecular structure, **h** amylose's molecular structure (0.1–1 nm). Reproduced with permission from Elsevier [52]

starch nanoparticles from various sources. Kim et al. prepared crystalline starch nanoparticles using cold acid hydrolysis and ultrasonication from waxy maize starch [56]. They used aqueous sulfuric acid solution (3.16 M, 14.7% solids) was hydrolyzed for 2–6 days, either isothermally at 40 °C or 4 °C, or at cycled temperatures of 4 and 40 °C (1 day each). They found that the yield of starch hydrolysates depended on the hydrolysis temperature and time, which varied from 6.8 to 78%. Scanning electron microscopy revealed that the starch nanoparticles had globular shapes with diameters ranging from 50 to 90 nm.

Shi et al. prepared sodium trimetaphosphate (STMP)-cross-linked starch nanoparticle through high-pressure homogenization combined with water-in-oil (w/o) miniemulsion cross-linking technique [57]. Dynamic light scattering (DLS) and transmission electron microscopy (TEM) revealed that starch nanoparticles had narrow size distribution, good dispersibility and spherical shape. They demonstrated that when judicious amount of surfactant in oil is used, sufficiently small size particles with good stability could be produced. Fabian et al. extracted rice bran starch from defatted rice bran using a wet-milling process [58]. Starch was extracted from the bran by first soaking it in water. The mixture was then subjected to blending and washing with water, alcohol and alkali solution. About 83% of the rice bran starch was recovered. They concluded that, the defatted rice bran can be a good source of starch that is suitable for applications in food and pharmaceutical industry and other new applications such as a potential material in the biomedical field.

7.6 Characterization of Nanoparticles from Agrowastes

7.6.1 FTIR (Fourier Transform Infrared Spectroscopy)

FTIR spectroscopy involves the interaction of energy and matter. By recording the magnitude of absorbed energy and comparing it with available database, the present functional groups can be identified, and chemical structure of the sample can be characterized. FTIR spectroscopy is an effective non-destructive method for studying the physico chemical properties of lignocellulosic materials. The plant fiber consists of three main materials which include cellulose, hemicellulose and lignin. These materials are mainly composed of alkanes, esters, aromatics, ketones and alcohols with different oxygen containing functional groups. In most literature, FTIR has been used to confirm the removal of the amorphous portions like lignin and hemicellulose from the fiber matrix after the chemical treatments such as alkali treatment and bleaching process or to find the extracted cellulose nanofiber contains any lignin or hemicellulose portion. The infrared spectra of cellulose, hemicellulose and lignin were studied in the literature [12, 15, 59–61]. Chirayil et al. reported the FTIR spectroscopic analysis of the untreated, alkali treated, bleached and acid treated isora fibers were shown in Fig. 7.12 [15]. They observed a peak at 3300 cm^{-1} in the spectra of all fibers which corresponds to the OH stretching vibrations of hydrogen bonded hydroxyl group and it showed the hydrophilic tendency of the fiber [9].

The peak at 1630 cm^{-1} was due to the bending mode vibration of the absorbed water. The peak at 1228 cm^{-1} in raw fiber was due to the aromatic skeletal vibration of lignin. Similar report was reported by Tibolla et al. on the FTIR studies of the cellulose nanofibers obtained by enzymatic treatment from banana peel [11]. They confirmed that, the enzymatic treatment effectively removed the lignin and

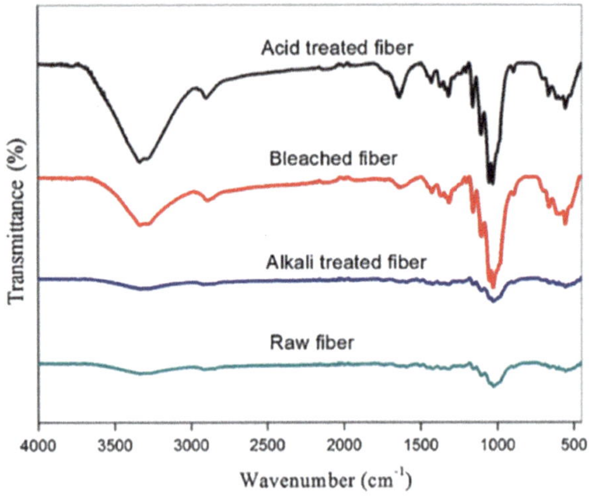

Fig. 7.12 FTIR spectra of fibers at different stages of preparation. Reproduced with permission from Elsevier Ltd. [15]

hemicellulose (amorphous portion) from fiber matrix. Pelissari et al. reported that the bleaching step removed most of the lignin from the cellulose nanofibers obtained from banana peels [19]. They observed the vibrations at 1525 cm^{-1} (aromatic ring vibrations), 1238 cm^{-1} (guaiaryl ring breathing with stretching C=O) and 761 cm^{-1} (C–H deformations) disappeared after bleaching. Similar work was done by Karimi et al. and they investigated the FTIR of the unbleached (UBNF) and bleached (BNF) kenaf bast cellulosic fiber samples and finding that there were no remaining traces of lignin in the obtained UBNF and BNF samples [62].

Wang et al. studied the FTIR spectra of the acetylated chitin nanocrystals: prepared through surface modification. From Fourier transform infrared spectroscopy, they found that the hydroxyl groups were partly replaced by the acetyl groups on the surface of chitin nanocrystals [63]. They showed that, compared to the chitin nanocrystals, the acetylated chitin nanocrystals show an obvious ester signal at 1742 cm^{-1}, and the intensity of the ester bands gradually become stronger with the increase of reaction time. In addition, the band at 1385 cm^{-1}, which was undoubtedly the CH_3 symmetrical deformational mode, was greatly strengthened in the spectra of the acetylated chitin nanocrystals, the results above indicate that the acetyl groups were introduced into the chitin nanocrystals as shown in Fig. 7.13. Finally, they concluded that the acetylated chitin nanocrystals were formed during the chemical modification process.

Ifuku et al. also conducted FTIR studies of chitin nanofibers from crab shells [50]. They studied FT-IR spectra of commercially available pure R-chitin chemical derived from crab shell, newly prepared chitin nanofibers from crab shell by removal matrix components, and dried crab shell powder derived from red king crab without purification. They found that, the spectrum of the dried crab shell powder was very different from the other two spectra because of the matrix components included in the shell. On the other hand, spectrum of newly prepare chitin nanofibers was in excellent agreement with the spectrum of commercial pure R-chitin, indicating that the matrix, protein, and minerals were well-removed by conventional processing. Ma et al. studied the FTIR spectra of corn starch, starch

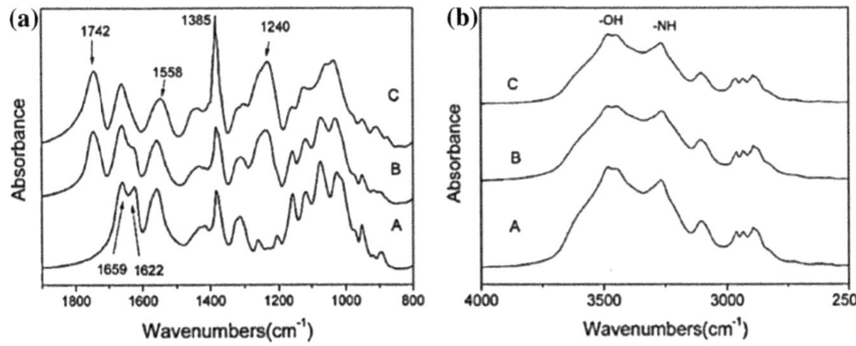

Fig. 7.13 FTIR spectra of chitin nanocrystals (**a**); acetylated chitin nanocrystals (**b**). Reproduced from Elsevier Ltd. [63]

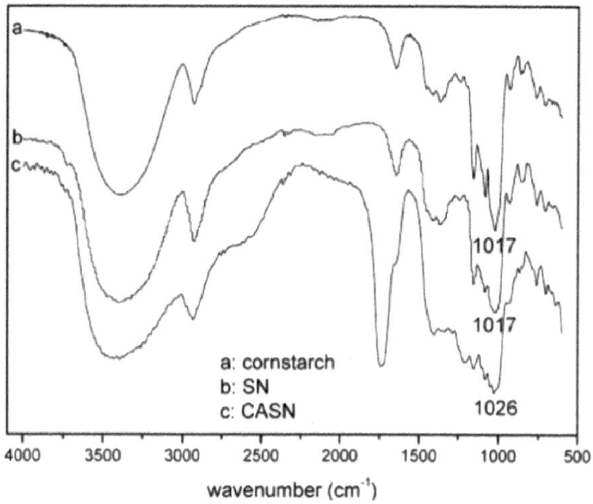

Fig. 7.14 FTIR spectra of corn starch, SN, and CASN. Reprinted with permission from American Chemical Society [64]

nanoparticle (SN) and citric acid modified starch nanoparticles (CASN) [64]. Figure 7.14 showed the FTIR spectra of corn starch, SN and CASN powders. The characteristic peak occurred at 1650 cm^{-1}, was due to the presence of water in starch. The absorption bands between 1000 and 1200 cm^{-1} were characteristic of the –C–O– stretching on polysaccharide skeleton. Corn starch and SN exhibited similar FTIR spectra. For CASN powders, a new peak at 1738 cm^{-1} was characteristic of an ester group. The other peak at 1017 cm^{-1} was attributed to C–O bond stretching of the C–O–C group in the anhydroglucose ring.

7.6.2 XPS (X-Ray Photoelectron Spectroscopy)

X-ray photoelectron spectroscopy (XPS) also known as Electron spectroscopy for chemical analysis (ESCA). XPS has been applied to a broad range of materials for surface analysis. XPS provides valuable quantitative and chemical state information from the surface of the material being studied. A very few works have been reported the XPS studies of cellulose nanofibers. Ahola et al. studied the chemical state information of the cellulose nanofibril model films by XPS [65]. They prepared thin and smooth native cellulose model films by spin coating of cellulose nanofibril dispersions onto silica substrates. They found out the relative atomic concentrations of oxygen, carbon, nitrogen, silicon and carboxyl groups for the model films and their references shown in Table 7.2. They observed in both model films the amount of detected substrate material (silicon and nitrogen) increases compared to reference, indicating that the surface is covered with the cellulose nanofibrils. Aulin et al. also studied the surface chemical composition of nanofibrillated cellulose aerogels and films before and after perfluorodecyl trichlorosilane (PFOTS) coating by XPS [66].

Table 7.2 Relative atomic concentrations of oxygen, carbon, nitrogen, silicon and carboxylic groups for low and high charge cellulose nanofibril model films on silica and their references

Material	{O}	{C}	{Si}	{N}	{COO}	D
APTS (3-amino propyl trimethoxy silane)	23.9	45.9	21.8	8.4	0.28	12.3
LCcellulose films	28.7	52.8	13.2	5.3	0.05	10.7
SiO_2	37.4	6.3	56.3		0.36	16.9
HCcellulose films	40.4	42.0	17.7		0.97	11.4

Reproduced with permission from American Chemical Society [65]

They found that the carbon, oxygen and very small amounts of Na as counterions of the carboxymethyl cellulose fibrils on the surface of the cellulose samples, whereas the modified samples clearly showed the presence of fluorine and silicon. They concluded that the PFOTS had reacted with the cellulose. Wang et al. studied the X-ray photoelectron spectroscopy of the acetylated chitin nanocrystals prepared via surface modification with acetic anhydride to investigate changes of elemental composition on the surface of the acetylated chitin nanocrystals [63]. They studied full XPS spectra of both chitin nanocrystals and acetylated chitin nanocrystals is shown in Fig. 7.15. They observed the signals around binding energies of 532, 398 and 286 eV correspond to the 1s orbital electrons of oxygen, nitrogen and carbon, respectively. Carbon, nitrogen and oxygen atoms were main components in both chitin nanocrystals and acetylated chitin nanocrystals. The ratio of oxygen-to-carbon for both chitin nanocrystals and acetylated chitin nanocrystals were almost 0.5, because the ratio of oxygen-to-carbon for acetyl groups, similar to the unmodified chitin nanocrystals, was also 0.5. Moreover, they observed that, the nitrogen content gradually decreased after modification, while both the oxygen and carbon contents increased which was due to the acetyl groups replacing the hydrogen atoms in the acetylated modification process.

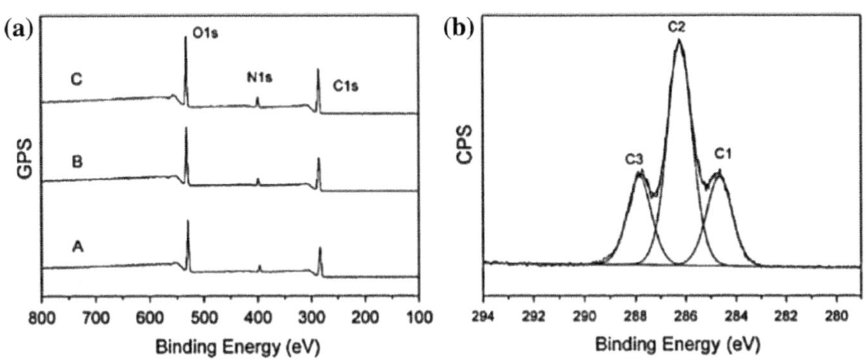

Fig. 7.15 Full XPS spectra (**a**) of the chitin nanocrystals (A), acetylated chitin nanocrystals-1 (B) and acetylated chitin nanocrystals-2 (C) and (**b**) the deconvolutions of the C 1s signal into its constituent contributions for chitin nanocrystals. Reproduced with permission from Elsevier Ltd. [63]

Wang et al. studied X-ray photoelectron spectroscopy of chitin nanocrystals unmodified and modified with poly (3-hydroxybutyrate-co-3-hydroxyvalerate) PHBV. They prepared chemically modified chitin nanocrystals via grafting of poly (3-hydroxybutyrate-co-3-hydroxyvalerate) (PHBV) onto chitin backbone via chlorination. From the XPS analysis they confirmed the PHBV grafting on the chitin nanocrystals [67]. Thielemans et al. studied the X-ray photon electro spectroscopy of the Nanoscale monocrystalline starch particles to investigate the surface elemental composition and modification [68]. They modified the starch nanoparticles using stearic acid chloride and polyethylene glycol methyl ether. From XPS studies, they found that the Stearate modification was quite effective due to the increased carbon content to 92%. Decomposition of the C 1s signal confirmed the replacement of the oxygen-rich starch surface by an aliphatic surface. The C 1s signal was almost completely composed of the C–C/C–H contribution (C1). Oxygen-carbon linkages (contributions C2 to C5) were found in only small quantities. Compared to the unmodified starch decomposition results, the ester carbon contribution (contribution C5) increases relative to the ether and alcohol carbon contributions (C4 and C3, respectively). They concluded that, this provides an additional proof of successful grafting of stearate to starch.

7.6.3 Contact Angle Studies

The contact angle is the angle, conventionally measured through the liquid, where a liquid–vapor interface meets a solid surface. It quantifies the wettability of a solid surface by a liquid via the Young's equation. Nowadays contact angle technique has been extensively used worldwide in the entire scientific community to investigate hydrophilicity and hydrophobicity of the materials. There is numerous work that have been reported on the contact angle studies of the agro-based nanoparticles like nanocellulose, nanochitin and starch nanoparticles.

The nanoscale dimensions and properties of cellulose nanofibers promoted their use as good alternatives to conventional fibers, because it provides nanocomposites with enhanced mechanical properties even at very low loadings. These enhanced mechanical properties are due to its OH functionality that enables good interactions with compatible polymer matrices i.e., polar polymers because of hydrogen bond formation; whereas the hydrophobic matrices such as polyolefin have low interfacial compatibility due to the high polar character of the cellulose nanofibers. Therefore, their incorporation in hydrophobic matrices has been achieved by specific chemical modifications that decrease the hydrophilic character of the cellulose nanofibers. Very few works have reported about the hydrophilicity and free energy determination of cellulose nanofiber and its composites. Ashori et al. studied the hydrophilicity of the acetylated cellulose nanofibers and native cellulose fibers by contact angle measurement [69]. They found that the surface characteristics of acetylated cellulose nanofibers were changed from hydrophilic to more

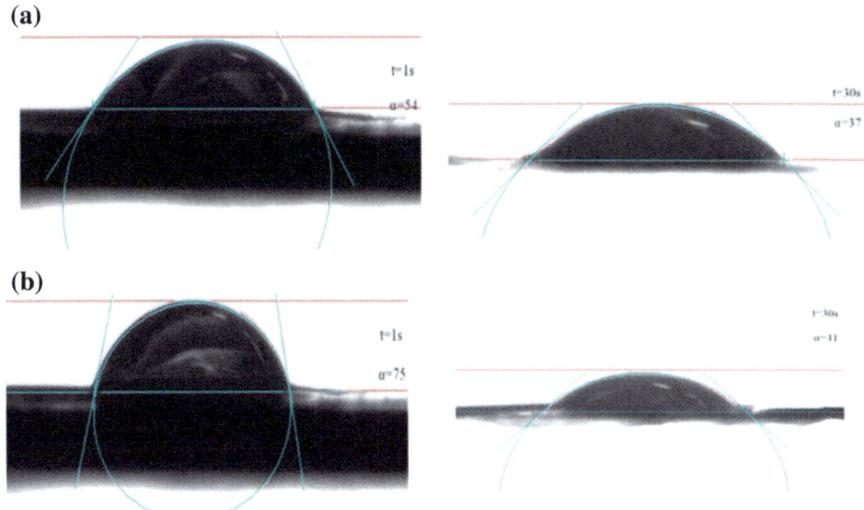

Fig. 7.16 Images of water drops used to determine contact angles of native (**a**) and acetylated (**b**) CNFs after 1 s (left) and 30 s (right). Copyright 2015. Reproduced with permission from Elsevier Ltd. [69]

hydrophobic as shown in Fig. 7.16. They concluded that the obtained acetylated cellulose nanofibers showed more hydrophobic surface, which was in favor of enhancing the interactions in the hydrophobic non-polar mediums.

Trovatti et al. studied contact angle measurements of bacterial cellulose nanofiber coated with polystyrene (BCPS) Nanocomposites [70]. They found a considerable enhancement of the hydrophobic nature of the BCPS surface through the respective increase in contact angle compared to native bacterial cellulose nanofibers (BC), which were 43.2 ($\pm$1) and 59.5° ($\pm$0.9) for Bacterial cellulose nanofibers and BCPS respectively. Contact angle measurement is also used to evaluate the surface hydrophilicity of the ultrafiltration membranes. Kong et al. studied the surface hydrophilicity of the cellulose triacetate (CTA) ultrafiltration membrane coated with TEMPO-oxidized cellulose nanofibers (TOCNS) by measuring static water contact angle [71]. They observed that the water contact angles of the blend membranes were in the range of 47.1°–72.6° which was lower than that of (CTA) membrane surface which was 84.9°. They concluded that the addition of TOCNS enhanced the membrane surface hydrophilicity, which increases the water permeation flux during the water treatment (reference) due to the hydrophilic group of TOCNS (–COOH): this was confirmed by the water contact angle studies.

Wang et al. studied the contact angle measurements of the chitin nanocrystals and acetylated chitin nanocrystals [63]. They studied contact angle measurements to investigate the surface tension and wettability of the chitin nanocrystals using water and hydrophobic diiodomethane liquid droplets. They observed that contact angle (θ) value of chitin nanocrystals with water is 31°, while that of the acetylated chitin

Table 7.3 Contact angle (θ) values of chitin nanocrystals and acetylated chitin nanocrystals

Sample	Water contact angle (θ) (°)	Diiodomethane contact angle (θ) (°)
Chitin nanocrystals	31	51
Acetylated chitin nanocrystals-1	61	39
Acetylated chitin nanocrystals-2	68	36

Reproduced with permission from Elsevier Ltd. [63]

nanocrystals increased to 61° and 68°, which was due to the partial replacement of hydroxyl groups by acetyl groups on the surface of acetylated chitin nanocrystals. Correspondingly, the contact angle (θ) value with diiodomethane decreased from 51° to 39° and 36° after the modification as shown in Table 7.3. Moreover, they concluded that the increased surface hydrophobicity of the acetylated chitin nanocrystals confirmed the acetylated modification.

Goetz et al. studied the hydrophilicity of the electrospun cellulose acetate membrane coated with chitin nanocrystals via contact angle measurements [72]. From contact angle measurements they found that the chitin nanocrystal surface coating significantly impacted the surface properties of the membranes, producing a super hydrophilic membrane (contact angle 0°) from the original hydrophobic CA mats (contact angle 132°) as shown in Fig. 7.17. Moreover, they demonstrated that the coated membranes showed significant reduction in bio fouling and biofilm formation.

Feng et al. studied the contact angle measurements of chitin whisker-graft-polycaprolactone (CHW-g-PCL) to investigate the hydrophilicity of the demonstrated composite [73]. They observed that the CHW-g-PCL with grafted PCL chains showed a distinct hydrophobicity. As shown in Fig. 7.17, the contact angles of water (θ) on the CHW-g-PCLs' surfaces were higher than 86° and the θ water values gradually increased with an increase of the PCL content in CHW-g-PCL. The CHW-g-PCL (I) and CHWg-PCL (II) had the similar water contact angle (θ) values as 86° and 87°, respectively, while the CHW-g-PCL (III) with the highest PCL content showed the maximum θ as 107°. They concluded that the higher the PCL content in CHW-g-PCLs was, more hydrophobic the bio nanocomposites became.

Nafchi et al. investigated the hydrophilicity of the sago starch films filled with nanorod-rich zinc oxide using contact angle measurements [74]. They observed that the films showed higher contact angle with the incorporation of zinc oxide nanorods as shown in Fig. 7.18. They found that by the incorporation of zinc oxide nanorods into the sago starch films, the composite showed decreased tendency to absorb water and the surface of the films became more hydrophobic.

Li et al. studied the water contact angles of rice, waxy maize, wheat and potato starch in order to investigate the hydrophilicity of the samples [75]. They observed the contact angle of water on rice, waxy maize, wheat and potato starch were 48°,

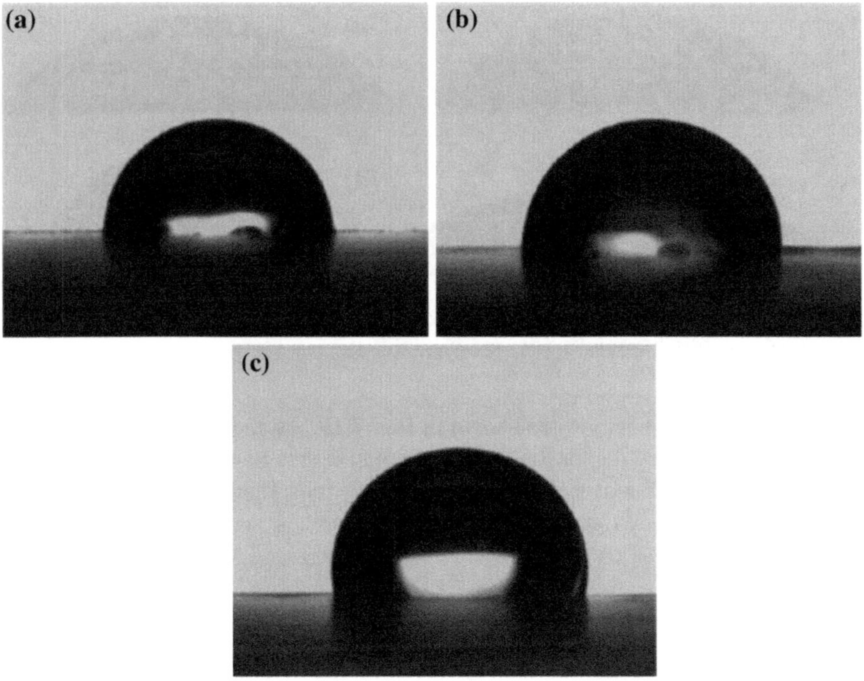

Fig. 7.17 Photographs of a water drop on the surface of thermoforming CHW-g-PCL sheets with various CHW content (**a**: CHW-g-PCL (I), **b**: CHW-g-PCL (II), **c**: CHW-g-PCL (III)). Reproduced with permission from John Wiley and Sons [73]

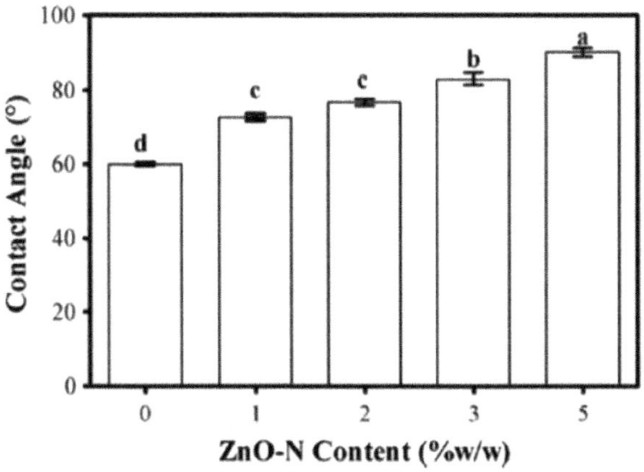

Fig. 7.18 Water contact angle on sago starch films surfaces as a function of ZnO-N content. Reproduced with permission from Elsevier Ltd. [74]

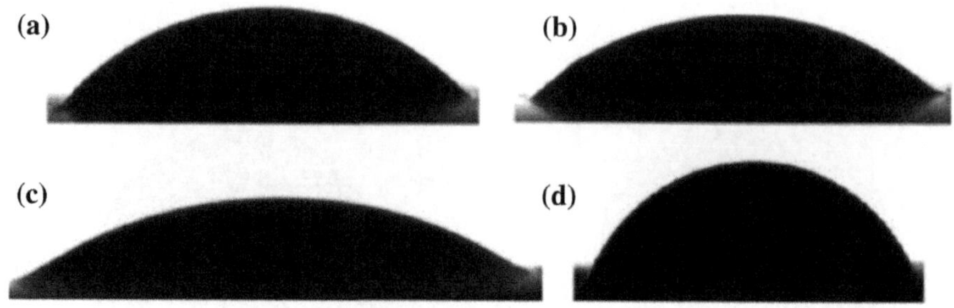

Fig. 7.19 Contact angles of **a** rice starch, **b** waxy maize starch, **c** wheat starch and **d** potato starch. Reproduced with permission from Elsevier Ltd. [75]

45°, 29° and 63°, respectively as shown in Fig. 7.19. All these contact angles were smaller than 90°. They demonstrated that, the native starch was hydrophilic, but the hydrophilicity of different starch granules was varied. Their results indicated that potato starch was more hydrophobic than other starches, and wheat starch was the most hydrophilic.

7.6.4 Zeta Potential Measurements of Agro-Based Nanoparticles

The zeta potential is a key indicator of the stability of the cellulose nanofiber suspensions. It indicates the degree of repulsion between adjacent, similarly charged particles in dispersion. For molecules and particles that are small enough, a high zeta potential denotes stability; i.e., the solution or dispersion will resist aggregation. When the potential is low, attraction exceeds repulsion and the dispersion will break a flocculate [76]. Nanofiber suspensions with high zeta potential (absolute value) are electrically more stable, whereas those with low zeta potentials tend to aggregate, compromising their ability to act as reinforcing agents in polymer matrices. According to Bondeson and Oksman hydrolysis with sulfuric acid causes electrostatic repulsion between nanofibers in suspension—this process introduces natural sulfate groups in the nanofiber surface [77]. The performance of the nanofiber as reinforcing agent is also associated with other features such as surface charge (zeta potential).

Pelissari et al. reported the zeta potential measurements of cellulose nanofiber suspensions extracted from banana peels [19]. They increased the number of passages of cellulose nanofibers in the high pressure homogenizer and obtained high stable aqueous suspensions of cellulose nanofibers with zeta potential values ranging from −16.1 to −44.1 mV. They reported corresponding dimensions and zeta potential values of cellulose nanofibers obtained by different number of passages in the high pressure homogenizer. They concluded that the cellulose nanofibers with high numbers of passages showed high stable dispersions which was due to the fact that the greater homogenization promoted higher electrostatic repulsion between the fibers, which provided enough surface charges to stabilize the suspension. They also observed that, in neutral water the nanofiber suspensions exhibited negative zeta potential. Similar values have been reported for cellulose nanofiber isolated from banana plant wastes [78]. They reported the zeta potential values ranging between −20 and −41.32 mV. Tibolla et al. reported the nanofibers obtained by chemical treatment and enzymatic treatment exhibited negative zeta potential were −16.1 and 25.7 mV respectively [11]. They found that the enzymatic treatment led to cellulose nanofibers with highest zeta potential and electrically stable than the cellulose nanofibers were obtained by chemical treatment.

Tzoumaki et al. investigated the surface charge of the chitin nanocrystals extracted from crude chitin via zeta potential measurements [79]. They have measured zeta-potential of the samples at pH values between 3.0 and 8.0 in order to investigate the effect of pH of the aqueous medium on the chitin nanocrystals charge density. They observed that, at acidic pH (3.0) the chitin nanocrystals showed strong positive zeta potential charge value of 50 mV. This was due to the protonation of amino groups to form $–NH_3^+$ in acidic media. It was also observed that, the charge of the dispersion was neutralized as the pH increased to values of 7.0 and showed zeta potential value of 20 mV. They concluded that, the enhancement in zeta potential value of chitin nanocrystals from negative to highly positive were due to the hetero polyelectrolyte property of both the amino groups (positively charged at acidic pH), and carboxylate groups (negatively charged at alkali pH) generated from the oxidation of alcoholic groups during bleaching.

Goetz et al. investigated the surface charges of the electrospun cellulose acetate membrane and electrospun cellulose acetate membrane coated with chitin nanocrystals via zeta potential measurements [72]. They observed that cellulose acetate (CA) membranes showed a surface potential of −30.2 mV whereas the cellulose acetate membrane coated with chitin nanocrystals (CA-ChNC) membranes displayed a ζ-potential of −4.7 mV at pH 7.5. They concluded that the reason for the enhancement of the zeta potential value from −30.2 to −4 mV was due to the chemical structure of the chitin nanocrystals on the surface of cellulose acetate that changed the surface properties of the CA membrane, making them less negative. Li et al. studied the water contact angles of rice, waxy maize, wheat and potato starch in order to investigate the hydrophilicity of the samples [75]. They obtained the zeta-potential of starch granules which were −20.6, −19.0, −20.3, and −4.0 mV for rice, waxy maize, wheat and potato starch, respectively.

Wei et al. studied the effect of different dispersion pHs on zeta potentials and agglomeration behavior of starch nanocrystals (SNC) prepared via sulfuric acid hydrolysis [80]. They showed that the Zeta potentials of all the samples were negative, indicating that SNC had negative surface charges as shown in Fig. 7.20. This was due to the presence of carboxyl and sulfate esters on the surface of SNC after H_2SO_4 hydrolysis. Their zeta potential measurements showed that, the zeta

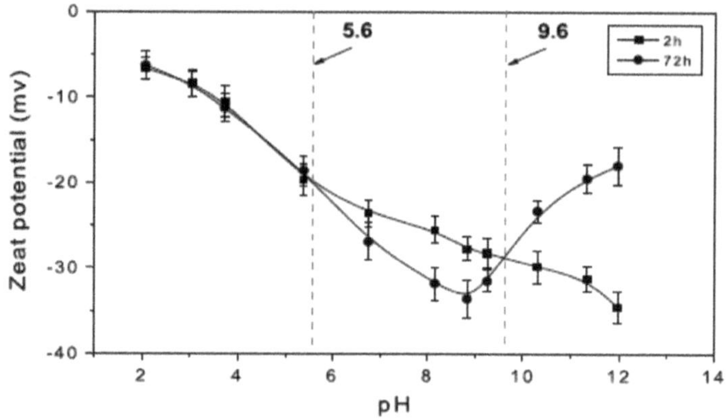

Fig. 7.20 Zeta potentials of SNC suspensions at different dispersion pH placed for 2 and 72 h. Reproduced with permission from Elsevier Ltd. [80]

potential of starch nanocrystals was decreased from −6.7 to −34.5 mV as the dispersion pH increased from 2.07 to 11.96. Moreover, they concluded that the smaller the starch nanocrystals the wider their distribution peaks with the increase of dispersion pH.

7.6.5 Morphological Characterization of Agrowaste Based Nanoparticles

Morphology investigates the details of structures commonly encountered in materials into the types of overall shapes to understand how different materials interact with one another. Morphology is also related to macroscopic properties and both conformation and morphology are related to polymer microstructure. There are a number of characterization techniques which can help to determine the size, size distribution, and shape of the bio-based nanomaterials. While in the case of cellulose the prepared nanocellulose exists in varied length, width, and shape. Properties depend on a number of factors such as the cellulose sources, method and time of preparation, etc. While native starch granules contain more or less concentric "growth rings" that are readily visible by optical or electronic microscopy whereas in the case of chitin, it existed as a fibrillar crystalline material [81]. Microscopy provides direct images of individual particles and allows characterization of its morphology and size (length, width, and thickness). The common microscopic techniques include transmission electron microscopy (TEM), atomic force microscopy (AFM), field emission gun scanning electron microscopy (FEG-SEM). Figure 7.21 shows the hierarchical structure of polysaccharides, which indicates the importance of morphological analysis to view their structural details.

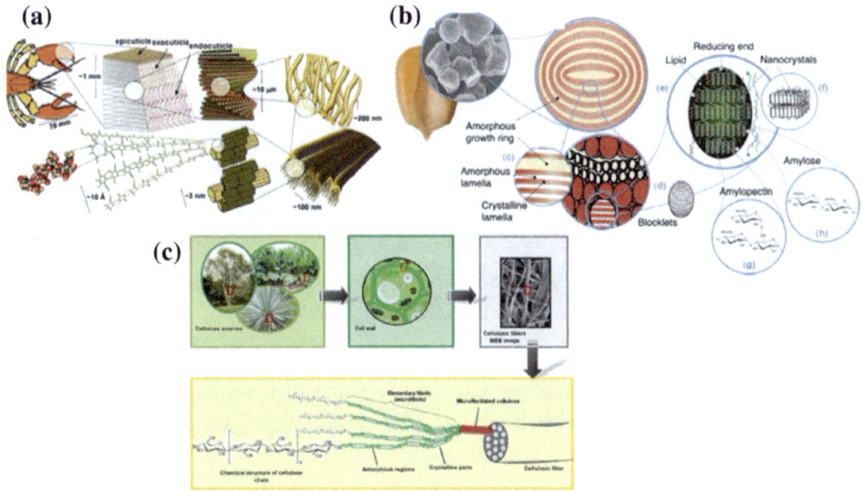

Fig. 7.21 Hierarchical structure of **a** chitin. Reproduced with permission from Springer Nature [44]. **b** Starch. Reproduced with permission from Elsevier Ltd. [3]. **c** Cellulose. Reproduced with permission from Elsevier Ltd. [47, 52, 82]

7.6.5.1 Atomic Force Microscopy (AFM)

In atomic force microscope the physical properties of a sample are detected by passing a sharp probe on a cantilever over the sample and detecting deflections in the cantilever using reflections from a laser beam. Figure 7.22 shows the schematic representation of AFM. AFM provides accurate and precise information about the

Fig. 7.22 Schematic illustration of AFM

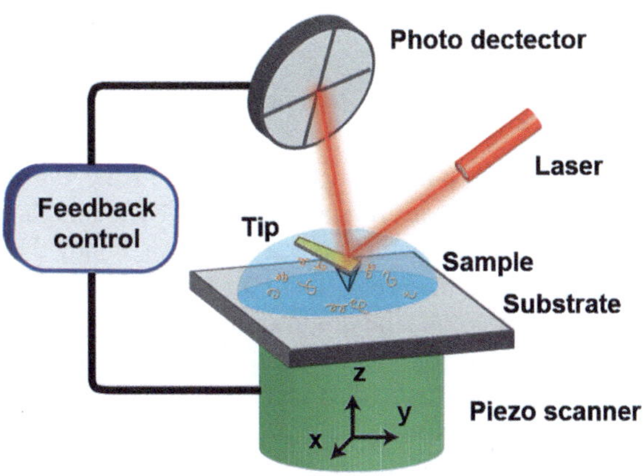

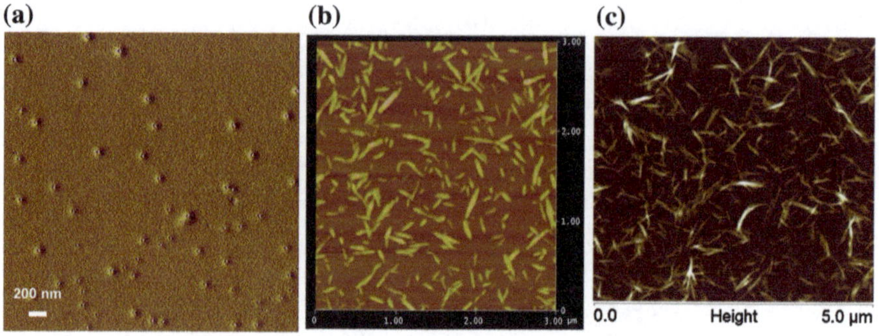

Fig. 7.23 AFM images of **a** starch nanocrystals. Reproduced from Elsevier Ltd. [88]. **b** Cellulose nanocrystals. Reproduced from Elsevier Ltd. [89]. **c** Chitin nanowhiskers. Reproduced with permission from Elsevier Ltd. [90]

dimensions of cellulose nanofibrils on morphology, surface topography, mechanical properties, and adhesion of nanocellulose under ambient conditions [83–86]. It can provide the thickness of the nanocellulose particles, nanoscale morphology and roughness, lateral associations that occur between the nanofibrils, porous network structure and gives an idea regarding its mechanical, optical, and permeability properties and shape parameter of individual particles. Surface roughness of the cellulose influences the flow properties, local mass transfer, the contact angle, and membrane hydrophilicity etc. AFM can also support in determining the particle size distribution in the nanorange of cellulose fibers [87].

Atomic force microscope images of nanocellulose from various sources and the AFM can be useful in finding the changes that is found by the influence of different factors. Figure 7.23 gives an example of atomic force microscopy (AFM) images of (a) starch nanocrystals, (b) cellulose nanocrystals and (c) chitin nanowhiskers.

7.6.5.2 Transmission Electron Microscopy (TEM)

TEM imaging plays a significant role in the morphological studies of cellulose microfibrils and nanocrystals. TEM has been a method usually preferred to envisage the shape and structure of individual or bundled CNCs [87]. The schematic of image formation (Bright Field Imaging) in transmission electron microscopy is given in Fig. 7.24. The origin of the image contrast is the variation of intensities of transmitted and diffracted beams due to the differences in diffraction conditions depending on the microstructural features on the electron path. Figure 7.25 shows the TEM images of the (a) starch nanocrystals, (b) chitin nanocrystals and (c) cellulose nanocrystals. However, there are some points that need to be noticed in the characterization of nanoparticles by TEM. It is important that while characterizing TEM, it needs to be checked whether a satisfying dispersion of nanoparticles has been obtained.

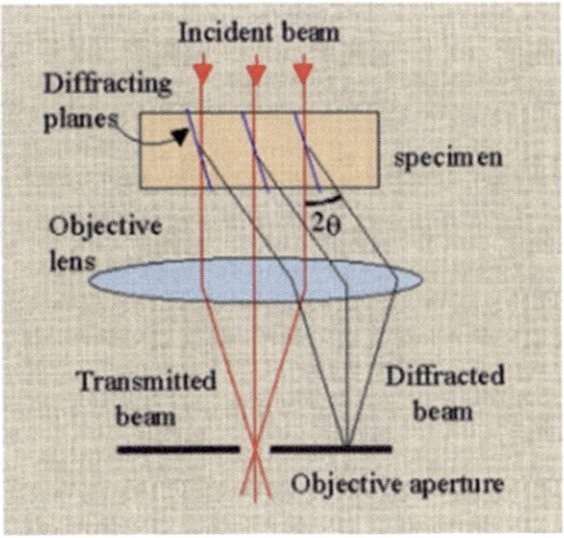

Fig. 7.24 Schematic of image formation (Bright Field Imaging) in transmission electron microscopy

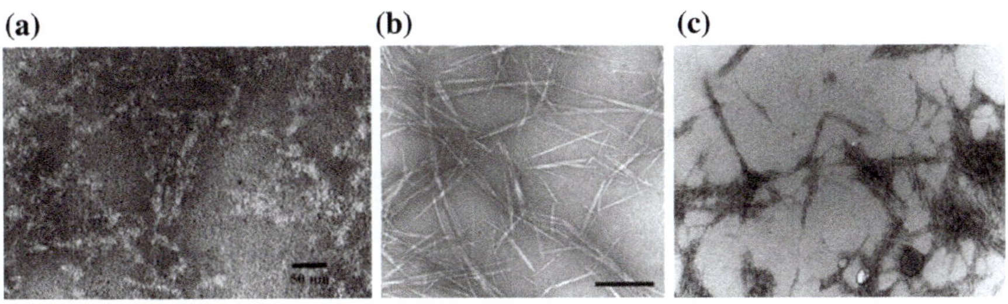

Fig. 7.25 TEM images of **a** starch nanocrystals. Reproduced with permission from Elsevier Ltd. [53]. **b** Chitin nanocrystals. Reprinted with permission from American Chemical Society [91]. **c** Cellulose nanocrystals. Reproduced with permission from Elsevier Ltd. [84]

7.6.6 Thermal Analysis of Agro-Based Nanoparticles

Thermal analysis is a technique used to identify chemical transformations such as decomposition, melting, polymerization, crystallization or oxidation of a material under heating conditions. Different techniques such as Thermogravimetric Analysis (TGA), Differential Thermal Analysis (DTA) and Differential Scanning Calorimetry (DSC) are available and different equipments and their principle are described in detail in Chap. 2. In this section, examples of results obtained for micro cellulose, cellulose nanofibrils, and cellulose nanocrystals are presented. Thermal analyses are techniques allowed to identify thermal transformation of agro waste nanocrystals and allow to calculate the kinetics of the thermal decomposition which are important data to define operating conditions to use these materials in different fields as chemical or environmental.

7.6.6.1 Nanocellulose and Micro Cellulose from Agrowaste

The micro cellulose was extracted from raw pineapple leaves by alkali treatment, bleaching, acid treatment and neutralization. The nanocellulose was obtained from the micro cellulose by ultraturrax used. Thermal analysis (Thermogravimetric and calorimetric) were carried out in a TG-DSC 111 of Setaram. Measurements were made under air atmosphere with a flow of 50 mL/min. The temperature program applied is: an isotherm of 20 min at 30 °C is applied to get uniform temperature throughout the oven and a temperature ramp from 30 to 800 °C with a heating rate of 10 °C/min and a cooling until 30 °C. The samples of 10 mg were analyzed in platinum crucible. Figure 7.26 shows the weight loss of micro cellulose and nanocellulose during the heating and Fig. 7.27 present derivative of the weight loss as a function of the temperature. The decomposition of the micro and nanocellulose occurs in three steps (Fig. 7.26) and reaches 90 wt%. The first step (until 150 °C) corresponds to the dehydration of the sample and is of the same order in the two cases. The second step corresponding to a temperature range from 140 to 370 °C corresponds to the decomposition of hemicellulose followed by the last step which corresponds to degradation of cellulose and lignin content. The derivative of weight loss allows to identify the temperatures decomposition during the thermal treatment. From Fig. 7.28, the temperature of the maximum rate of the decomposition of the hemicellulose is situated at 290 °C. For the last step, the thermal degradation of

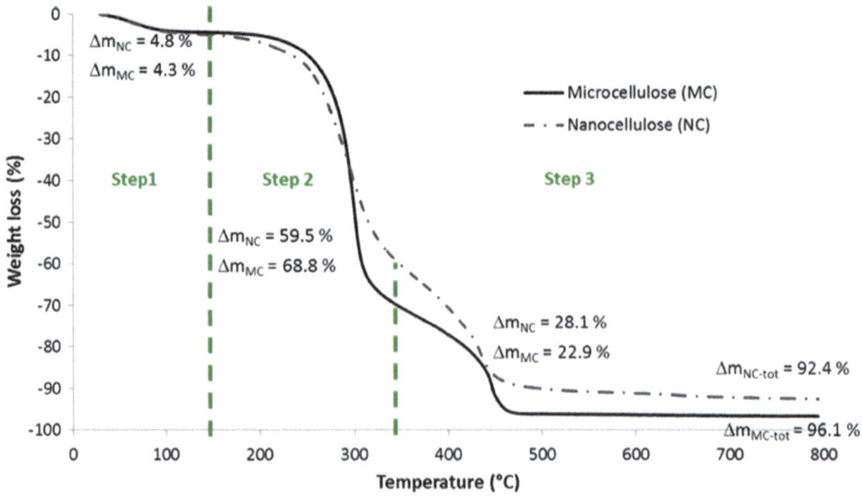

Fig. 7.26 Thermogravimetric curves of micro cellulose and nanocellulose

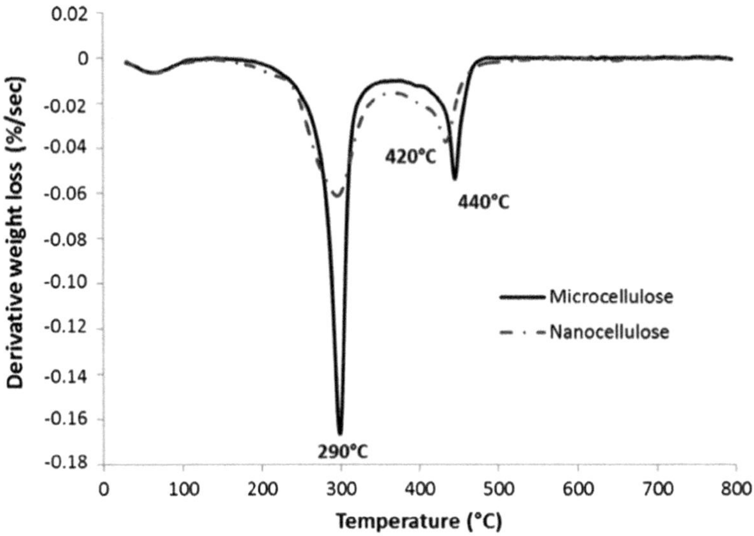

Fig. 7.27 Calorimetric curves of micro cellulose and nanocellulose

Fig. 7.28 Weight loss curves of (a) untreated, (b) CNF-1, (c) CNF-2 and (d) CNF-3 [94]

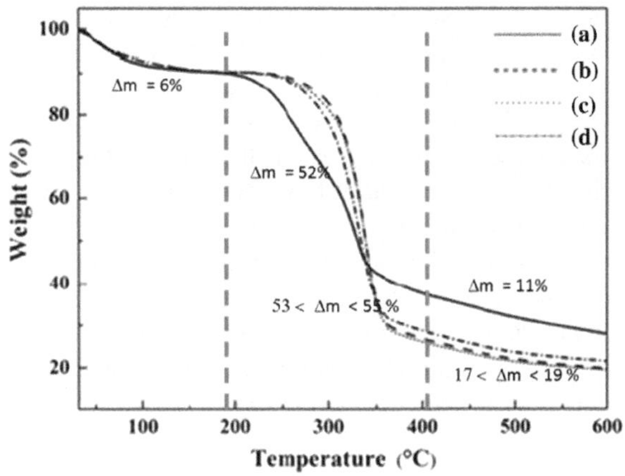

cellulose and lignin is of 425 °C for nanocellulose and 440 °C for micro cellulose. The decrease of 15 °C of the maximum rate of temperature decomposition for nanocellulose can be explained by the shorter structure dissociation due to the use of ultraturrax. The identification of cellulose and lignin content is difficult due to their temperature of thermal decomposition very close. An experimental method to separate these two decompositions is to decrease the heating rate during thermogravimetric analyses of 5 °C/min. Ogah et al. studied thermal stability of different agro nanofibers (corncob, rice hull, walnut shell, flax shive) to identify the decomposition of hemicellulose, cellulose and lignin [92]. Thermogravimetric were made with a heating rate ranging from 5 to 20 °C/min. From the weight loss and derivative weight loss curves of the agro fibers they confirmed the first step corresponds to the evaporation of the moisture then a second step to hemicellulose decomposition between 180 and 300 °C due to its weak branched-chain structure. From 290 to 360 °C the cellulose decomposition occurs due to its thermally stable linear chain structure. The last decomposition from 300 to 500 °C corresponds to lignin. The broad range of its temperature decomposition is due to complex and highly cross-linked structure. In the study of Romero Millán et al., the thermal decomposition of cellulose, hemicellulose and lignin of tropical lignocellulosic agro waste were quantified [93]. Authors used Gaussian Lorentzian functions to fit symmetric curves and Fraser-Suzuki function to fit the derivative weight loss. These deconvolutions allowed to calculate a pseudo quantity of each component in the agro waste analyzed.

7.6.6.2 Cellulose Nanofibrils (CNF)

Thermal analysis was used to evaluate the degradation of cellulose nanofibrils (CNF) extracted from coconut palm petioles residues by chemical and mechanical treatments [93]. Thermogravimetric analyses were carried out with a Thermo gravimetric Analyzer of Netzsch, 209F1 model. The procedure used is a heating from room temperature to 600 °C with a heating rate of 10 °C/min under nitrogen atmosphere which the flow is 30 ml/min. The mass of the sample used for analysis is of 6 mg. Figure 7.28 show curves obtained for weight loss and derivative weight loss for the sample untreated, the CNF1 obtained by acid treatment, CNF2 after grinding and CNF3 after homogenization. For all samples, the mass loss occurs in three steps as shown in Fig. 7.28. The first step from the room temperature to 130 °C corresponds to the moisture contain in the samples, the second step (up to 315 °C) corresponds to the thermal decomposition of hemicellulose. For higher temperature, the last step occurs and corresponds to the decomposition of lignin and cellulose.

Xie et al. studied the properties of cellulose nanofibrils extracted from bamboo using microwave liquefaction [95]. 2 mg of sample were heated until 800 °C with a heating rate of 20 °C/min and under nitrogen atmosphere. The weight loss curves obtained shows typical results of the literature with different steps as described above in the study of Xu et al. [94]. From weight loss curve, the derivative weight loss can be calculated as the derivative from the time and gives information about the maximum decomposition rate temperature (Tmax). As shown in Fig. 7.29, the Tmax corresponding to the decomposition of cellulose in original bamboo is 333 °C. The derivative weight loss curve allows to observe the decrease of this temperature for the microwave liquefied residue sample of 25 °C. This decrease can be explained by the dissociated structure of the bamboo during its microwave treatment. So, it is important to use weight loss curves but also the corresponding derivative weight loss curve allows to explain the phenomena which can occur during thermal treatment.

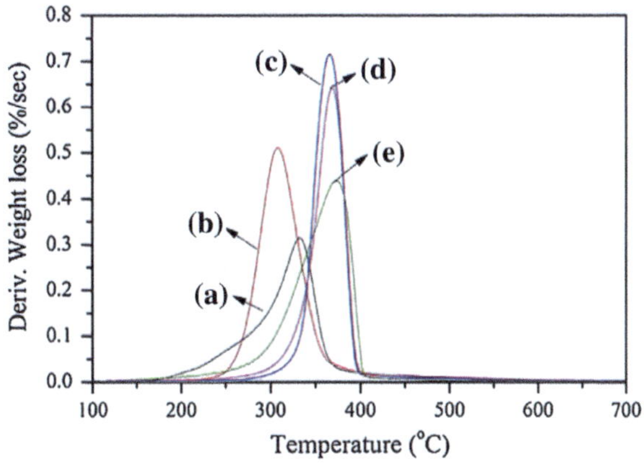

Fig. 7.29 Derivative weight loss curves of (a) original bamboo, (b) microwave liquefied residue, (c) bleached residue, (d) alkali treated residue, (e) cellulose nanofiber. Reprinted with permission from Elsevier [95]

7.6.6.3 Nanocrystals and Nanofibers from Agrowastes

An example obtained for chitin and cellulose nanocrystals is presented by Robles et al. [96]. As shown in Fig. 7.30, the thermal resistance of composites increases and can be seen by a shift of the top of the peak of derivative weight loss to higher temperatures degradation. Thermal analyses are used to characterize nanocrystals and nanofibers from agro wastes to identify decomposition steps and the corresponding temperatures and numerous studies are about the cellulose and/or starch

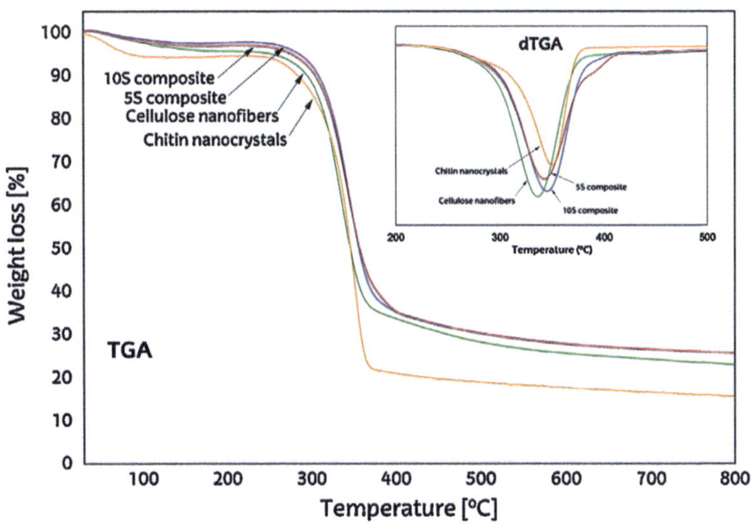

Fig. 7.30 Weight loss and derivative weight loss curves for different composites and nanomaterials. Reprinted with permission from Elsevier [96]

decomposition [97–100] but also to determine the temperature of thermal transition as shown by Sun et al. [101].

7.6.6.4 Kinetics Parameters

From thermal analysis kinetics parameters of the thermal decomposition of agro wastes nanoparticles materials can be calculated as shown by numerous authors [100, 102–104]. The rate of the thermal decomposition can be described by Eq. (7.1).

$$d\alpha/dt = k\,(T)f\,(\alpha) \tag{7.1}$$

where t is the reaction time, k (T) is a constant and f (α) is a function depending on the conversion of the material α.

The conversion α at the time is expressed by:

$$\alpha = (w_0 - w_t)/(w_0 - w_f) \tag{7.2}$$

With w_0 the initial sample weight, w_t the mass sample weight at the time t and w_f the final sample weight. All sample weights are expressed in mg.

The reaction rate constant depends on the Arrhenius equation and can be described by the following equation:

$$d\alpha/dt = A\exp(Ea/RT)f(\alpha) \tag{7.3}$$

where A is the pre exponential factor, Ea the activation energy and R the constant of ideal gas.

Equation (7.3) can be transformed for a non-isothermal rate expression relating the relation rate as a function of temperature at a constant heating rate and after equation can be integrated in the temperature ranging of the analysis. From Eqs. (7.1), (7.2) and (7.3) and some integration as described by Henrique et al., kinetic parameters can be determined [102]. Authors used two methods, Flymn-Ozawa and Kissinger to calculate the conversion and activation energy of cellulose nanocrystals (CN) extracted from purified mango seeds, *Eucalyptus urograndis* and cellophane. An example obtained for cellulose nanocrystals extracted from purified mango seeds is presented in Fig. 7.31 and shows the activation energy can be very different for a same conversion rate and depends on the treatment used for the extraction.

Numerous studies used modelling to determines thermal decomposition kinetics, activation energy of agrowastes [92, 93, 105]. Thermogravimetric curves of agro fibers with different heating temperature were modelling by Fly-Wall-Ozawa and Coats Redfern methods to determine activation energies [92]. A study of tropical lignocellulosic agro wastes determine the Arrhenius parameters and activation energy by using three isoconversional methods, Flynn-Wall-Ozawa (FWO),

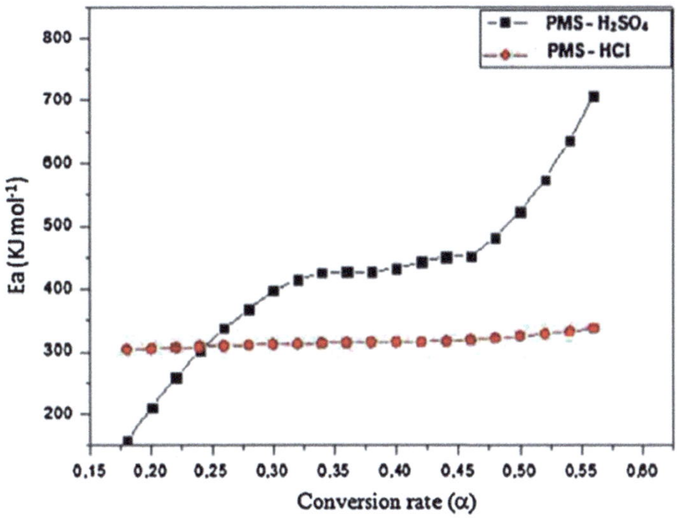

Fig. 7.31 Activation energy (Ea) versus conversion rate (α) calculated from thermal decomposition of CN extracted by two methods. Reproduced with permission from Elsevier [102]

Kissinger-Akahira-Sunose (KAS) and Friedman [93]. It was also shown that the decomposition mechanisms for hemicellulose, cellulose and lignin were a second order kinetic model. The use of these models was described and also applied to wood decomposition during pyro gasification [105].

7.6.7 *Small Angle Neutron and X-Ray Analysis of Agro-Based Nanoparticles*

7.6.7.1 Small Angle Scattering Definition

As a function of the particle size, techniques used to characterize the structure can be different as presented in Fig. 7.32. Small angle scattering (Small angle neutron and X-ray) corresponds to nanoscale structure as for nanoparticles, nanocomposites or nanotubes.

The first study of small angle scattering was proposed by Guinier in 1938 [106]. Then, different authors developed this method with theoretical and experimental considerations to improve the characterization of the structure of materials [107–109].

Small-angle scattering (SAS, angle ~0–2°) is used to determine the structure of materials and it is based on the small deflection of collimated radiation away from the straight path after it interacts with structures that are larger than the wavelength of the radiation. The "angle" in SAS refers to the angle of radiation deflection. SAS

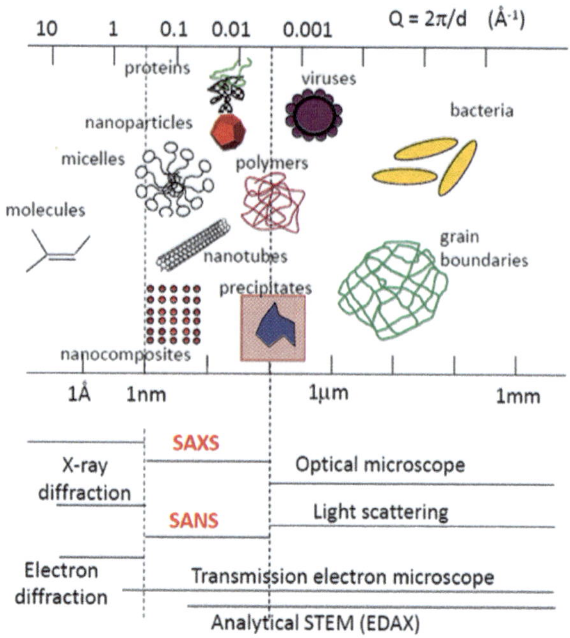

Fig. 7.32 Techniques of analysis as a function of particle size [109]

techniques can give information about the shape and size of structures in a sample. Small-angle X-ray scattering (SAXS) and small-angle neutron scattering (SANS) are two SAS techniques which are extensively used for nanomaterials. In Small-Angle scattering, the scattered intensity is collected very close to the direct beam that passes through the sample without deviation. At small angles, signal comes from the atoms and molecules organization in a scale from a hundred nanometers to a micrometer. All samples which haven't the same electronic density conduct to a signal at small angles. Neutron beam can be produced in a range of wavelengths from 0.01 to 3 nm: the fundamental difference between neutron and electromagnetic radiation (X-ray) is the mechanism by which incident radiation interacts with the matter.

SANS and SAXS allow the study of the structure at nanoscale and are nondestructive techniques.

For small-angle scattering measurement, as presented in Fig. 7.33, an incident beam (monochromatic wave of wavelength λ) interacts with the sample and the wave vector of the scattered radiation is noted $\overrightarrow{k_i} = 2\pi/\lambda$ while the wave vector of the incident radiation is $\overrightarrow{k_f}$. The resulting scattering vector is $\vec{q} = \overrightarrow{k_f} - \overrightarrow{k_i}$ and can be expressed by the following equation:

$$\|\vec{q}\| = \mathrm{q} = 4\pi/\lambda\ \sin(\theta) \tag{7.4}$$

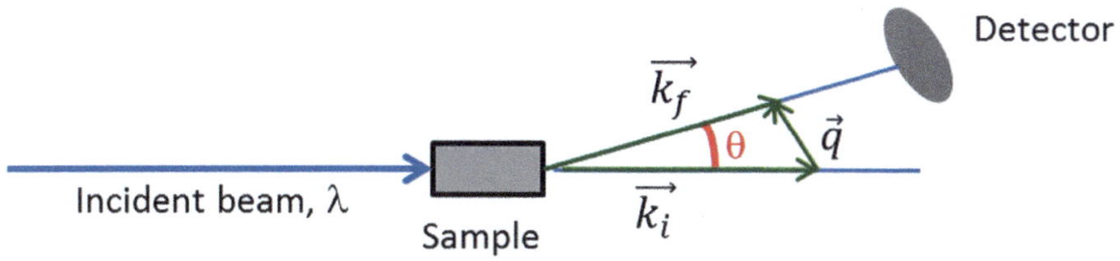

Fig. 7.33 Schematic small scattering angle measurement

Small-angle scattering measurement is an elastic scattering of radiation at small angles with:

$$0.3 < 2\theta > 5^{\circ} \text{ or } Q < 0.5\text{Å}^{-1} \quad q = \frac{4\pi}{\lambda}\sin\theta = \frac{2\pi}{d}\left(\text{Å}^{-1}\right) \tag{7.5}$$

The technique of SANS uses elastic neutron scattering at small scattering angle and allows investigating the structure of matter at a scale of 1–100 nm. A beam of neutron goes to the sample (powder, crystal or aqueous solution). Neutrons are elastically scattered by nuclear interaction with the nuclei or interaction with magnetic momentum of unpaired electron. SANS used collimation of the neutron beam to determine the scattering angle of a neutron which results in an ever lower signal to noise ratio for data that contains information on the properties of a sample at relatively long strength beyond 1 μm [110].

The X-ray scattering technique is based on the principle that a sample is illuminated by X-rays and the scattered radiation is registered by a detector. As measurements are made at small angle which is very close to the primary beam, the technique uses the brilliance of X-ray photon beams provided by particle accelerators known as synchrotrons [107]. The angular range contains information about the shape and size and can give information on nanoparticles ranging between 5 and 25 nm.

Some approximations and laws have to be used to obtain information of the particles structure from the scattering measurement. Data which can be obtained from measurements and data-processing methods are presented in Fig. 7.34 [111]. Three zones can be distinguished in this graph. The first is useful for the determination of the particle size using the Guinier equation. The second zone could be used for the determination of the shape and using the Fourier equation while the surface could be determined the last zone with the Porod equation.

The Porod law is applied for scattering at high value of q which can occurs when aggregates are present in the sample and is expressed by the following equation:

$$\text{I}(\text{q}) \propto \text{q}^{-4} \tag{7.6}$$

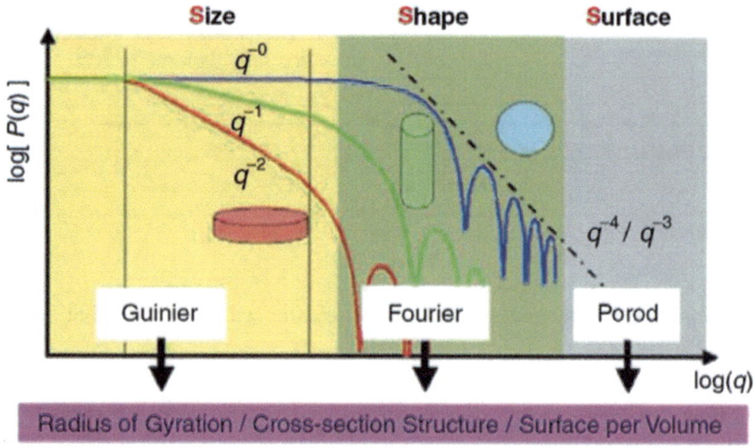

Fig. 7.34 Schematic information obtainable from various data processing. Reproduced with permission from Anton Paar [111]

For low scattering, the Guinier's approximation is used and depends on the radius of gyration Rg as shown in Eq. (7.7).

$$I(q) = I(0)\exp\left[-(q\,Rg)^2/3\right] \tag{7.7}$$

$$\ln(I(q)) = \ln(I(0)) - \left[q^2\,Rg^2\right]/3 \tag{7.8}$$

The radius of gyration can be extracted from the slope of the plot of ln(I(q)) as a function of q^2.

The radius of gyration is different as a function of the shape considered. For example, for a sphere $Rg = 3/5\ R^2$, for a cylinder $Rg^2 = R^2/2 + L^2/12$.

To extract information of the structure of nanoparticle from experimental data, different approximations and modeling have to be use. The modelling equations and the software corresponding are in constant improvement and are described by numerous authors [112–116].

7.6.7.2 Examples of SAXS and SANS Measurements

Examples of SAXS and SANS techniques used to determine characteristics of nanocellulose and chitin extracted from agro waste are presented in this section.

Mao et al. showed in their study the different characteristics of two types of nanocellulose extracted from wood: the one obtained by TEMPO-oxidized cellulose nanofiber (TOCN) and one sulfuric acid hydrolyzed cellulose nanocrystals (SACN) using SAXS and SANS techniques [116]. Experiments of SANS were carried out with a NG7-30m SANS beam line in NIST Center for Neutron research with different sample-to-detector distances [117]. For analysis, the suspensions of TOCN (0.1 and 0.2 wt%) were prepared. For SAXS measurements, a beam line 7.3.3 in Advance Light source, Lawrence Berkeley National Laboratory was used [118]. The shape and size of nanocellulose were modeled by different models considering the polydispersity. The different models are developed as a function of the particles shaped cross section considered. From three sections presented in Fig. 7.35 authors developed a parallelepiped model (Fig. 7.35a) which considers the three dimensions, the ribbon model. Figure 7.35b and a Gaussian-approximated parallelepiped model (GAP) (Fig. 7.35c). Results obtained by using GAP model were the best and are shown in Fig. 7.36. From this model, dimensions of the TOCN particles were obtained, as shown in Table 7.4.

As it is shown above, SANS and SAXS techniques can be used to determine the dimensions of particles. They can also be used to identify the structure changes during a treatment as shown by Penttilä et al. and presented in Fig. 7.37 [119]. Authors used these techniques to follow the polymeric structure formed by

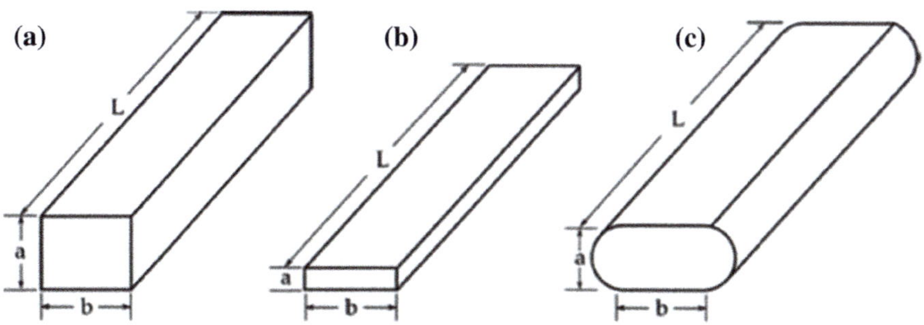

Fig. 7.35 Schematic rectangular shaped cross sections **a** parallelepiped model, **b** ribbon model where b ≫ a, **c** cross section of the rectangle with semicircle ends. Reproduced with permission from American Chemical Society [116]

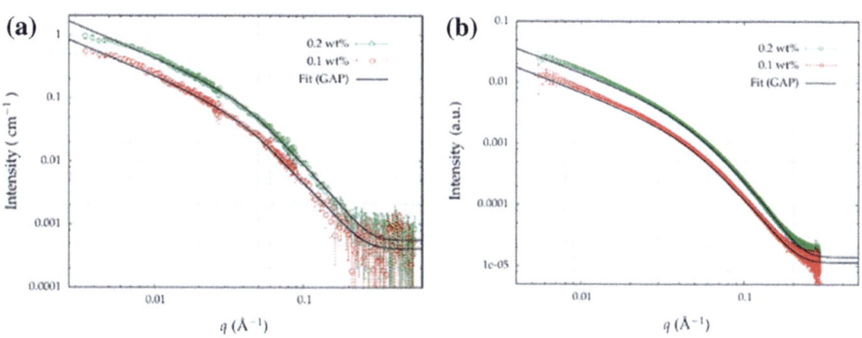

Fig. 7.36 SANS (**a**) and SAXS (**b**) curves of TOCN suspensions. Reproduced with permission from American Chemical Society [116]

Table 7.4 Cross sectional dimensions of TOCN particles

Mass fraction (%)	a (nm)/SANS	a (nm)/SAXS	b (nm)/SANS	b (nm)/SAXS
0.1	1.7 (0.15)	1.7 ($1.0 * 10^{-2}$)	8.2 (0.17)	7.7 ($4.4 * 10^{-2}$)
0.2	1.7 ($7.8 * 10^{-2}$)	2.3 ($7.3 * 10^{-2}$)	8.0 ($9.1 * 10^{-2}$)	7.8 ($1.3 * 10^{-2}$)

Reproduced with permission from American Chemical Society [116]

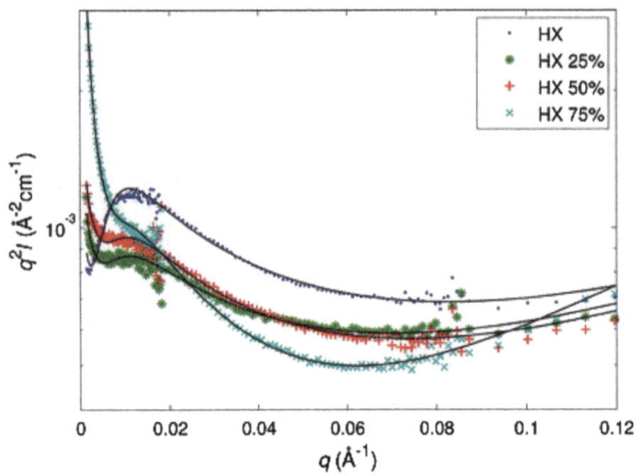

Fig. 7.37 Kratky plot of fits. Reproduced with permission from Springer Nature [119]

individual cellulose fibril during the hydrolysis of nanofibrillated cellulose samples. To identify changes they used a function (Eq. 7.9) to fit data.

$$I(q) = \frac{A}{q^n} + \frac{C}{1 + (\xi q)^m} + B \tag{7.9}$$

where A/q^n corresponds to Porod type behavior of the scattering intensities, the second term characterizes the fibril network, n and m are power law exponents independent of each other, A and C are scaling constants and B accounts for a constant background.

The fitting of the experimental data allows identifying the different parameters of Eq. (7.9). The Kratky plot (Fig. 7.37) is also used to see the structure changes. Authors described how to relate the Kratky plot behavior (peak position and the evolution of intensities, weakened for example) with the fibril network correlation term of Eq. (7.9) and so they identified structure change during hydrolysis.

Chitin nanofibers were prepared from chemical and mechanical treatment of shrimp shells and suspension of 0.05–0.07 wt% were characterized by SAXS to analyze the shape size and size distribution [120]. Data obtained were fitted by two polydisperse models, the cylinder and the ribbon model described in a previous study of Su et al. [121]. Figure 7.38 shows the comparison of experimental data and their fitting with the two models. The Ribbon model appears as the better of the two

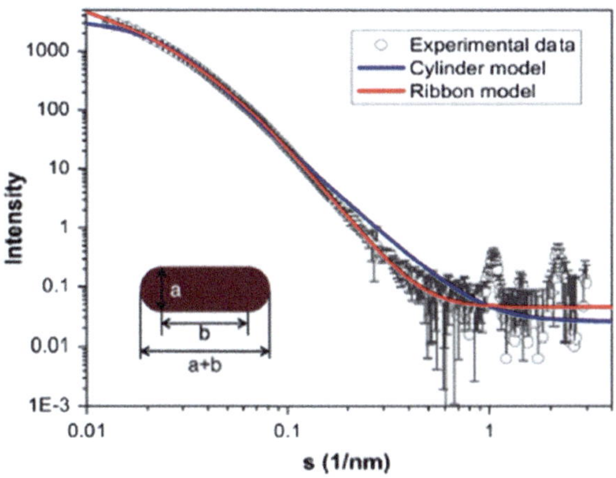

Fig. 7.38 Comparison between polydisperse ribbon and cylinder models and experimental data. Reproduced with permission from Elsevier [120]

and the average width calculated with this model is of about 23 nm. This result was in agreement with results obtained by TEM images of single nanofibers of about 20 nm.

SAXS allows to determine cellulose nanofibers with a size of around 7 nm and thickness of about 15 nm produced by TEMPO-mediated oxidation of raw jute fiber and was described using ribbon model and confirmed by TEM analysis. Reproduced with permission from Elsevier [122].

7.6.7.3 Dynamic Light Scattering Techniques of Agro-Based Nanoparticles

Dynamic Light Scattering (DLS) is also called as Photon correlation Spectroscopy (PCS) or Quasi-elastic light scattering (QELS). This technique is largely used in chemistry, biochemistry and physics fields for nanoparticles, polymers, emulsions or colloids applications. Dynamic Light Scattering is a spectroscopy method which is non-invasive and allows to measure the particles or molecules size distribution in solution or suspension [123, 124]. The particle size measurements are ranging between 0.3 and 10 nm. A standard specifies the application of dynamic light scattering (DLS) to the measurement of average hydrodynamic particle size and the measurement of the size distribution of mainly submicrometric-sized particles using Dynamic light scattering techniques is available [125].

7.6.7.4 Dynamic Light Scattering Theory

A schematic graph of DLS apparatus principle is presented in Fig. 7.39 and the system is composed with a laser which provides a light source to illuminate the sample contained in a cell. The sample is composed by a suspension of solid

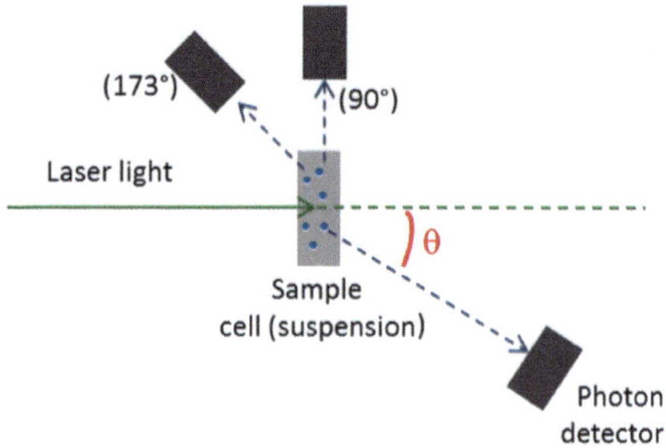

Fig. 7.39 Schematic graph of DLS technique

nanoparticles. The sample is illuminated by the laser beam and the fluctuations of the scattered light are detected at a known scattering angle θ by a fast photon detector. The detector measures the scattered light at different scattering angles θ which depends on its position and can vary between 20 and 173°. Most used scattering angles are 90 and 173°. The signal shows random changes due to the changing relative position of the particles, due to their Brownian motion.

The principle is based on a monochromatic incident light which goes through a solution where particles are in suspension and particles diffract the incident light in all directions with different intensities as a function of time. When particles are in suspension, they are moving randomly due to collisions with the solvent by the Brownian motion. The DLS measurement consists in the measurement of the speed at which the particle is diffusing due to Brownian motion. Speed of diffusion is measured by measuring the rate at which the intensity of the scattered light fluctuates. Small particles cause more fluctuations than larger particles and inversely as shown in Fig. 7.40. During Brownian motion, the distance between particles changes thus creating a shift between the frequencies of the incident light and scattered light.

The DLS technique measures a diameter which corresponds to how particles diffuse within a fluid, so it corresponds to a hydrodynamic diameter (D_h). The velocity of the Brownian motion depends on the translational diffusion coefficient (D_t) and on the size of particles (D_h). The translational diffusion coefficient depends on the size, the structure of the surface particles and on the concentration and the type of ions present in the medium.

By this method, the hydrodynamic diameter measured is considered as the diameter of a sphere which has the same translational diffusion coefficient as the particle. So, the Brownian motion of a particle depends on the diffusion coefficient and is related to the particle size by the Stokes-Einstein's equation (Eq. 7.10).

$$D_h = kBT/3\pi\,\eta\,D_t \tag{7.10}$$

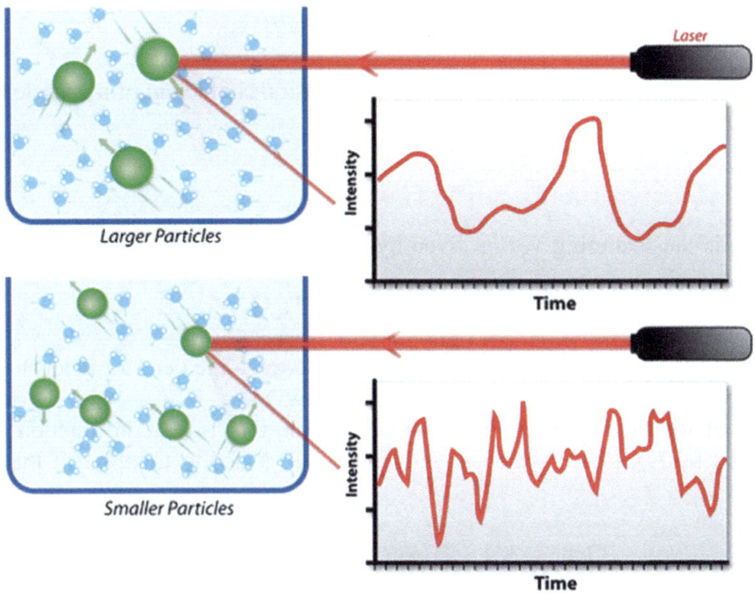

Fig. 7.40 Intensity fluctuations as a function of the size of particle in suspension. *Mike Jones*

where: D_h is the hydrodynamic diameter, D_t the translational diffusion coefficient, k the Boltzman's constant (= 1.3806 10^{-23} J/K), T the thermodynamic temperature and η the viscosity of the medium.

The intensity fluctuations arising from the Brownian motion of the particles can be measured directly but this measure is not sufficient and it is necessary to use a correlator function. The demonstration and use of the correlator function have been described by some authors [126–128].

Percora et al. explained that the intensity of light scattered from a suspension of particle into a scattering angle is the result of the interference on the square law detector between light scattered from different particles in the medium [128]. The light scattered from different particles depend on relative positions of particles relative to the direction of the incoming and scattered light beams. At a given time, the total scattered intensity given at a scattering angle depend on the position of the particle. These scattered intensities are a stochastic signal linked to the Brownian motion and using an autocorrelation the scattered intensity time correlation can be calculated.

For monodispersed particle in Brownian motion, the correlation function can be expressed by Eq. (7.11) which is an exponential decaying function of the decay time τ.

$$G(\tau) = A[1 + B\exp(-2\Gamma\,\tau)] \tag{7.11}$$

where Γ is the delay constant related to the diffusion coefficient of a particle of size D_t by Eq. (7.12).

$$\Gamma = D_t q^2 \tag{7.12}$$

where q is the scattering vector given by the equation:

$$q = [4\,\Pi\,n_o \sin(\theta/2)]/\lambda$$

where n_o is the refractive index, θ is the angle at which the detector is located from the sample cell and λ the incident laser wavelength.

For polydispersed particle in Brownian motion, the correlation function can be expressed by Eq. (7.13) which is an exponential decaying function of the decay time τ.

$$G(\tau) = A[1 + B\exp(-2(\Gamma'\tau + a/2 + \ldots))] \tag{7.13}$$

All correlation functions are used and are available in the different software used for dynamic light scattering measurement and allows to determine the hydrodynamic diameter of particle and the particle size distribution in suspensions.

7.6.7.5 Examples of DLS Measurements

In this section, literature examples of DLS measurement used to determine the particle size and size distribution of agro wastes nanocrystals and nanofiber are presented.

In their study, Rahimi Kord Sofla et al. extracted cellulose from bagasse and after treatment they obtained cellulose nanoparticles [98]. The cellulose nanocrystals (CNC) were obtained from an acid hydrolysis of cellulose and the cellulose nanofibrils (CNF) were obtained from milling of cellulose. Measurements were carried out with a Malvern 3000 Zetasizer nano-ZS. Figure 7.41 presents the particles size distribution obtained for CNC and CNF. For CNC particles, a main peak is present and shows a nano-scale range for more than 90% vol. of the particle population and a minimum of particle size about 100 nm. For CNF particles, the size distribution is polymodal and shows two particle size families, one in the range of the nanometer and one in the micrometer range.

Cellulose nanofibrils can be extracted from eucalyptus fibers by TEMPO treatment and homogenization at room temperature as shown by Gamelas et al. [129]. Cellulose nanofibrils size measurements were carried out with a Zetasize NanoZS of Malvern Instruments which allows size measurements in the range from 0.6 nm to 6 μm. The sample analyzed contains 0.2 wt% of cellulose nanofibrils. Figure 7.42 presents results of intensity as a function of size obtained for the two

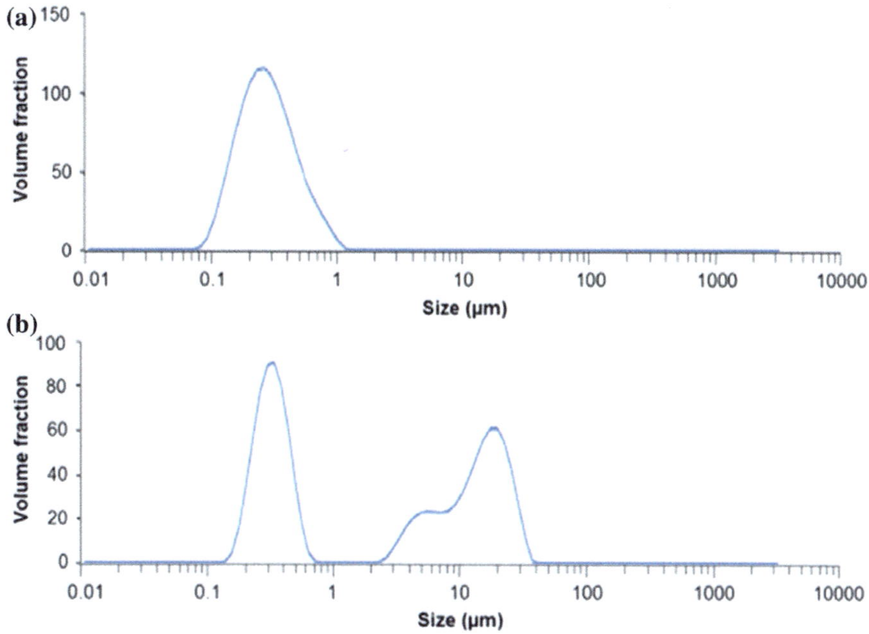

Fig. 7.41 Size distribution **a** of CNC particle and **b** of CNF particle [98]

Fig. 7.42 Intensity distribution for CNF-5p (A) and CNF-15p (B). Reproduced with permission from Elsevier [129]

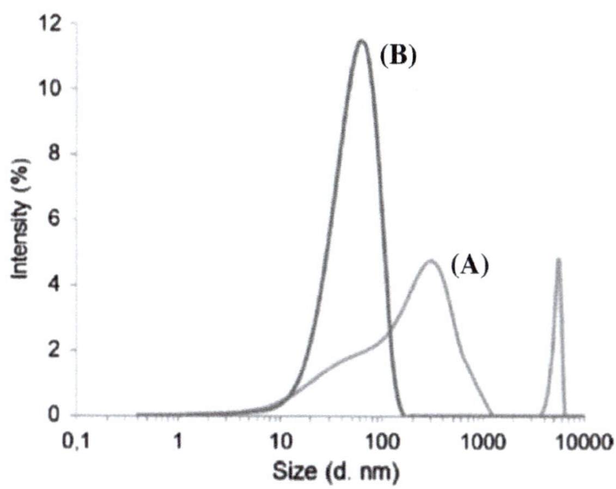

samples, the CNF-5p corresponding to a homogenization 5 times at 300 bar and the CNF-15p corresponding to a homogenization 15 times. The intensity as a function of the size is monomodal for the sample CNF-15p with the main peak centered around 55 nm. For the sample CNF-5p the range of the size is broad and presents two main peaks, one with a particle range from around 5 to 1000 nm and the second around 5000 nm. As the size distribution of CNF-15p is uniform, authors described a method to coupled DLS and AFM (Atomic Force Microscopy) analysis to calculate the length of the nanofibrils from data obtained by these two techniques. For this, they consider the hydrodynamic diameter determined by DLS (DLS) and the average width measured by AFM (WAFM). From equations and assumptions described in the paper of Gamelas et al., the relationship obtained between the length (L) and the hydrodynamic diameter is the following [129]:

$$L = \frac{2}{3}\frac{(D_{LS})^3}{(X_{AFM})^2} \tag{7.14}$$

DLS measurement can be also used to determine the size of chitin nanofibers. Wang et al. prepared chitin nanofibers from commercial chitin powder which were transformed by esterification and ultrasound treatment [130]. Analyses were carried out with a Malvern Zetasizer Nano-ZS 90 apparatus. The sample preparation was a suspension of chitin nanofiber of 1 wt% into purified water. From this measurement, authors presented equivalent size measuring from the Brownian motion of the nanofibers in the suspension and size distribution (PDI) of chitin nanofibers. All calculations were made with the Zetasizer software of Malvern Instruments version 7.01 [131]. Figure 7.43 shows the particles size distribution obtained and in Table 7.5 are presented the mean equivalent size of the particle determined and the corresponding polydispersity index (PDI). The chitin nanofibers not treated present a particle size distribution with three main peaks corresponding to a mean size of 1909 nm and a PDI of 0.84 as shown in Fig. 7.43. The ultrasound treatment allows to obtain smaller particles and a particle size distribution more uniform. For example, after 45 min of treatment with ultrasound, the particle size distribution is monomodal and the mean size is of 203 nm and the PDI of 0.25.

Chakraborty et al. measured the particle size of starch nanoparticles in solutions via DLS [132]. The apparatus used is a N4 plus Particle Sizer of Beckman-Coulter with a He–Ne laser as a light source. The autocorrelation (g2(t)) function of light

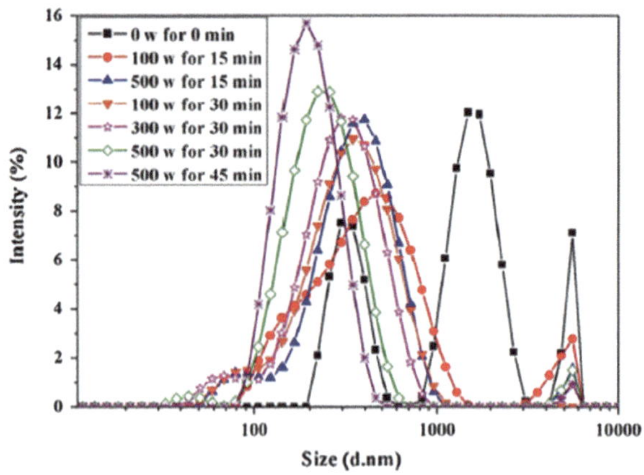

Fig. 7.43 Size distribution of chitin nanofibers after different ultrasound treatment. Reproduced with permission from Elsevier [130]

Table 7.5 Mean equivalent size and PDI of chitin nanofibers after different ultrasound treatment

Ultrasound treatment condition	Mean size (nm)	PDI
0 W for 0 min	1909 ± 565.9	0.84 ± 0.19
100 W for 15 min	372.6 ± 6.3	0.37 ± 0.04
500 W for 15 min	321.5 ± 12.5	0.32 ± 0.03
100 W for 30 min	280.1 ± 4.8	0.32 ± 0.04
300 W for 30 min	364.7 ± 6.4	0.31 ± 0.02
500 W for 30 min	223.6 ± 2.8	0.27 ± 0.03
500 W for 45 min	203.1 ± 2.2	0.25 ± 0.01

Reproduced with permission from Elsevier [130]

scattering is measured and converted into a distribution of the apparent diameter assuming that the light scattering comes from spherical particles independently diffusional motion. The normalized autocorrelation functions of aqueous solutions of starch nanoparticles (1 wt%) obtained for different scattering angles θ are plotted as a function of the delay time as shown in Fig. 7.44. The results indicate diffusional behavior for starch nanoparticles. From these normalized autocorrelation functions (g2(t)), the size distribution was calculated and is presented in Fig. 7.45. Two main peaks are present and correspond to a particle size of 50 and 300 nm. Electron micrograph measurement combining with these results confirmed that the peak of 300 nm corresponds to aggregates. The peak situated around 300 nm moves to smaller diameter when the scattering angle increases, this example confirms the presence of aggregates distribution. In this paper, authors show also the concentration and temperature effects on the particles size measurement by dynamic light scattering. These measurements allow showing particles aggregation when the concentration increases and a variation of the temperature from 25 to 65 °C do not lead to significant changes in the size distributions.

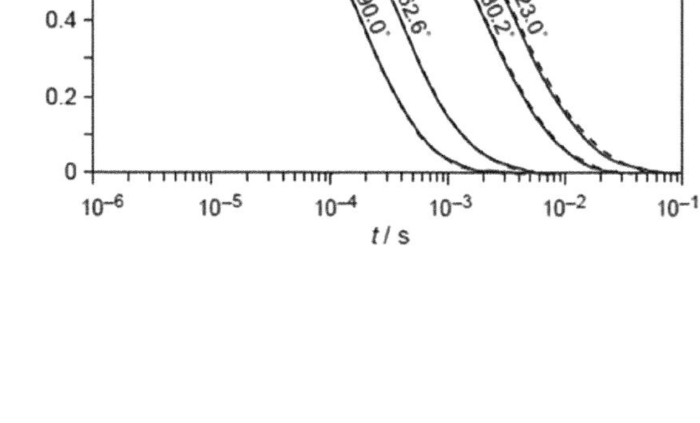

Fig. 7.44 Normalized autocorrelation function as a function of delay time. Reproduced with permission from Elsevier [132]

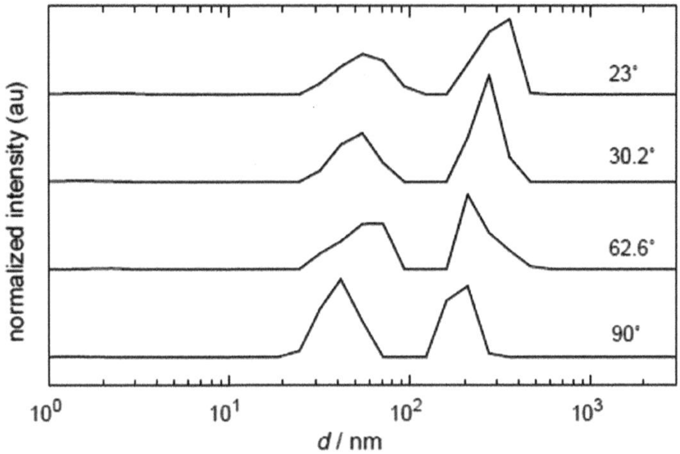

Fig. 7.45 Particle size distribution of starch nanoparticles in 1 wt% solution in water at 25 °C. Reproduced with permission from Elsevier [132]

7.7 Conclusion

Agro-industrial and other solid waste disposal is another serious issue of concern in developing countries. The present chapter explores how we can use these agrowaste biomasses for the extraction of various highly engineered nanoparticles that can be used for various technological applications. Based on the availability of agrowaste materials, we can extract various nanomaterials like nanocellulose, nanochitin and starch nanoparticles for the various engineering and technological applications. The effective utilization of agrowastes for the extraction of the various nanoparticles with excellent properties high strength, surface area, aspect ratio etc. provides a solution for reducing the agrowaste. In this context, this review summarizes the extraction procedures, characterization techniques and properties of various agro-based nanoparticles like cellulose nanofibers, cellulose nanocrystals, chitin nanofibers, chitin nanocrystals and starch nanoparticles. In this review, an attempt has been made to increase the understanding of the importance and characteristics of the agro-based nanoparticles like nanocellulose, nanochitin and starch nanoparticles by describing various aspects including their properties, characterization techniques and extraction procedures.

References

1. Rajinipriya, M., Nagalakshmaiah, M., Robert, M., Elkoun, S.: Importance of agricultural and industrial waste in the field of nanocellulose and recent industrial developments of wood based nanocellulose: a review. ACS Sustain. Chem. Eng. **6**, 2807–2828 (2018)

2. Wang, B., Sain, M.: Isolation of nanofibers from soybean source and their reinforcing capability on synthetic polymers. Compos. Sci. Technol. **67**, 2521–2527 (2007)
3. Alemdar, A., Sain, M.: Isolation and characterization of nanofibers from agricultural residues—wheat straw and soy hulls. Bioresour. Technol. **99**, 1664–1671 (2008)
4. Chakraborty, A., Sain, M., Kortschot, M.: Cellulose microfibrils: a novel method of preparation using high shear refining and cryocrushing. Holzforschung **59**, 102–107 (2005)
5. Kalia, S., Boufi, S., Celli, A., Kango, S.: Nanofibrillated cellulose: surface modification and potential applications. Colloid Polym. Sci. **292**, 5–31 (2014)
6. Abe, K., Iwamoto, S., Yano, H.: Obtaining cellulose nanofibers with a uniform width of 15 nm from wood. Biomacromolecules **8**, 3276–3278 (2007)
7. Iwamoto, S., Nakagaito, A.N., Yano, H., Nogi, M.: Optically transparent composites reinforced with plant fiber-based nanofibers. Appl. Phys. A Mater. Sci. Process. **81**, 1109–1112 (2005)
8. Ma, Z., Kotaki, M., Ramakrishna, S.: Electrospun cellulose nanofiber as affinity membrane. J. Membr. Sci. **265**, 115–123 (2005)
9. Henriksson, M., Henriksson, G., Berglund, L.A., Lindström, T.: An environmentally friendly method for enzyme-assisted preparation of microfibrillated cellulose (MFC) nanofibers. Eur. Polym. J. **43**, 3434–3441 (2007)
10. Pääkko, M., Ankerfors, M., Kosonen, H., Nykänen, A., Ahola, S., Österberg, M., Ruokolainen, J., Laine, J., Larsson, P.T., Ikkala, O., Lindström, T.: Enzymatic hydrolysis combined with mechanical shearing and high-pressure homogenization for nanoscale cellulose fibrils and strong gels. Biomacromolecules **8**, 1934–1941 (2007)
11. Tibolla, H., Pelissari, F.M., Menegalli, F.C.: Cellulose nanofibers produced from banana peel by chemical and enzymatic treatment. LWT Food Sci. Technol. **59**, 1311–1318 (2014)
12. Li, M., Wang, L.J., Li, D., Cheng, Y.-L., Adhikari, B.: Preparation and characterization of cellulose nanofibers from de-pectinated sugar beet pulp. Carbohydr. Polym. **102**, 136–143 (2014)
13. Chen, W., Abe, K., Uetani, K., Yu, H., Liu, Y., Yano, H.: Individual cotton cellulose nanofibers: pretreatment and fibrillation technique. Cellulose **21**, 1517–1528 (2014)
14. Deepa, B., Abraham, E., Cherian, B.M., Bismarck, A., Blaker, J.J., Pothan, L.A., Leao, A.L., de Souza, S.F., Kottaisamy, M.: Structure, morphology and thermal characteristics of banana nano fibers obtained by steam explosion. Bioresour. Technol. **102**, 1988–1997 (2011)
15. Chirayil, C.J., Joy, J., Mathew, L., Mozetic, M., Koetz, J., Thomas, S.: Isolation and characterization of cellulose nanofibrils from *Helicteres isora* plant. Ind. Crops Prod. **59**, 27–34 (2014)
16. Cherian, B.M., Leão, A.L., de Souza, S.F., Thomas, S., Pothan, L.A., Kottaisamy, M.: Isolation of nanocellulose from pineapple leaf fibres by steam explosion. Carbohydr. Polym. **81**, 720–725 (2010)
17. Fortunati, E., Luzi, F., Jiménez, A., Gopakumar, D.A., Puglia, D., Thomas, S., Kenny, J.M., Chiralt, A., Torre, L.: Revalorization of sunflower stalks as novel sources of cellulose nanofibrils and nanocrystals and their effect on wheat gluten bionanocomposite properties. Carbohydr. Polym. **149**, 357–368 (2016)
18. Turbak, A.F.A., Snyder, F., Fred, W., Sandberg, K., Albin, F.T., Snyder, K., Sandberg, R.: Suspensions containing microfibrillated cellulose. J. Appl. Polym. Sci. Appl. Polym. Symp. **37**, 1–5 (1983)
19. Pelissari, F.M., Sobral, P.J.D.A., Menegalli, F.C.: Isolation and characterization of cellulose nanofibers from banana peels. Cellulose **21**, 417–432 (2014)
20. Floros, M., Hojabri, L., Abraham, E., Jose, J., Thomas, S., Pothan, L., Leao, A.L., Narine, S.: Enhancement of thermal stability, strength and extensibility of lipid-based polyurethanes with cellulose-based nanofibers. Polym. Degrad. Stab. **97**, 1970–1978 (2012)
21. Siqueira, G., Bras, J., Dufresne, A.: Cellulosic bionanocomposites: a review of preparation, properties and applications. Polymers **2**, 728–765 (2010)

22. Ng, H.M., Sin, L.T., Tee, T.T., Bee, S.T., Hui, D., Low, C.Y., Rahmat, A.R.: Extraction of cellulose nanocrystals from plant sources for application as reinforcing agent in polymers. Compos. Part B Eng. **75**, 176–200 (2015)
23. Savadekar, N.R., Mhaske, S.T.: Synthesis of nano cellulose fibers and effect on thermoplastics starch based films. Carbohydr. Polym. **89**, 146–151 (2012)
24. Zainuddin, S.Y.Z., Ahmad, I., Kargarzadeh, H., Abdullah, I., Dufresne, A.: Potential of using multiscale kenaf fibers as reinforcing filler in cassava starch-kenaf biocomposites. Carbohydr. Polym. **92**, 2299–2305 (2013)
25. Majeed, K., Jawaid, M., Hassan, A., Abu Bakar, A., Abdul Khalil, H.P.S., Salema, A.A., Inuwa, I.: Potential materials for food packaging from nanoclay/natural fibres filled hybrid composites. Mater. Des. **46**, 391–410 (2013)
26. Shi, J., Shi, S.Q., Barnes, H.M., Pittman, C.U.: A chemical process for preparing cellulosic fibers hierarchically from kenaf bast fibers. BioResources **6**, 879–890 (2011)
27. Shin, H.K., Pyo Jeun, J., Bin Kim, H., Hyun Kang, P.: Isolation of cellulose fibers from kenaf using electron beam. Radiat. Phys. Chem. **81**, 936–940 (2012)
28. Panaitescu, D.M., Frone, A.N., Nicolae, C.: Micro- and nano-mechanical characterization of polyamide 11 and its composites containing cellulose nanofibers. Eur. Polym. J. **49**, 3857–3866 (2013)
29. Vilay, V., Mariatti, M., Mat Taib, R., Todo, M.: Effect of fiber surface treatment and fiber loading on the properties of bagasse fiber-reinforced unsaturated polyester composites. Compos. Sci. Technol. **68**, 631–638 (2008)
30. Tee, T.T., Sin, L.T., Gobinath, R., Bee, S.T., Hui, D., Rahmat, A.R., Kong, I., Fang, Q.: Investigation of nano-size montmorillonite on enhancing polyvinyl alcohol-starch blends prepared via solution cast approach. Compos. Part B Eng. **47**, 238–247 (2013)
31. Alemdar, A., Sain, M.: Biocomposites from wheat straw nanofibers: morphology, thermal and mechanical properties. Compos. Sci. Technol. **68**, 557–565 (2008)
32. Maiti, S., Jayaramudu, J., Das, K., Reddy, S.M., Sadiku, R., Ray, S.S., Liu, D.: Preparation and characterization of nano-cellulose with new shape from different precursor. Carbohydr. Polym. **98**, 562–567 (2013)
33. Sonia, A., Priya Dasan, K., Alex, R.: Celluloses microfibres (CMF) reinforced poly (ethylene-co-vinyl acetate) (EVA) composites: dynamic mechanical, gamma and thermal ageing studies. Chem. Eng. J. **228**, 1214–1222 (2013)
34. Lee, S.Y., Mohan, D.J., Kang, I.A., Doh, G.H., Lee, S., Han, S.O.: Nanocellulose reinforced PVA composite films: effects of acid treatment and filler loading. Fibers Polym. **10**, 77–82 (2009)
35. Danial, W.H., Abdul Majid, Z., Mohd Muhid, M.N., Triwahyono, S., Bakar, M.B., Ramli, Z.: The reuse of wastepaper for the extraction of cellulose nanocrystals. Carbohydr. Polym. **118**, 165–169 (2015)
36. Kargarzadeh, H., Ahmad, I., Abdullah, I., Dufresne, A., Zainudin, S.Y., Sheltami, R.M.: Effects of hydrolysis conditions on the morphology, crystallinity, and thermal stability of cellulose nanocrystals extracted from kenaf bast fibers. Cellulose **19**, 855–866 (2012)
37. Jiang, F., Hsieh, Y.-L.: Cellulose nanocrystal isolation from tomato peels and assembled nanofibers. Carbohydr. Polym. **122**, 60–68 (2015)
38. Lamaming, J., Hashim, R., Sulaiman, O., Leh, C.P., Sugimoto, T., Nordin, N.A.: Cellulose nanocrystals isolated from oil palm trunk. Carbohydr. Polym. **127**, 202–208 (2015)
39. Bettaieb, F., Khiari, R., Hassan, M.L., Belgacem, M.N., Bras, J., Dufresne, A., Mhenni, M. F.: Preparation and characterization of new cellulose nanocrystals from marine biomass *Posidonia oceanica*. Ind. Crops Prod. **72**, 175–182 (2015)
40. Fortunati, E., Luzi, F., Puglia, D., Petrucci, R., Kenny, J.M., Torre, L.: Processing of PLA nanocomposites with cellulose nanocrystals extracted from *Posidonia oceanica* waste: innovative reuse of coastal plant. Ind. Crops Prod. **67**, 439–447 (2015)
41. Mohamed, M.A., Salleh, W.N.W., Jaafar, J., Asri, S.E.A.M., Ismail, A.F.: Physicochemical properties of "green" nanocrystalline cellulose isolated from recycled newspaper. RSC Adv. **5**, 29842–29849 (2015)

42. Camarero Espinosa, S., Kuhnt, T., Foster, E.J., Weder, C.: Isolation of thermally stable cellulose nanocrystals by phosphoric acid hydrolysis. Biomacromolecules **14**, 1223–1230 (2013)
43. Tang, Y., Shen, X., Zhang, J., Guo, D., Kong, F., Zhang, N.: Extraction of cellulose nano-crystals from old corrugated container fiber using phosphoric acid and enzymatic hydrolysis followed by sonication. Carbohydr. Polym. **125**, 360–366 (2015)
44. Liu, Y., Wang, H., Yu, G., Yu, Q., Li, B., Mu, X.: A novel approach for the preparation of nanocrystalline cellulose by using phosphotungstic acid. Carbohydr. Polym. **110**, 415–422 (2014)
45. Yu, H., Qin, Z., Liang, B., Liu, N., Zhou, Z., Chen, L.: Facile extraction of thermally stable cellulose nanocrystals with a high yield of 93% through hydrochloric acid hydrolysis under hydrothermal conditions. J. Mater. Chem. A **1**, 3938 (2013)
46. Tibolla, H., Pelissari, F.M., Rodrigues, M.I., Menegalli, F.C.: Cellulose nanofibers produced from banana peel by enzymatic treatment: study of process conditions. Ind. Crops Prod. **95**, 664–674 (2017)
47. Barikani, M., Oliaei, E., Seddiqi, H., Honarkar, H.: Preparation and application of chitin and its derivatives: a review. J. Iran. Polym. **23**, 307–326 (2014)
48. Zeng, J.B., He, Y.S., Li, S.L., Wang, Y.Z.: Chitin whiskers: an overview. Biomacromolecules **13**, 1–11 (2012)
49. Paillet, M., Dufresne, A.: Chitin whisker reinforced thermoplastic nanocomposites. Macromolecules **34**, 6527–6530 (2001)
50. Ifuku, S., Nogi, M., Abe, K., Yoshioka, M., Morimoto, M., Saimoto, H., Yano, H.: Preparation of chitin nanofibers with a uniform width as chitin from crab shells. Biomacromolecules **10**, 1584–1588 (2009)
51. Kadokawa, J., Takegawa, A., Mine, S., Prasad, K.: Preparation of chitin nanowhiskers using an ionic liquid and their composite materials with poly(vinyl alcohol). Carbohydr. Polym. **84**, 1408–1412 (2011)
52. Le Corre, D., Angellier-Coussy, H.: Preparation and application of starch nanoparticles for nanocomposites: a review. React. Funct. Polym. **85**, 97–120 (2014)
53. Chen, G., Wei, M., Chen, J., Huang, J., Dufresne, A., Chang, P.R.: Simultaneous reinforcing and toughening: new nanocomposites of waterborne polyurethane filled with low loading level of starch nanocrystals. Polymer **49**, 1860–1870 (2008)
54. Hebeish, A., El-Rafie, M.H., EL-Sheikh, M.A., El-Naggar, M.E.: Ultra-fine characteristics of starch nanoparticles prepared using native starch with and without surfactant. J. Inorg. Organomet. Polym. Mater. **24**, 515–524 (2014)
55. Gonçalves, P.M., Noreña, C.P.Z., da Silveira, N.P., Brandelli, A.: Characterization of starch nanoparticles obtained from *Araucaria angustifolia* seeds by acid hydrolysis and ultrasound. LWT Food Sci. Technol. **58**, 21–27 (2014)
56. Kim, H.Y., Park, D.J., Kim, J.Y., Lim, S.T.: Preparation of crystalline starch nanoparticles using cold acid hydrolysis and ultrasonication. Carbohydr. Polym. **98**, 295–301 (2013)
57. Shi, A.M., Li, D., Wang, L.J., Li, B.Z., Adhikari, B.: Preparation of starch-based nanoparticles through high-pressure homogenization and miniemulsion cross-linking: influence of various process parameters on particle size and stability. Carbohydr. Polym. **83**, 1604–1610 (2011)
58. Fabian, C., Ayucitra, A., Ismadji, S., Ju, Y.H.: Isolation and characterization of starch from defatted rice bran. J. Taiwan Inst. Chem. Eng. **42**, 86–91 (2011)
59. Mondragon, G., Fernandes, S., Retegi, A., Peña, C., Algar, I., Eceiza, A., Arbelaiz, A.: A common strategy to extracting cellulose nanoentities from different plants. Ind. Crops Prod. **55**, 140–148 (2014)
60. Yang, H., Yan, R., Chen, H., Lee, D.H., Zheng, C.: Characteristics of hemicellulose, cellulose and lignin pyrolysis. Fuel **86**, 1781–1788 (2007)
61. Morán, J.I., Alvarez, V.A., Cyras, V.P., Vázquez, A.: Extraction of cellulose and preparation of nanocellulose from sisal fibers. Cellulose **15**, 149–159 (2008)

62. Karimi, S., Tahir, P.M., Karimi, A., Dufresne, A., Abdulkhani, A.: Kenaf bast cellulosic fibers hierarchy: a comprehensive approach from micro to nano. Carbohydr. Polym. **101**, 878–885 (2014)
63. Wang, B., Li, J., Zhang, J., Li, H., Chen, P., Gu, Q., Wang, Z.: Thermo-mechanical properties of the composite made of poly (3-hydroxybutyrate-co-3-hydroxyvalerate) and acetylated chitin nanocrystals. Carbohydr. Polym. **95**, 100–106 (2013)
64. Ma, X., Jian, R., Chang, P.R., Yu, J.: Fabrication and characterization of citric acid-modified starch nanoparticles/plasticized-starch composites. Biomacromolecules **9**, 3314–3320 (2008)
65. Ahola, S., Salmi, J., Johansson, L.S., Laine, J., Österberg, M.: Model films from native cellulose nanofibrils. Preparation, swelling, and surface interactions. Biomacromolecules **9**, 1273–1282 (2008)
66. Aulin, C., Netrval, J., Wågberg, L., Lindström, T.: Aerogels from nanofibrillated cellulose with tunable oleophobicity. Soft Matter **6**, 3298 (2010)
67. Wang, J., Wang, Z., Li, J., Wang, B., Liu, J., Chen, P., Miao, M., Gu, Q.: Chitin nanocrystals grafted with poly(3-hydroxybutyrate-co-3-hydroxyvalerate) and their effects on thermal behavior of PHBV. Carbohydr. Polym. **87**, 784–789 (2012)
68. Thielemans, W., Belgacem, M.N., Dufresne, A.: Starch nanocrystals with large chain surface modifications. Langmuir **22**, 4804–4810 (2006)
69. Ashori, A., Babaee, M., Jonoobi, M., Hamzeh, Y.: Solvent-free acetylation of cellulose nanofibers for improving compatibility and dispersion. Carbohydr. Polym. **102**, 369–375 (2014)
70. Trovatti, E., Carvalho, A.J.F., Ribeiro, S.J.L., Gandini, A.: Simple green approach to reinforce natural rubber with bacterial cellulose nanofibers. Biomacromolecules **14**, 2667–2674 (2013)
71. Kong, L., Zhang, D., Shao, Z., Han, B., Lv, Y., Gao, K., Peng, X.: Superior effect of TEMPO-oxidized cellulose nanofibrils (TOCNs) on the performance of cellulose triacetate (CTA) ultrafiltration membrane. Desalination **332**, 117–125 (2014)
72. Goetz, L.A., Jalvo, B., Rosal, R., Mathew, A.P.: Superhydrophilic anti-fouling electrospun cellulose acetate membranes coated with chitin nanocrystals for water filtration. J. Membr. Sci. **510**, 238–248 (2016)
73. Feng, L., Zhou, Z., Dufresne, A., Huang, J., Wei, M., An, L.: Structure and properties of new thermoforming bionanocomposites based ort chitin whisker-graft-polycaprolactone. J. Appl. Polym. Sci. **112**, 2830–2837 (2009)
74. Nafchi, A.M., Alias, A.K., Mahmud, S., Robal, M.: Antimicrobial, rheological, and physicochemical properties of sago starch films filled with nanorod-rich zinc oxide. J. Food Eng. **113**, 511–519 (2012)
75. Li, C., Li, Y., Sun, P., Yang, C.: Pickering emulsions stabilized by native starch granules. Colloids Surf. A Physicochem. Eng. Asp. **431**, 142–149 (2013)
76. Tholstrup Sejersen, M., Salomonsen, T., Ipsen, R., Clark, R., Rolin, C., Balling Engelsen, S.: Zeta potential of pectin-stabilised casein aggregates in acidified milk drinks. Int. Dairy J. **17**, 302–307 (2007)
77. Bondeson, D., Oksman, K.: Dispersion and characteristics of surfactant modified cellulose whiskers nanocomposites. Compos. Interfaces **14**, 617–630 (2007)
78. Elanthikkal, S., Gopalakrishnapanicker, U., Varghese, S., Guthrie, J.T.: Cellulose microfibres produced from banana plant wastes: isolation and characterization. Carbohydr. Polym. **80**, 852–859 (2010)
79. Tzoumaki, M.V., Karefyllakis, D., Moschakis, T., Biliaderis, C.G., Scholten, E.: Aqueous foams stabilized by chitin nanocrystals. Soft Matter **11**, 6245–6253 (2015)
80. Wei, B., Hu, X., Li, H., Wu, C., Xu, X., Jin, Z., Tian, Y.: Effect of pHs on dispersity of maize starch nanocrystals in aqueous medium. Food Hydrocoll. **36**, 369–373 (2014)
81. Pearson, F.G., Marchessault, R.H., Liang, C.Y.: Infrared spectra of crystalline polysaccharides. V. Chitin. J. Polym. Sci. **43**, 101–116 (1960)

82. Lavoine, N., Desloges, I., Dufresne, A., Bras, J.: Microfibrillated cellulose—its barrier properties and applications in cellulosic materials: a review. Carbohydr. Polym. **90**, 735–764 (2012)
83. Le Normand, M., Moriana, R., Ek, M.: Isolation and characterization of cellulose nanocrystals from spruce bark in a biorefinery perspective. Carbohydr. Polym. **111**, 979–987 (2014)
84. Flauzino Neto, W.P., Silvério, H.A., Dantas, N.O., Pasquini, D.: Extraction and characterization of cellulose nanocrystals from agro-industrial residue—soy hulls. Ind. Crops Prod. **42**, 480–488 (2013)
85. dos Santos, R.M., Flauzino Neto, W.P., Silvério, H.A., Martins, D.F., Dantas, N.O., Pasquini, D.: Cellulose nanocrystals from pineapple leaf, a new approach for the reuse of this agrowaste. Ind. Crops Prod. **50**, 707–714 (2013)
86. Sacui, I.A., Nieuwendaal, R.C., Burnett, D.J., Stranick, S.J., Jorfi, M., Weder, C., Foster, E. J., Olsson, R.T., Gilman, J.W.: Comparison of the properties of cellulose nanocrystals and cellulose nanofibrils isolated from bacteria, tunicate, and wood processed using acid, enzymatic, mechanical, and oxidative methods. ACS Appl. Mater. Interfaces **6**, 6127–6138 (2014)
87. Silvério, H.A., Flauzino Neto, W.P., Dantas, N.O., Pasquini, D.: Extraction and characterization of cellulose nanocrystals from corncob for application as reinforcing agent in nanocomposites. Ind. Crops Prod. **44**, 427–436 (2013)
88. Espino-Pérez, E., Gilbert, R.G., Domenek, S., Brochier-Salon, M.C., Belgacem, M.N., Bras, J.: Nanocomposites with functionalised polysaccharide nanocrystals through aqueous free radical polymerisation promoted by ozonolysis. Carbohydr. Polym. **135**, 256–266 (2016)
89. Liu, H., Cui, S., Shang, S., Wang, D., Song, J.: Properties of rosin-based waterborne polyurethanes/cellulose nanocrystals composites. Carbohydr. Polym. **96**, 510–515 (2013)
90. Salaberria, A.M., Diaz, R.H., Labidi, J., Fernandes, S.C.M.: Role of chitin nanocrystals and nanofibers on physical, mechanical and functional properties in thermoplastic starch films. Food Hydrocoll. **46**, 93–102 (2015)
91. Fan, Y., Saito, T., Isogai, A.: Chitin nanocrystals prepared by TEMPO-mediated oxidation of α-chitin. Biomacromolecules **9**, 192–198 (2008)
92. Ogah, A.O., Afiukwa, J.N., Englund, K.: Characterization and comparison of thermal stability of agro waste fibers in bio-composites application. J. Chem. Eng. Chem. Res. **1**, 84–93 (2014)
93. Romero Millán, L.M., Sierra Vargas, F.E., Nzihou, A.: Kinetic analysis of tropical lignocellulosic agrowaste pyrolysis. Bioenergy Res. **10**, 832–845 (2017)
94. Xu, C., Zhu, S., Xing, C., Li, D., Zhu, N., Zhou, H.: Isolation and properties of cellulose nanofibrils from coconut palm petioles by different mechanical process. PLoS One **10** (2015)
95. Xie, J., Hse, C.-Y., De Hoop, C.F., Hu, T., Qi, J., Shupe, T.F.: Isolation and characterization of cellulose nanofibers from bamboo using microwave liquefaction combined with chemical treatment and ultrasonication. Carbohydr. Polym. **151**, 725–734 (2016)
96. Robles, E., Salaberria, A.M., Herrera, R., Fernandes, S.C.M., Labidi, J.: Self-bonded composite films based on cellulose nanofibers and chitin nanocrystals as antifungal materials. Carbohydr. Polym. **144**, 41–49 (2016)
97. Costa, L.A.D.S., Fonseca, A.F., Pereira, F.V., Druzian, J.I.: Extraction and characterization of cellulose nanocrystals from corn strover. Cellul. Chem. Technol. **49**, 127–133 (2015)
98. Rahimi Kord Sofla, M., Brown, R.J., Tsuzuki, T., Rainey, T.J.: A comparison of cellulose nanocrystals and cellulose nanofibres extracted from bagasse using acid and ball milling methods. Adv. Nat. Sci. Nanosci. Nanotechnol. **7**, 035004 (2016). https://doi.org/10.1088/2043-6262/7/3/035004
99. Aggarwal, P., Dollimore, D.: A comparative study of the degradation of different starches using thermal analysis. Talanta **43**, 1527–1530 (1996)
100. Aggarwal, P., Dollimore, D.: The combustion of starch, cellulose and cationically modified products of these compounds investigated using thermal analysis. Thermochim. Acta **291**, 65–72 (1997)

101. Sun, Q., Fan, H., Xiong, L.: Preparation and characterization of starch nanoparticles through ultrasonic-assisted oxidation methods. Carbohydr. Polym. **106**, 359–364 (2014)
102. Henrique, M.A., Flauzino Neto, W.P., Silvério, H.A., Martins, D.F., Gurgel, L.V.A., da Silva Barud, H., de Morais, L.C., Pasquini, D.: Kinetic study of the thermal decomposition of cellulose nanocrystals with different polymorphs, cellulose I and II, extracted from different sources and using different types of acids. Ind. Crops Prod. **76**, 128–140 (2015)
103. Stawski, D., Rabiej, S., Herczyńska, L., Draczyński, Z.: Thermogravimetric analysis of chitins of different origin. J. Therm. Anal. Calorim. **93**, 489–494 (2008)
104. Jiménez, A., Berenguer, V., López, J., Sánchez, A.: Thermal degradation study of poly(vinyl chloride): kinetic analysis of thermogravimetric data. J. Appl. Polym. Sci. **50**, 1565–1573 (1993)
105. Lopez, G., Artetxe, M., Amutio, M., Alvarez, J., Bilbao, J., Olazar, M.: Recent advances in the gasification of waste plastics. A critical overview. Renew. Sustain. Energy Rev. **82**, 576–596 (2018)
106. Guinier, A., Fournet, G.: Small angle scattering of X-rays. J. Polym. Sci. **1**, 268 (1955)
107. Porod, G., Glatter, O., Kratky, O.: Small Angle X-ray Scattering. Academic Press, London (1982)
108. Feigin, L.A., Svergun, D.I.: Structure Analysis by Small-Angle X-Ray and Neutron Scattering. Plenum Press, London (1989)
109. Windsor, C.G.: An introduction to small angle scattering. J. Appl. Cryst. **21**, 582–588 (1988)
110. Jackson, A.J.: Introduction to Small-Angle Neutron Scattering and Neutron Reflectometry, pp. 1–24. NIST Center for Neutron Research (2008)
111. Boldon, L., Laliberte, F., Liu, L.: Review of the fundamental theories behind small angle X-ray scattering, molecular dynamics simulations, and relevant integrated application. Nano Rev. **6**, 25661 (2015)
112. Franke, D., Petoukhov, M.V., Konarev, P.V., Panjkovich, A., Tuukkanen, A., Mertens, H.D. T., Kikhney, A.G., Hajizadeh, N.R., Franklin, J.M., Jeffries, C.M., Svergun, D.I.: ATSAS 2.8 : a comprehensive data analysis suite for small-angle scattering from macromolecular solutions. J. Appl. Crystallogr. **50**, 1212–1225 (2017)
113. Kline, S.R.: Reduction and analysis of SANS and USANS data using IGOR Pro. J. Appl. Crystallogr. **39**, 895–900 (2006)
114. Ilavsky, J., Jemian, P.R.: Irena: tool suite for modeling and analysis of small-angle scattering. J. Appl. Crystallogr. **42**, 347–353 (2009)
115. Spinozzi, F., Beltramini, M.: QUAFIT: a novel method for the quaternary structure determination from small-angle scattering data. Biophys. J. **103**, 511–521 (2012)
116. Mao, Y., Liu, K., Zhan, C., Geng, L., Chu, B., Hsiao, B.S.: Characterization of nanocellulose using small-angle neutron, X-ray, and dynamic light scattering techniques. J. Phys. Chem. B **121**, 1340–1351 (2017)
117. Glinka, C.J., Barker, J.G., Hammouda, B., Krueger, S., Moyer, J.J., Orts, W.J.: The 30 m small-angle neutron scattering instruments at the National Institute of Standards and Technology. J. Appl. Crystallogr. **31**, 430–445 (1998)
118. Hexemer, A., Bras, W., Glossinger, J., Schaible, E., Gann, E., Kirian, R., MacDowell, A., Church, M., Rude, B., Padmore, H.: A SAXS/WAXS/GISAXS beamline with multilayer monochromator. J. Phys. **247**, 012007 (2010)
119. Penttilä, P.A., Várnai, A., Fernández, M., Kontro, I., Liljeström, V., Lindner, P., Siika-aho, M., Viikari, L., Serimaa, R.: Small-angle scattering study of structural changes in the microfibril network of nanocellulose during enzymatic hydrolysis. Cellulose **20**, 1031–1040 (2013)
120. Yang, R., Su, Y., Aubrecht, K.B., Wang, X., Ma, H., Grubbs, R.B., Hsiao, B.S., Chu, B.: Thiol-functionalized chitin nanofibers for As (III) adsorption. Polymer **60**, 9–17 (2015)
121. Su, Y., Burger, C., Hsiao, B.S., Chu, B.: Characterization of TEMPO-oxidized cellulose nanofibers in aqueous suspension by small-angle X-ray scattering. J. Appl. Crystallogr. **47**, 788–798 (2014)

122. Geng, L., Peng, X., Zhan, C., Naderi, A., Sharma, P.R., Mao, Y., Hsiao, B.S.: Structure characterization of cellulose nanofiber hydrogel as functions of concentration and ionic strength. Cellulose **24**, 5417–5429 (2017)
123. Berne, B.J., Pecora, R.: Dynamic Light Scattering: With Applications to Chemistry, Biology, and Physics (2000)
124. Kato, H., Nakamura, A., Takahashi, K., Kinugasa, S.: Accurate size and size-distribution determination of polystyrene latex nanoparticles in aqueous medium using dynamic light scattering and asymmetrical flow field flow fractionation with multi-angle light scattering. Nanomaterials **2**, 15–30 (2012)
125. ISO 22412, Particle Size Analysis—Dynamic Light Scattering (DLS). Available: https://www.iso.org/obp/ui/#iso:std:iso:22412:ed-2:v1:en. Accessed Feb 2017
126. Brown, W.: Dynamic Light Scattering: The Method and Some Applications. Oxford University Press, USA (1993)
127. Goldburg, W.I.: Dynamic light scattering. Am. J. Phys. **67**, 1152–1160 (1999)
128. Zhang, S., Yang, H., Singh, L.: Dynamic light scattering measurement of nanometer particles in liquids. CEUR Workshop Proc. **1225**, 41–42 (2014)
129. Gamelas, J.A.F., Pedrosa, J., Lourenço, A.F., Mutjé, P., González, I., Chinga-Carrasco, G., Singh, G., Ferreira, P.J.T.: On the morphology of cellulose nanofibrils obtained by TEMPO-mediated oxidation and mechanical treatment. Micron **72**, 28–33 (2015)
130. Wang, Q., Yan, X., Chang, Y., Ren, L., Zhou, J.: Fabrication and characterization of chitin nanofibers through esterification and ultrasound treatment. Carbohydr. Polym. **180**, 81–87 (2018)
131. Malvern Instruments: Electrophoretic Mobility and the Zetasizer Quality Factor. Malvern Tech. Note 1–2 (2011)
132. Chakraborty, S., Sahoo, B., Teraoka, I., Gross, R.A.: Solution properties of starch nanoparticles in water and DMSO as studied by dynamic light scattering. Carbohydr. Polym. **60**, 475–481 (2005)

Chapter 8
Food Waste and Manure

Carol Sze Ki Lin, Muthupandian Ashokkumar, Guneet Kaur, Chong Li, Xiaotong Li, Khai Lun Ong and Daniel Pleissner

Abstract Investigation of composition of biomass is an important pre-requisite for determining its suitability for various downstream applications. In recent times, food waste has emerged as a valuable biomass feedstock which could be valorized for production of fuels, chemicals and materials. Prior to use as a feedstock, it is pertinent to perform a detailed composition analysis and gather critical information about nutrient content including carbon, nitrogen and lipid. Additionally, analysis of minor constituents in food waste is significant to understand the possibility of their toxic or inhibitory effects during biotechnological conversions. Another abundant biomass source is manure which primarily arises from animal feeding operations. The major application of manure is land-applied fertilizer besides the recent investigations for fuel and energy. These intended applications demand complete characterization of nutrient content and quality. Thus, this chapter is focused on characterization and analysis of food waste and manure. Methods for sampling, handling and pretreatment of food waste and manure are discussed. Furthermore, a detailed nutrient analysis and equipment used for analysis is

C. S. K. Lin (✉) · G. Kaur · C. Li · X. Li · K. L. Ong
School of Energy and Environment, City University of Hong Kong, Tat Chee Avenue, Kowloon, Hong Kong
e-mail: carollin@cityu.edu.hk

G. Kaur
Sino-Forest Applied Research Centre for Pearl River Delta Environment, Department of Biology, Hong Kong Baptist University, Kowloon Tong, Hong Kong

D. Pleissner
Faculty of Sustainability, Sustainable Chemistry (Resource Efficiency), Institute for Sustainable and Environmental Chemistry, Leuphana University Lüneburg, Universitätsallee 1/C13.203, 21335 Lüneburg, Germany

M. Ashokkumar
Faculty of Science, School of Chemistry, The University of Melbourne, Melbourne, VIC 3010, Australia

C. Li
Agricultural Genomics Institute at Shenzhen, Chinese Academy of Agricultural Sciences, Shenzhen, Guangdong, People's Republic of China

A. Nzihou (ed.), *Handbook on Characterization of Biomass, Biowaste and Related By-products*, https://doi.org/10.1007/978-3-030-35020-8_8

described. We further discuss the application of food waste processing techniques which could facilitate the characterization, treatment and product recovery.

8.1 Generation of Food Waste and Manure and Need for Characterization

Food waste is defined as the food which is still able for human consumption has been removed from food supply chain and it is commonly but not completely occurring at both retail and consumer levels [1, 2]. Every year approximately 1.3 billion tonnes of food produced for human consumption in the world is wasted. According to the Food and Agricultural Organisation (FAO), approximately 40–50% of fruits, vegetables and root crops, 35% of fish, 30% of cereals and 20% of meats, dairy and oil seeds are wasted globally every year [3]. Conventionally, this type of food waste undergoes the first generation waste treatment including disposal in landfills and incineration. Both options are undesirable since landfilling may cause severe health and sustainability issues, whereas incineration of high moisture content of food waste results in release of dioxins that lead to environmental problems [4].

Food waste is a valuable resource for biofuels, biochemicals and biomaterials production [5]. Food waste mainly contains carbohydrates, proteins, lipids and trace minerals. It also contains minor constituents such as food and colour additives. Nutrients in food waste can be used by microorganism to produce value added products. Food wastes can have various compositions, for example, food waste consisting of rice is rich in carbohydrate, and food waste consisting of meat is rich in proteins and lipids [4]. Proper procedures for food waste handling and storage is needed because food waste contains high content of moisture. It is easy to spoil and release bad odour. Besides that, accurate knowledge of physical, chemical and composition of food waste is important information for food waste valorization application. Food waste characterization can be done in several methods. Its nature, composition and nutritional value are the properties to determine the methods of food waste collection, storage, handling and utilization. Determination of chemical composition such as carbohydrate and amino acids of food waste is important because it affects hydrolysis and fermentation application [6]. The presence of preservatives and colorants in food waste also can cause inhibition of enzymatic hydrolysis and fermentation. Hence, knowledge of methods and standard procedures for proper handling and characterization of food waste is important. Such methods are introduced in this chapter, which include sampling and preconditioning of food waste, physico-chemical characterization and composition analysis, extraction techniques for recovery of useful components and equipment used for characterization.

Manure is a valuable source of nutrients for crops and it can improve soil productivity. It is an organic matter which is mostly composed of animal faeces and urine, and may contain livestock bedding, additional water and wasted feed.

It contains a wide range of nutrients such as nitrogen, phosphorous and potassium as well as secondary or micro-nutrients such as copper, manganese, zinc and sodium [7, 8]. The additional bedding in manures makes them an excellent source of organic matter which helps to improve soil quality when applied to land. However, the properties of manure depend on a broad array of factors including animal species, age, diet, digestibility, housing, environment and stage of production [9]. Furthermore, the handling and storage of manure before land application also influence the final quality of manure nutrients [10]. This makes manure management more difficult than synthetic fertilizers. Accurate knowledge of physical and chemical properties of manure is important to derive its full benefits as a fertilizer and ensures its efficient land application. An additional concern is the possible environmental contamination in the case of excessive manure application and/or leaching [11]. Manure characterization can be done in several ways. Its solid content, size and composition are the properties which determine manure collection, storage, handling and utilization. Determination of important nutrients such as nitrogen, phosphorous and potassium is essential since it affects land application rates and treatment techniques [12]. A detailed account of presence of secondary elements, trace (heavy) elements and other contaminants such as pesticides is also important to avoid environmental risks to surface and groundwater, as well as threats to human and livestock health through the food chain [13]. Therefore, there is a pressing need for efficient manure management through proper characterization to enable sustainable use of nutrients and mitigate environmental impacts.

8.2 Introduction of Sampling and Preconditioning of Food Waste

The main sources of food waste include manufacturers, retailers, catering business and consumers. Food waste from manufacturers, retailers and catering business is also known as 'industrial food waste', and another type of food waste is known as 'household food waste' [14], which includes meat and fish, dairy and eggs, bakery, cakes and desserts, fresh fruit, drinks, fresh vegetables and salads, meals and others [15]. Industrial food waste includes molasses and bagasse from sugar refining, fruit and vegetable residues, residues from wineries, breweries and dairy wastes [14]. Besides that, grocery retail store food waste is also categorized as industrial food waste.

8.2.1 Sampling Method

Sampling of food waste is an important and useful procedure because a quality estimate of food waste can be obtained quickly and accurately without the need for analysis of total population of food waste [16]. Probability sampling technique is used to ensure valid and reliable inferences from a sample. There are four types of methods which are commonly used in sampling. These include simple random sampling, stratified sampling, systematic sampling and cluster sampling. Every subject has an equal chance to be selected for the study in a simple random sampling method. In stratified sampling method, the sample is divided into subgroups and then sample is withdrawn from each subgroup using simple random sample method. In the systematic sampling method, the sample is selected with a fixed periodic interval. Sampling interval is calculated according the population size and divided by the desired sample size. Cluster sampling involves the random selection of the sample from a subset of the clusters wherein the population of food waste is divided into clusters. These clusters are normally based on geographic area and district [17]. Simple random sampling normally applies on homogenous food waste [18]. Stratified sampling should be applied when certain factors of the area was studied (e.g. seasonal, geographical and socio-economic factors), those factors affect the composition of food waste [19]. Systematic sampling is suitable for not stratified homogenous food waste [18]. Cluster sampling should be applied when all elements for sample frame is not suitable [20].

8.2.2 Sampling Conditioning Before Analysis

Food waste is rich in nutrients such as carbon, nitrogen and lipid sources. In addition, the moisture content of food waste is high. Therefore, it has to be treated immediately after collection as microbial growth will rapidly cause food waste spoilage and release of bad odour.

Different methods can be used for food waste treatment. According to our former study, food waste was blended after collection and kept at 4 °C until further processing [21]. Subsequently, the blended food waste was processed using a food waste treatment machine to produce food waste powder. The two processes involved in food waste treatment include shredding and dehydration at 100–150 °C. On the other hand, Yu et al. [22] used a freeze drying process to treat the food waste after collection [22]. The freeze-dried food waste was then ground and sieved through 0.2 mm mesh. Finally, the food waste powder was stored in cold room at 4 °C under dark conditions. Li et al. [23] collected fruits and vegetables waste from street market (Sham Shui Po, Hong Kong) [23]. After that, they were washed with deionized water and blended into pastes using a blender. The resultant pastes were kept at −20 °C until further use. Pagliaccia et al. [24] collected food waste, chopped and screened it. A knife was applied to shred the food waste [24]. Storage occurred at −20 °C. Different methods of pre-treatment have different impact of shelf life of storage. Freeze drying method enables longer shelf life for food waste storage as compared to dehydration and wet storage. Since the moisture content of food waste using freeze drying method is lower than dehydration and wet storage [25].

8.3 Physico-Chemical Characterization and Composition Analysis of Food Waste

In a sustainable waste management strategy, determination of nature, composition and nutritional value of food waste is an important step. Identification and quantification of food waste provides knowledge of food waste composition. This information is useful for selection of further processing methods. The food waste composition plays an important role in producing valuable products and waste treatment [26]. For example, high carbohydrate content is suitable for alcohol production. However, carbohydrate content could be too high thereby causing inhibitory effect on microbial growth and alcohol production. Hence, suitable sugar concentration is needed for fermentation and this critical information can be obtained through composition analysis [27].

Physical characterization of food waste includes determination of weight, moisture, total solid, ash and volume. The moisture content of a sample is important because it affects the food waste composition. Total solid content provides information of the age, origin or pretreatment. Chemical characterization of food waste includes proximate analysis i.e. total carbohydrates, proteins, lipids and pH. The pH of food waste represents freshness of the waste, pH changes upon hydrolysis, fermentation or storage [26].

8.3.1 Moisture Content

Determination of moisture content is necessary in order to calculate food constituents on a dry weight basis. Different types of equipment can be used to determine moisture content. This includes forced draft oven, vacuum oven, microwave drying, rapid moisture analyzer, toluene distillation, Karl Fisher and near infrared spectroscopy.

In the forced draft oven method, the moisture content of a sample is calculated on the basis of weight loss after heating at specified conditions (AOAC 2005, Method 930.29) [28]. Moisture content determination using the vacuum oven method is similar to forced draft oven method, except that it involves heating of sample at reduced pressure conditions for water removal (AOAC 2005, Method 926.12) [28]. In the microwave drying method, the moisture content of a sample is calculated based on weight loss after heating using microwave energy (AOAC 2005, Method 985.26) [28]. Rapid moisture analyzer method measures the percentage of moisture in a sample based on weight loss upon heating at controlled high heating condition [29]. Principle of toluene distillation method is the moisture content which is measured through volume of water removed from distillation, where moisture in sample is co-distilled with toluene (AOAC 2005, Method 925.04) [28]. Karl Fisher method is a titration-based method which is based on the principle that the amount of water in a sample is directly proportional to the volume of Karl Fisher reagent (KF) needed to achieve the endpoint of titration (AOAC 2005, Method 991.01, 984.20) [28]. Iodine is reduced by sulphur dioxide from KF reagent when it reacts with water. Principle of near infrared analyzer method is based on the determination of moisture content of a sample by measuring the energy that is transmitted by the sample, which is inversely proportional to the energy absorbed [29].

8.3.2 Carbohydrate Content (Total Carbohydrate, Starch, Free Sugar)

Carbohydrate content of food waste is important because it indicates the amount of total carbohydrate, starch and free sugar that can be recovered and reused as nutrient sources in microbial fermentation for production of value-added products.

In 1956, Dubois et al. developed a colorimetric method to determine the carbohydrate content. Free reducing groups from polysaccharides, oligosaccharides, simple sugars and derivatives react with phenol and concentrated sulfuric acid to form orange-yellow colour solution. This method could be used for measuring small quantities of sugars in samples which could be extracted with volatile solvent [30]. In 1972, Lever developed another colorimetric method for determination of carbohydrates. Reducing carbohydrates react with acid hydrazides in alkaline solution to give yellow anions. It can be used to detect glucose or monosaccharides with concentration less than 1 μg [31]. Phenol-Sulfuric method and UV-Vis spectrophotometer have been applied in determination of total carbohydrate in food waste by WRAP [32].

In 1969, Gur et al. used colorimetric method for starch determination based on the Nielsen method which includes colorimetric determination of starch with iodine, and the Morell method which includes precipitation of starch with iodine and its subsequent hydrolysis to release reducing sugar [33]. In a new analytical method, starch content was measured through glucose produced from enzymatic hydrolysis of starch. Starch was hydrolyzed by thermostable α-amylase into soluble maltodextrins. Then, it underwent hydrolysis to produce glucose using amyloglucosidase [34]. Megazyme's total starch kit was applied for starch content determination in food waste by Pleissner et al. and Ma et al. [35, 36].

Free sugar determination was performed by cold water extraction. Approximately 5 g of dried food waste powder was added with 50 mL deionized water and shaken for half an hour. After that, 2 mL of zinc sulphate solution (30%, w/w) and potassium ferrocyanide (15%, w/w) were added sequentially. After shaking, the filtrate was analyzed by HPLC [37]. Another free sugar determination was carried out by water extraction in which 1 g of dried food waste powder sample was mixed with 10 mL deionized water. Extraction of free sugar from food waste was then performed by sonication for 30 min. After that, centrifugation of food waste mixture was carried out at 10,000 g for 10 min. Supernatant was filtered and analyzed with HPLC [38].

8.3.3 Protein and Free Amino Nitrogen Content

Nitrogen is present in all organisms in the form of proteins, amino acids, nucleotides as well as heteroatom in molecules. Due to the ubiquitous occurrence of nitrogen, the investigation of quantification methods has a long history. For instance, the determination of proteins in blood was carried out using the so-called 'van Slyke method' just over a hundred years ago. The van Slyke method is based on the reaction of aliphatic amino groups with nitrous acid and quantification of released gaseous nitrogen [39]. This method was a highly recommended method among others for quantitative analyses of nitrogen-containing compounds. Examples include (i) the relative digestibility of proteins, (ii) proteolytic enzymes, (iii) amino acids and (iv) amino nitrogen in urine. Pleissner et al. and Ulusoy et al. built upon this method and used the reaction of aliphatic amino groups with nitrous acid in order to convert amino acids into their corresponding α-hydroxy acids [40, 41]. Their motivations were to develop a chromatographic method to analyze amino acids in fermentation broth using an Aminex HPX-87H column (Bio-Rad, CA, USA). This column was packed with a polymer-based matrix which contains 8% cross-linked resin in ionic hydrogen form. Therefore, it would exclude ionic molecules. Since sulfuric acid was used as an eluent, amino acids were in ionic form and therefore it was not separated. By converting amino acids into their corresponding α-hydroxy acids and determination of the latter, single amino acids could be analyzed indirectly even in complex solutions such as fermentation broths [40, 41]. Therefore, the application of this chromatographic method for the quantification and identification of amino acids in food waste hydrolysates can be considered as an alternative option.

In 1973, Lie published a method for determination of free α-amino nitrogen in malt and wort using ninhydrin [42]. Ninhydrin causes an oxidative decarboxylation of α-amino acids and production of CO_2, NH_3 and an aldehyde with one carbon atom less than the parent amino acid. Reduced ninhydrin reacts with unreduced ninhydrin and forms a blue complex. The intensity of the blue complex corresponds with the concentration of free α-amino nitrogen and it can be measured photometrically at a wavelength of 570 nm [42]. In the past, this method has not only been applied for the analysis of malt and wort, but also for food waste hydrolysates [36], wheat flour hydrolysate [43] and hydrolyzed rapeseed meal [44].

The van Slyke method, the modified methods by Pleissner et al. and Ulusoy et al. as well as the ninhydrin method of Lie work when amino acids are presented as monomers with a free amino nitrogen group [39–42]. However, proteins and peptides need to be hydrolyzed first. The simplest method for analyzing the content of proteins is by UV-spectroscopy. Aromatic amino acids such as tryptophan and tyrosine absorb at 280 nm. Nevertheless, the composition of aromatic amino acids can be different between proteins, and thus this method is only reliable for quantification of the same protein. Another option is the Bradford assay based on the binding of Coomassie Blue dye to proteins. Binding of the dye is non-covalent, which includes interaction with basic and hydrophobic surfaces. Higher quantity of proteins would lead to more bindings of dyes. The fraction of Coomassie dye interacting with protein can be quantified photometrically at 595 nm [45]. However, the interaction of Coomassie dye with protein might be disturbed when proteins are modified, for instance, by glycosylation.

Complex materials such as food waste and manure may contain compounds either interacting with dyes or also absorbing at 280 nm. Furthermore, the complexity and recalcitrance of solid biomass may hinder the quantification of nitrogen-containing compounds. Therefore, a tough pretreatment is needed in order to release all nitrogen-containing compounds for the analysis of nitrogen in complex substrates with a certain solid fraction. One of the approaches is the digestion of the compact structure using strong acids under heat, and the nitrogen measurement would be conducted using a total nitrogen analyzer [36]. However, the chemical standard method is the determination of total nitrogen according to Kjeldahl. Johan Kjeldahl developed this method in 1883 and even more than 130 years later, it is still one of the most widely used standard in many laboratories worldwide. Total nitrogen is in the forms of dissolved inorganic and organic nitrogen, as well as of particulate inorganic and organic nitrogen. An advantage of the Kjeldahl method is that it can be used for analyses of both liquid and solid environmental samples.

The Kjeldahl method is based on a complete breakdown of materials using hot sulfuric acid. Furthermore, the addition of salts, oxidative agents and catalysts such as mercury used in early days can improve the digestion [46]. Over the years, several improvements of the Kjeldahl method have been reported. One of the improved methods is the microscale Kjeldahl nitrogen determination [47]. Campins-Falco et al. aimed at reducing the scale of the conventional Kjeldahl method from 800 mL to 25 mL [47]. This reduction in scale results in the reduction of volumes of reactants, which is in line with effort associated to the recent development of green and sustainable chemistry approaches. For 20 mL of sample, 0.75 mL of concentrated sulfuric acid was added and heated at 370 °C for 25 min. After dilution of the digestion residue, 12.5 M NaOH was added and ammonium was distilled. Distilled ammonia was neutralized using 0.02 M sulfuric acid. Sodium hydroxide (10 M) was added to the distillation solution and the Kjeldahl nitrogen was measured using an ammonia selective electrode. According to Campins-Falco et al., almost 100% recoveries and good precision have been achieved, and the cost of analysis was significantly reduced in comparison to the conventional Kjeldahl method by a factor of 10 [47].

In addition, conversion factors have been established for estimation of protein content in biomass and food waste based on free amino nitrogen or Kjeldahl nitrogen determination. For instance, Mariotti et al. reviewed a couple of conversion factors related to food [48]. They proposed an average conversion of 5.6, which was more appropriate than the commonly used factor of 6.25. Since food waste consists of food, a factor of 5.6 can be applied to estimate its protein content. Based on the quantification of free amino nitrogen by Lie [42] and Pleissner et al. [49], a conversion factor of 5.7 can be used to estimate the protein content. This is similar to the value suggested by Merrill and Watt [50].

8.3.4 Lipid Content

Lipid content determination is important because lipids influence aerobic and anaerobic fermentation in various ways, including growth stimulation, improving the tolerance and as antifoam agents [26]. Free fatty acids in lipids have been responsible for controlling ethanol tolerance of yeast [51]. Furthermore, when molasses were used as substrate for fermentation, a low lipid content stimulated high production of ethanol, high lipid content of molasses reduced ethanol production [26]. There are many methods to determine the lipids content of food waste. Hexane/isopropanol (3:2) method was applied in determination of lipid content of food waste by Ma et al. [35]. Lipid content of food waste was determined using the chloroform methanol extraction method by WRAP [32].

Lipid content of food waste was determined by a method of Bligh and Dyer [52]. Principle of the lipid content determination is based on the fact that when a mixture of chloroform and methanol is homogenized with the tissue of food waste, it forms a monophasic solution. When the homogenate is diluted with water, it forms a biphasic solution system. The methanol-water layer contains non-lipids substance and the chloroforms layer contains the lipids. Then, the chloroform layer is isolated to obtain the lipid extract. Oven-dried food waste (100 mg) is re-suspended in 5 mL of CH_3Cl:CH_3OH (1:2 v/v) and shaken for 24 h at room temperature. After that, pellets and supernatant are separated by centrifugation at 5,000 g for 5 min. Pellets are re-suspended in CH_3Cl:CH_3OH (1:1 v/v), and shaken for 12 h at room temperature. Then, supernatant is collected by centrifugation at 5,000 g for 5 min. Supernatants from both extractions are combined. Distilled water in the ratio of 1:5 is added and shaken for 15 min to remove non-lipid components. The organic phase is collected and evaporated at 60 °C. Then, the mass of the lipid extract is measured [52].

8.3.5 Minor Constituent Analysis of Food Waste

The addition of food and colour ingredients into food is to improve food safety and maintain its freshness. For instance, preservatives slow down the rate of product spoilage by microbial growth in order to maintain the food quality. In addition, food and colour ingredients also improve and maintain the nutritional values. For example, fortification of food with fibre, vitamins and minerals is done to reduce malnutrition. Food and colour ingredients also improve taste, texture and appearance of foods. Food colorant is usually added to improve the appearance, while emulsifiers and thickeners are added to provide food texture [53].

Colour additive is any dye, pigment or any substance which is applied to food for the purpose of imparting colour. Colour additives are usually added to beverages, candies, snack foods and processed meat. Typical colour additives includes tartrazine, sunset yellow, ponceau-4R and indigo carmine [53]. The type of food ingredients includes preservatives, colour additives and food additives. Preservatives are normally applied to baked goods, beverages, cured meat, cereals etc. The most common preservatives include sodium benzoate, potassium sorbate and calcium propionate. Since many food preservatives and colour additives are added to food, determination of preservatives and colorants is needed for identification and quantification. Preservatives and colorants might have significant effects towards enzymatic hydrolysis and fermentation using food waste as substrate.

8.3.5.1 Determination of Preservatives

Preservatives are food additives that are added in different food products for prevention from microbial spoilage and extension of their shelf life [54]. Preservatives in food waste might have inhibition effects on enzymatic hydrolysis. Hence, determination of preservatives content is needed for identification and quantification. Janovitz-Klapp et al. [55] conducted inhibition studies to prove that apple polyphenol oxidase was inhibited by aromatic carboxylic acids including benzoic acid [55]. Further evidence of inhibition was provided through raw starch hydrolysis in which *Chalara paradoxa* glucoamylase was strongly inhibited by potassium metabisulfite [56].

Several methods can be used for determination of preservatives in food waste. Flow injection analysis has been used to detect preservatives in meat product by Ruiz-Capillas and Jimenez-Colmenero [57]. Determination of calcium propionate in breads has been carried out using colour complex formation method by Phechkrajang and Yooyong [58]. Preservatives determination in food waste was carried out according to Aubin [59]. Sample was prepared first by mixing 1 g of oven dried food waste with 10 mL of 30% acetonitrile. The mixture was sonicated and filtered for HPLC analysis. The HPLC analysis for preservatives and caffeine involves the use of two types of mobile phases in isocratic elution, which include 10 mmol sodium phosphate buffer at pH 6.0 (mobile phase A), and acetonitrile (mobile phase B). The flow rate was 0.9 mL/min with a solvent ratio of 98% mobile phase A and 2% mobile phase B. Preservatives were separated using ACQUITY UPLC BEH column C18 at 40 °C. The absorbance was detected at a wavelength of 214 nm using the photodiode array (PDA) detector.

8.3.5.2 Colorants Determination

Many analytical methods have been developed for synthetic colorant determination in food and beverages. The methods include Raman spectroscopic [60], capillary electrophoresis [61], voltammetry, HPLC with diode array detection [62–64], ultra-performance liquid chromatography (UPLC) with PDA detection [65] and spectrophotometry [66]. Colorant was extracted by adding 5 mL of 80:1:19 ethanol: ammonia: water solution to a food waste sample and sonication for 30 min and analyzed using UV/Vis spectrophotometer at different wavelengths corresponding to each colorant (tartrazine, sunset yellow, ponceau-4R, indigo carmine). Standard curve was obtained by measuring the absorbance of different concentrations of colorant standards [66]. Colorant of sample was further analyzed with Acquity UPLC BEH C18 column using UPLC equipped with PDA conditions. Three mobile phases were applied, which contained ammonium acetate, methanol and acetonitrile, respectively, and the flow rate was set at 0.45 mL/min [65]. Capillary electrophoresis with photodiode array detection (CE-PDA) was used for separation of synthetic dyes by López-Montes et al. [61]. Separations was carried out using a fused-silica capillary. Peak identification was done using UV spectrum and migration times.

8.4 Storage and Handling of Manure

Manures are cheap sources of valuable nutrients required for plant growth and contribute significantly to long term fertility of soil. These are important aspects of any farming system which does not rely on expensive fertilizers. However, much of the nutrient values contained in manures can be lost due to improper handling and storage of materials. Primarily, the system and length of storage affect the nitrogen content of manure. Long term storages often result in reduced organic nitrogen and increased ammonium nitrogen with the increased possibility of the latter being released into the atmosphere along with risk of nitrous oxide emissions, thus causing environmental damage [11]. Furthermore, poorly managed manure can harbour intestinal parasites and enter to the water cycle via surface runoff or leach into groundwater, and thereby cause environmental concerns. Efficient removal of manure from livestock production facility is essential for optimal livestock productivity [10]. Thus, correct storage and handling is necessary to preserve the quality of manure and maintain their utility for crop growth and soil improvement. Such manure management practices are discussed in this section.

The components of manure management system include collection, transfer, storage, treatment and utilization. Laws, regulations, guidelines and procedures for design and construction of manure handling and storage are provided by district, provincial or national regulatory authorities to meet the desired objectives. Some of these regulations are provided by for e.g. the United States Department of Agriculture (USDA) [67], Ministry of Agriculture in British Columbia, Agricultural Operations Act, Ministry of Agriculture and Forestry, Alberta in Canada, and Joint Food and Agricultural Organization and International Atomic Energy Agency (FAO/IAEA) [68] in Asian countries.

Prior to storage, the handling of manure, equipment selection and maintenance is governed by the manure type. The latter is classified as either a liquid, semi-solid (slurry) or solid, based on its solid content. Liquid manure contains less than 5% solids while semi-solid manure has 5–25% solids. Solid manure contains greater than 25% solids by addition of bedding or removal of liquid by draining or drying.

The type of storage structure and its operation affect the nutrient uniformity and losses upon recovery of manure from storage and thus it should be carefully designed [69].

8.4.1 Manure Handling Equipment

When handling manure, it is important to ensure that handling equipment is designed for that purpose and it is used according to manufacturer's instructions. The equipment must be capable of functioning reliably in a corrosive environment. It also requires proper maintenance if it is expected to have a long service life. Liquid manure systems are generally handled using pumps. The presence of solid clumps in liquid manure can clog the pumps. Problems occur when ear tags, hair, teeth, tails, and other objects enter the pump. Therefore, it is essential to adopt some method of screening or cutting solid material. For proper storage of liquid manures, agitation of solids is sometimes required to bring them back into suspension. In such circumstances, the use of chopper pumps is appropriate since these do not become easily clogged with solids and it can be operated at high speeds. Solid and semi-solid manures are transferred by conveyers, augers, piston pumps or front-end loaders to storage area [67].

8.4.2 Manure Storage Structure

Manures are usually stored in confinement, either in open lots or enclosed, roofed structures. Manure storage facilities are designed based on the type and consistency of manure to be handled in the system, location, size required and the methods of filling and emptying. In general, a good storage structure should fulfil the following purposes [70]:

(i) be sufficiently impervious to prevent leakages;
(ii) provide an appropriate level of odour control; and
(iii) provide appropriate flexibility for timing application operations.

The location of storage structure should be close to barns to allow convenient filling without limiting the expansion of facilities. It is also important that it is located sufficiently close to fields for application purposes and/or can easily be connected by long distance pipelines. Due to odour and unsightly nature of manure, the location of storage should be out of sight of roads and dwellings. Storages must not be constructed on banks of river or drainage channels to avoid collecting surface runoff and surface/groundwater should be protected from potential spills from storage structure.

The size of storage is determined by volume of manure produced and length of time the manure will be contained in it. The volume in turn depends on the size of operation, management practices and facilities design. Furthermore, the volume of manure can increase due to water spillage and washing. Overall, it is important to determine the manure production rates as accurately as possible, particularly for expensive concrete storage structures. Additional consideration should be paid to storage of solids manure which contains large amount of bedding and are often stored as stockpiles for subsequent spreading operations. In such cases, it should be ensured that storage has reserved capacity to accommodate precipitation and runoff.

The storage structures should be constructed to accommodate equipment for pumping and agitating the storage in sufficient locations to ensure that stored manure can be completely removed. Wet manure or liquid runoff is usually stored in three types of storages. These include earthen dykes, concrete tanks below ground and concrete or steel tanks above the ground. Variations of these storage structures are used in specific countries in accordance with the regulations to develop efficient manure management practices. While earthen storage ponds, concrete storage structures, large roofed buildings are usually used in the US, covered dumpsters, covered free-standing manure piles, enclosed truck bed, 3-walled structures roof or tarp cover are used in the Canadian region [67, 70]. The situation in Asian countries is, however, different and diverse methods are used for manure storage. Manual separation of excreta into liquid and solid fractions with storage on floor is quite common in small-sized farms. In some other small farms, the fattening pens are divided into two parts, including a living area with concrete floor and a lower lying channel area where excreta are collected by scraping the solid floor area, which is followed by mixing with straw. In some big farms, the solids are not separated from liquid manure and a slurry is produced which is stored on solid or slatted floors [68].

Earthen storage structures offer low capital cost relative to volume stored. However, the major disadvantage of this structure is that due to its anaerobic conditions and large exposed surface area, large quantities of odorous gases are released in the air. Additionally, high permeability of earthen dykes results in undesirable nutrient losses during storage. In contrast, concrete tanks below ground are more expensive than earthen dykes, however, their impermeable nature makes them suitable for maintaining valuable manure nutrients, especially in areas having sandy soils. The problem of odour release is also avoided in concrete tanks except during the agitation of manure for emptying the storage. The design and operation of concrete tanks needs additional consideration. Due to its closed design, sufficient access ports should be provided for the pump and compartmentalization may be necessitated for more effective agitation. Appropriate safety precautions should be taken against noxious and explosive gas (e.g. methane) hazards while inspecting or working in the storage tank [71]. Concrete tanks above the ground could be more expensive than tanks below the ground, depending on their sizes. These are either circular silo type with an open or closed-top or rectangular structure. Concrete staves, reinforced cast-in place concrete, glass lined steel panels or spiral wound coated steel may be considered for construction of these storage tanks [72]. The possibility of creation of a crust on top in such tank structures allows considerable odour reduction. However, due to their high construction costs, these structures might only be used in situations where space is limited or where soil conditions do not permit use of an in-ground storage.

Storage of manure as solids offers certain advantages. These include less volume, less odour due to less bacterial action at lower moisture content, less runoff potential and relatively high nutrient retention [73]. Conditions for storing solid manure differ for low-rainfall (arid) and humid areas. For arid areas, solid manure is usually stacked or stockpiled, with appropriate provisions for runoff collection and treatment. In more humid areas, concrete structures with walls are used to confine solids. These facilities may be provided with roofs to eliminate the effects of rainfall. Another widely operation in used solid manure storage system is composting. Composting manures further improves the biological, chemical and physical properties of soil by promoting the growth of beneficial microorganisms. Due to its high pH, composting results in a more favourable soil environment for plant growth as compared to fresh manure. Depending on the temperature used for composting, the process can also reduce or kill weed seeds, plant and animal pathogens present in manure to promote a safer plant growth.

8.5 Physico-Chemical Characterization and Composition Analysis of Manure

The properties and quality of manure is affected by several factors including animal species, diet, digestibility and storage conditions. To allow efficient land application of manure, a detailed and accurate knowledge of its physical and chemical properties is important. Moisture content, total solids/volatile solids ratio, macro-nutrient content in terms of nitrogen, phosphorous, potassium, and minor constituents such as sulphur, calcium, magnesium and trace metals are some of the nutritional components of manure which are required to be analyzed. Methods for determination of these physico-chemical properties are discussed in Sect. 8.5.1–8.5.6.

8.5.1 Dry Matter (DM) or Moisture Content

Manure is a valuable source of nutrients for crops and it can improve soil productivity. It is characterized in several ways. Important properties for manure collection, storage, handling and utilization include the solids content (the percentage of solids per unit of liquid), size and makeup of manure solids (fixed and volatile solids, suspended solids and dissolved solids) [74]. In general, manure can be classified based on how manure must be handled. However, the most suitable way to classify manure depends primarily on its dry matter content, as the manure handling characteristics vary as consistency changes from liquid to solid.

On one end of the spectrum is lagoon liquid with very low solids content (less than 1%) that can be used for irrigation by either big guns or center pivot irrigation systems with small nozzles. On the other end of the spectrum is solid manure, which normally has more than 25% solids. In between of these two conditions are the ones containing 5–25% solids. Sand is another challenging solid that is sometimes used as dairy bedding. Alternatively, the most typical manures, such as dairy, beef, and swine manure, may be either solid or liquid (with inclusion of urine or barn floor and milking parl or wash water), while horse and poultry manures are solid. Solid manures often include bedding materials [75]. In general, the nutrient values are related to solids concentration: the higher the solids concentration, the higher the nutrient concentration.

Nitrogen, phosphorus, and potassium are the major nutrients of manure. Nutrients are divided between soluble and insoluble states. While the major soluble nutrients are found in the liquid (urine) that are more readily available for crop usage, the remaining soluble and insoluble nutrients are found in the solids (faeces) of excreted manure. For example, 80% of the phosphorus is insoluble and exists in settled solids, but 80% of the potassium is highly soluble.

Normally, the moisture content of the manure is determined by dry weight method. After the collection of manure, samples will be homogenized in a blender to pass a 3 mm screen [76], and then put into an oven at 105 °C for 24 h [77].

8.5.2 Total Solids/Volatile Solid Ratio

As one of the major components in manure, total solids (TS) is the solids left after the water is excluded [78]. Generally, TS is comprised of total dissolved solids (TDS), total suspended solids (TSS), and settleable solids. TDS is composed of the substance in solution or suspension liquid that with the size smaller than 1.5 μm [79]. TSS is the solids in the liquid stream that exceed 2 μm and remains suspended in the solution indefinitely [80]. Settleable solids are the matters, with any size, which do not remain in suspension or dissolved form. These solids can impact the characteristics of manure, including the particle size, nutrient and salt concentrations. Normally, manure contains various nutrients such as nitrogen (N), phosphorous (P), potassium (K), calcium (Ca), magnesium (Mg), sulfur (S), manganese (Mn), copper (Cu), zinc (Zn), chlorine (Cl), boron (B), iron (Fe) and molybdenum (Mo) [81], but their contents vary depending on the species, manure handling practices and storage time. Table 8.1 shows the values of the major nutrients in 1 ton (or 241 gallons) of fresh manure and 1 ton of dry manure (solids) at 15% moisture for each animal type [82]. According to Table 8.1, N, P and K with relatively high concentrations are found in the manure. However, as phosphorus tends to form a precipitate or adhere to particles (solids) in manure, the concentration of phosphorus in the TS stream is often thought to be higher than in the

Table 8.1 Nutrient content for different sources of manures

Animal type	Animal size	Nutrient content						
		Ratio	Water	TS	VS	N	P	K
	lb.	*Wm/Wa*	*% wb*	*% db*	*% db*	*% db*	*% db*	*% db*
Dairy cattle								
Calf	150	–	88	12	85.7	3.2	0.6	2.6
Heifer	750	0.09	88	12	85.3	3	0.9	2.8
Dry cow	1000	0.08	88	12	85.3	3.7	0.6	2.4
Lactating cow	1400	0.09	88	12	85	5.5	0.7	2.5
Veal	250	0.04	96	4	43.8	11.1	8.3	16.7
Beef cattle								
Calf	450	0.06	92	8	84.7	6.7	4.8	5.3
High forage	1100	0.08	92	8	89.4	8.3	2.9	4.9
High energy	1100	0.07	92	8	91.9	8.4	3.3	5
Cow	1000	0.06	88	12	77.9	4.1	2.5	3.4
Swine								
Nursey	25	0.11	89	11	81.5	6.7	3.4	3.4
Grow-finish	150	0.06	89	11	80	7.7	4.8	3.8
Gestating	275	0.03	91	9	85.5	7.4	5.9	5.9
Lactating sow	375	0.06	90	10	90.2	8	5.8	6.2
Boar	350	0.02	91	9	89.4	7.7	6.2	6.2
Sheep								
Ewes	100	0.04	75	25	82.7	4	2	4
Poultry								
Layer	4	0.07	75	25	75.4	5.4	4.2	2.5
Broiler	2	0.09	74	26	72.3	4.9	3	2.4
Turkey	20	0.05	75	25	76	5.6	4.8	2.4
Duck	6	0.06	73	27	59.6	5.2	4.3	3.1
Equine								
Horse	1100	0.05	85	15	85	2.3	0.3	0.7

Reference Adapted from Bulletin 604 [82]. *Wm*: daily weight of manure, *Wa*: average weight of animal during this stage of production, *% wb*: percent wet basis, *% db*: percent dry basis

un-separated manure. Moller et al. indicated that dry matter (DM) fractions >0.5 mm will contain 5–7% of the total phosphorus [83].

Nevertheless, with the diminishing of the natural nutrients and fertilizer prices trending higher, the importance to retain and effectively apply manure nutrients in agriculture has been recognized. According to the prediction, world supply of phosphate fertilizer will reach a peak in 2033 and then decline [84]. In the meanwhile, another nutrient that receives special attention due to its decline from atmospheric concentrations is sulphur [85, 86].

On the other hand, TS can also be subdivided as volatile solids (VS) and fixed solids or ash. The VS is always determined by the ignition of dry solids at 550 °C and gives an approximation of the organic matter present in the waste [87, 88]. Normally, the VS is composed of bio-degradable volatile solids (BVS) including saccharine, starch, organic acids, cellulose, lipids and proteins that can be easily decomposed by microorganisms in most storages and lagoons [81], and the refractory volatile solids (RVS) which resists bacterial digestion.

During the degradation process, VS in manure is broken down by anaerobic bacteria. Normally, the organic nitrogen is converted to ammonium nitrogen, but sometimes this ammonium nitrogen would be lost to the atmosphere as ammonia. Thanks to this process, the unavailable organic nitrogen in plant can be converted to ammonium-N that is useful for plant. On the other hand, owing to the complicate reactions including anaerobic digestion, mineralization and volatilization during the production, collection and storage of manure, the VS of manure vary with the digestibility of the ration, animal age, the amount of feed wasted, manure handling methods and storage time [81]. Therefore, the determination of volatile solids is indispensable in the design and operation of manure digestion, vacuum filter and incineration plants.

8.5.3 Total Kjeldahl Nitrogen and Total Ammonium Content

All types of manure contain nitrogen source. Normally nitrogen found in manure is made up by organic nitrogen and ammonium nitrogen. The amount and form of nitrogen source in manure determine the quality of manure fertilizer. Nitrogen is an essential element of nucleic acids and proteins. However, crops required large amounts of nitrogen, which is always deficient in agricultural soils [89].

Measurement of total nitrogen in manure is to determine all type of nitrogen present. This includes organic nitrogen, ammonium nitrogen and any type of nitrate. The concentration of nitrogen in manure is variable and not all types of nitrogen are presented in manures. Quantification of ammonium content in manure is to determine the main inorganic form of nitrogen. Ammonium is obtained from the breakdown of organic nitrogen in faeces, and conversion of uric acid and urea in urine.

Liquid manures contain more ammonium nitrogen than solid manure. Ammonium nitrogen is very useful for plants because it can be readily used by plants [89].

Pan et al. [90] analyzed nitrogen in manure using Kjeldahl nitrogen method [90]. Test portions were digested with sulfuric acid, potassium sulphate and copper catalyst in an aluminum block at 420 °C. Subsequently, the digestate was neutralized by concentrated sodium hydroxide. Steam distillation was applied to liberate ammonia and trapped in weak boric acid. Colorimetric titration using hydrochloric acid was performed to obtain nitrogen content. Powell et al. [91] determined nitrogen content in manure using Elementar VarioMax CN analyzer

(Elementar, Germany) [91]. Liu et al. [23] determined ammonia nitrogen content using formaldehyde method in China National or Industrial Standards (GB/T3600-2000) [92]. Ammonium cation was detected by flow analysis and spectrometric detection [93].

8.5.4 Phosphorous Content

Similar to nitrogen, phosphorous is present in all organic materials. Among the nitrogen-containing base and five-carbon sugar, phosphorous forms the backbone of DNA and it is a part of the energy carrier adenosine triphosphate. Therefore, production of biomass would not be possible without phosphorous. Phosphorous is one of the essential elements for life, which cannot get lost and its recovery is absolutely necessary in order to cover the demand of phosphorous-containing fertilizers in agriculture. Phosphorous in biomass appears either as phosphate or polyphosphate. First developments regarding the determination of phosphate were made in medicine. In 1925, Fiske and Subbarow developed a method for determination of phosphate in blood and urine [94]. Methods of choice were colorimetric ones based on the reaction of phosphate and molybdic acid to form phosphomolybdic acid, and its subsequent reduction by hydroquinone resulting in formation of a blue colour. The intensity of colour was directly proportional to the concentration of phosphate. The determination was relatively simple where phosphate and molybdic acid could react. However, the application of such method would be challenging for solid or complex materials, or when phosphate is present in the form of polyphosphate, such as phytic acid, or when inorganic and organic phosphorous are present. Nevertheless, materials can be digested and free phosphate released by acidic treatment. Commonly used acids are sulfuric acid, nitric acid and hydrogen peroxide. However, when the quantity of sulfuric acid is too high, then the reduction of phosphomolybdic acid is disturbed. When the quantity is too low, then portion of phosphate is not released, and thus it was neither quantified.

The colorimetric phosphate determination method has been used for several biological feedstocks. Jakubus used this method for determination of total phosphorous in compost and its waste residues [95]. They incinerated 1 g sample at 550 °C for 3 h. Afterwards, 5 mL concentrated HCl was added to the ash and heated for 30 min at 180 °C. The concentration range was in g per kg. Lugo-Ospina et al. evaluated quick tests for determinations in dairy manure. Tests were based on the molybdic acid assay, but pretreatments were different [96]. For instance, potassium persulfate and 12 mL 5.5 M H_2SO_4 were added to 2 mL manure suspensions in one of the approaches. The mixture was heated for 30 min at 180 °C and digested at 350 °C for 30 min to reduce the overall volume to 4–5 mL. Afterwards, the solution was diluted with demineralized water to 35 mL and used in total phosphorous determination. Jastrzebska further determined phosphorous in meat samples [97]. Concentration of phosphorous was determined using colorimetric method after mineralization of meat using a Kjeldahl apparatus with nitric acid and hydrogen peroxide.

8.5.5 Potassium Content

Potassium from manure is good for plant growth. Potassium content in manure is variable and it is present in a water soluble and inorganic form [89]. Potassium is an essential macronutrient for plants growth and development. It controls enzyme activity, concentration of metabolites and transcription level of a number of genes in plants. It is important for crop yield as well as human nutrition [98]. Spectrochemical methods have been applied for long time to determine elemental concentrations in manures. Colorimetry, atomic absorption spectroscopy (AAS) and inductively coupled plasma-atomic emission spectroscopy (ICP-AES) are some methods which have been used to determine elemental concentrations in manure. Advantage of AAS is its high specificity for individual elements. ICP-AES is generally free from interferences [99]. Colorimetry method is a rapid and reliable method which is applicable to large number of samples [100].

In 1950, Whittles and Little [100] invented a colorimetric method to determine small amount of potassium [100]. The method involves sample extraction with Bray's extracting reagent and precipitation of potassium as cobalt nitrite. Nitrite group in cobalt nitrite precipitate undergoes diazotization of sulphanilic acid and subsequent coupling with dimethylaniline to give methyl orange whose colour is measured with a Spekker absorptiometer. Recently, Montégut et al. [93] measured mineral composition in manure including potassium oxide (K_2O) using inductively coupled plasma optical emission spectrometry (ICP-OES) [93]. Potassium ion (K) concentration was determined by a Varian A.A 240 FS atomic absorption spectrometer. Hollow cathode lamps of K was used as a radiation source. The wavelength and applied electric current were 766.5 nm and 5.0 mA, respectively. Potassium analyses used 13.5 of air-acetylene flame with flow rates of 2.0 L/min, respectively. According to Roa-Espinosa et al. [101], X-ray fluorescence (XRF) spectrometry is a rapid and precise method that can be used to determine quantitative elemental analysis of macro- and micro-nutrients in dairy manure [101]. It can be used to determine elements includes N, P, K, Ca, Mg and others. Sample was pelletized under 4 tons of pressure for 2 min using hydraulic press. After that, sample underwent elemental analysis using a PHILIPS MagiX apparatus to perform wavelength dispersive XRF.

8.5.6 Secondary Elements, Heavy Metals and Trace Contaminants Content

Due to the use of feed additives, manure from factory farms contains high considerable amount of heavy metals, especially the heavy metals such as copper and zinc. Their potential accumulation on land raises concerns regarding the environmental risks to surface and groundwater, in addition to threatening the food chain and agriculture. In addition to heavy metal contamination, manures might also contain trace amounts of pesticides, especially herbicides [102]. Pyridine carboxylic acid-based herbicides such as aminopyralid, fluroxypyr, picloram, clopyralid, and triclopyr are usually applied to hay fields or pastures which are considered to be safe for livestock consumption. Upon passage through animals' digestive tract, these are excreted in urine and manure, and it can actually remain active in manure for an unusually long period of time. The use of such contaminated manures in fields can cause devastating damage to crops. Therefore, it is essential to perform complete assessment of contaminants such as heavy metals, pesticides and mycotoxins in manure to ensure their safe application in agro-ecosystem.

The method for determination of dissolved anions and cations concentration (secondary elements) in manure is based on UNE-EN ISO 10304-1 and Standard Methods 21st ED [103, 104]. The latter is a joint publication of the American Public Health Association (APHA), the American Water Works Association (AWWA) and the Water Environment Federation (WEF). The analysis is carried out by HPLC using anion or cation-selective columns. The method is simple which involves sample filtration by 0.2 μm followed by *in situ* analysis. Position of the maximum in the obtained chromatographs indicates the presence of a specific ion, while area indicates the concentration of ion.

Trace metals in manures exist in different forms. These include water-soluble, exchangeable, linked to organic substances, co-precipitated with oxides, carbonates and phosphates and ions in crystalline lattices of primary minerals. The metals present in the first three forms are the most readily available for plant uptake [105]. Several methods have been developed for heavy metal assessment in manure. Phytoavailability of metals has been largely applied as a one-step soil extracting procedure to predict the size of pool that may be depleted by plant during the growth period. Although a simple method, phytoavailability does not indicate the total heavy metal content of sample. Another method is called the sequential extraction procedure which allows the determination of different chemical forms in which the element appears to be associated in a sample [106]. As opposed to phytoavailability, this method provides key information about the extent of retention of elements in sample and their ability to be released in soil. These sequential extraction methods are usually used to evaluate the actual and potential mobility of metals in the environment. This helps to understand the uptake of metals by plants in comparison with chemical pool available in soil. Heavy metal specification is a more sophisticated method for heavy metal assessment since it determines the availability of metals for plant uptake and potential for contamination of groundwater following the application of manure to agricultural lands. Sequential fractionation method categorizes metals which are associated with chemically homogenous fractions, which ultimately decide its availability for uptake. Therefore, the method can determine the amount and proportions of metals in different forms such as water-soluble, exchangeable or linked to organic substances in manure samples [107, 108]. Since the specific chemical form of metals in manures affects its mobility and bioavailability, evaluations based on chemical associations are considered to be important while analyzing manure samples.

For the determination of heavy metals, the manure samples are air-dried, crushed and sieved to ensure homogeneity. This is followed by digestion of samples in a mixture of acids such as HNO_3 and $HClO_4$. Specifically, 0.25 g sample is digested with 5 mL concentrated $HClO_4$ over a hot plate for 1 h after which it is dried. Subsequently, 20% HNO_3 is added to the sample with heating for an additional hour. The resulting solution is then diluted to 50 mL with water and passed through a 0.2 μm filter. Total elemental analysis for quantification of common metals such as nickel, iron, zinc, mercury, cadmium is performed using an atomic absorption spectrophotometer. Modified versions of sequential extraction method have also been employed to fractionate the solid-phase chemical forms of the metals. Sequential extraction has been performed using 0.01 M CH_3COOH for 16 h, 0.1 M $NH_2OH{\cdot}HCl$ for 16 h, H_2O_2 for 1 h, followed by heating at 85 °C for 1 h and final extraction with 1 M ammonium acetate for 16 h [109]. In another method, trace elements are fractionated into exchangeable, adsorbed, organically bound, carbonate precipitated and residual forms by sequential extractions with 25 mL of following reagents: 0.5 M KNO_3 for 16 h, 3 times extraction with de-ionized water for 2 h, 0.5 M NaOH for 16 h, 0.05 M Na_2EDTA for 6 h, and 4 M HNO_3 for 16 h at 80 °C [106]. After each extraction, the sample is centrifuged at 5,000 rpm for 15 min and the resultant supernatant is decanted, filtered and then analysed.

An European standard EN 15662: Determination of pesticide residues using gas chromatography-mass spectrometry GC-MS (Agilent 7890) and/or liquid chromatography-mass spectrometry/mass spectrometry LC-MS/MS (Agilent 6400), following acetonitrile extraction/partitioning and clean-up by dispersive SPE-QuEChERS-method (Solid Phase Extraction Quick-easy-Cheap-Effective-Robust-Safe) are used for determination of pesticides in manure [110]. Manure samples are extracted with 10 mL acetonitrile. Solid manure samples with low water content (<80%) are diluted with water to obtain approximately 10 g water sample. Magnesium sulphate, sodium chloride and buffering citrate salts are added to sample which is then shaken intensively for 1 min and centrifuged for 5 min at >3,000 g for phase separation. Organic phase is then cleaned up by dispersive solid phase extraction (DSPE) to remove the residual water. Formic acid (50 μL, 5%, w/v) is added to 5 mL extract to acidify and improve the storage solubility of base-sensitive pesticides. The final extract is then analysed by LC or GC-based determination analysis.

8.6 Advanced Analytical Methods for Food Waste and Manure Characterization

The section describes the advanced analytical methods which have been used for characterization of food waste and manure.

8.6.1 High Performance Liquid Chromatography (HPLC)

According to IUPAC definition, chromatography is a physical separation method in which the components to be separated are distributed between stationary and mobile phases [111]. HPLC is a fast technique, which separates mixtures into individual components with high precision and specificity. HPLC instruments include pumps, solvent reservoirs, injector, detector and a data acquisition system [112]. HPLC can be used in diverse research areas including environmental, food and flavour, pharmaceutical and forensics.

There are several kinds of HPLC, such as normal phase chromatography, reverse phase chromatography, ion exchange chromatography, size exclusion chromatography and their applications depending on the nature of compounds to be separated. For normal phase chromatography, the stationary phase is polar and mobile phase is nonpolar. Polar samples are retained for longer periods on polar surface of the column than less polar molecules. For reverse phase chromatography, the stationary phase is nonpolar compounds and mobile phase is a polar solvent. Here, nonpolar samples are retained for longer time in column than polar molecules. In the case of ion exchange chromatography, the stationary phase is an opposite ionically charged surface to the sample ions and mobile phase is an aqueous buffer. Strongly charged samples are retained for longer time on ionic surface. Size exclusion chromatography separates the samples on the basis of molecular size. The column is packed with materials with controlled pore size. Large molecules are washed out from the column first, then followed by small molecules [112].

Carbohydrates in food waste can be analysed by HPLC. Type of carbohydrates in food waste includes *monosaccharides,* such as glucose, fructose, galactose and sorbitol; *disaccharides,* such as sucrose, lactose and maltose; *trisaccharides* such as raffinose; and *polysaccharides,* such as starch [112]. Aminex HPX-87H column (Bio-Rad, USA) is commonly used for analysis of carbohydrates presented in biological fluids. Properties of this column include 8% of cross-linked resin, hydrogen ionic form, pH range from 1 to 3 and 300 $\times$ 7.8 mm column size. Recent research conducted by our groups applied HPX-87H column to analyze sugar and organic acids, with the use of H_2SO_4 as mobile phase [21, 113]. A C_{18} column can be used to separate organic acids in food waste including preservatives such a ssodium benzoate and potassium sorbate, caffeine. Kuprovskytë et al. [54] performed separation of preservatives includes benzoate sorbate, methyl, ethyl and propyl esters of p-hydroxybenzoic acid in beverages using the Separon SGX C_{18} column (TESSEK Ltd., Prague) [54]. Mobile phase contained 5 mmol/L aqueous acetate buffer (pH 5) with 40% (v/v) acetonitrile at a flow rate of 0.2 mL/min and UV detection at 254 nm. Similarly, Aubin [59] performed separation of preservatives and caffeine from soft drinks using ACQUITY UPLC BEH C_{18} Column [59]. Column temperature was set to 40 °C, mobile phase (Solvent A: 10 mmol sodium phosphate buffer, pH 6 and Solvent B: acetonitrile) with 0.9 mL/min of flow rate, and the wavelength of UV detector was set to 214 nm.

HPLC can be used to determine the concentrations of constituents of food waste, such as sugars, amino acids, preservatives and colourants concentration, and manure including secondary elements, heavy metals and trace contaminants concentration. The constituents of food waste and manure obtained from HPLC would provide useful information for further processing. The advantage of HPLC is that the compounds being analysed do not need to be volatile, unlike gas chromatography. However, the compounds have to solubilize in the mobile phase [114].

8.6.2 Gas Chromatography and Mass Spectrometry (GC-MS)

Gas chromatography (GC) is a common type of chromatography used for separation and analyses of compounds that can be vaporized without decomposition. GC is commonly used for determination of the purity of a particular substance, or for identification and quantification of compounds in a mixture via separations using a gas as the mobile phase [115]. The system is mainly composed of a carrier gas as the mobile phase, an inlet to deliver sample to a column, the column where separations occur, an oven as a thermostat for the column, a detector to register the presence of a chemical in the column, and a data system to record and display the chromatogram. These components of a gas chromatograph have been unchanged in function or purpose during the last 40 years although technology has been improved in design, materials, and methodology [116].

The carrier gas or mobile phase in GC is an essential, but limiting, facet in separations. Carrier gas is needed to carry constituents of a sample through the column and yet the choice of possible gases is restricted [116]. The inert gas of helium remains the most commonly used carrier gas in about 90% of instruments although nitrogen or hydrogen is preferred for improved separations [117]. The chromatographic process begins when sample is introduced into the column, ideally without disrupting flows in the column. In order to keep the results to be reproducible, suitable sample inlet methods are preferable for the least change in pressure or flow of the carrier gas or mobile phase. Currently, syringe injection is the commonly used method for placing samples to a GC column, since it is convenient and generally effective although the thermoplastic septum develops leaks after repeated injections. Other options such as pyrolysis GC and purge-and-trap methods are also popular in certain specific applications [116]. Samples either in liquids or solids form, must be converted to gaseous state before passing through the GC column. Therefore, most gas chromatographs are equipped with ovens to keep the column at temperatures from 40 to 350 °C. Conventional ovens consisted of a resistive wire coil that spread heat into the inner volume of the oven. Alternatives to conventional ovens were designed later and may be helpful for short columns or conditions where the space is limited [117, 118]. Effluent from the column enters a detector where the composition of the carrier gas stream is characterized.

The mainstays in GC have been the flame ionization detector (FID), the thermal conductivity detector (TCD) and the electron capture detector (ECD). Other commercially available detectors include the photoionization detector (PID), the nitrogen–phosphorus detector and the atomic emission detector, though these have been less prevalent historically than the FID, TCD, and ECD [119].

Currently, with the increasing complexity of analytes and consequent increase in the demand for analytical accuracy, GC becomes a technique with obvious limitation. Even though GC equipped with a traditional detector such as flame ionization detector can separate volatile and semi-volatile compounds with great resolution, it cannot identify them. In contrast, the specific detector of mass spectrum (MS) can provide detailed structural information on most compounds so that they can be exactly identified and quantified. However, it cannot readily separate them [120]. In this case, combining the two techniques, namely Gas Chromatography-Mass Spectrometry (GC-MS), can take the advantages of GC and MS to identify different substances within a test sample with high resolution [121]. During the GC-MS process, different molecules in a mixture are retained by the column of GC due to their chemical properties and then eluted from the column at different times, which allows the downstream MS to capture, ionize, accelerate, deflect and detect the ionized molecules separately. The mass spectrometer does this by breaking each molecule into ionized fragments and detecting these fragments using their mass-to-charge ratio [121].

Since GC-MS is highly effective and versatile, it is widely used for research purposes in diverse fields, such as monitoring and tracking organic pollutants in the environment, analysis of nutrients, harmful materials or aromatic compounds in food, beverage, flavour and fragrance, bio-analysis of blood and urine for biological and pesticides detections, and it is also emerging as an important technique in the field of food waste and manure characterization [120–124].

Food wastes are usually composed of highly complex nutrients and organic materials such as fats, sugars, proteins and vitamins, and inorganic material, such as water and minerals. Apart from natural constituents, xenobiotic compounds derived from a variety of sources could also be introduced to the waste. Similarly, manures contain various materials including carbohydrates, lipids, proteins, volatile compounds and minerals that can be determined by the GC-MS regardless of the species of the manures and their handling practices or storage time.

The analysis of food can vary in a scope. For example, a GC–MS method can be used for the qualitative/quantitative analyses of untargeted volatiles (for example, aroma) or targeted ones such as pesticides. Furthermore, a GC–MS method can also be exploited for the generation of a chromatography profile (fingerprinting), with the aim of distinguishing between food samples of the same type, for example, to determine geographical origin of samples. However, for any purpose, the extent to which one or more of the aforementioned features prevails is dependent on the initial analytical objective. The methods and equipment conditions for the composition determination of food waste [120, 125] and manure [123, 124] can be found from the related studies [44, 122].

8.6.3 Ultraviolet-Visible Spectrophotometer

Spectrophotometry is the equipment for the measurement of the refection or absorption properties of a material. A broad range of wavelength can be dealt by spectrophotometry, including near-ultraviolet, visible light and near-infrared. Ultraviolet and visible regions of the spectrum are most commonly used in spectrophotometers, in which the wavelength ranges from 190 to 1100 nm. Spectrophotometry is made up of two parts, a spectrometer and a photometer. The spectrometer is used for providing any wavelength of light (near-ultraviolet, visible light and near-infrared), while the function of photometer is to measure the intensity of light. The sample is placed between spectrometer and photometer. The intensity of light passing through the sample (I) is measured and compared to the intensity of light before passing through the sample (I_0). The ratio of light intensity after and before passing through sample (I/I_0) is called transmittance ($\%T$). The output result (absorbance) is calculated by the transmittance (Eq. 8.1):

$$\mathrm{A} = -\log\left(\frac{\%T}{100\%}\right) \tag{8.1}$$

A basic spectrophotometer includes the following parts: a light source, a holder for the sample, a monochromator and a detector [126]. A Tungsten filament is used for visible light region, and a deuterium arc lamp for ultraviolet region (190–400 nm). The typical parts of a detector consist of a photomultiplier tube, a photodiode, a photodiode array or a charge-coupled device (CCD) [127]. Single photodiode detectors and photomultiplier tubes are used with scanning monochromators, which can filter the light so that only a single wavelength reaches the detector at a time [128].

Samples for UV-V is spectrophotometry are most often in the form of liquids, although the gas and solid forms are also acceptable for measurement. In general, liquid samples are placed in a transparent cell, known as a cuvette. The shape of cuvettes is typically rectangular, commonly with a width of 1 cm. Glass test tubes can also be used as cuvettes in some instruments. The criterion of sample container is that radiation can pass over the spectral region of interest. The most widely used cuvettes are made of silica or quartz glass because they can allow the UV wavelength to pass without any absorption. Glass and plastic cuvettes are also common, but they can be only used for visible wavelengths since the glass and most of plastics can absorb UV wavelengths. UV-V is spectrophotometer can be used in various research fields, like materials, chemistry, biochemistry and biology. They are widely used in laboratories for various studies, for example, enzyme activity assay, determinations of nitrogen and protein concentration, measurements of phosphate component. Information about colorant impurities in food waste hydrolysate can also be obtained from the ultraviolet and visible regions of the spectrum. Any hydrolysate containing large amounts of organic matter may have absorption peaks in the UV-V is region of the spectrum. Most of the colour in the sample appears in the wavelength of 250–700 nm [129].

8.6.4 Total Nitrogen Unit

There are four methods to determine total nitrogen including Total Kjeldahl Nitrogen (TKN), Ammonia (NH_3^+), Nitrite (NO_2^-) and Nitrate (NO_3^{2-}). TKN is an analysis method for bound nitrogen. All these methods suffer from certain disadvantages, the most common one being time-consuming and laborious sample preparation technique. TKN method and method of ammonia determination also use hazardous chemicals such as mercuric oxide, sulfuric acid, boric acid and phenol which are harmful to both life and land. Nitrate analysis requires a separate sampling for acidification of the sample before analysis, while nitrite analysis is time consuming [130, 131].

With the aim to provide simple and non-hazardous sample preparation for nitrogen analysis, the Total Nitrogen Module (TNM-1) was introduced by Shimadzu Company. The advantage of TNM-1 is that it allows analysis of sample in less than 4 min, requires no chemicals for sample preparation and/or analysis and produces no hazardous wastes. In TNM-1, samples containing nitrogen are introduced into an oxygen-rich combustion tube with platinum catalyst at 720 °C. Bound nitrogen is converted to nitrogen monoxide (NO), further oxidized to nitrite (NO_2) in the presence of ozone, and then it is detected by the chemiluminescence detector [131]. The CN analyzer (Vario Max CN, Elementar, Hanau, Germany) was introduced in 1988, it was the first combustion type nitrogen analyzer. It applied Dumas combustion method for determination of nitrogen in different fields such as food and agricultural products. It is fast, safe and environmental friendly compared to wet chemical Kjeldahl method [132].

Nitrogen content in food waste and manure can be determined using the Total Nitrogen Module. Pleissner et al. [49] performed total nitrogen content using a TOC-V CSH total organic carbon analyzer (Shimadzu, Japan) equipped with a TNM-1 total nitrogen measuring unit (Shimadzu, Japan) [49]. Nitrogen compounds from food waste were extracted through acid hydrolysis, then samples were diluted and filtered prior to total nitrogen content determination. Xu et al. [133] measured total nitrogen concentration in manures using combustion oxidation on a CN analyzer (Vario Max CN, Elementar, Hanau, Germany) [133].

8.6.5 ^{31}P-NMR Spectroscopy

The assessment of phosphorous composition in animal manures is of great interest for their long-term application to land. This is because the manure phosphorous holding capacity of soil depends mainly on the forms and bioavailability of phosphorous in manure. The latter must be considered for manure application to land to avoid soil phosphorous accumulation and an acceleration of eutrophication [134]. Traditionally, four solution-based techniques have been used to analyze the phosphorous composition of animal manure. These include sequential fractionation,

ion-pair chromatography, enzymatic hydrolysis, and solution ^{31}-P nuclear magnetic resonance (^{31}P-NMR) spectroscopy [12, 134–136]. While fractionation procedure has been successfully used to evaluate manure phosphorous composition particularly in terms of phytic acid, (myo-inositol hexakisphosphate), the most dominant organic phosphorous compound in a wide range of manures, the analysis by chromatographic techniques becomes complicated and erroneous since phosphates exist in complexes with mineral and organic compounds in manures [137]. Several researchers have used ^{31}P-NMR to identify organic phosphorous in foods, animal feeds, digestate and manures. ^{31}P-NMR offers an advantage that an entire range of phosphorous compounds including both organic and inorganic can be identified with one simple extraction procedure to provide better characterization of phosphorous in the samples. Solution ^{31}P-NMR has been used for quantification of phytic acid in food, animal feed and sewage sludge. Orthophosphate monoesters and diesters in NaOH extracts of swine slurry were identified by ^{31}P-NMR method [138]. Crouse et al. [139] characterized functional phosphorous groups in NaOH-EDTA extracts of turkey litter using ^{31}P-NMR [139]. The value of solid-state ^{31}P-NMR had been limited due to poor spectral resolution. However, improvements in extraction procedure, signal identification and understanding of compound degradation during extraction and analyses have advanced this technique. Information on the forms of phosphorous in manure has been pivotal to determine its environmental fate.

For solution ^{31}P-NMR, 2 g manure is extracted with 40 mL NaOH-EDTA for 4 h at 20 °C. The inclusion of EDTA in alkaline extraction solution markedly improves phosphorous recovery [140]. The mixtures are shaken horizontally and then centrifuged at 10,000 g for 30 min. The aliquots are analysed for total phosphorous by inductively coupled plasma atomic emission spectroscopy (ICP-AES). Reactive phosphorous, which amounts to inorganic orthophosphate is determined by molybdate colorimetry. Unreactive phosphorous (organic phosphorous plus inorganic polyphosphates) is calculated as the difference between total and reactive phosphorous [141]. Remaining extract after centrifugation is frozen rapidly at −80 °C, lyophilized and ground to a fine powder. Immediately before ^{31}P-NMR, ~100 mg freeze dried extract is redissolved in 0.9 mL 1 M NaOH and 0.1 mL D_2O (for signal lock) and transferred to 5 mm NMR tube. Spectra are obtained using spectrometer operating at 200 MHz for ^{31}P and 500.134 MHz for ^{1}H. Individual phosphorous compounds or functional groups from ^{31}P-NMR spectra are identified by their chemical shifts of signals (ppm) relative to 85% H_3PO_4, and comparison with literature reports. Signal areas are calculated by integration and phosphorous concentrations are calculated by multiplying the proportion of total spectral area assigned to a specific signal by the total phosphorous concentration in the original extract [136]. In well-resolved spectra, sum of areas of four signals at approximately 5.95, 5.06, 4.70 and 4.54 ppm in the ratio 1:2:2:1 is used to quantify phytic acid concentrations. In the case of overlapping of phytic acid signal with those from other monoesters, the signal from the phosphate at C-2 position on the inositol ring (at ~5.95 ppm) is multiplied by 6 to calculate phytic acid concentration. This is considered suitable since the signal at this position is often well resolved compared to that from other orthophosphate and monoester signals. Usually for manure samples, analytical errors are estimated to be approximately 5% for larger signals and 10% for smaller signals [137, 138]. Statistical analysis is performed using Statistical Analysis System [142] and regression analysis is performed using the generalized linear models function in SAS.

8.6.6 X-Ray Absorption Near Edge Structure (XANES) Spectroscopy

Identification of the chemical phosphorous species which exist in manures is important for our understanding of the long-term potential of P loss to water when manure is land applied [7]. Forty to fifty percent of phosphorous in manures such as those from poultry waste is in the organic form and mainly consists of organic phosphorous compounds such as inositol phosphates, sugar phosphates, nucleic acids and phospholipids. Phosphates sorbed on the surfaces of clay minerals and iron or aluminum hydroxides are important inorganic phosphorous forms which are less bioavailable in water. Additionally, manures such as poultry litters also contain phosphorous in mineral-phase litters e.g. potassium phosphate, pyrophosphate and aluminum phosphate. Thus, it is important to characterize phosphorous into organic and inorganic forms during phosphorous speciation for correct prediction of their environmental mobility. Several fractionation techniques have been applied, where phosphorous is partitioned into operationally defined fractions based on its solubility in a series of extractants and provide information about organic and inorganic phosphorous species in manure [143]. While these techniques can assist in predicting the effects of land application of organic phosphorous sources on soil phosphorous, these cannot provide specific information about the exact chemical species of phosphorous in manure [144]. This does not allow prediction of long term effects of manure phosphorous on water quality due to non-availability of information on actual phases of phosphorous present in these materials. Sequential extraction procedures can be complemented with X-Ray absorption near edge structure (XANES) spectroscopy to characterize the different forms of phosphorous in soils and manures [145, 146].

XANES spectroscopy offers a means of analyzing solids at in situ moisture contents, ambient pressures and temperatures without extensive alteration of samples [147]. It is used to specify phosphorous in a series of manure with a proper detection of organic and inorganic phosphorous species. Analysis by XANES spectroscopy starts with sample characterization for P, Al and Ca using inductively coupled plasma atomic emission spectroscopy (ICP-AES). This is followed by phosphorous sequential extraction which determines the molybdate-reactive P (inorganic P), as the percentage of amount of phosphorous removal in each extraction is measured using the ascorbic acidphosphomolybednum blue method. The resulting data is compared with the total phosphorous estimated by ICP-AES to determine the relative distribution of phosphorous in organic and inorganic phases. For XANES spectroscopic analysis, the residue after each step of sequential fractionation is used. Several steps are followed for sample preparation. The samples are shock-frozen with liquid nitrogen and freeze-dried for 24 h before analysis. Dried manure samples are ground into a fine powder using an agate mortar and pestle, following which a thin layer is mounted on a Plexiglas sample holder using phosphorus-free cellophane tape. XANES spectroscopy is conducted at beamlines X-19 A and X-15 B. These two beamlines tender X-ray lines designed for light element X-ray analysis. Data is collected for both samples and standards on beamline X-19 A using a monochromator with Si (111) crystals and a specific flux. A helium-purged sample chamber with a fluorescence detector is used for data collection at this line. Standard spectra are also collected at beamline X-15 B using a monochromator at specific flux. The monochromator is calibrated to the maximum peak energy of the first derivative (E_0 2149 eV) using a standard at the start of every run and after every beam dump, usually every 12 h [146]. In some other protocols, the monochromator can also be calibrated using NaH_2PO_4 powder and maximum of the standard spectra is assigned an energy value of 2,158.5 eV. The energy scale is normalized by subtracting the calibration energy from all obtained spectra. P XANES-spectra from manure samples are compared with spectra collected for organic phosphorous standard using a linear combination fitting method, a tool available with XANES spectroscopy equipment. Linear combination fitting is conducted across the energy range of 2,142–2,179 eV. Linear relationships between the XANES linear combination data and chemical fractionation data are analysed using statistical analysis methods such as PROC REG [148].

8.6.7 *Fourier Transform Infrared Reflectance Photoacoustic (FTIRPAS) Spectroscopy*

Infrared techniques such as transmittance spectroscopy and reflectance spectroscopy in combination with multivariate analysis can be used in characterization of manure, particularly for rapid quantification of mineralizable or labile fraction of carbon in manure [8, 149]. Since near infrared region consists of highly overlapping overtones and combination bands, it possesses some limits in qualitative analysis about chemical components. Additionally, optical opaque samples such as manure make the quantitative analysis even more difficult. Diffuse reflectance mid-infrared Fourier transform spectroscopy (DRIFT) and attenuated total reflectance mid-infrared Fourier transform spectroscopy (ATR-FTIR) have been used in the past for characterization of soil particles, and other organic waste products [150, 151]. The existence of abundant literature for interpretation of spectra generated by FTIR techniques provides an advantage over NIRS. However, DRIFT strongly depends upon material particle size and therefore it requires time to prepare samples of appropriate size suitable for analysis. ATR-FTIR requires little sample preparation but the sample should be close to water saturation to ensure a good contact between the sample and ATR crystal. It also typically results in low resolution with very dark and opaque samples such as manures [152]. Sample dilution with potassium bromide powder is often used to increase reflectance and avoid spectral distortion and non-linearities.

Recently, Fourier transform mid-infrared photoacoustic spectroscopy (FTIR-PAS) has been applied to soil samples and shown to overcome the above limitations. It is considered to be very suitable for characterization of manure. FTIR-PAS is a combination of FTIR and a photoacoustic detector (PAS) and the analysis is based on the absorption of electromagnetic radiation by analyte molecules [153]. Development of highly sensitive microphones in the last decade has increased the sensitivity of PAS and facilitated to achieve better results than the traditional FTIR. FTIR-PAS generates a signal by interaction of infrared beam with the surface of sample. This leads to production of a heat wave that can be detected by a sensitive microphone and transformed into a regular absorption spectrum. The resulting spectrum differs from both the equivalent transmittance and reflectance spectra since the technique detects non-radiative transitions in the sample. The detected wave amplitude in PAS is proportional to sample concentration and therefore this technique allows measurement of infrared absorption of relatively dark samples, such as manure without any pre-treatment [154]. When coupled with multivariate statistical analysis such as partial least squares (PLS), FTIR-PAS can be used for quickly and effortlessly to obtain important information such as chemical composition and carbon mineralization of samples. This provides critical information to develop application strategies, such as time of application and application amounts of manure in the fields. The manure samples are air-dried at room temperature and passed through a 2 mm sieve for analysis. Photoacoustic spectra are recorded for samples using an infrared spectrophotometer which is equipped with a photoacoustic cell. Around 200 mg sample is placed in the cell holding cup and purged with dry helium at 10 mL/min for 10 s. Subsequently, the scans are conducted in the wavelength range of 500–4,000 cm^{-1} with a resolution of 4 cm^{-1} [155]. For data processing, data reduction is achieved by the principal component analysis (PCA), which is commonly used to reduce the dimensionality of infrared spectra by generating a small number of coefficients called PCA scores that retain most of the variability (information) present in the original data. Probabilistic neural networks (PNN) are used to make manure identification [156]. Number of PCA scores retained from PCA decomposition and number of training samples are used to determine the number of input and hidden nodes, respectively. PAS spectra are coupled to PLS model consisting of for example 75% samples randomly chosen as calibration set, and 25% samples used as validation set. This information is used for quantitative analysis of organic matter content in manure samples. Distinctive bands for different animal manure spectra are generated by FTIR-PAS which establish its application for fast characterization of animal manures.

8.6.8 Near Infrared (NIR) Spectroscopy

Presently, several methods are used to determine the constituents of interest such as nitrogen fractions, mineral and dry matter content. There are two main methods to obtain these analyses which include (i) book values based on average of manure samples from a particular region, and (ii) analyses in commercial testing laboratory. Manure nutrient contents can vary substantially depending on age, species, nutrition of animal, amount of bedding and storage practices [9]. In such a case, using book values to determine the nutrient content can be inaccurate. The lab-based methods are accurate but time consuming, expensive and inconvenient for most producers. Furthermore, in order to combine manure analyses with precision agriculture, a rapid and accurate method is needed to use the correct amount of manure to the correct field location. Spectroscopic methods such as near-infrared diffuse reflectance spectroscopy (NIRS) offers the possibility of rapid, non-destructive analyses of manure without generating chemical wastes as those obtained from traditional digestion based methods. NIRS requires little or no sample preparation and allows simultaneous measurement of several analytes, thus making it more suitable for real-time analysis during spreading of manure in the field [157].

NIRS was firstly developed in the early 1970s for rapid analysis of moisture content of cereal grains. However, over the past few decades, it has rapidly developed to be a fast and robust analytical tool for characterization of agricultural, pharmaceutical and food products. The technique is particularly useful for determination of moisture and organic (carbon, nitrogen) content of soils and other organic waste products [158]. NIRS combines applied spectroscopy and statistics. Its principle is based on the absorbance of covalent chemical bonds formed between light atoms such as carbon, nitrogen, oxygen and hydrogen. These bonds have primary absorbances in the infrared region (IR) and combination bands that absorb light in the NIR region i.e. 780–2500 nm. This region of the electromagnetic spectrum is particularly useful since a linear relationship between absorbance and concentration is exhibited by many biological and agricultural samples, in contrast to the case in the IR region [149]. Thus, light reflected by or transmitted from samples in the NIR region is amplified, digitized and recorded as absorbance. For samples under analysis, spectral data is collected statistically and converted to calibration equations. This is then used with results of conventional chemical analyses on the same samples to predict the concentrations of constituents in further samples of a similar type [159]. The NIRS technique utilizes samples in dry or ground or 'as it is' form. Samples are scanned using a transport module with the sample placed in a polyethylene bag, thus allowing a sample path $\sim$12 cm to be scanned. Though exact protocols can vary, usually the data is collected every 2 nm at a nominal bandwidth of 10 nm. A ceramic standard is used for background spectra and spectra are computed as log (1/R), where R is reflectance [160].

Several researchers have reported the use of NIRS to analyze nutrient concentrations in manure samples. Total nitrogen, total carbon and crude ash in cattle manure was analysed [161]. NIRS was used to predict the total solids (TS), total nitrogen (TN), ammonia-nitrogen and potassium in swine lagoon effluent and solid beef manure samples using spectral data [162]. It was also reported that NIRS could measure organic nitrogen, TN, and moisture but not the mineral content in poultry manure [9]. The use of NIRS for total dissolved nitrogen, suspended nitrogen, soluble reactive phosphorous, magnesium, potassium, and suspended phosphorous in hog manure was demonstrated [163]. In general, the application of NIRS for mineral content analysis of manures has been limited.

In addition to manure, NIRS also finds applications in food products to determine its constituents such as fat, moisture, proteins, acid and neutral detergent fibres [164]. In dairy industry, NIRS has been used for analyzing major components of milk, skim milk and fermented milk products. In cheese industry, NIRS has been successfully used to determine proteins, moisture, lactose and fat [165, 166]. Since food waste also comprises the same constituents, this opens up the possibility of using NIRS for food waste composition analysis as well.

Acknowledgements The authors acknowledge the Innovation and Technology Fundings (ITS/323/11), (ITS/353/12), (ITP/087/15FP) and (ITP/109/15TP) from Innovation and Technology Commission in Hong Kong. We are grateful to the industrial sponsors Starbucks Hong Kong, PepsiCo Inc. and Novozymes®. Daniel Pleissner acknowledges the Max Buchner Research Foundation in Frankfurt, Germany.

References

1. Parfitt, J., Barthel, M., Macnaughton, S.: Food waste within food supply chains: quantification and potential for change to 2050. Philos. Trans. R. Soc. Lond. B Biol. Sci. **365**, 3065–3081 (2010)
2. Rezaei, M., Liu, B.: Food loss and waste in the food supply chain. International Nut and Dried Fruit council, Reus (2017)
3. FAO: Global Food Losses and Food Waste: Extent, Causes and Prevention, Save Food! Rome. http://dx.doi.org/10.1098/rstb.2010.0126 (2011). Accessed 18 Dec 2017
4. Paritosh, K., Kushwaha, S.K., Yadav, M., Pareek, N., Chawade, A., Vivekanand, V.: Food waste to energy: an overview of sustainable approaches for food waste management and nutrient recycling. Biomed. Res. Int. **2017**, 19 (2017)
5. Lin, C.S.K., Pfaltzgraff, L.A., Herrero-Davila, L., Mubofu, E.B., Abderrahim, S., Clark, J. H., Koutinas, A.A., Kopsahelis, N., Stamatelatou, K., Dickson, F., Thankappan, S., Mohamed, Z., Brocklesby, R., Luque, R.: Food waste as a valuable resource for the production of chemicals, materials and fuels. Current situation and global perspective. Energy Environ. Sci. **6**(2), 426–464 (2013)
6. Waqas, M., Rehan, M., Khan, M.D., Nizami, A.-S.: Conversion of food waste to fermentation products. (2018). https://doi.org/10.1016/B978-0-08-100596-5.22294-4
7. Negassa, W., Leinweber, P.: How does the Hedley sequential phosphorus fractionation reflect impacts of land use and management on soil phosphorus: a review. J. Plant Nutr. Soil Sci. **172**, 305–325 (2009)
8. Peltre, C., Bruun, S., Du, C., Thomsen, I.K., Jensen, L.S.: Assessing soil constituents and labile soil organic carbon by mid-infrared photoacoustic spectroscopy. Soil Biol. Biochem. **77** (Supplement C), 41–50 (2014)
9. Reeves, J.B.: Near-infrared diffuse reflectance spectroscopy for the analysis of poultry manures. J. Agric. Food Chem. **49**, 2193–2197 (2001)

10. Ogejo, J. A.: Selection and Location of Poultry and Livestock Manure Storage, pp. 442–307. Virginia Cooperative Extension, VCE publication (2009)
11. Bicudo, J.R.: Animal solid manure: storage, handling and disposal. Encyclopedia of Life Support Systems (EOLSS) vol. II, (2009)
12. He, Z., Honeycutt, C.W., Griffin, T.S., Cade-Menun, B.J., Pellechia, P.J., Dou, Z.: Phosphorus forms in conventional and organic dairy manure identified by solution and solid state p-31 NMR spectroscopy. J. Environ. Qual. **38**, 1909–1918 (2009)
13. Zhu, W., Yao, W., Zhang, Z., Wu, Y.: Heavy metal behavior and dissolved organic matter (DOM) characterization of vermicomposted pig manure amended with rice straw. Environ. Sci. Pollut. Res. **21**, 12684–12692 (2014)
14. Kosseva, M.R.: Chapter 3—sources, characterization, and composition of food industry wastes. In: Food Industry Wastes, pp. 37–60. Academic Press, San Diego (2013)
15. WRAP: The Food We Waste, Banbury, UK. http://www.wrap.org.uk/downloads/The_Food_We_Waste_v2__2_.99cb5cae.5635.pdf (2008). Accessed 14 Dec 2017
16. Morawicki, R.O.: Sampling and sample preparation. In: Food Analysis, pp. 69–81. Springer US, Boston, MA (2010)
17. Suresh, K., Thomas, S., Suresh, G.: Design, data analysis and sampling techniques for clinical research. Ann. Indian Acad. Neurol. **14**, 287–290 (2011)
18. Swyngedouw, C., Crépin, J.M.: Sampling methods for site characterization. Environ. Geochem. 13–27 (2008)
19. Sahimaa, O., Hupponen, M., Horttanainen, M., Sorvari, J.: Method for residual household waste composition studies. Waste Manag. **46**, 3–14 (2015)
20. Elfil, M., Negida, A.: Sampling methods in clinical research; an educational review. Emerg. **5**, e52 (2017)
21. Kwan, T.H., Hu, Y., Lin, C.S.K.: Valorisation of food waste via fungal hydrolysis and lactic acid fermentation with *Lactobacillus casei* Shirota. Bioresour. Technol. **217**, 129–136 (2016)
22. Yu, I.K.M., Tsang, D.C.W., Yip, A.C.K., Chen, S.S., Wang, L., Ok, Y.S., Poon, C.S.: Catalytic valorization of starch-rich food waste into hydroxymethylfurfural (HMF): controlling relative kinetics for high productivity. Bioresour. Technol. **237**, 222–230 (2017)
23. Li, C., Yang, X., Gao, S., Chuh, A.H., Lin, C.S.K.: Hydrolysis of fruit and vegetable waste for efficient succinic acid production with engineered *Yarrowia lipolytica*. J. Clean. Prod. **179**, 151–159 (2018)
24. Pagliaccia, P., Gallipoli, A., Gianico, A., Montecchio, D., Braguglia, C.M.: Single stage anaerobic bioconversion of food waste in mono and co-digestion with olive husks: impact of thermal pretreatment on hydrogen and methane production. Int. J. Hydrog. Energy **41**(2), 905–915 (2016)
25. Ciurzyńska, A., Lenart, A.: Freeze-drying—application in food processing and biotechnology. Pol. J. Food Nutr. Sci. **3**, 165–171 (2011)
26. Hegde, S., Lodge, J.S., Trabold, T.A.: Characteristics of food processing wastes and their use in sustainable alcohol production. Renew. Sust. Energ. Rev. **81**, 510–523 (2018)
27. Meghwal, M., Goyal, M.R.: Developing technologies in food science: status, applications, and challenges. CRC Press Taylor & Francis Group (2017)
28. AOAC: Official Methods of Analysis of the Association of Analytical Chemists International, 18th edn, AOAC International, Gathersburg, MD U.S.A. (2005)
29. Nielsen, S.S.: Determination of moisture content. In: Nielsen, S.S. (eds) Food Analysis Laboratory Manual, pp 17–27. Springer US, Boston, MA (2010)
30. DuBois, M., Gilles, K.A., Hamilton, J.K., Rebers, P.T., Smith, F.: Colorimetric method for determination of sugars and related substances. Anal. Chem. **28**(3), 350–356 (1956)
31. Lever, M.: A new reaction for colorimetric determination of carbohydrates. Anal. Biochem. **47**(1), 273–279 (1972)
32. WRAP: Food Waste Chemical Analysis. http://www.wrapcymru.org.uk/sites/files/wrap/Technical_report_food_waste_characterisation_Wales_2009x2.9086.pdf (2010). Accessed 14 Dec 2017

33. Gur, A., Cohen, A., Bravdo, B.-A.: Colorimetric method for starch determination. J. Agric. Food Chem. **17**, 347–351 (1969)
34. Megazyme: Total Starch Assay Procedure (Amyloglucosidase/ α-Amylase method). https://secure.megazyme.com/files/Booklet/K-TSTA_DATA.pdf (2017). Accessed 20 Dec 2017
35. Ma, Y., Cai, W., Liu, Y.: An integrated engineering system for maximizing bioenergy production from food waste. Appl. Energy **206**, 83–89 (2017)
36. Pleissner, D., Kwan, T.H., Lin, C.S.K.: Fungal hydrolysis in submerged fermentation for food waste treatment and fermentation feedstock preparation. Bioresour. Technol. **158** (Supplement C), 48–54 (2014)
37. Pleissner, D., Demichelis, F., Mariano, S., Fiore, S., Navarro Gutiérrez, I.M., Schneider, R., Venus, J.: Direct production of lactic acid based on simultaneous saccharification and fermentation of mixed restaurant food waste. J. Clean. Prod. **143**, 615–623 (2017)
38. Karkacier, M., Erbas, M., Uslu, M.K., Aksu, M.: Comparison of different extraction and detection methods for sugars using amino-bonded phase HPLC. J. Chromatogr. Sci. **41**, 331–333 (2003)
39. Van Slyke, D.D.: A method for quantitative determination of aliphatic amino groups applications to the study of proteolysis and proteolytic products. J. Biol. Chem. **9**, 185–204 (1911)
40. Pleissner, D., Wimmer, R., Eriksen, N.T.: Quantification of amino acids in fermentation media by isocratic HPLC analysis of their α-hydroxy acid derivatives. Anal. Chem. **83**(1), 175–181 (2011)
41. Ulusoy, S., Ulusoy, H.I., Pleissner, D., Eriksen, N.T.: Nitrosation and analysis of amino acid derivatives by isocratic HPLC. RSC Adv. **6**(16), 13120–13128 (2016)
42. Lie, S.: The EBC-ninhydrin method for dertermination of free alpha amino nitrogen. J. Inst. Brew. **79**(1), 37–41 (1973)
43. Du, C., Lin, S.K.C., Koutinas, A., Wang, R., Webb, C.: Succinic acid production from wheat using a biorefining strategy. Appl. Microbiol. Biotechnol. **76**(6), 1263–1270 (2007)
44. Wang, R., Shaarani, S.M., Godoy, L.C., Melikoglu, M., Vergara, C.S., Koutinas, A., Webb, C.: Bioconversion of rapeseed meal for the production of a generic microbial feedstock. Enzyme Microb. Technol. **47**(3), 77–83 (2010)
45. Bradford, M.M.: A rapid and sensitive method for the quantitation of microgram quantities of protein utilizing the principle of protein-dye binding. Anal. Biochem. **72**(1), 248–254 (1976)
46. Kirk, P.L.: Kjeldahl method for total nitrogen. Anal. Chem. **22**(2), 354–358 (1950)
47. Campins-Falco, P., Meseguer-Lloret, S., Climent-Santamaria, T., Molins-Legua, C.: A microscale Kjeldahl nitrogen determination for environmental waters. Talanta **75**(4), 1123–1126 (2008)
48. Mariotti, F., Tomé, D., Mirand, P.P.: Converting nitrogen into protein—beyond 6.25 and Jones' factors. Crit. Rev. Food Sci. Nutr. **48**(2), 177–184 (2008)
49. Pleissner, D., Lam, W.C., Sun, Z., Lin, C.S.K.: Food waste as nutrient source in heterotrophic microalgae cultivation. Bioresour. Technol. **137**(Supplement C), 139–146 (2013)
50. Merrill, A.L., Watt, B.K.: Energy value of foods: basis and derivation. United States Department of Agriculture, Washington, DC (1973)
51. Ghareib, M., Youssef, K.A., Khalil, A.A.: Ethanol tolerance of *Saccharomyces cerevisiae* and its relationship to lipid content and composition. Folia Microbiol. **33**(6), 447–452 (1988)
52. Bligh, E.G., Dyer, W.J.: A rapid method of total lipid extraction and purification. Can. J. Biochem. Phys. **37**(8), 911–917 (1959)
53. FDA: Overview of Food Ingredients, Additives & Colors. https://www.fda.gov/Food/IngredientsPackagingLabeling/FoodAdditivesIngredients/ucm094211.htm#introduction (2010). Accessed 19 Dec 2017
54. Kuprovskytė, K., Pranaitytė, B., Padarauskas, A.: Isocratic HPLC determination of preservatives in beverages. Chemija **13**, 160–163 (2002)

55. Janovitz-Klapp, A.H., Richard, F.C., Goupy, P.M., Nicolas, J.J.: Inhibition studies on apple polyphenol oxidase. J. Agr. Food Chem. **38**(4), 926–931 (1990)
56. Noda, T., Ohtani, T., Shiina, T., Nawa, Y.: Semi-continuous hydrolysis of sweet potato raw starch by *Chalara paradoxa* glucoamylase. J. Food Sci. **57**(6), 1348–1352 (1992)
57. Ruiz-Capillas, C., Jimenez-Colmenero, F.: Determination of preservatives in meat products by flow injection analysis (FIA). Food Addit. Contam. Part A **25**(10), 1167–1178 (2008)
58. Phechkrajang, C.M., Yooyong, S.: Fast and simple method for semiquantitative determination of calcium propionate in bread samples. J. Food Drug Anal. **25**(2), 254–259 (2017)
59. Aubin, A.: Acquity UPLC for the rapid analysis of soft drinks. https://www.waters.com/webassets/cms/library/docs/720001053en.pdf (2014). Accessed 16 Nov 2017
60. Uhlemann, U., Strelau, K.K., Weber, K., Da Costa Filho, P.A., Rösch, P., Popp, J.: Raman spectroscopic determination of norbixin and tartrazine in sugar. Food Addit. Contam. Part A **29**(8), 1244–1255 (2012)
61. López-Montes, A.M., Dupont, A.-L., Desmazières, B., Lavédrine, B.: Identification of synthetic dyes in early colour photographs using capillary electrophoresis and electrospray ionisation–mass spectrometry. Talanta **114**(Supplement C), 217–226 (2013)
62. de Araújo Siqueira Bento, W., Lima, B.P., and Paim, A.P.S.: Simultaneous determination of synthetic colorants in yogurt by HPLC. Food Chem. **183**, 154–160 (2015)
63. González, M., Gallego, M., Valcárcel, M.: Determination of natural and synthetic colorants in prescreened dairy samples using liquid chromatography-diode array detection. Anal. Chem. **75**(3), 685–693 (2003)
64. Karanikolopoulos, G., Gerakis, A., Papadopoulou, K., Mastrantoni, I.: Determination of synthetic food colorants in fish products by an HPLC-DAD method. Food Chem. **177** (Supplement C), 197–203 (2015)
65. Brady, E., Burgess, J.: Analysis of artificial food dyes using ultraperformance liquid chromatography and an extended wavelength Photo Diode Array detector. http://www.waters.com/webassets/cms/library/docs/720005217en.pdf (2014). Accessed 10 Jan 2018
66. Sigmann, S.B., Wheeler, D.E.: The quantitative determination of food dyes in powdered drink mixes. A high school or general science experiment. J. Chem. Edu. **81**(10), 1475 (2004)
67. USDA: National Engineering Handbook Part 651 (NEH 651), Agricultural waste management field handbook, NRCS, Iowa (2012)
68. FAO/IAEA: Division of Nuclear Techniques in Food and Agriculture, International Atomic Energy Agency (IAEA), Guidelines for sustainable manure management in asian livestock production systems. Vienna, Austria https://www-pub.iaea.org/MTCD/publications/PDF/TE_1582_web.pdf (2008). Accessed 12 Jan 2018
69. Ogejo, J.A.: Poultry and livestock manure storage: management and safety. (2009)
70. Ministry of Agriculture, B.C., Canada. Farm structures fact sheet, manure storage structure (2015)
71. Beaver, R.L., Field, W.E.: Summary of documented fatalities in livestock manure storage and handling facilities 1975–2004. J. agromedicine **12**(2), 3–23 (2007)
72. Po, M.: Manure handling and storage- farm practices guidelines for pig producers in Manitoba. (2007)
73. Forestry.: Agricultural Operation Practices Act. http://www.agriculture.alberta.ca/app21/ (2002). Accessed 19 Jan 2018
74. Lorimor, J., Powers, W., Sutton, A.: Manure management systems series, section one-manure characteristics. In: Manure Management Systems Series-18, P. 24. Iowa State University (2004)
75. Murphy, S.: Manure sampling and analysis. Rutgers cooperative research & extension, (NJAES), The State University of New Jersey, Rutgers. https://www.researchgate.net/file.PostFileLoader.html?id=572d78ad615e272a5d37a987&asseKey = AS%3A358975368384513%401462597805690 (2006). Accessed 20 Feb 2018
76. Dou, Z., Galligan, D., Allshouse, R., Toth, J., Ramberg, C., Ferguson, J.: Manure sampling for nutrient analysis. J. Environ. Qual. **30**(4), 1432–1437 (2001)

77. Zhu, N.: Composting of high moisture content swine manure with corncob in a pilot-scale aerated static bin system. Bioresour. Technol. **97**, 1870–1875 (2006)
78. Flotats Ripoll, X., Foged, H., Bonmatí Blasi, A., Palatsi Civit, J., Magrí Aloy, A., Schelde, K.M.: Manure processing technologies. http://agro-technology-atlas.eu/docs/21010_technical_report_II_manure_processing_technologies.pdf (2012). Accessed 10 Jan 2018
79. Hamilton, D., Zhang, H.: Solids content of wastewater and manure. Oklahoma Cooperative Extension Service. Publication BAE-1759. https://shareok.org/bitstream/handle/11244/49599/oksd_bae_1759_2011-10.pdf?sequence=1 (2011). Accessed 14 Nov 2019
80. Møller, H.B., Sommer, S.G., Ahring, B.K.: Separation efficiency and particle size distribution in relation to manure type and storage conditions. Bioresour. Technol. **85**(2), 189–196 (2002)
81. Chastain, J.P., Camberato, J.J., Albrecht, J.E., Adams, J.: Swine manure production and nutrient content, South Carolina confined animal manure managers certification program, pp. 1–17. Clemson University, SC (1999)
82. James, R., Eastridge, M., Brown, L., Elder, K., Foster, S., Hoorman, J., Joyce, M., Keener, H., Mancl, K., Monnin, M.: Ohio livestock manure management guide: Bulletin 604. The Ohio State University Extension, Columbus (2006)
83. Møller, H.B., Lund, I., Sommer, S.G.: Solid–liquid separation of livestock slurry: efficiency and cost. Bioresour. Technol. **74**(3), 223–229 (2000)
84. Cordell, D., Drangert, J.-O., White, S.: The story of phosphorus: global food security and food for thought. Glob. Environ. Change **19**(2), 292–305 (2009)
85. Camberato, J., Maloney, S., Casteel, S.: Sulfur deficiency in corn, department of agronomy soil fertility update. Purdue University, West Lafayette, IN (2012)
86. Sawyer, J.E., Lang, B., Barker, D.W.: Sulfur fertilization response in Iowa corn and soybean production. In: Proceedings of the 2012 Winconsin Crop Management Conference, vol. 51, 39–48 2012
87. Eastman, J.A., Ferguson, J.F.: Solubilization of particulate organic carbon during the acid phase of anaerobic digestion. J. Water Pollut. Control Fed., 352–366 (1981)
88. Peces, M., Astals, S., Mata-Alvarez, J.: Assessing total and volatile solids in municipal solid waste samples. Environ. Technol. **35**, 3041–3046 (2014)
89. Manitoba Agriculture Food and Rural Development: Properties of Manure. https://www.gov.mb.ca/agriculture/environment/nutrient-management/pubs/properties-of-manure.pdf (2015). Accessed 20 Dec 2017
90. Pan, L., Shang, Q.H., Ma, X.K., Wu, Y., Long, S.F., Wang, Q.Q., Piao, X.S.: Coated compound proteases improve nitrogen utilization by decreasing manure nitrogen output for growing pigs fed sorghum soybean meal based diets. Anim. Feed Sci. Tech. **230** (Supplement C), 136–142 (2017)
91. Powell, J.M., Barros, T., Danes, M., Aguerre, M., Wattiaux, M., Reed, K.: Nitrogen use efficiencies to grow, feed, and recycle manure from the major diet components fed to dairy cows in the USA. Agric. Ecosyst. Environ. **239** (Supplement C), 274–282 (2017)
92. Liu, C., Guo, T., Chen, Y., Meng, Q., Zhu, C., Huang, H.: Physicochemical characteristics of stored cattle manure affect methane emissions by inducing divergence of methanogens that have different interactions with bacteria. Agri. Ecosyst. Environ. **253** (Supplement C), 38–47 (2018)
93. Montégut, G., Michelin, L., Brendlé, J., Lebeau, B., Patarin, J.: Ammonium and potassium removal from swine liquid manure using clinoptilolite, chabazite and faujasite zeolites. J. Environ. Manag. **167**, 147–155 (2016)
94. Fiske, C.H., Subbarow, Y.: The colorimetric determination of phosphorus. J. Biol. Chem. **66** (2), 375–400 (1925)
95. Jakubus, M.: Estimation of phosphorus bioavailability from composted organic wastes. Chem. Speciation Bioavailability **28**(1–4), 189–198 (2016)
96. Lugo-Ospina, A., Dao, T.H., Van Kessel, J.A., Reeves, J.B.: Evaluation of quick tests for phosphorus determination in dairy manures. Environ. Pollut. **135**(1), 155–162 (2005)

97. Jastrzębska, A.: Modifications of spectrophotometric methods for total phosphorus determination in meat samples. Chem. Pap. **63**(1), 47 (2008)
98. Amtmann, A., Rubio, F.: Potassium in Plants, In: *eLS*, Wiley, Ltd. (2001) https://doi.org/10.1002/9780470015902.a0023737
99. Peters, J.: Recommended Methods of Manure Analysis. http://learningstore.uwex.edu/assets/pdfs/A3769.pdf (2003). Accessed 4 Dec 2017
100. Whittles, C.L., Little, R.C.: A colorimetric method for the determination of potassium and its application to the analysis of soil extracts. J. Sci. Food Agric. **1**(11), 323–326 (1950)
101. Roa-Espinosa, A., Markley, J.L., Vu, T.T., Filatov, E., Roa-Lauby, S.W.: Elemental analysis of nutrients in dairy manure by automated X-ray fluorescence spectrometry. J. Environ. Anal. Chem. **3**(1), 1–6 (2016)
102. Chen, W.-C., Hsu, F.-Y., Yen, J.-H.: Effect of green manure amendment on herbicide pendimethalin on soil. J. Environ. Sci. Health B, 1–8 (2017)
103. APHA: Standard Methods For The Examination of Water and Wastewater. 21st edn., Washington. DC. (2005)
104. Standardization TSAf: UNE-EN ISO 10304-1: 2009 water quality—determination of dissolved anions by liquid chromatography of ions—Part 1: determination of bromide, chloride, fluoride, nitrate, nitrite, phosphate and sulfate (2009)
105. Iwegbue C.M.A, N.G., Osakwe S.A.: Recycling waste in agriculture: efficacy of composting in ameliorating trace metal availabilityand soil borne pathogens. Eur. J. Sci. Res. **11**, 572–577 (2005)
106. Irshad, M., Malik, A.H., Shaukat, S., Mushtaq, S., Ashraf, M.: Characterization of heavy metals in livestock manures. Pol. J. Environ. Stud. **22**(4), 1257–1262 (2013)
107. Bolan, N., Adriano, D., Mahimairaja, S.: Distribution and bioavailability of trace elements in livestock and poultry manure by-products. Crit. Rev. Environ. Sci. Technol. **34**(3), 291–338 (2004)
108. He, Z., Fortuna, A.-M., Senwo, Z.N., Tazisong, I.A., Honeycutt, C.W., Griffin, T.S.: Hydrochloric fractions in Hedley fractio-nation may contain inorganic and organic phosphorus. Soil Sci. Soc. Am. J. **72**(5), 893–899 (2006)
109. Achiba, W.B., Lakhdar, A., Gabteni, N., Laing, G.D., Verloo, M., Boeckx, P., Van Cleemput, O., Jedidi, N., Gallali, T.: Accumulation and fractionation of trace metals in a Tunisian calcareous soil amended with farmyard manure and municipal solid waste compost. J. Hazard. Mater. **176**(1), 99–108 (2010)
110. Standardization, E.C.f. EN 15662 Foods of plant origin—determination of pesticide residues using GC-MS and/or LC-MS/MS following acetonitrile extraction/partitioning and cleanup by dispersive SPE—QuEChERS-method, Brussels. (2008)
111. Ettre, L.S.: Nomenclature for chromatography (IUPAC recommendations 1993). Pure Appl. Chem. **65**(4), 819 (1993)
112. Nollet, L.M.L.: Food Analysis by HPLC, 2nd edn, Taylor & Francis (2000)
113. Yu, I.K.M., Ong, K.L., Tsang, D.C.W., Haque, M.A., Kwan, T.H., Chen, S.S., Uisan, K., Kulkarni, S., Lin, C.S.K.: Chemical transformation of food and beverage waste-derived fructose to hydroxymethylfurfural as a value-added product. Catal. Today **314**, 70–77 (2018)
114. Talcott, S.: High performance liquid chromatography. In: Nielsen, S.S. (ed.) Food Analysis Laboratory Manual, pp. 145–154. Springer US, Boston, MA (2010)
115. Harris, D.C.: Quantitative Chemical Anslysis, 8th edn. Freemand Palgrave Macmillan International Edition. Royal Society of Chemistry, London (2010)
116. Eiceman, G.A.: Instrumentation of gas chromatography. In: Meyers, R.A. (ed.) Encyclopedia of Analytical Chemistry. Wiley Ltd., Chichester (2006)
117. Snyder, A.P., Harden, C.S., Brittain, A.H., Kim, M.G., Arnold, N.S., Meuzelaar, H.L.C.: Portable hand-held gas chromatography/ion mobility spectrometry device. Anal. Chem. **65**(3), 299–306 (1993)
118. Jain, V., Phillips, J.B.: Fast temperature programming on fused-silica open-tubular capillary columns by direct resistive heating. J. Chromatogr. Sci. **33**(1), 55–59 (1995)

119. Eiceman, G.A., Hill, H.H., Gardea-Torresdey, J.: Gas chromatography. Anal. Chem. **72**(12), 137–144 (2000)
120. Hussain, S.Z., Maqbool, K.: GC-MS: principle, technique and its application in food science. J. CURR Sci. **13**, 116–126 (2014)
121. Sparkman, O.D., Penton, Z., Kitson, F.G.: Gas chromatography and mass spectrometry: a practical guide, Academic Press (2011)
122. Schnitzer, M.I., Monreal, C.M., Jandl, G., Leinweber, P., Fransham, P.B.: The conversion of chicken manure to biooil by fast pyrolysis II. Analysis of chicken manure, biooils, and char by curie-point pyrolysis-gas chromatography/mass spectrometry (Cp Py-GC/MS). J. Environ. Sci. Health Part B **42**, 79–95 (2007)
123. Yasuhara, A.: Identification of volatile compounds in poultry manure by gas chromatography—mass spectrometry. J. Chromatogr. A **387**, 371–378 (1987)
124. Yuan, M., Xiao, H., Lu, Y., Huang, H., Jiang, J., Zhao, Y., Mao, T.: Measuring volatile organic compounds by headspace trap GC-MS in the Beijing food laboratory. Am. Lab., On-Line Ed. (2009)
125. Lehotay, S.J., Hajšlová, J.: Application of gas chromatography in food analysis. Trends Anal. Chem. **21**(9), 686–697 (2002)
126. Chen, Z., Deutsch, T.G., Dinh, H.N., Domen, K., Emery, K., Forman, A.J., Gaillard, N., Garland, R., Heske, C., Jaramillo, T.F., Kleiman-Shwarsctein, A., Miller, E., Takanabe, K., Turner, J.: UV-Vis spectroscopy. In: Photoelectrochemical Water Splitting: Standards, Experimental Methods, and Protocols, pp. 49–62. Springer, New York (2013)
127. Perkampus, H.-H., Grinter, H.-C.: UV-VIS Spectroscopy and its Applications. Springer-Verlag, Berlin Heidelberg (1992)
128. Förster, H.: UV/VIS spectroscopy. In: Karge, H.G., Weitkamp, J. (eds.) Characterization I, pp. 337–426. Springer, Berlin, Heidelberg (2004)
129. Haque, M.A., Yang, X., Ong, K.L., Tang, W.-T., Kwan, T.H., Kulkarni, S., Lin, C.S.K.: Bioconversion of beverage waste to high fructose syrup as a value-added product. Food Bioprod. Process. **105**, 179–187 (2017)
130. APHA, AWWA, and WEF.: Standard Methods for the Examination of Water and Wastewater: 4500-N Nitrogen (Organic), 20th edn, (1999)
131. Goldman, M.S., Clifford, R.H.: Shimadzu's total nitrogen module eliminates environmentally un-friendly environmental methods. https://www.ssi.shimadzu.com/products/literature/env/Shimadzu_TN_Module_Eliminates_EnvMethods.pdf (2017). Accessed 9 Jan 2018
132. Elementar: Maximum performance on a micro foot print. https://www.elementar.de/en/products/organic-elemental-analysis/vario-max-cube.html (2018). Accessed 14 Jan 2018
133. Xu, F., Li, Y., Ge, X., Yang, L., Li, Y.: Anaerobic digestion of food waste—challenges and opportunities. Bioresour. Technol. (2017)
134. Leytem, A.B., Kwanyuen, P., Plumstead, P.W., Maguire, R.O., Brake, J.: Evaluation of phosphorus characterization in broiler ileal digesta, manure, and litter samples: 31P-NMR vs HPLC. J. Environ. Qual. **37**(2), 494–500 (2008)
135. He, Z., Honeycutt, C.W., Cade-Menun, B.J., Senwo, Z.N., Tazisong, I.A.: Phosphorus in poultry litter and soil: enzymatic and nuclear magnetic resonance characterization. Soil Sci. Soc. Am. J. **72**(5), 1425–1433 (2008)
136. Li, G., Li, H., Leffelaar, P.A., Shen, J., Zhang, F.: Characterization of phosphorus in animal manures collected from three (dairy, swine, and broiler) farms in China. PLoS ONE **9**, e102698 (2014). https://doi.org/10.1371/journal.pone.0102698
137. Kemme, P.A., Lommen, A., De Jonge, L.H., Van der Klis, J.D., Jongbloed, A.W., Mroz, Z., Beynen, A.C.: Quantification of inositol phosphates using (31)P nuclear magnetic resonance spectroscopy in animal nutrition. J. Agric. Food Chem. **47**, 5116–5121 (1999)
138. Leinweber, P., Haumaier, L., Zech, W.: Sequential extractions and 31P-NMR spectroscopy of phosphorus forms in animal manures, whole soils and particle-size separates from a densely populated livestock area in northwest Germany. Biol. Fertil. Soils **25**(1), 89–94 (1997)

139. Crouse, D.A., Sierzputowska-Gracz, H., Mikkelsen, R.L.: Optimization of sample pH and temperature for phosphorus-31 nuclear magnetic resonance spectroscopy of poultry manure extracts. Commun. Soil Sci. Plant Anal. **31**(1–2), 229–240 (2000)
140. Bowman, R.A., Moir, J.O.: Basic EDTA as an extractant for soil organic phosphorus. Soil Sci. Soc. Am. J. **57**(6), 1516–1518 (1993)
141. Murphy, J., Riley, J.P.: A modified single solution method for the determination of phosphate in natural waters. Anal. Chim. Acta **27**, 31–36 (1962)
142. SAS Institute Inc., S.O. Version 8 edn., Cary, NC (1999)
143. Condron, L.M., Newman, S.: Revisiting the fundamentals of phosphorus fractionation of sediments and soils. J. Soil. Sediment **11**(5), 830–840 (2011)
144. Kar, G., Hundal, L.S., Schoenau, J.J., Peak, D.: Direct chemical speciation of P in sequential chemical extraction residues using P K-Edge X-Ray absorption near-edge structure spectroscopy. Soil Sci. **176**(11), 589–595 (2011)
145. Ajiboye, B., Akinremi, O.O., Hu, Y., Flaten, D.N.: Phosphorus speciation of sequential extracts of organic amendments using nuclear magnetic resonance and X-ray absorption near-edge structure spectroscopies. J. Environ. Qual. **36**, 1563–1576 (2007)
146. Khare, N., Hesterberg, D., Beauchemin, S., Wang, S.-L.: XANES Determination of adsorbed phosphate distribution between Ferrihydrite and Boehmite in mixtures. Soil Sci. Soc. Am. J. **68**(2), 460–469 (2004)
147. Liu, J., Yang, J., Cade-Menun, B.J., Liang, X., Hu, Y., Liu, C.W., Zhao, Y., Li, L., Shi, J.: Complementary phosphorus speciation in agricultural soils by sequential fractionation, solution 31P nuclear magnetic resonance, and phosphorus K-edge X-ray absorption near-edge structure spectroscopy. J. Environ. Qual. **42**(6), 1763–1770 (2013)
148. SAS Institute: Statistical Analysis Software. Release 9, Cary, NC. (2003)
149. Peltre, C., Thuriès, L., Barthès, B., Brunet, D., Morvan, T., Nicolardot, B., Parnaudeau, V., Houot, S.: Near infrared reflectance spectroscopy: a tool to characterize the composition of different types of exogenous organic matter and their behaviour in soil. Soil. Biol. and Biochem. **43**(1), 197–205 (2011)
150. Calderón, F., Haddix, M., Conant, R., Magrini-Bair, K., Paul, E.: Diffuse-reflectance fourier-transform mid-infrared spectroscopy as a method of characterizing changes in soil organic matter. Soil Sci. Soc. Am. J. **77**(5), 1591–1600 (2013)
151. Ellerbrock, R.H., Gerke, H.H., Böhm, C.: In situ DRIFT characterization of organic matter composition on soil structural surfaces. Soil Sci. Soc. Am. J. **73**(2), 531–540 (2009)
152. Smidt, E., Meissl, K., Schwanninger, M., Lechner, P.: Classification of waste materials using Fourier transform infrared spectroscopy and soft independent modeling of class analogy. Waste Manage. **28**(10), 1699–1710 (2008)
153. Du, C., Zhou, J., Wang, H., Chen, X., Zhu, A., Zhang, J.: Determination of soil properties using Fourier transform mid-infrared photoacoustic spectroscopy. Vib. Spectro. **49**(1), 32–37 (2009)
154. Changwen, D., Guiqin, Z., Jianmin, Z., Huoyan, W., Xiaoqin, C., Yuanhua, D., Hui, W.: Characterization of animal manures using mid-infrared photoacoustic spectroscopy. Bioresour. Technol. **101**, 6273–6277 (2010)
155. Bekiaris, G., Bruun, S., Peltre, C., Houot, S., Jensen, L.S.: FTIR-PAS: A powerful tool for characterising the chemical composition and predicting the labile C fraction of various organic waste products. Waste manag. (New York, N.Y.) **39**, 45–56 (2015)
156. Specht, D.F.: Probabilistic neural networks. Neural Netw. **3**(1), 109–118 (1990)
157. Malley, D.F., McClure, C., Martin, P.D., Buckley, K., McCaughey, W.P.: Compositional analysis of cattle manure during composting using a field-portable near-infrared spectrometer. Commun. Soil Sci. Plant Anal. **36**(4–6), 455–475 (2005)
158. McGrath, J.M., Sims, J.T., Maguire, R.O., Saylor, W.W., Angel, C.R., Turner, B.L.: Broiler diet modification and litter storage. J. Environ. Qual. **34**(5), 1896–1909 (2005)
159. García-Sánchez, F., Galvez-Sola, L., Nicolás, J.J.M., Muelas-Domingo, R., Nieves, M.: Using near-infrared spectroscopy in agricultural systems. In: Kyprianidis, K. G., and Skvaril, J. (eds) Developments in Near-Infrared Spectroscopy, p. Ch. 05, InTech, Rijeka (2017)

160. Chen, L., Xing, L., Han, L.: Review of the application of near-infrared spectroscopy technology to determine the chemical composition of animal manure. J. Environ. Qual. **42** (4), 1015–1028 (2013)
161. Asai, T., Shimizu, S., Koga, T., Sato, M.: Quick determination of total nitrogen, total carbon, and crude ash in cattle manure using near-infrared reflectance spectroscopy. Nippon Dojo Hiryogaku Zasshi **64**, 669–675 (1993)
162. Millmier, A., Lorimor, J., Hurburgh Jr., C., Fulhage, C., Hattey, J., Zhang, H.: Nearinfrared sensing of manure nutrients. Trans. ASAE **43**, 903–908 (2000)
163. Malley, D.F., Yesmin, L., Eilers, R.G.: Rapid analysis of hog manure and manure-amended soils using near-infrared spectroscopy sponsoring organization: PDK projects. Inc. Soil Sci. Soc. Am. J. **66**(5), 1677–1686 (2002)
164. Blanco, M., Villarroya, I.: NIR spectroscopy: a rapid-response analytical tool. Trends Anal. Chem. **21**(4), 240–250 (2002)
165. Laporte, M.F., Paquin, P.: Near-infrared analysis of fat, protein, and casein in cow's milk. J. Agric. Food Chem. **47**, 2600–2605 (1999)
166. Lee, S.J., Jeon, I.J., Harbers, L.H.: Near-infrared reflectance spectroscopy for rapid analysis of curds during cheddar cheese making. J. Food Sci. **62**(1), 53–56 (1997)

Chapter 9
Sludge

Patricia Arlabosse, Ange Nzihou, Stewart Oakley, Martial Sauceau, Christelle Tribout, Fei Wang and Yaqian Zhao

Abstract Water treatment residuals, sewage sludge and dredged sediments, hereinafter referred to as sludge, share common features, like inherent high moisture content, high organic and mineral pollutant load and above all a pasty consistency. Several recovery opportunities exist. However, sludge composition is closely correlated with human activity and the design of the recovery processes requires knowledge of physical, thermal or biological characteristics of the sludge that will be processed, often even before the specific sludge exists. Therefore, sludge characterization is a key challenge for the process design, control and optimization and further valorization. The present chapter is divided into four parts dedicated to the composition, the material properties, the dynamic processing properties and the environmental assessments, respectively. Standard tests methods for the characterization of solids and water distribution in sludge are introduced first. Then, the fate of organic constituents, inorganic nonmetallic and metallic constituents as well as pathogenic organism. In the second part, standard and state-of-the-art methods

P. Arlabosse (✉) · A. Nzihou · M. Sauceau
RAPSODEE CNRS, UMR-5302, Université de Toulouse, IMT Mines Albi, Campus Jarlard, 81013 Albi Cedex, 09, France
e-mail: Patricia.Arlabosse@mines-albi.fr

S. Oakley
Department of Civil Engineering, California State University, Chico, CA 95929-0930, USA

C. Tribout
UPS, INSA, LMDC, Université de Toulouse, 135 Avenue de Rangueil, 31077 Toulouse Cedex 4, France

F. Wang
College of Energy Engineering, Zhejiang University, Yuquan Campus, 38 Zheda Road, Hangzhou 310027, China

Y. Zhao
School of Civil Engineering, UCD Dooge Centre for Water Resource Research, University College Dublin, Belfield, Dublin 4, Ireland

State Key Laboratory of Eco-Hydraulic Engineering in Arid Area, Xi'an University of Technology, Xi'an 710048, People's Republic of China

A. Nzihou (ed.), *Handbook on Characterization of Biomass, Biowaste and Related By-products*, https://doi.org/10.1007/978-3-030-35020-8_9

for the characterization of chemical, physical and thermal properties of sludge are described. Dynamic tests, required for the selection and design of conventional processes for sludge concentration and conveying, are detailed in the third part. Finally, methods for assessing the environment impacts of sludge and secondary raw materials derived from sludge are addressed.

9.1 Introduction

Water is essential to human life. Treatment and purification are indispensable to produce potable water for human consumption (referred to as drinking water) or to fulfill the requirements of medical, pharmacological, chemical and industrial applications (referred to as process water). However, the reliable and safe water production for human beings is at the cost of the pollutants transference into and the generation of water treatment residuals (WTRs).

When water is of no further immediate value to the purpose for which it was used or in the pursuit of which it was produced, it has to be collected and treated in wastewater treatment plants (WWTPs) prior to its release to the environment. Again, health and environment protection is at the cost of pollutants transference into and the generation of municipal and industrial sewage sludge.

The treated wastewater is then discharged into a water recipient (mostly in rivers). As water flows toward the ocean, sediments derived from erosion of the watershed area and banks, surface runoff or industrial and human discharges are carried by the draught and end up settling in slow-moving area and ports. Large volumes of sediments are dredged each year in order to maintain river traffic and harbor activities.

In the following, water treatment residuals, sewage sludge and dredged sediments will be collectively referred to as sludge.

Sludge processing and disposal represent one of the most complex technical problems facing engineers. A variety of recovery opportunities exist. However, the inherent high moisture content and high pollutant load of sludge require appropriate chemical, biological or thermal treatments to minimize its impact on the ecosystem. Knowing the composition and basic physicochemical and thermal properties of the material that will be processed is critical for process design. Unfortunately, sludge composition is closely correlated with human activity and varies depending on location, origin, time of the year, implemented processes (if any), aging etc. Therefore, sludge characterization is a key concern for process design, control and optimization and for sludge valorization.

Considering the biological activity of the microorganisms inherently present in sludge, preserving the integrity of the sample from the time of sampling to the time of analysis is critical. Recommendations for sample collection, handling, storage and subsampling are provided in the first paragraph.

Reducing sludge volume to reduce transportation costs and promote a better valorization is often the primary concern. To this end, characterizing the various categories of solids and the water distribution in sludge is frequently the starting point of any recovery strategy. A further concern is related to potential health risks of disposal options with regard to the high pollutant load of sludge (including nonmetallic and metallic constituents and pathogenic organisms). Standard tests methods for characterizing the composition are addressed in the second part of the chapter.

The third paragraph is dedicated to standard and state-of-the-art methods for the characterization of basic intrinsic physical, chemical and thermal material properties, the knowledge of which is essential for proper process design and operation. Most of the material properties depend only on the composition while others can be influenced by the physical consistency and structure of the material (like, for instance, the thermal conductivity of partially dried samples).

Sludge consistency may vary from liquid, to semi-solid and solid state. Besides, shearing can enhance sludge stickiness. This raises severe technical issues for handling, pumping and conveying the sludge. Available dynamic tests for measuring sludge's flowability, solidity and plasticity are described in the fourth section. New methods under development for characterizing sludge cohesion and adhesion are also introduced.

Finally, standard and state-of-the-art methods for assessing the environment impacts of secondary raw materials derived from dredged sediments and sludge are outlined.

9.2 Sampling, Handling, Storage and Subsampling Procedures

Preparing a small test sample that is representative of the original material is required for all the tests described in the following. Reference standards for sample collection, handling, storage and subsampling are listed in Table 9.1.

Sludge may contain pathogenic organisms such as leptospirosis, fungal spores and chemical contaminants. Appropriate personal protection equipment (respiratory protective masks, safety glasses, protective gloves …) should be worn to reduce human exposure, especially eyes and mucous membranes, to these contaminants. The use of the ventilation system is recommended, for instance when drying the samples.

Table 9.1 Standards related to collection, handling, storage and subsampling

Titles	References
ISO 5667-13:2011 Water quality—Sampling—Part 13: Guidance on sampling of sludges	[1]
ISO 5667-15:2009 Water quality—Sampling—Part 15: Guidance on preservation and handling of sludge and sediment samples	[2]
EN 16179 Sludge, treated biowaste and soil—Guidance for sample pretreatment	[3]

Sludge might biodegrade, volatilize, oxidize or be reduced. Care should be taken to retain the integrity of the collected sample. Preserve the sample immediately upon collection, in suitable containers and at low temperature. Most of the time, polyethylene, polypropylene, polycarbonate and glass containers can be used. However, care should be taken to physical/chemical interaction between the container and the analyte, for instance between polyethylene flasks and some heavy metals such as mercury. Biological activity, leading to methane production and potential pressure build-up, is almost unavoidable. Filling the container to a maximum of 80% of its capacity is strongly advised, except if analyses of volatile compounds are required. In such a case, to minimize losses by volatilization, the sample should be collected directly into the laboratory container with no headspace, the container sealed, refrigerated and analyzed as soon as possible. Transportation within 24 h at 5 ± 3 °C and manual release of pressure are recommended. Frozen at −18 °C can also be used but fast frozen can cause cellular disruption, which might bias the results of some analyses. Furthermore, microbial activity can only be determined without fixation by drying or frozen [2].

Storage duration and conditions depends on the analyte(s) to be characterized. A temperature of 3 ± 2 °C in air tight and dark containers is advised. A storage period of 4–14 days is often the maximum recommended, except for pH (1 day). For volatile organic compounds, storage time should be kept to minimum before the extraction but extracts may then be stored for up to one month at 3 ± 2 °C. As a general rule, the authors recommend to periodically repeat one or two experiments using the same operating conditions to check the influence of aging on the investigated property.

When preparing a small test sample, pretreatments are often required [3]. Extraneous materials shall be removed and recorded. Homogenization by mixing is advised for subsampling a liquid sludge. If demixing is likely to occur, subsample should be taken during the mixing process. For solid sludge, careful mixing and quartering is recommended. For some analytes such as trace elements, moderately volatile compounds and AOX, the required minimum mass of the sample to be used depends on the maximum particle size (expressed as the diameter at which 95% of the population has a diameter less than this value, D_{95}—please refer to Sect. 9.4.1.1 for the description of the analytical method).

9.3 Constituents of Sludge

Sludge is a heterogeneous mixture of microorganisms, mineral particles, colloids, organic polymers, cations, fibers, etc. with a liquid, semi-solid or pasty consistency. Characterization of the solids is often the starting point for any analytical investigation and is essential to select the appropriate recovery option.

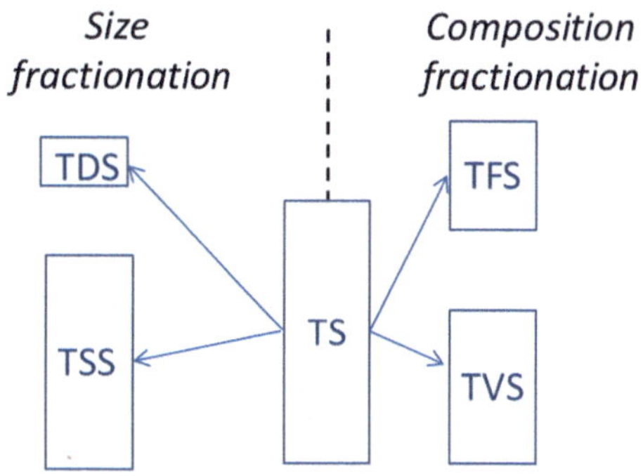

Fig. 9.1 Categories of solids in wastewater and sludge (TS: Total solids, TSS: Total suspended solids, TDS: Total dissolved solids, TVS: Total volatile solids, TFS, Total fixed solids)

9.3.1 Solids

Solids in sludge can be classified as floating/settleable, dissolved/suspended or volatile/fixed [4]. Figure 9.1 illustrates the interrelationships between the various fractions. Analytical methods for their characterization are described below.

9.3.1.1 Total Solids

General Discussion

Total solids (TS) describe the residue remaining after a wet sample has been dried to a constant weight in an oven at 103–105 °C. The difference in mass between the wet sample and the dried residue is used to compute the total solids content (TS), as follows [5, 6]:

$$TS = \frac{(m_{r+d} - m_d)}{(m_{ws+d} - m_d)} \times 100 \quad (9.1)$$

where

m_{ws+d} is the weight of the dish containing the wet sample, in grams
m_d is the weight of the empty dish, in grams
m_{r+d} is the weight of the dish containing the dried residue, in grams
TS is expressed in weight percent. All values shall be rounded to the nearest 0.1%.

For the wet basis moisture content (W_{wb}), expressed as a percentage, and the dry basis moisture (W_{db}), expressed in g of water per g of total solids, the following relationships apply:

$$W_{wb} = 1 - TS \quad (9.2)$$

Table 9.2 Standards related to TS characterization

Titles	References
EN 12880:2000 Characterization of sludges—Determination of dry residue and water content	[5]
Standard Methods for the Examination of Water and Wastewater—Method 2540.G Total, Fixed, and Volatile Solids in Solid and Semisolid Samples	[6]

$$W_{db} = \frac{1}{TS} - 1 \quad (9.3)$$

Reference standards for TS characterization are listed in Table 9.2.

Apparatus

Measurement [5, 6] requires:

- An evaporating dish made of porcelain, platinum or high-silica glass
- A drying oven ventilated, thermostatically controlled, for operation at 105 °C
- An analytical balance
- A desiccator provided with a fresh and powerful desiccant (silica gel or phosphorus pentoxide for instance)
- A steam bath, for fluid samples only.

Procedure

The two steps procedure requires the preparation of the dish before the analysis of the samples as follows:

Preparation of the evaporating dish

- Heat a clean and empty dish to the oven temperature, T_{oven}, for a given duration, $t_{heating}$
- Store the dish and cool it down to room temperature in a desiccator until needed
- Weigh the dish immediately before use (m_d).

Samples analysis

- Weigh an appropriate sample mass in the evaporating dish (m_{ws+d})
- Put the dish containing the wet sample in the oven at T_{oven} and dry the sample overnight
- Cool the dish in a desiccator down to the balance temperature
- Weigh the dish and its contents for the first time

Table 9.3 Specific conditions according to standards

	Reference [5]	Reference [6]
T_{oven}	105 ± 5 °C	103–105 °C
$t_{heating}$	At least 30 min	1 h
Appropriate sample mass	Amount that will yield a dry residue with a targeted weight of at least 0.5 g	25–50 g of wet sludge
Equilibrium criteria	Weight change for 1 extra hour of drying less than 0.5% of previous weigh or 2 mg, whichever is the largest	Weight change for 1 extra hour of drying less than 4% of previous weigh or 50 mg, whichever is less
Scale's accuracy	At least 1 mg	At least 10 mg

- Repeat cycle of drying for 1 h, cooling and weighing until a constant mass (m_{r+d}) is obtained for the container and its contents.

The specific requirements of the standards are reported in Table 9.3.

Remarks:

1 If volatile solids (please refer to Sect. 9.3.1.3) are to be subsequently measured, the dish must be heated to 550 °C for $t_{heating}$ in a muffle furnace.
2 When the moisture content of the sludge sample is high (i.e. for fluid samples able to flow more or less readily), as much water as possible should be evaporated to dryness using a steam bath to avoid matter losses through splattering. Then the remaining sample has to be dried in the oven for 1 h.

Interferences

Weight losses due to ammonium carbonate degradation and organic matter volatilization may occur, depending on the sludge origin, the temperature and the heating time, and skew the results.

Water-trapping crust may appear during drying. A glass rod can be weighted with the evaporating dish and used to mix the sludge and prevent crust formation during the drying process.

Unavoidably, evaporation occurs at the surface of the wet samples during weighing. Make all weighings quickly to limit this weight loss.

Dry residues are often hygroscopic. Limit the contact time with air to avoid moisture absorption on the solids.

Table 9.4 Typical TS content after conventional sewage sludge treatments

Unit operation	Typical TS content (wt%)
Clarification	0.5–3
Thickening	2–10
Anaerobic digestion	2–5
Mechanical dewatering	18–35
Thermal drying	65–95

Comment

Typical TS values after common solids treatments (see [4]) are reported in Table 9.4.

9.3.1.2 Suspended Solids

In regards to size, total solids in liquid sludge can be classified as dissolved, colloidal and suspended [7]. Total suspended solids (TSS) refer to total solids in a separate phase from the solution, namely the portion of total solids that can be trapped by a filter with a specified pore size, and the total filterable solids (TFS), the portion that passes through the filter. The TSS test quantifies the particulate solids, excluding soluble and part of the colloidal particles (depending on the pore size of the filter). Filtration through glass fibre filters is the reference method. However, in case of filter plugging or for samples with high total suspended solids (TSS) content, the filtration duration increases. When this duration exceeds 30 min, the centrifugation method is recommended [8].

Filtration Through Glass Fibre Filters [9, 10]

General discussion

The liquid sample is filtered through a glass fibre filter under vacuum. The filter with the residue trapped is dried in an oven at 105 °C and the residue mass is determined by weighing. The total suspended solids (TSS) content, expressed in mg/L, is computed as follows [9, 10]:

$$TSS = \frac{(m_{r+f} - m_f)}{V_s} \times 1000 \tag{9.4}$$

where

m_{r+f} is the weight of the filter with the dried trapped residue, in grams
m_f is the weight of the filter, in grams
V_s is the volume of the sample, in mL.

Table 9.5 Standards related to TSS characterization by filtration

Titles	References
EN 872:2005 Water quality—Determination of suspended solids. Method by filtration through glass fibre filters	[8]
Standard Methods for the Examination of Water and Wastewater—Method 2540.D Total Suspended solids dried at 103–105 °C	[9]
ISO 11923:1997 Water quality—Determination of suspended solids by filtration through glass-fibre filters	[10]

If the sample has been weighted, 1 g is considered to be equal to 1 mL.

The total dissolved solids content (TDS), also called total filterable solids, is given by [9, 10]:

$$TDS = TS - TSS \tag{9.5}$$

TDS can be experimentally determined by drying the filtrate volume in the oven at 105 °C until a steady mass is achieved.

Reference standards are listed in Table 9.5.

Apparatus

Experiments require [9, 10]:

- A suction filtration apparatus
- A circular glass fiber filter, without binder, having a mass per unit area between 50 and 100 g/m^2
- A drying oven ventilated, thermostatically controlled, for operation at 105 °C
- A drying rack or Petri dish
- An analytical balance with an accuracy of at least 0.1 mg.

Procedure

The two steps procedure requires the preparation of the filter before the analysis of the samples as follows:

Filter preparation

- Insert the glass fiber filter paper in the filtration apparatus
- Apply vacuum and wash the filter with distilled water
- Continue suction to remove all traces of water
- Turn off vacuum
- Remove the filter and dry it for 1 h at 105 °C in an inert weighing dish
- Cool in desiccator to room temperature and weigh
- Repeat cycle of drying, cooling and weighing until a constant weight is obtained
- Store it in desiccator until use.

Sample analysis

- Weigh the filter (m_f) and place it in the funnel, the smooth side down
- Connect the filtration flask to the vacuum line
- Wet the filter with a small volume of distilled water to seal it
- Homogenize the sample, measure an appropriate sample volume (V_s)
- Filter the sample
- Rinse
- Continue suction until the filter looks dry
- Turn off vacuum and remove the filter
- Dry it in an oven at 105 °C, cool down to room temperature in a desiccator and weigh the filter with the dried trapped residue, until a constant mass (m_{r+f}) is obtained.

The specific requirements of the standards are reported in Table 9.6.

Remarks:

1 Read the sample volume with an accuracy of at least 2%. If a measuring cylinder is used, weigh sample having a volume less than 25 mL. Usually, the sample volume poured in the measuring cylinder ranges from 25 to 100 mL, depending on the concentration.
2 The lower limit of determination for TSS is 2 mg/L. No upper limit is specified. However, centrifugation is the most appropriate method for high TSS values, usually beyond 10 g/L.

Table 9.6 Specific conditions according to standards

	Reference [8]	Reference [9]	Reference [10]
T_{oven}	105 ± 2 °C	103–105 °C	105 ± 2 °C
Volume for filter preparation	150 mL	3 successive 20 mL washes	150 mL
Appropriate sample volume	Volume that will yield a dry trapped residue with a targeted weight of 5–50 mg	Volume that will yield a dry trapped residue with a targeted weight of 2.5–200 mg	Volume that will yield a dry trapped residue with a targeted weight of 5–50 mg
Volume for the rinsing process	20 mL for filter washing + 20 mL for funnel rinsing	3 successive 10 mL washes for filter washing	20 mL for filter washing + 20 mL for funnel rinsing
Samples with dissolved solids	For TDS > 1 g/L, use 3 additional 50 mL volumes to rinse the filter	Additional washings may be required	For TDS > 1 g/L, use 3 additional 50 mL volumes to rinse the filter
t_{drying}	At least 1 h with a maximum of 14 to 16 h	1 h	1 or 2 h
Equilibrium criteria	–	Weight change less than 4% of the previous weighing or 0.5 mg, whichever is less	–

3 As the filter characteristics depends on its origin, provide the filter references and the name of the manufacturer in the test report.

Interferences

Values of TSS depend, amongst other, on the type of the filter paper and its pore size. Glass fiber filters with nominal pore size varying from 0.45 to 2.0 μm are commonly used [4]. The lower the pore size, the higher the TSS content.

Characteristics of the particles (size and nature), storage duration and conditions as well as pH may have an influence on the TSS values. It is recommended to quantify TSS within four hours following material sampling. Otherwise the sample has to be stored away from light at 1–5 °C for a recommended maximum of two days. Samples should not be frozen.

The presence of oil or any other non-miscible organic liquid, likely to adsorb on the filter and to partially desorb during drying, can disturb the test result. It is advisable to wash the sample with ethanol and then hexane prior to drying it at 105 °C. This pre-treatment shall be mentioned in the test report.

Even if it has no fundamental significances, the TSS test is used routinely to assess the performances of conventional treatment processes like anaerobic digestion for instance.

Comment

A check test with a microcrystalline cellulose suspension, used as reference material, is recommended [8, 10]. Prepare a 500 mg/L reference suspension, using oven dried TLC grade microcrystalline cellulose and distilled water. From this suspension, prepare on a daily basis a 50 mg/L working suspension. Use 200 mL of the working suspension as a sample. The recovery shall be between 90 and 110%.

Centrifugation

General discussion

Liquid samples are centrifuged at high rotational speeds for several minutes. The supernatant liquor is discarded and the sludge pellets are collected and dried in an oven at 105 °C and the dried mass is determined by weighing. The total suspended solids (TSS) content, expressed in mg/L, is computed as follows [11]:

$$TSS = \frac{(m_{r+d} - m_d)}{V_s} \times 1000 \quad (9.6)$$

where

m_{r+d} is the weight of the evaporating dish with its content, in grams
m_d is the weight of the evaporating dish, in grams
V_s is the volume of the sample, in mL.

Table 9.7 Standard related to TSS characterization by centrifugation

Title	Reference
NF T90-105-2 Water quality—Determination of suspended solids. Method by centrifugation	[11]

Reference standard for TSS characterization by centrifugation is listed in Table 9.7.

Apparatus

Experiments require:

- Laboratory centrifuge
- Centrifuge bottles with capacity from 100 to 500 mL
- Drying oven ventilated, thermostatically controlled, for operation at 105 ± 2 °C
- Analytical balance, with an accuracy of at least 0.1 mg
- Inert weighing dish.

Procedure

The experimental procedure is as follows:

- Homogenize the sample and quickly pour the sample into the centrifuge bottles
- Fill around 80% of the bottles
- Centrifuge the samples at 4000–5000 rpm for 10 min [12]
- Remove the supernatant liquor and collect carefully all the sludge pellets
- Weigh an inert weighing dish (m_d) and dry all the sludge pellets at 105 °C in this dish overnight
- Cool in desiccator to room temperature and weigh
- Repeat cycle of drying, cooling and weighing until a constant weight is obtained
- Wash the bottles with 2 successive 20 mL washes
- Collect the washing water in the evaporating dish and dry the dish with its content in the oven at 105 °C
- Cool in desiccator to room temperature and weigh (m_{r+d})
- Repeat cycle of washing, drying, cooling and weighing until the weigh change does not exceed 0.5 mg [11].

Interferences

It is recommended to quantify TSS within four hours following material sampling. Otherwise the sample has to be stored away from light at 1–5 °C for a recommended maximum of one day. Samples should not be frozen.

The amount of interstitial water trapped in the sludge pellet may depend on the applied mechanical strain.

Refrigerated centrifuge or short centrifugation runs are recommended to limit the temperature rise and fat solubilization.

Comment

Without polymer conditioning, sludge pellet does not have a homogenous structure (a lower and fairly hard zone and an upper and extremely pasty zone with a high organic matter content), because the particles are graded according to their density. Collecting carefully all the sludge pellets is essential.

9.3.1.3 Volatile and Fixed Solids

Total volatile solids (TVS) describe those solids that can be volatilized and burned off when total solids (TS) are ignited at 550 °C and total fixed solids (TFS) the residue on ignition.

General Discussion

A dry sample is heated to a constant weight in a furnace at 550 °C. The mass loss resulting from sample heating, called the loss on ignition (LOI) in standards or more commonly referred to as total volatile solids, is expressed as a weight percentage of TS. The total volatiles solids (TVS) content, expressed as a percentage of TS, is computed as follows:

$$TVS = \frac{(m_{ds+d} - m_{ir+d})}{(m_{ds+d} - m_{d})} \times 100 \tag{9.7}$$

where

m_{ds+d} is the weight of the crucible containing the dried sludge, in grams
m_d is the weight of the empty crucible, in grams
m_{ir+d} is the weight of the crucible containing the residue on ignition, in grams.

All values shall be rounded to the nearest 0.1%.

The total fixed solids content (TFS), the residue on ignition, of the dry sample, expressed as a percentage of TS, is given by:

$$TFS = 100 - TVS \tag{9.8}$$

The reference standard is listed in Table 9.8.

Table 9.8 Standard related to TVS and TFS characterization

Title	Reference
Standard Methods for the Examination of Water and Wastewater—Method 2540. G Total, Fixed, and Volatile Solids in Solid and Semisolid Samples	[6]

Apparatus

Experiments require [6]:

- Flat bottom crucible made of porcelain, nickel, platinum or silica, 50 to 70 mm in diameter
- Muffle furnace or equivalent, for operation at 550 ± 25 °C
- Analytical balance
- Desiccator with a fresh and powerful desiccant (silica gel for instance)
- Spatulas.

Procedure

The two steps procedure requires the preparation of the dish before the analysis of the samples as follows:

Preparation of the crucible

- If the determination of TVS is carried out in the same crucible as the determination of TS, refer to Sect. 9.3.1.1 for the crucible preparation.
- If not, heat a clean and empty crucible at 550 °C for a given duration, $t_{crucible, heating}$
- Store the crucible and cool it down to room temperature in a desiccator until needed.
- Weigh the crucible immediately before use (m_d).

Sample analysis

- Weigh into the crucible a dried sludge sample (m_{ds+d})
- Put the crucible with the dried sludge in the muffle furnace at 550 °C for 1 h
- Cool the container and its content in a desiccator down to the balance temperature and weigh the container and its contents for the first time
- Repeat igniting for 30 min, cooling and weighing until a constant mass (m_{ir+d}) is obtained for the container and its contents.

The specific requirements of the standard are reported in Table 9.9.

Table 9.9 Specific conditions according to standards

	Reference [6]
$T_{crucible, heating}$	At least 1 h
Sample	Dried residue from the TS test (i.e. fraction recovered after drying 25–50 g of wet sludge)
Equilibrium criteria	Weight change for 1 extra hour of drying less than 4% or 50 mg, whichever is less
Scale's accuracy	At least 10 mg

Interferences

Unavoidable loss of volatile matter occurs during drying, reducing TVS. Chemically bound water can be released at 550 °C, contributing to TVS. Dry residues are often hygroscopic. Appropriate storage conditions are required to prevent moisture uptake on the dried sludge before weighing. Limit the contact time with moist air to avoid moisture absorption on the solids during weighing.

For a number of reasons, TVS is considered as a rough approximation of the amount of organic matter (OM) in the sample. Loss of volatile inorganic substances and salts (sulphides, iron or other metals) can occur at 550 °C while, at the same time, oxidation of organocalcium or organometallic complexes, for instance, requires temperature well above 550 °C. For limed sludge, carbon dioxide or sulphuric oxides formed during the combustion may react with calcium oxide or hydroxide to yield calcium carbonate that breaks down at higher temperature. Better characterization of organic matter can be made by such tests as total organic carbon, BOD and COD.

Comment

Between 550 and 900 °C, volatile solids mainly consist of CO_2 produced by the decomposition of carbonates. To roughly quantify the carbonate content, total solids are sometimes ignited at 900 °C, the difference with TFS remaining after ignition at 550 °C being the carbonate content.

9.3.1.4 Typical Data

Fifteen sewage sludge samples were collected in fourteen French WWTPs, which all have a capacity exceeding 30,000 population equivalents. This sampling represents a large variety of wastewater treatment processes and covers the main types of sewage sludge that can be encountered in French WWTP: from highly offensive sludge coming from the primary settling tank (S1 and S2) to stabilized sludge stemming from the mesophilic anaerobic digester (S10–S12) or from alkaline stabilization with lime (S13–S15), by way of chemical (S3) and extended aeration (S4–S9) sludge. All the samples were taken after the mechanical dewatering stage and analyzed according to the Method 2540.G [6] in the RAPSODEE Research Centre.

TS, TVS and TFS contents as well as the mass loss on TFS between 550 and 900 °C, Δ(TFS), are presented in Table 9.10. As expected, both the wastewater treatments and the dewatering processes implemented have an influence on the sludge composition. For instance, higher dewatering performances (TS) are achieved with filter presses than with belt-filter presses and centrifuges. Liming also improves the solid/liquid separation. And finally, stabilization treatments decrease TVS, the alkaline stabilization being much effective than the mesophilic anaerobic digestion. It is therefore difficult to make general statements about an average TVS/TFS composition for all type of sludge. However the orders of magnitude of TS, TVS and TFS contents reported in Table 9.10 are quite conventional [4].

Δ(TFS) is usually lower than 3%, except here for S10 and S11. Both sludge samples have undergone an alkaline stabilization with lime. The higher mass losses observed for S10 and S11 result from the decomposition of carbonates under the influence of temperature, as emphasized previously. These carbonates are produced

Table 9.10 Influence of wastewater treatment processes on sewage sludge composition

Sample	Type of sewage sludge	Dewatering equipment	TS (wt %)	TVS (wt% TS)	TFS (wt% TS)	ΔTFS (wt% TS)
S1	Primary sludge	Belt-filter press	29.3	66.9	30.2	2.8
S2	Primary sludge	Centrifuge	26.7	79.1	19.8	1.1
S3	Mixture of chemical sludge (70%) with biological sludge (30%)	Centrifuge	28.4	68.0	30.2	1.8
S4	Extended aeration sludge without primary settling	Centrifuge	18.7	75.3	23.6	1.1
S5	Extended aeration sludge without primary settling	Centrifuge	21.8	75.9	23.5	0.6
S6	Extended aeration sludge without primary settling	Centrifuge	17.4	69.5	28.8	1.7
S7	Extended aeration sludge without primary settling	Centrifuge	19.8	54.0	44.1	1.9
S8	Extended aeration sludge without primary settling	Centrifuge	20.5	75.4	23.6	1.0
S9	Extended aeration sludge without primary settling	Centrifuge	17.0	75.6	23.2	1.2
S10	Mixture of primary and biological sludge anaerobically digested	Filter press	25.1	59.2	38.4	2.4
S11	Mixture of primary and biological sludge anaerobically digested	Belt-filter press	19.0	67.6	30.7	1.8
S12	Anaerobically digested extended aeration sludge without primary settling	Centrifuge	17.4	66.7	30.9	2.5
S13	Sludge S10 alkalinally stabilized	Filter press	33.4	46.3	46.4	7.3
S14	Alkalinally stabilized activated sludge with primary settling	Filter press	28.4	45.8	47.5	6.6
S15	Alkalinally stabilized extended aeration sludge without primary settling	Filter press	33.1	34.0	48.2	17.9

by the carbonation reaction between the slaked lime ($Ca(OH)_2$) and the carbon dioxide in the atmosphere during the stabilization and maturation treatments. Thus, this mass loss must be taken into account when determining TFS, especially for alkalinally stabilized sludge.

9.3.2 *Moisture Distribution*

Even with modern optical means, it is quite difficult to get an accurate picture of sludge structure. Sludge is made up of diverse solids suspended or dissolved in a water continuum. Proximity to the solids, especially during dewatering and drying, changes the behavior of the water molecules. As a result, all the water within the sludge does not have the same properties in terms of viscosity, density, vapor pressure etc. All of this tremendously affects the performances of the dewatering and drying facilities.

Two types of water were initially introduced: the *free* water refers to the water fraction which is not influenced by the solid particles while the *bound* water is related to the water fraction whose properties are modified by the presence of the solids. More complex discrete water classifications appeared in the mid-1990s [13–16]. However, no consensus has yet emerged in the scientific literature. Besides, as several properties changes simultaneously, different techniques have been used to characterize the moisture in sludge (Table 9.11) and data are often contradictory. More information on the most commonly used analytical methods is available in [16].

Rather than a discrete water classification with unclear thresholds, the approach based on the measurement of the water binding energy, using sorption isotherms tests [16, 17], seems more appropriate. These tests rest on the thermodynamic

Table 9.11 Techniques available to characterize the moisture in sludge

Techniques
Drying methods
Vacuum drying
Atmospheric drying
Freezing method
Dilatometry
Mechanical methods
Centrifugal settling test
Expression test
Filtration test
Thermal analysis
Thermogravimetry (TGA)
Differential scanning calorimetry (DSC)
Combined TGA/DSC
Others
Water vapor sorption
Nuclear magnetic resonance

concept of water activity, as desorption isotherms represent the change in moisture content of a sample at equilibrium and at a specified temperature versus the change in water activity.

For liquid-vapor equilibrium, thermodynamics defines the activity a_i of a species i in a mixture as follows:

$$a_i = \gamma_i x_i = \frac{\varphi_i y_i P}{f_{i L.pure}} \tag{9.9}$$

where

γ_i is the activity coefficient (−)
x_i is the molar fraction of the species i in the liquid phase, in mol/mol
φ_i is the fugacity coefficient (−)
P is the total pressure, in Pa
$f_{i.\ Lpure}$ is the fugacity of the species i as a pure liquid in equilibrium with its vapor, in Pa.

Assuming that the species i adsorbed on a solid is pure, the corresponding liquid molar fraction x_i is equal to 1 and thus the activity is equal to the activity coefficient γ_i. As the vapor phase is usually assumed to behave like an ideal gas, the fugacity coefficient φ_i is then taken equal to 1. Furthermore, at moderate pressure, a conventional assumption is that the fugacity of the pure liquid in equilibrium with its vapor is equal to the saturation pressure of the component at the operating temperature, $P_{sat.i.}$. Thus, the activity of the species i may be written:

$$a_i = \frac{y_i P}{P_{sat.i}} = \frac{P_i}{P_{sat.i}} \tag{9.10}$$

The water activity, a_w, can be defined as the ratio of partial pressure of water in the surrounding air to the saturation pressure of water at a given temperature. It should be noted that this is also the definition of the relative humidity.

Two main categories of methods [18, 19] exist for the measurement of water activity: the *manometric* methods in which the pressure above a known mass of sample is measured and the *gravimetric* methods in which the pressure above the sample is fixed and the mass measured. Ritzi [20] asserts that manometric methods were not suitable for materials containing large amounts of volatiles or those undergoing a respiration process, like sludge.

9.3.2.1 Saturated Salts Method

Among gravimetric methods, the saturated salts method is widely adopted as a standard [21].

General Discussion

The sample is put into contact with an atmosphere at controlled temperature and steady-state water vapor pressure. The sample is manually weighed until its mass remains constant. Once the equilibrium is reached, the equilibrium amount of water sorbed is determined. Having characterizing several equilibriums at constant temperature, the sorption curve $W_{db} = f(a_w)$ can be plotted.

Apparatus

Experiments require [21]:

- 25 * 25 mm weighing bottles with ground-in stoppers
- Petri dishes, able to provide sufficient place to accommodate 5 weighing bottles
- Trivets
- 1L glass jars sealed with standard glass covers and rubber seal rings
- A thermostatically controlled water bath with space for five sorption containers
- A desiccator
- An analytical balance with an accuracy of at least 0.1 mg
- Laboratory drying oven
- Salts of analytical grade purity
- Spatulas.

Procedure

The two steps procedure requires the preparation of the salts solutions before the analysis of the samples as follows:

Preparation of 10 aqueous solutions of pure salts

- Add the quantity of salt indicated in Table 9.12 to a glass jar (called the sorption container in the following)
- Add while stirring the reported volume of distilled water
- Close the sorption container and wait for one week
- Stir the solutions briefly once a day
- Repeat the procedure for all the salts.

Sample analysis

- Wash five weighing bottles with distilled water
- Dry them for 3 h at 105 °C in the drying oven and then cool them down to ambient temperature in the desiccator
- Weigh the empty bottles (m_b) to the nearest 0.1 mg
- Place one sample in the first weighing bottle and leave the lid open. Prepare four others weighing bottles, in order to have 5 replicates

Table 9.12 Nominal water activities for the 10 salts

Salt	Quantities		Nominal water activity at		
	Salt (g)	Distilled water (mL)	15 °C	25 °C	35 °C
Lithium chloride	150	85	0.1194 ± 0.0005	0.1105 ± 0.0005	0.1117 ± 0.0005
Potassium acetate	200	65	0.234 ± 0.003	0.2245 ± 0.0005	0.2161 (at 30 °C)
Magnesium chloride	200	25	0.3330 ± 0.0005	0.3300 ± 0.0005	0.3200 ± 0.0005
Potassium carbonate	200	90	0.43	0.4276 ± 0.0005	0.4317 (at 30 °C)
Magnesium nitrate	200	30	0.5587 ± 0.0005	0.5286 ± 0.0005	0.4972 ± 0.0005
Sodium bromide	200	80	0.61	0.58	0.5455 ± 0.004
Strontium chloride	200	50	0.7413 ± 0.0005	0.7083 ± 0.0005	0.6911 (at 30 °C)
Sodium chloride	200	60	0.76	0.75	0.75
Potassium chloride	200	80	0.8598 ± 0.0005	0.8426 ± 0.0005	0.83
Barium chloride	250	70	0.9107 ± 0.0005	0.9019 ± 0.0002	0.89

- Put the five bottles in a given sorption container
- Close the sorption container hermetically
- Put it in the thermostatically controlled water bath
- Repeat this protocol for the four remaining sorption containers.
- After a certain period of time, open the sorption containers, close the lids of the weighing bottles, cool them in a desiccator down to the balance temperature and weigh them
- Repeat contact with the atmosphere at controlled temperature and established constant vapor pressure, cooling and weighing until a constant mass is reached (m_{b+eq}). The weight should be considered as constant when the mass change over a given period (24 h for instance) is lower than 0.5 mg
- Dry the weighing bottles with their contents at 105 °C
- Cool them in the desiccator down to room temperature and weigh
- Repeat drying, cooling and weighing until a constant mass is reached (m_{b+d})
- Compute the dry basis moisture content (W_{db}).

Interferences

Sludge has to be stored away from light at 1–5 °C for a short period. Samples should not be frozen.

Long equilibration period during sorption measurement may lead to bacterial growth, especially in the lower range of moisture content, which consequently invalidates the results [22].

Radiative heat transfer between the jars and the environment, which would affect the temperature of the jar contents, shall be limited. For instance, the tops of the jars in the water-bath could be covered by a reflecting heat shield.

Comment

The experimental procedure was established in the four-year European COST 90 project on physical properties of food. Microcrystalline cellulose (MCC) has been selected as reference material. Results of the 10 laboratories, that applied the COST 90 procedure to MCC, were collected for the certification of the MCC water content when in equilibrium with the atmosphere above 10 specified salt solutions at 25 °C. Accordingly, MCC is often used as a standard. The mean adsorption isotherm curve determined for the ten water activities as well as the standard deviations for each point can be found in [23].

9.3.2.2 Dynamic Vapor System

As several days, or even weeks, may be required to achieve the equilibrium for materials having high initial moisture content with the saturated salts method, commercial water activity meters, such as the Dynamic Vapor System (DVS) described in Chap. 2, can also be used.

General Discussion

The principle is similar to that of the saturated salts method: the sample is put into contact with an atmosphere at controlled temperature and well-established water vapor pressure. However, the implementation is different. In the Dynamic Vapor System (DVS), the same sludge sample is exposed to a series of step changes in relative humidity and the mass change is monitored as a function of time with a Cahn type microbalance. The sample mass must be allowed to reach gravimetric equilibrium at each step change in humidity before progressing to the next humidity level. The desired desorption cycle as well as the equilibrium criteria are defined by the user.

Apparatus

The apparatus is described in Chap. 2.

Procedure

- Carefully check that the baseline drift, when recording the mass as a function of time, is lower than three times the resolution of the microbalance over a period of 20 s. This preliminary stage is useful for checking the dryness of the crucibles and the absence of static electricity
- Define the desired sorption cycle (desorption only or desorption/adsorption cycles) and the relative humidity values. Set the 2 equilibrium criteria
- The first criterion is a limiting value of the percentage of mass variation over a certain time, dm/dt, expressed in %/min
- The second criterion is the stage duration, which includes a minimum time during which the mass should remain stable and a maximum duration on the plateau at constant relative humidity
- Select an appropriate wet sample mass that will yield a dried residue with a targeted weight of 20 mg.

For sludge, typical values for the equilibrium criteria are twenty minutes for the minimum time and 2×10^{-3} %/min for the dm/dt criterion, which means that the equilibrium is considered when the mass variation will be less than 0.04 mg/min during at least 20 min. Due to the high moisture content of the sludge, a recommended value for the maximum duration on each plateau at each constant relative humidity is 500 min. However, above 80% relative humidity, longer equilibration times are requested and the maximum duration should be increased.

The commercial software, supplied with the apparatus, provides a flexible and powerful environment for quantitative analysis of data and rapid plotting of isotherm and mass plots, as illustrated in Fig. 9.2.

Interferences

Sludge has to be stored away from light at 1–5 °C for a short period. Samples should not be frozen.

Comment

After mechanical dewatering, the moisture content is usually high, typically around 80% (w.b) for centrifuged sewage sludge. Even with a maximum duration set to 2000 min, i.e. the maximum value allowed on the DVS2 instrument provided by

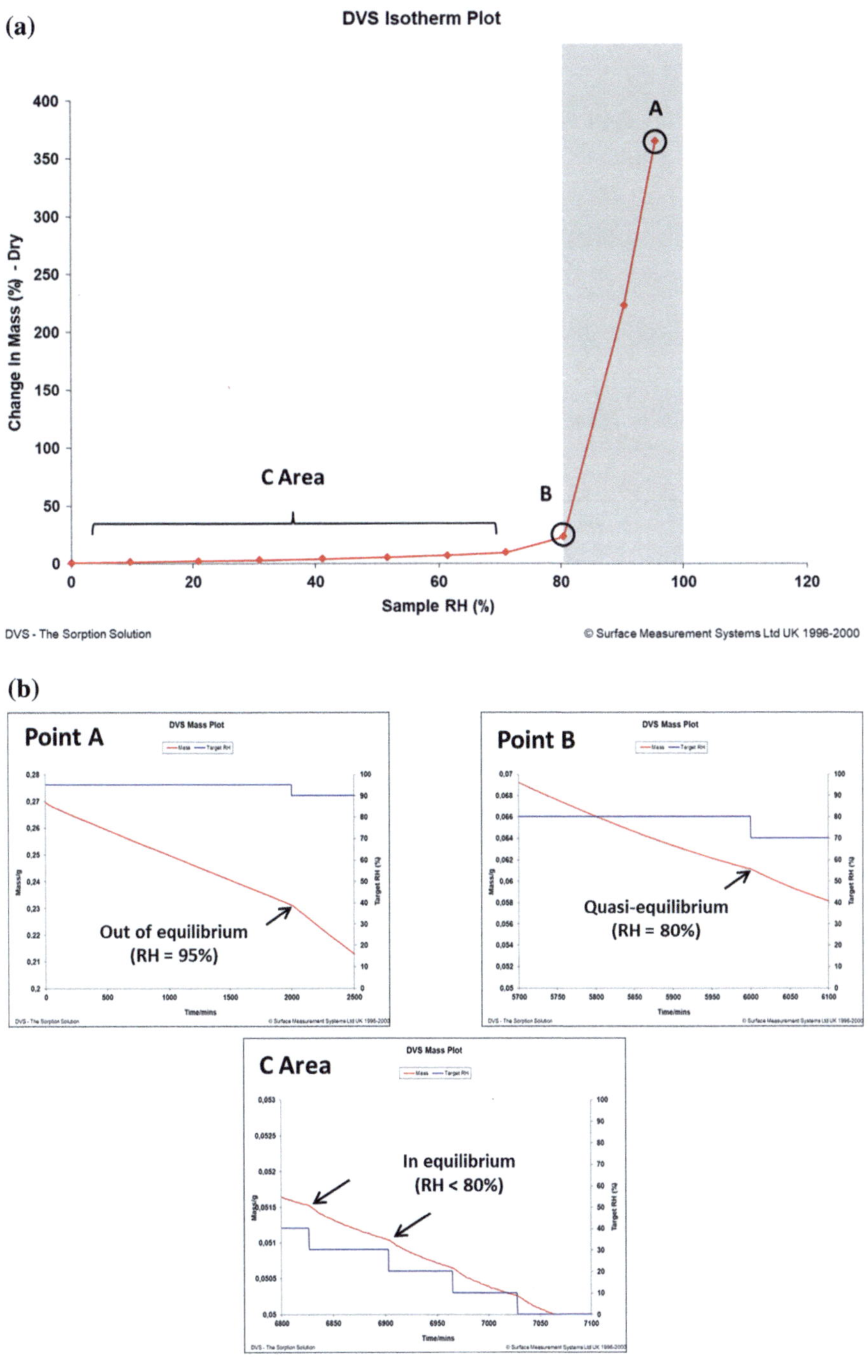

Fig. 9.2 Typical isotherm (**a**) and mass plots (**b**) for a dewatered sewage sludge (dm/dt = 2×10^{-3} %/min; t_{max} = 2000 min at 95, 90 and 80% relative humidity and t_{max} = 500 min for 0 < RH < 80%; t_{min} = 20 min)

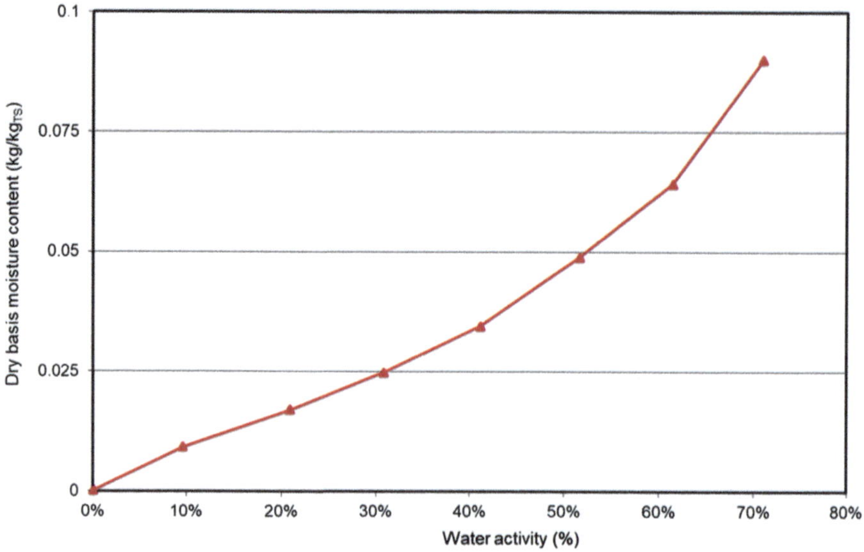

Fig. 9.3 Typical S-shape of a sludge desorption isotherm

Surface Measurement Systems, reaching a gravimetric equilibrium is not possible for the first two steps at 95 (point A) and 90% relative humidity, as can be seen clearly on the mass plots in Fig. 9.2b. At the end of these first two steps, the percentages of mass variation over a certain time were 7.1×10^{-3} and 1.2×10^{-2} %/min, respectively. Accordingly, these two apparent equilibrium points should be disregarded. However, setting these long steps at high relative humidity is essential for the rest of the analysis. The long drying time (6000 min in this case) allows reaching quasi-equilibrium at 80% relative humidity (point B) and true equilibrium at lower relative humidities (C area).

Usually desorption isotherms describe S-shape (Fig. 9.3), corresponding to type II in the classification of Brunauer et al. [24]. A large number of models are available in the literature for fitting the experimental data [17].

9.3.2.3 Heat of Sorption

Heat of sorption, ΔH_s, can be determined from sorption isotherm data, using the Clausius Clapeyron equation:

$$\frac{dP_v}{dT} = \frac{\Delta H_v + \Delta H_s}{T(V_v - V_l)} \tag{9.11}$$

where

P_v is the vapor pressure, in Pa
T is the temperature, in K
ΔH_v is the heat of evaporation of pure water, in J/mol
ΔH_s is the heat of sorption, in J/mol
V_v and V_l are the molar volumes of vapor and liquid respectively, in m³/mol.

Neglecting the volume of liquid, considering one mole of ideal gas and assuming ΔH_s constant over a small range of temperature, it comes:

$$ln(a_w) = -\frac{\Delta H_s}{R}\frac{1}{T} + cte \tag{9.12}$$

Accordingly, $\Delta H_s/R$ is the slope of the $ln(a_w) = f(1/T)$ line.

9.3.3 CHNSO Analysis

The predominant elements in the carbohydrates, fats and proteins composing the volatile solids of dry sludge are carbon (C), oxygen (O), hydrogen (H) and nitrogen (N). The approximate percentage of these elements can be determined by a technique known as ultimate (elemental or CHNSO) analysis.

9.3.3.1 General Discussion

CHNS analysis requires a complete combustion of the dried sample at high temperature (~ 1800 °C). After removal of the oxygen and conversion of nitrogen oxides to nitrogen gas, the evolved gas (CO_2, N_2, H_2O and SO_2) are analyzed. The mass percentage of carbon, hydrogen, nitrogen and sulphur contained in the sample is deduced from the concentration of each individual gas species. The oxygen content is either measured, knowing the CO concentration of the gases recovered after sample pyrolysis in a helium atmosphere at around 1000 °C, or calculated from a mass balance, as follows:

$$O = 100 - C - H - N - S - TFS \tag{9.13}$$

where

C, H, O, N, S and TFS are mass fractions expressed in weight percent on a dry basis (wt% TS).

9.3.3.2 Apparatus and Procedure

The apparatus is described in Chap. 2.

Protocols of both liquid and solid samples are detailed in Chap. 2. Equations to compute the results on 'as received basis' are also given in Chapter 2.

9.3.3.3 Interferences

Most of the time, moisture-free samples are used. However, organic matter volatilization may occur during drying and skew the results.

9.3.3.4 Comment

CHNS analyses were performed for the fifteen sewage sludge samples introduced in Sect. 9.3.1.4, using a NA 2100N/Protein analyzer (Thermoquest, Milan, Italy). The oxygen fraction was calculated from the mass balance, using Eq. (9.13).

Sewage sludge composition depends strongly on the origin of the wastewater, the type of processing to which they have been subjected, the aging of the solids, the location of the plant. Therefore, Table 9.13 provides typical CHNSO values. From this ultimate analysis, an empirical chemical formula of the volatile fraction can be determined. Considering 100 g of the sample, grams of each atom can be

Table 9.13 Typical CHNSO values in a sewage sludge

Sludge	C (wt% TS)		H (wt% TS)		N (wt% TS)		S (wt% TS)		O (wt% TS)	
	Mean	SD	Mean	SD	Mean	SD	Mean	SD	Mean	SD
S1	46.0	4.2	6.7	0.6	4.6	1.7	0.2	0.4	12.4	10.0
S2	43.7	1.2	6.5	0.2	5.0	3.2	0.3	0.1	24.8	6.9
S3	35.0	1.0	6.1	0.5	3.3	0.5	0.4	0.1	25.0	3.1
S4	38.7	0.1	5.7	0.1	6.3	0.2	0.4	0.0	25.4	0.7
S5	40.5	0.3	6.0	0.1	6.7	0.0	0.6	0.0	22.7	0.8
S6	35.3	0.3	5.2	0.1	6.2	0.3	0.0	0.0	24.5	1.1
S7	26.3	0.9	4.0	0.4	4.1	0.0	0.3	0.1	21.3	2.3
S8	49.0	1.8	5.0	0.4	7.2	0.1	0.2	0.2	15.1	3.7
S9	43.8	1.4	5.1	0.1	7.2	0.1	0.5	0.2	20.2	2.5
S10	27.0	0.4	3.7	0.1	2.9	0.1	0.9	0.2	27.1	1.4
S11	35.9	0.0	5.3	0.0	5.5	0.0	0.0	0.0	22.7	0.3
S12	31.0	1.0	4.9	0.1	4.7	0.2	0.4	0.7	28.2	3.0
S13	27.7	0.4	4.0	0.1	3.1	0.1	0.9	0.2	17.9	1.5
S14	33.9	16.8	3.6	0.4	5.0	2.1	0.1	0.2	9.8	28.0
S15	23.6	1.1	4.1	0.1	3.8	0.4	0.0	0.1	20.3	2.6

converted in molar amount, using the molecular weight. As the chemical formula is related to the ratios between the quantities of the elements, the molar ratios need then to be reduced to their lowest whole numbers.

9.3.4 Organic Contaminants

Organic contaminants in sludge include [25]:

- polycyclic aromatic hydrocarbons (PAHs),
- dioxin-like compounds, i.e. polychlorinated biphenyls (PCBs), polychlorinated dibenzo-p-dioxins (PCDDs) and polychlorinated dibenzofurans (PCDFs). PCDDs and PCDFs are collectively referred to as PCDD/Fs in the following,
- nonylphenole and nonylphenole ethoxylates with 1 or 2 ethoxy groups (NP/NPE)
- di(2-ethylhexyl)phthalate (DEHP)
- linear alkylbenzene sulphonates (LAS)
- and 'AOX', the so-called 'sum of halogenated organic compounds'.

Table 9.14 provides a classification of these organic substances according to their behaviour in soils. Limit values for concentrations of some organic

Table 9.14 Overview on the behaviour of organic substances of concern in soils

Substance	Human/mammalian toxicity	Ecotoxicity	Water solubility	Persistence	Concentration levels
AOX	–	–	–	–	H
LAS	M	Aquatic: H Terrestrial: M Bioaccumulation: H	H	M	H
DEHP	L	Aquatic: M to H Terrestrial: L Bioaccumulation: H	L	M	H
Nonylphenole	M	Aquatic: H Terrestrial: M Bioaccumulation: H	H	M	H
PAHs	H	Aquatic: H Terrestrial: H Bioaccumulation: H	L	H	H
PCBs	M	Aquatic: H Terrestrial: H Bioaccumulation: H	L	H	L
PCDD/Fs	H	Aquatic: H Terrestrial: H Bioaccumulation: H	L	H	L
TBT	H	Aquatic: H Bioaccumulation: H	M	H	H

Reprinted with permission from Andreas Prüeß [25]

Table 9.15 Maximum permissible limits in sewage sludge for land application according to French, German and Chinese legislations

Pollutant	Ceiling concentration (mg/kg$_{TS}$)		
	French regulation [26]	German regulation [27]	Chinese regulation [28]
7 PCBs[a]	0.8 for the sum of the 7 congeners	0.2 per congener	0.2 per congener
Fluoranthene	5	–	–
Benzo(b) fluoranthene	2.5	–	–
Benzo(a)pyrene	2	–	3
PCDD/Fs	–	10^{-4}	10^{-4}
AOX	–	500	500

[a]PCB 28, 52, 101, 138, 153, 180 + PCB 118 for France

compounds in sludge are already established in some Member States of the European Union for land application (Table 9.15). However, no restrictions apply under the EU and US regulatory frameworks.

Introduction of limiting values for dioxin-like compounds [29] and others substance classes [25] was considered in several countries but, up to now, the proposed thresholds were not implemented, land application of sewage sludge being not likely to pose a problem.

9.3.4.1 General Discussion

Characterization of organic contaminants in sludge requires a robust and sensitive three-step analytical procedure. Sample preparation includes extraction and clean-up. A variety of standard and novel extraction techniques (Table 9.16) are available [30].

The main purification and concentration treatments include sulphur removal using potassium hydroxide in ethanol or tetrabutylammonium hydrogen sulfate saturated with sodium sulfite and copper, solid phase extraction (SPE) using normal phase and gel permeation chromatography (GPC) for high-molecular weight compounds elimination. Gas chromatography (GC), coupled with mass spectrometry (MS), and liquid chromatography (LC) with mass selective detection (MS) or fluorescence detection (FD) are the two main instrumental techniques used for sample analysis. Reference standards related to organic contaminants characterization are listed in Table 9.17.

Table 9.16 Main techniques for organic compounds extraction in sludge

State of the art	Soxhlet Soxtec Ultrasound assisted extraction
Novel	Focused ultrasound assisted liquid extraction Pressurized liquid extraction Pressurized hot water extraction Microwave-assisted extraction Supercritical fluid extraction Matrix-solid phase dispersion

Table 9.17 Standards related to organic contaminants characterization

Titles	References
EN 15527:2008 Determination of polycyclic aromatic hydrocarbons (PAHs) in waste using gas chromatography mass spectrometry	[31]
Standard Methods for the Examination of Water and Wastewater—Method 8275A. Semi volatile organic compounds (PAHs and PCBs) in soils/sludge and solid wastes using thermal extraction/gas chromatography/mass spectroscopy (TE/GC/MS)	[32]
ISO 13914:2013 Soil quality—Determination of dioxins and furans and dioxin-like polychlorinated biphenyls by gas chromatography with high-resolution mass selective detection (GC/HRMS)	[33]
ISO/TS 13907:2012 Soil quality—Determination of nonylphenols (NP) and nonylphenol-mono- and diethoxylates—Method by gas chromatography with mass selective detection (GC-MS)	[34]
ISO 13913:2014 Soil quality—Determination of selected phthalates using capillary gas chromatography with mass spectrometric detection (GC/MS)	[35]
ISO/TS 13896:2012 Soil quality—Determination of linear alkylbenzene sulfonate (LAS)—Method by HPLC with fluorescence detection (LC-FLD) and mass selective detection (LC-MSD)	[36]
EN 16166:2012 Sludge, treated biowaste and soil—Determination of adsorbable organically bound halogens (AOX)	[37]
Standard Methods for the Examination of Water and Wastewater—Method 1650. Adsorbable organic halides by adsorption and coulometric titration	[38]

9.3.4.2 Apparatus and Procedure

Chapter 8 is dedicated to the characterization of dioxins and furans. Extraction, clean-up and detection techniques for the determination of priority and emerging contaminants in sewage sludge have been recently reviewed and compared by Zuloaga et al. [30]. State-of-the art extraction techniques involve long analytical and manual manipulation of the extracts. Pressurized liquid extraction is a popular method whose efficiency has been proven. For the detection, gas or liquid chromatography coupled to mass spectrometry is preferred.

9.3.4.3 Comments

Typical values of PAHs, NP/NPEs, DEHPs, dioxin-like congeners and LAS are reported in Tables 9.18, 9.19, 9.20, 9.21, 9.22.

9.3.5 Alkaline, Alkali Earth Metals and Transition Metals

Metallic constituents in sludge include alkaline metals (Na, K), alkali earth metals (Ca, Mg), transition metals (As, Cr, Cu, Co, Fe, Mn, Mo, and Ni, for example) and post-transition metals (B, Cd, Hg, Pb and Zn, for instance), according to the classification provided in Chap. 2. Some of them are essential to plant growth (Table 9.23) while others may adversely affect the receiving environment.

Heavy metals limits for land application of sewage sludge are reported in Table 9.24.

Table 9.18 Concentrations of PAHs (mg/kg_{TS}) in sewage sludge of various countries

Country	Year	Number of compounds summed	Number of samples	Range	Median
Denmark	1995	18	20	<0.01–8.5	–
Denmark	1993/1994	18	9	0.42–2.4	–
Norway	–	–	36	0.7–30	3.9
Sweden	–	6	–	–	1.6
Part of Germany	–	6	124	0.4–12.83	–
Part of Germany	–	16	88	0.25–16.28	–

Reprinted with permission from Andreas Prüeß [25]

Table 9.19 Overview of concentrations of nonylphenole and ethoxylates (in mg/kg_{TS}) in Scandinavian sewage sludge

Country	Year	Number of samples	Range	Median
Norway	–	36	22–650	136
Norway	1989	19	25–2298	189
Sweden	1993	23	23–171	82
Sweden	1989/1991	27	44–7214	825
Denmark	1995	20	0.3–67	8
Denmark	1993/1994	9	55–537	–

Reprinted with permission from Andreas Prüeß [25]

Table 9.20 Concentrations of DEHP (in mg/kg_{TS}) in sewage sludge of various countries

Country	Year	Number of samples	Range	Median
Norway	–	36	<1–140	58
Norway	1989	19	27–1115	83
Sweden	1989/1991	27	25–661	170
Denmark	1995	20	3.9–170	24.5
Denmark	1993/1994	9	17–120	38

Reprinted with permission from Andreas Prüeß [25]

9.3.5.1 General Discussion

A variety of methods can be used to characterize metals in sludge, including atomic absorption spectrometry, flame photometry, inductively coupled plasma emission spectrometry. The choice often depends on the precision and the sensitivity required. Among all techniques applied, ICP emission techniques are applicable over a broad linear range and are especially sensitive to refractory elements. Inductively coupled plasma optimal emission spectroscopy (ICP-OES) is now widely used for the accurate elemental analysis in complex matrix, such as sludge. Prior to the analysis, solid samples are digested using reagents. The resulting sample solution is nebulized and then vaporized into the core of inductively coupled argon plasma. The analyte species are atomized, ionized and thermally excited. An optical emission spectrometer (OES) measures the intensity of radiation emitted at the element-specific, characteristic wavelength from the analyte atoms. Intensity measurements are converted to elemental concentration by comparison with calibration standards, commercially available.

Reference standards related to metals characterization are listed in Table 9.25.

Table 9.26 provides a list of elements that are commonly characterized using ICP-OES.

9.3.5.2 Apparatus and Procedure

The apparatus is described in Chap. 2.

Protocols are detailed in Chap. 2 for sample preparation and analysis. Reagents for sludge digestion are nitric acid and aqua regia (a solution obtained by mixing concentrated nitric acid with concentrated hydrochloric acid) [43, 46, 47]. Operating conditions for the digestion process are summarized in Table 9.27.

Table 9.21 Concentration of dioxin-like congeners (ng/kg_{TS}) in 14 sewage sludge collected from Australian sources

Congeners	Wastewater treatment plant														Mean	SD
	A	B	C	D	E	F	G	H	I	J	K	L	M	N		
2,3,7,8-TCDF	2.6	3.9	2.1	0.83	1.5	1.3	1.1	1.2	<0.4	0.6	11	0.775	<0.6	1.3	2.4	2.9
2,3,7,8-TCDD	1.6	0.72	<0.3	<0.4	<.5	<0.5	<0.6	<0.5	<0.4	<0.4	<1	<0.3	<0.3	<0.3	1.2	0.6
1,2,3,7.8-PeCDF	0.75	1.4	<0.3	<0.3	0.43	<0.3	<0.5	0.66	<0.1	<0.2	0.65	0.35	<0.1	0.77	0.7	0.3
2,3,4,7,8-PeCDF	0.8	2.1	<0.4	<0.6	<0.7	<0.4	0.89	<0.4	<0.4	<0.7	2	>0.3	<0.3	1.4	1.4	0.6
1,2,3,7,8-PeCDD	0.91	1.7	<0.4	0.72	<2	<0.6	<0.6	1.2	<0.9	<1	1.2	1	0.56	<1	1.0	0.4
1,2,3,4,7,8-HxCDF	2.9	2.4	<0.3	<0.5	1.1	<0.6	1	0.84	<0.5	<0.6	1.4	0.83	0 61	<2	1.4	0.8
2,3.4,6,7,8-HxCDF	2.1	2.9	<0.4	0.82	1.9	0.82	1.3	1	0.6	0.67	2.9	0.97	0.43	1.6	1.4	0.8
1,2,3,6,7.8-HxCDF	1.7	4.1	<0.3	1.5	1.6	1.5	2	1.5	<1	<1	1.1	0.62	<0.4	2.2	1.8	0.9
1,2,3,7,8.9-HxCDF	<0.2	1.3	<.2	<0.2	<0.3	<0.2	<0.2	0.15	<0.4	<0.3	<0.1	<0.08	<0.3	<0.7	0.7	0.8
1,2,3,4,7,8-HxCDD	1.2	2.5	0.22	0.72	<2	<0.7	0.64	0.51	<0.4	0.5	I.I	0.63	<0.4	<0.7	0.9	0.7
1,2,3.6.7,8-HxCDD	4	7.4	1	3.3	2.3	2.1	2.6	3	<2	3.1	6.4	5	3.7	3.3	3.6	1.8
1,2,3.7,8,9-HxCDD	1.7	5.2	<0.5	1.1	1.3	<2	1.3	2.7	<0.6	<1	3.5	1.8	1.3	<2	2.2	1.4
1,2,3.4,6.7.8-HpCDF	42	53	8.6	15	27	28	32	32	16	18	24	18	14	48	27	13
1,2,3,4,7,8,9-HpCDF	1.6	1.6	<0.3	0.46	1.2	0.76	0.99	0.97	<0.7	0.39	1.8	1.5	<0.2	0.9	1.1	0.5
1,2,3,4.6,7,8-HpCDD	150	230	30	26	61	59	68	75	50	67	130	150	63	95	90	57
OCDF	400	160	34	23	76	77	68	96	19	36	79	84	22	170	96	99
OCDD	20,900	7820	540	190	1290	1940	690	1180	420	640	2560	1410	500	1990	3005	5495
WHO_{98} TEQ ΣPCDD/Fs	**8.6**	**10.2**	**0.8**	**2.0**	**2.0**	**1.7**	**2.5**	**3.5**	**0.8**	**1.4**	**6.8**	**3.9**	**2.0**	**3.2**	**3.5**	**2.9**
WHO_{05} TEQ ΣPCDD/Fs	**12.7**	**11.3**	**0.9**	**2.0**	**2.3**	**2.1**	**2.5**	**3.8**	**0.9**	**1.5**	**6.9**	**4.2**	**2.1**	**3.4**	**4.0**	**3.7**
PCB 77	19	37	2.9	4.7	18	7.3	27	13	6.1	9.2	18	5	3.4	24	14	10
PCB 81	<1	<0.6	<0.5	2.3	1.8	<1	3	0.76	<0.9	<0.8	1.5	0.48	<0.2	2.6	2	1
PCB 126	1650	2400	210	110	1270	280	950	790	560	670	1210	260	250	1920	895	712
PCB 169	110	130	12	21	160	29	57	100	45	53	90	28	19	150	72	51

(continued)

Table 9.21 (continued)

Congeners	Wastewater treatment plant														Mean	SD
	A	B	C	D	E	F	G	H	I	J	K	L	M	N		
PCB 105	4480	5790	540	410	3250	730	2840	2150	1390	1640	3010	700	660	4850	2317	1772
PCB 114	81	120	8.6	6.3	56	2	42	41	29	32	70	12	8	77	42	35
PCB 118	900	850	88	100	400	130	340	210	98	140	360	84	87	860	332	311
PCB 123	150	240	23	22	90	27	80	52	23	33	85	19	17	200	76	72
PCB 156	280	320	26	30	130	37	120	75	31	51	130	31	28	260	111	103
PCB 157	100	45	5.4	8.2	29	10	20	14	9.6	13	20	6.6	<4	49	25	27
PCB 167	8.6	10.2	0.8	2.0	2.0	1.7	2.5	3.5	0.8	1.4	6.8	3.9	2.0	3.2	3.5	2.9
PCB 189	12.7	11.3	0.9	2.0	2.3	2.1	2.5	3.8	0.9	1.5	6.9	4.2	2.1	3.4	4.0	3.7
WHO_{98} TEQ ΣPCBs	**3.1**	**5.2**	**0.4**	**0.6**	**2.6**	**0.9**	**3.4**	**1.8**	**0.9**	**1.3**	**2.5**	**0.7**	**0.5**	**3.7**	**2.0**	**1.5**
WHO_{m} TEQ ΣPCBs	**2.2**	**4.0**	**0.3**	**0.6**	**2.0**	**0.8**	**3.0**	**1.5**	**0.7**	**1.0**	**2.0**	**0.6**	**0.4**	**2.8**	**1.6**	**1.1**
I-TEQ	**27.3**	**16.5**	**1.3**	**1.8**	**3.2**	**3.5**	**3.2**	**4.1**	**1.2**	**2**	**8.6**	**4.8**	**2.2**	**5.2**	**6.1**	**7.3**
WHO_{98} TEQ (ΣPCDD/Fs + ΣPCBs)	**11.8**	**15.3**	**1.2**	**2.6**	**4.7**	**2.6**	**5.9**	**5.3**	**1.7**	**2.7**	**9.3**	**4.6**	**2.5**	**7**	**5.5**	**4.1**
WHO_{05} TEQ (ΣPCDD/Fs + ΣPCBs)	**14.9**	**15.3**	**1.2**	**2.6**	**4.3**	**2.8**	**5.4**	**5.2**	**1.6**	**2.6**	**8.9**	**4.8**	**2.5**	**6.1**	**5.6**	**4.5**
Total TCDD	18	29	9.7	58	39.5	16	41	16	10	10	7.9	7.8	97	10	>26	26
Total TCDF	42	70	7.7	37	32	9.7	87	39	3.2	13	110	16	31	5.8	36	33
Total PeCDD	19	31	13	26	13	12	21	15	42	81	14	6.1	16	140	32	36
Total PeCDF	66	45	4.6	34	17	7	43	41	5.4	8	260	14	27	7.3	41	66
Total HxCDD	35	86	9.3	23	22	13	28	28	14	23	60	38	37	21	31	20
Total HxCDF	45	46	4.6	23	15	29	21	20	8.5	11	66	16	24	6.2	24	18
Total HpCDD	280	530	56	50	120	130	30	160	87	120	260	270	200	110	172	132
Total HpCDF	78	98	8.6	16	41	29	53	56	16	19	69	58	49	14	43	27
Total PCDD/Fs	21,900	8920	690	480	1640	2260	1180	1650	630	960	3490	1920	2640	840	3514	5709

Reprinted with permission from Elsevier [29]

Table 9.22 Concentrations of LAS (in mg/kg_{TS}) in sewage sludge from selected countries

Country	Number of WWTPs	Type of sewage sludge	Range
Denmark	19	–	11–16,100
Germany	8	Anaerobically digested	1600–11,800
Germany	10	Aerobic	182–432
Italy	1	Anaerobically digested	11,500–14,000
Spain	3	Anaerobically digested	12,100–17,800
Spain	2	Non-treated	400–700
Switzerland	10	Anaerobically digested	2900–11,900
United Kingdom	5	Anaerobically digested	9300–18,800

Reprinted with permission from Andreas Prüeß [25]

Table 9.23 List of macro and micro-nutrients for plants

Macro-nutrients	C, H, O, N, P, K, S, Ca, Mg
Micro-nutrients	Metallic elements: Fe, Mo, B, Cu, Mn, Na, Zn, Ni, Co Halogens: Cl Inorganic nonmetallic constituents: Se, Si

Table 9.24 Maximum permissible limits in sewage sludge for land application according to EU, US, Chinese and French legislations

Heavy metal	Ceiling concentration (mg/kg_{TS})				
	EU regulation [39]	French legislation [26]	US regulation [40]	Chinese regulation [41]	
				pH < 6.5	pH $\geq$ 6.5
As	–	–	75	75	75
Cd	20–40	10	85	5	20
Cr	–	1000	3000	600	1000
Cu	1000–1750	1000	4300	250	500
Hg	16–25	10	57	5	15
Ni	300–400	200	420	100	200
Pb	750–1200	800	840	300	1000
Se		–	100	–	–
Zn	25–4000	3000	7500	500	1000
Cr + Cu + Ni + Zn	–	4000	–	–	–
B	–	–	–	150	150

Table 9.25 Standards related to metals characterization by ICP-EOS

Titles	References
Standard Methods for the Examination of Water and Wastewater—Method 3120.B Inductively Coupled Plasma (ICP) Method	[42]
Standard Methods for the Examination of Water and Wastewater—Method 3030.F. Nitric Acid-Hydrochloric Acid Digestion	[43]
Standard Methods for the Examination of Water and Wastewater—Method 3030.K. Microwave-Assisted Digestion	[44]
EN 16170:2016 Sludge, treated biowaste and soil—Determination of elements using inductively coupled plasma optical emission spectrometry (ICP-OES)	[45]
EN 16173: Sludge, treated biowaste and soil—Digestion of nitric acid soluble fractions of elements	[46]
EN 16174: Sludge, treated biowaste and soil—Digestion of aqua regia soluble fractions of elements	[47]

Table 9.26 Non-exhaustive list of elements in digest solutions of sludge that can be characterized using ICP-OES

Elements	As, B, Ca, Cd, Co, Cr, Cu, Fe, Hg, K, Mg, Mn, Mo, Na, Ni, P, Pb, S, Se[a], Si[a], Zn

[a]Non-metallic elements

Table 9.27 Digestion methods

Digestion system	Standard	Reagent	Temperature (°C)	Time (min)
Opened	[45]	HNO_3/HCl	Boiling under reflux	–
	[47]	HNO_3/HCl	Boiling under reflux	120
Closed (microwave system)	[44]	HNO_3	160 °C for 10 min and then 165–170 °C for 10 min	
	[46]	HNO_3	175 °C	10
	[47]	HNO_3/HCl	175 °C	10

9.3.5.3 Interferences

Closed digestion systems or open systems with absorption vessel are recommended to limit the risk of volatilization. As digestion with nitric acid does not necessarily lead to a complete element breakdown, microwave digestion procedure may be used as an alternative.

9.3.5.4 Remark

X-ray fluorescence spectrometry (see Chap. 2 for the description of the technique) can be used as a screening method for sludge. Based on matrix-independent calibration curves, this method can be used for the analysis of the total element characterization at a semi-quantitative level according to EN 15309 standard [48].

9.3.5.5 Comments

Typical heavy metals contents of sewage sludge and water treatment residuals (WTRs) are reported in Tables 9.28 and 9.29, respectively. It is reasonable to understand that the heavy metals content of sludge varies considerably depending on the place and the origin. Al and Fe are of course the major elements in WTRs, as aluminum and ferric salts are used as primary coagulants. However, others heavy metals are also reported in WTRs analyses, although their concentrations are usually lower than in sewage sludge. Heavy metals in WTRs are carried along with raw/source water and/or present as impurities. They are also the indication of source water quality or pollution although they were concentrated into WTRs during treatment process of coagulation/flocculation.

The composition variety of sludge is a concern that the variety is not only from region to region but also from timeline. As such, the compositions of WTRs extracted from dozens of literature in a 10-year interval (three groups) are presented in Fig. 9.4. The WTRs data in 1996 was averaged from the available published papers/reports from 1990 to 1999 [58–60] while the data in 2007 (2005–2008) is derived from Ippolito et al. [53]. The composition data in 2017 is averaged from 9 published papers from 2014–2017 [61–69].

As all the data scattered in a world dimension, the standard deviations (SD) are very high within each group. Many studies have highlighted the increase of the natural organic matter in the surface water as a result of climate change, which will pose increasing pressure on the water treatment works and may result in high coagulant and volatile solids content in WTRs [70]. However, the possible volatile content increase is not reflected in WTRs of 2017 group. In terms of the main

Table 9.28 Typical heavy metals content (in mg/kg_{TS}) of sewage sludge

Country	Cu	Ni	Zn	Cr	Cd	Pb
Canada	180–2300	37–179	354–640	66–2021	2.3–10	26–465
China	131.2	49.3–95.5	783.4–3096	45.8–78.4	5.9–13	57.5–109.3
Germany	394.5	23.3	834	50	1.5	67.7
Hong Kong	275	44.5–622	1009–2823	663	–	52.5–57
India	112–255	192–293	870–1510	102–8110	41–54	91–129
Italy	280–543	19	1500	–	2.1	72
Spain	370	23.2–36.5	871–1626	54.4–3809	2.37–18.3	167–223
UK	204–337	58.5	778	159.5	3.5	221.5
USA	562–616	71	1285	178	25	170

Table 9.29 Physicochemical composition of WTRs sludge

Parameter	Units	WTRs conditioned with alum salts	WTRs conditioned with ferric salts	References
pH		6.5 ± 0.3[a]	7.0 ± 1.3	Inputs from [51–57]
Total solids	mg/L	2500–52,345	2132–5074	
Al	mg/kg_{TS}	118,700 ± 24,260	61,390 ± 35,920	
Fe		37,000 ± 19,740	220,900 ± 32,200	
Ca		10,360 ± 4299	nd	
Mg		2407 ± 572	nd	
Na		355 ± 142	nd	
K		3547 ± 582	nd	
S		6763 ± 2955	nd	
Mn		2998 ± 1122	1088 ± 178	
Zn		98 ± 31	36 ± 4	
Cu		624 ± 581	46 ± 12	
Ni		28 ± 10	64 ± 14	
Pb		22 ± 12	47 ± 1	
Cr		20 ± 7	38 ± 4	
Cd		0.12 ± 0.02	nd	
Hg		0.46	nd	

Reprinted with permission from Elsevier [50]
nd not determined
[a]Numbers are means ± SD

constituents, the volatile solids and the Al/Fe (as coagulant) contents remained at 200–300 g/kg_{TS} and ~100 g/kg_{TS}, respectively. The Ca contents in the three groups are also within the same level.

Regarding the plant nutrients content in WTRs, the macronutrients (N, P, K, and Mg) in the three groups varied marginally in Fig. 9.4. The N contents were about 5 g/kg_{TS} on average and the others were located at an average level of 2.5 g/kg_{TS}. In terms of the micronutrients in the WTRs, the Mn is outstanding over others in 2017 group with a level of 12.0 ± 18.7 g/kg_{TS} which is significantly higher than 2007 and 1996 group. Mn is an essential plant mineral nutrient, playing a key role in several physiological processes, particularly photosynthesis [71]. Its deficiency is a widespread problem, most often occurring in sandy soils, organic soils with a pH above 6 and heavily weathered, tropical soils [72]. There is an average of 5–14 g-Mn/kg-soil content in the sandy soils. With this kind of Mn-rich WTR application in the soil, it could probably benefit the plant growth when applied in Mn-deficient soils. On the other hand, Mn phytotoxicity can be unwelcomely introduced into the soil [63]. Thus, Mn concentration in the WTRs should be given special attention before considering the land application. Regarding the heavy metals content in the WTRs, most of the metals contents in 2017 group are similar

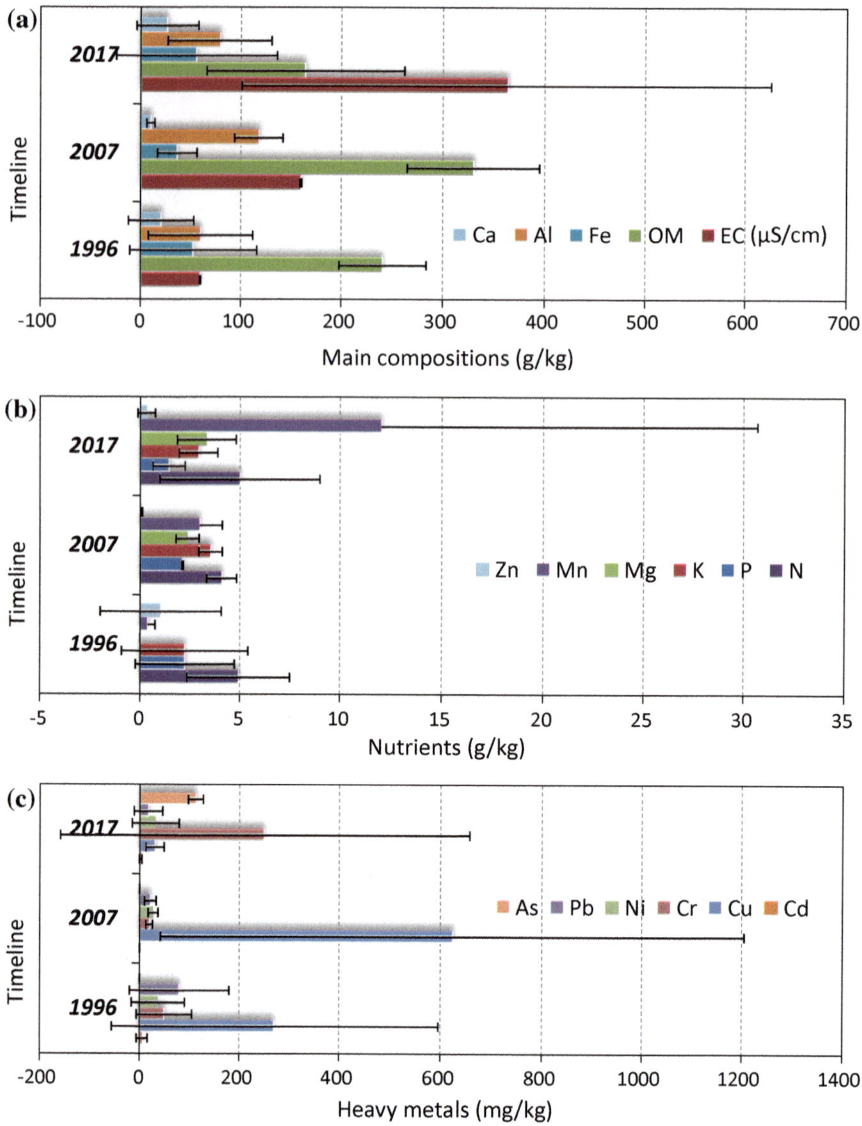

Fig. 9.4 The compositions of the WTR in a decade-interval [OM—organic matter; EC—electrical conductivity (μS/cm)]

to previous samples except for the As and Cr. The As and Cr contents in group 2017 are considerably higher than other two groups, particularly the Cr content which is as high as 720 mg/kg_{TS} in some cases [68]. With a comprehensive comparison, the compositions of the WTRs did not show a significant change during the past 20 years except for several elements.

9.3.6 Pathogenic Microorganisms

Sludge has long been valorized as a source of nutrients and as a soil conditioner for use in agriculture. Indeed, the Sewage Sludge Directive of the European Union seeks to encourage the use of sewage sludge in agriculture [39] and the USEPA promotes beneficial use of properly treated sludge in land application [73]. The World Health Organization has also developed detailed guidelines for the use of treated excreta and sludge in agriculture in developing countries as an important beneficial resource that can improve nutrition and food security in impoverished communities [74]. The valorization and beneficial use of sewage sludge, however, requires the control of pathogenic organisms that are concentrated in the sludge as a result of conventional wastewater treatment processes.

The majority of the pathogens found in sludge come from human sources, although animal pathogens can enter the sewer systems from slaughterhouses, food processing plants and stormwater runoff. The concentrations and types of pathogens found in wastewater and sludge reflect the health status and sanitation level in a given region. Pathogen concentrations are highest in low income countries and lowest in high income countries, and some pathogens, such as *Ascaris* species, may be very common in large cities in income countries, with 30–60% prevalence, and essentially absent in industrialized cities in Europe or the US [75].

If sewage sludge is to be used as a valuable renewable resource in agriculture, pathogenic must be reduced to levels that pose minimal risks to the population and the environment. Conventional sludge treatment, however, was developed to stabilize the putrescible fraction of the sludge and not to reduce pathogens. Consequently, the engineering technologies and management strategies that have been developed specifically for pathogen removal in sludge for use in agriculture will be discussed below.

9.3.6.1 Characteristics of Pathogens in Sludge

Table 9.30 presents the principal groups of pathogens of concern in sewage sludge [73, 76]. The pathogens are divided into four main groups: bacteria, viruses, protozoa, and helminths. All of the pathogens are transmitted principally by the fecal-oral route, and only the helminths have a latency period in the environment when they are not in the infective stage.

Only bacterial pathogens have the ability to grow in the environment under certain conditions, and this is a concern in sludge management that will be discussed below. While there is some immunity to bacterial and viral infections, there is no immunity to protozoan and helminth infections and persons who have had a protozoan or helminth disease can become re-infected as long as the environment is contaminated with these pathogens.

The minimum infective dose is the minimum number of pathogens that must be ingested by the fecal-oral route in order to become infected. Table 9.31 lists the

Table 9.30 Principal pathogens of concern in sewage sludge

Bacteria	Protozoa
Salmonella sp *Shigella* sp. *Yersinia* sp. *Vibrio cholerae* *Campylobacter jejuni* *Escherichia coli* (pathogenic strain)	*Cryptosporidium* sp. *Giardia lamblia* *Entamoeba histolytica* *Balantidium coli* *Toxoplasma gondii*

Viruses	Helminths
Hepatitis A	*Ascaris lumbricoides*
Norwalk	*Trichuris trichiura*
Rotaviruses	*Ancylostoma duodenale*
Polioviruses	*Necator americanus*
Coxsackieviruses	*Taenia saginata*
Echoviruses	*Taenia solium*
Reovirus	*Hymenolepis nana*
Astroviruses	*Toxocara canis*
Caliciviruses	

Table 9.31 Minimum Infective Dose (MID) for pathogen group

Pathogen	Minimum infective dose
Bacteria	10^2–10^6
Viruses	10^2
Protozoa (oo)cysts	10^0–10^2
Helminth eggs[a]	10^0–10^1

[a]Eggs must be in the infective stage

ranges of minimum infective doses (MID) for the four general groups of pathogens. Generally, bacteria have the highest MID, followed by viruses, protozoa and helminth eggs. While hundreds to thousands of bacteria or viruses must be ingested to be infected, as few as one protozoan (oo)cyst or one helminth egg [77] can cause an infection in humans. Protozoan (oo)cysts and helminth eggs are thus the principal pathogens to be controlled in the use of sludge in agriculture.

Persistence of sludge pathogens in the environment is shown in Tables 9.32 and 9.33. Protozoan (oo)cysts generally have the shortest survival time of less than 30 days, with bacteria and viruses surviving up to several months. Helminth eggs are the most persistent of all pathogens and can survive months to years in the environment. Survival of the pathogen groups on crops is lower than in the soil but still significant on the order of several weeks to several months (Table 9.33).

Table 9.34 presents the concentration ranges of pathogens in sewage sludge from various conventional treatment processes. Anaerobic digestion was designed historically to stabilize the sludge by degrading putrescible organic matter (volatile solids) and not to remove pathogens.

The final conventional sludge treatment process, sludge dewatering, was developed to remove water and reduce volume for final disposal without

Table 9.32 Survival times of pathogens in soil [74]

Pathogen	Survival in Soil at 20–30 °C	
	Normal range	Absolute maximum[a]
Bacteria: Fecal coliforms *Salmonella*	<70 days usually < 20 days	1 year
Viruses	<100 days usually < 20 days	1 year
Protozoa (oo)cysts	<20 days usually < 10 days	>2 months
Helminth eggs	Months to years	7 years

Reprinted with permission from WHO
[a]This occurs only under very optimum environmental conditions for the pathogen

Table 9.33 Survival times of pathogens on crops [74]

Pathogen	Survival on Crops at 20–30 °C
Bacteria	
Fecal coliforms	<30, usually <15
Salmonella	<30, usually <15
Shigella	<10, usually <5
Vibrio cholerae	<5, usually <2
Viruses	
Enterovirus	<60, usually <15
Protozoa (oo)cysts	
Entamoeba histolytica	<10, usually <2
Cryptosporidium	<3, usually <2
Helminth eggs	
Ascaris	<60, usually <30
Taenia	<60, usually <30

Reprinted with permission from WHO

considering pathogen content for subsequent reuse. Thus conventional wastewater treatment practice is insufficient to provide a sewage sludge free of pathogens and advanced treatment of sludge for agricultural reuse is necessary. Alternatively, in developing countries a simpler approach using sludge storage for long time periods at ambient temperatures may be more feasible.

9.3.6.2 Design of Pathogen Reduction Processes in Sludge

The survival of pathogens in the environment is a function of various factors, including temperature, moisture, time of exposure, type of pathogen, etc. In liquid wastewater treatment systems the inactivation of pathogens is related by a first order kinetics inactivation equation such as following equations:

Table 9.34 Range of concentrations of pathogens in sewage sludge

Pathogen	Type of sludge	Concentration range number (kg_{TS})
Helminth eggs	Primary sludge	10^3–10^4
	Digested sludge	10^2–10^3
	Partially-dewatered sludge	10^1–10^3
	Partially-dewatered sludge from aerobic treatment	10^2–7.5×10^4
	Anaerobic sludge	6.3×10^3–1.5×10^4
Protozoan (oo)cysts	Primary sludge	7.7×10^4–3.0×10^6
	Digested sludge	3.0×10^4–4.1×10^6
	Dewatered sludge	7.0×10^1–10^2
Bacteria	Primary sludge	10^1–8.8×10^6
	Extended aeration sludge	10^8
Viruses	Primary sludge	10–10^6

Reprinted with permission from IWA Publishing [77]

$$\frac{dN}{dt} = -kN \tag{9.14}$$

$$\ln \frac{N_t}{N_0} = -kt \tag{9.15}$$

where

N_t is the pathogen concentration at time t,
N_O is the original pathogen concentration,
t is the time, in minutes, hours or days
k is the inactivation constant.

The value of k is a function of pathogen type, temperature, and various environmental conditions.

Because of the heterogeneous nature of sludge, which can range from liquids, and semi-solids to solids, empirical approaches have been developed that relate treatment technique to pathogen reduction at levels below the limits of detection. This approach implies that the resulting treated sludge should be free of pathogens and can be used in agriculture without additional controls. Periodic monitoring of sludge for select pathogens, however, should be performed and will be discussed below.

Time and Temperature (Batch and Continuous/Intermittent Feed Systems)

It has long been known that exposure to heat is an effective method of pathogen inactivation if the temperature is far above optimum growth temperatures [75]. Figure 9.5 shows the original time-temperature data presented by Feachem et al. [75] that is still used as a guide in the design of pathogen reduction systems for

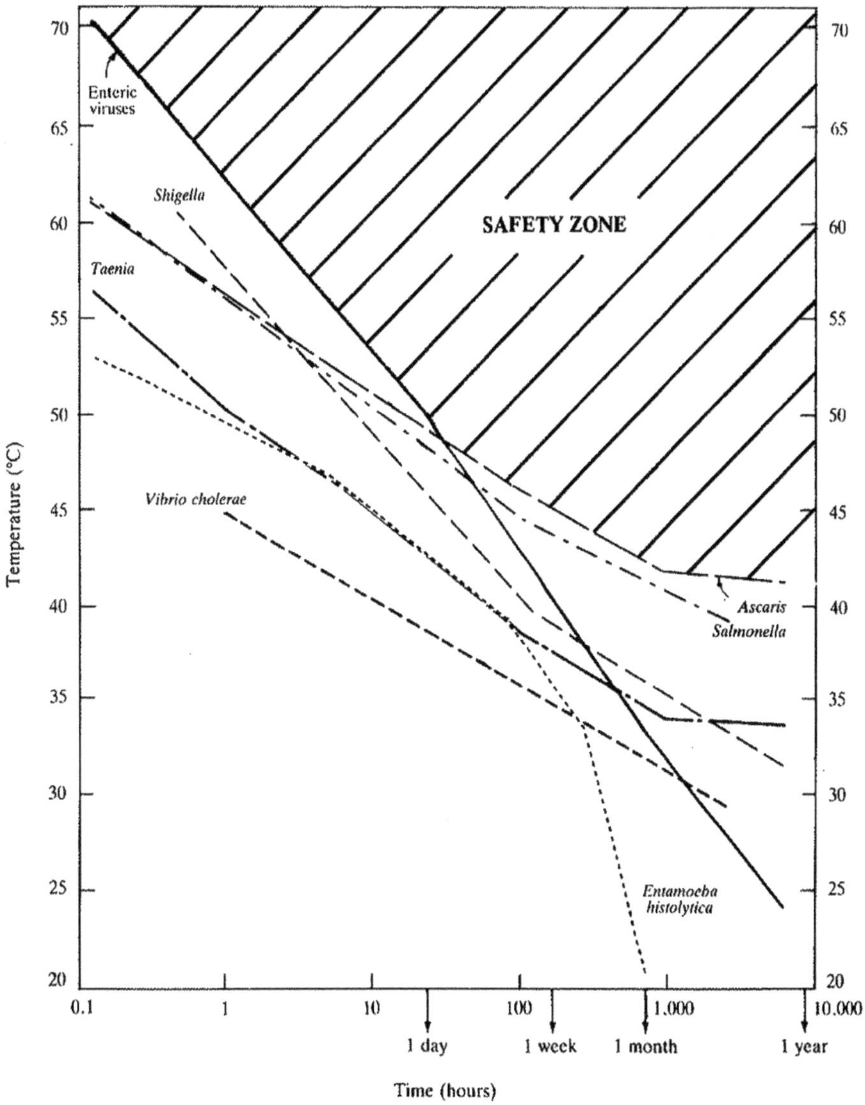

Fig. 9.5 Influence on time and temperature for selected pathogens in excreta and sludge. Reprinted with permission from World Bank [77]

sludge. These data suggest that batch systems of sludge, whether dewatered or not, could inactivate all pathogens if, for example, a temperature of 70 °C was maintained for as short as 7 min, or a temperature of 50 °C for approximately 3 days. Empirical data from full scale plants, however, show that in practice times need to be longer.

Thermal drying of sludge at high temperatures is currently one of the most common methods of pathogen reduction in Europe and the US. The USEPA [73] has developed the following equations for time-temperature regimes for thermally treated sludge, both liquid and solid, that were developed from studies of reduction of *Ascaris*, Enterovirus, and *Salmonella* spp. as well as the original data from Feachem et al. [75]:

Sludge with TS > 7% (liquid and solid sludge):

$$D = \left(1.63 \times 10^{9}\right)/10^{0.14T} \tag{9.16}$$

where

D is the time, in minutes
T is the temperature, in °C.

Sludge with TS < 7% (liquid sludge):

$$D = \left(50.07 \times 10^{10}\right)/10^{0.14T} \tag{9.17}$$

Figure 9.6 shows a plot of the equation for sludge with TS greater than 7%.

Figure 9.6 suggests that all pathogens could be eliminated at 70 °C at a time of only 0.26 min. In engineering practice, however, a safety factor must be applied

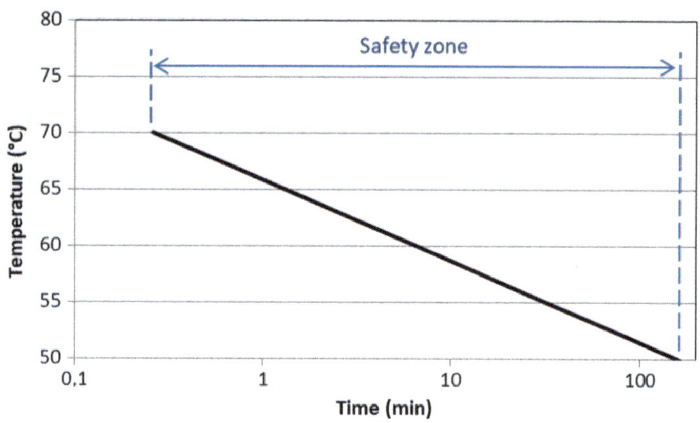

Fig. 9.6 Influence on time and temperature for all pathogens in liquid or solid sludge with TS > 7%, as formulated by Eq. 9.16

and typically the retention time for thermal dryers at 70 °C is designed for at least 30 min and at 80 °C for at least 10 min [77].

Thermophilic biological processes, such as thermophilic anaerobic or aerobic digestion, or aerobic composting, can easily reach temperatures above 50 °C and thus sludge digestion could include pathogen reduction if it were incorporated into the design of the treatment plant. Continuous or intermittent feeding digesters, however, must be operated to take into account the short-circuiting of the influent sludge as shown in the following equation [77]:

$$\frac{C}{C_0} = \frac{R}{(10^{DR_\theta} - 1 + R)} \tag{9.18}$$

where

C/C_0 is the fraction of pathogens surviving,
R is the fraction of the reactor content replaced each day,
R_0 is the interval between feeds
D is the decimal decay rate, t^{-1}.

As an example, Table 9.35 shows the effect of feed interval on the theoretical reduction for a digester with a design hydraulic retention time (HRT) equal to 15 days. It is for this reason that batch thermal treatment, if possible, is preferable to continuous or intermittent feed unless the digester is carefully monitored and the feeding interval controlled.

Dewatered Sludge Treatment with CaO (Quicklime)

When CaO is added to dewatered sludge it reacts with the moisture causing an exothermic reaction that raises the temperature of the sludge up from 50 to 70 °C. The lime and sludge have to be very well mixed so the temperature is reached throughout for pathogen reduction. This method has been proven to destroy *Ascaris* eggs and various viruses within 24 h.

Summary of Pathogen Reduction Processes

Table 9.36 presents a summary of the principal pathogen reduction processes for agricultural reuse of sludge as practiced in the EU and USA. These processes are

Table 9.35 Reduction of pathogens at different feed intervals, $D = 1.0\ d^{-1}$, HRT = 15 d

Interval between feeds (h)	$\log_{10}$ Reduction
2.4	0.69
24	2.10
72	4.18

Table 9.36 Summary of additional pathogen reduction processes[a] for agricultural reuse of sludge

Process[a]	Design parameters
Thermal drying	Dewatered sludge should be heated to at least 80 °C for 10 min and moisture content reduced to <10%
Windrow composting	Batches of sludge kept at 55 °C for 4 h between each of 3 turnings, followed the maturation period to complete the composting process
Aerated pile/in-vessel composting	Batch to be kept at a minimum of 40 °C for at least 5 days and for at least 4 h at a minimum of 55 °C followed the maturation period to complete the composting process
Thermophilic digestion (Anaerobic or Aerobic)	Sludge should achieve a temperature of at least 55 for a minimum of 4 h after the last feed and before the next withdrawal. Digester should be designed to operate at a temperature of at least 55 °C with a mean hydraulic retention time to stabilize the sludge
Stabilization with CaO	Dewatered sludge should be thoroughly mixed with quicklime to reach a pH $\geq$ 12.0 and a minimum temperature of 55 °C for 2 h after mixing

Reprinted with permission from the European Commission [76]
[a]These processes are additional to conventional sludge treatment as historically practiced without consideration of pathogen reduction for agricultural reuse

Table 9.37 Summary of natural pathogen reduction processes for agricultural reuse of sludge

Process	Design parameters
Long-term storage 2–20 °C	Sludge must be dewatered by air drying, covered and stored for 1.5–2 years. All pathogens should be reduced to low levels. Long-term storage of dewatered sludge can inactivate all pathogens
Long-term storage 20–35 °C	Sludge must be dewatered by air drying, covered and stored for >1 year. Substantial to total inactivation of viruses, bacteria, and protozoa. More or less complete inactivation of helminths after 1 year

Reprinted with permission from WHO [74]

considered additional to conventional sludge treatment and processes that historically did not consider pathogen reduction for agricultural reuse of sludge.

Table 9.37 presents the natural pathogen reduction process, long-term storage at ambient temperatures, proposed by the World Health Organization as an appropriate solution to pathogen reduction in developing countries.

9.3.6.3 Monitoring of Pathogens

Pathogens in treated sludge need be monitored for the following reasons:

- to check the performance of a specific treatment process (see Tables 9.36 and 9.37)

- to check the quality of the final product before land application.

Unfortunately it is not possible to analyze the numerous pathogens that may be present in sludge and indicator species need to be chosen. Indicator organisms should satisfy the following criteria [78]:

- to be quantified by simple, reliable, precise and inexpensive analytical techniques,
- present high resistance to pathogen reduction processes such as *Ascaris*,
- concentrations of non-pathogenic indicators need to correlate with specific pathogens of concern,
- to be present in sufficient numbers to ensure precise quantitative analysis,
- to be able to withstand reduction processes at the same level as potentially present pathogens.

As an example from the US, the indicator organisms selected by the USEPA include microorganisms in the groups of bacteria, viruses, and helminths [73]:

- Fecal coliforms
- *Salmonella* species
- Total culturable viruses (enteroviruses and reoviruses)
- *Ascaris* ova (detection, enumeration, viability determination)

Fecal coliforms are indicators of fecal material from humans and warm blooded animals and their decline in pathogen reduction processes is the same proportion as enteric bacterial pathogens [73]. *Salmonella* species are bacterial pathogens of concern that can be routinely monitored; their reduction should be comparable with other bacterial pathogens. Enteroviruses and reoviruses are viral pathogens and their reduction in treatment processes should be comparable among other groups of viral pathogens. Finally, *Ascaris* is the hardiest of all helminths and is the only helminth in which viability can be measured; if *Ascaris* species are not present in sludge, or if their ova are not viable, it can be assumed that no other type of helminth could survive [73]. The analyses for these organisms require specialized personnel and equipment.

Analytical monitoring must take into account:

- Sampling onsite in a heterogeneous matrix
- Sample conservation and transport to laboratory
- Sampling at laboratory
- Extraction/purification steps
- Analytical method to detect organisms

The most frequently encountered problems include representative sampling, conservation of samples, differences in culture media, existence of a background in samples [78]. There are numerous standards for sampling sludge that vary from country to country; the only international standards for sampling and microbiological analysis in sludge, however, are guidance documents that do not report data on precision [78]. The development of standard methods of sampling and microbiological analysis in sludge is an area of much research and is urgently needed in the future.

9.4 Physical, Chemical and Thermal Characteristics

Varieties of technologies are available to reduce the volume (thickening, conditioning, mechanical dewatering, drying, incineration) and to improve the quality (anaerobic or aerobic digestion, composting or liming) of the sludge. A number of physical, chemical and thermal properties affect the efficiency of these processes and the reuse potential of the sludge.

9.4.1 Physical Properties

Among the most influent physical properties, particle size and distribution, sludge density and specific surface area strongly affect solid/liquid separation. There are of special concern in sludge conditioning, whose aim is to increase particle size by combining the smallest particles into large aggregates.

9.4.1.1 Particle Size Distribution

Supra-colloidal particles with $1 < d < 100$ μm strongly influence the efficiency of the settling, thickening and dewatering processes. Thus, knowing the relative amounts of particles present in a sample sorted according to the size [79], i.e. characterizing the particle size distribution (PSD), is fundamental. A variety of methods are available, including microscopy and optical image analysis, sieving, photocentrifuge, sedimentation or permeability tests and optical sizing techniques. Among optical methods, laser diffraction allows to simultaneously characterize both the size and the shape of the particles in a large size range ($0.1 < d < 2$ mm), in a wide concentration range and with a high reproducibility.

General Discussion

The particle size distribution is determined by measuring the angular variation in intensity of light scattered as a laser beam passes through the dispersed particulate sample. Large particles scatter light at small angles relative to the laser beam and small particles scatter light at large angles. The angular scattering intensity data is then analyzed to calculate the size of the particles responsible for creating the scattering pattern, using the Fraunhofer approximation for particles bigger than 50

Table 9.38 Standards related to particle size distribution

Titles	References
ISO 13320:2009—Particle size analysis—Laser diffraction methods	[80]
GB/T 19077-2016—Particle size analysis—Laser diffraction methods	[81]

microns or the Mie theory of light scattering for smaller ones according to the ISO 13320 standard [80]. The particle size is reported as a volume equivalent sphere diameter. The reader can refer to Chap. 2 for a comprehensive presentation of the technique.

Results are usually reported as a volume weighted distribution, with several characteristic diameters:

- Percentile D_{vx}, the diameter at which x% of the population has a diameter less than this value. It is standard practice to report D_{v10}, D_{v50} and D_{v90};
- Volume moment mean, also called De Brouckere mean diameter, D [3, 4];
- Surface area moment mean, known as Sauter mean diameter, D [2, 3].

According to the ISO 13320 standard [80], a minimum of three independent measurements is required and the coefficients of variation must be less than 3% for D_{v50} and less than 5% for D_{v10} and D_{v90}.

Another way to define the width of the distribution is to compute the span, as follows:

$$SPAN = \frac{D_{v90} - D_{v10}}{D_{v50}} \tag{9.19}$$

Reference standards are listed in Table 9.38.

Apparatus

Measurement requires [80]:

- A laser analyzer
- A stirrer
- An ultrasonic probe
- Bakers.

Procedure

Preparation of the samples. The wet processing method is usually applied. Homogenizing the batch and taking a sample representative of the batch is fundamental. Most of the time, dispersion and dilution of the sludge samples in the dispersion cell, either using water or centrifuged mixed liquor, will be needed to achieve an optimal obscuration value during the measurement. Avoiding decantation in the dispersion cell is critical. Stirring should be efficient enough to transport the biggest particles. Ultrasonication can be also be used to facilitate proper dispersion of particles in the sludge. With intermediate ultrasonication, the measured size distribution should not change significantly if the sample is well dispersed and the particles are neither fragile nor soluble. A preliminary check on the dispersion quality by visual/microscopic inspection of the suspension is recommended.

The minimum volume of the sample, required for repeatable measurement, increases as the width of the size distribution becomes greater [82]. This allows a sufficient number of large particles to be present. Accordingly, the volume of the dispersion sludge required to suspend these samples also increases if the limits of optical concentration are to be observed.

Sample analysis. The reader can refer to Chap. 2 for a comprehensive presentation of the protocol. It includes three stages [80]: instrument setup, measurement of the background noise (blank experiment with a particle-free dispersion medium) and final measurement of the scattering pattern of dispersed sample(s). Values of light refraction index and the absorption coefficient for the dispersed phase being unknown, it is recommended to leave the defaults values provided by the manufacturer. Recommended values are usually $n_{refraction} = 1.52$ and $n_{absorption} = 0.1$. The fractional difference between a dispersed sample and a blank experiment is given as an obscuration value which is indicative of the total amount of scattered light and the particle concentration. To have a large quantity to analyze, an obscuration of 15–25% is recommended. Care must be taken not to exceed the limit of 30%, in order to avoid multi-diffraction phenomena. The operator must ensure that the stirrer speed and the volume flowrate of the pump are high enough to transport all the particles by changing the set point values. Laser measurement results must be in accordance with the visual observations.

Interferences

Systematic measurement errors (bias) may arise from improper sample preparation, departure from the theoretical assumptions for the particulate material and/or improper operation or functioning of the instrument.

Comments

Figure 9.7 highlights agglomeration/desagglomeration phenomena likely to occur during the measurements. Two WTRs, referred to as A and B in the following, were sampled in the same drinking water treatment plant. The only difference between the two samples arises from the coagulant used for conditioning. Experiments were conducted without sonification. Water was used as the dispersant ($n_{refraction} = 1.33$). Obscuration and the stirrer speed were set to $20 \pm 5\%$ and 2600 rpm, respectively.

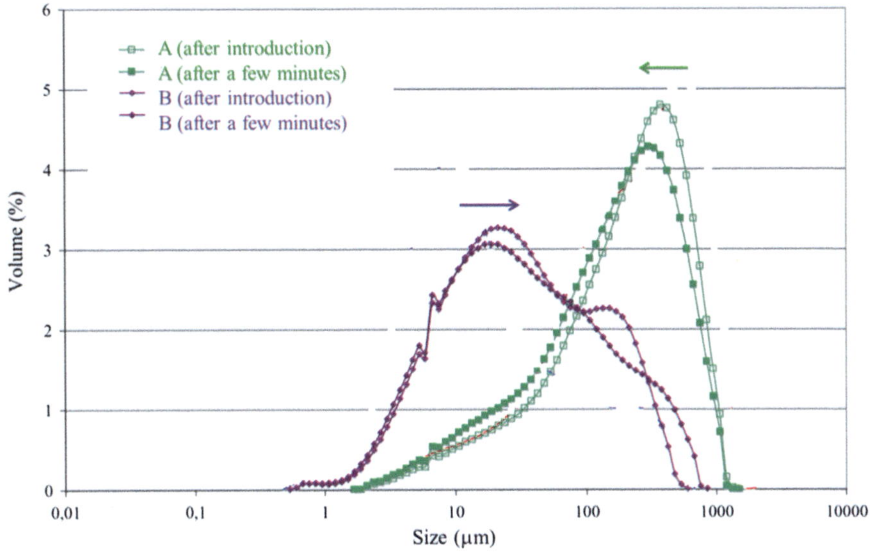

Fig. 9.7 Agglomeration/deagglomeration phenomena during the PSD characterization of two WTRs

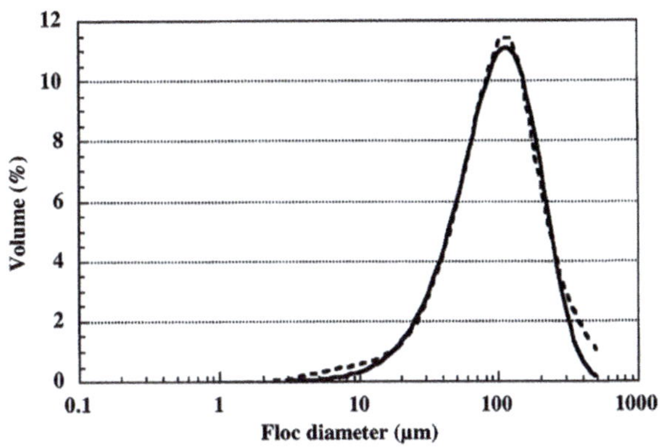

Fig. 9.8 Typical PSD curves for liquid sewage sludge. Dilution with centrifuged mixed liquor, without sonication, stirring speed = 100 rpm. Reprinted with permission from Elsevier [84]

Default values were used for the refraction and absorption indexes of the sludge particles. The residuals obtained during the analyses were around 1%, thereby validating the choice of these coefficients. As emphasized, the PSD changes over time. After a few minutes, particles of sample B tend to agglomerate while the biggest particles in sample A deagglomerate.

Typical particle size distributions of activated sewage sludge collected from the aeration tanks are shown in Fig. 9.8. Monomodal distributions are usually observed [83, 84].

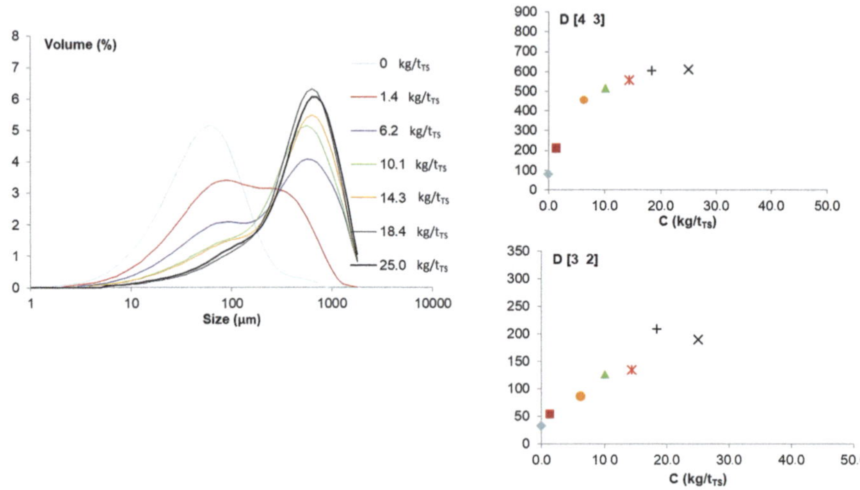

Fig. 9.9 Influence of polymer concentration on the particle size distribution

This method is thus sensitive enough to closely characterize the influence of polymer conditioning on the PSD [85]. For example, on Fig. 9.9, the PSD of an activated sewage sludge collected from the anaerobic co-digestion tank and flocculated with rising concentrations of Zetag® 9018 BASF flocculant are highlighted. The addition of flocculant progressively forms larger agglomerates, or flocs, until the flocculant concentration exceeds a certain limit. Above this limit, corresponding to the optimal flocculant dosage, flocculant addition does not significantly alter the PSD.

9.4.1.2 Sludge Density

General Discussion

The absolute density, ρ, is defined as mass per unit volume, the SI unit is kg/m^3. Specific gravity, SG, is the ratio of the material's density to the density of a standard substance under specified conditions, usually water at 4 °C. It's a relative quantity with no units. A common device for measuring density is a pycnometer. The reader can refer to Chap. 2 for a comprehensive description of the apparatus and the recommended protocol. A simple gravimetric procedure can also be used to estimate the absolute density. It requires the knowledge of the sample mass weighted in air, $m_{s,\text{in air}}$, and the sample mass weighted in liquid, $m_{s,\text{in water}}$, for the same given volume. The absolute density (ρ) is computed as follows:

$$\rho = \rho_w \frac{m_{s,\,in\,air}}{m_{water} - m_{s,in\,water}} \tag{9.20}$$

$$\rho = \rho_w \frac{(m_{c+s} - \mathrm{m}_c)}{(m_{c+w} - m_c) - (m_{c+s+w} - m_{c+s})} \tag{9.21}$$

where

m_c is the weight of empty container (with lid), in grams
m_{c+w} is the weight of water-filled container (with lid), in grams
m_{c+s} is the weight of sample-filled container (with lid), in grams
m_{c+s+w} is the weight of sample and water-filled container, in grams
ρ_w is the water density at 4 °C, i.e. 1 000 kg/m^3.

SG is derived as follows:

$$\mathrm{SG} = \frac{\rho}{\rho_{ref}} \tag{9.22}$$

where

ρ_{ref} is the density of a standard substance under specified conditions.

Reference standard related to density measurement is listed in Table 9.39.

Apparatus

Measurement requires [86]:

- Weighing Container
- Balance, with a sensitivity of 0.01 g.

Procedure

The procedure is as follows:

- Weigh the empty container with lid on, m_c.
- Fill the container with reagent water, place lid on container, wipe off excess water and weigh the water-filled container, m_{c+w}.
- Empty the container and dry before filling with sample.
- Fill the calibrated weighing container with as much sample as possible without exerting pressure on the container. Place the lid on the container and weigh the container with sludge and lid in air, m_{c+s}.

Table 9.39 Standard related to density measurement

Title	Reference
ASTM D5057-17 Standard Test Method for Screening Apparent Specific Gravity and Bulk Density of Waste	[86]

- Fill remaining space in the container with water, taking care that air bubbles are not trapped in sludge or the container. Place the lid on and weigh the container with sludge, water and lid, $\dot{m}_{c+s+w}$.

All the mass should be weighted to the nearest 0.01 g.

Interferences

Excessive temperatures causing loss of sample components due to vaporization could result in erroneous readings.

Large, obvious void spaces interfere in this test method and will give inaccurate results because of the false volume measured.

Comments

Mass-volume relationships are crucial for the design of reactors and storage facilities. Sludge volume depends mainly on the moisture content and to a lesser degree on the mineral or organic character of the solid matter.

When the dry basis moisture content of the sludge (W_{db}) is known, the density (ρ) of the wet sludge, considered as a two phase (water and total solids) mixture, may be computed from the density of the total solids (ρ_{TS}) and that of the water (ρ_w) according to the following equation:

$$\frac{1}{\rho} = \frac{W_{db}}{1 + W_{db}} \frac{1}{\rho_w} + \frac{1}{1 + W_{db}} \frac{1}{\rho_{TS}} \tag{9.23}$$

where

ρ, ρ_{TS} and ρ_w are expressed in kg/m^3
W_{db} is given in kg of water/kg of total solids.

In the same way, knowing the total solid (TS) content, the specific gravity of the wet sludge (SG) may be computed from the specific gravity of the total solids (SG_{TS}) and that of the water (SG_w) according to:

$$\frac{1}{SG} = \frac{TS}{SG_{TS}} + \frac{1 - TS}{SG_w} \tag{9.24}$$

where

TS is the weight percent of total solids.

Even if it is of limited usefulness, the specific gravities of the total solids (TS), the total volatile solids (TVS) and the total fixed solids (TFS) are linked by the following relationship:

$$\frac{1}{SG_{TS}} = \frac{TVS}{SG_{TVS}} + \frac{TFS}{SG_{TFS}} \tag{9.25}$$

Density measurements were performed for the 15 sewage sludge samples introduced in Sect. 9.3.1.4, using an Accupyc 1330TC Pycnometer (Micromeritics, Creil, France). Before the analyses, dry sludge samples were ground with a Martar Grinder Retsch 9001 (Retsch, Haan, Germany) for 30 min. Since the density affects specifically the storage and the conveyance of the sludge, all the experiments were performed at 25 °C and the temperature-dependence of this property was not investigated. Mean values are reported in Table 9.40. The measure deviation from the mean value is small, except for S1–S3 samples. Indeed, the composition of primary sludge is more diverse than that of biological sludge, which causes problems of sampling. If the specific gravity of the water is taken to be 1.0, the specific gravity of a dewatered sludge having 20% of total solids (i.e. $W_{db} = 4$ kg/kg TS) is around 1.08. Higher values can be achieved when the TVS content of the sludge is low (Fig. 9.10). However the moisture content has a greater influence on the sludge density than the mineral or organic character of the total solids.

As expected (Fig. 9.11), the specific gravity of the total solids is linearly correlated with the total volatile solids content of the sludge. From this, it can be deduced that the specific gravity of the total volatile solids, SG_{TVS}, is around 1.296 and that of the total fixed solids, SG_{TVS}, around 2.634.

9.4.1.3 Specific Surface Area

General Discussion

Specific surface area (a) is defined as the total surface area of the solid per gram of solid material. It is expressed in m^2/g. A common method for measuring the specific surface area of dried sludge is gas adsorption. The reader can refer to Chap. 2 for a comprehensive description of the current apparatus and the recommended protocols. The specific surface area of wet sludge can be determined by dye adsorption [87, 88], a simple and rapid method requiring a colorimetric analysis. The method requires the determination of the adsorption isotherm that describes the amount of solute adsorbed on per unit amount of adsorbent and concentration of solute in solution at equilibrium. Rhodamine B molecule, methylene blue or Lissamine Scarlet 4R are frequently used as solute. The amount of dye adsorbed as a monolayer is given by the plateau value on the adsorption isotherm. Once this amount is known, the specific surface area can be deduced as follows:

$$a = Y\,N_A\,A \tag{9.26}$$

where

a is the specific surface area, in m^2/g
Y is the amount of adsorbed dye, in mol/g
N_A is the Avogadro's number (6.023×10^{23} molecules/mol)
A is the area covered by each molecule, in m^2 per molecule.

Table 9.40 Density of the total solids

Sludge	S1	S2	S3	S4	S5	S6	S7	S8	S9	S10	S11	S12	S13	S14	S15
ρ_{TS} (kg m^{-3})	1590	1377	1510	1502	1481	1547	1668	1492	1485	1668	1551	1575	1811	1879	1839
RSD (%)	1.1	1.1	1.2	0.5	0.0	0.2	0.2	0.1	0.3	0.0	0.5	0.3	0.1	0.4	0.1
ρ for TS = 20% (kg m^{-3})	1078	1055	1070	1069	1067	1073	1084	1068	1067	1084	1074	1076	1095	1100	1097

Values are averages of seven replicates

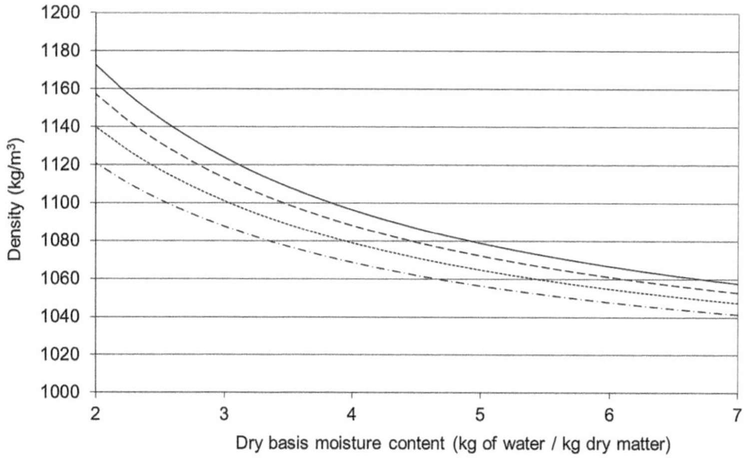

Fig. 9.10 Influence of the dry basis moisture content on the sludge density of four different total volatile solids content (- TVS = 45%, - - TVS = 55%, … TVS = 65% and -.-TVS = 75%)

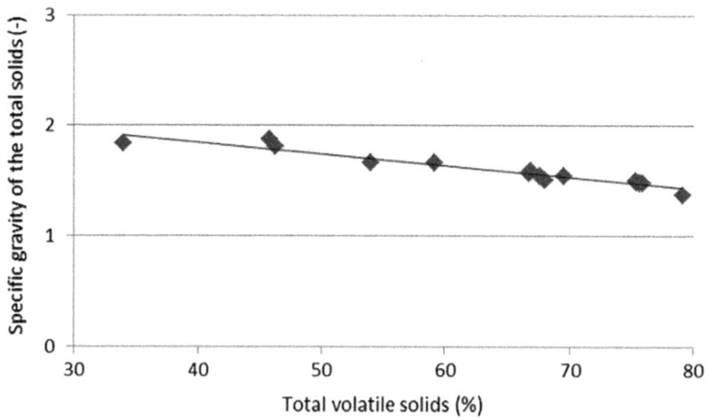

Fig. 9.11 Evolution of the total solids density with the total volatile solids content

For this method to be sound, orientation of the adsorbed molecules (parallel or perpendicular to the surface) must be known with confidence, as it has a direct impact on the value of the area covered by each molecule, A.

Apparatus

Experiments require:

- Flasks
- Flask shaker

- Centrifuge
- UV-Vis spectro-photometer
- Dye.

Procedure

The procedure is as follows:

- Partially fill the flasks with sludge and aqueous solutions with different dye concentrations.
- Make up the volume with acidified water, to maintain a constant pH.
- Shake vigorously the flasks to ensure that equilibrium conditions are reached and complete adsorption established. Duration of at least 30 min is recommended.
- Centrifuge the flasks.
- Collect and, if necessary, dilute the supernatants.
- Determine the concentrations of the dye remaining in the supernatants with the UV-Vis photo-spectrometer.
- Draw the adsorption isotherm.

Prior calibration using solutions with different known dyes concentration is needed.

Interferences

The presence of impurities in the adsorbates and the pH may have influences on the adsorption.

Comments

Figure 9.12 displays typical examples of dye adsorption curves for activated sludge using Lissamine Scarlet 4R dye [88]. Maximum dye adsorption is usually between 70 and 180 mmol/kg (range determined from more than 50 dye adsorption curves). The area covered by each Lissamine Scarlet 4R dye molecules is 196 $\dot{A}^2$. Assuming a mean value of 120 mmol/kg, the specific surface area of activated sludge is then equal to 142 m^2/g.

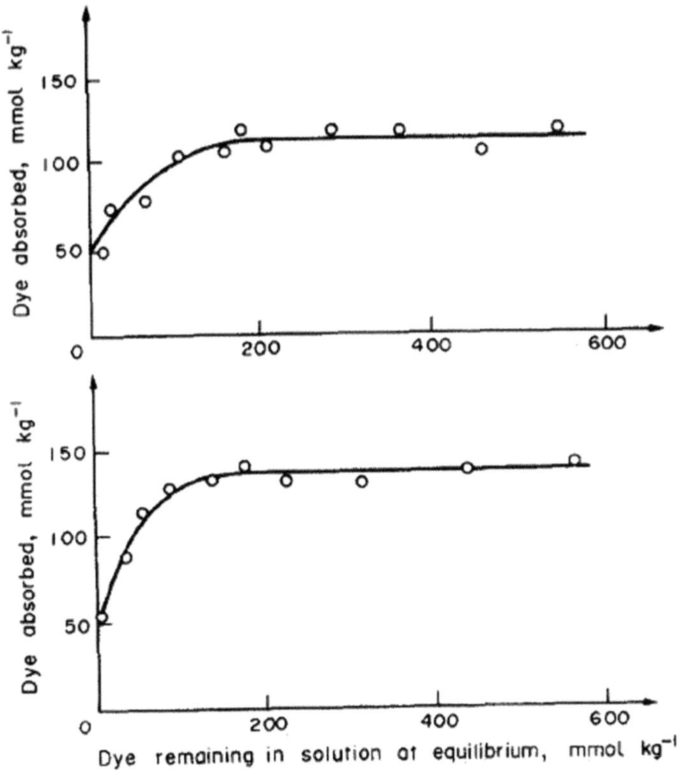

Fig. 9.12 Examples of dye adsorption curves for activated sludge using Lissamine Scarlet 4R dye. Reprinted with permission from Elsevier [88]

9.4.2 Chemical Characteristics

Among the most influent chemical properties, pH and alkalinity affect sludge conditioning or anaerobic digestion, for instance, while speciation and bioavailability of heavy metal fraction are of specific concern for sludge reuse.

9.4.2.1 pH

General Discussion

pH, also called the hydrogen ion activity index, is a scale of hydrogen ion activity in the solution, which reflects the degree of acidity of the solution. pH is defined as the negative logarithm of the hydrogen-ion concentration.

pH of aqueous samples is usually measured with pH meter. Prior to the analysis, the instrument must be calibrated by using at least two standard buffer solutions.

Table 9.41 Standard related to pH characterization

Title	Reference
ASTM D4980-89 Standard Test Methods for Screening of pH in Waste	[89]

The results should be reported to the nearest 0.1 pH unit. Though less accurate, pH papers and indicator solutions, that change color at definite pH values, are also used.

Reference standard related to pH characterization is listed in Table 9.41.

Apparatus

Experiments require:

- pH meter with means for temperature compensation. A wide variety of instruments are commercially available with various specifications and optional equipment.
- Glass Electrode.
- Reference Electrode. A silver-silver chloride or other reference electrode of constant potential may be used.
- Beaker.
- Magnetic Stirrer and TFE-Fluorocarbon-Coated Stirring Bar.
- Thermometer or Temperature Sensor for automatic compensation.

Procedure

Calibration and Standardization. Each instrument/electrode system must be calibrated at two points that bracket the expected pH of the samples and are approximately three pH units or more apart. Adjustments must be repeated until readings are within 0.05 pH units of the buffer solution values.

Sample analysis. The procedure for sample analysis is as follows:

- Add reagent water to the sludge to prepare a 10% slurry of sludge in reagent water.
- Stir continuously for 5 min and then let the solution stand for 15 min to allow most of the suspended solids to settle out from the suspension.
- Immerse the electrode into the clear supernatant and take measurement.
- Thoroughly rinse and gently wipe the electrode prior to measuring pH of each sample.
- Note and record sample pH and temperature. If the sample's temperature differs by more than 2 °C from the buffer solution, correct the measured pH values.
- Repeat measurement on successive volumes of sample until values differ by less than 0.1 pH unit. Two or three volume changes are usually sufficient.

Interferences

Sodium error at pH level > 10 can be reduced or eliminated by using a low sodium error electrode.

Coatings of oily material or particulate matter can impair electrode response. These coatings can usually be removed by gentle wiping or detergent washing, followed by dilute water rinsing. An additional treatment with hydrochloric acid 1 + 9, (1 volume HCl added to 9 volumes of H_2O) may be necessary to remove any remaining film.

Comments

Range suitable for the existence of most biological active life is narrow, typically from 6 to 9, but optimum pH for bacterial growth is between 6.5 and 7.5. Accordingly, the pH of sludge is usually around 7, but may vary from 5 to 8 [90, 91]. Anaerobic digestion process is especially sensitive to pH, as the methanogenic activity is inhibited below 6.8.

9.4.2.2 Alkalinity and Buffering Capacity

General Discussion

Alkalinity indicates the ability of water to absorb protons and thus represents the capacity of the water to resist changes in pH. Usually, alkalinity is measured by titrating the solution with a strong acid (HCl–0.02 N), until its pH changes abruptly. To develop the titration curve, standard acid is added to the sample in small increments and the pH is read after each addition. The cumulative volume of solution added is plotted against the observed pH values. Alkalinity related to a particular pH is computed as follows:

$$Alk\,(as\ meq/L) = \frac{AN}{B} * C \tag{9.27}$$

where

C is a constant equal to 1000 mL/L
A is the volume of the standard acid required for the titration, in mL
N is the normality of the standard solution (−)
B is the volume of the sample titrated, in mL.

Alkalinity is expressed in milliequivalents per liter (meq/L), which corresponds to the amount of acid added as a titrant in millimoles per liter.

Table 9.42 Standard related to alkalinity characterization

Title	Reference
ASTM International. ASTM D1067-16 Standard Test Methods for Acidity or Alkalinity of Water	[92]

Results are often reported in terms of calcium carbonate, mg/L as $CaCO_3$. The following relationship can be used to convert the value:

$$Alk(as\,CaCO_3) = Alk\,(as\,meq/L) * 50 \tag{9.28}$$

Reference standard is listed in Table 9.42.

Apparatus

Experiments require:

- Any commercial pH meter. Calibration is required prior to the analysis
- Glass and reference electrodes
- Magnetic stirrer
- Beaker
- Pipets.

Procedure

The procedure is as follows:

- Pipette 100 mL of the sample into the beaker.
- Measure the pH of the sample.
- Add 0.02 N HCl solution in increments of 0.5 mL or less.
- After each addition, stir the solution thoroughly.
- Determine the pH when the mixture has reached equilibrium (indicated by a constant reading).
- Continue the titration. The inflection point corresponding to the complete titration of carbonic acid salts is around pH 3.9.
- Once the necessary data for the titration curve have been obtained, plot the cumulative millilitres of standard solution added to the sample aliquot against the observed pH values and compute the alkalinity relative to a particular.

If the sample requires appreciably more than 25 mL of standard solution for its titration, a smaller aliquot or a 0.1 N solution, prepared and standardized in the same manner, must be used.

Interferences

Oily matter or suspended solids may interfere with pH measurements. However these materials may not be removed to increase precision, because they might affect the acid-consuming property of the sample.

Development of a precipitate during titration may make the glass electrode sluggish and skew the results.

Comments

CO_2 is the main consumer of alkalinity in an anaerobic digester. A well-established digester has a total alkalinity of 2000–5000 mg/L as $CaCO_3$ [4].

The measurement carried out to determine the alkalinity can also serve to determine the buffering capacity of the sample [92]. Buffering capacity is a quantitative measure of the resistance of a buffer solution to pH change on addition of acids (or bases). It is defined as the moles/L of strong acid (or strong bases) needed for a change in one pH unit of a solution. Buffering capacity depends on the concentration of the buffer and on the type and concentration of the acid (or base) to be added to the buffer solution. Sludge shows a certain buffering capacity as indicated by weak inflexion points [93].

9.4.2.3 Mobility and Bioavailability of Heavy Metal Fractions

General Discussion

According to the waste management hierarchy, agricultural reuse is widespread worldwide. As sludge is frequently loaded with heavy metals, mobility and bioavailability of the metals have been of great concern. The slightly modified Community Bureau of Reference (BCR) three-step sequential extraction procedure [94–96] is usually applied to study the partitioning of heavy metals in four fractions (arranged in relation to the mobility and bioavailability of the metals):

- exchangeable metal and carbonate-associated fraction (F1),
- reducible fraction (F2),
- oxidizable fraction (F3),
- residual fraction (F4).

Reference standard is listed in Table 9.43.

Table 9.43 Standard related to speciation characterization

Title	Reference
ISO 12782:2012 Soil quality—Parameters for geochemical modelling of leaching and speciation of constituents in soils and materials	[97]

Apparatus

- Flame atomic absorption spectrometry (FAAS), electrothermal atomic absorption spectrometry (ETAAS), inductively coupled plasma atomic emission spectrometry (ICP-AES) or inductively coupled plasma optimal emission spectroscopy (ICP-OES)
- pH meter
- Centrifuge
- A suction filtration apparatus.

Procedure

The procedure is as follows:

- Determine the total concentration of metals in the original sample (refer to Sect. 9.3.5) first.
- Then apply sequential extraction procedure described in Fig. 9.13 and analyze the heavy metals in each extract immediately.

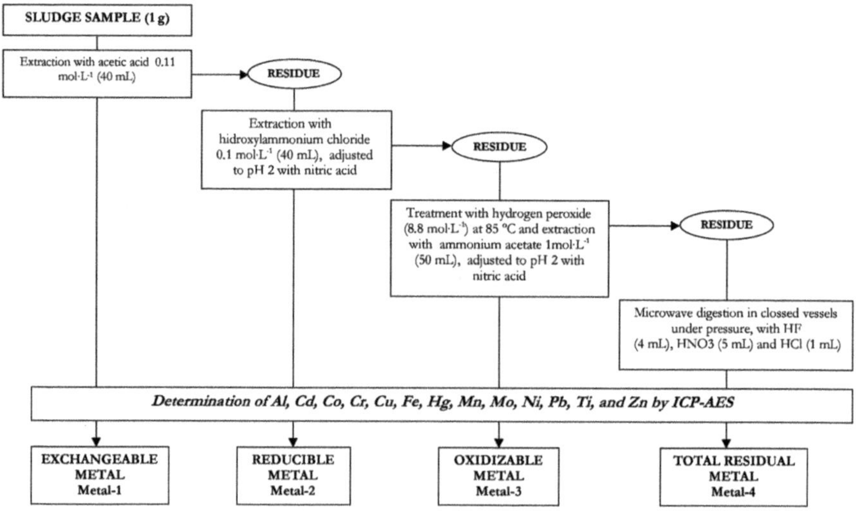

Fig. 9.13 Sequential extraction procedure to determine the different metal fractions. Reprinted with permission from Elsevier [98]

Comment

Figures 9.14 and 9.15 show the distribution of heavy metals in the four different fractions for 9 textile sludge sampled in China [96] and 10 municipal sewage sludge sampled in Spain [98], respectively. Heavy metals are mainly associated with the

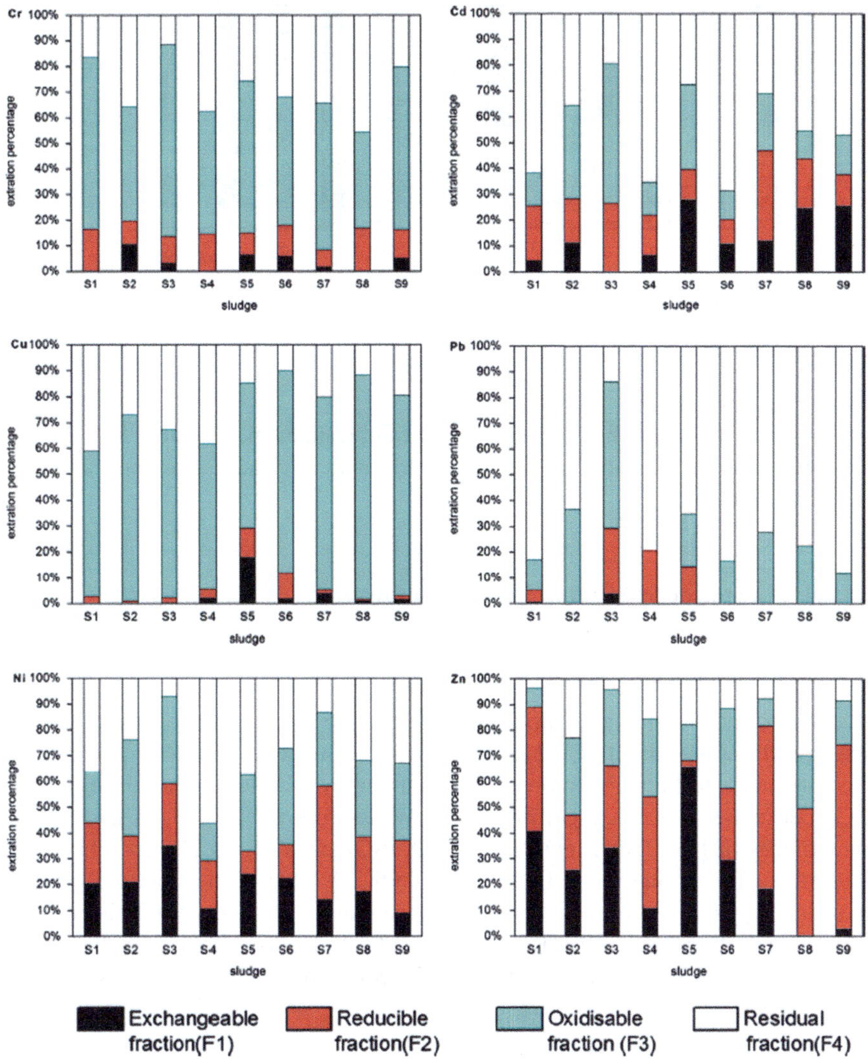

Fig. 9.14 Distribution of fractions for selected metals in 9 different sludge from China. Reprinted with permission from Elsevier [96]

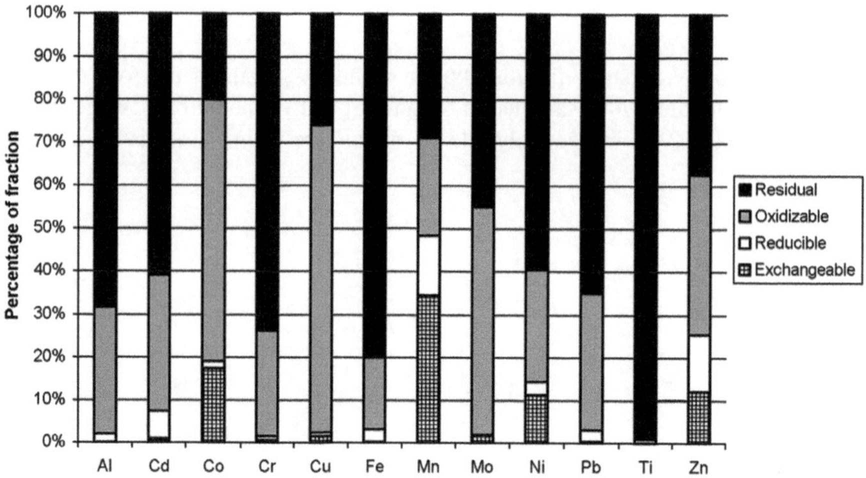

Fig. 9.15 Mean percentages, in relation to the total concentration, of each of the metal fractions obtained after fractionation of 10 samples collected in different WWTPs in Spain. Reprinted with permission from Elsevier [98]

two less available fractions (oxidizable metal and residual metal). Only Mn, Zn and, to a lesser extent, Ni show the highest share of the available (exchangeable and reducible) fractions.

9.4.2.4 Chemical Speciation, Phase Identification

General Discussion

The composition of sludge is varied according to the nature of their production. For instance, the composition of WTRs differ with the quality of raw water, coagulants used, treatment technology involved and final quality of water produced. In general, SiO_2 constitutes the major portion of the WTRs followed by Al_2O_3 and Fe_2O_3, other oxides such as CaO, MgO, Na_2O, K_2O, P_2O_5 and TiO_2 are also found in small percentage. The amount of Al_2O_3 or Fe_2O_3 in the WTRs is also associated with the coagulant applied (Al or Fe salts) and concentration of these metals in the raw water. X-ray diffraction (XRD) provides rapid information about crystallographic structure and can be used to determine the specific mineral phases in a sample.

Apparatus [99–101]

Experiments require:

- X-ray powder diffractometer

- Evaporating dish
- Oven
- Balance.

Procedure

Operating principles of X-ray powder diffractometers (XRD) were described in details in Chap. 2. Sample preparation is a critical stage. Indeed, the X-ray beam penetration in the sample is around 100 μm. For dried sludge samples, grinding is recommended for reducing the particle size and ensuring that sufficient grains contribute to the reflection of the beam. Additional details were available in Chap. 2. The method for phase identification will be explained in Chap. 15.

Comments

Al and Fe hydroxides of the WTRs were found to exhibit amorphous nature. WTRs are often called alum sludge or ferric sludge, depending on the chemicals (aluminium-based or ferric-based) used during the coagulation process. X-ray diffraction (XRD) analysis of alum sludge showed no sharp characteristic diffraction peaks, indicating poorly ordered particles, having no distinct shape or form and suggesting that the crystalline $Al(OH)_3$ phase is absent within alum sludge, although pure aluminium hydroxide exhibited a regular crystalline structure [99, 100]. The XRD pattern shown in Fig. 9.16 illustrates silica (SiO_2) as the major crystalline phase present in the amorphous alum sludge. It is important to note that ageing (18 months) seems no significant effect on alum sludge structure. Even the

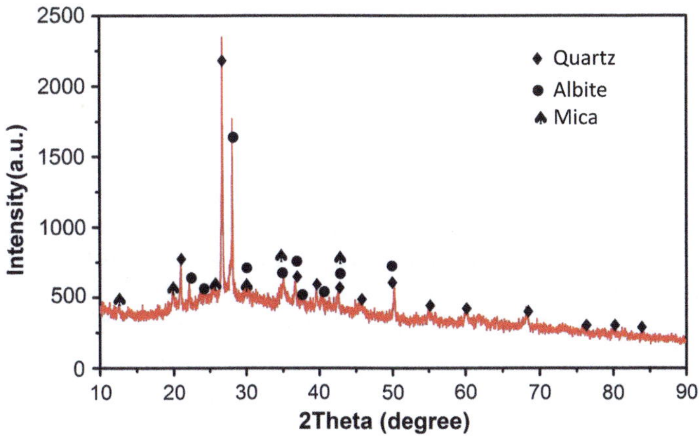

Fig. 9.16 XRD image of a typical alum sludge

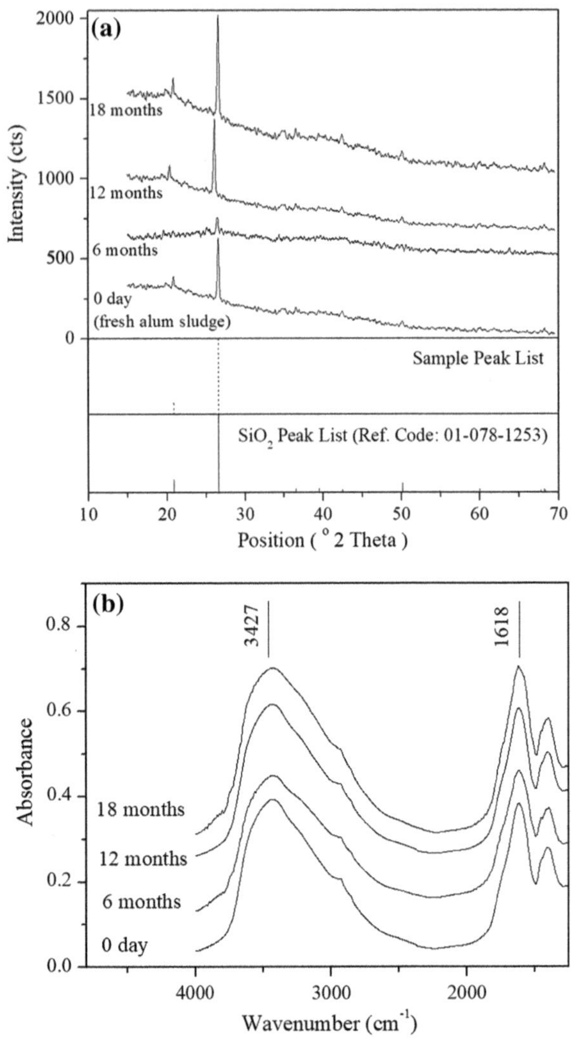

Fig. 9.17 XRD (**a**) and ATR-FTIR (**b**) spectra of alum sludge samples with ageing times. Reprinted with permission from Elsevier [101]

aged sludge exhibited similar XRD and ATR-FTIR characteristics as that of fresh sludge (Fig. 9.17), indicating that the crystallization did not occur during the ageing process and XRD pattern illustrates the amorphous structure of alum sludge [101].

Figure 9.18 shows the scanning electron microscope (SEM) observation of WTRs. There is no well-crystalline appearance being found on sludge surface and therefore it was reported that WTRs has amorphous structure [100, 102]. Babatunde et al. [103] further pointed the highly porous structure of the sludge surface. Makris et al. [104] reported that the total micropore volumes for alum and ferric sludge were 0.042 and 0.012 cm^3/g, respectively. The amorphous and porous nature of Al and Fe hydroxides makes the sludge an adsorption site for majority of anions [105].

Fig. 9.18 SEM images of alum sludge with magnitude of 5000 (upper) and 10,000 (lower)

9.4.3 Thermal Properties

Among the most influent thermal properties, heating values are by far the most important ones for energy recovery from sludge. However, the specific heat and thermal conductivity are also important material properties for any thermal processes, especially during the transient operation.

9.4.3.1 Heating Values

General Discussion

The heating value is the amount of energy released by the complete combustion of the substance. It is measured as a unit of energy per unit mass or volume of substance. As a basis for determining the performance of an energy system, it is necessary to differentiate between higher heating value (HHV) and lower heating value (LHV), as already mentioned in Chap. 2.

HHV includes the latent heat of condensation of the water in the combustion flue gas. Usually, the temperature of the exhaust flue gas of an industrial boiler is high enough to prevent water condensation. As a consequence, the effective heat available for use in the boiler is a lower amount called LHV. LHV is the amount of heat released by fully combusting a specified quantity of fuel less the heat of condensation of water in the combustion product (i.e. the water formed during the combustion from the hydrogen in the sludge and the water originally present in the raw sludge). Even though it is frequently quoted in the technical literature, HHV is of limited practical use with regard to LHV. However, existing methods cannot directly determine LHV. The methodology consists of determining HHV using a bomb calorimeter and then calculating LHV from the measured HHV value. Calculation requires information on the intrinsic moisture and hydrogen content of the sample, which transform into water vapor during combustion of the sample.

Usually, the sample is dried prior to the analysis. Thus, $\mathrm{HHV_{TS}}$, the higher heating value of the total solids, will be directly calculated by the calorimeter and LHV of the dry sample, $\mathrm{LHV_{TS}}$, is computed according to:

$$\mathrm{LHV}_{TS} = \mathrm{HHV}_{TS} - L_v \frac{9H}{100} \tag{9.29}$$

where

H is the mass fractions of hydrogen in the total solids, expressed in weight percent on a dry basis.
L_v is the latent heat of vaporization of the water, expressed in the same units as $\mathrm{HHV_{TS}}$ and $\mathrm{LHV_{TS}}$ (usually in MJ/kg).

Table 9.44 Standards related to heating value

Titles	References
CEN/TS 14918:2005 Solid Biofuels—Method for the determination of calorific value	[106]
EN 15400 Solid recovered fuels—Determination of calorific value	[107]

Then, the heating values of the wet sludge, HHV and LHV, may be calculated knowing the initial moisture content of the sample, as follows:

$$HHV = \left(1 - \frac{W_{db}}{1 + W_{db}}\right) HHV_{TS} \quad (9.30)$$

$$LHV = \left(1 - \frac{W_{db}}{1 + W_{db}}\right) LHV_{TS} - \frac{W_{db}}{1 + W_{db}} L_v \quad (9.31)$$

where

W_{db} is expressed in kg of water per kg of dry solids.

Reference standards are listed in Table 9.44.

Apparatus

Experiments require:

- Bomb calorimeter
- Evaporating dish
- Oven
- Balance.

The bomb calorimeter system is described in details in Chap. 2.

Procedure

Calibration and measurement methods were detailed in Chap. 2. Due to the high moisture content of sludge, the sample has to be dried to a constant weight in an oven at 105 ± 5 °C prior to the analysis.

Interferences

When using a bomb calorimeter, correction for extraneous energy is necessary. The detailed method can be referred in Chap. 2.

Table 9.45 Typical HHV_{TS} values for several types of sewage sludge

Type of sludge	Heating value (MJ/Kg of DS) range	Typical
Raw sludge	23–29	25.5
Activated sludge	16–23	21
Anaerobically digested primary sludge	9–13	11
Raw chemically precipitated primary sludge	14–18	16
Biological tilter sludge	16–23	19.5

Reprinted with permission from Elsevier [91]

Comments

Typical higher heating values of the total solids are given in Table 9.45. Stabilized sludge have significantly lower heating values than raw sludge. Generally, the heating value is equivalent to that of some of the lower grades of coal.

Many correlations are available in the literature [108–111] to estimate the HHV of biomass fuels from the proximate or ultimate analyses or from thermochemical and stoichiometric considerations. The accuracy of some of the empirical formulae was statistically evaluated for biomass fuels [109, 111] and for sewage sludge from Thailand [108]. These studies emphasized that correlations based on the ultimate analysis are generally more accurate than those based on proximate analysis while popular expressions, such as Dulong or Steuer equation, derived from thermo-chemical concepts cannot be generalized beyond the specific material for which they were established. Parikh et al. [111] introduced a general correlation, based on proximate analysis of solid fuels, to calculate HHV. The entire spectrum of solid carbonaceous materials like coals, lignite, all types of biomass material, and char to residue-derived fuels, over 450 data points from the literature and 100 additional data points for further validation, have been considered in derivation of the following correlation:

$$HHV_{TS} = 0.3536\,\Delta TFS + 0.1559\,TVS - 0.0078\,TFS \tag{9.32}$$

whereas Thipkhunthod et al. [108] proposed the following relation for sewage sludge:

$$HHV_{TS} = 0.2839\,\Delta TFS + 0.2557\,TVS - 2.3864 \tag{9.33}$$

In both equations, ΔTFS, TVS and TFS are the mass fraction expressed in weight percent on dry basis (see Sect. 9.3.1.4) and HHV_{TS} is the heating value of the total solids in MJ/kg.

The Milne formula is often recommended for computing the HHV_{TS} of solid fuels from the ultimate analysis whereas US EPA [112] advises the use of Dulong equation. Milne equation is as follows:

$$HHV_{TS} = 0.341\,C + 1.322\,H - 0.12\,(O + N) + 0.0686\,S - 0.0153\,TFS \quad (9.34)$$

and Dulong equation is written as:

$$HHV_{TS} = 0.3386\,C + 1.4317\left(H - \frac{O}{8}\right) + 0.0941\,S \quad (9.35)$$

In both equations, C, H, O, N, S (see Sect. 9.3.3) and TFS are the mass fraction expressed in weight percent on dry base and HHV_{TS} the heating value of the total solids in MJ/kg.

To assess the accuracy of these correlations, a comparison is carried out on the basis of average absolute (AAE) and bias (ABE) errors, computed as follows:

$$AAE = \frac{1}{n}\sum_{1}^{n}\left|\frac{HHV_{TS,calc} - HHV_{TS,exp}}{HHV_{TS,exp}}\right| \quad (9.36)$$

$$ABE = \frac{1}{n}\sum_{1}^{n}\left(\frac{HHV_{TS,calc} - HHV_{TS,exp}}{HHV_{TS,exp}}\right) \quad (9.37)$$

Data collected in the literature [108, 113] have been supplemented by authors' data. Therefore, HHV experiments were conducted for the fifteen sewage sludge samples described in Sect. 9.3.1.4. A IKA-C5000 bomb calorimeter (IKA Labotechnik, Stanfen, Germany) was used for these analyses. Experimental HHV_{TS} values are reported in Table 9.46.

The comparison between experimental data and higher heating values of the total solids computed from correlations (9.32)–(9.33) or (9.34)–(9.35) are plotted in Figs. 9.19 and 9.20, respectively.

Average absolute (AAE) and bias (ABE) errors are given in Table 9.47.

The general correlation introduced by Parikh et al. [111] for any fuel does not account for the experimental results. Satisfactory results can be given by Milne and Dulong equations after considering the precision of elemental analyses but the lowest average absolute and bias errors can be obtained by the correlation suggested by Thipkhunthod et al. [108]. Only based on proximate analysis data, this equation provides a rapid, simple and economical estimation of the HHV_{TS}. To further simplify the characterization step, the higher heating value of the total solids is significantly correlated (R^2 = 0.8627) with the total volatile solids (TVS) content, as can been seen in Fig. 9.21. The average absolute error of the correlation given below is 10.4% and bias error is 2.2% with respect to the measured HHV_{TS} values.

$$HHV_{TS} = 0.2817\,TVS - 2.1645 \quad (9.38)$$

Table 9.46 Measured higher heating values of the total solids

Sludge	S1	S2	S3	S4	S5	S6	S7	S8	S9	S10	S11	S12	S13	S14	S15
HHV_{TS} (MJ/kg)	18.0	19.4	17.7	17.3	18.4	15.9	11.6	17.2	16.6	11.0	16.2	13.2	11.1	6.5	9.2
RSD (%)	8.2	0.5	0.5	1.2	0.6	0.1	3.1	0.2	0.7	2.1	0.2	1.1	1.9	5.3	1.1

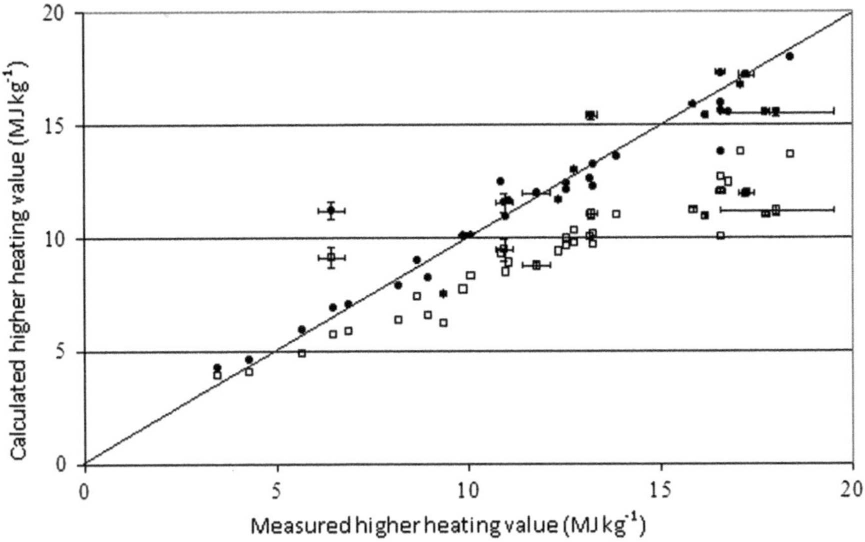

Fig. 9.19 Comparison between experimental data and higher heating values of the total solids computed from correlations (9.32) and (9.33) (empty square box Parikh's correlation [111], filled circle Thipkhunthod's correlation [108])

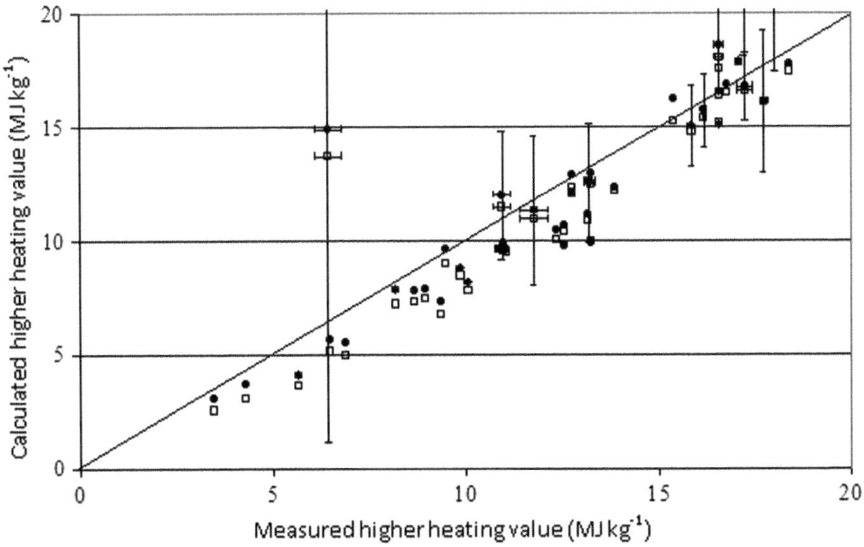

Fig. 9.20 Comparison between experimental data and higher heating values of the total solids computed from correlations (9.34) and (9.35) (empty square box Milne formula, filled circle Dulong equation [112])

Table 9.47 Errors generated when applying the correlations to the experimental data

	AAE (%)	ABE (%)
Parikh's correlation [111], Eq. (9.32)	23.73	−21.15
Thipkhunthod's correlation [108], Eq. (9.33)	7.47	0.77
Milne formula, Eq. (9.34)	15.41	−7.26
Dulong equation [112], Eq. (9.35)	10.99	−6.78

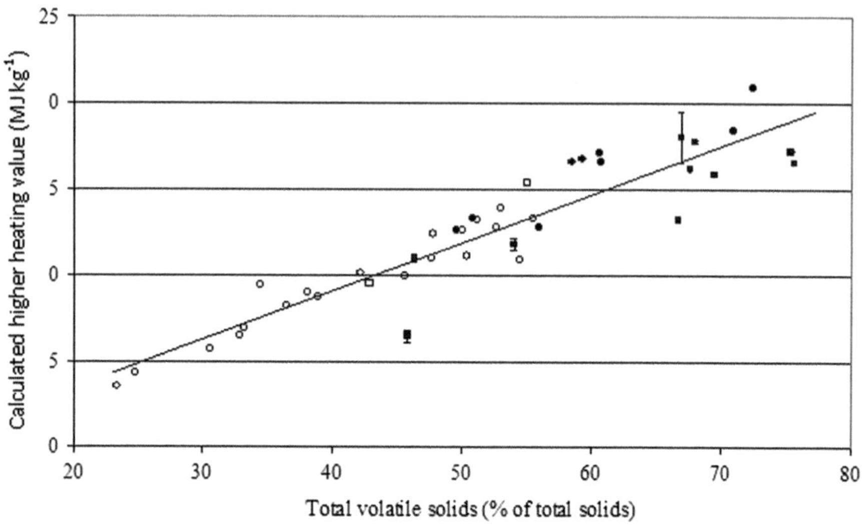

Fig. 9.21 Evolution of the higher heating value of total solids with the total volatile solids content (empty square box authors' data, empty circle Thipkhunthod et al. [108], filled circle Calvo et al. [113])

9.4.3.2 Specific Heat

General Discussion

The specific heat, C_p, is an intensive thermodynamic property, which states the amount of heat per unit mass of the substance required to raise its temperature by 1 °C. C_p is expressed in kJ/(kg·K).

C_p is temperature dependent, particularly at low temperatures. In the engineering practice, the specific heat of the material used is one of the decisive factors under conditions of transient heat flow, especially in sizing sludge cooling and heating equipment.

Scanning calorimeters are conventionally used to determine specific heat of liquids, gases and powder. C_p is deduced from heat measurement as a function of temperature while the sample is subjected to a controlled temperature program (usually a linear variation of the temperature with time). The testing procedure provides an instantaneous C_p over a temperature range.

C_p depends upon chemical composition but the test procedure is limited to dry sludge sample. The specific heat of the wet sludge is computed from the specific heat of the water, $C_{p,w}$, and that of the total solids, $C_{p,TS}$, as follows:

$$C_p(W_{db}, T) = \frac{W_{db}}{1 + W_{db}} C_{p,w}(T) + \frac{1}{1 + W_{db}} C_{p,TS}(T) \tag{9.39}$$

where

W_{db} is the dry basis moisture content, in kg of water per kg of dry solids.

Apparatus

Experiments require:

- Scanning calorimeter
- Evaporating dish
- Oven
- Balance.

Calorimetry systems were described in Chap. 2.

Procedure

Methods are detailed in Chap. 2. Prior to the analysis, the sample has to be dried to a constant weight in an oven at 105 ± 5 °C and ground to a fine powder using a hammer mill.

Interference

When a material is heated, physico-chemical phenomena such as chemical reactions, glass transition etc., can take place at a given temperature. These transformations imply heat absorption or release and skew the results. The investigated temperature range should be limited to 25–100 °C to avoid chemical decomposition of the samples.

Comments

The C_p change with temperature depends on the composition of the sludge. Furthermore, the sludge morphology strongly affects the C_p value. To highlight the influence of those factors, the specific heat of the total solid, $C_{p,TS}$, were measured for the fifteen sewage sludge samples described in Sect. 9.3.1.4. A Tian-Calvet type

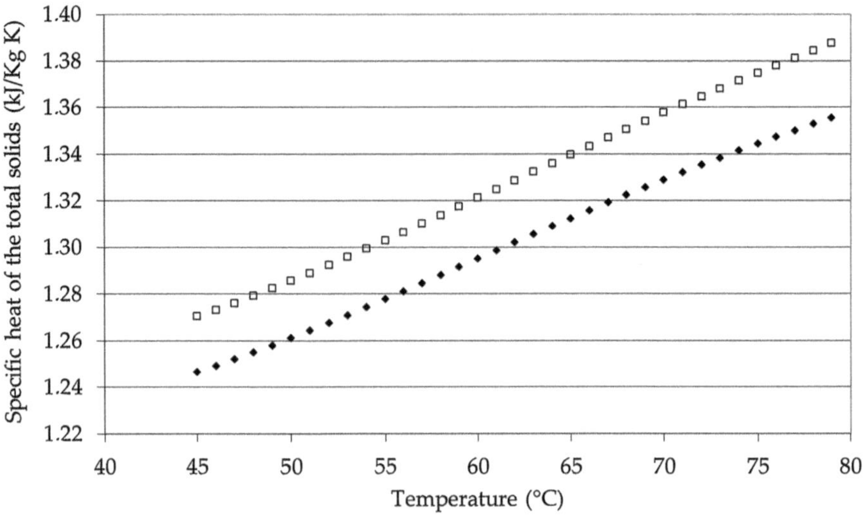

Fig. 9.22 Influence of the sludge morphology (sample S3) on the specific heat of the total solids (filled circle Powder, empty square box Pellets)

microcalorimeter, model C80 II calorimeter (Setaram, Caluire, France) was used for these analyses. The temperature program was to increase the temperature from ambient to 95 °C at the rate of 0.1 °C/ min.

As the dried sludge is a porous material, the specific heat of the total solids depends strongly on the sludge morphology (particle size distribution, particle shape and thus sample internal porosity). This phenomenon is highlighted in Fig. 9.22. Depending what is the intended purpose, this influence has to be taken into account.

The evolution of the specific heat with the temperature is plotted in Fig. 9.23 for eleven of the 15 tested samples. Except for samples S6 and S10 and to a certain extent for sample S7, $C_{p,TS}$ slightly increases with the temperature within the temperature range investigated.

As a first approximation, linear regressions, based on the least squares method, were performed to fit the experimental results with the following model:

$$C_{p,TS} = aT + b \tag{9.40}$$

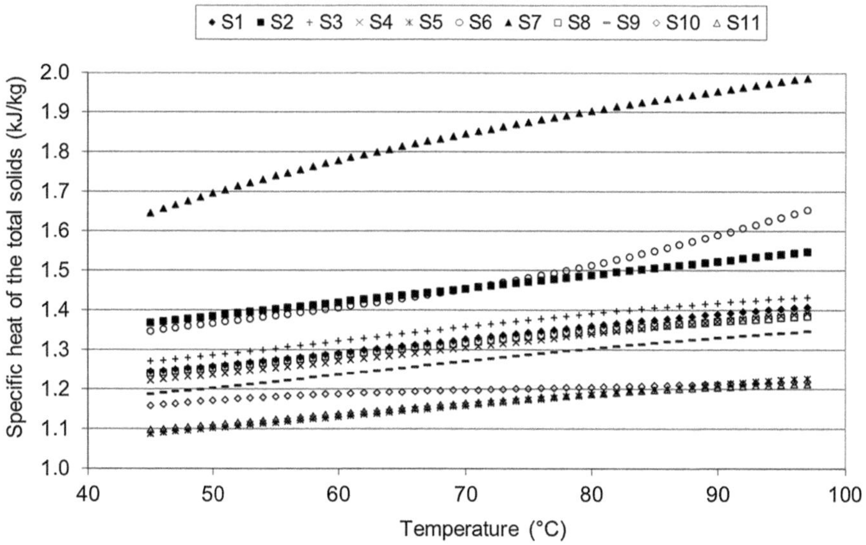

Fig. 9.23 Evolution of the total solids specific heat with the temperature

where

T is the temperature, in °C
a and b are the two regression coefficients, expressed in kJ/(kg.K) and kJ/kg respectively.

Values of the estimated regression coefficients are reported in Table 9.48. Constant a, which depends on the sludge porosity and the particle size distribution, cannot be correlated with the sludge composition, as long as the porosity is unknown. Conversely, constant b, which is independent of the sludge morphology, is significantly correlated ($R^2 = 0.826$) with the total volatile solids content as follows:

$$b = 0.0368\, TVS + 0.6851 \tag{9.41}$$

where

TVS is the total volatile solids content, expressed in weight percent of the total solids.

9.4.3.3 Thermal Conductivity

General Discussion

The thermal conductivity (λ) indicates the ability of a material to transfer heat by conduction. It is evaluated primarily in terms of the Fourier's Law for heat

Table 9.48 Estimated regression coefficients

Sample	a (J/kg K)	b (J/kg)	R^2
S1	1096.3	3.3	0.9988
S2	1189.1	3.4	0.9975
S3	1213	3.44	0.9888
S4	1119.4	2.9	0.9889
S5	1114.1	3.3	0.9982
S6	1070.6	3.3	0.9996
S7	967.1	2.7	0.9991
S8	1068.8	5.7	0.9771
S9	1387.8	6.4	0.9877
S10	1047.1	2.7	0.9955
S11	1076.2	3.3	0.9976
S12	1047.2	3.2	0.9971
S13	1123.3	1	0.9538
S14	996.4	2.3	0.9874
S15	1090.6	2.1	0.991

conduction. In general, thermal conductivity is a tensor property, expressing the anisotropy of the property.

Several direct methods are available in the literature for determining λ. All these methods are based on transient heat transfer experiments, using a probe inserted in the sample initially in thermal equilibrium. The method consists of measuring the temperature response of a heat step imposed on the probe by Joule effect. Consequently, the probe shall include a heating element and an integrated temperature sensor. Thermal conductivity is obtained from an analysis of the temperature time series data. The temperature decay with time after the cessation of heating can also be included in the analysis to minimize effects of temperature drift during measurement.

The thermal conductivity should be calculated as follows:

$$\lambda = \frac{Q}{4\pi\Delta T}\ln(t), 0 < t \leq t_1 \tag{9.42}$$

$$\lambda = \frac{Q}{4\pi\Delta T}\ln\left(\frac{t}{t - t_1}\right), t > t_1 \tag{9.43}$$

where

Q is the heat input per unit length of heater, in W/m,
ΔT is the change in temperature from time zero, in K,
t_1 is the heating time, in s,
t is the time from the beginning of heating, in s,
λ is the thermal conductivity, expressed in W/(m K).

Reference standard is listed in Table 9.49.

Table 9.49 Standard text related to thermal conductivity

Title	Reference
ASTM D5334-14 Standard Test Method for Determination of Thermal Conductivity of Soil and Soft Rock by Thermal Needle Probe Procedure	[114]

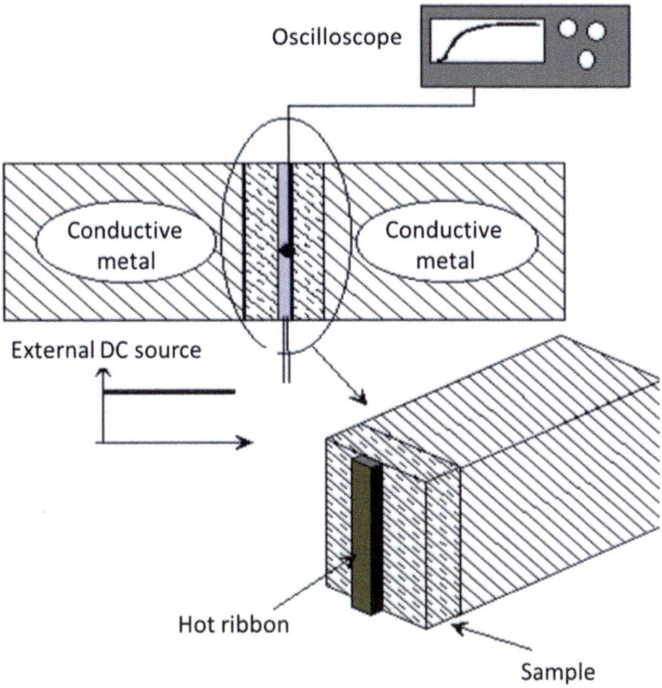

Fig. 9.24 Basic configuration for the hot ribbon method. Reprinted with permission from Ferrasse [116]

Apparatus

Experiments require:

- Probe
- External DC Source
- Temperature Recorder
- Volt/Ohm meter or oscilloscope.

A needle probe having a large length to diameter ratio to simulate conditions for an infinitely long and thin heating source can be used [114]. Alternative methods use a hot ribbon inserted between two sample layers contained into two hollowed out metallic mass [115] (Fig. 9.24). A thermally conductive material is recommended: it promotes heat diffusion through the sample layer and allows the control of the boundary conditions (need information of solving the inverse problem and estimating λ). The ribbon width should be equal to the thickness of the sludge layer.

Procedure

The procedure is as follows:

- Bring the sample to equilibrium at the selected testing temperature.
- Insert the probe into the sample and apply a known constant electrical power.
- Record the increase of temperature during an appropriate duration.
- Turn off the constant current source.
- If cooling data are to be included in the analysis, record the decrease of temperature throughout a cooling period equivalent in duration to the heating cycle.
- Use a suitable inverse method (graphical or statistical) to determine thermal conductivity.
- Measure the moisture content of the sample.

Interference

Redistribution of water can occur when heating the probe results in thermal or hydraulic gradients and phase change of water. These errors can be minimized by adding less total heat to the sample through either minimizing power applied to the probe and/or minimizing the heating duration. If experiments are conducted at high temperature, it is recommended to check that water evaporation during the experiments was low by measuring both the initial and final moisture contents of the sample.

For the hot wire method, care must be taken not to choose a too long heating duration. Indeed, this method is valid only when the thermal pulse does not encounter the boundaries of the sample.

Comments

Thermal conductivity decreases significantly with increasing total solids (Table 9.50). Compared to moisture content, the influence of the temperature on the thermal conductivity is weak. The data reported in Table 9.50 were achieved with the hot ribbon method, using a 6.1 mm thick sludge layer. As long as the moisture content is high, thermal conductivity is close to the thermal conductivity of the water (λ_{water} = 0.6 W/(m K)). The internal porosity of the dried matrix can influence significantly the thermal conductivity of the dried sludge. In this case, the material should be oven-dried to constant mass at 105 °C.

Table 9.50 Influence of the dry basis moisture and sample temperature on the thermal conductivity of a sewage sludge (λ expressed in W/(m.K)) [116]

	$W_{d.b}$ = 5.2 kg/kg	$W_{d.b}$ = 4 kg/kg	$W_{d.b}$ = 1 kg/kg	$W_{d.b}$ = 0 kg/kg
T_{sample} = 25 °C	0.57	0.55	0.28	0.13
T_{sample} = 60 °C	0.55	–	0.28	0.13
T_{sample} = 95 °C	–	–	–	0.13

Reprinted with permission from J. H. Ferrasse

9.5 Processing Characteristics

9.5.1 Introduction

The physical consistency, which is a characteristic strictly linked to the rheological properties, is a parameter of fundamental importance in sludge characterization, as it strongly affects almost all treatments, utilizations and disposal operations [117].

The assessment of physical and mechanical characterization methods and tests firstly requires the definition of the different physical states of sludge. The following three consistency categories have been proposed [118, 119]:

- Liquid: sludge flowing under the effect of gravity or low pressure;
- Paste-like: sludge capable of continuous flow under the effect of pressure above a certain threshold and having a shear resistance below a certain threshold;
- Solid: sludge having a high shear resistance.

This involves the need to set up laboratory methodologies to define the boundary limit between liquid and paste-like behaviors (flowability) and that between solid and paste-like behaviors (solidity). The definition of these limits could be slightly different for other types of materials. For instance, the Atterberg limits comprising a liquid and a plastic limits are defined on the basis of a plastic state instead of a paste-like one [120–122].

The consistency of a material describes its ability to support vertical and lateral loads. This ability is governed to a large extent by the shear strength of the material. Shear may be defined as the tendency of one part of a material to slide with respect to the other. This process of attachment of a substance to an identical or a similar substance is called the cohesion, while the adhesion refers to the attachment of a substance to the surface of another substance. Both adhesion and cohesion contribute to the stickiness of sludge and are also essential for many key unit operations in sludge treatments and valorization, such as for instance pumping, dewatering and drying.

To access to flowability, solidity, plasticity and stickiness, many apparatuses exist. Several of them, which can be applied in the field or in the laboratory, are suggested in standard methods for the characterization of various materials. Besides these standards, some non-standardized methods are also used in practice.

However, different conclusions are usually obtained by using different apparatuses and procedures and there are no standardized procedures to obtain comparable and reliable results for all materials. The available apparatuses can be classified into four main categories:

- tube type, often referred to as capillary viscometer,
- penetrometer,
- rotating type,
- other types.

In capillary viscometers, the viscosity is generally determined by measuring the time for a fixed volume of fluid to flow through the capillary calibrated viscometers. In a penetrometer, the resistance to tip penetration is measured. In rotating viscometer, the product studied fills the space between two elements. One of these two elements is rotated at a controlled rotation rate (torque respectively) in order to shear the sample contained in the gap and shear stress (rate respectively) is determined by measuring the resistant torque (rotation rate respectively) on this element. The others types of apparatuses also implement a flow of the material but with a specific geometry which can be more or less complex.

Reference [123] provides a comprehensive assessment and comparison of the different apparatuses for different materials. In the following, a description of the most commonly used apparatuses for sludge characterization, with the corresponding experimental methods, is proposed for each basic processing characteristic.

9.5.2 *Flowability*

9.5.2.1 General Discussion

Flowability is closely connected to rheological properties, rheology studying the relationship between shear stress (force per unit area required to produce the shearing action [124]) and shear rate (measure of the spatial change in speed at which the intermediate layers move with respect to each other [124]). The rheological behavior of dilute sludge (TS < 3 wt%) is Newtonian, like water, i.e. the dynamic viscosity (ratio of the applied shear stress to the shear rate in a rotating viscometer) is independent of the flow rate and no initial resistance is shown if a force is applied at rest (yield stress, defined as the minimum shear stress to initiate flow [124]). At higher concentration, the behavior becomes non-Newtonian, as sludge may exhibit a yield stress and the viscosity may vary with the shear rate [125]. It means that the whole plot of shear stress versus shear rate, also called rheogram, has to be determined [126].

Two main categories of apparatuses are usually used to evaluate this property [125, 126]: tube type (often referred as capillary viscometer) and rotating type.

Table 9.51 Summary of the recommended apparatuses for flowability

Apparatus/method
Rotational viscometer
Flow cone
Penetration cone
Kasumeter
Inclined plane
Modified slump test

However, several empirical apparatuses also exist. The recommended equipment and methods for characterizing sludge flowability [123] are listed in Table 9.51.

Flow cone is a simple, inexpensive and maneuverable apparatus but it is only suitable for Newtonian fluids. Moreover, sludge and treated bio-waste can contain coarse particles, likely to occlude the discharge orifice, and thus require a sieving pre-treatment [123].

Inclined plane and modified slump test methods are not standardized. Inclined plane method can be used to calculate the yield stress of visco-plastic fluids by determining the depth of the sludge remaining at the rest on the inclined channel [127]. It allows to measure the yield stress of coarse suspension without any pre-treatment but it has been tested only in a narrow yield stress range (from 35 to 90 Pa) [123].

Modified slump test is adapted for measuring the yield stress of a pasty sludge at different solid concentrations by determining the height loss of a cylindrical sample left flowing on a horizontal plane [128]. It has the advantage to involve a small volume of materials and to not be influenced by the presence of coarse particles, but the flow is not easy to be controlled for all materials [123].

Rotational viscometers have proven to be efficient tools to study rheological properties of sludge as they allow the determination of absolute viscosity of Newtonian and non-Newtonian substances, for every flow curve of the fluid in a wide range of viscosity values [125, 126]. There are no standardized methods for sludge but several exist for other materials, such as bitumen [129–131] and paint [132, 133].

Cone penetrometer is a simple apparatus easy to use, which allows measuring the resistance to tip penetration. The tip's shape determines the value of the calibration constant. The obtained yield stress is a measurement of the forces required to initiate and maintain flow from a storage vessel. This apparatus is suitable for testing both in laboratory and field, meanwhile the measure is very fast and the absolute value of the yield stress obtained can be compared with results of the reference laboratory method. The method described here is standardized for another type of materials [134].

The use of Kasumeter is a good approach, with in particular the mechanical simplicity (associated to a low cost), the geometric similarity with pipelines, the possibility of measurement at high shear rates and the diameter dependent effects [126]. However, a quite complicate procedure is often needed and only a poor

Table 9.52 Standards related to flowability measurements

Titles	References
ASTM D2884-93 Standard test method for yield stress of heterogeneous propellants by cone penetration method	[134]
EN 16720-1 Characterization of sludges—Physical consistency—Part 1: Determination of flowability—Method by extrusion tube apparatus	[135]

control of operative parameters is possible [123]. The following described apparatus has been identified as a promising method for field determination of the yield stress of high concentrated slurries [123] and has been finally standardized for sludge [135].

Reference standards related to flowability measurements are listed in Table 9.52.

9.5.2.2 Coaxial Rotational Viscometer

Apparatus

Firstly, the appropriate geometry, among the available ones, must be selected in accordance to the sludge consistency. Four main geometries exist (Fig. 9.25):

- coaxial cylinders with moving inner or outer cylinder (configuration b and e),
- rotating blades (configuration a),
- cone-plate (configuration c),
- plate-plate (configuration d).

The imposed range of torque and rotation speed of the rheometer motor implies to adapt the geometry to the range of materials properties; the higher the contact surface, the lower the viscosity.

In most cases, rheometric devices of small gap are used, in order to minimize the spatial variations in shear rate and shear stress [126]. In this case, the sheared geometry is well known and a calibration constant determined by using the

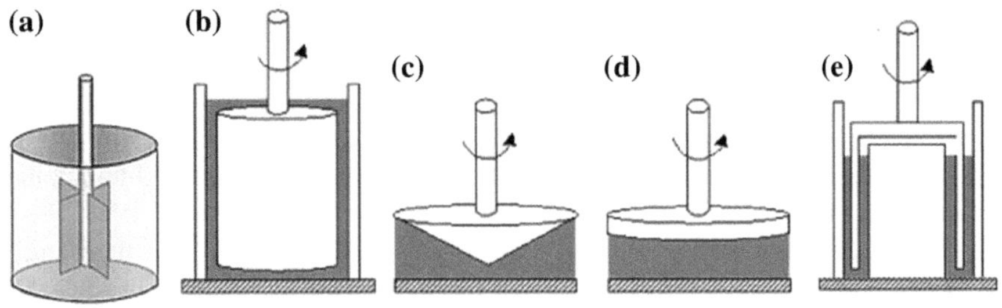

Fig. 9.25 Scheme of rotational viscometer: **a** vane, **b** concentric, **c** cone-plate, **d** parallel plate, **e** double concentric. Reprinted with permission from Elsevier [125]

Metzner-Otto calibration can be used directly to calculate the viscosity of the sample [136]. The cone-plate geometry is generally excluded on the basis of the large size of sludge particles relative to the gap [123]. On the contrary, the plate-plate geometry exhibits the possibility to adapt the gap but with an associated increase of the spatial variation of the flow.

If a specific rheometric device (vane apparatus or systemic rheology) is used, the velocity gradient is less well-defined, so only a mean value based on the mechanical energy dissipated in the medium can be obtained [126].

Procedure

There is no specific standard for sludge. However, the experimental procedure described hereafter is generic, whatever the material [129–133]. The sample is placed between the rotor and the stator. In regard of the calibration constant, the contact surface has to be respected what implies to remove any surplus samples.

In general, the measurement is implemented after a pre-shearing or after a fixed time to avoid stationary regime and obtain repeatable results.

The viscosity is then determined at one or several fixed shear rates. For a more complete measurement, a sweep of shear stress or strain is applied. An increasing sweep followed by a decreasing one allows checking the presence of hysteresis and the eventual thixotropic behaviour.

9.5.2.3 Magnesium Penetration Cone

Apparatus

The apparatus is described on the Fig. 9.26. It is composed mainly of a cup and a penetrometer. The cup should be made of a material compatible with the materials tested. The cone should fall, when released, without appreciable friction for at least 42 mm but not more than 60 mm so the cone cannot hit the bottom of the container. A spatula with a stiff blade 30 mm wide and at least 150 mm long is also necessary.

The material temperature has to be regulated prior to the trial at 298 ± 0.5 K. For unworked penetrations, means must be provided for protecting the material surface from atmosphere moisture [134].

Procedure

The material has to be transferred in the cup to overfill it, packed and leveled by using the spatula. The apparatus is set in the zero position, so that the tip of the cone just touches the surface of the test sample (the shadow of the cone tip is an aid to

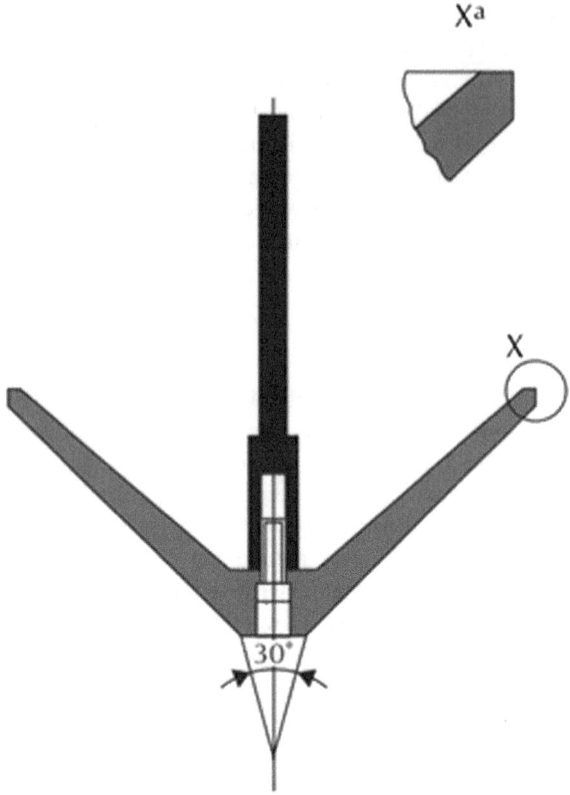

Fig. 9.26 Magnesium penetrometer cone. Reprinted with permission from IRSTEA [137]

accurate setting). The test rod is then released rapidly and allowed to drop for 5.0 ± 0.5 s and the penetration depth is read. The yield stress τ_0 (Pa) is calculated as follows:

$$\tau_0 = \frac{1}{10}\frac{98.1(30.0 - \left(1.047 \times 10^{-6}\right)\left(P - 107.4\right)^3\rho - 0.198\rho\right)}{4.44 \times 10^{-4}(P - 107.4)^2 + 1.260} \tag{9.44}$$

where

P is the penetration, in tenths of a mm
ρ is the density of the suspension, in kg/m^3.

If the penetration depth is over 20 mm, the sample can be used for only one test. If this penetration depth is lower than 20 mm, 3 tests have to be performed in a single container at different positions (spacing on three radii 120° apart and midway between the center and side of the container) so that the cone neither strikes the side of the container nor impinges on the disturbed area made in a previous test. The three test results should be averaged. If the penetration exceeds 42 mm, the sample should be reported as too soft.

The material of which the cone is made can be changed to determine an apparent yield stress in relation to a specific structural material or lining [134].

9.5.2.4 Kasumeter

Apparatus

The extrusion apparatus (also called Kasumeter) consists of a cylindrical acrylic container (100 mm internal diameter and 350 mm deep) to which calibrated pipes of length $l_d = 0.2$ m and different diameters ϕ_d (5, 10, and 20 mm) can be fitted at the bottom (Fig. 9.27) [135].

Procedure

A sufficient volume of the test sample is required (about 3 L to fill up the apparatus). The measurement procedure includes the following steps: (i) the Kasumeter is filled up with the suspension by keeping closed the end-side of the calibrated tube to avoid the presence of air bubbles in the calibrated tube, (ii) the end-side of the calibrated tube is opened, (iii) and the time is measured until the flow is

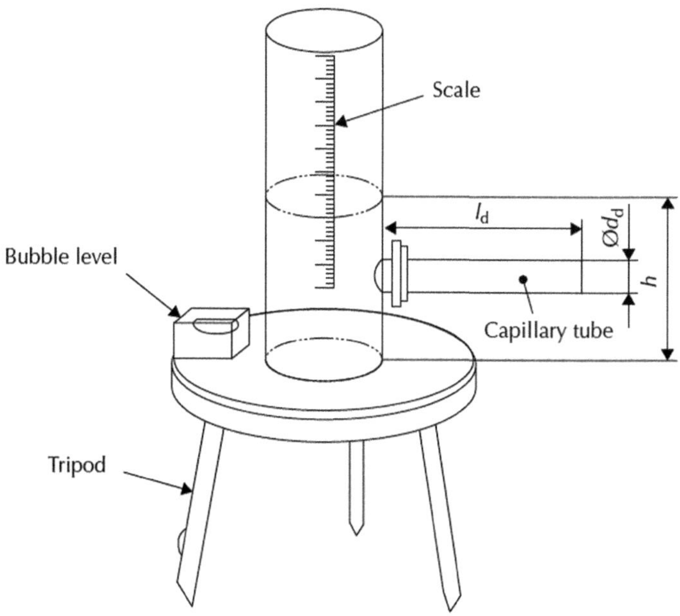

Fig. 9.27 Scheme of extrusion tube apparatus (Kasumeter). Reprinted with permission from IRSTEA [137]

discontinuous (reduced to one drop falling per second), and (iv) the height of the suspension remaining in the cylinder is measured. The flowability index (F_i) can be then calculated through the following formula:

$$F_i = \frac{3}{8}\frac{\rho g h r_d}{l_d} \tag{9.45}$$

where

ρ is the sludge density, in kg/m^3
g is the gravity acceleration, in m/s^2
h is the height of the suspension remaining in the cylinder, in m
r_d is the pipe radius, in m
l_d is the pipe length, in m.

Measurements are repeated with the same pipe but also with the 3 pipes of different diameters. If one of two of these 3 mean values differs by more than 20% from the calculated mean value, F_i has to be considered only as "indicative" and "not valid", respectively.

Limits exist also on the height h. For h values higher than 30 cm, results are "not significant". When h values are higher than 8 cm with the 20 mm diameter pipe, and/or higher than 16 cm with the 5 mm diameter pipe, "not valid" or "indicative" results have to be expected.

9.5.2.5 Comments

Rotational rheometers have been widely used to examine the influence of operating conditions and physicochemical properties on the rheology of sewage sludge [125, 126].

Kasumeter apparatus has been applied to two synthetic suspensions of kaolin and quartz sand in water, to avoid problems connected to the complex and changing nature of real sludge, and to one sewage sludge at different solid concentrations [138]. Results (with five pipe diameters) have provided yield stress results comparable to those obtained by a conventional rotational viscometer, thus allowing evaluating the limit of flowability in a very simple way (Fig. 9.28).

The standard states that the sludge could be considered in a liquid physical state when the flowability index (F_i) is lower than 25 Pa [135]. Validation tests were carried out by ten operators (from 8 institutions/laboratories of 3 countries) on three different sludge: a thickened sewage sludge (TS = 5.6 wt%), an inorganic sludge from a marble manufacture (TS = 66 wt%) and a mixture of thickened and dewatered sludge (TS = 8.4 wt%). The overall means of F_i were respectively 4.2, 5, and 22.5. The results for the first two sludge lead to a similar F_i value for nevertheless different TS, what clearly evidence that solids' concentration cannot be used

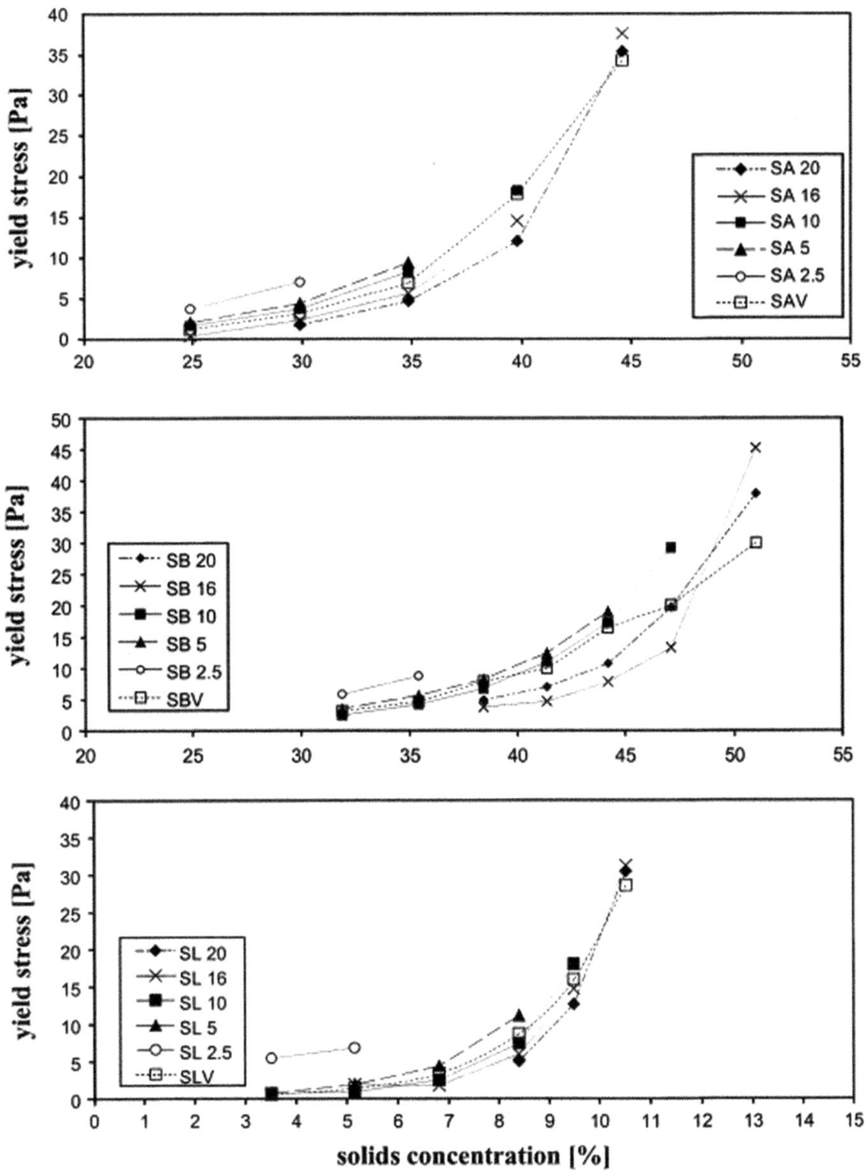

Fig. 9.28 Yield stress versus solids concentration (TS) for three sludge (SA: 90/10% kaolin/quartz sand; SB:75/25% kaolin/quartz sand; SL: sewage sludge; number: pipe diameter in mm; V: viscometer). Reprinted with permission from Elsevier [138]

as an indicator of a liquid physical state. For the third one, only 60% of the results show that the material consistency is liquid, so the F_i value could only be considered as indicative.

9.5.3 *Solidity*

9.5.3.1 General Discussion

A list of the devices recommended for assessing sludge solidity is provided in Table 9.53.

"Vane" shear apparatus is one of the most commonly used equipment in the field of soil mechanics and procedure are standardized [139, 140]. To obtain reliable measurements, the studied materials have to be cohesive.

Penetrometers [141, 142] are suitable to obtain reliable results of the measurements both in laboratory and field. The penetrometer tip is proposed in a large range and can be thus adapted to the material studied. However, whatever the tip is, the test execution has to be implemented precisely as the measurement has to be carrying out within a fixed duration. An example of procedure for a penetrometer is presented below.

The Vicat needle apparatus is a specific penetrometer, with a tip in the shape of a needle, used for cements or plastics [143–146]. In consequence, it could be adapted only for solid materials like dewatered sludge, particular bio-waste and soil. The measuring procedure is based on a fixed penetration depth of the needle instead of a fixed duration, which can be more comfortable during the measurements.

Pocket penetrometers are well suited for field tests, thanks to low measurement duration, very good handling and reproducibility of test values. The specific measured properties can sometimes be converted in another parameter as the "vane shear strength" by using a linear regression.

Reference standards related to solidity measurements are listed in Table 9.54.

Table 9.53 Summary of the recommended apparatuses for solidity

Method/apparatus	Physical property
Penetrometer	Shear strength
Vicat needle apparatus	Shear strength
Vane shear apparatus	Vane shear strength
Pocket penetrometers	Bearing capacity or unconfined compression strength

Table 9.54 Standards related to solidity measurements

Titles	References
ASTM D217-17 Standard test methods for cone penetration of lubricating grease	[141]
ISO 2137 Petroleum products and lubricants—Determination of cone penetration of lubricating grease and petrolatum	[142]

9.5.3.2 Penetration Cone

Apparatus

The apparatus is similar to the one described previously for the liquid limit. The major differences lie in the shape and the mass of the cone [142].

Procedure

The procedure is also the same as previously. The material has to be transferred in the cup to overfill it, packed and leveled by using the spatula. The apparatus is set in the zero position, and the test rod is then released rapidly and allowed to drop for 5.0 ± 0.5 s and the penetration depth is read. The sludge cohesion (undrained shear strength τ_u) can be derived from the cone penetration depth using the following empirical equation [147]:

$$\tau_u = \frac{Kmg}{P^2} \tag{9.46}$$

where

K is a function of the apex angle of the cone
m is the mass of the cone, in g
g is the acceleration of gravity, in m/s^2
P is the depth of penetration, in tenths of a mm.

9.5.3.3 Comments

It has to be noticed that, in literature, a limited attention was paid to pasty and highly concentrated sludge (i.e. TS > 15%). In this TS range, the rheological characterization is difficult. The reproducibility of the measurements is often poor, due to cracks and fractures [125, 126], and the experimental results achieved may not be fully representative of the effective material properties. However, new methodologies have been recently developed to characterize pasty sludge with high shear resistance using conventional rheometric apparatuses.

Concerning rotational viscometer, a well-controlled experimental procedure [148] has been proposed to accurately determine the surface effectively sheared, even in the presence of fractures. This surface is calculated by scattering a classical stress sweep with measurements at a reference strain value. The implementation of this procedure allows the correct determination of solid-like characteristics for TS ranging from 20 to 45 wt% but also shows that pasty and highly concentrated sludge highlight normal forces caused by dilatancy.

Another specific methodology has been developed based on uniaxial single and cyclic compression tests [149]. It has allowed characterizing the rheological and textural properties of highly concentrated sludge (TS around 20%). The cyclic compression tests reveal that the sludge behaves as a viscoelastic body when the true strain is smaller than the yield stress and as a visco-elasto-plastic once exceeding this yield stress.

In any case, it is at least possible to determine the boundaries between the different behaviors.

9.5.4 *Plasticity*

9.5.4.1 General Discussion

The physical consistency of a material, or more precisely the transitions between the different physical states also called the Atterberg limits, can be characterized through basic measures at lab scale. Initially defined for soils, the Atterberg limits tests have also been applied to sludge [120–122]. Two methods, the penetration cone or the cup method, are used to assess the liquid limit w_L while the thread method is applied for measuring the plastic limit w_P. The plasticity index, calculated from w_L and w_P, gives a numerical measure of the sludge plasticity. Reference standards related to the measurements of the Atterberg limits are listed in Table 9.55.

9.5.4.2 Liquid Limit (Penetration Cone)

Apparatus

The apparatus is similar to the penetrometer described in Fig. 9.26 [134]. The cone apparatus should permit the cone to be held firmly initially and to be released instantaneously to fall freely in a vertical direction into the material. The cone should be made of stainless steel or duralumin material of 80 g mass with an apex angle of 30° [122], but it can also be a cone of 60 g mass with an apex angle of 60° [121]. The height of the conical section of the cone should be 20 mm or greater.

Table 9.55 Standards related to the Atterberg limits

Titles	References
NF P94-051 Sols: Reconnaissance et essais—Détermination des limites d'Atterberg—Limite de liquidité à la coupelle—Limite de plasticité au rouleau	[120]
XP CEN ISO/TS 17892-12 Geotechnical investigation and testing—Laboratory testing of soil—Part 12: Determination of Atterberg limits	[121]
ASTM D4318 Standard Test Methods for Liquid Limit, Plastic Limit, and Plasticity Index of Soils	[122]

To ensure that the point remains sufficiently sharp for the purposes of the test, the cone should be replaced if the point can no longer be felt when brushed lightly with the tip of the finger when the tip is pushed through a hole of a fixed diameter.

The mixing cups should be made of steel, porcelain or plastic, of cylindrical or semi-spherical shape, with a diameter greater than 55 mm and depth of 30 mm if the 60 g/60° cone is used and a depth of 40 mm if the 80 g/30° cone is used.

Procedure

A sample of the material of sufficient size has to be mixed with distilled water to obtain a thick homogeneous paste. Water is added until the first cone penetrometer reading is about a fixed value (7 mm for the 60 g/60° cone and 15 mm for the 80 g/30° cone).

A portion of the mixed paste is placed into the cup, taking care not to trap air, and excess material is removed to give a smooth level surface. With the penetration cone locked, the cone is lowered so that it just touches the surface of the paste and the initial position is recorded. Then, the cone is released for a period of 5 ± 1 s and the final position should be recorded. The penetration of the cone should be calculated as the difference between the initial and final position of the cone shaft. The test should be repeated until subsequent penetrations are repeatable. The average of the penetrations should be then calculated.

The penetration test should be repeated using the same specimen of material but at different moisture contents. Once the moisture contents of the different samples of material have been determined, the relationship between water content and the cone penetration should be plotted with the water contents on a linear scale as abscissa and the cone penetrations on a log scale as ordinate (linear scale can also be used for the 80 g/30° cone). The best straight line fitting the plotted points should be drawn. From this graph, the water content corresponding to a defined cone penetration should be read and corresponds to the liquidity w_L (10 mm for the 60 g/60° cone and 20 mm for the 80 g/30° cone).

9.5.4.3 Liquid Limit (Cup Method)

Apparatus

The device [122] consists of a brass cup suspended from a carriage designed to control its drop onto the surface of a block, which serves as the base of the device. A flat grooving tool, made of plastic or non-corroding-metal, is also necessary. The design of the tool may vary as long as the essential dimensions are maintained.

Procedure

A portion of the material is placed in the cup at an imposed level. Then, a groove is formed in the material by drawing the tool on a line joining the highest point to the lowest point on the rim of the cup. Then the cup is lifted and dropped by turning the crank at a rate of 2 drops per second until the two halves of the material come in contact at the bottom of the groove along a certain distance (10 mm [120] or 13 mm [122]). The closing of the groove has to occur because of the slump of the material and not the sliding on the cup surface. The number of drops, N, required to close the groove is recorded.

The procedure should be repeated using the same material but at different water contents to have a number of drops from about 15–35. Then the relationship between the water contents and the corresponding numbers of drops of the cup is plotted on a semi-logarithmic graph with the water contents as ordinate on the linear scale, and the number of drops as abscissa on a logarithmic scale. The best straight line is drawn through the plotted points. The water content corresponding to the intersection of the line with the 25-drop abscissa is considered as the liquid limit, LL, of the material and round to the nearest whole number.

If the number of drops required to close the groove is always less than 25, the liquid limit could not be determined, and the sludge is reported as non-plastic.

9.5.4.4 Plastic Limit

Apparatus

A mixing plate is necessary. It should be flat, clean, smooth and free from scratches. A convenient size of plate is about 10 mm thick and 300 mm square [122].

Procedure

A sample should be taken from the soil paste prepared previously for the determination of the liquid limit. It should be allowed to partially dry until it becomes plastic enough to be shaped into a ball.

The material should be moulded between the fingers to equalize the distribution of moisture. Then, the thread or the ball should be rolled to reduce the diameter of the thread to about 3 mm by using regular forward and back movements of the hand at a rate of about 1 complete movement per second. It is important to maintain a uniform rolling pressure.

Previous steps should be repeated until the thread shears both longitudinally (at least 10 cm of length) and transversely when it has been rolled to about 3 mm diameter. The water content of the first crumbling point corresponds to the plastic limit.

9.5.4.5 Plasticity Index

The plasticity index, I_P, can be calculated according to the following equation [122]:

$$I_p = w_L - w_P \tag{9.47}$$

9.5.4.6 Comments

These limits have been determined to evaluate the dewatering potential of sludge [150–152]. For paste-like sludge sampled after the mechanical dewatering stage, the more organic the sludge, the higher the liquid and the plastic limits and the more water it retains (Fig. 9.29) [150]. Whatever the organic fraction, the moisture content after mechanical dewatering (i.e. w_{init}) is between the Atterberg limits, w_L and w_P. Moreover, the rheological and textural characteristics of the sludge are correlated with its drying and shrinkage aptitudes [150, 151] and also with mechanical dewatering aptitude [152]. This approach confirms the validity and interest of standardized consistency tests.

The determination of these limits can be used to define the application range of a test. For instance, to carry out uniaxial compression tests, it has been shown that, the sludge has to behave as a plastic solid, with solids content between the liquid limit (to ensure that the sludge sample cannot flow under its own weight by gravity) and the plastic limit (to ensure that it can be considered as a continuous medium) [149]. The studied digested sludge has a total solids mass content of 20.4% (i.e. W_{db} = 3.9 kg/kg$_{TS}$), which is higher than the liquid limits, and the plastic limit has been measured at around TS = 37% [(i.e. W_{db} = 1.7 kg/kg$_{TS}$)] [153].

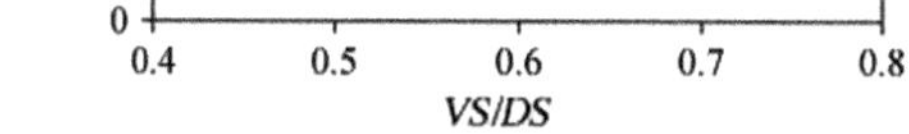

Fig. 9.29 Variations in Atterberg limits (liquid, w_L and plastic, w_P) and initial water content (w_{init}) with organic fraction (DS: dry solids mass%; VS: volatile solids mass%; water content expressed, here, in % on a dry base). Reprinted with permission from Elsevier [150]

9.5.5 *Stickiness*

9.5.5.1 General Discussion

The process of attachment of a substance to the surface of another substance is called the adhesion, while the cohesion refers to the attachment of a substance to an identical or a similar substance. Both adhesion and cohesion contribute to the stickiness of sludge. This sticking ability may lead to a significant reduction of dehydration capacity and efficiency [154]. Therefore, it is necessary to develop a reliable method to determine the adhesion and cohesion of sludge and then identify the essential factors that dominate the sticky properties.

Peeters et al. [155] employed a simplified version of the Jenike Shear Test [156] to map the sticky phase of an activated sewage sludge. Influence of the solid contents on the adhesion force was investigated in the range of 9–99%. Li et al. [157] also developed adhesion/cohesion test devices, which can measure independently the adhesion and the cohesion stress under an imposed normal stress. The adhesion test is nearly the same as the Peeters' shear test with the difference that it is equipped with a heating table to measure the adhesion stress at different temperatures. For cohesion measurements, the device consists of two open cylinders of the same diameter, superposed one on top of the other. More recently, the adhesive/cohesive behaviour of highly concentrated sewage sludge has been characterized by comparing a cylindrical Jenike-like shear test unit and a square Casagrande-like geometry shear test unit [153, 158].

9.5.5.2 Shear Test

Apparatus

A schematic diagram of the shear testing device is presented in Fig. 9.30 and the photograph of Fig. 9.31 depicts the main components. In the following, the numbers between parentheses refer to Fig. 9.30. The open (hollow) steel cylinder (3) has an internal diameter of 5.5 cm and a height of 10 cm, and is further characterized by a weight of 388 g and a wall thickness of 2.5 mm. The massive cylinder (4) has a weight of 2.2 kg, with an external diameter of 5.3 cm, which fits closely in the open cylinder. The open cylinder is connected by the flexible steel cable (6) and the small wheel (7) to the recipient (8), which is used to increase the shear stress on the sludge. The purpose of the wheel is to redirect the flexible steel cable in a vertical direction. A flexible plastic bag (5) is arranged over the massive consolidation cylinder—this further avoids disturbing the intentionally created adhesion between the sludge and the steel surface when this massive cylinder is removed.

To study the stickiness of sludge during the drying process, the surface on which are placed the geometry can also be heated [157].

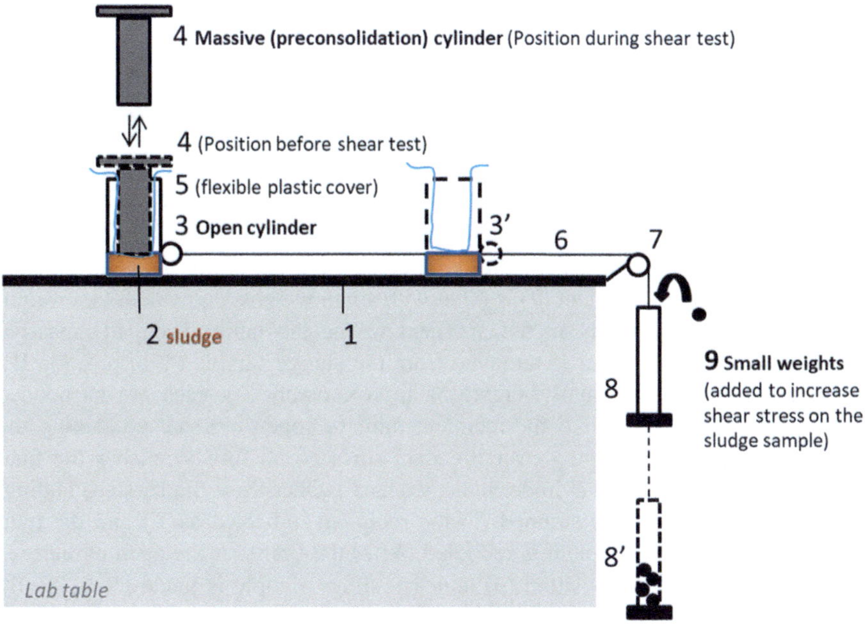

Fig. 9.30 Scheme of the shear test. (1) lab table, (2) sludge sample, (3) open (hollow) steel cylinder, (4) massive steel cylinder, (5) flexible plastic cover, (6) flexible steel wire, (7) wheel, (8) recipient and (9) steel balls. Adapted and reprinted with permission from Peeters [155]

Fig. 9.31 Photo of the main components. Adapted and reprinted with permission from Peeters [155]

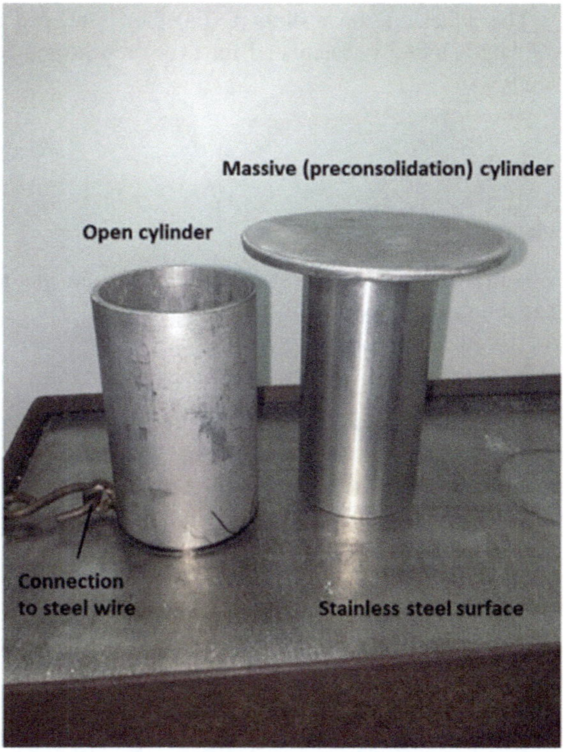

Procedure

First, conditioned sludge (2) is put in the open (hollow) steel cylinder (3) which is placed on the left side of the stainless steel surface (1), i.e., the starting position of the cylinder for the test. Sludge (approximately 15 g) is brought into this open cylinder and uniformly distributed over the circular surface. Next, the sludge is consolidated during 1 min on the stainless steel surface by means of the massive cylinder (4, in position A). In this way a normal stress of 9.1 kPa is applied which presses the sludge onto the steel surface, inducing adhesion of the sludge to the steel surface. Initially, the recipient (7) is filled with small weights equivalent to a weight of 100 g and hangs freely in the air, next to the lab table. Then, the massive (pre-consolidation) cylinder is removed from the sludge sample (4, in position B). From that moment on, small weights of approximately 3 g each are introduced approximately every 5 s into the recipient until the open cylinder containing the activated sludge starts sliding over the steel surface and finally reaches the final mark on this surface (open cylinder indicated as 3′). Once the cylinder starts sliding, the addition of weights is stopped (in the recipient indicated as 7′) and the total weight M (kg) of the recipient is recorded. After the first test, the open cylinder is placed back in the start position (3) and the sludge sample is made loose, gently mixed with a laboratory spoon in the open cylinder itself (typically during 20 s), and again distributed uniformly over the circular contact surface as was done during the first trial with this sample. The procedure is repeated again as described above, for the second time. A third trial is also performed to get 3 repeats of the shear test for the same sample.

The adhesion shear stress τ (Pa) that is needed to cause slipping of the consolidated sludge is calculated for every repeat and plotted as function of the total solids:

$$\tau = \frac{M \times g}{A} \tag{9.48}$$

where

M is the weight needed to cause slipping of the sludge over the steel surface, in kg
g is the gravity acceleration, in m/s^2
A is the circular contact surface between sludge and steel surface (2.38×10^{-3} m^2).

For a cohesion test, the mobile open geometry containing the sludge is superimposed on another similar geometry. This other part is also filled with sludge but it is fixed on the surface. The similar procedure as for adhesion test allows measuring the shear stress required to cause slipping of the consolidated sludge in contact with other sludge [153, 157, 158].

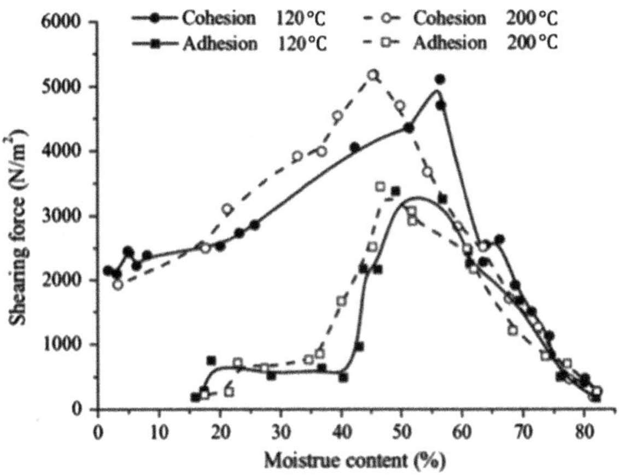

Fig. 9.32 Shearing forces versus moisture content. Reprinted with permission from Taylor and Francis [157]

9.5.5.3 Comments

It has been showed that the most significant sludge adhesion and cohesion emerges at total solids content of approximately 30–55% (i.e. $45 < W_{wb} < 70$ wt%) and 40–70% (i.e. $30 < W_{wb} < 60$ wt%), respectively, and at a heating temperature of 120 ° C (Fig. 9.32) [157]. This confirms the online measurement of the torque on the stirring paddles during a drying process which shows a sharp increase in torque indicates the start of the so-called "plastic region', while the rapid decrease means its end [159]. An increase of the heating temperature to 200 °C only brings a small change to the sticky range location. The adhesion and cohesion nature of sludge exhibit a large effect on its drying characteristics.

9.5.6 Interferences

In most of apparatus, the shear rate and shear stress are spatially averaged over the fluid volume, what can lead to important errors.

Whenever possible, the tests should be carried out on material from its natural state.

However, if the material contains coarse materials (grains of sand, fibres and other solid materials), there is a risk of measurement disturbance or apparatus obstruction (in particular in the case of apparatus with a small gap). This coarse fraction should be thus excluded from the sample (for instance, particles larger than 2.5 mm in the Kasumeter apparatus). If it cannot be removed by hand, the particles should be removed using a reproducible method (for instance a wet sieving), in general described in the standard.

In several devices, the material has to be packed to remove the entrapped air and/or leveled (for instance for a penetrometer). Moreover, the sludge has to be homogenized just before the test to obtain a repeatable measurement, whatever the history of the material. This can be done by a manual mixing or by implementing a pre-shearing step before the measurement. But this pre-treatment results in a modification of the raw sludge, as it alters the microstructure of the sludge and the moisture distribution [158]. This has thus to be done with a minimum of manipulation to limit the modifications of the sludge.

Several experiments have also to be carried out on the same material at different water contents, which requires a preliminary preparation step of the material. This step can modify the structure of the material and has thus to be in a reproducible way, in general described in the standard.

Another drawback is the separation of particles due to the gravitational and centrifugal fields. It can be sometimes solved by the use of a specific rheometric device (vane apparatus or systemic rheology) leading to a reduction of sedimentation but, as previously evoked, also to the measurement of only bulk properties [126].

The phenomenon of slip occurring at the geometry/material interface has to be avoided and can be overcome by artificially increasing the roughness of the device surfaces.

It is also necessary to ensure that the geometry of the device is still correct (for instance, a point of a cone sufficiently sharp) to be able to use calibration constant or to obtain a valid result. Several standards also describe methods to check the state of the used device.

9.5.7 Conclusion

A large variety of apparatus exists to measure the rheological properties of pasty materials such sludge. For adaptation of methods to a wider range of materials or for a more exact measurement, it is possible to vary the dimensions of the apparatuses (rotor, tube diameter, weight and size of cone, etc.) and the procedures.

Other important aspects are the sampling and the preparation of the sample. A pre-treatment can be implemented to carry out the measurement and to obtain repeatable results (sieving, compaction, pre-shearing, etc.). However, it leads to a modification of the materials which can result in a modification of the measured properties. Thus, these steps have to be controlled, repeatable, and standardized. It has to be noticed that some detailed procedures exist in several standards.

At laboratory scales, for developing new formulations, a combination of methods should be used. In the industry, where fast methods and low cost are required, the automated methods will be preferred for tests of raw materials or process control parameters.

9.6 Environmental Assessments

9.6.1 Introduction

The environmental assessment methods presented in this section are proposed for waste material in general and are relevant for the evaluation of the behaviour of sludge.

Current regulations define three modes of management: immersion, re-use and disposal to landfill. The most widely used solutions, involving nearly 90% of marine sediments, are immersion and putting back into suspension. Indeed, sustainable development logic makes it a priority to return sediments to their original habitats unless they are contaminated. The 2016 Law on the Blue Economy introduced stricter environmental regulations including a ban, that will come into force on 1 January 2025, on dumping polluted dredging sludge into the sea (Article 19) [160], thus fulfilling a commitment made at the "Grenelle de l'environnement" process of the sea. Immersion will remain authorised for dredging sediments "considered to be environmentally neutral" and the law advocates the establishment of an industry for sludge treatment and recuperation of associated macro-waste

If their characteristics allow, sediments may be recycled for different uses, listed in Table 9.56 (extracted and adapted from [161]).

Several methods exist for environmental assessment of the potential impacts of the pollutants containing waste material; one of these is the EN 12920 standard [162] which is part of the Detailed Risk Evaluation and uses a "matrix" type methodology, developed for the evaluation of the risks of a solid matrix [163]. This type of methodology is particularly well-adapted to evaluate the environmental

Table 9.56 Recycling applications for sediments

Use	Conditions[a]
Road-building	Inert or non-hazardous sediment
Agriculture	Inert or non-hazardous sediment
Landscaping	Inert sediment or specific study in accordance with EN 12920 + A1 [162]
Strengthening river banks	Inert sediment or acceptable for immersion
Filling in quarries	Inert sediment
Covering waste storage installations	Inert or non-hazardous sediment depending on type of use
Maritime works	Inert sediment or specific study in accordance with EN 12920 + A1 [162]
Construction products	Inert sediment or specific study in accordance with EN 12920 + A1 [162]

[a]The qualification of sediments as "inert", "non-hazardous" or "hazardous" should be taken in the regulatory meaning. Sediment is inert if its pollution content is lower than the limit values defined by the dedicated regulation

acceptability of recycling mineral residues. Release[1] of soluble components in contact with water is considered to be one of the main processes that can lead to potential environmental risk during the life cycle of the waste material. Leaching tests are the most widely used methods for evaluating the polluting potential of raw or stabilized waste materials.

European standard EN 12920 [162] is a seven-stage methodology that aims to "characterize the leaching behaviour of waste materials under specified conditions". Here, the term "leaching" is understood as solid/liquid contact, whether the waste material is percolated or washed. This methodology has the advantage of taking into account the specific properties of the waste material and the conditions of the scenario considered to be as representative as possible. Its objective is to describe in details the source term in as much detail as possible and evaluate the polluting potential of the waste materials intended to be recycled.

Stage 1 of EN 12920 [162] is to describe the problem and the envisaged solutions.

Stage 2 consists of describing the proposed scenario, enabling the best-adapted leaching test(s) to be selected. The environmental characterization must be carried out using tests that recreate the conditions to which the alternative material will be exposed. The environmental compatibility with the different stages of its use must be ensured. This includes the storage before use, the introduction into the environment after possibly having undergone stabilisation/solidification, and the end of its life. Since conditions may vary considerably during its life cycle, validation must be carried out under different test conditions. When sediment is recycled for landscaping, the material will be in granular form, for example to create an embankment. On the other hand, a stabilized/solidified waste material used as a substrate in road-building will be in the form of blocks. In both cases, the materials will be subject to rainwater infiltration. The chosen tests must take into account this difference in structure depending on the envisaged scenario. The phenomena that occur when granular materials come into contact with a liquid are different from those that occur with a block material. On the other hand, tests must also be adapted to the constraints of industrialists, which imply that they must enable the desired characteristics to be determined within a "reasonable" timescale.

Stage 3 consists of complete characterization of the waste material: total chemical composition but also the speciation of chemical species, physical and geotechnical properties etc. (see Sects. 9.4.2.3 and 9.4.2.4). However, it is recognised that the environmental performance of a material must be based on release rather than on the total content of potentially hazardous components. A priori, there is no simple relationship that enables the release potential of contaminants generated by contact with rainwater to be determined from the chemical composition of the waste material. This is exacerbated by the fact that, in order to verify whether the amount of pollutants released exceeds the threshold setup by the given

[1]In this document "release" is used to indicate the transfer of an ion from one environment to another.

regulation, a maximum release is not as important as changes in concentrations released over time [164]. Moreover, changes in the source term will lead to changes in the resulting release from the “target”.

This implies that considering the underlying mechanisms that control the release [165], which are determined in stage 4 of the EN 12920 methodology [162]. Many factors influence the release; they may be internal, either chemical (pH, chemical form, temperature, presence of organic matter, ionic force etc.) or physical (porosity, permeability, tortuosity, particle sizes etc.) but external factors (temperature and pH of the environment etc.) also have an effect. The state of the material (granular or in blocks) also influences the release. Parametric tests will bring to light the main phenomena and the effects of those parameters on the release of the components of the waste material.

Stage 5 involves modelling leaching behaviour, the objective being to provide a simulation and a prediction corresponding to the scenario envisaged in stage 1.

Stage 6 involves verifying the results in situ in order to evaluate the proposed model and the last stage provides the conclusion and enables a decision to be taken with respect to the problem stated and the envisaged scenario.

The objective of this environmental characterization stage is to establish the link between the definition of the proposed scenario (stage 2) and the understanding of the mechanisms (stage 4), which both require the use of leaching tests. This will facilitate the choice of the most suitable leaching test(s) to obtain the data that will be inserted later into a numerical model.

9.6.2 Tests and Parameters

9.6.2.1 Leaching Tests

One of the major environmental risks is the potential release of contaminants due to contact with water. Many tests exist to evaluate release and they are mostly built on the same basic principle, but with characteristics adapted to the specific conditions of use [164]. The main tests and their operating conditions are presented in Table 9.57.

The hierarchy proposed by the CEN TC 292, splits the tests in three categories and is taken from the Directive relating to disposal to landfill (Council of Europe Directive 1999/31/CE dated 26 April 1999, [166]) and from the Decision conforming to Annex II of that Directive (Council of Europe Decision 2033/33/CE dated 19 December 2002, [167]) relating to the elimination of waste materials.

It distinguishes:

- ***“basic characterization tests”*** (category a) to obtain information about short- and long-term leaching behaviour but also about the characteristic properties of waste materials;

Table 9.57 Main leaching protocols

	Name of the test	Year	Country	Type of test	Type of materia	Leaching solution	pH	Number of steps	L/S (L/kg) or L/A	Duration	Stirring
Basic charactersation test (category a)	NEN 7371 [168]	2004	Netherlands	Batch	Granular <125 μm	dm water	7 and 4	2	50	3 h	yes
	NEN 7375 Tank Leaching Test [169]	2004	Netherlands	Dynamic batch	Block	dm water	7	8	depends on sample	64 days	no
	EN 14405[a] [170]	2017	Europe	Percolation — continuous upward flow	Granular (crushing possible)	dm water	7	1	Up to 10 (v = 15 cm/day)	72 h (equilibrium) + test	no
	EN 14429 [171]	2015	Europe	Batch	granular <1 mm	Water + acid/base	From 2 to 12 (8 values)	1	10	48 h	yes
	EN 14997[a] [172]	2015	Europe	Batch	Granular <1 mm	Water + arid/base	from 2 to 12(8 values)	1	10 (in 2 stages)	48 h (in 2 periods)	yes
	EN 15863 [173]	2015	Europe	Dynamic batch—periodic renewal	Block (min 40 * 40 * 40 mm)	dm water	NC	8	L/A = 8 mL/cm^2	64 days	no
	Noidtest method availability test	1995	Scandinavia	Batch	granular <125 μm	dm water	7 and 4	2	100	3 and 18 h	yes

(continued)

Table 9.57 (continued)

	Name of the test	Year	Country	Type of test	Type of materia	Leaching solution	pH	Number of steps	L/S (L/kg) or L/A	Duration	Stirring
	NTENVIR 003 [174]										
	XP CEN/TS 15864 [175]	2016	Europe	Dynamic batch—continuous renewal	Block (min 40 * 40 * 40 mm)	dm water	NC	8	L/A between 2 and 10 mL/cm^2	36 days	yes
Compliance tests category b)	EN 12457 (1–4) [176–179]	2002	Europe	Batch	Granular < 4 or 10 mm	dm water	NC	1 or 2	2 or 10/2 and 10	24 h/6and 18 h	yes
	NF X 31-211 [180]	2012	France	Batch	Block Ø4, h8 cm	dm water	NC	1or 2	10	24 h or 16 h/ leaching	yes
	TCLP method 1311[a] [181]	1994	USA	Batch	<9.5 mm	Water + acids	3 (initial) or 5	1	20	24 h	yes
	XP CEN/TS 15862 [182]	2012	Europe	Static batch	Block (min 40 * 40 * 40 mm)	dm water	NC	1	L/ A = 12 cm^3/ cm^2	24 h	yes

dm demineralised, *NC* not checked

[a]possibility to measure organics and inorganics

- ***"compliance tests"*** (category b) to determine whether the waste material conforms to threshold regulatory behaviour or values based on the leaching behaviour;
- ***"in situ verification tests"*** (category c) which are rapid checks to confirm that the waste material is similar to that which has undergone characterization or compliance tests.

In situ tests (category c) include lysimeters or the construction of a full-scale embankment or road incorporating the waste material under consideration; they are as close as possible to the conditions of the proposed scenario because they represent, at medium or large scale, the material as it will be used in real conditions. The issue is that these options require very long testing times and are very expensive; moreover, it is impossible to carry out in situ tests for all types of waste materials to be evaluated.

Basic characterization tests (category a) are more detailed than compliance tests (category b), the latter being more suitable for daily use. Basic characterization tests provide a description of the material release behaviour, enabling the latter to be estimated [165]. The simplicity of compliance tests offers practical and financial advantages but they do not characterize the behaviour of a waste material.

Consequently, once leaching has been defined by the characterization test, a simple measurement using a compliance test is sufficient to evaluate the behaviour of the materials. It is also important to link the compliance test to the standard characterization test previously carried out [183].

This hierarchy has been proved for a wide variety of alternative materials [165]. The combining of characterization of the materials behaviour with the use of simpler compliance tests provides the necessary understanding and at the same time limits the need for numerous useless tests.

Another classification, complementary to the previous, is based on the technique used during the leaching test [184]:

- ***static***, also known as "batch". The tests are carried out in a closed reaction chamber with a constant liquid-to-solid ratio and provide information on the long-term behaviour (cumulated emissions) if one considers that equilibrium is attained at the end of the leaching test. The solution may be renewed or not;
- ***dynamic***. Dynamic leaching, also known as percolation, consists of passing a constantly renewed liquid through the material.

Batch Tests

The batch leaching test, or extraction test, lasts only a short time and is easy to carry out. Several parameters are required:

- Composition and granulometry of the solid sample;
- pH of the solution, adjusted during the test or not;

- Liquid/solid ratio (L/S);
- Contact time;
- Number of extractions;
- Stirring speed and procedure;
- Open or closed system.

For granular waste materials, the test is generally carried out with stirring. On the other hand, for blocks, the test is carried out either in static (one extraction) or dynamic (several extractions) mode. The main reason that could justify the use of this test of dynamic leaching tests is the fact that release on blocks is related to the surface and not, as is the case for granular materials. For the latter, percolation and the equilibrium between the solid and the leachate [173, 175].

Because they are rapid and easy to set up, batch tests are very often used in guides for the acceptability tests as "checking tests" for which the results could be compared with regulatory thresholds [164]. On the other hand, they do not predict behaviour in the very long term because the leaching methods in the pH range 4–7 enable the dissolution of the most soluble solid phases but not aluminosilicate phases, which are more stable [185]. Indeed, silicates in general are hard materials. However, at more extreme pH, the alteration of these phases could happened with the release of alkalis.

Dynamic Tests

Dynamic tests, also called *column leaching tests* or *percolation tests,* consist of passing a continuous flow of liquid through the material. Several parameters are involved:

- Composition and granulometry of the solid sample;
- pH of the solution, adjusted during the test or not;
- Flow rate;
- Temperature;
- Contact time;
- Direction of flow.

This type of test is of great interest. Unlike batch tests, in which the material is in prolonged contact with a liquid, percolation tests come close to the real conditions of the considered scenario and risks associated, because the infiltration of rainwater which is a dynamic problem. Moreover, dynamic tests enable the user to evaluate the risk of exceeding a threshold at a given moment. Batch tests only provide an estimation of cumulative release whereas dynamic tests provide a continuous record of the phenomenon. Despite these advantages, column leaching tests are less used than batch tests due to repeatability issues. A descending flow, which appears to be the most natural, often leads to the creation of preferential pathways, even if the samples have been prepared properly. There is a risk that the whole sample will not be traversed by the liquid and, if this is the case, the concentrations of recovered percolates will be biased. It is possible to apply an ascending flow to overcome this problem and thus ensure better reproducibility of the results. The resulting mechanical and thermodynamic scenarios are not very realistic [185].

To summarize, batch tests are cheap, simple and provide chemical data for mechanical applications. Column tests are more expensive and more complex to set up but provide more realistic results in terms of flow and transport phenomena [186].

It is also important to note that leaching tests are often used for inorganic components. The processes that regulate the leachability of organic pollutants are fundamentally different to those for inorganic pollutants [177]. The role of organic matter is recognised to be a key factor in the transfer of inorganic and organic contaminants into the aqueous phase and consequently their transfer into the ground and groundwater. Organic matter also has an important role in the speciation and the bioavailability of contaminants in sludge or soils for instance [187]. Moreover, the presence of organic matter can be an issue when stabilisation of a waste material is considered for further recycling; indeed, in the case of stabilisation using cement, the organic matter disrupts the setting of the cement and leads to the reduction of mechanical strength. As a conclusion, it is worth mentioning hare that in most of the tests mentioned above the released behaviour of organic components could be assessed.

9.6.2.2 Other Tests

The list of tests in Table 9.57 is not exhaustive indeed, leaching tests exist in similar conditions to the ones presented in Table 9.57.

Among the batch leaching tests, the most widely-used are American tests in addition to the Toxicity Characteristic Leaching Procedure (TCLP) [181] are the Synthetic Precipitation Leaching Procedure (SPLP), ASTM-D3987 and Waste Extraction Test (WET), as it can be seen in Table 9.58 [186, 188]. It should be noted that these American tests are aggressive because of the low pH used.

For tests that investigate the behaviour of waste materials as a function of pH, we can cite, in addition to CEN/TS 14429 [171] and CEN/TS 14997 [172] (see Table 9.57), the XP CEN ISO/TS 21268-4 test [189] which is under revision, and the American EPA Method 1313 test [190], which are based on the same principles.

Table 9.58 Principal characteristics of the leaching protocols in the main American tests

Characteristics	SPLP EPA method 1312	ASTM D 3987	WET
particle size	10 mm	As in environment	2 mm
L/S ratio	20	20/4	10
Leaching time	18 h	18 h	48 h
pH	4.2	2.88 or 4.93	5
Leaching fluid	Acetic acid mixture, nitric and sulphuric acids	ASTM Type IV reagent water	Citric acid

It should be noted that the CEN/TS 14429 [171] is only used for studying the release of inorganic components unlike the other tests where both organics and inorganics can be measured.

Column tests include the ISO/AWI 21268-3 test, currently being developed by ISO TC 190/SC7/WG6 (to replace the XP CEN ISO/TS 21268-3 test from 2009 [191]) which has identical conditions to the CEN/TS 14405 but which will also be applicable to organic contaminants (unlike the CEN/TS 14405 which is only intended for use with inorganic contaminants) in order to determine the ecological properties of the soil with respect to micro-organisms, flora and fauna [186, 191].

The American ASTM D4874-95 test [192] is a column test applicable to both inorganic compounds and semi- and non-volatile organic compounds.

Other methods such as sequential extraction, multiple extractions and the measurement of the CEC can also be used to identify and evaluate the mobility and availability of heavy metals in the environment (see Sect. 9.4.2.3).

9.6.2.3 Influence of Experimental Conditions

In the EN 12920 methodology [162], the leaching tests must be chosen in relation to the proposed recycling scenario. Thanks to their different experimental protocols, the existing tests offer the possibility of choosing the most suitable for the proposed scenario or to get the desired information. Moreover, the conditions imposed by the leaching tests will influence the release that is obtained. Indeed, the mechanisms of retention or release depend on the physical and chemical conditions and the concentrations measured in the leachates depend on the applied parameters.

Tests vary as a function of two criteria:

- the **physical state** of the material: granular or block; granular materials present a much greater specific surface, thus increasing the waste materials reactivity capacity for a reaction;
- the **leaching reagent** and the **conditions of contact**: aggressive or otherwise depending on the value of the pH, a greater or lesser degree of dilution represented by the value of L/S, contact time.

Depending on the chosen environment, the chemical equilibria could be modified, and the resulting reactional kinetics would influence the quantities sorted out. Three main parameters control the release of pollutants [164]:

- transfer kinetics from the solid to the liquid phase;
- the solubility of the phases linked to the pollutants, which depends, among other parameters, on the pH and the redox potential;
- complexation and sorption phenomena which can increase or decrease the mobility of leachable species.

A single leaching test does not provide enough information to understand the behaviour of the waste material and the complexity of the phenomena inherent to the test. The identification of the species to which the pollutants are bound and the evaluation of the importance of the parameters influencing the release will enable changes in the waste material in a given context to be described. The Network Harmonisation of Leaching Extraction Tests [184, 193, 194] has identified pH, redox potential and the presence of organic and inorganic complexes as the important parameters. In the case of sludge and soils, the main factors controlling the release are the dissolved and particulate organic matter and the pH [187]. Their influence is detailed below, as well as that of other parameters which can become dominant, particularly when studying inorganic species.

PH and Redox Potential

Several fractions may be distinguished in a waste material:

- an inert fraction;
- a mobile soluble and available fraction;
- a potentially mobile fraction, which is partly active.

Depending on the leaching agent used, these three levels of mobilization (see Table 9.59) can be recovered in the leachate [195] cited in [164]. Strong acids dissolve part of the waste material which would not have been mobilised with water; this leads to an over-estimation of the quantities of pollutants in the leachates.

The pH of the material is crucial to determine the release of the components. This is true for all types of material (blocks, granular, soils, waste materials, sediment etc.). The species are sensitive to the environment in which they are found, and the conditions imposed by the fluid surrounding the sub-product determine the passage into the solution of the components and their concentration (see Fig. 9.33, based on [203]).

Figure 9.33 presents the release of cadmium for a soil modified with sludge [187]. The different pH values found in the environment are represented by rectangles. This demonstrates the interest of studying the influence of pH on release because the pH at which leaching starts depends on the pH of the material itself, on that of the surrounding environment and on the buffering capacity of the material. It

Table 9.59 Leaching solutions used in relation to the type of fraction to be mobilised

Soil fraction	Characteristic	Commonly used leaching solutions
Immobile	Non-reactive	Strong acid solutions (aqua regia, HCl, HNO_3)
Potentially mobile	Potentially active + active	Chelating solution, reagents buffered or not (NH_4OAc + EDTA, EDTA, acetic acid, DTPA + $CaCl_2$)
Mobile	Active	Solutions of unbuffered, neutral salts ($NaNO_3$, $CaCl_2$, NH_4NO_3)
Total		Solutions of strong acid (aqua regia, HCl, HNO_3)

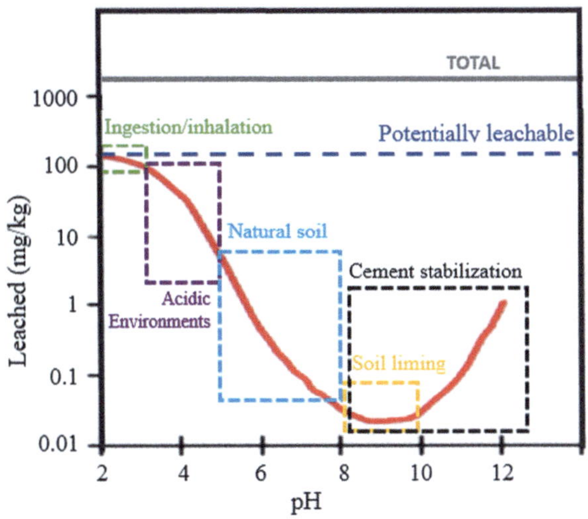

Fig. 9.33 Release of cadmium using a pH dependence test on a soil amended with sludge—effects on release depending on the environment

could be seen that a range of low pH enables the case of inhalation/ingestion to help to understand the modelling of absorption mechanism. When the amended soil is exposed to acid conditions, the leachability is very sensitive to small variations in pH.

It is also important to note the difference between the total content of a product (shown using a thick, solid line) and the leachable amount (dot lines). The difference can be small in the case of soluble salts (Na, K, Cl) or big (one or several orders of magnitude) for metals trapped in a mineral phase [164]. This curve can then vary depending on different factors.

Depending on the elements being studied, one can generally observe two types of behaviour:

- the solubilisation of the species is pH-independent (Na, K etc.);
- the solubilisation is pH-dependent (Ca, Mg etc.). This is also the case for amphoteric metals (Pb, Zn etc.) whose solubility increases with varying acid or alkaline pHs.

Thus, release curves are similar for different groups of elements, only the absolute level may vary between different materials. This implies that the phases controlling solubility are the same; only the relative significance of the influencing factors may differ from one material to another.

Figure 9.34 presents the general leaching behaviour of three groups of components as a function of pH: soluble salts, cations and anions. It may be seen that the three groups show different trends during leaching, thus showing the great amplitude of release as a function of pH.

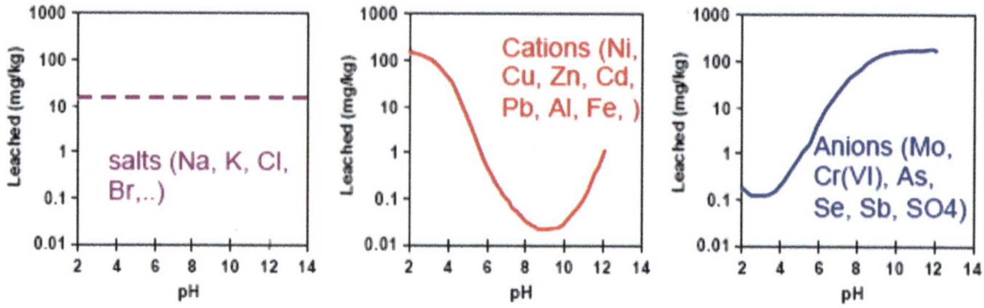

Fig. 9.34 General leaching behaviour of salts, cations and anions. Reprinted with permission from van der Sloot [181]

These differences in behaviour may be explained by their chemical speciation. The form in which the contaminant is found, influenced particularly by whether it is in an oxidised or a reduced state, determines its leaching behaviour. The oxidation of a material initially in a reduced form leads to leaching of heavy metals whereas reduction has the opposite effect [165].

The curves in Fig. 9.34 show that [165, 196]:

- the solubility of salts does not depend on pH. Upon contact with water, they dissolve instantaneously. They are governed by their availability and not by their solubility[2];
- when elements are stable in a cationic form (which is the case for most metals), solubility initially decreases when pH increases; it then reaches a minimum and begins to increase again when the element becomes anionic;
- when elements are stable in an anionic form, solubility generally increases when pH increases.

For each material considered, the quantities can differ under the influence of several parameters but the leaching models for the different groups of elements are systematic.

Standardised tests enable the behaviour of waste material to be determined in different environments and underline the influence of pH on the dissolution of species; this is particularly the case with the EN 14429 [171] and EN 14997 [172] tests which make it possible in particular to measure the acid-basic neutralisation capacity of the material. A detailed example is given in Sect. 9.6.3.

Redox potential is an important parameter for checking the mobility of metallic elements but because it is difficult to verify experimentally, the protocols rarely include it. Significant variations in redox potential may occur in real conditions due to changes in weather conditions (alternating periods of flood and drought).

---

[2]It should be remembered that *solubility* indicates the tendency of substances to be mobilised by leaching; *availability* is the release potential in relation to total content.

Generally speaking, the mobility of metallic elements increases in a reducing environment and increasing acidity accentuates this. This is due to the destabilisation of phases like iron oxide which carry metallic and oxyanionic trace elements [164].

Characteristics of Organic Matter

Dissolved (DOC) or in suspension (known as particulate, POC) organic matter containing waste plays an important role in the mobilisation or sorption of components and influences the concentration, availability and transport of metal ions in the environment [203].

The soluble fraction of the organic matter (DOC) can lead to significant mobility of metal ions if their activity in solution is weak. The degradability of the organic matter in sludge, compost, soil or sediment is an important aspect that determines to what extent DOC may be formed. The nature of the DOC is also very dependent on the stage of degradation. In the initial phases of degradation of materials rich in organic matter (e.g. sewage sludge and compost), low molecular weight organic acids and sugars are formed. At an extended degradation level, fulvic and humic substances are generated with significantly different properties in terms of metal interaction. It is thus necessary to know more about the organic matter that is present in order to better understand its behaviour with respect to release.

Figure 9.35 shows for instance, the comparison between the release of a sewage sludge and a soil amended with that sludge. It highlights the importance of the organic matter and changes in release with respect to surrounding pH value. The figure shows that the quantities of leached metal are higher at low pH. The much greater quantity of dissolved organic matter (DOC) present in the initial sewage sludge means that, even at low pH, it retains the metals, leading to a lower quantity of release compared to the soil amended with this sludge for this pH range.

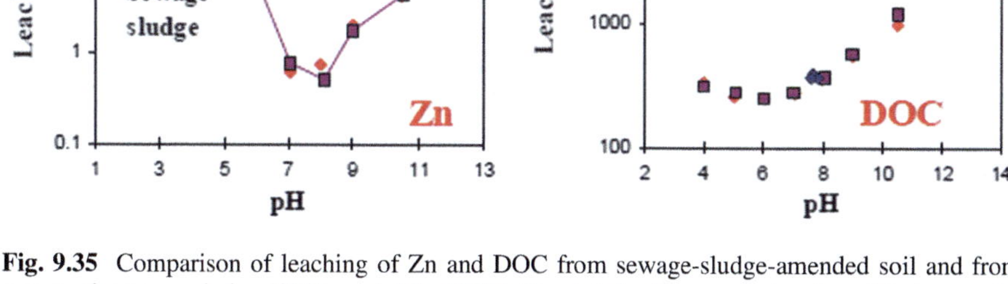

Fig. 9.35 Comparison of leaching of Zn and DOC from sewage-sludge-amended soil and from raw rural sewage sludge [209] quoted by [203]. Reprinted with permission from H. A. van der Sloot

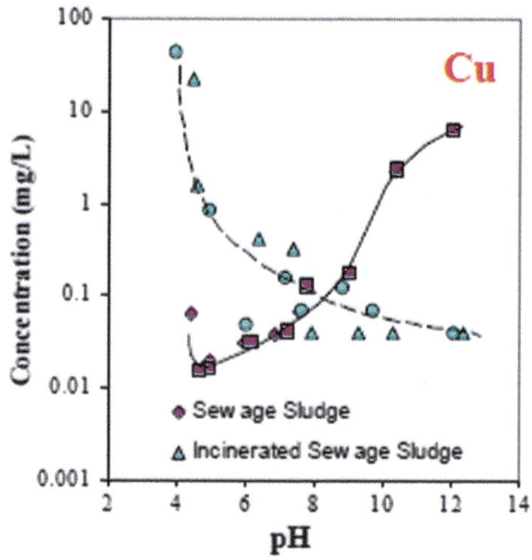

Fig. 9.36 Comparison of Cu leaching behaviour of raw sewage sludge and incinerated sewage sludge. Reprinted with permission from H. A. van der Sloot [187]

Another example, highlighting the effects of incineration treatment on sewage sludge, is presented below. The heat treatment has radically changed the release behaviour of the metals, for example copper (see Fig. 9.36). The pH of the sewage sludge has increased from an initial value of 7.3–11.9 after treatment. This shows the importance of understanding the relationship between release and pH. The very high leachability of metals at high pH due to the complexation of the DOC in the initial sludge is totally eliminated after the incineration stage. Moreover, the disappearance of the organic matter after heat treatment leads to an increase in the release of copper at low pH because the bonds between the metal and the organic matter have been destroyed.

These two examples clearly show the importance of considering the scenario and the future recycling environment, and not only the characterization of the waste material.

Size of Particles and Samples

The majority of tests on granular materials require the particle size to be reduced. This reduces testing time by accelerating release phenomena and reaching chemical equilibrium more quickly [197]. However, it also leads to a greater degree of release because the surface in contact with the liquid is increased. The finer the particles, the greater the mobilisation of the metallic elements [164].

For materials in blocks, sample shape and size are also important. Shape is critical because the results are expressed by unit surface area and this parameter must thus be measurable. Samples with irregular surfaces are more difficult to deal with and coring is necessary to determine the surface area in contact with the leaching agent, in order to be able to express the results and interpret the release. The size is also important with respect to impoverishment, which is a function of porosity, tortuosity and the interaction of the components being studied with the solid matrix over a given time period. If the dimensions are too small, there is a risk of insufficient mobile species to measure [173].

The Liquid to Solid Ratio L/S

This is an essential parameter in understanding leaching tests. The Liquid to Solid ratio, expressed in L/kg, describes the quantity of the leaching agent in contact with the total solids of the waste material. The value of this ratio, which varies depending on which test is considered (2, 5, 10, 20 or even 100) (see Table 9.57), may be linked to leaching time in a real scenario. Hjelmar [198] cited by [164] proposes a simplified equation involving certain hypotheses relating the value of L/S to time:

$$\frac{L}{S} = \frac{I \times t}{\rho \times h} \tag{9.49}$$

where

L/S is the liquid to solid ration, in L/kg
I is the infiltration, in mm/year
t is the time, in years
ρ is the apparent density of the material, in kg/m^3
h is the height of the column of material subject to leaching, in m.

It may then be seen that the L/S ratio is proportional to time (see Fig. 9.37).

To obtain an order of magnitude, if one applies the equation to an infiltration of about 300 mm/year in France, for sewage sludge with an apparent density of 1000 kg/m^3 [199], an L/S of 10 L/kg corresponds to a time of 333 years for a height of 10 m, in a storage hypothesis for example. In the case of re-use and recycling of a sample of 1 m thick, an L/S ratio of 10 L/kg only corresponds to 33 years for the same results. This shows that batch tests with an L/S of 10 L/kg simulate long-term release tests.

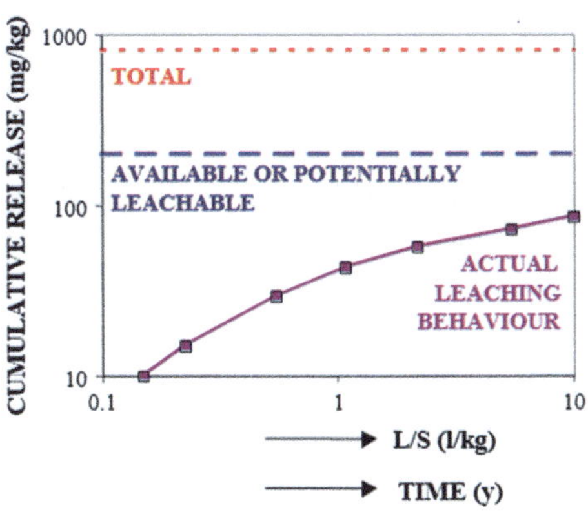

Fig. 9.37 Leaching as a function of L/S (and time) versus total composition. Reprinted with permission from H. A. van der Sloot [183]

The L/S ratio is an important parameter in the measurement of release because values that are too high or too low can prejudice the estimation of the quantities of pollutant sorted out. Indeed, the standard EN ISO 18772 [200] raises the question of the advantages and disadvantages of the L/S ratio. A low L/S ratio has the advantage of being closer to the reality but can pose experimental problems because it is difficult to extract leachates in very small quantities, particularly for soils rich in clay or organic matter. In addition, the small volume of leachate recovered can be difficult to carry out later characterizations. Moreover, a problem of over-saturation can also occur. On the contrary, a big L/S ratio is easier to obtain experimentally, and the volumes obtained are sufficient for their characterization, but in this case, a dilution problem could appear and bias the results if the concentrations are below the detection thresholds of the measuring instruments.

It is also necessary to remain aware that, of course, the cumulative release obtained with batch tests provides an interesting indication but does not enable the behaviour to be qualified in terms of released kinetics or exceeding a threshold at a given moment.

Contact Time

The contact time between the leaching reagent and the waste material is defined by the standards, with the objective of reaching chemical equilibrium. This state depends on other parameters such as the value of the L/S ratio (see preceding section) and the sample granulometry. A test involving a granular sample with a small particle size will reach equilibrium more quickly.

A study carried out on waste materials and soils [201] cited by Colombel [164] showed that, for the majority of materials with a particle size of less than 2 mm, a contact time of 48 h is sufficient to reach equilibrium, except in cases where the diffusion of species seems to be the predominant mechanism, in which case a more in-depth study is necessary. In the same way, for other materials it has been shown that for the release of inorganic elements during column leaching of granular materials, increasing the contact time had little influence on the measurements [202].

A test determining the salting-out kinetics of species could allow checking an equilibrium state has been reached.

9.6.3 Examples

Several tests among those presented can provide important and useful information for the characterization of the environmental behaviour of sludge and soils under different conditions of exposure. Leaching test results may be expressed either in leachate concentration (mg/L) or as the release rate of an element (mg/kg of total solids). The choice of units depends on the data with which the results are to be compared. The results of regulation tests are often presented in mg/L. This enables a comparison to be made between the solubility of the contaminants, reflecting the chemical speciation of the elements, and the parameters of the leaching agent (e.g. pH). On the other hand, in order to estimate availability, the measurements must be converted into release rates, i.e. from mg/L to mg/kg.

To summarize, results presented in mg/L show the impact on solubility whereas results in mg/kg show the impact on availability [203].

Three examples of tests frequently used to characterize sludge or waste materials are detailed below: the pH dependence test EN 14429 [171], the American TCLP test [181] and the EN 12457-2 test [177].

9.6.3.1 The pH Dependence Test

This is one of the main characterization tests which demonstrates the mobilisation of a species in relation to pH and is an important tool, serving as a reference for studies of release in soils and sediments [203]. It is highly recommended [165, 203–206] as it covers a wide range of pH and allow to determine the buffering power of the material. It gives an idea of the sensitivity of the materials to the external conditions imposed upon them. This test takes into account variation in pH, an essential parameter as discussed previously. All the contaminants that are controlled by solubility or sorption show pH-dependent release. The leaching model obtained with the "pH dependence test" is the result of the combination of chemical factors specific to the material that control release, such as the presence of salts in the products, redox properties, buffering capacity, ionic force and chemical speciation [165]. The test results can also be extrapolated to relevant field conditions and thus provide a lot of information about the quantities likely to be salted out under different conditions of exposure. It is thus possible to estimate the release from the waste material at the different stages of its life cycle. Finally, it also provides an indication of total dissolved concentrations as a function of pH, which is necessary in order to estimate absorption by plants and their ingestion (bioavailability) [187].

It should be noted that the principle is to leach different samples to obtain pre-determined stationary pH values at the end of the extraction period, with an L/S ratio of 10 L/kg with leaching reagents containing predefined quantities of acid or base. The reagents used are nitric acid and/or caustic soda. Eight pH values ranging from 2 to 12, including the natural pH, should be obtained in such a way that the difference between two consecutive values does not exceed 1.5 pH units. This requires a preliminary test to determine the quantities of acid or base necessary to obtain the desired pH while respecting the imposed constraint. At least eight test samples must thus be prepared with the material crushed to 1 mm. The leaching agent is introduced; the previously established quantities of acid or base are then introduced in three stages: at t_o, t_o + 30 min, t_o + 2 h. The samples are stirred for 48 h, to make sure that the variations in pH are limited as it is considered that equilibrium has been reached. The solutions are then filtered and acidified while awaiting analysis.

In addition to providing information on release as a function of pH for the species studied, this test also gives the Acid Neutralising Capacity (ANC). It is thus useful for:

- qualifying and quantifying the resistance of the material to acid or basic attacks as a function of pH and the quantity of acid or base necessary to achieve a given final pH;
- identifying the trends of chemical behaviour and the levels of availability of the components at different pH values; it can serve as a basis for modelling chemical behaviour;
- providing the data to produce a model of dynamic behaviour.

On the other hand, the standard states clearly that the test is not appropriate for:

- quantifying the maximum removable fraction; indeed, in this test, it is considered that a state of stationary equilibrium close to chemical equilibrium is reached;
- simulating actual situations in specific scenarios, information about low L/S value will be needed.

This test characterizes two parameters of the behaviour of the waste material:

- its acid neutralising capacity (ANC);
- the availability of pollutants as a function of pH.

Acid Neutralising Capacity (ANC)

The ANC shows how the material studied will behave with respect to chemical attack by representing pH as a function of the quantity of H^+ ions added. A typical curve includes plateaux corresponding to the different buffering capacities of the material, and points of inflexion, the last point corresponding to the acid neutralising capacity.

Figure 9.38 (based on [171, 207]) presents the curves obtained for three sediments: a lake sediment, a contaminated river sediment [171] and a contaminated river sediment that has been treated (phosphatisation and calcination) to stabilize the metal [207]. The neutralisation curves of the first two sediments come from the standard of the pH dependence test. Fang et al. [208] obtained similar results with sewage sludge composts and soils. The curves first give information about the natural pH of the sediments (no acid added). The natural pH of the treated sediments is 12.2 whereas the "raw" sediments have a natural pH of between 7 and 8. The subsequent release of metals will thus be different. Indeed, the relevant pH range lies between 4 (acidic conditions) and 8 for soils and sludge [187]. It has been shown that treatment by phosphatisation and calcination does lead to a very significant increase in pH [209].

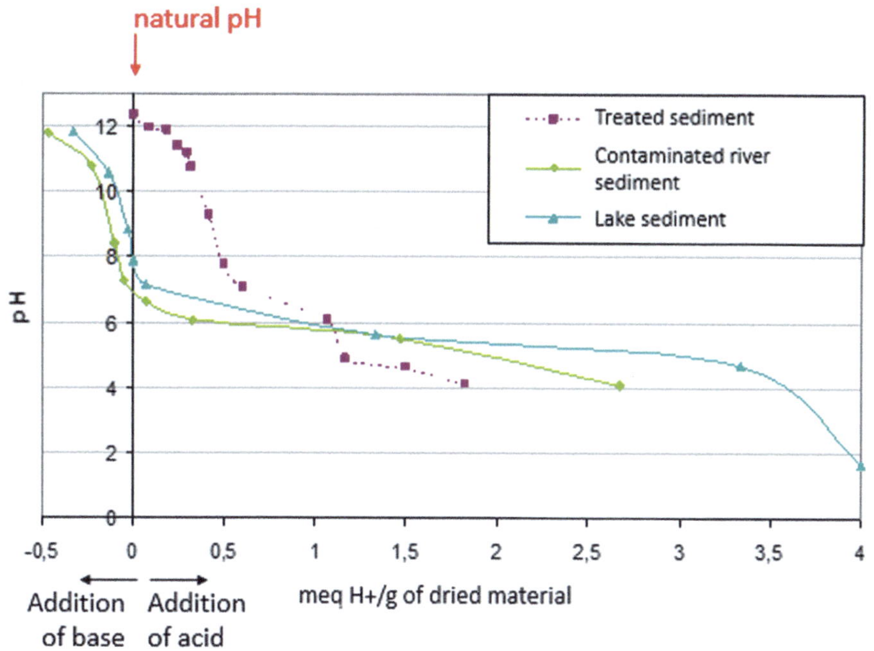

Fig. 9.38 Neutralisation curves for different types of sediment

Next, the curves are similar but offset along the X-axis. They do indeed present one or more plateaux and points of inflexion. It is particularly noteworthy that the plateau around pH 7, which corresponds to the buffering capacity of the sediments, is smaller for the treated sediment than that for the others. A greater concentration of carbonates in the untreated sediments could explain that the plateau remains despite the addition of acid [207]. The end of the acid neutralising capacity is reached with the addition of 1.15 meq H^+/g for the treated sediment, which is low when compared to that of the lake sediment, for example, requiring the addition of 3.7 meq H^+/g.

The neutralisation capacity of the treated sediment is thus lower than that of the other types of sediment. The treated sediment thus appears to be a material that is sensitive to external aggression. Nevertheless, everything depends on the context in which it is used [204]. If it is used in an acid environment, it could undergo undesirable leaching. On the other hand, if it is used in a basic environment such as a cement matrix, that will not cause any problems even though its acid neutralising capacity is low.

Availability of Pollutants as a Function of PH

The release from a raw, contaminated harbour dredging sediment (DB) was studied, as well as the effects of phosphatisation with different types of phosphoric acid (DA6P and DA7P) followed by calcination at 650 °C (DA6C and DA7C) [210]. The release of arsenic and chromium as a function of pH are presented in Fig. 9.39. The results are presented in mg/kg of total solids. The release curves of arsenic show that at the natural pH of the raw sediment, which is around 8, the release by the raw

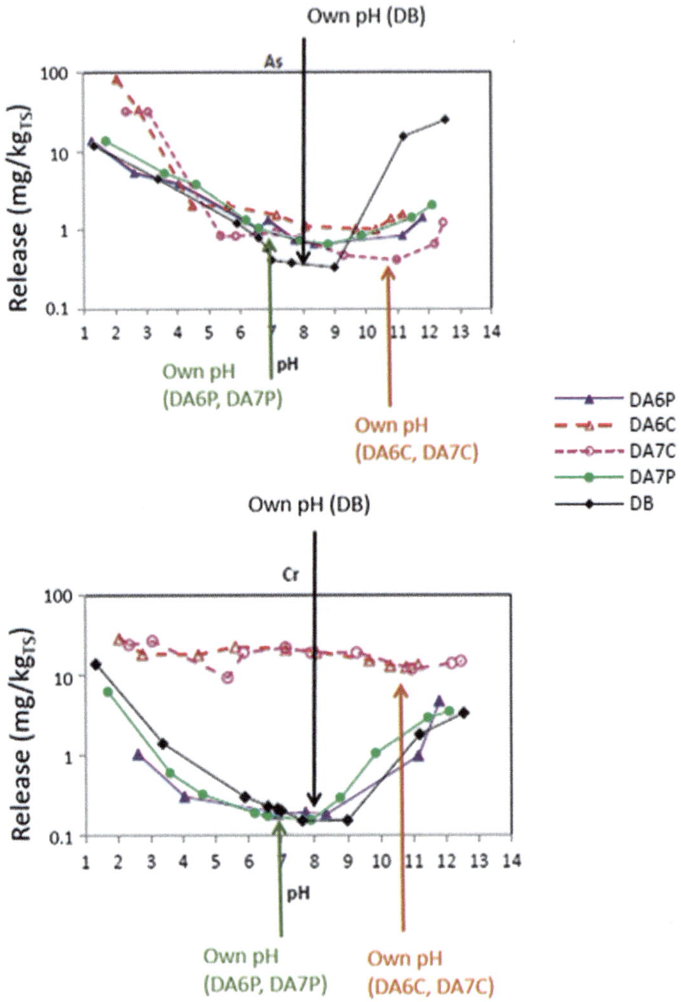

Fig. 9.39 Release of arsenic and chromium as a function of pH for raw dredging sludge (DB), phosphated (DA6P and DA7P) and phosphated and calcinated sludge (DA6C and DA7C). Reprinted with permission from Dia [210]

sediment (DB) is lower than that of the treated sediments (DA6P, DA7P, DA6C and DA7C). It has been shown for soils that the addition of phosphoric acid produces a slight increase in the mobilisable fraction of arsenic ([211] cited by [210]). The hypothesis of the chemical similarity between arsenate and phosphate ions has been suggested to explain this phenomenon. Indeed, the partial substitution of phosphates by arsenates leads to the formation of soluble salts such as $Pb_2(AsO_4)Cl$.

It may also be noted that at very low pH, the phosphated and calcinated sediments (DA6C and DA7C) show considerably greater release for arsenic. The decomposition and elimination of the organic matter after calcination could explain the increase in the quantities released (see Sect. 9.6.2.3 and Fig. 9.36). The opposite effect, expected for higher pH, is also present but the fact that the sediments that had only been phosphated, which still contain organic matter, also show lower release than the raw sediment, shows the impact of phosphatisation.

The release of chromium from the phosphated and calcinated sediments (DA6C and DA7C) show a trend that is clearly different to that of the raw sediments (DB) or those that have only been phosphated (DA6P and DA7P). Calcination leads to a significant increase in release, particularly at intermediate pH. This difference may be linked to the speciation of chromium [210]. The latter is often present in clays in the form of a trivalent compound (Cr III) and during heating of the basic mix (e.g. cement manufacture), it is oxidised to the more soluble and more toxic hexavalent chromium (Cr VI). In the same way, the disappearance of the organic matter from the completely treated sediment certainly has influence on the release of this species. Monitoring of the release from the organic matter could provide an understanding of the contribution of the latter to the observed release.

The release curves in Fig. 9.39 also show that if the raw sludge (DB) is to be re-used in contact with another soil (in the pH range relevant to soils and sludge, lying between 4 and 8 [187]), the pH range consideration would be that presenting the lowest release for these two species. The fact that total arsenic and chromium content are also shown could provide information about the leachable of species quantities.

9.6.3.2 Compliance Test: TCLP

This test is often used to check the compliance and to evaluate the release potential of waste material under severe conditions. Indeed, the TCLP uses acids whereas most of the leaching tests are performed with demineralized water as a leaching solution. With TCLP, the objective is not to mimic real conditions.

Before starting the TCLP, the leaching solution must be chosen. The choice of pH of the leaching solution for this American test depends on the alkalinity and the buffering capacity of waste material considered. Acetic acid is used as a week. The test is carried out directly in a buffered environment [185]. The test then uses crushed material of which the dimension of the biggest piece is less than 9.5 mm, in a PVC bottle where Ø = 10 cm and h = 20 cm. The quantity of weighed material is 75 g. The quantity of acetic acid solution added is such that the water/dry solid ratio

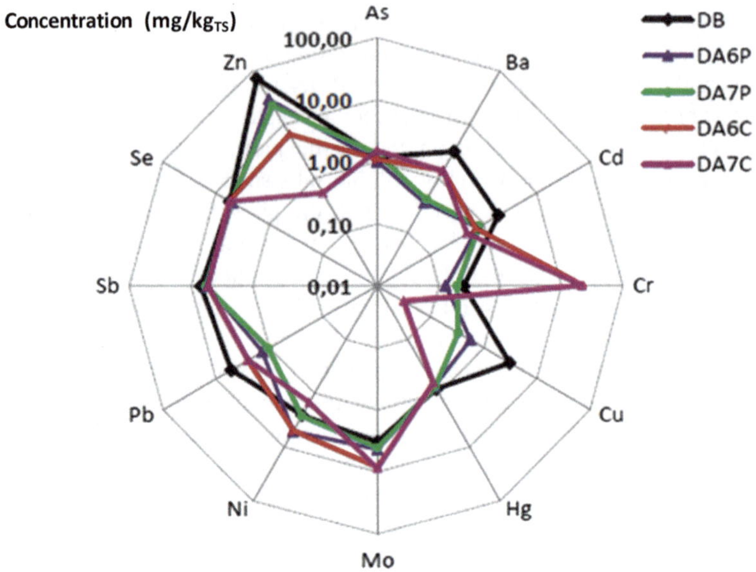

Fig. 9.40 Release of heavy metals using the TCLP test. Reprinted with permission from Dia [210]

is 20 L/S. The content is stirred for 18 h on a rotary stirrer. At the end of the experiment, the solution is filtered (Ø = 45 μm), the pH measured and the solution analysed.

Dia also carried out the TCLP test on raw, contaminated harbour dredging sediment (DB), on the same sediments phosphated with different phosphoric acid solutions (DA6P and DA7P) and on those that have in addition been calcinated at 650 °C (DA6C and DA7C) [210] (see Sect. 10.6.3.1). The results are presented in Fig. 9.40 and show that the treatment applied (phosphatisation and/or calcination) to the harbour sediment, contaminated mainly with zinc, led to a significant reduction in the availability of this pollutant. The quantities of barium, cadmium, copper, lead and zinc that salted out are higher for the raw sediment (DB) than for the treated sediments, whichever treatment is applied. The mobilisation of arsenic, mercury, antimony and selenium are of the same order of magnitude for all the sediments. However, molybdenum and above all chromium are more available in the treated sediments. This result has already been highlighted by other studies [207, 209] and has been explained in terms of the oxidation of chromium during calcination, changing its valency from (III) to the more soluble (VI).

9.6.3.3 Compliance Test: EN 12457-2

EN 12457-2 [177] is a compliance test carried out on material crushed or sieved to 4 mm, with demineralised water as the leaching reagent with an L/S ratio of 10 L/kg for 24 h. This European standard was mainly designed to study the behaviour of inorganic components of waste materials. It does not take into account the specific characteristics of non-polar organic components nor the consequences of microbiological change in degradable organic waste materials [177].

It is specified that the material subjected to this test will be under standardised conditions that are not considered to be representative of a real exposure scenario. It is a test that validates compliance to reference thresholds and is commonly used to estimate the polluting potential of a material. It is moreover part of the qualification procedure for waste materials going to landfill, in accordance with the European Council decision no. 2003/33/CE, 19 December 2002 [167].

In order to continue with the same materials as those presented in the previous tests, Fig. 9.41 presents the release results on the raw and/or treated sediments used by Dia [210]. It should be recalled that the tests were carried out on raw, contaminated harbour dredging sediment (DB), on the same sediments phosphated (for heavy metal stabilization purpose) with different grades of phosphoric acid (DA6P and DA7P) and on those that had in addition been calcinated at 650 °C (DA6C and DA7C).

First, it may be noted that this test is less aggressive than the TCLP, given the difference in the leaching solutions, because the quantities salted out are globally

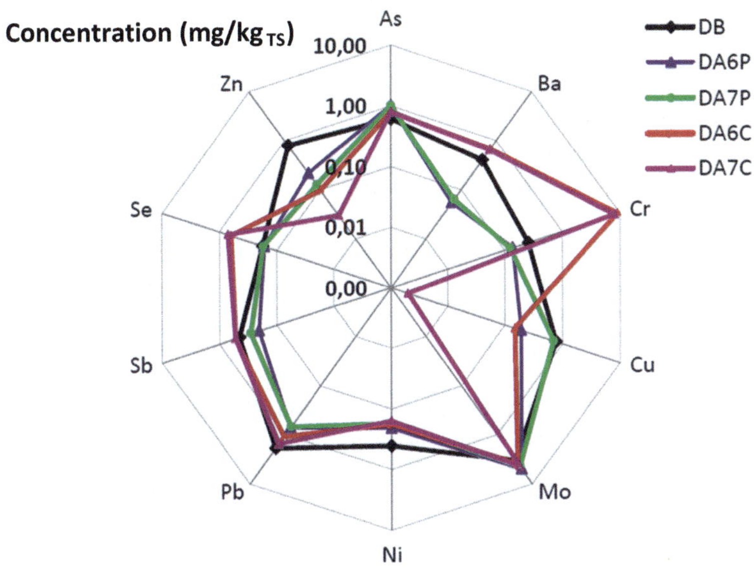

Fig. 9.41 Release of heavy metals determined using the EN 12457-2 test. Reprinted with permission from Dia [210]

lower. The quantities of certain pollutants salted out (cadmium and mercury) were even below the detection threshold. The remarks made for the TCLP for the release of chromium and zinc remain valid:

- The phosphatisation and calcination treatments considerably reduce the release of zinc;
- The release of chromium is amplified when the whole treatment is applied.

Some excess release from treated sediments compared to raw sediment fall within the range of measurement uncertainty for molybdenum and selenium [210].

9.6.3.4 Correspondence Between the Results of the Different Tests

The hierarchy of leaching tests recommends carrying out a basic characterization test in order to understand the main release mechanisms of the waste material being studied, then, in a second stage, to use of routine compliance tests that are experimentally easier.

The conditions suggested for most of the test protocols often enable results to be compared. An example is given in Fig. 9.42, comparing the results of basic characterization and compliance tests carried out as a function of pH or L/S.

The relationship between characterization and compliance is highlighted.

Hierarchy in leaching tests with a distinction between characterization, compliance and on-site verification is useful where leaching tests are relevant. Reporting compliance test data in relation to previously determined characterization level test data enhances the potential use of the short compliance leaching test results. Minimum requirements need to be defined to make optimal use of compliance test data [183].

Among the examples presented, the pH dependence test is a basic characterization test while the TCLP and the EN 12457-2 are compliance tests (see Table 9.57). It is important to note that the L/S ratio used is not identical to tests: L/S of 10 L/kg for the pH dependence test and EN 12457-2 are considered while the value of L/S for TCLP is 20. However, the presentation of the results in terms of release (in mg/kg_{TS}) and not in terms of concentration (mg/L) overcomes this difference. Using mg/kg as a unit for release is necessary in order to compare data from tests carried out at different L/S ratios [203]. It is then necessary to plot the pH values measured during the compliance tests onto the pH dependence curve to see the relationship.

Many studies have shown this close correlation. We can cite, for example, the work of Slo et al. [187] on contaminated harbour dredging sediments using several tests: EN 12457 and other extraction tests ($CaCl_2$ extraction and sequential chemical extraction) to estimate the contamination of soils. The results from the sequential chemical extraction are calculated as a cumulative leached quantity at the pH corresponding to the subsequent more aggressive extraction step. The pH dependence test was carried out over the full pH range covered by all other test methods and showed very good agreement (Fig. 9.43).

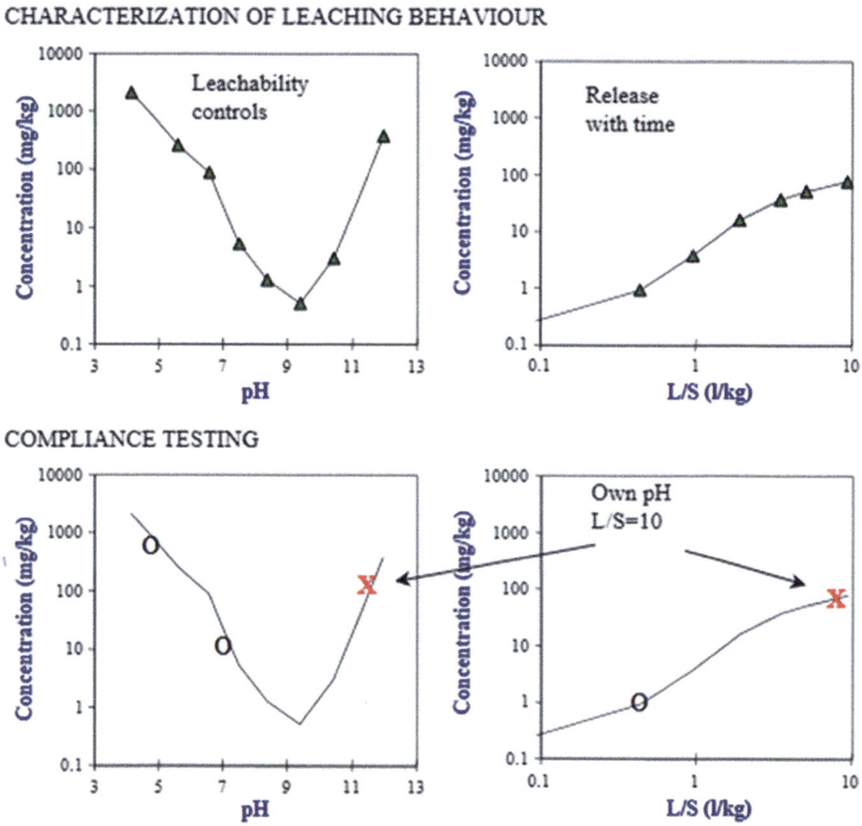

Fig. 9.42 Illustration of data used in a testing hierarchy. Compliance test data are presented in relation to previous characterization data. Reprinted with permission from van der Sloot [183]

For comparison purposes, the total composition is presented in the horizontal broken line. The curve obtained with the pH dependence test can thus be seen as a material characteristic. This applies to metals, oxyanions, major and minor elements and DOC. Here only a limited number of parameters are shown, but the same observation could be made for a wide range of components.

The compliance between the results of the tests is valid whichever element is considered. This justifies the use of a routine compliance test which is easier to set up. The interest of the pH dependence test is obvious because it enables many results from other leaching tests to be confirmed; this is why it appears to be an essential test for environmental characterization. This also shows that choosing tests that fit in with the hierarchy suggested by the CEN TC 292 and the requirements of the study enables the number of tests to be reduced while obtaining in-depth knowledge of waste material behaviour.

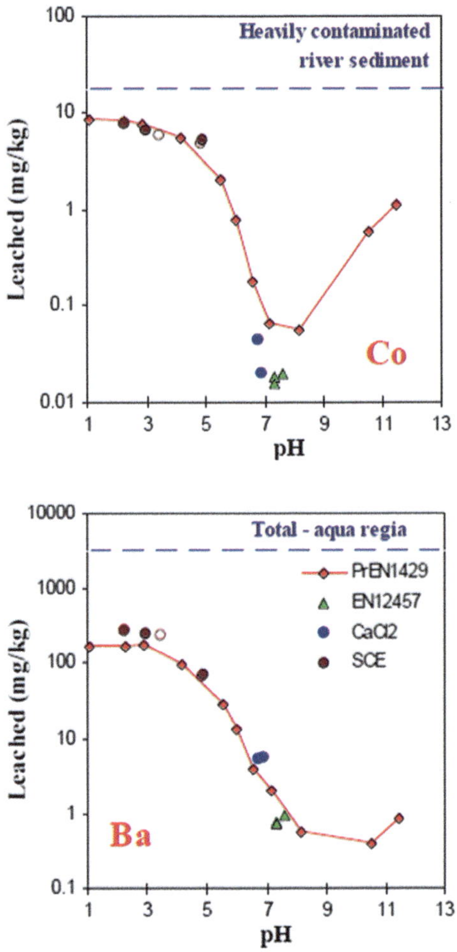

Fig. 9.43 Leaching of Co and Ba from a heavily contaminated river sediment using a range of leaching test procedures: pH dependence test, EN 12457-2, $CaCl_2$ extraction and sequential chemical extraction. Reprinted with permission from van der Sloot [187]

9.6.4 *Regulatory Aspects of Managing Sludge on Land*

The results obtained using leaching tests to characterize environmental behaviour are then compared with the threshold values set by the regulations. The main guidelines giving thresholds are presented in Tables 9.60, 9.61, 9.62, 9.63, 9.64, 9.65 and 9.66 and refer to the leaching tests presented in the previous section. It is also important to keep in mind that the regulation in this field is changing rapidly; new laws or documents may thus appear and modify or complement the regulatory aspects developed in this section.

Table 9.60 Thresholds (mg/kg_{TS}) recommended in France—EN 12 457-2 leaching tests

Parameter	Limit value (mg/kg_{TS})
As	0.5
Ba	20
Cd	0.04
Total Cr	0.5
Cu	2
Hg	0.01
Mo	0.5
Ni	0.4
Pb	0.5
Sb	0.06
Se	0.1
Zn	4
Chloride[a]	800
Fluoride	10
Sulphate[a]	1000[b]
Phenol index	1
TOC (total organic carbon) on eluate[c]	500
SF (soluble fraction)[a]	4000

[a]If waste material exceeds at least one of the values fixed for chloride, sulphate or the soluble fraction, the waste material may also be requested to comply with the admission criteria if it respects either the values for chloride and sulphate, or that for the soluble fraction

[b]If the waste material exceeds this value for sulphate, it can also be judged to comply with the admission criteria if leaching does not exceed the following values: 1500 mg/L with an L/S ratio of 0.1 l/kg and 6000 mg/kg of total solids with an L/S ratio of 10 L/kg. It is necessary to use the CEN/TS 14405 percolation test to determine the value when L/S = 0.1 L/kg under conditions of initial equilibrium; the value corresponding to L/S = 10 L/kg may be determined using an EN 12457-2 leaching test or using an CEN/TS 14405 percolation test under conditions approaching local equilibrium

[c]If the waste material does not meet the limit value indicated for total organic carbon on eluate at its own pH value, it can also be subjected to an EN 12457-2 leaching test with a pH of between 7.5 and 8.0. The waste material may also be requested to comply with the admission criteria for total organic carbon on eluate if the result of this test does not exceed 500 mg/kg of total solids

Table 9.61 Threshold values (mg/kg_{TS}) recommended in France—analyses of total content

Parameter	Limit value (mg/kg_{TS})
TOC (total organic carbon)	30,000[a]
BTEX (benzene, toluene, ethylbenzene and xylenes)	6
PCBs (polychlorobiphenyls 7 congeners)	1
Hydrocarbons (C10–C40)	500
PAHs (polycyclic aromatic hydrocarbons)	50

[a]For soils, a higher limit value can be accepted, it the limit value of 500 mg/kg total solids is respected for total organic carbon on eluate, either at the pH of the soil, or at a pH between 7.5 and 8.0

Table 9.62 Leaching and percolation limit values for granular inert materials in landfill

Parameter	Leaching (EN 12457)	Leaching (EN 12457)	Percolation (EN 14405)
	L/S = 2 L/kg	L/S = 10 L/kg	
	Value (mg/kg_{TS})	Value (mg/kg_{TS})	C_0 (mg/L)
As	0.1	0.5	0.06
Ba	7	20	4
Cd	0.03	0.04	0.02
Total Cr	0.2	0.5	0.1
Cu	0.9	2	0.6
Hg	0.003	0.01	0.002
Mo	0.3	0.5	0.2
Ni	0.2	0.4	0.12
Pb	0.2	0.5	0.15
Sb	0.02	0.06	0.1
Se	0.06	0.1	0.04
Zn	2	4	1.2
Fluorides	4	10	2.5
Chlorides	550	800	460
Sulphates	560[a]	1000[a]	1500
Phenol index	0.5	1	0.3
TOC in eluate[b]	240	500	160
Soluble fraction[c]	2500	4000	–

[a]If the waste material does not meet the values for sulphate, it may still be considered to comply with the admission criteria if the leaching does not exceed the following values: 1500 mg/L of C0 where the L/S *ratio* = 0.1 L/kg and 6000 mg/kg with an L/S *ratio* = 10 L/kg. It is necessary to use a percolation test in order to determine the limit value when L/S = 0.1 L/kg under conditions of initial equilibrium; the corresponding value at L/S = 10 L/kg can be determined by a batch leaching test or by a percolation test under conditions approaching local equilibrium

[b]If the waste material does not meet the indicated values for total organic carbon in the eluate at its own pH value, it can also be subjected to a test with an L/S ratio = 10 L/kg and a pH of between 7.5 and 8. The waste material may be considered to comply with the admission criteria for TOC in the eluate if the result of this evaluation does not exceed 500 mg/kg (a draft method based on European pre-Standard no. 14429 is available)

[c]The values corresponding to the soluble fraction (SF) may be used instead of the fixed values for sulphate and chloride

9.6.4.1 Regulatory Classification of Sludge on Land

When, for environmental or health reasons, it is not possible or desirable to immerse sediments managing them on land is one option often considered.

Table 9.63 Other criteria to be checked for inert materials to be landfilled

Parameter	Value (mg/kg)
TOC (total organic carbon)	30,000[a]
BTEX (benzene, toluene, ethylbenzene and xylenes)	6
PCBs (polychlorobiphenyls, 7 congeners)	1
Total hydrocarbons (C10–C40)	500
PAHs (polycyclic aromatic hydrocarbons)	Member States fix the limit value

[a]For soils, a higher limit value may be admitted by the relevant authorities, on condition that the limit value of 500 mg/kg is respected for the TOC in the eluate for L/S = 10 L/kg, either at the pH of the soil, or at a pH of between 7.5 and 8.0

Table 9.64 Leaching and percolation thresholds for granular non-hazardous materials to be landfilled

Parameter	Leaching (EN 12457)	Leaching (EN 12457)	Percolation (EN 14405)
	L/S = 2 L/kg	L/S = 10 L/kg	
	Value (mg/kg$_{TS}$)	Value (mg/kg$_{TS}$)	C_0 (mg/L)
As	0.4	2	0.3
Ba	30	100	20
Cd	0.6	1	0.3
Total Cr	4	10	2.5
Cu	25	50	30
Hg	0.05	0.2	0.03
Mo	5	10	3.5
Ni	5	10	3
Pb	5	10	3
Sb	0.2	0.7	0.15
Se	0.3	0.5	0.2
Zn	25	50	15
Fluorides	60	150	40
Chlorides	10,000	15,000	8500
Sulphates	10,000	20,000	7000
TOC in the eluate[a]	380	800	250
Soluble fraction[b]	40,000	60,000	–

[a]If the waste material does not meet the indicated values for total organic carbon in the eluate at its own pH value, it can also be subjected to a test with an L/S ratio = 10 L/kg and a pH of between 7.5 and 8. The waste material may be considered to comply with the admission criteria for TOC in the eluate if the result of this evaluation does not exceed 800 mg/kg (a draft method based on European pre-Standard no. 14429 is available)

[b]The values corresponding to the SF may be used instead of the fixed values for sulphate and chloride

Table 9.65 Leaching and percolation thresholds for granular hazardous materials for landfill

Parameter	Leaching (EN 12457)	Leaching (EN 12457)	Percolation (EN 14405)
	L/S = 2L/kg	L/S = 10 L/kg	
	Value (mg/kg_{TS})	Value (mg/kg_{TS})	C_0 (mg/L)
As	6	25	3
Ba	100	300	60
Cd	3	5	1.7
Total Cr	25	70	15
Cu	50	100	60
Hg	0.5	2	0.3
Mo	20	30	10
Ni	20	40	12
Pb	25	50	15
Sb	2	5	1
Se	4	7	3
Zn	90	200	60
Fluorides	200	500	120
Chlorides	17,000	25,000	15,000
Sulphates	25,000	50,000	17,000
TOC in the eluate[a]	480	1000	320
Soluble fraction[b]	70,000	100,000	–

[a]If the waste material does not meet the indicated values for total organic carbon in the eluate at its own pH value, it can also be subjected to a test with an L/S ratio = 10 L/kg and a pH of between 7.5 and 8. The waste material may be considered to comply with the admission criteria for TOC in the eluate if the result of this evaluation does not exceed 1000 mg/kg (a draft method based on European pre-Standard no. 14429 is available)
[b]The values corresponding to the SF may be used instead of the fixed values for sulphate and chloride

Table 9.66 Other criteria to be checked for waste hazardous materials to be landfilled

Parameter	Values
Loss on ignition[a]	10%
TOC[b]	6%[b]
ANC (acid neutralising capacity)	To be evaluated

[a]Either Loss on Ignition or TOC should be used
[b]If this value is exceeded, a higher limit value can be accepted by the relevant authority on condition that the limit value of 1000 mg/kg for the total organic carbon in the eluate is respected for L/S = 10 L/kg, either at the pH of the waste material, or at a pH of between 7.5 and 8.0

There is a common framework at EU level. According to the CEAMaS,[3] the legislation for the treatment of dredged materials is complex and based on the interpretation of the European Water Framework Directive (WFD), waste materials and the strategy for the marine environment. Despite this common framework, countries are free to choose the methods they use to achieve the European objectives, which has led to different regulatory choices (Laws on water vs laws on waste materials). As a consequence, different calculation methods and regulatory thresholds as used leading to a different interpretation of the results. Finally, the transposition into national law can take more or less time depending on the country, leading to variable situations [212].

The EN 12457-2 leaching test [177] is thus used to evaluate the sediment behaviour, as well as to analyse the total content. In France, for instance the thresholds are presented in Tables 9.60 and 9.61 extracted from [213].

9.6.4.2 Thresholds for Waste Disposal [183]

When no other solution is possible, the sediments should be directed to landfill (ISD):

- Site for the Disposal of Hazardous Waste Materials (ISDD);
- Site for the Disposal of Non-hazardous Waste Materials (ISDND);
- Site for the Disposal of Inert Materials (ISDI).

Leaching tests must be carried out before waste materials can be admitted to landfill sites to enable the compatibility of pollutant emissions to be checked with respect to the environment. Decision no. 2003/33/CE dated 19/12/02 [167] sets out the criteria and the procedures for the admission of waste materials to landfill, in accordance with Article 16 and Annex II of Directive 1999/31/CE [166].

Leaching thresholds are fixed by leaching tests EN 12457 (1–4) [176–179] with an L/S of 2 or 10 L/kg depending on the test; percolation test EN 14405 [170] is also proposed with the measurement of C_0 (first eluate of the percolation test, with L/S = 0.1 L/kg), directly expressed in mg/L. The thresholds, extracted from [167], are presented in:

- Tables 9.62 and 9.63 for admission of **inert** waste materials to landfill;
- Table 9.64 for admission of **non-hazardous** waste materials to landfill;
- Tables 9.65 and 9.66 for admission of **hazardous** waste materials to landfill.

[3]The CEAMaS project is an international project promoted and financed by the European Union aiming to encourage knowledge and a consensus to bring forward new solutions for the re-use of marine sediments to be applicable throughout Europe. (http://www.ceamas.eu/fr).

9.6.5 Conclusion

Sludge understood as sewage sludge and dredged sediments could be considered has hazardous materials when their content of the toxic elements such as heavy metals or some organics are above the emission thresholds. For non-polluted sludge, various options are relevant for their use and recycling as secondary materials after ensuring their compliance with the regulation criteria.

The EN 12920 methodological standard [162] underlines the importance of characterising the sludge but also taking into account the proposed recycling scenario and the conditions that it imposes.

Different tests are available to evaluate the potential for pollutant release. They have different conditions that take account the intended objectives (release in extreme conditions, conditions close to reality etc.). It is possible to obtain the results of a compliance test from a basic characterization test.

Several factors govern the leaching of contaminants, but pH is one of the most important. Indeed, the pH of the material and its environment are crucial in determining the release of the components. This is valid for all sorts of materials (block, granular, cement, soils, waste materials, sediments etc.). The conditions imposed by the fluid surrounding the by-product determine the concentration of the different elements and each material has its own pH dependent release curve. The strong influence of pH on release is due to the fact that the dissolution of the majority of minerals is pH dependent.

It is important to characterize the raw sludge but also the sludge recycled as secondary materials.

Different guidelines have been produced containing proposals for thresholds to be respected; it is necessary to be aware that this is a constantly developing field and that these guidelines keep evolving.

Moreover, biological aspects as ecotoxicity have not been treated in this study, but it is important to keep in mind that such tests are encouraging (despite the difficulties to perform them) since they can bring some other information from an environmental point of view.

We also have seen the complexity of obtaining test results that are representative of the reality of the proposed scenario: any test carried out will only provide a better approximation to real behaviour. The introduction of experimental data into a calculation model could enable more precise prediction of the effective release of the species and provide a better understanding of the chemical speciation, the long-term control release. In order to evaluate the release in a given scenario, modelling the dynamic release of a material under specific exposure conditions would be relevant, using models that combine chemical reactions and transport [187]. The choice of a calculation code for interpreting and modelling more or less complex geochemical systems is not necessarily evident [201]. One can cite, for example, work by Fang et al. [208] who used the ORCHESTRA modelling framework embedded in LeachXS to determine the chemical speciation of the heavy metals in the solid and liquid phases. It is based on the results of the pH-dependent leaching test [214, 215]. The results of the pH dependence test can also be used as input to chemical speciation codes such as MINTEQA2 and ECOSAT to identify the solubility controlling mineral phases, the role of sorption onto iron oxide phases or complexation with dissolved organic matter [167].

References

1. ISO: Water Quality—Sampling—Part 13: Guidance on Sampling of Sludges. 5667-13:2011 (2011)
2. ISO: Water Quality—Sampling—Part 15: Guidance on Preservation and Handling of Sludge and Sediment Samples. 5667-15:2009 (2009)
3. AFNOR: Sludge, Treated Biowaste and Soil—Guidance for Sample Pretreatment. NF EN 16179 (2012)
4. Tchobanoglous, G., Burton, F.: Wastewater engineering. In: Treatment and Reuse, 4th edn. McGrawHill, New York (2003)
5. AFNOR: Characterization of Sludges—Determination of Dry Residue and Water Content. FR EN 12880:2000 (2000)
6. Rice, E.W., Baird, R.B., Eaton, A.D., Clesceri, L.S.: Standard Methods for the Examination of Water and Wastewater—Method 2540.G Total, Fixed, and Volatile Solids in Solid and Semisolid Samples. American Public Health Association, Washington DC (2012)
7. Spellman, F.R.: Handbook of Water and Wastewater Treatment Plant Operations. CRC Press, Boca Raton (2014)
8. AFNOR: Water Quality—Determination of Suspended Solids. Method by Filtration Through Glass Fibre Filters. NF EN 872 (2005)
9. Rice, E.W., Baird, R.B., Eaton, A.D., Clesceri, L.S.: Standard Methods for the Examination of Water and Wastewater—Method 2540.D Total Suspended solids dried at 103–105°C. American Public Health Association, Washington DC (2012)
10. AFNOR: Water Quality—Determination of Suspended Solids by Filtration Through Glass Filters. ISO 11923:1997 (1997)
11. AFNOR: Water Quality—Determination of Suspended Solids. Method by Centrifugation. NF T90-105-2 (1997). (in French)
12. Degrémont: Water Treatment Handbook, 7th edn. Lavoisier SAS, Cachan (2007)
13. Vesilind, A.: The role of water in sludge dewatering. Water Environ. Res. **66**(1), 4–11 (1994)
14. Colin, F., Gazbar, S.: Distribution of water in sludges in relation to their mechanical dewatering. Water Res. **29**(8), 2000–2005 (1995)
15. Legrand, V.: Study of deswelling and flocculation of polyelectrolyte gel networks: application to wastewater sludge dewatering. PhD dissertation. Université Paris VI, France (1997). (in French)
16. Vaxelaire, J., Cezac, P.: Moisture distribution in activated sludges: a review. Water Res. **38** (9), 2215–2230 (2004)
17. Vaxelaire, J.: Moisture sorption characteristics of waste activated sludge. J. Chem. Technol. Biotechnol. **76**(4), 377–382 (2001)
18. Leung, H.K.: Water Activity and Other Colligate Properties of Foods. ASAE Paper No. 83–6508, Chicago (1983)
19. Rouquerol, F., Rouquerol, J., Sing, K.S.W., Llewellyn, P., Maurin, G.: Adsorption by Powders and Porous Solids: Principles, Methodology and Applications. Academic Press, Oxford (1999)
20. Ritzi, S.S.H.: Thermodynamic properties of foods in dehydration. In: Dekker, Marcel (ed.) Engineering Properties of Foods, pp. 359–436. Marcel Dekker, New-York (1986)

21. Jowitt, R., Wagstaffe, P.J.: The Certification of Water Content of Microcrystalline Cellulose (MCC) at 10 Water Activities: CRM 302. Commission of the European Communities, Luxembourg (1989)
22. Rahman, S.: Food Properties Handbook. CRC Press, Boca Raton (1995)
23. Wolf, W., Spiess, W.E.L., Jung, G., Wiesser, H., Bizot, H., Duckworth, R.B.: The water-vapour sorption isotherms of microcrystalline cellulose (MCC) and of purified potato starch. Results of a collaborative study. J. Food Eng. **3**(1), 51–73 (1984)
24. Brunauer, S., Emmett, P.H., Teller E.: The adsorption of gases in multimolecular layers. J.A. C.S. **60**, 309–319 (1938)
25. Erhardt, W., Prüeß, A.: Organic Contaminants in Sewage Sludge for Agricultural Use. European Commission Joint Research Centre. http://ec.europa.eu/environment/archives/waste/sludge/pdf/organics_in_sludge.pdf (2001). Accessed 15 Mar 2018
26. Arrêté du 8 janvier 1998 fixant les prescriptions techniques applicables aux épandages de boues sur les sols agricoles pris en application du décret n°97–1133 du 8 décembre 1997 relatif à l'épandage des boues issues du traitement des eaux usées. (in French)
27. Wiechmann, B., Dienemann, C., Kabbe, C., Brandt S., Vogel, I., Roskosch, A.: Sewage Sludge Management in Germany. Umweltbundesamt. https://www.umweltbundesamt.de/sites/default/files/medien/378/publikationen/sewage_sludge_management_in_germany.pdf (2013). Accessed 15 Mar 2018
28. GB: Discharge Standard of Pollutants for Municipal Wastewater Treatment Plant. 18918-2002
29. Clarke, B., Porter, N., Symons, R., Blackbeard, J., Ades, P., Marriott, P.: Dioxin-like compounds in Australian sewage-sludge—review and national survey. Chemosphere **72**, 1215–1228 (2008)
30. Zuloaga, O., Navarro, P., Bizkarguenaga, E., Iparraguirre, A., Vallejo, A., Olivares, M., Prieto, A.: Overview of extraction, clean-up and detection techniques for the determination of organic pollutants in sewage sludge: a review. Anal. Chim. Acta **736**, 7–29 (2012)
31. AFNOR: Determination of Polycyclic Aromatic Hydrocarbons (PAHs) in Waste using Gas Chromatography Mass Spectrometry. NF EN 15527:2008 (2008)
32. US EPA: Standard Methods for the Examination of Water and Wastewater—Method 8275A. Semi Volatile Organic Compounds (PAHs and PCBs) in Soils/Sludge and Solid Wastes Using Thermal Extraction/Gas Chromatography/Mass Spectroscopy (TE/GC/MS). https://www.epa.gov/sites/production/files/2015-12/documents/8275a.pdf (1996). Accessed 18 May 2018
33. ISO: Soil Quality—Determination of Dioxins and Furans and Dioxin-like Polychlorinated Biphenyls by Gas Chromatography with High-Resolution Mass Selective Detection (GC/HRMS). 13914:2013 (2013)
34. ISO: Soil Quality—Determination of Nonylphenols (NP) and Nonylphenol-mono- and Diethoxylates—Method by Gas Chromatography with Mass Selective Detection (GC-MS). 13907:2012 (2012)
35. ISO: Soil Quality—Determination of Selected Phthalates Using Capillary Gas Chromatography with Mass Spectrometric Detection (GC/MS). 13913:2014 (2014)
36. ISO: Soil Quality—Determination of Linear Alkylbenzene Sulfonate (LAS)—Method by HPLC with Fluorescence Detection (LC-FLD) and Mass Selective Detection (LC-MSD). 13896:2012 (2012)
37. AFNOR: Sludge, Treated Biowaste and Soil—Determination of Adsorbable Organically Bound Halogens (AOX). NF EN 16166:2012 (2012)
38. US EPA Standard Methods for the Examination of Water and Wastewater—Method 1650. Adsorbable organic halides by adsorption and coulometric titration. https://www.epa.gov/sites/production/files/2015-10/documents/method_1650c_1997.pdf (1997). Accessed 18 May 2018
39. Commission of the European Communities: Council Directive 86/278/EEC of 12 June 1986 on the Protection of the Environment, and in Particular of the Soil, When Sewage Sludge is Used in Agriculture (1986)

40. Us, E.P.A.: Land Application of Sewage Sludge—A Guide for Land Appliers on the Requirements of the Federal Standards for the use or Disposal of Sewage Sludge, 40 CFR Part 503. United States Environmental Protection Agency, Washington DC (1994)
41. GB: Control Standards for Pollutants in Sludge from Agricultural Use. 4284-84 (1984)
42. Rice, E.W., Baird, R.B., Eaton, A.D., Clesceri, L.S.: Standard Methods for the Examination of Water and Wastewater—Method 3120.B Inductively Coupled Plasma (ICP) Method. American Public Health Association, Washington DC (2012)
43. Rice, E.W., Baird, R.B., Eaton, A.D., Clesceri, L.S.: Standard Methods for the Examination of Water and Wastewater—Method 3030.F. Nitric Acid-Hydrochloric Acid Digestion. American Public Health Association, Washington DC (2012)
44. Rice, E.W., Baird, R.B., Eaton, A.D., Clesceri, L.S.: Standard Methods for the Examination of Water and Wastewater—Method 3030.K. Microwave Assisted Digestion American Public Health Association, Washington DC (2012)
45. AFNOR: Sludge, Treated Biowaste and Soil—Determination of Elements Using Inductively Coupled Plasma Optical Emission Spectrometry (ICP-OES). FR EN 16170:2016 (2016)
46. AFNOR: Sludge, Treated Biowaste and Soil—Digestion of Nitric Acid Soluble Fractions of Elements. FR EN 16173:2012 (2012)
47. AFNOR: Sludge, Treated Biowaste and Soil—Digestion of Aqua Regia Soluble Fractions of Elements. FR EN 16174:2012 (2012)
48. AFNOR: Determination of Elemental Composition by X-ray Fluorescence. FR EN 15309:2009 (2009)
49. Pathak, A., Dastidar, M.G., Sreekrishnan, T.R.: Bioleaching of heavy metals from sewage sludge: a review. J. Environ. Manage. **90**, 2343–2353 (2009)
50. Ahmad, T., Ahmad, K., Alam, M.: Sustainable management of water treatment sludge through 3'R' concept. J. Clean. Prod. **124**(Supplement C), 1–13 (2016)
51. Babatunde, A.O., Zhao, Y.Q.: Constructive approaches toward water treatment works sludge management: an international review of beneficial reuses. Crit. Rev. Environ. Sci. Technol. **37**, 129–164 (2007)
52. Chiang, K.-Y., Chou, P.H., Hua, C.R., Chien, K.L., Cheeseman, C.: Light weight bricks manufactured from water treatment sludge and rice husks. J. Hazard. Mater. **171**, 76–82 (2009)
53. Ippolito, J.A., Barbarick, K.A., Elliot, H.A.: Drinking water treatment residuals: a review of recent uses. J. Environ. Qual. **2**, 1–12 (2011)
54. Razali, M., Zhao, Y.Q., Bruen, M.: Effectiveness of a drinking water treatment sludge in removing different phosphorus species from aqueous solution. Sep. Purif. Technol. **55**, 300–306 (2007)
55. Sales, A., Rodrigues de Souza, F., do Cousto Rosa Almeida, F.: Mechanical properties of concrete produced with a composite of water treatment sludge and sawdust. Constr. Build. Mater. **25**(6), 2793–2798 (2011)
56. Sotero-Santos, R.B., Rocha, O., Povinelli, J.: Evaluation of watertreatment sludges toxicity using the Daphnia bioassay. Water Res. **39**(16), 3909–3917 (2005)
57. Zou, J.L., Xu, G.R., Li, G.B.: Ceramsite obtained from water and wastewater sludge and its characteristics affected by Fe_2O_3, CaO, and MgO. J. Hazard. Mater. **165**, 995–1001 (2009)
58. Ippolito, J.A., Barbarick, K.A., Redente, E.F.: Co-application effects of water treatment residuals and biosoilds on two range grasses. J. Environ. Qual. **28**, 1644–1650 (1999)
59. Heil, D.M., Barbarick, K.A.: Water treatment sludge influence on the growth of Sorghum-Sudangrass. J. Environ. Qual. **19**, 292–298 (1989)
60. Elliott, H.A., Singer, L.M.: Effect of water treatment sludge on growth and element composition of tomato shoots. Commun. Soil Sci. Plant Anal. **19**, 345–354 (1988)
61. Diquattro, S., Garau, G., Lauro, G.P., Silvetti, M., Deiana, S., Castaldi, P.: Municipal solid waste compost as a novel sorbent for antimony (V): adsorption and release trials at acidic pH. Environ. Sci. Pollut. Res. **25**(6), 5603–5615 (2018)

62. Elkhatib, E., Moharem, M., Mahdy, A., Mesalem, M.: Sorption, release and forms of mercury in contaminated soils stabilized with water treatment residual nanoparticles. Land Degrad. Dev. **28**, 752–761 (2017)
63. Tay, D.Y.Y., Fujinuma, R., Wendling, L.A.: Drinking water treatment residual use in urban soils: balancing metal immobilization and phosphorus availability. Geoderma **305**, 113–121 (2017)
64. Ociński, D., Jacukowicz-Sobala, I., Mazur, P., Raczyk, J., Kociolek-Balawejder, E.: Water treatment residuals containing iron and manganese oxides for arsenic removal from water—characterization of physicochemical properties and adsorption studies. Chem. Eng. J. **294**, 210–221 (2016)
65. Elkhatib, E.A., Moharem, M.L.: Immobilization of copper, lead, and nickel in two arid soils amended with biosolids: effect of drinking water treatment residuals. J.S.S. **15**,1937–1946 (2015)
66. Ippolito J.A.: Aluminum-based water treatment residual use in a constructed wetland for capturing urban runoff phosphorus: column study. Water Air Soil Pollut. **226**, 334 (2015)
67. Silvetti, M., Castaldi, P., Garau, G., Demurtas, D., Deiana, S.: Sorption of Cadmium (II) and Zinc (II) from aqueous solution by water treatment residuals at different pH values. Water, Air Soil Pollut. **226**, 313 (2015)
68. Wang, C., Pei, Y., Zhao, Y.: Comparison of metal lability in air-dried and fresh dewatered drinking water treatment residuals. J. Environ. Sci. Health A **50**, 135–143 (2015)
69. Elliott, H.A., Taylor, M.: Phosphorus partitioning in co-dewatering biosolids and water treatment residuals. Wat. Sci. Technol. **70**, 422–430 (2014)
70. Bhatnagar, A., Sillanpää, M.: Removal of natural organic matter (NOM) and its constituents from water by adsorption—a review. Chemosphere **166**, 497–510 (2017)
71. Schmidt, S.B., Jensen, P.E., Husted, S.: Manganese deficiency in plants: the impact on photosystem II. Trends Plant Sci. **21**, 622–632 (2016)
72. Crop Nutrition: Manganese in Crop Production. http://www.cropnutrition.com/manganese-in-crop-production (2018). Accessed 20 Feb 2018
73. US EPA Control of pathogens and vector attraction in sewage sludge. EPA/625/R-92/013. https://www.epa.gov/biosolids/control-pathogens-and-vector-attraction-sewage-sludge (1999). Accessed 18 May 2018
74. WHO: Guidelines for the Safe use of Wastewater, Excreta, and Greywater, vol. 4. Excreta and greywater use in agriculture. Geneva (2006)
75. Feachem, R.G., Bradley, D.J., Garelick, H., Mara, D.D.: Sanitation and Disease. Health Aspects of Excreta and Wastewater Management. Wiley, Chichester (1983)
76. Carrington, E.G.: Evaluation of Sludge Treatments for Pathogen Reduction-Final Report, Report No. CO 5026/1, European Commission (2001)
77. Andreoli, C.V., Von Sperling, M., Fernandes, F.: Sludge Treatment and Disposal, Biological Wastewater Treatment Series, vol. 4. IWA Publishing, London (2007)
78. Lepeuple, A.S., Gaval, G., Jovic, M., de Roulin, M.R.: Literature review on levels of pathogens and their abatement in sludges, soil and treated biowaste. Horizontal Project 6 (2004)
79. Jillavenkatesa, A., Lum, L.H., Dapkunas, S.: Particle Size Characterization. NIST Special Publication 960-1 (2001). https://dx.doi.org/10.6028/NBS.SP.960-1
80. ISO Particle Size Analysis—Laser Diffraction Methods. 13320:2009 (2009)
81. GB/T Particle Size Analysis—Laser diffraction methods. 19077-2016 (2016)
82. ISO Particle Size Analysis-Laser Diffraction Methods—Part 1: General Principles. 13220:1999-1 (1999)
83. Bieganowski, A., Lagod, G., Ryzak, M., Montusiewicz, A., Chomczynska, M., Sochan, A.: Measurement of activated sludge particle diameters using laser diffraction method. Ecol. Chem. Eng. S **19**(4), 597–608 (2012)
84. Chaignon, V., Lartiges, B.S., El Samrani, A., Mustin, C.: Evolution of size distribution and transfer of mineral particles between flocs in activated sludges: an insight into floc exchange dynamics. Water Res. **36**(3), 676–684 (2002)

85. Charlou, C.: Caractérisation et modéisation de l'écoulement de boues résiduaires dans un sécheur à palettes, PhD dissertation, Université de Toulouse, 200 p. (2014). (in French)
86. ASTM Standard Test Method for Screening Apparent Specific Gravity and Bulk Density of Waste. D5057-17 (2017)
87. Sørensen, B.L., Wakeman, R.J.: Filtration characterization and specific surface area measurement of activated sludge by Rhodamine B adsorption. Water Res. **30**(1), 115–121 (1996)
88. Smith, P.G., Coackley, P.: A method for determining specific surface area of activated sludge by dye adsorption. Water Res. **17**(5), 595–598 (1983)
89. ASTM Standard Test Methods for Screening of pH in Waste. D4980-89 (1989)
90. Tan, G.D.: Analysis of characteristics of sludge in Beijing wastewater treatment plant. Sci. Technol. Inf. **7**, 340 (2011)
91. Fytili, D., Zabaniotou, A.: Utilization of sewage sludge in EU application of old and new methods—a review. Renew. Sust. Energ. Rev. **12**(1), 116–140 (2008)
92. ASTM Standard Test Methods for Acidity or Alkalinity of Water. D1067-16 (2016)
93. Sun, J.H., Zhang, S.T., Yao, Y.: On the buffering performance of activated sludge. J. Saf. Environ. **1**, 199–202 (2015)
94. Quevauviller, P., Donard, O.F.X., Maier, E.A., Griepink, B.: Improvements of speciation analyses in environmental matrices. Microchim. Acta **109**(1–4), 169–190 (1992)
95. Pustišek, N., Milačič, R., Veber, M.: Use of the BCR three-step sequential extraction procedure for the study of the partitioning of Cd, Pb and Zn in various soil samples. J. Soils Sed. **1**(1), 25–29 (2001)
96. Liang, X., Ning, X., Chen, G., Lin, M., Liu, J., Wang, Y.: Concentrations and speciation of heavy metals in sludge from nine textile dyeing plants. Ecotoxicol. Environ. Saf. **98**, 128–134 (2013)
97. ISO Soil Quality—Parameters for Geochemical Modelling of Leaching and Speciation of Constituents in Soils and Materials. 12782:2012 (2012)
98. Alonso, E., Villar, P., Santos, A., Aparicio, I.: Fractionation of heavy metals in sludge from anaerobic wastewater stabilization ponds in Southern Spain. Waste Manage. **26**, 1270–1276 (2006)
99. Yang, Y., Tomlinson, D., Kennedy, S., Zhao, Y.Q.: Dewatered alum sludge: a potential adsorbent for phosphorus removal. Water Sci. Technol. **54**, 207–213 (2006)
100. Yang, Y., Zhao, Y.Q., Babatunde, A.O., Wang, L., Ren, Y.X., Han, Y.: Characteristics and mechanisms of phosphate adsorption on dewatered alum sludge. Sep. Purif. Technol. **51**, 193–200 (2006)
101. Yang, Y., Zhao, Y.Q., Kearney, P.: Influence of ageing on the structure and phosphate adsorption capacity of dewatered alum sludge. Chem. Eng. J. **145**, 276–284 (2008)
102. Wang, Y., Ren, B., Zhao, Y.Q., English, A., Cannon, M.: A comparison of alum sludge with peat for aqueous glyphosate removal for maximizing their value for practical use. Water Sci. Technol. (2018). https://doi.org/10.2166/wst.2018.165
103. Babatunde, A.O., Zhao, Y.Q., Yang, Y., Kearney, P.: Reuse of dewatered aluminium-coagulated water treatment residual to immobilize phosphorus: batch and column trials using a condensed phosphate. Chem. Eng. J. **136**, 108–115 (2008)
104. Makris, K.C., Harris, W.G., O'Conno, G.A., Obreza, T.A.: Phosphorus immobilization in micropores of drinking-water treatment residuals: implications for long-term stability. Environ. Sci. Technol. **38**, 6590–6596 (2004)
105. Makris, K.C., Harris, W.G., O'Conno, G.A., Obreza, T.A., Helliot, H.A.: Physicochemical properties related to long-term phosphorus retention by drinking water treatment residuals. Environ. Sci. Technol. **39**, 4280–4289 (2005)
106. AFNOR Solid Biofuels—Method for the Determination of Calorific Value, CEN/TS 14918:2005 (2005)
107. AFNOR Solid recovered fuels—Determination of Calorific Value, EN 15400 (2011)

108. Thipkhunthod, P., Meeyoo, V., Rangsunvigit, P., Kitiyanan, B., Siemanond, K., Rirksomboon, T.: Predicting the heating value of sewage sludges in Thailand from proximate and ultimate analyses. Fuel **84**, 849–857 (2005)
109. Demirbas, A.: Calculation of higher heating values of biomass fuels. Fuel **76**, 431–434 (1997)
110. Cordero, T., Marquez, F., Rodriguez-Mirasol, J., Rodriguez, J.J.: Predicting heating value of lignocellulosic and carbonaceous materials from proximate analysis. Fuel **80**, 1567–1571 (2001)
111. Parikh, J., Channiwala, S.A., Ghosal, G.K.: A correlation for calculating HHV from proximate analysis of solid fuels. Fuel **84**, 487–494 (2005)
112. US EPA Process Design Manual Sludge Treatment and Disposal, US Environmental Protection Agency (1979)
113. Calvo, L.F., Otero, M., Jenkins, B.M., Garcia, A.I., Moran, A.: Heating process characteristics and kinetics of sewage sludge in different atmospheres. Thermochim. Acta **409**, 127–135 (2004)
114. ASTM D5334-14 Standard Test Method for Determination of Thermal Conductivity of Soil and Soft Rock by Thermal Needle Probe Procedure. ASTM International, West Conshohocken
115. Ladevie, B., Fudym, O., Batsale, J.C.: A new simple device to estimate the thermophysical properties of insulating materials. Int. Commun. Heat Mass Transf. **27**(4), 473–484 (2000)
116. Ferrasse, J.H.: Développement d'outils expérimentaux pour le dimensionnement de procédés de séchage conductif avec agitation: application à des boues de stations d'épuration urbaines. PhD Dissertation, Université de Toulouse, p. 207 (2000). (in French)
117. Spinosa, L., Wichmann, K.: Sludge characterization: the role of physical consistency. Water Sci. Technol. **49**(10), 59–65 (2004)
118. AFNOR Usefulness of Rheological Characterization in Sludge Management. CEN/TC308 Report CEN/TC308/WG2-N25 (1995)
119. Spinosa, L., Lotito, V.: Rheological Measurements and Their Application in Biosolids Management. Water **21**, 28–29 (1999)
120. AFNOR: Sols: reconnaissance et essais - Détermination des limites d'Atterberg - Limite de liquidité à la coupelle - Limite de plasticité au rouleau, NF P94-051 (1993)
121. AFNOR: Geotechnical Investigation and Testing—Laboratory Testing of Soil—Part 12: Determination of Atterberg limits, XP CEN ISO/TS 17892-12 (2006)
122. ASTM: Standard Test Methods for Liquid Limit, Plastic Limit, and Plasticity Index of Soils, D4318 (2017)
123. AFNOR: Characterization of Sludges—Physical Consistency—Thixotropic Behaviour and Piling Behavior, CEN/TR 15463 (2007)
124. AFNOR, Glossary of Wastewater Engineering Terms, EN 16323 (2014)
125. Eshtiaghi, N., Markis, F., Yap, S.D., Baudez, J.C., Slatter, P.: Rheological characterisation of municipal sludge: a review. Water Res. **47**(15), 5493–5510 (2013)
126. Seyssiecq, I., Ferrasse, J.H., Roche, N.: State-of-the-art: rheological characterisation of wastewater treatment sludge. Biochem. Eng. J. **16**(1), 41–56 (2003)
127. Coussot, P.: Steady, laminar, flow of concentrated mud suspensions in open-channel. J. Hydraul. Res. **32**(4), 535–559 (1994)
128. Baudez, J.C., Chabot, F., Coussot, P.: Rheological interpretation of the slump test. Appl. Rheol. **12**(3), 133–141 (2002)
129. ASTM: Standard Test Method for Determining the Viscosity of Emulsified Asphalts Using a Rotational Paddle Viscometer, D7226-13 (2017)
130. AFNOR: Determination of Dynamic Viscosity of Bituminous Binder Using a Rotating Spindle Apparatus, NF EN 13302 (2010)
131. AFNOR: Bitumen and Bituminous Binders Determination of Dynamic Viscosity of Modified Bitumen by Cone and Plate Method, NF EN 13702 (2010)
132. ASTM: Standard Test Method for High-Shear Viscosity Using a Cone/Plate Viscometer, D4287 (2014)

133. AFNOR: Determination of Viscosity Using Rotary Viscometers. Part 1: Cone-and-Plate Viscometer Operated at a High Rate of Shear, NF EN ISO 2884-1 (2006)
134. ASTM: Standard Test Method for Yield Stress of Heterogeneous Propellants by Cone Penetration Method, D2884-93 (2002)
135. AFNOR: Characterization of Sludges—Physical Consistency—Part 1: Determination of Flowability—Method by Extrusion Tube Apparatus, NF EN 16720-1 (2016)
136. Ratkovich, N., Horn, W., Helmus, F.P., Rosenberger, S., Naessens, W., Nopens, I., Bentzen, T.R.: Activated sludge rheology: a critical review on data collection and modelling. Water Res. **47**(2), 463–482 (2013)
137. Spinosa, L., Wichmann, K.: Physical consistency: a useful tool for sludge characterization to the implementation of European standards. Ingénieries Eau-Agriculture-Territoires, Numéro spécial "Les boues résiduaires: quelle caractérisation et quels impacts environnementaux pour l'épandage agricole?", pp. 9-22 (2008)
138. Spinosa, L., Lotito, V.: A simple method for evaluating sludge yield stress. Adv. Environ. Res. **7**(3), 655–659 (2003)
139. ASTM: Standard Test Methods for Laboratory Miniature Vane Shear Test for Saturated Fine-Grained Clayey Soil, D 4648 (2016)
140. AFNOR: Sols: reconnaissance et essais - Essai scissométrique en laboratoire, NF P94-072 (1995)
141. ASTM: Standard Test Methods for Cone Penetration of Lubricating Grease, D217-17 (2017)
142. AFNOR: Petroleum Products and Lubricants—Determination of Cone Penetration of Lubricating Grease and Petrolatum, NF ISO 2137 (1985)
143. AFNOR: Adjuvants pour béton, mortier et coulis—Méthodes d'essai—Partie 2: détermination du temps de prise, NF EN 480-2 (2006)
144. ASTM: Standard Test Methods for Time of Setting of Hydraulic Cement by Vicat Needle, C191 (2013)
145. AFNOR: Plastiques - Matières thermoplastiques - Détermination de la température de ramollissement Vicat, NF EN ISO 306 (1014)
146. ASTM: Standard Test Method for Vicat Softening Temperature of Plastics, D1525 (2017)
147. Hansbo, S.: A new approach to the determination of the shear strength of clays by the fall-cone test. Proc. Roy. SGI **14**, 7–48 (1957)
148. Mouzaoui, M., Baudez, J.C., Sauceau, M., Arlabosse, P.: Experimental rheological procedure adapted to dewatered sludge up to 45% dry matter. Water Res. **133**, 1–7 (2018)
149. Liang, F., Sauceau, M., Dusserre, G., Arlabosse, P.: A uniaxial cyclic compression method for characterizing the rheological and textural behaviors of mechanically dewatered sewage sludge. Water Res. **113**, 171–180 (2017)
150. Ruiz, T., Wisniewski, C., Kaosol, T., Persin, F.: Influence of organic content in dewatering and shrinkage of urban residual sludge under controlled atmospheric drying. Process Saf. Environ. **85**(B1), 104–110 (2007)
151. Ruiz, T., Wisniewski, C.: Correlation between dewatering and hydro-textural characteristics of sewage sludge during drying. Sep. Purif. Technol. **61**(2), 204–210 (2008)
152. Ruiz, T., Kaosol, T., Wisniewski, C.: Dewatering of urban residual sludges: filterability and hydro-textural characteristics of conditioned sludge. Sep. Purif. Technol. **72**(3), 275–281 (2010)
153. Liang, F.: Caractérisation et modélisation du comportement rhéologique des boues résiduaires. PhD Dissertation, Ecole des Mines d'Albi-Carmaux (2016). (in French)
154. Kudra, T.: Sticky region in drying—definition and identification. Drying Technol. **21**(8), 1457–1469 (2003)
155. Peeters, B., Dewil, R., Van Impe, J.F., Vernimmen, L., Smets, I.Y.: Using a shear test-based lab protocol to map the sticky phase of activated sludge. Environ. Eng. Sci. **28**(1), 81–85 (2010)
156. ASTM: Standard Test Method for Shear Testing of Bulk Solids Using the Jenike Shear Tester, ASTM D 6128 (2016)

157. Li, B., Wang, F., Chi, Y., Yan, J.H.: Adhesion and cohesion characteristics of sewage sludge during drying. Drying Technol. **32**(13), 1598–1607 (2014)
158. Liang, F., Chen, X., Mao, H., Sauceau, M., Arlabosse, P., Wang, F., Chi, Y.: The microstructure: a critical factor in characterising the sticky properties of highly concentrated sewage sludge. In: 6th International Conference on Engineering for Waste and Biomass Valorisation, Albi (2016)
159. Arlabosse, P., Chavez, S., Lecomte, D.: Method for thermal design of paddle dryers: application to municipal sewage sludge. Drying Technol. **22**(10), 2375–2393 (2004)
160. Loi n° 2016-816 du 20 juin 2016 pour l'économie bleue. Journal Officiel de la République Française n°0143 du 21 Juin 2016. https://www.legifrance.gouv.fr/afichTexte.do?cidTexte=JORFTEXT000032728685&categorie (2016). Accessed 20 May 2018
161. Ministère de l'écologie, du développement durable, des transports et du logement: Sédiments – Trois mode de gestion – Mode d'emploi. DREAL Nord Pas de Calais. https://www.hauts-de-france.developpement-durable.gouv.fr/IMG/pdf/troisvolets_pageparpage_les_sediments_2011.pdf (2011). Accessed 20 May 2018
162. AFNOR: Characterization of Waste. Methodology for the Determination of the Leaching Behaviour of Waste Under Specified Conditions, NF EN 12920 + A1 (2008)
163. Crest, M.: Etude multi-échelle de percolations instationnaires à travers un résidu minéral réactif et modélisations hydrodynamiques associées – Applicabilité de la norme NF EN 12920, PhD thesis, Institut National des Sciences Appliquées de Lyon (2007). (in French)
164. Colombel, L.: Caractérisation du comportement à la lixiviation: quels essais pour quels objectifs? BRGM. http://infoterre.brgm.fr/rapports/RP-60088-FR.pdf (2011). Accessed 20 May 2018
165. van der Sloot, H.A., Dijkstra, J.J.: Development of horizontally standardized leaching test for construction materials: a material based or release based approach? In: Identical Leaching Mechanisms for Different Materials. ECN. https://www.ecn.nl/docs/society/leaching/C04060.pdf (2004). Accessed 20 May 2018
166. Council of the European Union: Council Directive 1999/31/EC of 26 April 1999 on the landfill of waste. Official Journal L 182 published on 16/07/1999, 1-19 (1999)
167. Council of the European Union: Council Decision DC 2003/33/EC of 19 December 2002 establishing criteria and procedures for the acceptance of waste at landfills pursuant to Article 16 of and Annex II to Directive 1999/31/EC. Official Journal L 011 published on 16/01/2003, 27-49 (2003)
168. EA NEN: Leaching Characteristics—Determination of the Availability of Inorganic Components for Leaching—Solid Earthy and Stony Materials. NEN 7371 (2004)
169. EA NEN: Leaching Characteristics of Moulded or Monolithic Building and Waste Materials—Determination of Leaching of Inorganic Components with the Diffusion Test—Tank Leaching Test. NEN 7375 (2004)
170. AFNOR: Characterization of Waste—Leaching Behaviour Test—Up-Flow Percolation Test (under specified conditions). NF EN 14405 (2017)
171. AFNOR: Characterization of Waste—Leaching Behaviour Test—Influence of pH on Leaching with Initial Acid/Base Addition. NF EN 14429 (2015)
172. AFNOR: Characterization of Waste—Leaching Behaviour Test—Influence of pH on Leaching with Continuous pH Control. NF EN 14997 (2015)
173. AFNOR: Characterization of Waste—Leaching Behaviour Test for Basic Characterization—Dynamic Monolithic Leaching Test with Periodic Leachant Renewal, Under Fixed Conditions. NF EN 15863 (2015)
174. Nordtest: Solid Waste, Granular Inorganic Material: Availability Test (NT ENVIR 003). Nordtest. http://www.nordtest.info/index.php/methods/item/solid-waste-granular-inorganic-material-availability-test-nt-envir-003.html (1995)
175. AFNOR: Characterization of Waste—Leaching Behaviour Test for Basic Characterization—Dynamic Monolithic Leaching Test with Continuous Leachant Renewal Under Conditions Relevant for Specified Scenario(s). XP CEN/TS 15864 (2016)

176. AFNOR: Caractérisation des déchets - Lixiviation - Essai de conformité pour lixiviation des déchets fragmentés et des boues - Partie 1: essai en bâchée unique avec un rapport liquide-solide de 2 l/kg et une granularité inférieure à 4 mm (sans ou avec réduction de la granularité). NF EN 12457-1 (2002)
177. AFNOR: Caractérisation des déchets - Lixiviation - essai de conformité pour lixiviation des déchets fragmentés et des boues - Partie 2: Essai en bâchée unique avec un rapport L/S de 10L/kg et une granularité inférieure à 4 mm (sans ou avec réduction de la granularité). NF EN 12457-2 (2002)
178. AFNOR: Caractérisation des déchets - Lixiviation - Essai de conformité pour lixiviation des déchets fragmentés et des boues - Partie 3: essai en bâchée double avec un rapport liquide-solide de 2 l/kg et de 8 l/kg pour des matériaux à forte teneur en solides et une granularité inférieure à 4 mm (sans ou avec réduction de la granularité). NF EN 12457-3 (2002)
179. AFNOR: Caractérisation des déchets - Lixiviation - Essai de conformité pour lixiviation des déchets fragmentés et des boues - Partie 4: essai en bâchée unique avec un rapport liquide/ solide de 10 l/kg et une granularité inférieure à 10 mm (sans ou avec réduction de la granularité). NF EN 12457-4 (2002)
180. AFNOR: Déchets – Essai de lixiviation d'un déchet solide initialement massif ou généré par un procédé de solidification. NF X 31-211 (2012)
181. US EPA: Method 1311. Toxicity Characteristic Leaching Procedure, SW-846: Test Methods for Evaluating Solid Waste—Physical/Chemical Methods (1994)
182. AFNOR: Caractérisation des déchets - Essai de lixiviation de conformité - Essai de lixiviation en bâchée unique pour des monolithes avec un rapport liquide/ surface (L/A) fixe, pour des prises d'essai de dimensions minimales fixes. XP CEN/TS 15862 (2012)
183. van der Sloot, H.A.: European Activities on Harmonisation of Leaching/Extraction Tests and Standardisation in Relation to the Use of Alternative Materials in Construction. ECN. ftp://ecn.nl/pub/www/library/report/2001/rx01045.pdf (2001)
184. van der Sloot, H.A., Heasman, L., Quevauviller, Ph.: Harmonization of leaching/extraction tests. Stud. Environ. Sci. **70** (1997)
185. Lassin, A., Bodénan, F., Piantone, P., Blanc, P.: Essais de comportement des déchets à la lixiviation et modélisation des processus hydro-physico-chimiques associés, Etude bibliographique. BRGM. http://infoterre.brgm.fr/rapports/RP-51518-FR.pdf (2002). Accessed 20 May 2018
186. Tiwari, M.K., Bajpai, S., Dewangan, U.K., Tamrakar, R.K.: Suitability of leaching test methods for fly ash ans slag: a review. J. Radiat. Res. Appl. Sci. **8**, 523–537 (2015)
187. van der Sloot, H.A., Comans, R.N.J., Meeussen, J.C.L., Dijkstra, J.J.: Leaching Methods for Soil, Sludge and Treated Biowaste. ECN. https://www.ecn.nl/docs/society/horizontal/hor_desk_23_leaching2.pdf (2003) Accessed 20 May 2018
188. Waste Extraction Test, California Code Regulations, Sections 208 and 25141, Health and Safety Code
189. AFNOR: Qualité du sol - Modes opératoires de lixiviation en vue d'essais chimiques et écotoxicologiques ultérieurs des sols et matériaux du sol - Partie 4: essai de dépendance au pH avec ajout initial d'acide/base. XP CEN ISO/TS 21268-4 (2009)
190. US EPA: Method 1313. Liquid-solid partitioning as a function of extract pH using parallel batch extraction procedure, revision (2017)
191. AFNOR: Qualité du sol - Modes opératoires de lixiviation en vue d'essais chimiques et écotoxicologiques ultérieurs des sols et matériaux du sol - Partie 3: Essai de percolation à écoulement ascendant. XP CEN ISO/TS 21268-3 (2009)
192. ASTM: Standard test method for leaching solid material in a column apparatus. ASTM D4874-95 (2001)
193. van der Sloot, H.A.: Harmonisation of leaching/extraction procedures for sludge, compost, soil and sediment analyses. In: Quevauviller, P. (ed) Methodologies for Soil and Sediment Fractionation Studies, pp. 142–170. Royal Society of Chemistry, Cambridge (2002)

194. Commission, European: Technical Work in Support of the Network "Harmonisation of Leaching/Extraction Tests", Contract SMT4-CT96-2066. European Commission, Brussels (2004)
195. Gupta, S.K., Vollmer, M.K., Krebs, R.: The importance of mobile, mobilisable and pseudo total heavy metal fractions in soil for three-level risk assessment ans risk management. Sci. Total Environ. **178**, 11–20 (1996)
196. BRGM: Guide sur le comportement des pollutants dans les sols et les nappes – Application dans un contexte d'Evaluation Détaillée des Risques pour les resources en eau. BRGM. http://infoterre.brgm.fr/rapports/RP-50662-FR.pdf (2001). Accessed 20 May 2018
197. Kosson, D.S., van der Sloot, H.A., Sanchez, F., Garrabrants, A.C.: An integrated framework for evaluating leaching in waste management and utilization of secondary materials. Environ. Eng. Sci. **19**, 159–204 (2002)
198. Hjelmar, O.: Leachate from land disposal of coal fly-ash. Waste Manage. Res. **8**, 429–449 (1990)
199. Tribout, C.: Valorisation de sédiments traités en techniques routières: contribution à la mise en place d'un protocole d'acceptabilité, PhD Thesis, Université de Toulouse (2010). (in French)
200. AFNOR: Soil Quality—Guidance on Leaching Procedures for Subsequent Chemical and Ecotoxicological Testing of Soils and Soil Materials. NF EN ISO 18772 (2014)
201. Garrabrants A.C.: Development and Application of Fundamental Leaching Property Protocols for Evaluating Inorganic Release from Wastes and Soil, Master Thesis, The State University of New Jersey (1998)
202. Meza, S.L., Kalbe, U., Berger, W., Simon, F.-G.: Effect of contact time on the release of contaminants from granular waste materials during column leaching experiments. Waste Manage. **30**, 565–571 (2010)
203. van der Sloot, H.A.: Developments in evaluating environmental impact from utilization of bulk inert wastes using laboratory leaching tests and field verification. Waste Manage. **16**(1–3), 65–81 (1996)
204. van der Sloot, H.A., Hoede, D., Cresswell, D.J.F., Barton, J.R.: Leaching behaviour of synthetic aggregates. Waste Manage. **21**, 221–228 (2001)
205. van der SLoot, H.A.: Horizontal Standardization and Harmonization of Leaching Test Methods for Waste, Secondary Raw Materials, Construction Materials and (contaminated) soil. ECN. https://www.ecn.nl/docs/society/leaching/wascon_005.pdf (2003)
206. Quevauviller, P., van der Sloot, H.A., Urec, A., Muntaud, H., Gomeze, A., Rauret, G.: Conclusions of the workshop: harmonization of leaching/extraction tests for environmental risk assessment. Sci. Total Environ. **178**, 133–139 (1996)
207. Tribout, C., Husson, B., Nzihou, A.: Use of treated dredged sediments as road base materials: environmental assessment. Waste Biomass Valorization **2**(3), 337–346 (2011)
208. Fang, W., Wei, Y., Liu, J.: Comparative characterization of sewage sludge compost and soil: heavy metal leaching characteristics. J. Hazard. Mater. **310**, 1–10 (2016)
209. Kribi, S.: Décomposition des matières organiques et stabilisation des métaux lourds dans les sédiments de dragage, PhD Thesis, Institut National des Sciences Appliquées Lyon (2005). (in French)
210. Dia, M.: Traitement et valorisation de sédiments de dragage phosphatés en technique routière, PhD Thesis, Ecole des Mines de Douai & Université d'Artois (2013). (in French)
211. Theodoratos, P., Papassiopi, N., Xenidis, A.: Evaluation of monobasic calcium phosphate for the immobilization of heavy metals in contaminated soils from Lavrion. J. Hazard. Mater. **94**(2), 135–146 (2002)
212. RECORD: Impacts écologiques de sédiments pollués extraits et déposés en milieux terrestres. Etat des connaissances et évaluation des risques pour les écosystèmes. RECORD, n°14-1023/1A (2017). (in French)

213. Ministère de l'écologie, du développement durable et de l'énergie: Arrêté du 12 décembre 2014. Journal Officiel de la République Française n° 0289 du 14 décembre 2014 page 21032, texte n°11. https://www.legifrance.gouv.fr/affichTexte.do?cidTexte=JORFTEXT000029893828&categorieLien=id (2014). Accessed 20 May 2018
214. van der Sloot, H.A., van Zomeren, A., Meeussen, J.C.L., Seignette, P., Bleijerveld, R.: Test method selection, validation against field data, and predictive modelling for impact evaluation of stabilised waste disposal. J. Hazard. Mater. **141**(2), 354–369 (2007)
215. van der Sloot, H.A., Kosson, D.S.: Use of characterisation leaching tests and associated modelling tools in assessing the hazardous nature of wastes. J. Hazard. Mater. **207**, 36–43 (2012)

Chapter 10
Biogas

Hélène Métivier, Hassen Benbelkacem, Vincent Chatain, Lucy Culleton and Nathalie Dumont

Abstract Current waste management policies favor the valorization of biogas (from digester or landfill), usually used in combined heat and power plants, but also upgraded to produce biomethane. The characterization of biogas is a key challenge for the control and the optimization of treatment processes and further valorization. Raw biogas contains mainly methane and carbon dioxide. Trace components such as water, hydrogen sulfide, siloxanes, halogenated hydrocarbons, ammonia, oxygen and nitrogen can be present and might be problematic if not reduced and/or eliminated. The choice of adapted treatments and the assessment of their efficiencies depending on the selected valorization are still an issue. Indeed the biogas analysis with numerous components in large and trace quantities is difficult. The choice of the most appropriate analytical technique depends on the molecules of interest, their concentration, the desired precision and the gaseous matrix. A wide range of analytical techniques is described in the form of data sheets.

10.1 Introduction: What Are the Molecules to Analyze and Why?

10.1.1 Context and Need for Characterization

Biogas is a promising source of renewable energy, which recovery should be promoted, and even more when it proceeds from the residual potential of organic waste and the reduction of the greenhouse effect.

According to the report EurObserv'ER [1], in 2016 the total production of biogas in Europe was estimated to be about 16 million ton of oil equivalent, divided

---

H. Métivier (✉) · H. Benbelkacem · V. Chatain · N. Dumont
Laboratoire DEEP—EA 7429 (Déchets Eaux Environnement Pollutions—Waste Water Environment Pollutions), Université de Lyon—INSA Lyon, Villeurbanne, France
e-mail: helene.metivier@insa-lyon.fr

L. Culleton
National Physical Laboratory, Teddington, UK

A. Nzihou (ed.), *Handbook on Characterization of Biomass, Biowaste and Related By-products*, https://doi.org/10.1007/978-3-030-35020-8_10

in landfill, sewage sludge and other produced by anaerobic fermentation (such as farms, co-digestion sites, industrial food-processing plants, or waste methanization plants) as follows: 17.2%, 8.7% and 74.1%, respectively. Synthetic gas (Syngas, by thermal processes) appears still negligible.

The biogas produced in these anaerobic processes is raw renewable energy, consisting on average about 50–75% methane and 25–50% carbon dioxide; depending on the substrate, as well as, water vapor (1–5%) and other gases such as nitrogen (0–16%), oxygen (1–4%), hydrocarbons (1%), hydrogen sulfide (0.5%), ammonia (0.05%) and Volatile Organic Silicon Compounds (VOSiC's) (0–50 mg Si/Nm^3) [2]. The highest contents in O_2 and N_2 are a specificity of landfill biogas. The term biogas thus refers to a very wide variety of gases resulting from specific processes (industrial or not), feeding from different types of raw materials (crops, green waste, sewage sludge, food-processing industry waste, household waste …). Indeed, as shown in the Table 10.1, the amount, typology and specific composition of components present in the raw biogas is highly depending on the feedstocks that undergo methanization processes.

Depending on its final use, the biogas needs to be cleaned, compressed, stored and conveyed, as illustrated in the Fig. 10.1. The local needs and regulations have to be considered to ensure the feasibility of a given energy recovery way.

As will be developed in the following paragraphs, the presence of some of trace compounds in gas, such as hydrogen sulfide and VOSiC's, are particularly problematic and may affect the durability of certain recovery methods.

It is therefore essential to have tools and methods providing a reliable knowledge of biogas composition whether to evaluate the risks incurred by the conversion equipment and/or to pilot the treatment processes implemented before in order to

Table 10.1 Examples of biogas compositions according to waste undergoing anaerobic processes [2–4]

Compound		Household waste	Wastewater treatment plant	Agricultural waste	Waste of agri-food sector (distillery)	NHIW, Landfill
(% vol.)	CH_4	50–60	60–75	60–75	68	41–45
	CO_2	34–38	19–33	19–33	26	38–43
	N_2	0–5	0–1	0–1	1	4–12
	O_2	0–1	<0.5	0	–	1–3
	H_2O^a	6	6	6	5	1–3
(mg Nm^{-3})	H_2S	100–900	100–4000	3000–10,000	400	3000–4000
	NH_3)	0–5	–	50–100	–	–
	VOSiCs	0–50	0–20		–	

[a]at 40 °C
NHIW non-hazardous industrial waste

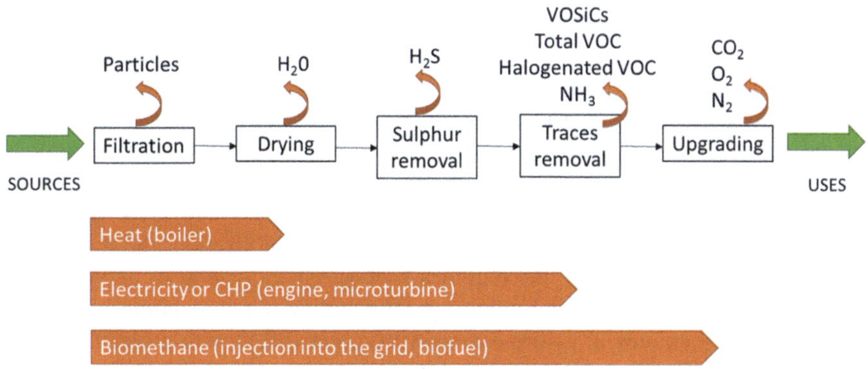

Fig. 10.1 Process treatment needs according to the type of energy conversion system

protect this equipment. In addition, analysis is also crucial to the development of the new biomethane sector. The European standard EN 16723 for the quality of biomethane and bioNGV includes indeed specifications [5, 6].

10.1.2 Biogas Valorization

Biogas has many energy uses, depending on the nature of biogas source and the local demand. As can be seen from Fig. 10.1, biogas can be used for heat production, electricity production but also as a fuel for vehicles or directly injected into the gas grid.

The simplest way to produce energy from biogas is the direct burning in burners for heating purpose. But, the standard use of biogas near the anaerobic digestion facility is the electricity generation, alone or with Combined Heat Power (CHP). The electricity produced is then consumed locally or, most of the time, exported to the electrical grid thanks to the preferential tariff for purchase of electricity. Before its burning in motor engines or micro-turbines, raw biogas has to be cleaned to remove harmful compounds like water, H_2S and VOSiC [7]. Water removal is usually done through refrigeration to a dew point of 4 °C. Several techniques are applied for H_2S removal:

- During anaerobic digestion by air/oxygen addition in the gaseous roof or by addition of iron chloride into the digester.
- After anaerobic digestion thanks to dry solutions like adsorption on impregnated activated carbon or using iron oxide or hydroxide; or wet solutions using chemical scrubbers or biological filters.

VOSiC removal is usually carried out using adsorption on activated carbon or polymer beds.

Biogas can also be upgraded to biomethane, gas sufficiently close to the properties of fossil-derived natural gas with more than 95% of methane and a key Wobbe index close to 13 kWh Nm^{-3}. The biomethane produced can then be directly injected into the gas grid and used for the same purposes as natural gas, or it can be compressed and used as Natural Gas Vehicle (NGV) fuel. But, prior to injection into the gas grid or use as a fuel for transportation, raw biogas must undergo upgrading process [8]. It is a multi-step procedure where, after removal of water, H_2S and VOSiC at lower levels than for CHP valorisation, the removal of CO_2 is done in order to reach the required Wobbe index. Several technologies are developed as commercial solutions: pressurized water scrubbing, chemical absorption using amines, Pressure Swing Adsorption (PSA) with molecular sieve, membrane separation or, more recently, cryogenic separation where nitrogen and oxygen can also be removed. Each of the mentioned technologies, except cryogenic separation, is in operation in large scale and provides biomethane with the required standards for injection into the gas grid or vehicle fuel. No real differences are observed regarding the investment costs or the energy consumption between the different technologies [9].

10.1.3 Energetic Value of Biogas: CH_4, CO_2, N_2, O_2, H_2O (Almost Major Compounds)

The energetic value of biogas relies on the methane content which depends on the abundance of other major compounds, mainly CO_2. Monitoring these compounds is important to run energy conversion systems. For example, engines used to produce electricity from biogas operate with lean gas, so that CH_4 content (and consequently CO_2 content too) may be within the range of operating conditions of engines: a minimum content of CH_4 is required but a maximum value should not be exceeded.

Other compounds like N_2, O_2 and H_2O, even present in smaller quantity than CH_4 or CO_2, can also be considered as major compounds since their content reaches several percent and hence are involved in the assessment of the low heating value (LHV) of biogas.

With this respect, H_2O is targeted by drying process because it has an adverse effect on the energetic value of biogas and H_2O measurements are helpful to monitor treatments' efficiency. Downstream of a dryer, H_2O usually remains as a minor compound, the effects of which will be discussed in the following Sect. (10.1.4.3).

Aside from being involved in the LHV of biogas, the oxygen content may be interesting to monitor, especially if adsorption onto impregnated activated carbon (targeting H_2S removal) is included in treatment facilities. Oxygen is indeed involved in the related adsorption mechanism [10] so that treatment efficiency may be drastically reduced if oxygen content is too low, becoming almost inoperative.

Nitrogen is not measured but its amount is calculated through a balance based on other major compounds. This is not always relevant, especially when biogas includes high content of various minor compounds the total of which exceeding 1%.

When biogas is upgraded into biomethane, the grid injection requires a minimum CH_4 content ranging from 90 to 99% according to the grade of the natural gas in the grid. For example the CH_4 content must be higher than 96.65% to reach a high heating value of 10.7 kWh/Nm^3 [11]. The analytical technique implemented must be able to certify that the targeted threshold is reached, the related uncertainty has to be well-known and as low as possible.

10.1.4 Harmful Compounds and Low Content VOCs (Almost Minor Compounds): Including H_2S, VOSiCs, Total VOCs, Halogenated VOCs, NH_3, H_2O

Some minor compounds like H_2S and VOSiCs are especially harmful to energy conversion equipment and hence need to be removed from biogas prior to energy conversion. Some others are not cause of concern for energy conversion but can be detrimental to biogas treatment processes. After a detailed presentation of these compounds, information is summarized in tabular form at the end of the section (see Sect. 10.1.4.7).

10.1.4.1 Hydrogen Sulphide (H_2S)

First of all, H_2S is a toxic gas even at low concentrations and it may induce acute or chronic intoxication. For example, an acute intoxication may result in loss of consciousness in the case of exposure to 500 ppm, in death if exposure is higher than 1000 ppm. So, for the sake of workers' health, it is important to know that H_2S is often present in biogas (concentrations ranging from 10 to 15,000 mg Nm^{-3} have been reported).

With regard to energy conversion, H_2S is targeted by treatment processes prior to combustion for two main reasons. On one hand, biogas is a humid gas and if it is not properly dried, water can be formed on cold points where H_2S may solubilized forming sulphuric acid and hence leading to equipment corrosion, especially when H_2S content of the biogas is high. On the other hand, H_2S is oxidized during biogas combustion leading to the formation of SO_x (sulphur oxides) that can be measured in exhaust gas. Since SO_x emissions are regulated, it is important to remove H_2S from biogas prior to energy conversion.

So, the efficiency of biogas treatment dedicated to H_2S removal has to be monitored to ensure good conditions for energy conversion, that is, safety of equipment and regulatory compliance.

Table 10.2 Why are minor compounds targeted in biogas characterization?

Minor compounds	Why are they targeted?
H_2S (hydrogen sulphide)	– toxic even at low concentrations, endanger the health/life of workers – responsible for the corrosion of equipment, especially in the presence of moisture: it can damage stainless steel under these conditions – responsible for the formation of SO_x (Sulphur oxides) during gas combustion and hence for SO_x emission – can damage membranes used for biogas treatment prior to energy conversion
VOSiC (volatile organic silicon compounds) or siloxanes	– cause silica deposits in engines during energy conversion – potentially harmful: responsible for the formation of silica nanoparticles in exhaust gas
H_2O	– responsible for the corrosion of equipment in the presence of H_2S – may have a negative impact on biogas treatment efficiency depending on the process – must be low in the case of biomethane production: dew point recommended below −5 °C at 7 kPa
Total VOC (volatile organic compounds)	– responsible for competitive adsorption in the case of adsorption treatment of other compounds (H_2S or VOSiCs) – some of them (terpenes, mercaptans) may damage membranes used for biogas treatment prior to energy conversion
Halogen-containing VOC	– may lead to acids formation during combustion and hence responsible for acidity of exhaust gas
NH_3 (ammonia)	– responsible for the corrosion of equipment – responsible for the formation of NO_x (Nitrogen oxides) during gas combustion and hence for NO_x emission

10.1.4.2 Volatile Organic Silicon Compounds (VOSiC)

VOSiC include especially methyl siloxanes (linear or cyclic molecules, with—O–Si–O—skeleton, commonly named siloxanes) but also silanes (Si_nH_{2n+2}, and substituted molecules) such as tetramethylsilane ($Si(CH_3)_4$, also referred as TMS), and silanols (with a Si–OH group) such as trimethylsilanol ($(CH_3)_3SiOH$), also referred as TMSol, typically found in landfill biogas). These VOSiC are especially known as harmful compounds towards energy conversion of biogas. Their oxidation during combustion leads indeed to the formation of deposits mainly composed of silica [4, 12, 13]. When biogas is used to produce electricity or heat, silica deposits affect several parts of combustion engines: wear of valve seat, of cylinder, fouling of pistons, of combustion chamber for example. The frequency of maintenance operations has to be increased leading to higher maintenance costs and losses of energy production namely losses of incomes. In worse cases, engine or turbine failure have been reported. In the case of biomethane (injection into the gas grid or use as NGV fuel), it can be assumed that VOSiCs may be cause of concern where gas will be burned.

The removal of VOSiCs prior to energy conversion is thus a mean to ensure the safety of combustion and energy conversion equipment. The monitoring of treatment processes is therefore essential.

In addition, the combustion of VOSiCs may give rise to a public health issue since silica nanoparticles are formed and released in the environment through exhaust gas [14]. In that respect, the need of upstream removal of Silicon-containing compounds and its monitoring is strengthened. Regarding health's workers, silica deposits include also heavy metals [4] so that the use of personal protective equipment, and especially respiratory protection, is compulsory during maintenance operations.

10.1.4.3 Water

Water has already been discussed in the section dedicated to the energetic value of biogas since up to several percent it can be viewed as a major compound. Nevertheless, the drying of biogas is usually incomplete so that water remains in the biogas as a minor compound. In this range of content, water may have two undesirable effects: as already mentioned (see Sect. 10.1.4.1), the presence of water in biogas is a key factor in the corrosive action of H_2S; moreover, water may affect the efficiency of treatment processes like adsorption (according to the adsorption media and the targeted compound, an optimal water content may be required for effective adsorption, excess of water is usually detrimental).

10.1.4.4 Total VOC

The presence of VOC [15] is mainly problematic with regard to treatment processes.

First, activated carbon adsorption is a very common treatment process aimed at H_2S and VOSiC removal, but VOC are adsorbed as well. Since competitive adsorption occurs it is important to assess the content of VOC with the view to properly design adsorption processes. Then, analysing targeted compounds can be sufficient to monitor the treatment efficiency and to manage the process. Monitoring VOC could be beneficial to the prediction of adsorbent's replacement need but is not essential. This also applies to other adsorbents.

Secondly, some VOC (terpenes, mercaptans) may react with and damage membranes used as separation process to upgrade biomethane. In this case it becomes essential to remove those compounds prior to membrane separation so that they are considered as targeted compounds for adsorption. With this respect, their monitoring becomes a topic of interest, especially if the breakthrough of these VOC occurs before the breakthrough of other targeted compounds (H_2S and VOSiC) at the outlet of the adsorbent filter. Up till now no studies related to that issue have been reported but the current development of biomethane production from wastewater treatment plant and even landfill biogas should arouse interest for that point.

10.1.4.5 Halogen-Containing VOC

During combustion these compounds are responsible for acids formation (hydrochloric acid from chlorinated VOC, hydrofluoric acid from fluorinated VOC …) which may be released in the environment with exhaust gas and/or be involved in the corrosion of equipment, in the acidification of engine oils (increasing the oil changes frequency). Fortunately, their content in biogas is usually low (traces) and they are adsorbed together with other adsorbable compounds. Special care is not required unless if high content is expected in biogas.

10.1.4.6 Ammonia (NH_3)

Ammonia is targeted for reasons similar to what has been described for H_2S: it may be corrosive for equipment and is converted into nitrogen oxides (NO_x) during combustion, NO_x emissions being regulated.

10.1.4.7 Summary

Information about minor compounds is summarized in Table 10.2.

10.2 Sampling Methods

Before any analysis, sampling is a critical step. If sampling is not properly conducted, analytical results will not be reliable, as relevant as the analytical method may be.

10.2.1 General Recommendations

Sampling concerns the analytical methods that work in the extractive mode. The gaseous sample is drawn from his pipe and brought to the analyzer by means of a sampling line the length of which should be kept to a minimum. This sampling line must be designed so as to first avoid interactions between gaseous compounds and the material the line is made of, or at least to limit them, and secondly to ensure that there is no cold spot where water and other corrosive gas or even analytes may condense. At last, it is usually necessary to include a filter (made of inert material such as passivated quartz wool 2 μm) to stop particles before the analyser. One may refer to the standard ISO 6712:1982 entitled "Gas analysis—Sampling and transfer equipment for gases supplying an analytical unit" [16].

Inert materials recommended for sampling lines are polytetrafluoroethylene (PTFE) or perfluoroalkoxy (PFA). Stainless steel 316 may also be used especially if passivated. For example, Silconert 2000® has been shown to be very effective passivation coating for analysing trace impurities (e.g. siloxanes). Monel® which is a nickel-based alloy, is the most corrosion-resistant material. When on-line analytical methods are used for spot measurements, it is always a good option to passivate the sampling line and any wetted connections in the entire sample path.

Preventing any condensation is important first because droplet (water and/or corrosive gas) may damage analysers or interfere in the analysis system, secondly because gaseous compounds may dissolve changing the composition of the gaseous sample. To preserve analyzers, it is recommended to incline the sampling line so as to allow condensate drain under gravity to a catch flask placed before the analyzer. According to the analytical technique and the composition of gas a heated line can be required to prevent condensation.

At last, if a pump is used it should be placed after the analyzer.

These general recommendations are usually sufficient for on-line analytical methods. Others may be considered in the case of offline analytical methods as developed below.

Whatever the sampling method, it is important to allow enough time for the purging of sampling lines with the gas sample before the analysis starts.

10.2.2 *Offline Analytical Methods*

In the case of laboratory analysis, samples are withdrawn on site, transported and potentially stored before analysis. To this end, sampling can be conducted without or with pre-concentration, the latest being especially indicated for trace or low concentration compounds.

10.2.2.1 Sampling Without Pre-concentration

Three types of container can be used: bags, canisters and gas collection tubes. Canisters and gas collection tubes are not widely used for the sampling of biogas. Canisters (made of stainless steel with inert internal surface) are useful to collect large volumes of gas (1−15 L) but are cumbersome and thus hardly transportable. Canisters are previously evacuated, the sampling being then achieved by suction. Gas collection tubes are usually made of glass so that they are fragile and not appropriate for the sampling of compounds containing silicon (like VOSiC). They are filled either by gas circulation or thanks to a pump following them.

Bags are most widely used, especially when samples are sent to external laboratory. They are indeed lightweight, resistant and ensure a satisfactory sealing. They are most often made of fluoropolymer, Tedlar© (polyvinyl fluoride) being the most commonly used.

Nevertheless, it is very important to pay attention to interactions which may exist between targeted compounds and the internal surface of the container (whatever the container may be) or even the connexion device, especially when low concentration compounds are to be analyzed. A simple precaution is to passivate the container with the sample, for example by filling and emptying a bag several times before sampling.

Despite these precautions, the sampling of VOSiC remains difficult. Even if Tedlar© bags seem to be the most appropriate container for the sampling of VOSiC without pre-concentration, several studies have pointed out problems of VOSiC losses during storage (even for short term storage) [17–19]. VOSiC may adsorb on fittings (especially in the case of stainless steel fittings, but not only), the use of silicone grease may induce bias in analysis (as well as any silicon containing element), a poorer stability is observed for VOSiC of highest molecular weights (with this regard, temperature is a key point and heating bags before analysis is beneficial). For these reasons, some laboratories stopped using bags and prefer carrying out absorption in suitable liquid solutions.

10.2.2.2 Sampling with Pre-concentration

Sampling with pre-concentration is especially intended when targeted compounds are present at low concentration, since it improves detection limits with respect to biogas content. It is carried out by absorption of the targeted compounds in a suitable liquid with the help of impingers or by adsorption onto a solid phase packed in a tube. Cryogenic trapping tests were conducted for analytical purposes [20] but this method remains marginal in terms of number of publications.

Concerning absorption, the gaseous phase bubbles in an absorbent solution having an affinity for the targeted compounds (comparable polarity for example). Several impingers are placed in series to ensure the complete entrapment of targeted compounds (the last impinger has to be free from targeted compounds). It is possible to use impingers filled with different solvents to absorb compounds with different physical or chemical properties. Besides the ability of absorbent solution to entrap targeted compounds, three other factors must be taken into account to ensure a good absorption and hence a satisfactory sampling. One is the flow rate of the gaseous phase in impingers which must enables the mass transfer to the liquid phase. Another factor is the volatility of the absorbent solution which should be low: if it evaporates too much during the bubbling, the concentration in solution may be overestimated. The last factor is the bubbling duration. If it is too short the concentration of targeted compounds in solutions may be too low to ensure a good quantification. If it is too long stripping of targeted compounds or saturation of absorbent solutions may occur leading to underestimation. Furthermore, a long bubbling step favours the evaporation of absorbent solutions. So, it is important to optimize the sampling duration. One of the advantages of this sampling technique is that the quantity of absorbent solution implemented enables to conduct as many analyses as necessary to obtain results with satisfactory quality. This technique provides an example of how a large uncertainty in the sampling step, can lead to variable results (that is, if flow rate and timing are not accurately monitored, the results may not be representative, despite having an accurate lab-based validation method), and this is true whatever the sampling method.

Concerning adsorption, the gaseous phase flows through the adsorption cartridge where targeted compounds are trapped. Various types of solid adsorbents are commercially available (resins, activated carbons, organic or inorganic polymers …), the properties of which varying according to targeted compounds. Adsorption capacity depends on specific surface area, pore size and surface chemistry. As for absorption, the duration and the flow rate of sampling are key parameters: the targeted compounds must be entirely retained on the adsorbent and the quantity trapped must be sufficient to enable the following quantification. After sampling, desorption may be achieved either by liquid extraction with a solvent, or thermally (thermodesorption). The choice is driven by the subsequent analytical technique used.

Disadvantages of sampling with pre-concentration are:

- The selectivity of absorbents or adsorbents: there is no universal absorbent/adsorbent so that it may be necessary to use several in series according to the targeted compounds. In the same time, competitive adsorption may occur in the case of a complex matrix like biogas.
- Absorption, adsorption and desorption are not always quantitative.

Hence, recovery rate has to be determined and taken into account to assess the uncertainty related to analytical results.

As far as biogas characterization is concerned, sampling with pre-concentration is more specifically used for VOSiC quantification. When absorption is implemented, various absorbent solutions may be used [21]: methanol, ethanol, acetone, hexane, aqueous solution. It is possible to perform a pseudo-speciation of VOSiC (polar and low polar) as a function of the affinity of the solvent, in particular its own polarity or functionalization. Regarding adsorption, the most commonly used are Tenax, Carbotrap and Carbosieve [21].

10.3 Characterization Methods and Equipment

The characterization of biogas quality is needed to monitor treatment processes that ensure the good conditions of energy conversion or to comply with European Standard EN16723 [5, 6]. All the more so as the composition of biogas may vary in time according to both inputs and conditions under which anaerobic digestion occurs.

With regard to on-site energy conversion of biogas (heat, electricity, CHP), regulatory requirements deal only with exhaust gas and there is no regulatory requirement related to biogas quality (prior to energy conversion). Biogas characterization is only conducted for operational control purpose but no characterization standards are available. The main difficulty with analyzing compounds in biogas, and especially trace compounds, is the complexity of this gas including up to several hundreds of minor compounds in a CH_4/CO_2 matrix, and in humid conditions.

As biomethane is concerned, the quality objective is to meet natural gas standards at the injection point. Standards specifically dedicated to biomethane are under discussion, especially including a requirement to total silicon content (related to VOSiC) which is not targeted in natural gas. Today there is no standardized characterization method for biomethane but related works are ongoing (a working group dedicated to standardization of test methods for biomethane (ISO/TC 193/SC 1/WG 25), and a European project dedicated to producing and validating the methods (EMPIR 'Metrology for biomethane'). Characterization methods used for natural gas may be used or transposed according to the considered compound. After upgrading, biomethane is indeed less complex than biogas and its quality is close to natural gas' one. Nevertheless characterization methods used for natural gas are not suitable in all cases, they have not yet been validated for use with biomethane. Furthermore, there is a need for standardized characterization method(s) to determine total silicon content. Regarding characterization of biogas along the treatment chain (before upgrading completion), it is the same issue than previously discussed with regard to on-site energy conversion.

During the early years of biogas conversion to energy, measurements have been mainly conducted periodically with low-cost portable analyzers, in both small and big energy conversion units. With the development of the biogas sector, on-line multi-gas analyzers are on the upward trend especially on big plants (higher costs).

Methods described in this section include electrochemistry, chromatography and spectrometry. Techniques will be presented in the form of data sheets including: electrochemical sensors, Non Dispersive InfraRed (NDIR), Absorption followed by laboratory analysis i.e. Gas Chromatography (GC) or Inductively Coupled Plasma (ICP), Flame ionization, Fourier Transform InfraRed spectroscopy (FTIR), GC coupled with various detectors, Optical Feedback Cavity Enhanced Absorption spectroscopy (OFCEAS), Ion Mobility Spectrometry (IMS), photoacoustic gas monitor, Tunable diode laser absorption spectroscopy (TDLAS), Direct Mass Spectroscopy. Items presented in data sheets are the following: targeted compounds, sampling methods, dimensions and weight of the analytical devices, application (qualitative, quantitative or semi-quantitative analysis), range, precision, principle, advantages and limitations. For the sake of an easy access to information, the Table 10.3 summarizes analytical techniques according to the compounds they can target.

Table 10.3 Available analytical techniques regarding targeted compounds in biogas

	Cost[a]	CH_4	CO_2	H_2O	H_2S	N_2	NH_3	O_2	VOCNM	VOSiC
Conventional and low cost techniques										
Electrochemical gas sensor	+				×		×	×		
NDIR	+(+)	×	×							×
Absorption and external analyze	+						×			×
Flame ionization	+(+)	×							(×)	
Techniques in the current trend										
FTIR	+++	×	×	×	×		×		×	×
GC	++ (+)	×	×		×	×	×	×	×	
OFCEAS	+++	×	×	×	×		×	×		
A-IMS and GC-IMS	++									×
GC-ICP-MS[b]	++++									×
Interesting techniques/non proven for biogas/biomethane										
Photoacoustic gas monitor	+++	×	×	×	×		×			
OA-ICOS	++	×	×	×	×		×			
TDLAS	++	×	×	×	×		×	×		
Direct MS	++++		×		×		×		×	×

[a]Cost range: + < 25 k€; 25 < ++ < 50 k€; 50 < +++ < 80 k€; ++++ >150 k€
[b]GC-ICP-MS is a technique still in development (NPL) but especially designed for biomethane

Regarding sampling methods:

- "on-line" will name an analyzer continuously connected to the process with an external line (heated or not),
- "in-line" will indicate a sensor that can be placed in a process vessel or stream of flowing gas,
- "in situ" or "portable" will refer to an analyzer connected to the process with an external line (heated or not) for one or more measurements.

10.3.1 Conventional and Low Cost Techniques

10.3.1.1 Electrochemical Gas Sensor

The electrochemical gas sensors are widely used because they are cheap and enable to quickly estimate the biogas composition. But the precision is not adequate for a

Table 10.4 Data sheet related to electrochemical gas sensors

Compounds	H_2S	NH_3	O_2
Sampling	Portable or on-line		
Dimensions/Weight	Multi-gas: 26×21×14 cm/<1 kg		
Application	Qualitative analysis		
Instrument range	1 ppb-100%	0–500 ppm 0–2000 ppm 0–10,000 ppm	0–5000 ppm
Precision	±10%		
Principle	Detect gases by producing a chemical reaction between the gas and oxygen contained in the sensor. Reactions occur on two electrodes (specific sensing electrode and counter electrode) and produce a small current, which is proportional to the concentration of the targeted gas		
Advantages	– ATEX – Cheap – Ease of use – Fast (few seconds)		
Instrument limitations	– Drift in time: it's often necessary to calibrate – Cross-sensitivity: e.g. SO_2 and NO for H_2S, CO_2 for O_2, H_2S, NO_2 and SO_2 for NH_3 – Memory effect: purge the system before each analyze with ambient air (until 15 min) – Temperature effect: an air-conditioned room or sealed cabinet with temperature controlled is preferable		

quantitative analysis because cross-sensitivities and drift time occur. So, it's important to be careful with the results obtained from this kind of analyzers (Table 10.4).

From a practical point of view:

- system must be purged with ambient air before use and in between analyses,
- measurements must be done under controlled temperature,
- sensors must be calibrated with standard bottles several times a year, depending on the use.

Researchers of Loughborough University in the UK are working on developing electrochemical gas sensor to quantify VOSiC, however it is in the early stages.

10.3.1.2 Non Dispersive InfraRed (NDIR)

The NDIR analyzers for CH_4 and CO_2 are widely used (Table 10.5), such as the electrochemical gas sensors. But the precision is not sufficient for a quantitative analysis because of cross-sensitivities. So, it's important to be careful with the results.

From a practical point of view, no water must enter in the analyzer so that a water filter is necessary and must be often replaced.

Only one VOSiC NDIR analyzer dedicated to biogas analysis exists but no feedbacks have been published yet.

10.3.1.3 Absorption and External Analysis

Apart from on-site analysis, another way to analyze ammonia and VOSiC consists in sampling by means of absorption in specific solution and subsequent analysis in laboratory with an appropriate analyzer. It's cheap but the results are not immediate and the absorption step on the field requires a technician (Table 10.6).

10.3.1.4 Flame Ionization

It is important to note that this data sheet (Table 10.7) is dedicated to simple portable instruments based on flame ionization, not coupled with a gas chromatography system (for GC-FID please refer to GC data sheet Sect. 10.3.2.2, Table 10.9).

The flame ionization technique is standardized for CH_4 analysis but this system can only analyze methane in biogas while multi-gas analysis is more suitable for biogas applications.

10.3.2 Techniques in the Current Trend

10.3.2.1 Fourier Transform InfraRed Spectroscopy (FTIR)

FTIR is a commonly used technique to analyze gases (Table 10.8). Biogas analysis is complicated but suppliers have optimized systems to minimize interferences. Many systems are installed on-site in sealed cabinet for "classical" compounds. FTIR analysis of VOSiC has been more recently developed. Only one instrument especially dedicated to VOSiC analysis is currently available-for-sale despite attempts of several suppliers. This system has been implemented to analyze biogas and assessed in a study for L2, L3, D4 and D5 [24]. Developments are still ongoing for VOSiC analysis. Given the difficulty to deal with interferences, one interesting approach is to couple FTIR with a pre-separation thanks to gas chromatography. A proof of concept has been published [25].

From a practical point of view:

- a heated line is required for the sampling,
- a nitrogen cylinder (>4.8) is required for baseline checking and optical system cleaning,
- the analyzer must be installed in a controlled temperature environment.

Table 10.5 Data sheet related to non dispersive infrared analyzers

Compounds	CH_4	CO_2	VOSiC
Sampling	Portable or on-line		On-line
Dimensions/Weight	220 × 155 × 60 mm/ 1.5–2 kg Rack/25 kg		Rack 475 × 170 × 350 mm/12 kg
Application	Qualitative analysis		Quantitative analysis
Instrument range	0–100%		1–400 mg/m^3
Precision	±10%		No data
Principle	Infrared gas analyzers detect gases based on the absorption of IR light by heteroatomic molecules or hydrocarbon. Each molecule vibrate and rotate at frequencies characteristic of internal bonds. The amount of IR light absorbed is proportional to the concentration. The term non-dispersive refers to the fact that all the light passes through the gas sample and is only filtered immediately before the detector		
Advantages	– ATEX – Cheap – Ease of use – Fast (few seconds) – Operating Temperature: 0-50°C		– Ease of use – Measurement on-line of total Siloxanes once an hour – Operating Temperature: ambient to 50 °C
Instrument limitations	– Cross-sensitivity for CH_4 with HCT – Low precision – Water filter		– No ATEX – Water trap – Calibrate once a year – Require an heated line – Requires a nitrogen supply from a cylinder

Table 10.6 Data sheet related to absorption and external analysis

Compounds	NH_3	VOSiC
Sampling	In situ	
Application	Quantitative analysis	
Instrument range	Depend on the time of absorption and the limit of detection of the analyzer	
Precision	5%	
Principle	Absorption in a specific solution with controlled flow and analysis in a laboratory with an ionic chromatography or by molecular absorption spectrometry (see NF X43-303 [22])	Absorption in a specific solution with controlled flow and analysis in a laboratory by ICP-OES [23]
Advantages	Cheap	
Instrument limitations	– No ATEX – Discontinuous analysis – Not immediate results – Require a technician to operate	

Table 10.7 Data sheet related to flame ionization analyzers

Compound	CH_4
Sampling	Portable or on-line or in-line
Dimensions/weight	330 × 305 × 76.2 mm/5 kg or 260 × 340 × 500 mm/15 kg 19 in. Rack
Application	Quantitative analysis
Instrument range	0.1–100,000 ppm
Precision	1 ppm
Principle	A flame ionization detector converts a hydrocarbon gas concentration into an electrical signal. The flame is hydrogen fueled and burns in hydrocarbon free air. Either side of the flame are two electrodes between which an electric field is generated by means of a polarization voltage. The sample is pre-mixed with the fuel gas and heated in the flame to a temperature at which the hydrocarbon molecules 'crack' and CH fragments appear. These CH fragments are oxidized by the oxygen in the combustion air resulting in the formation of CHO+ ions. The ion current then passes to the electrodes and the resulting current measured is proportional to the quantity of C atoms derived from organic compounds. A column can be added to separate methane and no methane
Advantages	– Automatic calibration – CH_4 analysis normalized ISO 25140:2010 – Ease of use – Fast: continuous analysis – Some instruments are ATEX – Operating temperature: 10–40 °C
Instrument limitations	– Requires an heated line – Requires H_2 supply from a (micro)cylinder – The instrument can analyze VOC but not in the biogas where CH_4 concentration is too high

Table 10.8 Data sheet related to FTIR analyzer

Compounds	CH_4	CO_2	H_2O	H_2S	NH_3	VOC	VOSiC
Sampling	On-line and some systems in-line						
Dimensions/ weight	19 in. rack/20–30 kg						
Application	Quantitative analysis						
Instrument range	0–1000 ppm	0–30%		Traces in the air	0–15 mg/m^3	0–20 mg (TOC)/m^3	No data
Precision	No data						
Principle	Infrared gas analyzers detect gases based on the absorption of IR light by heteroatomic molecules or hydrocarbon. The gas molecules absorb this radiation when the electromagnetic field of the molecule vibrates at the same frequency as the incoming IR radiation so this absorption is representative of the types of bonds present. The amount of IR light absorbed is proportional to the concentration. An interferometer modulates IR radiation and produces an interferogram (energy variation function of time) treated by the mathematical algorithm Fourier Transformation (FT) to obtain a spectrum (energy variation function of frequency)						
Advantages	– Low influence of pressure and temperature – Low detectable limit: 2% full scale – Internal calibration with He–Ne laser – Relatively fast (few minutes) – Screening of some VOC; only 1 instrument for VOSiC						
Instrument limitations	– For each gas, it's necessary to optimize the system to minimize the interferences – It's necessary to dilute a concentrate gas sample – No ATEX except if the analyzer is placed in a sealed cabinet with overpressure purge system – Require air-conditioned room or sealed cabinet with temperature controlled – Require a heated line for on-line system – Requires a nitrogen supply from a cylinder to do the zero point once a day						

10.3.2.2 Gas Chromatography

Gas chromatography is an efficient way to analyze gases even complex gases like biogas. There are a range of detectors used with GC according to the compounds targeted (Table10.9).

For industrial applications, suppliers propose microGC (except for MSD). GC-ICP-MS is being developed (see dedicated date sheet Sect. 10.3.2.5, Table 10.12).

Nanotechnology involving GC has been recently developed but no studies for biogas analysis have been published yet.

Regarding VOSiC, a pre-concentration step is usually required especially for low concentrations (outlet of treatment processes/biomethane). To date, GC analyses of VOSiC are commonly conducted in the laboratory, after sampling (see Sect. 10.2.2).

Table 10.9 Data sheet related to gas chromatography analyzers including different detectors

Compounds	CH_4	CO_2	H_2S	N_2	NH_3	O_2	VOC[a]
Detector	TCD (thermal conductivity detector) and/or MSD (mass detector), or FID (flame ionization detector for hydrocarbons), or SCD (Sulphur Chemiluminescence Detector for Sulphur species)						
Sampling	GC coupled with any detector except MS: On-line				GC-MSD: portable with a suitable vehicle		
Dimensions/weight	19 in. Rack/20–30 kg				Approx. 120 × 60 × 80 cm/ 50 kg		
Application	Quantitative analysis						
Instrument range	1 ppm–100%						
Precision	±2%						
Principle	Gas chromatography module separates the compounds. An instrument contains 1–3 analytical modules. Each module is a complete chromatograph with micro injector, analytical column (where the compounds are separated) and thermal conductivity detector. This TCD, non-destructive, can be coupling to a quadruple mass spectrometer which make possible the identification of each separated compounds						
Advantages	– CH_4 GC analysis is normalized (ISO 25139:2011) – Linear response over a wide dynamic range (6 decades) – No interferences – Relatively fast for the GC-TCD (few minutes) – Operating Temperature: ambient						
Instrument limitations	– No ATEX – Calibration once a week or more – Not really fast for the GC-MSD (30 min) – Require a water filter, a particle filter – In some case, require a technician to operate – Require a heated line – Requires an He supply like carrier gas						

[a]including VOSiC

From a practical point of view, suppliers develop simple software to be used by a non-scientist. The system requires precautions (water and particle filters, heated line), a gas to calibrate the system once a week or more and high purity He cylinder as carrier gas.

10.3.2.3 Optical Feedback Cavity Enhanced Absorption Spectroscopy (OFCEAS)

The OF-CEAS device is a simple and efficient system to analyze biogas (Table 10.10). The biogas application was validated in a study published by Chatain et al. [2]. However, each system must be optimized by the provider for one application.

Table 10.10 Data sheet related to OFCEAS analyzer

Compounds	CH_4	CO_2	H_2O	H_2S	NH_3	O_2
Sampling	On-line					
Dimensions/weight	19 in. rack/20 kg					
Application	Quantitative analysis					
Instrument range	50 ppm to 100%	1–100%	1000 ppm to 100%	50 ppm to 10%	50 ppm to 100%	1000 ppm to 100%
Precision	2%					
Principle	The system use the interaction between light and the molecules in a gas stream to determine the concentration of a given substance. With CEAS method, the pulse of the laser light is injected into high finesse optical cavity equipped with spherical and high reflective mirrors. Furthermore, the laser beam is feeding back onto itself to improve wavelength accuracy, reducing noise and increasing signal intensity (optical feedback). OFCEAS provides extremely long paths (>20 km)					
Advantages	– ATEX – Ease of use – Linear response over a wide dynamic range (6 decades) – No calibration in time (no instrumental response drift) – Very low detectable limit: ppb – Very good selectivity – Very fast (real time measurement) – Robust – Uses a patented low-pressure sampling system: heated lines not required with ambient T > 10 °C and reduced maintenance – Operating Temperature: 15–35 °C (standard) or 10–40 °C (optional)					
Instrument limitations	– Each system is optimized by the supplier for one application (adjustable range on request) – Requires dry Air supply from a cylinder (oil free)					

In practice, the use of a low-pressure sampling system is an advantage: it prevents water condensation in the analytical line and it reduces the exposure of the analytical device to corrosive compounds. The system doesn't need to be calibrated.

This multi-gas analyzer is able to measure all the compounds of interest in a biogas except VOSiC. This is the only transportable analyzer able to perform on-line NH_3 measurements.

10.3.2.4 Ion Mobility Spectrometry (IMS)

Two analyzers based on IMS have been developed: Analytical IMS (A-IMS), which is without pre-separation step, and GC-IMS which couples pre-separation thanks to gas chromatography and IMS detection. They were first developed to analyze VOC but application to VOSiC is especially interesting with the view to characterize biogas or biomethane (Table 10.11).

Table 10.11 Data sheet related to IMS analyzers

Compounds	VOC	VOSiC
Sampling	On-line or portable	
Dimensions/Weight	449 × 435 × 177 mm/15.5 kg	
Application	Quantitative analysis/trace analyzer	
Instrument range	0.1–5 mg(VOSiC)/m^3	
Precision	<10%	
Principle	The IMS (ion mobility spectrometry) is based on the specific drift times that ionized compounds need to pass a fixed distance (drift tube) in a defined electric field. For detection, the resulting ion current is measured by an electrometer as a function of time. With the GC model, the volatile compounds of samples under testing are pre-separated in time by a GC column. The discrete compounds elute directly into the IMS ionization chamber	
Advantages	– Ease of use – Fast (analysis in continuous) – Low detection limit: 1 μg/m^3 – Operating Temperature: ambient	
Instrument limitations	– No ATEX except if the analyzer is placed in a sealed cabinet with overpressure purge system – Factory calibration (periodicity unknown) – Short linear response (few mg/m^3) – Requires a N_2 supply from a cylinder as carrier gas or a circular gas flow unit (circulating air filtered by the CGFU to reach a 5.0 quality)	

A-IMS and GC-IMS are trace VOC analyzers. A-IMS provides only a qualitative monitoring of targeted compounds whereas GC-IMS enables quantitative measurements. Recently, GC-IMS was implemented to analyze VOSiC in biogas with success [21]. It has been proven to be appropriate to biomethane analysis and also to treated biogas analysis. It enables to quantify VOSiC at low concentrations, but only in a narrow range: if VOSiC content is too high, quantification is not possible anymore. So, this technique is not suitable to monitor the treatment efficiency by upstream and downstream measurements in the case of high VOSiC content, such as for landfill biogas for example. Fortunately, the management of treatment processes like the adsorbent's replacement can rely on downstream monitoring only.

This technique is more recent and less often used than FTIR or GC in the biogas sector.

In practice, the system requires high purity N_2 cylinder as carrier gas. According to the supplier, only one factory calibration is required. This statement is still to be confirmed through long time implementation. Details about the analyzer and the biogas application are available [26].

10.3.2.5 Gas Chromatography Inductively Coupled Plasma Mass Spectrometry (GC-ICP-MS)

The GC-ICP-MS technique (Table 10.12) is a flexible lab based technique that can be adapted to suit compositional variations caused by feedstock. It requires a technician to operate due to its complexity, but provides accurate and traceable quantification over a large dynamic range without the need for sample pre-concentration. The technique is still under development by NPL (National Physical Laboratory, UK), and will be optimized in order to further improve the limit of detection for total silicon.

Table 10.12 Data sheet related to GC-ICP-MS technology

Compounds	VOSiC
Sampling	Collection into evacuated passivated canisters
Dimensions/ weight	Approx. 99 × 117 × 73 cm/295 kg
Application	Quantitative analysis/trace analyzer
Instrument range	ppb to %
Precision	1–5% (range dependent)
Principle	The sample is introduced directly via a heated gas sample loop within the GC. An analytical column then separates the analytes from the matrix using a temperature program and valve switching system. With the matrix removed, the stream enters the front end of the ICP-MS and is exposed to a high temperature plasma, which breaks down the sample into its elemental constituents and performs ionization. The silicon ions are then directed via a variable magnetic field into an electron multiplier tube, which is used to perform a quantification of total silicon amount via comparison with a gas reference standard
Advantages	– Large linear dynamic range – High precision and accuracy – Low detection limit: ppb – Quantifies both total silicon and individual siloxanes – Flexible technique
Instrument limitations	– Lab based – Expensive (> 200 k€) – Requires a He and Ar supply from cylinders – Requires periodic maintenance

10.3.3 *Interesting Techniques but Non Proven for Biogas or Biomethane Applications*

10.3.3.1 Photoacoustic Gas Monitor

The photoacoustic gas monitor is a trace analyzer used for ambient air (Table 10.13). Nowadays no applications for biogas analysis are published but it will be interesting to test this instrument for biogas applications.

Table 10.13 Data sheet related to photoacoustic gas monitor

Compounds	CH_4	CO_2	H_2O	H_2S	NH_3
Sampling	On-line				
Dimensions/Weight	19 in. Rack/20–30 kg				
Application	Quantitative analysis for ambient air/trace analyzer				
Instrument range	100 ppmv				
Precision	No data available				
Principle	Any gas that absorbs infrared light can be analyzed. Light from an infrared light source is reflected off a mirror, passes through a mechanical chopper, which pulsates it, and then through one of the optical filters in the filter wheel. The gas selectively absorbs the light transmitted by the optical filter. Because the light is pulsating, the gas temperature increases and decreases, causing an equivalent increase and decrease in the pressure of the gas (an acoustic signal) in the closed cell Two or more microphones mounted in the cell wall measure this acoustic signal, which is directly proportional to the concentration of the monitored gas present in the cell				
Advantages	– Fast (continuous analysis) – Low detectable limit: ppb – Linear response over a wide dynamic range – Compensates for temperature and pressure fluctuations, water vapor interference and interference from other known gases				
Instrument limitations	– For each gas, it's necessary to optimize the system to minimize the interferences (water and other gases) – No ATEX – Two calibrations a year – Requires air-conditioned room or sealed cabinet with temperature controlled – Requires a particle filter – No published applications for biogas				

10.3.3.2 Off Axis Integrated Cavity Output Spectroscopy (OA-ICOS)

The OA-ICOS system is an efficient analyzer for impurities in air. No data was found for biogas analysis except for methane (Table 10.14).

10.3.3.3 Tunable Diode Laser Absorption Spectroscopy (TDLAS)

The TDLAS analyzer is easy to use on-line or in-line and is able to analyze trace or concentrated components (Table 10.15). This system has proven its efficiency in many industrials applications. However, this kind of analyzer is rarely implemented for biogas analysis.

Table 10.14 Data sheet related to OA-ICOS

Compounds	CH_4	CO_2	H_2O	H_2S	NH_3
Sampling	Portable or on-line				
Dimensions/ weight	13.4″ × 11.6″ × 6″ for the portable instrument/15 kg Rack 8.75″ × 19″ × 24″/29 kg				
Application	Quantitative analysis/trace analyzer				
Instrument range (industrial model)	0–10%	0.1–15%	0.1–98%	0.04–500 ppm	0.04–400ppm
Precision	No data available				
Principle	The system uses the interaction between light and the molecules in a gas stream to determine the concentration of a given substance. A laser is directed through a sample and a detector located after the sample cell measures the change in light, or absorption, due to the presence of known chemical components. Since different molecules absorb light at different wavelengths, mixing ratios of multiple gases can be determined by tuning the laser over a predetermined wavelength range. The beam enters the high fineness optical cavity by passing through the input mirror and is later collected as it passes through the output mirror. The light trapped inside can make a great number of round-trips between the cavity mirrors. Furthermore, the Off-Axis approach generates multiple reflections spatially separated. OA-ICOS provides very long paths (2–20 km, typical)				
Advantages	– Ease of use – No calibration – Very low detectable limit: ppb – Very good selectivity – Very fast (real time measurement) – Robust – Operating temperature: 5–45 °C				
Instrument limitations	Only 2 or 3 gas analyzed by one instrument				

Table 10.15 Data sheet related to Tunable Diode Laser Absorption Spectroscopy

Compounds	H_2O	NH_3	CH_4	CO_2	H_2S	O_2
Sampling	On-line or in-line					
Dimensions/weight	19″ in. Rack (13 kg) + sensors (5 and 6 kg)					
Application	Quantitative analysis/trace analyzer					
Instrument range	0–30%		0–100%			
Precision	2–5%					
Principle	The system uses the interaction between light and the molecules in a gas stream to determine the concentration of a given substance. A laser is precisely tunable to emit light at the specific resonant wavelength of the compound of interest in the near or mid-infrared region of the electromagnetic spectrum. The light passes through a multi-pass cavity or through a pipe process where the sample circulates. The laser energy causes the molecules of interest to vibrate that absorb energy					
Advantages	– ATEX – Ease of use – Good selectivity – Low detectable limit: subppm – The instrument range depends on the length of the multi-pass cavity or the length of the pipe – Very fast (real time measurement) – Robust – Operating temperature: −20 to +55 °C					
Instrument limitations	– Interfered by dust – One instrument for one or two target gas – Requires an N_2 or Air supply from a cylinder to clean the sensors surface					

In practice, one analyzer is devoted to the analysis of only one or two target compounds.

10.3.3.4 Direct Mass Spectrometry

Several technologies based on direct mass spectrometry have been developed and some of them seem to be interesting for biogas application: Selected Ion Flow Tube Mass Spectrometry (SIFT-MS), Proton Transfer Reaction—Time Of Flight MS (PTR-TOFMS) and Selective Reagent Ionization—TOFMS (SRI-TOMS).

The direct MS technique (Table 10.16) is commonly used to analyze many trace components in air or gases but it's still expensive. The biogas analysis is possible only if the biogas is first diluted in a Tedlar™ sampling bag. Considering the adsorption of some components on the bag, the analysis must be done immediately. This analyzer is cumbersome despite the efforts of suppliers. The calibration uses an internal standard.

Table 10.16 Data sheet related to technologies based on direct mass spectrometry

Compounds	CO_2	H_2S	NH_3	VOC	VOSiC
Sampling	On-line or portable with a suitable vehicle				
Dimensions/weight	Approx. 120 × 80 × 80 cm/200 kg				
Application	Quantitative analysis/trace analyzer				
Instrument range	ppt to 50 ppm				
Precision	No data available for biogas/biomethane applications				
Principle	With SIFT-MS the reagent ions are generated in the ion source by a microwave discharge operating in low pressure moist air and selected by an upstream quadrupole (ion source). Then the selected reagent ion is injected into a flow tube where they are carried along the tube in a stream of helium. The sample is then introduced at a known flow rate and the reactive compounds it contains are ionized by the reagent ion to form well-characterized product ions (reaction chamber). Product ions and unreacted reagent ions are sampled into a second quadrupole mass spectrometer (m/z analyzer).With a PTR-TOFMS, soft ionization technology where by proton transfer from H_3O^+, all compounds with a higher proton affinity (PA) than water are ionized. With SRI-TOFMS, the reagent ions are NO^+ and O_2^+ or Kr^+ to improve inorganic compounds detection. Compared to a quadrupole MS, TOF-MS have a much higher mass resolving power and a better duty cycle. All ions are virtually detected at once. It is not necessary to sequentially filter and detect the ions according to their m/z ratios				
Advantages	– Ease of use – Fast – Low detection limit: pptv – No humidity influence – Operating Temperature: ambient				
Instrument limitations	– No ATEX – Expensive (>200 k€) – Requires an He supply from a cylinder – The biogas analysis is not easy: the high concentration of CH_4 is a problem. The biogas must be diluted in a Tedlar™ sampling bags.				

10.4 Conclusion

Biogas may contain several hundreds of compounds so that analysis in such a complex gaseous phase is difficult. Potential sources of interference are indeed numerous and this is one of the biggest potential problems for both industry and the metrological community. The presence of water vapor, being likely to condensate, is the main cause of analytical difficulties. Among others, it can be mentioned that CO_2 in raw biogas may be an interfering gas for several spectroscopic methods.

The characterization of biogas may be conducted with two different aims in view. The first is to ensure that biogas quality will enable the energy conversion to be operated in good conditions. In this regard, there are no standards of quality available and no related analytical standards for most ways of energy conversion. There are only specifications (recommended values or threshold) regarding some compounds (especially CH_4, CO_2, H_2S and VOSiC). These specifications may vary according to the energy conversion system considered. The strongest requirements are in the case of biomethane production for which a standardization procedure is ongoing. Quality of biomethane at injection point should actually be similar to the natural gas' one, meaning high methane content and precisely known amounts of impurities. Hence analytical techniques should provide results with high precision. When biogas is used to produce heat and/or electricity, the content of methane may vary in a broader range, the required precision of methane measurement is thus lower. The second aim of biogas characterization is the monitoring of treatment processes. This aspect becomes increasingly important with the biogas sector development. It is actually crucial to control costs. The recent development of on-line analyzers seeks to respond to these analytical needs.

In the case of low concentration compounds, it is important to highlight that most analyzers are not able to quantify compounds with a good accuracy on both raw biogas and treated biogas. In that respect, two analyzers may be required to measure only one compound (or one family of compounds) according to the concentration range, or at least strategic analytical choices must be made consciously.

More generally, there is no universal analytical solution to carry out the characterization of a biogas. It depends on the aim of measurements, the targeted compounds, the concentration range, the accuracy expected, and the cost. The analytical solution is always a compromise.

References

1. EurObserv'ER: Observatoire des énergies renouvelables. Le baromètre 2017 des énergies renouvelables électriques en France. http://www.energies-renouvelables.org/observ-er/html/energie_renouvelable_france/Barometre-2017-ENR-elec-France.pdf (2017). Accessed 26 Oct 2018
2. RECORD: Mise en œuvre et évaluation critique de l'analyse d'effluents gazeux complexes par méthode utilisant le principe de l'analyse spectrale (laser) en cavités, 154, n°13-0149/1A. https://www.record-net.org/catalogue (2014). Accessed 26 Oct 2018
3. ADEME: Le biogaz et sa valorisation—Guide méthodologique. ADEME, Angers (1999)
4. Chottier, C.: Composés Organiques Volatils du Silicium et sulfure d'hydrogène – Analyse – Traitement – Impact sur la valorisation des biogaz, Ph.D. thesis, Insa de Lyon, Lyon (2011)
5. AFNOR: NF EN 16723-1 - Gaz naturel et biométhane pour utilisation dans le transport et biométhane pour injection dans les réseaux de gaz naturel - Partie 1: spécifications du biométhane pour injection dans les réseaux de gaz naturel (2017)

6. AFNOR: NF EN 16723-2 - Gaz naturel et biométhane pour utilisation dans le transport et biométhane pour injection dans les réseaux de gaz naturel - Partie 2: spécifications du carburant pour véhicules automobiles (2017)
7. Rickebosch, E., Drouillon, M., Vervaeren, H.: Techniques for transformation of biogas to biomethane. Biomass Bioenergy **35**, 1633–1645 (2011)
8. Sun, Q., Li, H., Yan J., Liu, L., Yu, Z., Yu, X.: Selection of appropriate biogas upgranding technology—a review of biogas cleaning, upgrading and utilization. Renew. Sustain. Energy Rev. **51**, 521–532 (2015)
9. Hoyer, K., Hulteberg C., Svensson, M., Jernberg, J., Norregard, O.: Biogas Upgrading—Technical Review. https://www.vav.griffel.net/filer/C_Energiforsk2016-275.pdf (2016). Accessed June 2016
10. Klein, J., Henning, K.-D.: Catalytic oxidation of hydrogen sulphide on activated carbons. Fuel **63**, 1064–1067 (1984)
11. ADEME: Etude technique, économique et environnementale sur l'injection portée de biomethane dans le réseau de gaz, ADEME, Angers (2016)
12. Glindemann, D., Morgenstern, P., Wenrich, R., Stottmeister, U., Bergmann, A.: Toxic oxide deposits from the combustion of landfill gas and biogas. Environ. Sci. Pollut. Res. **3**, 75–77 (1996)
13. Piechota, G., Iglinski, B., Buczkowski, R.: Development of measurement techniques for determination main and hazardous components in biogas utilized for energy purposes. Energy Convers. Manag. **68**, 219–226 (2013)
14. Tansel, B., Surita, S.C.: Oxidation of siloxanes during biogas combustion and nanotoxicity of Si-based particles released to the atmosphere. Environ. Toxicol. Pharmacol. **37**, 166–173 (2014)
15. Allen, M.R., Braithwaite, A., Hills, C.C.: Trace organic compounds in landfill gas at 7 UK waste disposal sites. Enrion. Sci. Technol. **31**, 1054–1061 (1997)
16. ISO 6712: 1982 Gas Analysis—Sampling and Transfer Equipment for Gases Supplying an Analytical Unit (1982)
17. Ajhar, M., Wens, B., Stollenwerk, K.H., Spalding, G., Yuce, S., Melin, T.: Suitability of Tedlar® gas sampling bags for siloxane quantification in landfill gas. Talanta **82**, 92–98 (2010)
18. Clark, C., Zytner, R.G., McBean, E.: Analyzing volatile organic siloxanes in landfill biogas. Can. J. Civ. Eng. **39**, 667–673 (2012)
19. Mariné, S., Pedrouzo, M., Marce, R.M., Fonseca, I., Borrull, F.: Comparison between sampling and analytical methods in characterization of pollutants in biogas. Talanta **100**, 145–153 (2012)
20. Grümping, R., Mikolajczak, D., Hirner, A.V.: Determination of trimethylsilanol in the environment by LT-GC/ICP-OES and GC-MS. Fresen. J. Anal. Chem. **361**, 133–139 (1998)
21. RECORD: Etat des lieux et comparatif expérimental des méthodes de détermination du silicium total dans un biogaz et un biométhane, 208 p, n°15-0157/1A (2017)
22. NFX 43-303: Stationary Source Emissions—Determination of ammonia (2011)
23. Germain, P., Dumont, N., Chatain, V.: Quantification of Total Silicon in Biogas. International patent WO/2006/129007 (2006)
24. Culleton, L., Brewster, V.: Field trial of traceable online measurements of siloxanes in landfill gases, EMPIR 14SIP06 Siloxanes. https://www.euramet.org (2016). Accessed 26 Oct 2018
25. Prism Analytical Technologies: On-site Monitoring of Siloxanes in Wastewater Digester and Landfill Gas, Application note 704, AN704-0515: MAX™ Technology: Siloxanes. https://www.maxanalytical.com/max-biogas-petroleum-laboratory (2015). Accessed 26 Oct 2018
26. IMPSPEX Diagnostic Limited. http://imspex.com/products/siloxanemeasurementinbiogas/. Accessed 26 Oct 2018

Chapter 11
Syngas

Augustina Ephraim, Rajesh Munirathinam, Ange Nzihou, Doan Pham Minh and Yohan Richardson

Abstract Syngas produced from biomass and biowaste gasification is a versatile renewable fuel and feedstock for chemical synthesis or heat and power production. Bio-syngas can be produced from different biomass feedstock (e.g. lignocellulosic biomass, pyrolysis oil, bioethanol, biogas) through thermal and/or catalytic reforming and gasification processes. The typical components of raw syngas can be classified into three groups: non-condensable gases (e.g. H_2, CO, CH_4 and CO_2), condensable gases (e.g. H_2O and tars) and impurities (e.g. HCl, NH_3, H_2S and particulates). Depending on the application, raw syngas may need to be cleaned of impurities and conditioned to adjust the H_2/CO ratio in order to meet the requirements of environmental regulations and downstream processes respectively. Hence, an accurate characterization of the syngas chemical composition, at the different processing stages, is important to control and optimize the process efficiency. In this chapter, various methods and equipment for sampling, preconditioning and analyzing the three groups of syngas components will be discussed. Analyses of the equipment detection limits, gas matrix sensitivity, and overall accuracy will also be made.

11.1 Introduction

The production of synthesis gas (syngas) from biomass and biowaste is currently considered as an attractive and promising route to produce chemicals, hydrogen, biofuels, and electricity by both the industrial and scientific communities [1–3]. Syngas is a mixture of hydrogen (H_2) and carbon monoxide (CO) produced from

A. Ephraim (✉) · R. Munirathinam · A. Nzihou · D. Pham Minh
RAPSODEE CNRS, UMR-5302, Université de Toulouse, IMT Mines Albi, Campus Jarlard, 81013 Albi Cedex, 09, France
e-mail: aug.ephraim@gmail.com

Y. Richardson
Biomass Energy and Biofuels Laboratory 2iE-CIRAD (LBEB), International Institute for Water and Environmental Engineering (2iE Foundation), Ouagadougou, Burkina Faso

A. Nzihou (ed.), *Handbook on Characterization of Biomass, Biowaste and Related By-products*, https://doi.org/10.1007/978-3-030-35020-8_11

the gasification of carbonaceous feedstock. Raw syngas generated from gasification or reforming processes contains contaminants that must be mitigated to meet process requirements and pollution control regulations. Figure 11.1 highlights the numerous uses of syngas, which range from heat or power applications such as Integrated Gasification Combined Cycle (IGCC) to a variety of synthetic fuels and chemicals.

Bio-syngas can be produced from various types of biomass feedstock. Depending on the physical and chemical characteristics of the feedstock different technological pathways can be applied to produce syngas, as depicted in Fig. 11.2. When lignocellulosic biomass is used as a direct feedstock, the classical syngas production process includes biomass gasification, gas cleaning up and gas conditioning. Other biomass vectors derived from both lignocellulosic biomass and fermentable biomass have also been considered as a secondary source of syngas. Biogas and bioethanol respectively produced by mechanization and ethanol fermentation of fermentable organic matter can be converted into syngas through reforming technologies [5, 6]. Lignocellulosic biomass can also be converted into a more suitable and economically advantageous feedstock for further conversion into syngas depending on the syngas production technology. This strategy includes flash pyrolysis to produce bio-oils, carbonization to produce char, and torrefaction to produce torrefied biomass. Bio-oils resulting from biomass flash pyrolysis can be used as a feedstock in both gasification and catalytic steam reforming technologies [7].

Despite the improvement of the bulk density of the biomass feedstock, the pyrolysis oils are cleaner than the biomass feedstock because many of the minerals and metals found in the latter are concentrated in the product char, which is separated from the product syngas. Carbonization and torrefaction processes are also

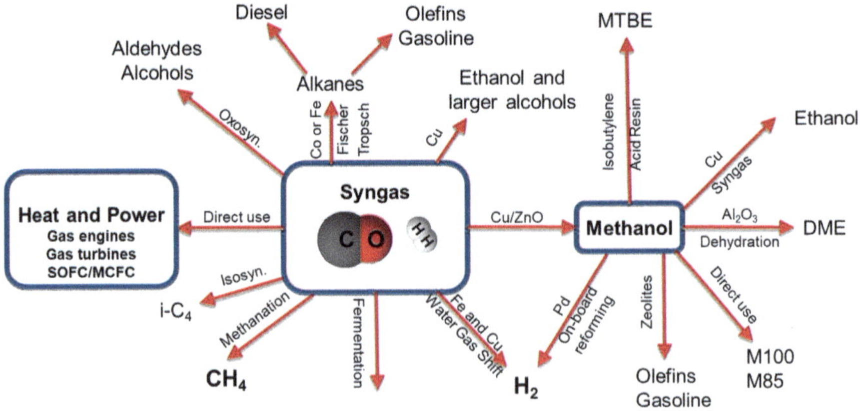

Fig. 11.1 Syngas conversion technologies [4]. Adapted with permission from American Chemical Society

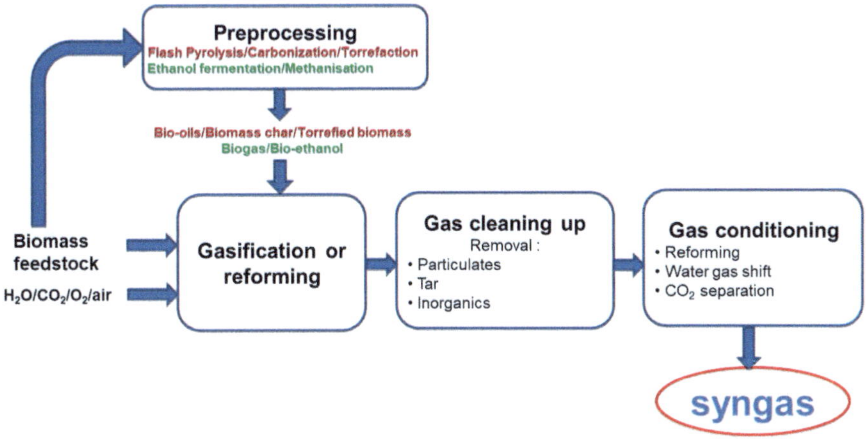

Fig. 11.2 Syngas production technologies from biomass feedstock

being considered in order to upgrade the biomass fuel for the subsequent gasification process [8].

Depending on the biomass feedstock used, the composition of the resulting raw syngas is expected to vary and therefore suitable gas clean-up technologies have to be implemented in order to remove the undesirable compounds.

The raw gas produced by gasification and reforming process is mainly composed of CO, H_2, CH_4, CO_2, H_2O, and N_2 (if air is used as the oxidizing agent). The syngas composition of the main products closely depends on the gasifier design and operating conditions, including the origin of the heat supply to the endothermic gasification reactions. Table 11.1 provides typical syngas composition from different gasifier designs.

In addition to these main products, syngas produced from biomass feedstock also contains secondary products such as tars, inorganic compounds (alkali metals, nitrogen, sulfur, and chlorine compounds), and solid particles. The tolerance of syngas end-use systems for these contaminants is highly variable and it is the role of

Table 11.1 Indicative composition of bio-syngas from different gasifier designs [9]

Syngas properties (vol.%, dry)	Allo-thermal	Auto-thermal	Entrained flow
H_2	40	26	39
CO	25	20	38
CO_2	21	35	20
CH_4	10	13	0.1
C_2H_4	2.5	3	0
N_2	1.5	3	3
H_2/CO	1.6	1.3	1.0
LHV (MJ/m^3)	14	12	10

Reprinted with permission from Elsevier

the clean-up and conditioning systems to fit the threshold required for each specific application. Table 11.2 provides the syngas specifications of impurities required for several applications. The use of external combustion in steam power stations is the least demanding application in terms of fuel gas purity. Gas quality requirements increase when internal combustion systems (e.g., gas engines and gas turbines) and more modern converters, such as fuel cells (e.g., solid oxide fuel cell (SOFC), molten carbonate fuel cell (MCFC), or proton exchange membrane fuel cell (PEMFC) are used. Syngas specifications become particularly restrictive for applications involving the catalytic conversion of syngas into chemicals or fuels, such as methanol and Fischer-Tropsch syntheses (F-T). Indeed, these contaminants could have a highly detrimental impact on downstream syngas processes. Depending on the end-use, the syngas that is cleaned up has to be further conditioned in order to fit the requirements of the desired application in terms of composition (H_2:CO ratio, H_2 purity), temperature, and pressure.

From an analytical point of view, the syngas species of interest can be lumped into three main categories: (i) Non-condensable gases, (ii) condensable gases, and (iii) impurities.

Non-condensable gases include mainly H_2, CH_4, CO, CO_2, N_2 and light hydrocarbons. These compounds remain in the gas phase and can be directly sampled and analyzed in the gas stream.

Condensable gases are composed of steam and condensable organic compounds called tars.

According to several definitions reported in the scientific community, tars refer to all the organic compounds produced under thermal or partial-oxidation regimes (gasification) of any organic material. Tars are generally assumed to be largely aromatic with higher molecular weights than benzene, although lighter compounds such as ethylene, cyclopentadiene, and benzene also have to be considered when specific applications of producer gas, such as fuel cells, are contemplated. Tars can be globally described as a complex mixture of organic compounds, which includes single-ring to multiple-ring aromatic compounds, along with numerous oxygen-containing hydrocarbons and complex polycyclic aromatic hydrocarbons (PAHs) [1, 11]. Among the contaminants of biomass gasification-derived raw gas, tars remain the most troublesome secondary products and tar removal still remains the subject of intense research and development [11]. Serious problems arise due both to their condensation/polymerization into more complex molecules and to aerosol formation, resulting in corrosion, clogging, and fouling of installations such as filters, heat exchangers, engines, or turbines where they reduce component performance and increase maintenance requirements. The presence of tar in syngas can also lead to the deactivation of catalysts used in downstream gasification processes such as Fischer-Tropsch or methanol syntheses and fuel cell applications as well [10, 11].

The impurities that can be found in bio-syngas include nitrogen-, sulfur-, and chlorine-containing compounds (e.g., NH_3, HCN, H_2S, COS and HCl), alkali and heavy metals and fine particles.

Table 11.2 Examples of syngas specifications on impurities required for specific applications [10]

Final syngas application	Particles	Tar	Sulfur (H_2S, COS)	Nitrogen (HCN, NH_3)	Alkali (primary K and Na)	Halides (primary HCl)	Heavy metals
Steam-cycle power station	Minimal requirements	Of no importance, but condensation must be avoided[a]	Final SO_x emissions limited by regulation	Final NO_x emissions limited by regulation	n.s	n.s	n.s
Gas engine	<50 mg m_N^{-3}	<100 mg m_N^{-3}	Final SO_x emissions limited by regulation	Final NO_x emissions limited by regulation	n.s	n.s	n.s
Gas turbine	<30 mg m_N^{-3}	<50 mg m_N^{-3}	<20 ppmV	<50 ppmV	<0.02 ppmV	<1 ppmV	n.s
Solid oxide fuel cell (SOFC)	<1 ppmW	Several tens to few hundred ppmV	Few ppmV		1 ppmV	Few ppmV	n.s
Molten carbonate fuel cell (MCFC)	<0.01 m	<2000 ppmW	<0.1 ppmV (H_2S)	<0.1 ppmW (HCN) <1%vol (NH_3)	n.s	<0.1 ppmW	n.s
Proton exchange membrane fuel cell (PEMFC)	n.s	<100 ppmV	<1 ppm	n.s	n.s	n.s	n.s
Methanol synthesis	<0.02 mg m_N^{-3}	<0.1 mg m_N^{-3}	<1 mg m_N^{-3}	<0.1 mg m_N^{-3}	n.s	<0.1 mg m_N^{-3}	n.s
Fischer-Tropsch synthesis (F-T)	<0.5 mg m_N^{-3}	<1 ppmV	<0.01 ppmV	<0.02 ppmV	<0.01 ppmV	<0.01 ppmV	<0.001 ppmV

Reprinted with permission from Elsevier

Nitrogen-, sulfur-, and chlorine-containing compounds are of concern due to their corrosive nature and because they are both poisons for metal-based catalysts such as nickel or iron-based catalysts and precursors of regulated emission products such as NO_x and SO_2. Compared to coal, biomass often contains 100 times less sulfur (typically in the form of hydrogen sulfide (H_2S) or carbonyl sulfide (COS)) [10], which usually does not exceed 0.5 g kg^{-1}. Nitrogen contaminants occur as cyanide (HCN) and ammonia (NH_3), which can decompose to molecular nitrogen (N_2) during biomass gasification [12]. Alkali metals in the ash material are also of concern for both gasifiers and end-use systems as they can lead to the agglomeration phenomenon in fluidized bed reactors [13], hot corrosion of metal surfaces, and catalysts deactivation [1]. Particle matter elutriated from a gasifier can be made of residual solid carbon and inorganic compounds from both the biomass and the elutriation of bed materials depending on the feedstock and process. Most of syngas applications require more than 99% particle removal owing to fouling, corrosion, and erosion issues that cause efficiency and safety concerns if they are not addressed [12].

Apart from the alkali metals, other metallic elements such as mercury, arsenic, cadmium, zinc and selenium can be found in trace amount in biomass feedstocks. Their fractions are generally less than 0.1% but can be higher if contaminated biomass wastes are used as biomass feedstocks. However, because they are highly toxic, their emissions are heavily controlled. Among all trace elements, mercury is the most highlighted toxin due to health concerns and its high corrosive reaction to aluminum [2].

11.2 Permanent Gas Characterization

As previously summarized in the Sect. 11.1, syngas is the main gaseous product of the pyrolysis and/or gasification process using various feedstocks such as charcoal, natural gas, heavy hydrocarbons originated from petroleum refineries, plastic wastes, organic wastes, agricultural wastes, wood wastes, and virgin biomass.

A common feature of raw syngas, which was also specified in the previous section, is its complexity with the simultaneous presence of different components: particulate matter, tars (condensable compounds), water vapor, permanent gas, acid inorganic gas, and even metals at high temperature. The analysis of the composition of permanent gas (CO, H_2, CH_4, CO_2, N_2, light hydrocarbons C1–C4) requires generally a cleaning step of raw syngas in order to avoid analyzer disturbance due to fine particles, tar and acid gas.

11.2.1 Sampling and Preconditioning

The main objective of this step is to separate particulate matter and tar from raw syngas, thus the resulting permanent gas can be analyzed without risk of contamination or damage to the analyzer. The most common method used for syngas preconditioning is described in the CEN BT/TF 143 Technical Specification [14]. More details on this Technical Specification can be found in Sect. 11.3 for tar analysis. According to this Technical Specification, raw syngas from gasifier is sampled by a heated probe equipped with a heated particle filter to remove solid matter. Then, this syngas passes through heated or chilled impringer bottles containing liquid absorbents, wherein tar molecules, water, and even acid gas can be absorbed. Syngas from the impringer bottles outlet can be eventually dried with a silicagel tube. The sampled gas volume is measured under dry conditions by means of a gas meter. Then, the cleaned syngas can be analyzed or sampled and stored for analysis later.

From this principle of syngas sampling and preconditioning, different sampling trains can be used for permanent gas analysis. The difference can be found at the syngas flow rate sampled for gas analysis, the number, form and size of impringer bottles used, the nature of solvent used, the temperature of impringer bottles etc. Figure 11.3 shows an example of the syngas sampling train available. It consists of: (1) one heated probe, (2) two empty impringer bottles kept at room temperature, (3) two impringer bottles containing iso-propanol as organic solvent and thermostated at −20 °C, (4) one empty impringer bottle thermostated at −20 °C, (5) one impringer bottle containing silicagel at room temperature, (6) one thermostatically-controlled bath, (7) a gas pump, (8) a gas flowmeter, (9) a gas counter, and (10) a gas analyzer and computer.

Cleaned syngas can be directly analyzed by adapted equipment or sampled in a special bag, bulb or bottle (or equivalent equipment) for storage and ulterior analysis. For a given equipment for gas sampling, the following criteria are commonly required for the bag:

- Equipped with at least a valve, allowing the sampling and the injection of gas for analysis.
- Light and easy to use.
- Made by inert material, without interaction with gas.
- Having adapted tear resistance.
- Adapted to sampling temperature.
- Having low gas permeability.

In the case of gas sampling bag, available commercial products can be related to Tedlar bag or Supel-Inert foil gas sampling bags [15, 16]. Tedlar bag includes a large range of volume from 0.5 to 100 L. The temperature and pressure limit for utilization is 204 °C and 560 bars, respectively. They are made of polyvinyl fluoride (thickness of 50 μm), with high chemical inert property. They have either a valve made of polypropylene or stainless steel and can be composed also by a septum which allows withdrawing or doping the gas inside the bag with a syringe.

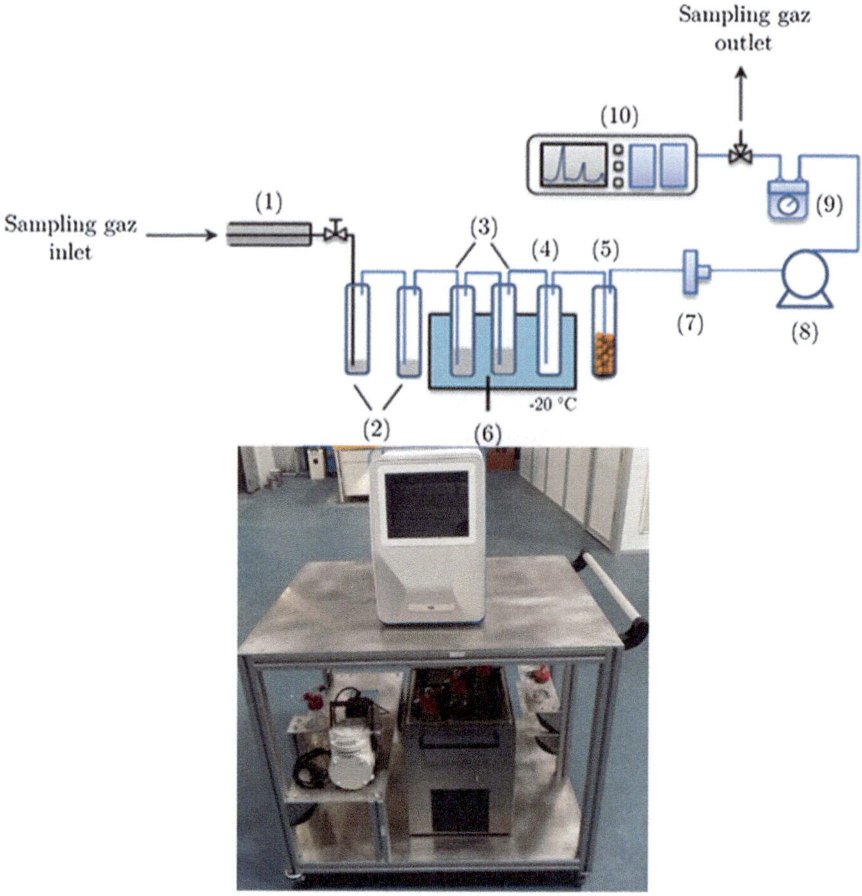

Fig. 11.3 Scheme (top) and photo (bottom) of a syngas sampling train

The permeability of oxygen and carbon dioxide is 50 and 172 cc/m^2/day, respectively, and that of water vapor is 9–57 g/m^2/day. Before gas sampling, the bag must be purged several times with inert gas such as nitrogen, helium or argon. The bag is firstly emptied with a pump, and then filled with an inert gas, and this procedure is repeated at least 3 times. The final empty bag is used for gas sampling.

A supel-Inert foil gas sampling bag is similar to a Tedlar bag but it is made of a multilayer structure bag. It is composed of two outer layers of aluminum film providing great barrier to gases permeating through the walls of the bag. This is the ideal solution for sampling low molecular weight compounds. Also, only foil bag material can hold hydrogen sulfide for long periods, from 5 to 7 days. However, it is not adapted for low ppm VOC's due to background levels.

Gas sampling bulb or bottle can be an alternative of Tedlar bag. They have the similar technical criteria compared to Tedlar bag. However, they are made of hard

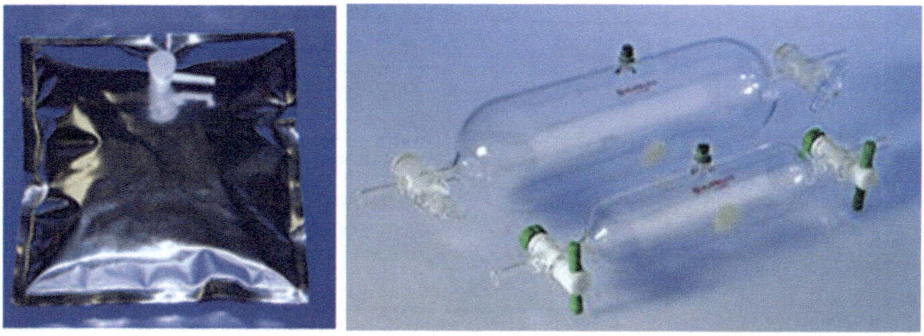

Fig. 11.4 Example of a Tedlar bag (left-hand-side) and a gas sampling bulb (right-hand-side)

materials such as glass, polypropylene etc. Thus, their form does not change during the utilization for gas sampling. Figure 11.4 shows examples of Tedlar bag and bulbs.

11.2.2 Analysis and Equipment

Cleaned syngas can be analyzed by different methods. The most used method is the gas chromatograph (GC). This separate analysis method is well described elsewhere [17]. Briefly, a GC is principally composed of an injector to receipt gas (or liquid) sample, a column where chemical compounds of sample is separated (called also stationary phase), a source of carrier gas (called also mobile phase), a detector, a heated oven for the control of the temperature of the injector, detector and column, a computer for the final treatment of results.

A conventional GC is generally equipped with only one column. Thus, the analysis time is usually long (typically more than 10 min), in particular for the separation of a complex gas mixture containing several molecules. Also, it cannot separate or detect all compounds of a gas mixture containing molecules of similar (for the separation) or very different (for the detection) properties. It is worth noting that GC analysis generally requires an internal standard because of the variable injection volume. Figure 11.5 shows the scheme of a conventional GC.

In order to improve the performance of the conventional GC having only 1 column, multicolumn GC systems have been developed. They can be equipped with different gas inlets, columns, injectors and detectors. For example, the model "Agilent 7890B" from Agilent is composed of two gas inlets and can receive several columns and detectors as described elsewhere [18]. These systems allow analyzing simultaneously complex gas mixtures with short analysis times. An example of a multicolumn GC is described thereafter in Sect. 11.2.3.

There are also multicolumn GC systems which are much more compact compared to a conventional GC and have very short analysis time of less than 3 min for

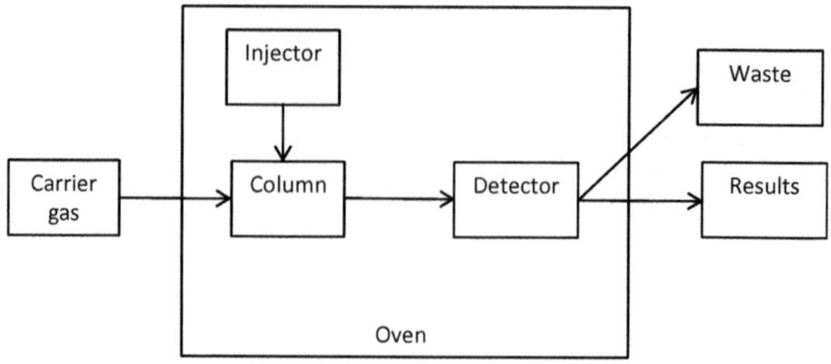

Fig. 11.5 Scheme of a conventional GC

complex gas mixtures. They are commonly called μ-GC. Equipped with an integrated computer, μ-GC can be also transportable for on-site analysis. The volume of gas sampled by the μ-GC can be kept constant, so no internal standard is needed. In addition, a μ-GC is generally equipped with an automatic injector, which can be programmed and controlled by a computer so that the sampling and injection can be periodically done. An example of μ-GC is described thereafter in the Sect. 11.2.3.

11.2.3 Examples of Syngas Analysis

11.2.3.1 Off-Line Measurement

Off-line measurement is related to punctual analysis. Permanent gas is sampled and analyzed with a given time interval between analyses.

Off-line analysis can answer to the need of different configurations of gasification. Research on the gasification process at lab scale is usually carried out in batch mode. Small quantity of biomass or wastes sample, going from some mg (TG-DSC analysis) to some g is generally used. Heating rate usually varies from 1 to 20 °C per min, and particularly up to around 50 °C per second in the case of the gasification with concentrated solar energy. The temperature range studied is typically from room temperature to 600–1200 °C, even at higher temperature in some cases. There are two cases of study: non-isothermal and isothermal processes.

Batch experiment at lab scale under isothermal regime

To study the thermal behavior of a sample at a given temperature and avoid all transformations of sample during the heating step, the isothermal regime is needed. In this case, the oven is firstly heated without sample to the desired temperature. At this temperature, sample is quickly introduced into the oven. Outlet syngas is cleaned and collected only during the plateau time at the studied temperature for ultimate analysis. We describe below an example of isothermal gasification of a Solid Recovered Fuel (SRF) under nitrogen atmosphere. This SRF was provided by a producer in Occitanie region in France and is currently used in cement factory. Table 11.3 shows the principal characteristics of this RDF.

The experimental apparatus (Fig. 11.6) is mainly composed of a reactor tube (20 mm internal diameter) made by inert ceramic and heated by an electrical furnace, and a gas trap system. For a given experiment, 700 mg of fine RDF particles was used. The furnace was firstly heated to 800 °C without sample. At this temperature, RDF sample contained in an inert ceramic crucible was quickly introduced into the center of the reactor tube (<15 s) under a nitrogen flow rate of 200 NmL/min. The syngas from the reactor outlet passed through the gas trap system and was collected in a Tedlar bag during a plateau time of 20 min. When the gas collection step was finished, its composition was analyzed by a μ-GC apparatus.

The gas trap system was composed of two impingers, the first one contained 100 mL of pure water, the second one contained 100 mL of iso-propanol, which were connected in cascade. This configuration was chosen by taking into account the low gas flow rate (around 200 NmL/min) and the low concentration of pollutants in this syngas due to high dilution of the syngas generated from RDF sample in the nitrogen flux. With these two impringers, all pollutants could be trapped: eventual fine particles, tar and acid gas such as HCl or H_2S. The syngas collected in Tedlar bag contained only permanent gas and N_2. The volume of the Tedlar bag used (5L) was calculated taking into account the nitrogen flow rate and the weight of RDF treated.

The μ-GC apparatus used in this example for permanent gas analysis was purchased from Agilent (MyGC model) via SRA Instrument (distributor in France). This machine is equipped with 4 modules. Each module has a backflush injector, a column and an enhanced/standard TCD detector which has high sensitivity for gas analysis, up to around 1 ppmv. The columns selected allow analyzing simultaneously all permanent gases of a typical syngas from gasification process for a short analysis time (<180 s): CO, H_2, CH_4, CO_2, Ar, N_2, H_2S, NH_3, and light hydrocarbons from C1 to C4. Particularly, this machine can analyze N_2 and Ar, which are

Table 11.3 Principal characteristics of the studied RDF

	Parameters	SRF 1
Moisture content (wt%, wet basis)		7.3
Proximate analysis (wt%, dry basis)	Volatiles	73.3
	Fixed carbon	9.4
	Ash	17.3
Elemental composition (wt%, dry basis)	C	46.2
	H	6.2
	N	0.6
	S	0.1
	Cl	0.3
LHV (MJ/kg, dry basis)		23.2

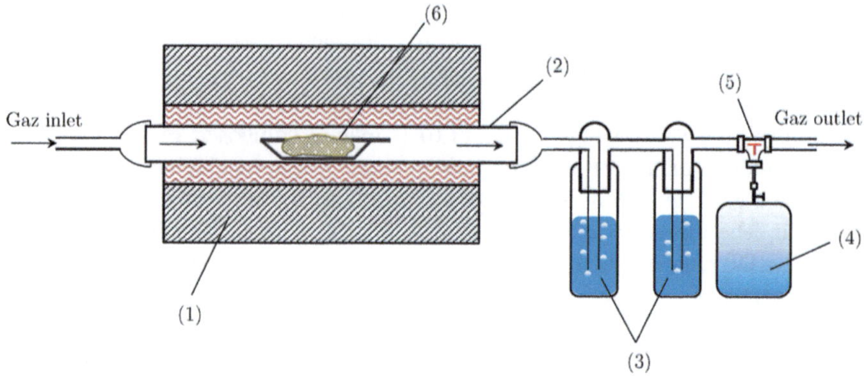

Fig. 11.6 Diagram of experimental device. (1) Furnace; (2) reactor; (3) impingers; (4) gas sampling bag; (5) three-way valve; (6) crucible with SRF sample

usually used as internal standard in GC analysis. Table 11.4 shows the operational conditions of the µ-GC analysis.

For the analysis, the Tedlar bag containing cleaned syngas was connected to the gas inlet of the µ-GC, as illustrated in Fig. 11.7. After fixing the analysis conditions as shown in Table 11.4, the analysis is started. An internal pump inside the µ-GC sends the syngas to the columns for the separation, detection and quantification of each gas.

Figure 11.8 shows the results obtained for the composition of the permanent gas produced by the gasification experiment described above by µ-GC analysis. As expected, H_2, CO, CO_2 and CH_4 were the main components of this syngas. C_2H_4 was formed at high concentration, probably due to high content of plastic such as PE and PP in SRF.

Batch experiment at lab scale under non-isothermal regime

This experiment was done with a vertical cylinder-shaped quartz tube reactor, heated by an electrical furnace [19]. A 65 mL quartz crucible containing 10 g of

Table 11.4 Example of operational conditions of the µ-GC analysis using MyGC model analyzed

Parameter	Column A	Column B	Column C	Column D
Name of the column-length	MS5A-10 m	MS5A-30 m	PLOTU-14 m	Alumina-10 m
Injection temperature (°C)	50	50	50	90
Column temperature (°C)	60	45	75	90
Sampling time (s)	20	20	20	20
Injection time(ms)	0	0	50	0
Back flush (s)	5.3	5.3	–	6
Pressure (psi)	28	50	28	28
Sensitivity	Standard	Standard	High	High
Carrier gas	Ar	He	He	He

Standard sensitivity: detection limit around 10 ppmv; High sensitivity: detection limit even <1 ppmv

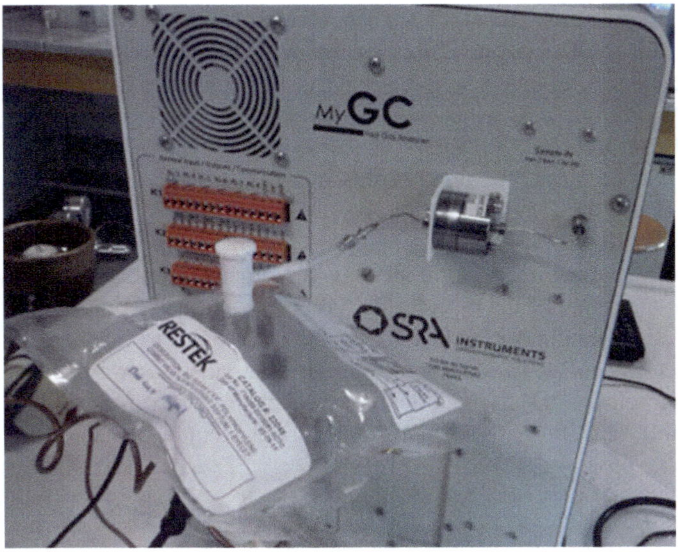

Fig. 11.7 Photo of the connection of a Tedlar bag to the gas inlet of a μ-GC for analysis

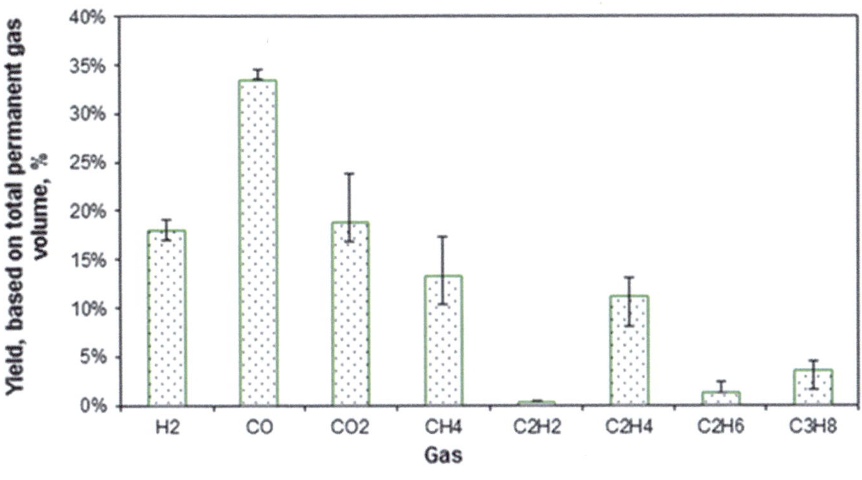

Fig. 11.8 Composition of the permanent gas from the gasification of a SRF at under nitrogen atmosphere at 800 °C with an isothermal regime

mixture of wood/plastic was placed at the center of the quartz tube. A nitrogen flux passed through the reactor from the bottom with a controlled gas flow rate (33 NmL/min). At the outlet of the reactor, tar molecules in the syngas were trapped by an impinger bottle (250 mL) filled with cotton wool and cooled to 0 °C. Then, the syngas was still cleaned by two consecutive impringer bottles containing each 100 mL of KOH (0.2 mol/L). Then, cleaned syngas was collected in a Tedlar bag of 5L for μ-GC analysis. This experimental setup is presented in Fig. 11.9.

Gasification experiments were carried out with mixtures of a poplar wood and different virgin plastics which were polyethylene (PE), polystyrene (PS) and polyvinyl chloride (PVC). Table 11.5 shows the ultimate analysis of these materials. Each wood/plastic mixture was made with fine particles of this poplar wood with one plastic for a total weight of 10 g, wherein the weight percentage of plastic reached 0, 5, 10, 30, 50 and 100%.

For a given experiment, after placing the crucible containing a wood/plastic mixture at the center of the reactor, the furnace was heated from room temperature to 750 at 20 °C/min under N_2 flux. Then, the reactor was maintained at 750 °C for 30 min before cooling to room temperature. The outlet gas was all collected in a Tedlar bag before analysis. The analysis was also done with the μ-GC described above. Table 11.6 shows the composition of permanent gas obtained with different wood/plastic mixtures.

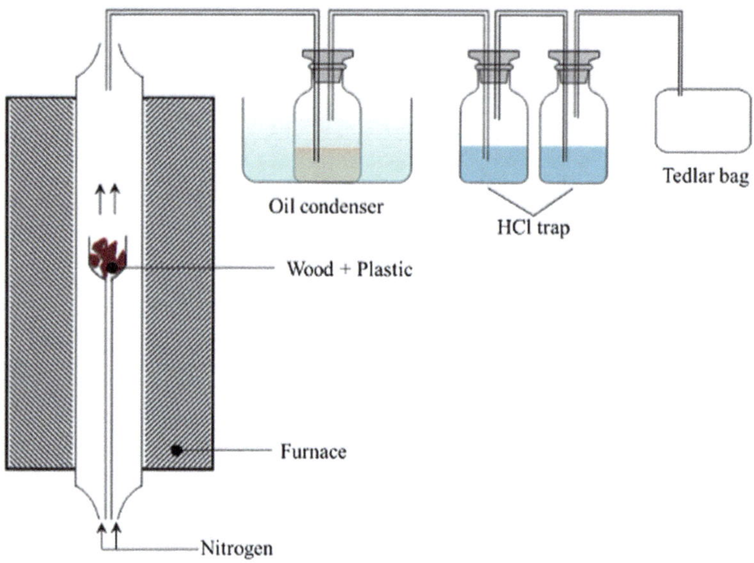

Fig. 11.9 Experimental setup for the gasification of poplar wood/plastic mixture. Reprinted with permission from E. Ephraim [19]

Table 11.5 Ultimate analysis (dry ash-free basis) of wood and plastic samples

	C (%)	H (%)	O (%)	N (%)	S (%)	CI (%)
PW	49.9	6.4	42.7	1.0	0.0	0.01
HDPE	86.1	14.5	0.0	0.0	0.0	0.0
PS	92.4	7.8	0.0	0.0	0.0	0.0
PVC	38.7	4.8	0.0	0.0	0.0	56.5

Reprinted with permission from E. Ephraim [19]

Table 11.6 μ-GC analysis of the syngas generated from the gasification of poplar wood/plastic mixtures

Plastic Content (wt%)	HDPE				PVC					PS			
	H_2 (wt %)	CH_4 (wt %)	CO (wt %)	CO_2 (wt %)	H_2 (wt %)	CH_4 (wt %)	CO (wt %)	CO_2 (wt %)	HCl (wt %)	H_2 (wt %)	CH_4 (wt %)	CO (wt %)	CO_2 (wt %)
0 (100 wt% PW)	0.16	1.54	6.10	7.17	0.16	1.54	6.10	7.17	0.10	0.16	1.54	6.10	7.17
5	0.15	1.40	5.00	6.00	0.20	1.36	4.80	6.15	0.48	0.18	1.59	6.66	6.89
10	0.13	1.16	4.20	4.97	0.19	1.12	3.85	4.97	0.56	0.18	1.59	6.52	6.95
30	0.11	1.04	3.16	3.97	0.25	1.24	3.57	3.62	1.19	0.21	1.52	6.20	6.53
50	0.21	1.095	4.51	3.82	0.26	1.31	3.73	3.66	11.90	0.15	0.87	3.46	3.34
100	0.16	4.07	0.00	0.05	0.36	2.50	0.07	0.05	54.79	0.03	0.18	0.15	0.03

Plastic content (wt%)	HOPE			PVC			PS		
	C_2H_4 (wt%)	C_2H_6 (wt%)	C_3H_8 (wt%)	C_2H_4 (wt%)	C_2H_6 (wt%)	C_3H_8 (wt%)	C_2H_4 (wt%)	C_2H_6 (wt%)	C_3H_8 (wt%)
0 (100 wt% PW)	0.26	0.20	0.19	0.26	0.20	0.19	0.26	0.20	0.19
5	0.15	0.08	0.09	0.19	0.12	0.13	0.38	0.18	0.22
10	0.39	0.17	0.24	0.13	0.12	0.11	0.34	0.18	0.24
30	0.47	0.23	0.44	0.18	0.17	0.13	0.34	0.15	0.15
50	0.68	0.33	0.38	0.23	0.29	0.17	0.28	0.10	0.09
100	3.80	0.01	3.90	0.69	0.45	0.24	0.09	0.01	0.00

Reprinted with permission from E. Ephraim [19]

11.2.3.2 On-Line Measurement

On-line analysis of syngas composition is done in the case of a continuous gasification process. In this case, raw syngas is continuously generated, and its composition is analyzed along the process. Raw syngas has to be cleaned before any analysis.

On-line analysis of syngas is usually carried out with μ-GC because of the following reasons:

- μ-GC is perfectly adapted for permanent gas analysis, with high accuracy.
- Analysis time is usually short (<3 min) and this is enough for a continuous monitoring of the gasification process.
- In the case of the μ-GC described above (MyGC model from Agilent), its utilization for continuous monitoring of cleaned syngas is easy. In fact, an analytical method is built by the constructor for continuous analysis. It is necessary to connect the inlet of analyzer to the cleaned syngas source by an adapted tube (stainless steel tube). This is applicable when the pressure of the cleaned syngas is smaller than 2 bars. If the syngas pressure is higher than 2 bars, a pressure regulator is needed.

Figure 11.10 shows an example of the connection of syngas source to the gas inlet of a μ-GC using a metal tube. This type of connection can be also used for the coupling of a μ-GC with another analyzer such as TG-DSC, TPR, TPD or TPO etc.

As example of syngas on-line analysis, Andersson et al. [18] worked on the synthesis of alcohols from syngas, wherein they showed the on-line analysis of syngas and others chemical compounds along the reaction. The chemical reactor scheme is presented in Fig. 11.11. The reactor was fed with a syngas wherein chemical reaction took place. Alcohols as products and unreacted CO and H_2 (syngas) were analyzed by the first GC1, while H_2S as a pollutant was analyzed by the second GC2.

GC1 is an Agilent 7890 allowing the analysis of all the products of the reaction except H_2S which was analyzed by the GC2. A scheme of this GC1 is shown in Fig. 11.12. This machine is composed of two gas inlets, which consist of two automated VICI-Valco gas valves, a thermal conductivity detector (TCD) and two flame ionization detectors (FID), and 6 columns. The detector TCD and the associated columns allowed the analysis of CO, H_2, CO_2, CH_4, N_2 and C2, while organic compounds (alcohols, esters, hydrocarbons…) were analyzed by two FID detectors and other columns. The automated VICI-Valco gas valves allowed the periodic injection of gas samples for the continuous monitoring of the reaction. More information about these analytic systems can be found elsewhere [18].

Figure 11.13 shows the analysis results of the catalytic process at various experimental conditions. From the outlet concentration of CH_4, CO and other hydrocarbons, the authors could determine the CO conversion, CH_4 selectivity and carbon balance. It is worth noting that CH_4 selectivity by TCD and FID detectors was similar to each other and carbon balance was well buckled along the reaction.

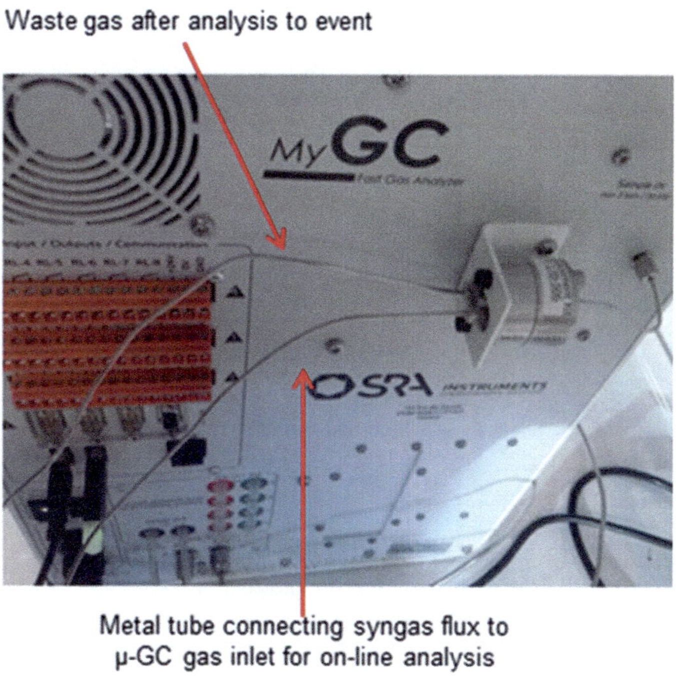

Fig. 11.10 Example of connection of syngas flux to µ-GC gas inlet for on-line analysis

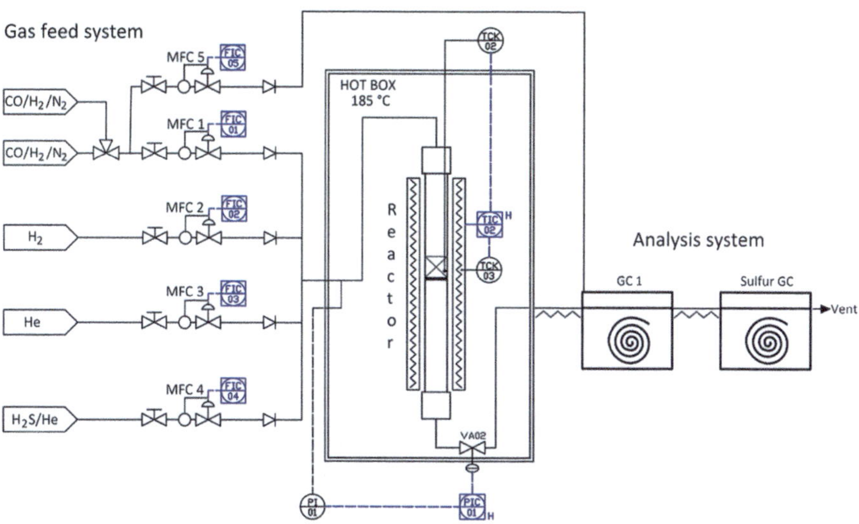

Fig. 11.11 Scheme of the alcohols synthesis process from syngaS. Reprinted with permission from Elsevier [18]

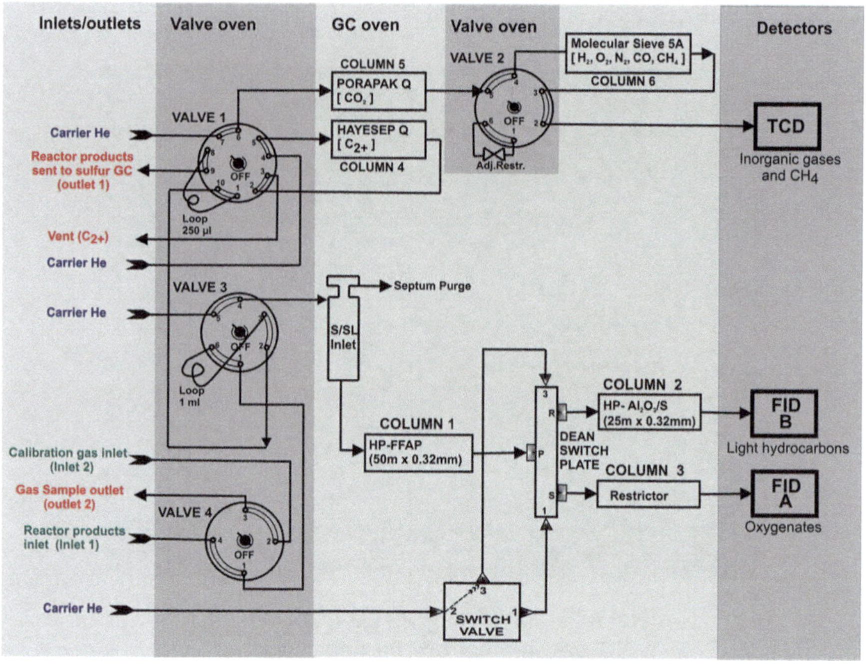

Fig. 11.12 Schematic overview of GC1. Reprinted with permission from Elsevier [18]

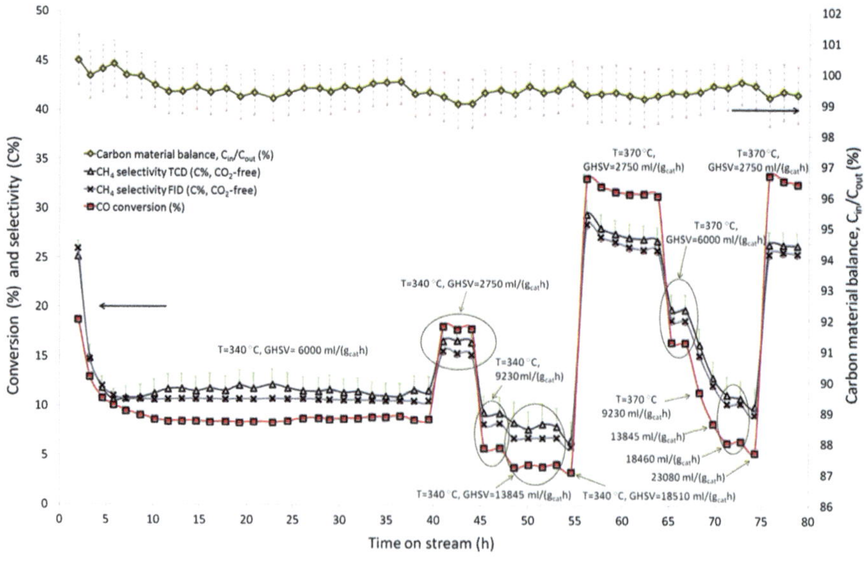

Fig. 11.13 Analysis of products from the transformation of syngas to alcohols [18]. Reprinted with permission from Elsevier

Figure 11.14 is the flowchart of the biomass CHP Güssing [20]. The gas composition of Fig. 11.15 was analysed with an online GC Syntech Spectras GC955, using a gas sampling valve and takes every 20 min a sample. Sampling of the synthesis gas was done from the clean gas after gas treatment of the CHP at atmospheric conditions. The GC system is equipped with two packed columns: the first one is a Porapak R 80/100 (1/8″ and 2 m) and the second one is a molsieve

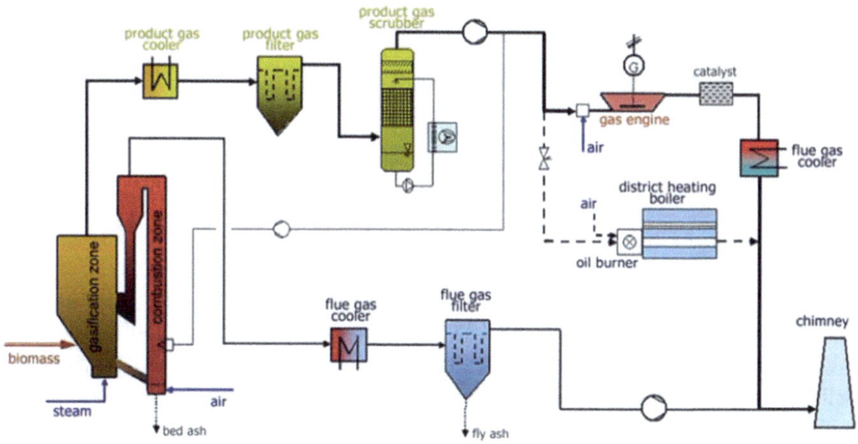

Fig. 11.14 Flow chart of biomass combined heat-power (CHP) Guessing [20]

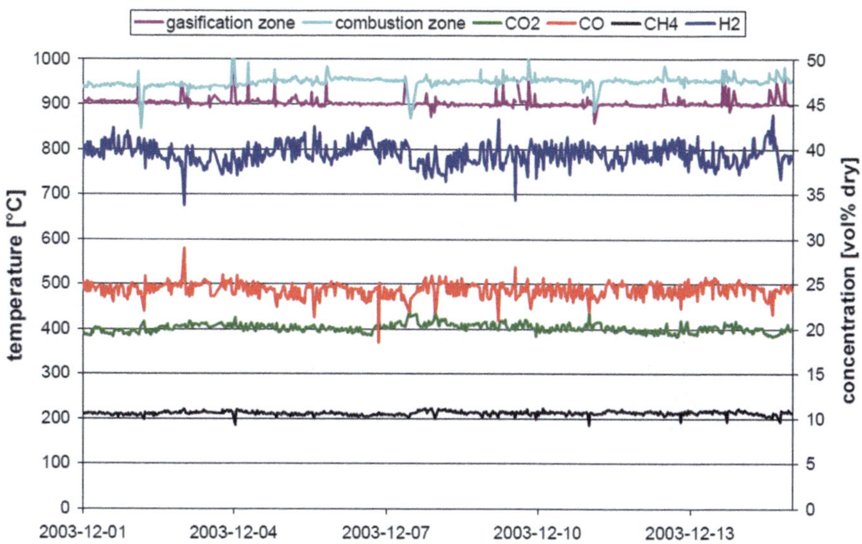

Fig. 11.15 Syngas composition from biomass combined heat-power (CHP) Guessing [20]

80/100 (1.8″ and 1.6 m). The temperature of injector and column were set at 80 °C and detector at 200 °C. The detector used was a TCD-detector (micro—TCD Vici). The carrier gas was He.

11.3 Condensable Gas Characterization

11.3.1 Offline Tar Analysis

11.3.1.1 Tar Protocol

Measuring Principle

The CEN/TS 15439 standard, commonly known as the tar protocol (TP), uses a solvent to sample and analyze tars in biomass-derived syngas [14]. The tar sampling system consists of a heated probe, a heated particle filter, a condenser and a series of impinger bottles containing a solvent for tar absorption. The sampling lines including the filter are heated (100–125 °C for updraft gasification; 300–350 °C for downdraft and fluidized –bed gasification) in order to prevent tar condensation as well as to avoid thermal decomposition of the organic compounds. Furthermore, the impinger bottles are placed in a temperature controlled bath so that staged warming and cooling of the sampled gas takes place. The gas is sampled for a specified period through the sampling line and filter and the flow rate is maintained with the aid of a pump.

Tar sampling device

As described in CEN/TS 15439 standard, syngas tar collection using the Tar Protocol is performed in a condenser containing six standard impinger bottles (100 or 250 mL) of which the first acts as a condenser for water [14]. 250 mL bottles are used when gas flow rates exceed 0.3 Nm^3/h. The first five bottles contain 50 mL solvent for tar absorption whereas the sixth bottle is kept empty. Bottles 1, 2 and 4 are heated to between 35 and 40 °C, whereas bottles 3, 5 and 6 are cooled to between −15 and −20 °C. In addition, bottles 2, 3, 5 and 6 are to be equipped with glass frits [14]. The reason for the interim temperature increase and the use of glass frits is to limit the formation of aerosols of tar and solvent, which can slip through the impinge bottles to the gas pump [21, 22]. After gas sampling, the solution is kept in storage bottles in a cool and dark environment until further analysis.

Regarding the choice of solvent, the CEN/TS 15439 standard recommends the use of isopropanol (i.e. 2-propanol) mainly due to its relatively high absorption capacity, low volatility and toxicity, as well as its ease of separation from tars [22]. Table 11.7 compares the performance of 2-propanol to other solvents that could be

Table 11.7 Comparison of solvents for tar absorption

Tar solvent	Advantages	Disadvantages	References
Acetone	High absorption capacity	High volatility. Loss of solvent may decrease mass transfer to the liquid phase because of lower liquid level and may lower the amount of tar dissolved until saturation is reached	[23]
Polar solvents [e.g. Dichloromethane (DCM), ethanol, 1-methoxy-2-propanol and isopropanol (2-propanol)]	Lower volatility than acetone and high solubility	–DCM is Toxic or carcinogenic, hence not recommended –1-methoxy-2-propanol is difficult to separate from Benzene and Toluene in GC analysis, and hard to evaporate in gravimetric analysis –Ethanol is subject to strict regulations in many countries	[23]
Hexane	Similar absorption capacity as 2-propanol	Higher loss through evaporation	[24]

used for tar absorption. According to the Tar Protocol, tar collection using isopropanol occurs both by condensation and absorption, and therefore, the cooling liquid used can either be a mixture of ice/salt/water or of isopropanol/dry ice or by cytostatic cooling of isopropanol [14].

Tar analysis

According to the tar protocol, two types of tar compounds can be determined—the gravimetric tar and a number of individual organic compounds (GC-detectable tars).

GC-detectable Tar

GC-detectable tar is tar that can be measured as a gaseous mixture of individual components using the standard gas chromatographic (GC) method. Standard GC analysis is not generally able to cover organic compounds that are larger than coronene (i.e. approximately 7 rings) [25].

Gravimetric tar

Gravimetric tar is defined as the residue obtained after the solvent is evaporated at specified conditions according to the tar protocol. In the gravimetric tar determination, all compounds larger than 3 rings are included. Although the individual species are not defined, those expected in biomass-derived syngas are listed in the tar protocol [25].

Both analysis of GC-detectable tar (individual compounds) and total gravimetric tars yield complementary information. The analyses can be performed separately and the user is free to select either one or both methods, depending on the type of information required.

Choice of analytical method

In deciding which of the methods for tar analysis to use, their efficiencies in capturing heavy and light tars should be considered. Analysis with GC is well suited for low-boiling small molecules (i.e. molecules with less than 7 rings), whereas larger molecules are harder to analyze with GC due to difficulties with evaporating and transporting such molecules through the chromatographic column. The opposite is true for gravimetric analysis, which captures heavy tars quite well but has low efficiencies for tars with low boiling points. In the CEN/TS 15439 standard [14], both GC and Gravimetric tar measurements are recommended in order to compensate for their respective limitations.

Calibration

Calibration using external standards without internal standards

External calibration standards of naphthalene is prepared at a minimum of five concentration levels. One of the calibration standards is at a concentration near the quantification limit, and the other concentrations correspond to the range of concentrations expected in the samples.

A table of values is drawn up which consist of the concentrations C_j in the standard solutions and of the associated measured values A_j (j = number of value pairs). The calibration line is determined from these values by linear regression using the equation below:

$$A_{Napthalene} = k * C_{Napthalene} + b \tag{11.1}$$

where

$A_{Napthalene}$: Measurement value for naphthalene as a function of $C_{Napthalene}$
$C_{Napthalene}$: Concentration of naphthalene in the calibration solution
k: Slope of the calibration line
b: Intercept of the calibration line

The linearity of the calibration function should be checked and must be better than $R^2 = 0.995$.

Calibration using internal standards (ISTD)

This follows a procedure similar to the calibration using external standards, except that each sample and working standard solutions are spiked with o-Terphenyl to attain the same end concentration for all working standard solutions. If the concentration of the internal standard is not the same in the sample and calibration standards a correction factor is required when calculating the results. It is important to use the same solvent composition and internal standard concentration for the working standard solutions and the samples.

To determine the calibration line, a series of measurement values is drawn up. It consists of the ratios of the concentrations C_j/C_{IS} of naphthalene to the internal standard and also of the ratios of the measured values A_j/A_{IS} of the naphthalene peak to the o-Terpenyl peak. From these pairs of values, the calibration line is determined by linear regression using the following equation:

$$(A_{Napthalene}/A_{IS}) = k * (C_{Napthalene}/C_{IS}) + b \tag{11.2}$$

where

$A_{Napthalene}$: Measurement value for naphthalene as a function of $C_{Napthalene}$
A_{IS}: Measurement value for o-Terphenyl as a function of C_{IS}
$C_{Napthalene}$: Concentration of naphthalene in the calibration solution
C_{IS}: Concentration of o-Terphenyl
k: Slope of the calibration line
b: Intercept of the calibration line

Calculation

Calculation using external standards without internal standards

If an external standard is used, the evaluation is performed using the following equation:

$$C_i = \frac{(A_i - b)}{k} \tag{11.3}$$

where

C_i: Concentration of organic molecule i in the isopropanol solution
A_i: Measurement value for organic molecule i as a function of C_i
k: Slope of the calibration line
b: Intercept of the calibration line

Calculation using internal standards (ISTD)

If an internal standard is used, the calculation is performed using the equation below:

$$C_i = \frac{\left(\frac{A_i}{A_{IS}} - b\right) C_{IS}}{k} \quad (11.4)$$

where

C_i: Concentration of organic molecule i in the isopropanol solution
A_i: Measurement value for organic molecule i as a function of C_i
C_{IS}: Concentration of o-Terphenyl
A_{IS}: Measurement value for o-Terphenyl as a function of C_{IS}
b: Intercept of the calibration line
k: Slope of the calibration line

Example of tar analysis using the 'Tar Protocol'

The analysis presented in this example was performed in accordance with the CEN/TS 15439 standard [14]. The syngas containing tar was produced from the air gasification of wood waste in a downdraft gasifier.

Tar sampling

Tar sampling in syngas was conducted using the same equipment setup as that defined by the Tar Protocol (Figs. 11.16 and 11.17). The syngas flow rate was 0.055 m^3/h and the sampling time was 45 min.

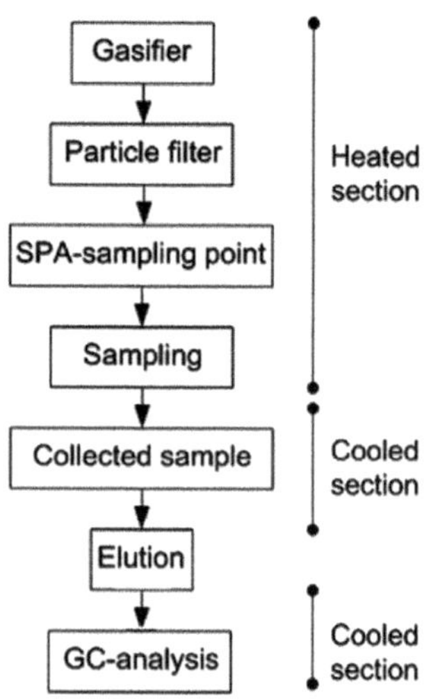

Fig. 11.16 Steps in the tar measurement process [27]. Adapted with permission from American Chemical Society

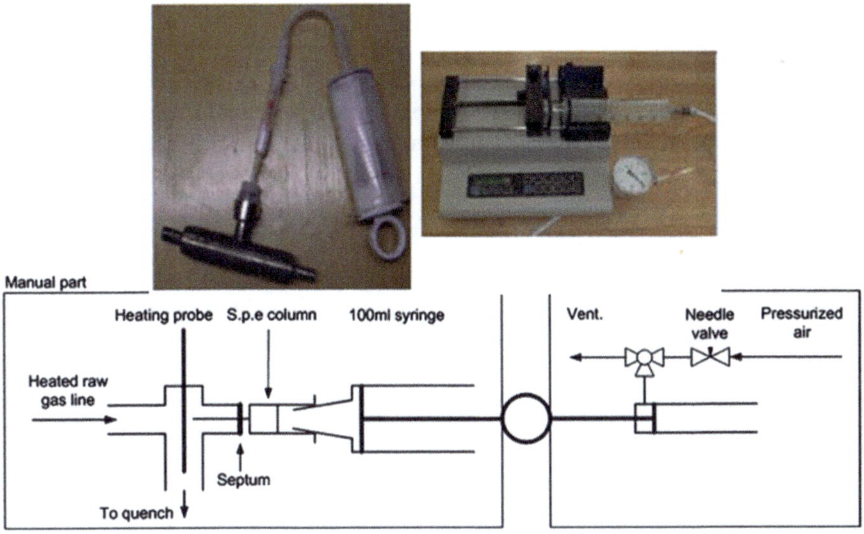

Fig. 11.17 Left panel: SPA sample point, spe column, and manual part of the extraction device. Right panel: pneumatic robot for consistent sample extraction [27]. Reprinted with permission from American Chemical Society

Tar analysis

A Shimadzu GCMS-TQ8030 system (Schimadzu Corporation, Kyoto, Japan) was used for the analysis. The following operating conditions applied are listed below:

Gas chromatograph (GC):

- Injection initial temperature: 250 °C
- Interface temperature: 250 °C
- Linear velocity: 45 cm/s
- Total flow: 23.2 mL/min
- Split ratio: 1:12
- Injection volume: 1 μL
- Oven temperature program held at 40 °C for 3 min, a ramp of 10 °C/min up to 270 °C, followed by a hold at 270 °C/min for 40 min
- Capillary column: 30 m length, 0.25 mm internal diameter and 0.25 μm film thickness (ZB-MS; phase reference—5% phenyl arylene and 95% dimethyl polysiloxane)
- Carrier gas: High purity helium (99.999 vol.%) with a flow rate of 1.6 mL/min.

Mass spectrometer (MS):

- Ion source temperature: 200 °C (70 eV electron ionization energy)
- Solvent cut time (delay): 1.5 min
- Scan range: 30–300 m/z

The detected peaks were identified using the spectral libraries NIST11, NIST11 s and PESTEI-3.

Calibration of the GCMS instrument was performed using four different concentrations of the external standards, Naphthalene, Acenaphthene, Fluorene, and Pyrene. These standards were chosen as they have different residence times in the capillary column and thus span the whole range of residence times of the identified tar compounds.

Results

The family of organic compounds identified in the tar samples using GC-MS are displayed in Table 11.8. The total tar content measured in the isopropanol solution was 269.8 ppm, which translates to 1162 mg/Nm^3 syngas. It is clear from the results that aromatic compounds, particularly Toluene, make up the majority of the tar components, followed by polycyclic aromatic hydrocarbons (PAHs) and nitrogen containing aromatics.

Table 11.8 List of tar compounds analyzed in syngas from wood waste gasification

Name	Conc (ppm)
Acids	**7.82**
Phosphonic acid, (p-hydroxyphenyl)-	6.43
Carbonic acid, phenyl propyl ester	1.10
Benzoic acid, diphenylmethyl	0.29
Phenols	**7.34**
Phenol, 2-methyl-	2.36
Phenol, 3-methyl-	4.03
Phenol, 2,5-dimethyl-	0.54
Phenol, 2,6-dimethyl-	0.12
p-Cresol	0.29
Furans	**1.55**
Benzofuran, 7-methyl-	0.27
Benzofuran, 2-methyl-	0.42
Dibenzofuran	0.34
Benzofuran, 2,3-dihydro-	0.35
Benzofuran, 4,7-dimethyl-	0.17
Aromatic compounds	**191.31**
Toluene	139.71
Ethylbenzene	3.04
o-Xylene	4.57
Benzene, 1-ethyl-2-methyl-	1.48
Benzene, 1-ethenyl-3-methyl-	0.85
Benzene, 1-ethenyl-4-methyl-	7.23
Benzene, 1,1′-(1-ethenyl-1,3-propanediyl) bis-	1.23
Benzene, cyclopropyl-	0.16
Indane	0.09
Benzene, 1-propynyl-	11.34

(continued)

Table 11.8 (continued)

Name	Conc (ppm)
1,3,5,7-Cyclooctatetraene	17.12
Benzene, 1,3-diethenyl-	0.40
Cycloprop[a]indene, 1,1a,6,6a-tetrahydro-	2.89
Biphenyl	0.96
4H-Cyclopenta[def]phenanthrene	0.14
1H-Cyclopropa[l]phenanthrene,1a,9b-dihydro-	0.09
PAHs	**36.76**
Phenylethyne	0.42
Acenaphthene	1.34
Naphthalene, 1,8-dimethyl-	0.71
Acenaphthylene	2.91
1-Isopropenylnaphthalene	0.83
2-Naphthalenol	0.48
1-Naphthalenol	0.22
Naphthalene, 2-methyl-	5.33
1,4-Methanonaphthalene, 1,4-dihydro-	0.37
Naphthalene, 2-ethyl-	0.39
1-Isopropenylnaphthalene	0.17
Fluorene	1.39
1H-Phenalene	0.20
Fluoranthene	0.32
Pyrene	0.33
Anthracene, 1,2-dihydro-	0.21
Phenanthrene	1.64
Anthracene, 9-methyl-	0.19
Azulene	19.11
5,16[1′,2′]:8,13[1″,2″]-Dibenzenodibenzo [a, g] cyclododecene,	0.19
Alicyclic compounds	**0.38**
Tetracyclo [5.3.0.0 < 2,6 > .0 < 3,10 >] deca-4,8-diene	0.38
Nitrogen containing aromatics	**23.90**
Benzonitrile	0.56
Benzenepropanenitrile, beta. -oxo-	0.51
5-Methylbenzimidazole	1.43
Iodoacetonitrile	0.07
Tetrahydro-2H-thiopyran-4-ylidenemalononitrile, 5,5-dioxide	0.27
5H-Cyclopentapyrazine, 6,7-dihydro-2,5-dimethyl-	0.72
Quinoline	0.90
1H-Indazole, 4-methyl-	0.23
2-Phenyl-4-methylimidazole	0.19
1,8-Diazacyclotetradecane-2,9-dione	19.02
Ketones	
5,9-Undecadien-2-one, 6,10-dimethyl-	0.75
Mixed oxygenates	
Triphenylphosphine oxide	1.99

11.3.1.2 Solid-Phase Adsorption (SPA)

Description

SPA is a dry sampling method, originally developed by KTH Sweden [26], by which tar components are condensed and adsorbed on a solid phase, and then desorbed for analysis. Figure 11.16 shows the steps through which the tar in syngas is sampled by SPA and analyzed by gas chromatography (GC). In this process, tar is produced in the gasifier (step 1) and is transported with the syngas through heated pipes and a particle filter (2) before it reaches the sampling point (3), where it is extracted (4). The collected sample (5) is eluted or desorbed (6) and then analyzed using a GC analyzer (7). SPA measures organic compounds between molecular weights of 104 g/mol (styrene) and 300 g/mol (coronene), with a reproducibility within 10% for most tar components [26].

A comparison of SPA to the Tar Protocol is presented in Table 11.9 [24, 27–30]. Factors that may influence the result of the total tar content measured or the concentration of individual components, include sampling method, extraction solutions, measurement intervals and analysis methods [22].

Table 11.9 Comparison of off-line tar measurement methods

Off-line method	Advantages	Disadvantages	Protocol exists?
Tar Protocol (TP)	More appropriate for measurements of gas with high concentration of tar species	Long sampling time (1–4 h) compared to SPA Long preparation period 2-propanol is harmful to health Formation of tar and solvent aerosol results in loss/waste of adsorbed tar Possible evaporation of solvent and tar Sampling system is bulky and complicated	YES
Solid-Phase Adsorption (SPA)	Recommended for tar sampling in gas with low tar concentration Short sampling time (1–10 min) Relatively small amount of solvent needed Low complexity Low error rate as no loss of tar by evaporation of solvent and formation of aerosol Relatively low costs High reproducibility	Desorption of BTX (Benzene, Toluene, Xylene) molecules during sample storage may result in inaccurate measurements of these molecules SPA with liquid extraction could become difficult to implement as gas flows with temperature higher than 300 °C	NO

Tar sampling

The sampling point for SPA measurements typically consists of a heated 'Tee' tube through which the syngas flows with a septum attached a short distance from the wall, to avoid melting, as shown in Fig. 11.17 [27]. The sampling point temperature is maintained at about 350 °C to avoid tar condensation. Sample extraction is usually performed with a syringe and pump to extract a set volume of raw syngas, usually 100 mL, through the Solid Phase Extraction (SPE) column [29]. During extraction of a sample from subatmospheric pressure systems, it is important to ensure that there are no leakages in the septum or the extraction device.

The sampling system can also be automated by fitting a pneumatic robot, which improves sampling reproducibility (right panel, Fig. 11.17). The robot consists of a pneumatic cylinder, connected to the syringe, which is filled with pressurized air at a flow rate regulated by a needle valve. This flow rate is calibrated to allow for set sample extraction times (e.g. 1 min). After collection of a sample, the pressure in the cylinder is released [27].

Examples of adsorbents used in SPE columns are as follows:

- SUPECLEAN LC-NH2 [31, 32]
- Sorbents made from graphitized carbon [33]
- CHROMABOND C18ec [34]
- Aminophases [27, 32, 35]
- Activated carbon [36].

Sample Storage

Ideally, the collected Solid Phase Extraction (SPE) columns should be eluted and analyzed directly after tar sampling in order to minimize desorption of volatile organic compounds. However, this is often impossible as laboratory facilities are not at the same location as the sampling site. In this case, the SPE tube should be capped at both ends after sampling, as shown in Fig. 11.18, and then stored in a fridge/freezer [29].

Previous studies have shown that storage at room temperature allows desorption BTEX compounds (benzene, toluene, ethylbenzene and xylenes), which thus compromises the accuracy of measurements of these compounds [26]. Such studies have also shown that storage at −20 °C for a period of 2 months results in complete desorption of BTEX compounds, as well as slight reductions in the amount of detected indene and phenol. Thus, it is important that the samples are analyzed as soon as possible after collection in order to minimize the measurement errors incurred from desorption of volatile tar compounds.

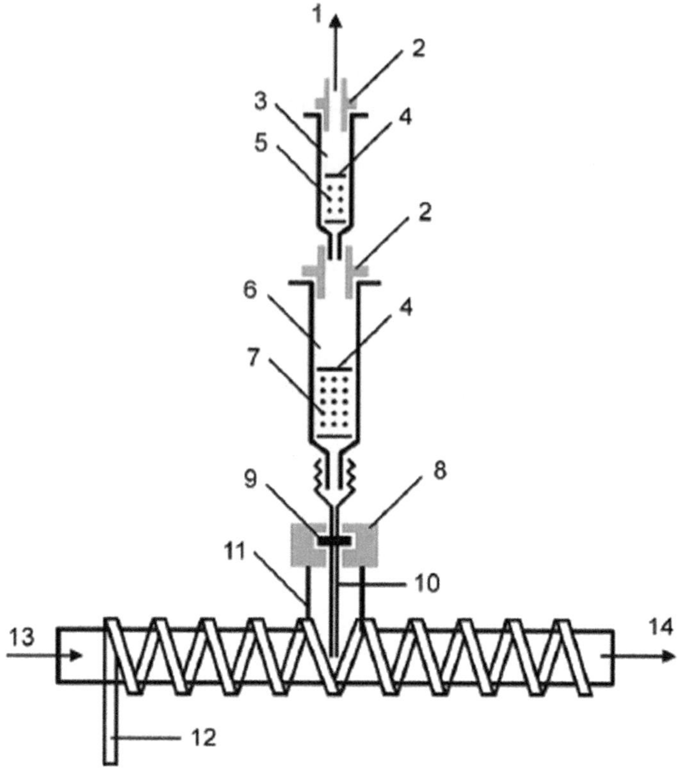

Fig. 11.18 A tar sampling device. 1—connection to a pump, 2—adapter (polypropylene), 3—sorbent column (polypropylene, 1 mL), 4—fritted disc (polyethylene), 5—activated coconut charcoal, 6—sorbent column (polypropylene, 4 mL), 7—amino-phase sorbent, 8—septum nut (polyethylene), 9—rubber/silicone septum, 10—hypodermic needle (stainless steel), 11—glass "Tee" tube, 12—heating tape, 13—product gas inlet, 14—connection to electric pump

Sample Preparation

Desorption or elution of tar compounds collected in SPE columns is performed with solvents such as dichloromethane, isopropanol and acetonitrile [29]. The solvents can be used individually or as a mixture, depending on the type of organic compounds targeted. Currently, there is no standardized method and procedure for elution. An example of an elution process is presented in Sect. 2.2.6.

GC analysis

Gas chromatography (GC) analysis of tar sampled by SPA is often used to determine compounds ranging in the molecular weight of 78 (benzene) and 300 (coronene). The instrument can be calibrated using external standards such as toluene or standards for all the major components (to reduce error in quantification). Moreover, internal standards may be used such as tert-butylcyclohexane (TBCH) or p-etoxyphenol (PEP), which allows for derivation of the phenolic fraction, thus improving the chromatographic performance [27].

Example of tar analysis using SPA method

Osipovs [28] has conducted a test on SPA method for determining the concentration of tar compounds in syngas produced from the gasification of peat biomass in a circulating fluidized bed (CFB) gasifier. The gas temperature at the gasifier outlet was 150 °C. The analysis results for the SPA method was compared with that of the cold solvent trapping (CST) method (i.e. 'Tar Protocol').

Sampling device

The tar sampling device used is shown in Fig. 11.20 and consists of the following sub-components:

- **2 columns joined consecutively**: Each column containing a different solid adsorbent. The first column was a 4 mL cartridge containing 500 mg loosely packed aminopropyl-bonded silica adsorbent. This sorbent is capable of adsorbing heavy tar compounds such as indene, naphthalene, acenaphthylene, acenaphthene, fluorine, phenanthrene, anthracene, fluoranthene and pyrene. The second column was a 1 mL adsorbent cartridge, in which 100 mg activated coconut charcoal was packed loosely. This adsorbent was meant to adsorb volatile organic compounds, mainly benzene and toluene.
- **"Tee" tube and air pump**: The "Tee" tube was heated to 250 °C using an insulated heating tape to minimize tar condensation. An air sampling pump was used to draw product gas through the heated tube.
- **Mass flow rate regulator** and a **gas vane-type pump**: For gas sampling at a constant flow rate of 100 mL/min.

Sample Storage

After sampling, the inlet and outlet sides of the cartridges were sealed with plastic stoppers.

Sample preparation

This involved two elution procedures for desorbing tar from the adsorbents in the two columns (or cartridges).

- 1st column tar desorption:
    i. A reservoir (an empty column) was connected to the inlet of the adsorbent cartridge, which in turn was positioned vertically above a vial (1.8 mL). A stand was used to hold the cartridge in position and the vial was placed on a laboratory jack.

ii. 3 × 600 μL of dichloromethane/acetonitrile (4:1 v/v) solvent mixture was added to the reservoir and was drawn through the amino-phase adsorbent by using a 100 mL syringe.
iii. The eluate (tar + solvent) was stored in the 1.8 mL vial.

- 2nd column tar desorption:
  i. Activated carbon sorbent was placed in separate vials, 1.8 mL each.
  ii. 1.0 mL of carbon disulphide solvent was added to each vial.
  iii. Crimp caps were immediately attached to each vial to prevent volatilization of organic compounds.
  iv. The vials were made to stand for at least 30 min with occasional agitation to allow for the desorption tar into the solvent solution.

Chemical Analysis

A Shimadzu GCMS-QP2010 equipment (Shimadzu Corporation, KYOTO, Japan) was used for tar analysis. The operating conditions for the gas chromatograph (GC) and mass spectrometer (MS) are as follows.

GC:

- Injection initial temperature: 250 °C
- Split ratio: 1:10
- Injection volume: 1 μL
- Oven temperature program held at 30 °C for 5 min, then 30–180 °C at the rate of 10 °C/min, 180–300 °C at the rate of 15 °C/min, and finally held at 300 °C for 5 min.
- Capillary column: 30 m length, 0.32 mm internal diameter and 0.25 μm film thickness (Rtx-5SIL-MS; phase reference—5% diphenyl—95% dimethyl polysiloxane; Restek)
- Carrier gas: High purity helium (99.999 vol.%) with a constant flow of 1.6 mL/min.
- Interface temperature: 310 °C

MS:

- Ion source temperature: 200 °C (70 eV electron ionization energy)
- Scan range: 35–300 m/z
- Analysis time: 31 min for dichloromethane/acetonitrile, and 15 min for carbon disulphide.

Calibration and quantification

The internal standards, Benzene-d_6, naphthalene-d_8, and phenol-d_6, were used to calibrate the GCMS equipment. Calibration curves obtained with five points, each repeated thrice. This lead to the identification of 48 tar compounds.

Results

Figure 11.19 compares the contents of some of the tar compounds identified using the SPA method with those obtained using the CST method (Tar Protocol). The results show that tar sampling using the SPA method yields significantly higher contents of Benzene, Toluene and Xylene (BTX). According to Osipovs [28], this is

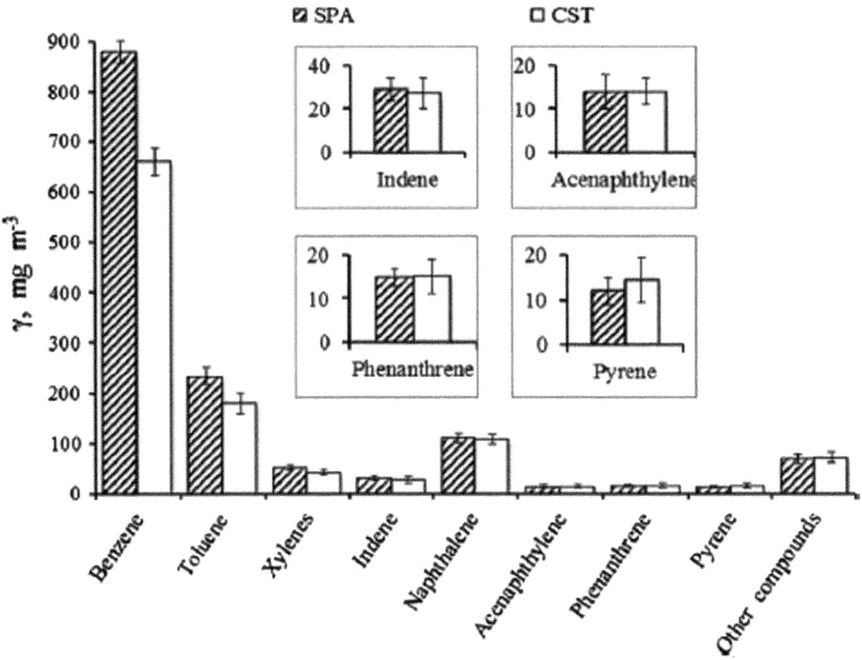

Fig. 11.19 Content of different compounds in mg/m^3 with standard deviation depending on the sampling method [28]. Reprinted with permission from Elsevier

due to the use of the second sorbent, activated coconut charcoal, which absorbs the BTX molecules and therefore enables their analysis. However, during the CST sampling process, the BTX molecules form aerosols with the solvent, which flows away from the sampling device uncontrolled. The author therefore demonstrates that this SPA method is more accurate for determining volatile organic compounds than the CST method, whereas the latter is more appropriate for analyzing heavy tar components due to the much larger volume of sampled gas.

11.3.2 Online Tar Analysis

11.3.2.1 Overview of Methods

Online methods for tar analysis are relatively new and most are still under development. A summary of the techniques under development is presented as follows:

- Laser-Inducted Fluorescence Measurement (LIF) by TU Berlin/TU Munich [37],
- Molecular Beam Mass Spectrometry by NREL [38],

- Time-Of-Flight Mass Spectrometry (TOF-MS) with laser ionization by the University of Rostock and KIT [39],
- Online Gas Chromatography/Mass Spectrometry (GC/MS) with electron and laser ionization by TU Berlin [40],
- Ion-Molecule Reactions Mass Spectrometry (IMR-MS) by CEA [41],
- Photo Ionization Detection (PID) by BTG Netherlands and KTH [42], and
- Flame Ionization Detection (FID) by IFK [43].

Generally, these tar online measurement devices are required to be robust, accurate, low-cost, non-intrusive and fast [43]. Table 11.10 compares the performance of PID, IMR-MS and FID in terms of their response time, quantitative and qualitative information they provide from the analysis, and their robustness ([41–43]).

11.3.2.2 Example of Online Tar Measurement

The work of Gredinger et al. [43] is discussed in this section, which focuses on the development and commissioning of the FID tar analyzer, carried out under the DEMTAR project at the Institute of Combustion and Power Technology (IFK). According to the authors, this online tar analyzer device is close to commercialization and will be manufactured by Ratfisch Analysenstysteme GmbH, located in Poing/Germany.

Measurement Principle

The general measurement principle is based on a differential measurement of the organically bound carbon in the sample gas of two sample loops with an equal volume (Figs. 11.20 and 11.21).

In the sampling phase (Fig. 11.20), the sample loops are consecutively filled with sample gas. In the analyzing phase (Fig. 11.21), nitrogen flushes the sample gas consecutively to the FID in two steps: Step 1: Analysis of sample loop 1 (solid line), step 2: Analysis of sample loop 2 (dashed line). To perform the differential measurement, the gas from loop 1 is passed over a suitable tar filter. The tar filter temperature can be controlled by Peltier elements in the range of 20–80 °C to condense only tars, whereas permanent gaseous hydrocarbons like C_1–C_4 and low-boiling substances like benzene (C_6) will therefore pass the tar filter. Furthermore, the gas from sample loop 2 is passed directly to the FID to determine the total hydrocarbon content. The entire analytical phase—except for the cooled tar filter—is heated to 300 °C to avoid tar condensation in the device.

FID Output Signal

The FID output signal is recorded and evaluated during the analyzing phase. Figure 11.22 shows the resulting peaks for each of the sample loops, which

Table 11.10 Comparison of some online tar measurement methods

Online method	Research group	Response time	Quantitate and qualitative information	Robustness	References
PID	KTH and BTG	(+) 60–90 s to detect a concentration change for gas flow of 20 mL/min (−) PID signal is stabilized within a minute (Compare SPA method: At least 40 min for sampling, sample-preparation and analysis)	(+) Accurate low tar concentration measurement (<200 mg/Nm3) (+) Good correlation between the total tar concentration and PID response (−) Unable to qualify and quantify individual compounds in tar	(−) The window of the excitation chamber needs periodic cleaning due to fouling, which causes the response to decrease. Cleaning cycle may range between a few hours to a day, depending on the gas composition	[42]
IMR-MS		(+) 25 s	(+) Can measure trace (<cm3/m3) and concentrated tar (+) Capable of measuring major gases (except H2), tar and inorganic pollutants in one apparatus (+) Capable of quantifying individual compounds in tar	(−) Signal overlapping for certain tar and inorganic compounds may need to be analyzed	[41]
FID	IFK	(+) < 60 s	(+) Accurate high tar concentration measurement (−) Currently unsuitable for low tar concentration measurement (−) Unable to qualify and quantify individual compounds in tar	(−) The analytical part of the device is heated to 300 °C, although 400 °C is the desired operating temperature to prevent tar condensation	[43]

resembles those obtained from other gas chromatography devices. The area of the peaks determines the respective hydrocarbon content, and the difference in the areas of peak 2 and 1 determines the tar content.

The analyzer is capable of performing three measurement ranges, from (1) 0–12, (2) 0–60 and (3) 0–120 g/m^3 total hydrocarbon. The change of the measurement

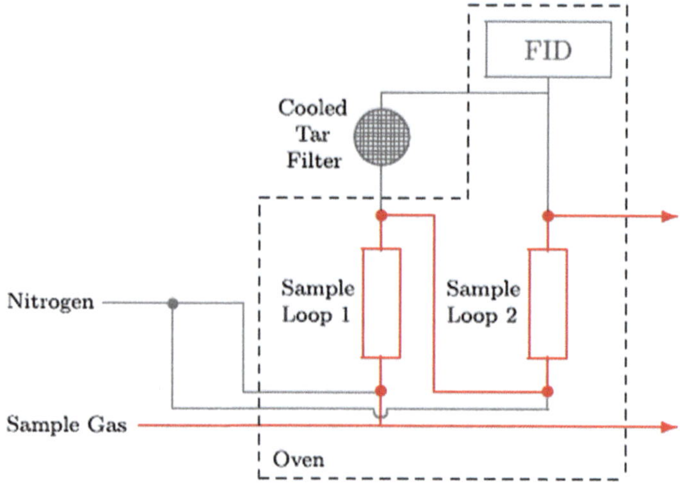

Fig. 11.20 Basic principle of the FID tar measuring system. Sampling phase [43]. Reprinted with permission from Elsevier

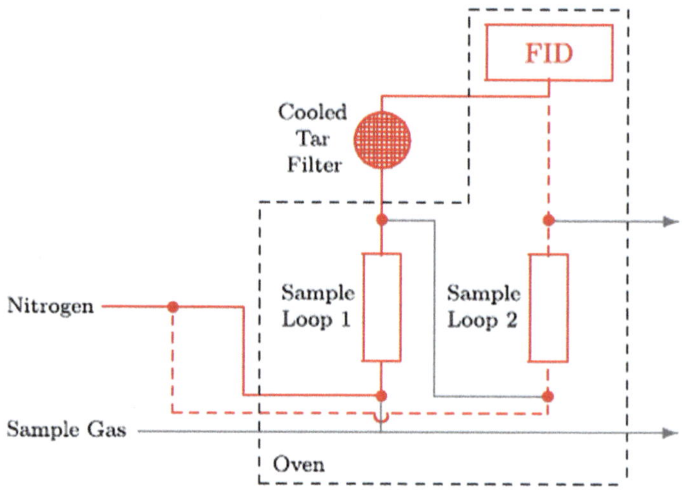

Fig. 11.21 Basic principle of the FID tar measuring system: Analyzing phase [43]. Reprinted with permission from Elsevier

ranges is done by different amplification of the FID current. According to Gredinger et al. [43], selecting a suitable measuring range allows for measuring low and high tar concentrations with high accuracy.

By selection of the suitable measuring range, this allows measuring small and high concentrations with high accuracy.

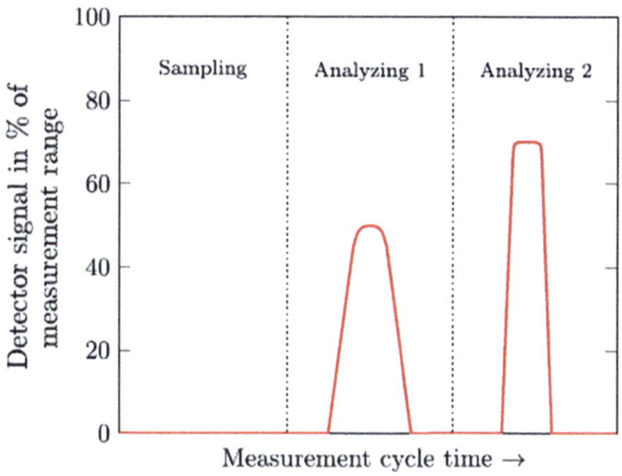

Fig. 11.22 FID-signal during sample and analysis cycle [43]. Reprinted with permission from Elsevier

Calibration

Because no permanent stable test gases with common tar species exist, calibration is conducted with propane with a concentration in the expected range of the total hydrocarbon concentration of the syngas. Measurements of the hydrocarbon mix (tars) then yield the carbon equivalent.

Results

The results of tests performed over 6 h during operation of a bubbling fluidized bed air gasification system, showed that the high tar concentrations measured using the online FID method had a small deviation compared to the results of the Tar Protocol. This method however is not yet suitable for measuring low tar concentrations, because this will require the use of a tar filter that does not absorb any benzene molecules, and thereby avoids a false recording of higher tar concentrations.

11.3.3 H_2O Analysis

11.3.3.1 Karl Fischer Principle

Karl Fischer analysis allows the determination of the quantity of water in tar collected during syngas sampling. This analytical method is based on the oxidation reaction of iodine with sulfur dioxide in the presence of water, as given by the equation:

$$I_2 + SO_2 + 2H_2O \leftrightarrow 2HI + H_2SO_4 \tag{11.5}$$

The analysis can be performed automatically by the titration apparatus Karl Fischer Mettler Toledo V20. Further description of the Karl Fisher method is presented in Chap. 2.

11.3.3.2 Procedure

Firstly, the 'Water Equivalent' (i.e. concentration) of the titrant (mixture of iodine and sulfur dioxide) has to be determined. To do this, a standard with a known water content is used, such as HYDRANAL Standard 10, whose water content is 10 mg/g. The Water Equivalent (WE) can therefore be determined using the equation:

$$WE = \frac{Std.M_{standard}}{V_{eq}} \tag{11.6}$$

WE: Water Equivalent (mg/mL)
Std: Water content of standard (mg_{H_2O}/g)
$M_{standard}$: Mass of standard introduced (g)
V_{eq}: Volume of titrant added (mL)

The titrant is introduced in the mixture of methanol (titration solvent) and the tar solution (containing isopropanol as solvent). While the added Iodine reacts with water in the tar, a significant potential difference is displayed by the electrode. Once all the water has reacted with the Iodine, a lowering of the potential is observed. When this measured potential is lower than 100 mV, we can assume that we are in a slight excess of Iodine and that the titration is complete.

We can therefore determine the water content as follows:

$$T_{KF} = \frac{V_q \cdot WE}{10 \cdot M_{sample}} \tag{11.7}$$

T_{KF}: Water content in mass percentage (% wt)
V_q: Volume of titrant used for sample measurement (mL)
M_{sample}: Mass of sample added (g)

However, the sample introduced also contains tar as well as isopropanol. The water content of pure tar is expressed in the following way:

Table 11.11 Water equivalent (WE) results for HYDRANAL Standard 10

Repeats	$M_{standard}$	V_{eq} (mL)	WE (mg/mL)
1	0.15	0.506	2.96
2	0.15	0.472	3.18
3	0.15	0.503	2.98
Average			3.04

Table 11.12 Measured water content in tar (T_{real}) using Karl Fischer method

Repeats	M_{sample}	V_q (mL)	T_{real} (% wt)
1	0.16	1.21	2.30
2	0.16	1.37	2.60
3	0.16	1.376	2.62
4	0.16	1.36	2.59
5	0.16	1.449	2.75
Average			2.57

$$T_{real} = T_{KF} \cdot \frac{m_{tar+isopropanol}}{m_{tar}} \tag{11.8}$$

T_{real}: Real water content in mass percentage (% wt)
$m_{tar+isopropanol}$: Mass of pyrolysis tar and isopropanol collected after sampling (g).
m_{tar}: Mass of pyrolysis tar recovered after sampling (g)

11.3.3.3 Example of H_2O Analysis Using Karl Fischer

The analysis of water content of syngas tar presented in this example was performed on a sample collected from the air gasification of wood waste in a downdraft gasifier. The tar sample was obtained using the Tar Protocol described in Sect. 11.3.1.1. Table 11.11 shows that the Water Equivalent of the HYDRANAL Standard 10 is 3.04 on average. Based on this, the water content in tar was determined to be 2.57 wt% as shown in Table 11.12.

11.4 Impurities Characterization

In the process of biomass gasification, apart from the formation of principal gas constituents such as CO and H_2 (syngas) together with CO_2, CH_4, lighter hydrocarbons (C_2 through C_6), H_2O, N_2 and tars, there is also a co-production of some impurities such as ammonia (NH_3), hydrogen cyanide (HCN), sulfur (e.g. H_2S, COS), chlorides (e.g. HCl), and particulate matter (fine particulates, soot, dust, etc.)

in low concentrations, alongside trace metals like As, Hg, Se, Cd, etc. [44, 45]. The presence of even small quantities of sulfur is detrimental to the downstream processes due to component fouling, catalytic poisoning and the environmental regulations. Of the different sulfur-containing compounds, H_2S is the dominant species formed during gasification process. Though the presence of nitrogen containing compounds depends on the process conditions and the nature of feedstock, they are present in minor quantities with NH_3 and HCN as principal components. Halogens are generally released as HCl or metal salts that are formed by binding to respective metals in biomass (e.g. alkali metal halides).

It is essential to characterize/quantify these impurities in the product gas from gasification process and subject them to cleaning process to set free of contaminants, so that the principal gas constituents can be used in different downstream processes [1, 12, 46]. Analytical techniques for quantifying these contaminants will be discussed in this section, as it is quintessential to assess the efficiency of both the biomass gasification unit and the gas cleaning systems. Most often the measurement techniques used for related processes such as coal gasification and coal combustion are applied directly or adopted with some modifications wherever required. Allowed level of contaminants in the syngas depends on the requirements set for downstream applications and environmental concerns, therefore, precise quantification of these impurities is to be established. The quantification of impurities can be done either by offline or online methods and will be discussed in the following sections.

11.4.1 Analysis of Ammonia

Ammonia and molecular nitrogen are found to be the dominant nitrogen species formed during biomass gasification [47]. NH_3 is generally trapped by using wet chemical sampling trains that consist of NH_3 absorbing solutions and then quantified by offline analysis. Since NH_3 is a basic compound, the absorbing solutions typically consist of acidic solutions, e.g., H_2SO_4 or HCl, which are effective in capturing NH_3 [48]. The concentration of adsorbing solutions depends on the gas composition and sampling time. It is recommended to keep the pH of the adsorbing solution below 1.0. Analysis of the adsorbed NH_3 in the acidic solution is usually performed using ion chromatography, ion selective electrodes, colorimetry, or titration method.

11.4.1.1 Sampling Method

The following points must be taken into account during NH_3 sampling for offline analysis [47].

- Tars present in the effluent should be removed without losing NH_3, otherwise, care must be taken to ovoid the condensation of tars in the NH_3 sampling unit.
- NH_3 loss caused by its interaction with components of the sample transport, for e.g., adsorption onto the walls of the sampling unit must be avoided.
- As NH_3 is highly water soluble, condensation of moisture in the sampling lines should be prevented.

An example for sampling method used for ammonia analysis formed during the gasification of corn/switch grass is described below [47].

- Syngas produced from the gasification unit is drawn into the sampling system (Fig. 11.23) at a flow rate of ca. 2 L/min (STP) for about 15 min per sample. This sampling time can be adjusted, depending on the concentration of NH_3 produced in the gas stream and the detection limit of the analytical methodology being employed.
- Particulate matter in the gas stream is removed by using a quartz-fiber thimble filter (25 mm × 90 mm). This filter assembly is also maintained at 450 °C. It is recommended to use filters that are nonreactive and possess less surface area that minimize analyte losses on the wall surfaces.
- Prior to the thimble filter assembly, highly porous, sintered metal filter is used for ash filtration.
- To prevent the condensation of tars the upstream lines leading to the tar condensation coil are maintained at 450 °C. The tar condensation coil (20–30 feet

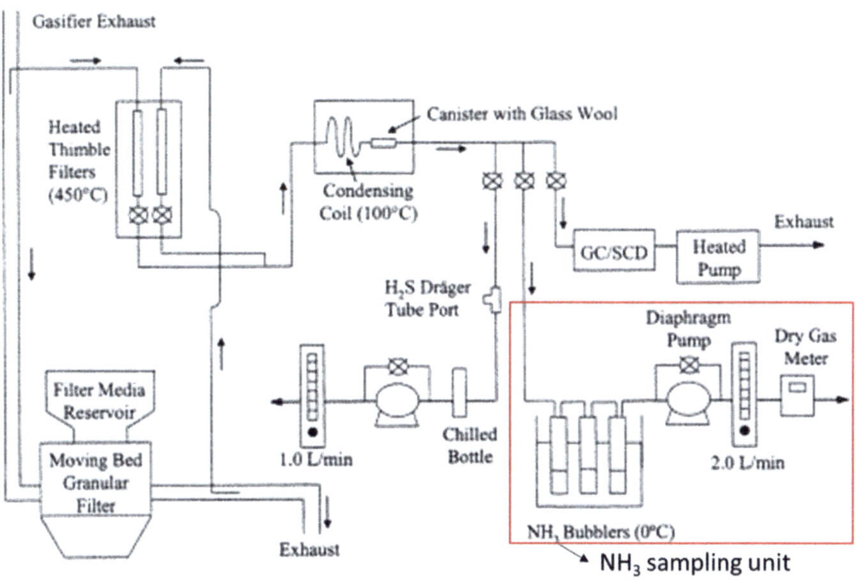

Fig. 11.23 Schematic diagram of the trace contaminant sampling system, with ammonia sampling unit outlined in red color [47]. Copyright © 2005, American Chemical Society

of 3/8 in. with 9.5 mm OD tubing) is maintained at 102 °C, aids in condensing the heavy tars by avoiding the moisture condensation. The larger diameter of the coil helps in avoiding the plugging and could be used for at least 6 h, after which the coil needs to be cleaned with dichloromethane or other suitable organic solvents. At this point, light hydrocarbons (that do not condense at 102 °C) still remains in the gas stream.

- At the exit of the tar condensation coil, a canister containing glass wool is placed to remove majority of heavy tar aerosols from the gas stream that are not removed in the condensation coil.
- The particulate-free and tar-free (except for non-condensable hydrocarbons) gases are then bubbled into a series of acidic absorbing solutions in the impinge bottles to quantitatively collect NH_3. NH_3 absorbs into the acidic solution by the addition of a proton to form NH_4^+ ions, which then analyzed in the ion chromatography. The sample line between the condenser coil and the NH_3 sampling train is heated at 110 °C to avoid moisture and light hydrocarbons condensation in the sample line. The impinge bottles are maintained at 0 °C by keeping them in an ice bath.
- The gas exciting the NH_3 sampling train passes through a dry gas meter, which measures the total volume of the gas collected.
- The NH_3 sampling train consists of a series of three precleaned 500-mL impingers (modified version of Greenburg-Smith impinge with no tapered tip), the first two of them contain 200 mL of 5% (v/v) aqueous H_2SO_4 (ACS reagent grade) absorbing solutions, whereas the third one contains silica gel or other suitable desiccant. HCl solutions can also be used in the place of H_2SO_4 solutions. Usage of greaseless impingers in the sampling train reduces the total number of liquid impingers needed for NH_3 sampling and also decreases the sampling time to about 10–20 min for raw gas.
- After sampling of raw syngas from the gasification process, the solutions in the impingers looked to be cloudy with a colloidal suspension. Analysis of this colloidal material by scanning electron microscopy-dispersive X-ray analysis (SEM-EDX) showed the presence of oxygenated and sulfonated organic compounds with particle sizes of less than 0.2 μm in diameter. To avoid colloidal suspension and the presence of sulfur and oxygen from H_2SO_4 absorbing solutions, HCl solutions are used instead. However, still the colloidal suspension is observed.
- The sampling solution for analytical purposes must be clear, therefore, the impinger solutions are filtered, and it is recommended to filter within 30 min of sample collection. Whatman No. 42 filters are used for this purpose. After filtration, the filter papers are thoroughly rinsed with aliquots of demineralized water to ensure no NH_4^+ ions are retained on the filter paper. The filtrate from each impinger is quantitatively recovered and diluted to 500 mL with deionized water, and this sample should be strictly clear. The pH of the solutions should be <1, if needed they can be acidified with H_2SO_4 and should be stored in a dark place at 2–5 °C prior to analysis. Thus prepared samples are expected to be stable for at least a month.

- The number of impinge bottles and quantity of solutions in each impinge bottles depends on the quantity of possible NH_3 formed during the gasification process and detection limit of the analysis method.

11.4.1.2 Analysis Method

The samples collected and prepared as described in the previous section can be analyzed using ion chromatography (IC), ion selective electrodes (ISE), colorimetry, or titration techniques [47] to quantify NH_4^+ ions formed during the gasification process. No NH_4^+ ions could be detected from these sampling solutions using IC analytical technique. In order to detect the source of this error, a series of liquid standards containing 1, 10, 100 ppm_w NH_4^+ ions in 5% (v/v) aqueous H_2SO_4 are prepared and analyzed in IC, still no NH_4^+ are detected. This strange behavior is attributed to the inability of the conductivity detector of IC to detect low levels of NH_4^+ ions in the presence of excessive ions originating from the acidic absorbing solutions. A series of 10 ppm_w NH_4^+ standards in aqueous H_2SO_4 solution with varying acid concentrations (from 5 to 0.1% (v/v)) analyzed using IC technique showed that the acid strength had a strong effect on the results (Table 11.13). Thus confirming the high ionic background signals associated with the concentrations of H_2SO_4 (any acidic solution) are responsible for unreliable results in detecting very low levels of NH_4^+ ions using IC technique. Therefore, this point must be taken into account while analyzing the NH_4^+ ions produced during the gasification process.

Analysis of 50 ppm_w NH_4^+ standards in 0.02, 0.5, and 5% (v/v) aqueous H_2SO_4 using different techniques, namely, colorimetry, titration, and ISE showed that consistent and reliable results are possible with colorimetry and titration method (Table 11.14). However, deviation in results from 50 ppm_w (NH_4^+ions concentration) were observed using ISE method and is attributed to the high ionic strength of H_2SO_4 solutions as observed in the case of IC analysis. Therefore, one among calorimetric or titrimetric method could be chosen to quantitatively analyze NH_4^+ ions. Results from these techniques can be used to calculate NH_3 concentrations in the syngas produced from biomass gasification by dividing the total amount of NH_3 (in micrograms) collected by the total volume of dry gas (at STP) sampled.

Table 11.13 Effect of acid strength on IC results obtained on a variety of 10 ppm_w NH_4^+ standards, table adopted from Ref. [47]

Acid strength of solution (%)	Reported NH_4^+ concentration (ppm_w)	
	Sample set 1	Sample set 2
5	0	0
2	3.6	0
1	1.0	0.8
0.5	8.6	2.2
0.2	7.4	3.0
0.1	4.4	1.5

Table 11.14 Effect of acid strength on results obtained on a variety of 50 ppm_w NH_4^+ standards using different analytical techniques, table adopted from ref. [47]

Acid strength of solution (%)	Reported NH_4^+ concentration (ppm_w)		
	Colorimetry	Titration	ISE
0.025	46	48	39
0.5	50	49	32
5	50	49	46

Table 11.15 Summary of NH_3 concentration determines in raw syngas from biomass gasification, table adopted from ref. [47]

Test no.	Feedstock	Duration of gasification (h)	Absorbing solution	NH_3 concentration (µg/at STP)		
				Range	Ave. ± σ	RSD (%)
1	Corn	4	5% H_2SO_4	5375–5542	5458 ± 119	2
2	Corn	3	5% H_2SO_4 and HCl	3309–4417	3775 ± 351	9
3	Corn	3	5% HCl	3132–3645	3330 ± 223	7
4	Switchgrass	4	5% HCl	816–2191	1422	40

The reproducibility and reliability of different analytical methods such as ion chromatography, ion selective electrode, colorimetry, and titration must be established before implementing a specific method to quantify the amount of NH_3 produced during gasification process. The NH_3 concentrations determined in raw syngas produced from corn/switchgrass gasification using titration method is presented in Table 11.15. The parameters such as gas meter calibration, daily variations in gasifier operating conditions, and feedstock homogeneity should be taken into consideration while comparing the results from different set of tests.

11.4.2 Analysis of HCN

Next to ammonia, HCN is formed in minor quantities among nitrogen-containing species during the biomass gasification. Wet chemical method is the most common method of measuring HCN in syngas [49]. In this case, syngas is bubbled through a basic solution (pH $\geq$ 12), e.g., dil. NaOH, wherein, HCN dissolves into the aqueous phase as cyanide ion. Thus sampled solution is then analyzed offline via an ion chromatography (IC) or ion selective electrodes. Fourier transform infrared spectroscopy (FTIR) [50] and gas chromatography with a nitrogen chemiluminescence detector [51] can also be used to measure HCN in the gas phase. Though online measurement techniques for HCN presents attractive aspects such as no requirement of tedious sampling system and avoids propagation of error arising from the multiple analytical steps with wet chemical methods, still most measurements of HCN in syngas are made using the later technique owing to its safety, high sensitivity, and robustness.

11.4.2.1 Sampling Method

The sampling method for HCN using wet chemical method is similar to that described for NH_3 in the previous section, however, HCN sampling requires basic absorption medium in contrast to the acidic medium required for NH_3 owing to their contrasting acid-basic character [52]. HCN sampling train can be prepared and collected separately in turns, i.e. one (HCN) after the other (NH_3). Collection in turns has some disadvantages such as complexity in operating two sampling train as it doubles the run time and also there can a drift in syngas quality between two collection sessions. In some cases simultaneous sample collection of NH_3 and HCN, respectively, can be done by arranging the impingers for respective gases in series as shown in Fig. 11.24 [49, 53]. However, in this case, there is a possibility of HCN being collected by the acidic solution in minor quantities that might lead to an underestimation of HCN concentration from the sample gas.

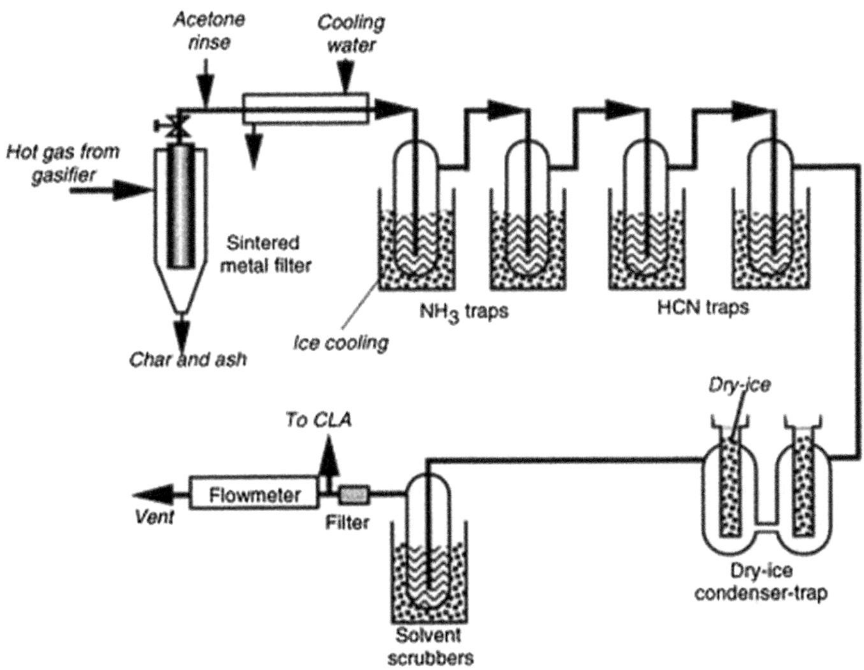

Fig. 11.24 Schematic diagram of NH_3/HCN sampling unit [49]. Reprinted with permission from American Chemical Society

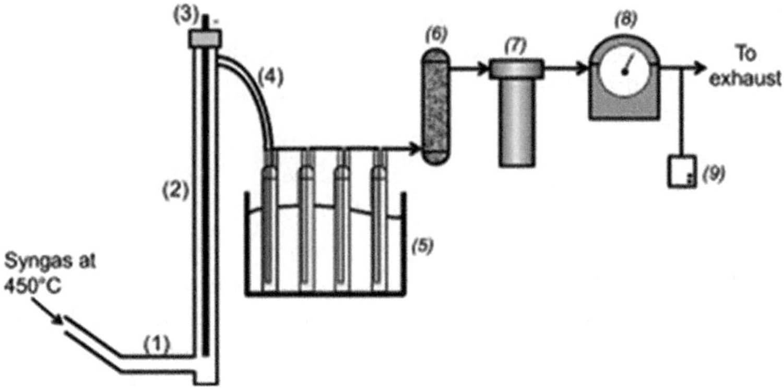

Fig. 11.25 Setup for testing HCN measurement techniques: (1) 22 mm ID stainless steel tubing at 150 °C, (2) 36 mm ID electrostatic precipitator case, electrically grounded, and heat traced to 110 °C, (3) ESP electrode mounted in an insulator, and operated at a negative voltage of 15–17 kV, (4) 13 mm ID Tygon tubing at room temperature, curving steeply downward to allow water condensation to run into impingers, (5) either four or five 500 mL glass impingers in series in water-ice bath, with each containing 200 mL of collection solution, (6) desiccant canister for water aerosol removal, (7) coalescing filter canister for final gas cleaning, (8) Ritter drum-type total flow gas meter, and (9) Varian CP-4900 micro gas chromatograph [52]. Reprinted with permission from Elsevier

An example for sampling method used to quantify HCN formed during the gasification of switch grass is described below and the sampling train setup is shown in Fig. 11.25 [52].

- The syngas stream from the gasifier is first subjected to cyclonic removal of particulate matter and then allowed to pass through the sampling system (Fig. 11.25).
- Tar from the gas stream is then removed with the electrostatic precipitator (ESP) whose outer wall is maintained at 110 °C to prevent moisture condensation (also no condensation of lightest hydrocarbons takes place).
- Gas stream coming out of the ESP passes through a Tygon tube with an inner diameter of 13 mm and a length of 380 mm which is connected to the HCN sampling train hosted in a water ice bath.
- The HCN sampling train consists of four 500 mL impinges filled with 200 mL of 100 mm NaOH solution in series and the Tygon tube is connected to the first impinger in such a way that all condensate is collected into the impinger.
- A desiccant canister and coalescence filter is fitted in the downstream of the impingers to remove the small amounts of aerosols that escapes the sampling train.
- At the end of the sampling system, a Ritter drum type gas flow meter is used to measure the dry syngas flowrate produced in the gasification process.

- The effect of connecting either acidic impinger or acetone scrubber before the HCN sampling train on the quantitative measurement of HCN in the gas stream is tested using the following two runs in a gasifier.
- Test run 1: In this run, an acidic impinger filled with 200 mL of 100 mm H_2SO_4 solution is installed in the upstream of HCN sampling train. In the control run, only the four impingers each containing 200 mL of 100 mm NaOH solution are employed without acidic impinger.
- Test run 2: In this run, acetone scrubber with 200 mL of acetone is installed in the upstream of HCN sampling train. The control run for this case is also similar to that of the test run 1.
- After the HCN collection time of 30 min, the impingers were instantaneously sealed with parafilm and stored on ice for 30–60 min until the disassembly of sampling system. After which the contents of each impinger is transferred into separate HDPE sample bottles with tight fitting lids. The acidic impinger bottle and its stopper are rinsed with small amounts of deionized water and transferred into its sample bottle. Similar protocol is followed for four HCN impingers and its stoppers, however, in this case 100 mm aqueous NaOH solution is used for rinsing. The H_2SO_4 sample is then diluted to 500 mL using deionized water. The other four HCN sample bottles are diluted to 500 mL using 100 mm aqueous NaOH solution. All these samples are maintained at 0–5 °C until analysis.

11.4.2.2 Analysis Method

The samples collected and prepared as described in the previous section are analyzed to quantify cyanide ions using ion chromatography (IC) with the following protocol [52].

- Each sample is agitated before sampling ca. 5 mL solution, and it is filtered using a Whatman 0.45 μm syringe tip filter. The filtrate is used to prepare 1500 μL of samples for analysis. Dilution of the samples with appropriate cyanide concentration must be carried out in accordance with the detection range of the analytical instrument. Three analysis vials are prepared in an identical manner for each sample (5 samples for each test run and 4 samples for control run) in order to replicate vial preparation.
- Thus prepared analysis vials are agitated and loaded into a Dionex ICS-3000 ion chromatograph (IC). Before these analyzes, the IC instrument should be calibrated from 1.0 to 10 ppm using standards prepared from commercial certified 1000 ppm aqueous cyanide solution.
- Each analysis vial (three such vials for each sample) is sampled three times in IC analysis, leading to a total of nine measurements per original collection impinger. In the case where the results from IC are out of the calibration range, new analysis vials should be prepared from the original 500 mL sample bottles

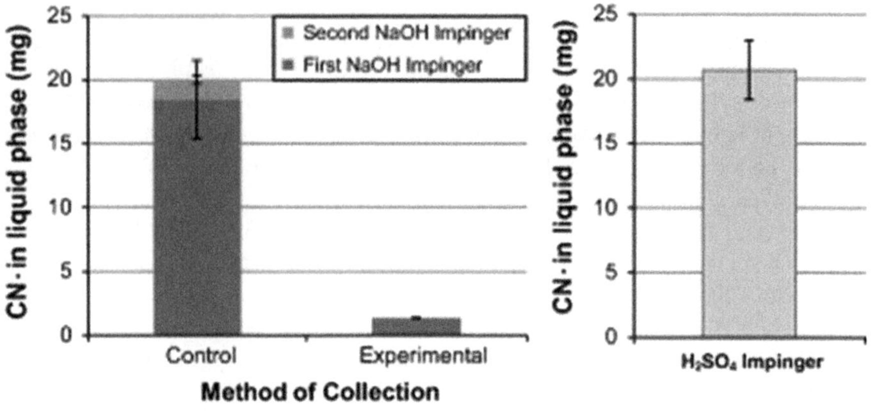

Fig. 11.26 When an H_2SO_4 impinger was added upstream of the series of 100 mm NaOH impingers, the total dissolved HCN concentration dropped 93% compared to the control. Dissolved HCN dropped from 19 mg to 1.4 mg in the first impinger, and from 1.6 mg to undetectable levels in the second NaOH impinger. Analysis of the H_2SO_4 itself found 21 mg, confirming that the missing HCN had been trapped by the H_2SO_4. The error bars in the graphed data signify 95% confidence intervals about the mean based on a two-tailed test using the Student-t distribution for eight degrees of freedom [52]. Reprinted with permission from Elsevier

with the revised dilution ration that fits within the calibration range. Thus prepared new vials should be reanalyzed.

- If the anticipated cyanide ion concentration in the sample bottles is less than 10 ppm, after filtration step, the sample can be introduced into IC vials undiluted. Cyanide ion concentration of less than 1 ppm falls below the detection range using this analytical technique. The HCN concentration result for each sample is calculated by taking the mean of nine IC readings.
- The results for test run 1 presented in Fig. 11.26 show that the addition of acidic impinger upstream of HCN sampling train underestimates the cyanide concentration, as it dissolves in acidic absorbing solution.
- The results of test run 2 presented in Fig. 11.27 also show that the acetone impinger added upstream of HCN sampling train underestimates the cyanide concentration, as it was captured by the acetone solution.

11.4.3 Analysis of H_2S and Other Sulfur Compounds

H_2S, carbonyl sulfide (COS), CS_2, mercaptans, thiophenic compounds such as thiophene, benzothiophene, etc. are the possible sulfur compounds formed based on the operating conditions in the gasification process [45, 54]. For instance, operating the gasifier at 1400 °C produces only gaseous molecules, namely, H_2S and COS. H_2S is the primary sulfur formed along with small amounts of COS, CS_2,

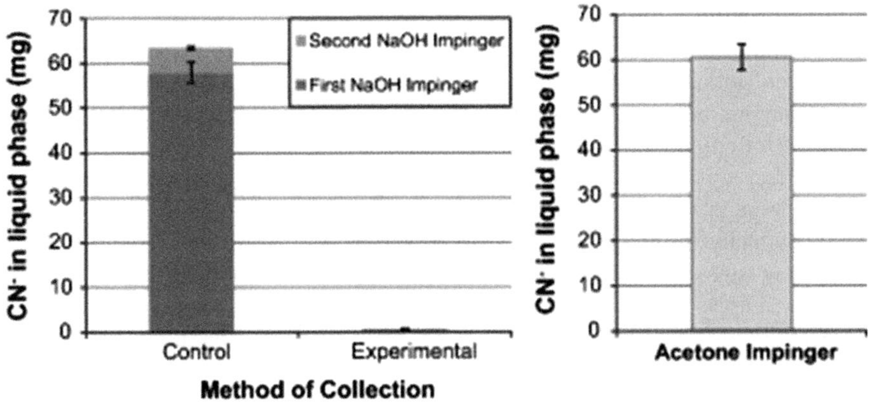

Fig. 11.27 When an acetone impinger was added upstream of the series of 100 mm NaOH impingers, the total dissolved HCN concentration dropped 99%. Dissolved HCN dropped from 58 mg to 0.58 mg in the first impinger, and from 5.5 mg to undetectable levels in the second NaOH impinger. Analysis of the acetone itself found 61 mg, confirming that the missing HCN was captured by the acetone. The error bars in the graphed data signify 95% confidence intervals about the mean based on a two-tailed test using the Student-t distribution for eight degrees of freedom [52]. Reprinted with permission from Elsevier

mercaptans and thiophene during typical gasification process operating at ca. 800 °C. H_2S and other gaseous sulfur compounds are mostly analyzed by gas chromatography either online or offline. The conventional wet chemical methods for H_2S capture using aqueous alkaline or cadmium absorbents are not widely used owing to the limitations such as sample instability, cadmium absorbent toxicity, and cumbersome sample preparation and analysis methods [48]. When trace amounts of H_2S is present in the gas stream, $AgNO_3$ impregnated filters can be used to trap H_2S gas in the form Ag_2S, which can be characterized using XRF spectroscopy to determine the amount of sulfur trapped on the filter [46].

11.4.3.1 Sampling Method

The following aspects needed to be considered while sampling gaseous sulfur compounds to be analyzed in gas chromatography [48, 54].

- The reactive and corrosive nature of H_2S and thiols containing SH group require careful attention to choose the materials for sampling. On glass or metal surfaces reactive adsorption of sulfur compounds take place even at low concentrations, therefore preference must be given to use inert materials like ceramics, PTFE, etc.
- When using stainless steel in the high temperature region of the sampling lines, maintaining sufficiently high flow rate, changing sampling lines upon noticing the variation in the usual concentration H_2S in the gas stream, and polishing the

tubing with chromium oxide surface might minimize the sulfur compounds adsorption problems.

- Particulate matter, tars and water vapors (moisture) should be removed from the gas stream, because their presence might lead to fluctuation in the quantification of sulfur compounds in the gas stream.
- It is recommended to remove particulate matter at high temperatures to ensure that there is no adsorption of H_2S.
- When analyzing the product gas stream offline, PTFE bags or aluminum-coated sample bags can be used to collect the gas and analyzed immediately. Storage of gas samples in these bags for long time period should be avoided. Aluminum-coated bags are safe and suitable for sampling sulfur and halogen compounds. Glass bulbs or other glass materials are recommended to be avoided to collect H_2S containing sample gas as it reacts with the surface of the glass.
- Online gas analysis using gas chromatography is very handy in quantifying the analyte more precisely, as it minuses the sampling time and a possible error during the sampling.
- In case the where tar-trapping solutions are used some of the sulfur species are absorbed, the sulfur analysis in these solutions are done by injecting these samples into the GC equipped with sulfur chemiluminescence detector (SCD).

A sampling technique to analyze H_2S and other gaseous sulfur molecules produced from biomass, peat or coal gasification process, was developed in Technical Research Centre of Finland (VTT) [48]. The principle is:

- The gas stream at the outlet of the gasifier is passed through a SiC sampling filter (ceramic filter) maintained at the process temperature and pressure to remove particulate matter. The product gas then enters into a short stainless steel tube at the beginning, which cools the gas to below 250 °C, and consequently passes through a Teflon-lined steel sampling tube.
- The secondary filter made of quartz is fitted in the sampling line to trap particulates that are escaped from the first filter and possibly damage the gas chromatograph.
- The process gas that exits the secondary filter is sent to analysis by online or offline gas chromatography. Offline analysis is handy when there is a condensation of tar in the product gas, as observed in wood gasification process. In this case, the product gas containing tar molecules condenses on the walls of PTFE bags evading the associated problem in GC analysis. It is recommended to analyze the samples collected bags within 5 h.

11.4.3.2 Analysis Method

The product gas sample from the gasifier passed through the filter and condenser, devoid of particles and tars is guided to an online GC (model 2014, Shimadzu, Kyoto, Japan) equipped with capillary column (Rtx-1, 60 m × 0.53 m × 7 μm,

Restek Corporation) and SCD detector (Sievers model 355. GE Analytical Instruments, Boulder, CO) to quantify the sulfur compounds [54].

- The sulfur response should be calibrated in GC before each test using 100 ppm H_2S in N_2 standard gas. Retention times for different sulfur compounds are identified before analyzing the product gas in GC, using a permeation tube gas generator (model CO395, Kin-Tek Laboratories, La Marque, TX) with tubes for thiophene, COS, benzothiophene and other appropriate sulfur compounds.
- The calibration of GC and retention times for various sulfur species are identified, after this step, in 10 min intervals, the samples are injected into GC by an automatic sampling.
- The diverse sulfur molecules separate inside the capillary column, based on their different affinity to the stationary phase coated onto the interior of the capillary column. Upon reaching the burner, all the sulfur species are converted to SO_2 molecules and are detected in the SCD detector.
- The sulfur species identified using GC-SCD is shown in Fig. 11.28.
- If the product gas is passed though tar-sampling impingers to eliminate tar molecules, the sample gas can be analyzed offline by GC-SCD.

GC (Varian GC 450) equipped with pulsed flame detector (PFPD) can also be used to detect gaseous sulfur molecules, namely, H_2S, COS and methyl mercaptan [55]. The calibration curves for three different sulfur species should be made before

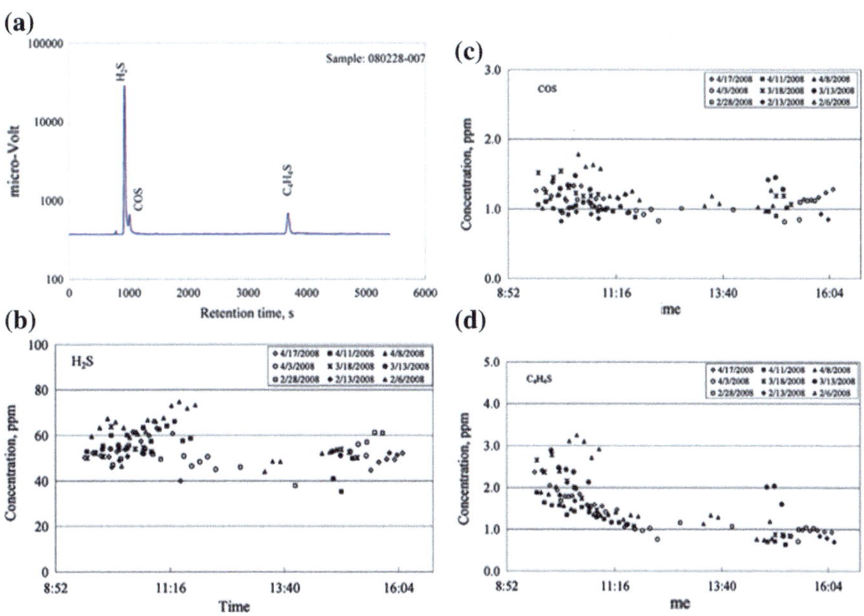

Fig. 11.28 Sulfur species identified by GC-SCD in the gas stream **a**, sulfur species concentration detected in gasification product gas (**b**, H_2S; c, COS; d, C_4H_4S; without N_2 deduction) [54]. Copyright © 2010, American Chemical Society

Table 11.16 Analysis method specification for GC-PFPD [55]

Item	Setup/specification
Column	CP-Sil 5CB, 50 m × 0.32 mm, df = 5 μm, CP7690
Temperature	Initial 50 °C for 2 min, ramping at 10 °C/min to 100 °C
Carrier gas	He, 2 mL/min, constant flow
Injector	Gas sample valve, sample loop 50 μL
Split ratio	1:50
Detector	PFPD, T = 250 °C
Combustion H_2	13 mL/min
Combustion air	17 mL/min (Air1), 10 mL/min (Air2)
Gate wide	10 ms
Gate delay	4 ms
Trigger level	200 millivolts

Reproduced with permission from X. Meng

performing the product gas analysis. For this purpose, gas cylinders (2000 ppm H_2S, 200 ppm COS, 50 ppm methyl mercaptan, N_2 to fill the balance) and (500 ppm H_2S, 200 ppm COS, N_2 to fill the balance) procured from Linde Gas BV Company are used. The analysis method with all the specifications for injector, column, and detector, in GC-PFPD to measure the sulfur species is given in Table 11.16.

11.4.4 Analysis of Halogens

Chloro compounds such as HCl, KCl and NaCl are the major halogenic species formed during the gasification process [44, 45]. HCl is extremely soluble and dissociate almost completely in aqueous solutions, which makes it easier to trap in acidic, neutral or basic solutions and these solutions can consequently be analyzed using ion chromatography to quantify halogenic species [48]. As HCl is an extremely corrosive gas and reacts easily with surfaces of gasifiers and downstream sampling components, therefore, it is quite challenging to precisely quantify HCl gas by online or offline analytical techniques. Small amounts of HCl gas in the gas stream can also be trapped using K_2CO_3-impregnated filters and consequently characterized using ion chromatography to determine the amount of chloride ions trapped on the filter [46].

11.4.4.1 Sampling Method

Some of the essential points to be considered while sampling halogenic (chloro) compounds are discussed below [48]. In general, the points discussed for sampling H_2S applies to HCl gas as well.

- Due to the extremely corrosive and reactive nature of HCl gas in the presence of moisture, it is recommended to use inert materials like quartz or PTFE in the sampling train.
- During the initial period of the gasification process, HCl formed may react with surfaces of the gasifier and downstream sampling components, therefore, before sampling the product gas, a stable operation of the reactor for long time (can be hours or days depending on various process parameters and feed nature) is usually required.
- Under the process conditions if both HCl and NH_3 are formed, it may combine to form NH_4Cl. To prevent the formation of such species, it is recommended to maintain the product gas at ca. 500 °C.
- If the HCl gas sampled is to be analyzed in IC, the most suitable solvents are distilled water and dilute H_2SO_4 (0.5% v/v). However, using NaOH as trapping solution requires the removal of Na^+ ions before performing IC analysis.

A sampling method to analyze halogenic species produced from biomass gasification process was developed in Technical Research Centre of Finland (VTT) [48]. The principle is:

- In this method, quartz liners are used inside the sampling lines including the parts that are under pressure. The only point where the product gas can encounter the bare metal surface is at the value chamber, which is the entry point for the product gas into the sampling system.
- The pressure is released at the value and sample gas is cooled and scrubbed by injecting purified halogen-free water into the opening of the quartz liner.
- After cooling, the sample gas with a flow rate of 5 dm_n^3/min is washed in four stages.
- The water injected into the liner is collected in the first washing stage which captures more than 98% of the total halogenic compounds from the sample gas, therefore, only the content from this stage is taken to analysis by a capillary electrophoresis system.

Another example for sampling method to analyze halogenic species produced from pyrolysis of wood and plastic is depicted in Fig. 11.29 [19]. The product gas leaving the pyrolysis reactor is bubbled through two impinger bottles containing 100 mL of 0.2 mol/L aqueous KOH solution for 30 min maintained at 0 °C, in order to trap halogenic species and taken to analysis in IC. The tar or particulate matter trapped along with chloride species in the KOH solution do not lead to any problem in IC analysis. More details concerning the sample preparation for IC analysis can be followed from sampling method for HCN (Sect. 1.2.1).

11.4.4.2 Analysis Method

The samples collected and prepared as described in the previous section can be analyzed using an Ion Chromatography System (ICS-3000 Dionex) enabled

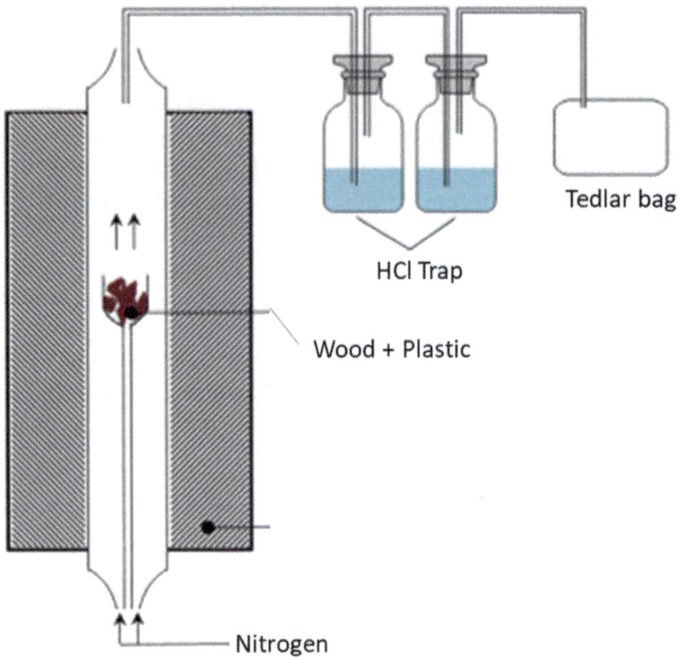

Fig. 11.29 Experimental setup of pyrolysis process, with HCl sampling train, adopted from thesis Ref. [19]. Reprinted with permission from A. Ephraim

quantification of HCl concentration in the aqueous phase [19]. More details about the IC analysis method can be found in the analysis method section of HCN (Sect. 1.2.2).

11.4.5 Solid Particulate Matter Collection

The product gas from the gasification process comprises of inorganic fuel residues (ash), residual solid carbon (non-reacted biomass also called as coalite), soot and other solid materials that are collectively represented as particulate matter (PM) [12, 56]. It has to be removed efficiently along with tar molecules to avoid associated problems such as abrasion, corrosion, and fouling of the sampling system as well as in the downstream applications. Cyclone separators, different types of filters, wet scrubbers, and electrostatic separators are used to trap the solid particles. PM is classified based on its aerodynamic diameter, for e.g., PM10 represents particles smaller than 10 μm and PM 2.5 represents particles smaller than 2.5 μm.

11.4.5.1 Sampling Method

The sampling method for particulate matter collection is discussed as follows [57, 58].

- The positioning of the components for particulate matter collection in the sampling system is depicted in module 2.
- The product gas coming out of the heated sampling probe (module 1) that comprises of PM are collected with cyclones and/or filters (module 2). Cyclones are useful in removing PM with diameter greater than 10 μm. Filters used are generally made of ceramic, polymeric, or natural fibers material.
- Quartz filters are used to collect the particles and majority of the particles (ca. 99.998%) should be trapped in these filters.
- The particles that bypass the filter are can be collected in the impinge bottles containing solvents (mostly isopropanol, others include water, aqueous acid/ base solutions) of the sampling train (module 3), however, tars are also dissolved in the solvent during the process.
- The type of the filters is chosen based on the particle concentration, for example, plate filters for particles with a minimum diameter of 90 mm for concentrations below 20 mg/m_n^3 and for concentrations above this range thimble filters are used.
- In case of unknown particle concentration thimble filters are recommended, because of the higher capacity and lower pressure drop, further, it is much easier to take a test sample and estimate the particle concentration.
- The choice of thimble filters also depends on the Soxhlet extraction performed thereafter. The recommended dimensions for the filter thimble are, diameter of ca. 30 mm and length of ca. 77 or 100 mm with the surface area of ca. 100 cm^2 that permits collection of several grams of particles. One must keep in mind the sample flow rate and nature of the tars (low temperature or high temperature) to decide about the filters choice.
- Preconditioning of plate or thimble filters involves drying them in an oven maintained at 110 °C under atmospheric pressure overnight. Later, the filters are removed from oven by wrapping it in aluminum foil and placed in a desiccator at room temperature. Upon cooling, the weight of the filters is measured (initial filter weight) using analytical balances with an accuracy of ±0.1 mg. This filter is now ready to be installed in the sampling to collect the particles.
- To avoid tar condensation on PM, the temperature at which PM are collected should be kept sufficiently high temperature (ca. 400 °C depending on the nature of tar formation during gasification).

11.4.5.2 Analysis Method

The particles collected in the filters are Soxhlet extracted prior to analysis, in order to remove tars that might have condensed. Thus extracted particles in the filters are dried and weighed (sampled filter weight) following the similar procedure that is described in the previous section. The mass of the particulate matter is calculated by subtracting the initial filter weight from the sampled filter weight. The particle concentration on the gasification product gas is reported in g/m_n^3, obtained by normalizing the volume of gas produced during the process [59]. Further these particles can be analyzed for its elemental composition and morphology by one or combination of the analytical techniques such as SEM-EDX, XRF, or ICP-MS/AES as described elsewhere [46, 58, 60].

11.5 Conclusion

In this chapter, different methods and equipment for sampling and analyzing syngas components have been presented. The syngas components have been broadly grouped as non-condensable gases (e.g. H_2, CO, CH_4 and CO_2), condensable gases (e.g. H_2O and tars) and impurities (e.g. halogens, NH_3, H_2S and HCN).

Both off-line and online techniques for syngas characterization have been discussed, and it is clear that standards that govern these techniques are largely inexistent, except for a few cases such as the 'Tar Protocol' for tar measurement. Thus, majority of the methods that have been presented here are from published works. Furthermore, the advantages of on-line techniques, including having real-time results, cost-savings and information about transient conditions, have accelerated their development by several research groups, as opposed to off-line methods. This trend is likely to continue in the foreseeable future due to the ever increasing pressures from emissions regulations and end-use device quality requirements.

For off-line sampling techniques, further research of effective and low-cost sorbents is required.

References

1. Richardson, Y., Blin, J., Julbe, A.: A short overview on purification and conditioning of syngas produced by biomass gasification: catalytic strategies, process intensification and new concepts. Prog. Energy Combust. Sci. **38**(6), 765–781 (2012)
2. Sikarwar, V.S., Zhao, M., Fennell, P.S., Shah, N., Anthony, E.J.: Progress in bio-fuel production from gasification. Prog. Energy Combust. Sci. **61**, 189–248 (2017)
3. Watson, J., Zhang, Y., Si, B., Chen, W.-T., de Souza, R.: Gasification of biowaste: a critical review and outlooks. Renew. Sustain. Energy Rev. **83**, 1–17 (2018)
4. Huber, G.W., Iborra, S., Corma, A.: Synthesis of transportation fuels from bio-mass: chemistry, catalysts, and engineering. Chem. Rev. **106**(9), 4044–4098 (2006)
5. Mota, N., Alvarez-Galvan, C., Navarro, R.M., Fierro, J.L.: Biogas as a source of renewable syngas production: advances and challenges. Biofuels **2**(3), 325–343 (2011)

6. Rossetti, I., Compagnoni, M., Finocchio, E., Ramis, G., Di Michele, A., Zucchini, A., Dzwigaj, S.: Syngas production via steam reforming of bioethanol over Ni–BEA catalysts: a BTL strategy. Int. J. Hydrog. Energy **41**(38), 16878–16889 (2016)
7. Zhang, Y., Brown, T.R., Hu, G., Brown, R.C.: Comparative techno-economic analysis of biohydrogen production via bio-oil gasification and bio-oil reforming. Biomass Bioenergy **51**, 99–108 (2013)
8. Chew, J.J., Doshi, V.: Recent advances in biomass pretreatment-Torrefaction fundamentals and technology. Renew. Sustain. Energy Rev. **15**(8), 4212–4222 (2011)
9. Zhang, W.: Automotive fuels from biomass via gasification. Fuel Process. Technol. **91**(8), 866–876 (2010)
10. Richardson, Y., Drobek, M., Julbe, A., Blin, J., Pinta, F.: Chapter 8—Biomass Gasification to Produce Syngas. In: Pandey, A., Bhaskar, T., Stöcker, M., Sukumaran, R.K. (eds.) Recent Advances in Thermo-Chemical Conversion of Biomass, pp. 213–250. Elsevier, Boston (2015)
11. Rios, M.L.V., González, A.M., Lora, E.E.S., del Olmo, O.A.A.: Reduction of tar generated during biomass gasification: a review. Biomass Bioenergy **108**, 345–370 (2018)
12. Woolcock, P.J., Brown, R.C.: A review of cleaning technologies for biomass-derived syngas. Biomass Bioenergy **52**, 54–84 (2013)
13. Bartels, M., Lin, W., Nijenhuis, J., Kapteijn, F., van Ommen, J.R.: Agglomeration in fluidized beds at high temperatures: mechanisms, detection and prevention. Prog. Energy Combust. Sci. **34**(5), 633–666 (2008)
14. DD CEN/TS 15439: Biomass gasification—Tar and particles in product gases—Sampling and analysis. https://shop.bsigroup.com/ProductDetail/?pid=000000000030163214 (2006) Accessed Mar 2007
15. Sigma-Aldrich: Gas sampling bags and samplers. https://www.sigmaaldrich.com/analytical-chromatography/air-monitoring/gas-sampling-bags.html#foilbags. Accessed Mar 2018
16. Restek: Sacs Tedlar. http://www.restek.fr/catalog/view//22052-3795. Accessed March 2018
17. Poole, C.F.: Gas Chromatography. 1st edn. Elsevier (2012)
18. Ephraim, A.: Valorization of wood and plastic waste by pyro-gasification and syn-gas cleaning. Ph.D. Thesis, Ecole nationale des Mines d'Albi-Carmaux, France (2016)
19. Andersson, R., Boutonnet, M., Järås, S.: On-line gas chromatographic analysis of higher alcohol synthesis products from syngas. J. Chromatogr. A **1247**, 134–145 (2012)
20. Rauch, R., Bosch, K., Hofbauer, H., Świerczyński, D., Courson, C., Kiennemann, A.: Comparison of different olivines for biomass steam gasification, pp. 799–809. CPL Press (2004)
21. Kübel, M.: Teerbildung und Teerkonversion bei der Biomassevergasung—Anwen-dung der nasschemischen Teerbestimmung nach CEN-Standard. Cuvillier Verlag (2007)
22. Neubert, M., Reil, S., Wolff, M., Pöcher, D., Stork, H., Ultsch, C., Meiler, M., Mes-ser, J., Kinzler, L., Dillig, M., Beer, S., Karl, J.: Experimental comparison of solid phase adsorption (SPA), activated carbon test tubes and tar protocol (DIN CEN/TS 15439) for tar analysis of biomass derived syngas. Biomass Bioenergy **105**, 443–452 (2017)
23. Brage, C., Sjöström, K.: An outline of R&D work supporting the tar guideline. Royal Institute of Technology (KTH). Technical report (2002)
24. Prando, D., Shivananda Ail, S., Chiaramonti, D., Baratieri, M., Dasappa, S.: Characterisation of the producer gas from an open top gasifier: assessment of different tar analysis approaches. Fuel **181**, 566–572 (2016)
25. van de Kamp, W., de Wild, P., Zielke, U., Suomalainen, M., Knoef, H., Good, J., Liliedahl, T., Unger, C., Whitehouse, M., Neeft, J., van de Hoek, H., Kiel, J.: Tar measurement standard for sampling and analysis of tars and particles in biomass gasification product gas. In: 4th European Biomass Conference and Exhibition, Paris, France (2015)
26. Grootjes, A. J.: Tar measurement by the Solid Phase Adsorption (SPA) method. In: 19th European Biomass Conference and Exhibition (2011)

27. Israelsson, M., Seemann, M., Thunman, H.: Assessment of the solid-phase adsorption method for sampling biomass-derived tar in industrial environments. Energy Fuels **27**(12), 7569–7578 (2013)
28. Osipovs, S.: Comparison of efficiency of two methods for tar sampling in the syn-gas. Fuel **103**, 387–392 (2013)
29. Engvall, K., Whitty, K.: Measurement and characterization of tars using the spa method: on-going developments on tar analysis. In: Workshop IEA Bioenergy Task 33, Lucerne University of Applied Sciences (2016)
30. Hasler, P., Nussbaumer, T.: Sampling and analysis of particles and tars from bio-mass gasifiers. Biomass Bioenergy **18**(1), 61–66 (2000)
31. Brage, C., Yu, Q.C., Chen, G., Sjöström, K.: Tar evolution profiles obtained from gasification of biomass and coal. Biomass Bioenergy **18**(1), 87–91 (2000)
32. Brage, C., Yu, Q.C., Chen, G., Sjöström, K.: Use of amino phase adsorbent for biomass tar sampling and separation. Fuel **76**(2), 137–142 (1997)
33. Dufour, A., Girods, P., Masson, E., Normand, S., Rogaume, Y., Zoulalian, A.: Comparison of two methods of measuring wood pyrolysis tar. J. Chromatogr. A **1164**(1), 240–247 (2007)
34. Reil, S., Karl, J., Beer, S.: Impact of gasification conditions on the tar yield of stratified downdraft gasifiers. In: Papers of the 22nd European Biomass Conference, 585–590 (2014)
35. Osipovs, S.: Sampling of benzene in tar matrices from biomass gasification using two different solid-phase sorbents. Anal. Bioanal. Chem. **391**(4), 1409–1417 (2008)
36. González, I.O., Pastor, R.M.P., Hervás, J.M.S.: Sampling of tar from sewage sludge gasification using solid phase adsorption. Anal. Bioanal. Chem. **403**(7), 2039–2046 (2012)
37. Sun, R., Zobel, N., Neubauer, Y., Cardenas Chavez, C., Behrendt, F.: Analysis of gas-phase polycyclic aromatic hydrocarbon mixtures by laser-induced fluorescence. Opt. Lasers Eng. **48** (12), 1231–1237 (2010)
38. Carpenter, D.L., Deutch, S.P., French, R.J.: Quantitative measurement of biomass gasifier tars using a molecular-beam mass spectrometer: comparison with traditional impinger sampling. Energy Fuels **21**(5), 3036–3043 (2007)
39. Fendt, A., Streibel, T., Sklorz, M., Richter, D., Dahmen, N., Zimmermann, R.: On-line process analysis of biomass flash pyrolysis gases enabled by soft photoionization mass spectrometry. Energy Fuels **26**(1), 701–711 (2012)
40. Neubauer, Y.: Online-analyse von teer aus der biomassevergasung mit lasermassen-spektrometrie. Ph.D. Thesis, Technische Universität Berlin, Berlin (2008)
41. Defoort, F., Thiery, S., Ravel, S.: A promising new on-line method of tar quantification by mass spectrometry during steam gasification of biomass. Biomass Bioenergy **65**, 64–71 (2014)
42. Ahmadi, M., Knoef, H., Van de Beld, B., Liliedahl, T., Engvall, K.: Development of a PID based on-line tar measurement method—Proof of concept. Fuel **113**, 113–121 (2013)
43. Gredinger, A., Spörl, R., Scheffknecht, G.: Comparison measurements of tar content in gasification systems between an online method and the tar protocol. Biomass Bioenergy **111** (2017)
44. Abdoulmoumine, N., Adhikari, S., Kulkarni, A., Chattanathan, S.: A review on biomass gasification syngas cleanup. Appl. Energy **155**, 294–307 (2015)
45. Chiche, D., Diverchy, C., Lucquin, A.-C., Porcheron, F., Defoort, F.: Synthesis Gas Purification. Oil Gas Sci. Technol. Rev. D'IFP Energ. Nouv **8**(4), 707–723 (2013)
46. Hoekman, S.K., Robbins, C., Wang, X., Zielinska, B., Schuetzle, D., Schuetzle, R.: Characterization of trace contaminants in syngas from the thermochemical conversion of biomass. Biomass Convers. Biorefinery **3**(4), 271–282 (2013)
47. Norton, G.A., Brown, R.C.: Wet chemical method for determining levels of ammonia in syngas from a biomass gasifier. Energy Fuels **19**(2), 618–624 (2005)
48. Stahlberg, P., Lappi, M., Kurkela, E., Simell, P., Oesch, P.: Sampling of Contaminants from Biomass Gasifiers. IEA Thermal Gasification of Biomass Task. Brussels, Belgium (1998)
49. Zhou, J., Masutani, S.M., Ishimura, D.M., Turn, S.Q., Kinoshita, C.M.: Release of fuel-bound nitrogen during biomass gasification. Ind. Eng. Chem. Res. **39**(3), 626–634 (2000)

50. De Jong, W., Ünal, Ö., Andries, J., Hein, K.R.G., Spliethoff, H.: Biomass and fossil fuel conversion by pressurised fluidised bed gasification using hot gas ceramic filters as gas cleaning. Biomass Bioenergy **25**(1), 59–83 (2003)
51. Agilent Technologies Inc.: Dual Plasma Sulfur and Nitrogen Chemiluminescence Detectors Dual Plasma Technology Dual Plasma Burner and Controller. https://www.agilent.com/cs/library/selectionguide/public/5989-6119EN.pdf (2006). Accessed 9 Jan 2018
52. Broer, K.M., Johnston, P.A., Haag, A., Brown, R.C.: Resolving inconsistencies in measurements of hydrogen cyanide in syngas. Fuel **140**, 97–101 (2015)
53. Vriesman, P., Heginuz, E., Sjöström, K.: Biomass gasification in a laboratory-scale AFBG: influence of the location of the feeding point on the fuel-N conversion. Fuel **79**(11), 1371–1378 (2000)
54. Cui, H., Turn, S.Q., Keffer, V., Evans, D., Tran, T., Foley, M.: Contaminant estimates and removal in product gas from biomass steam gasification. Energy Fuels **24**(2), 1222–1233 (2010)
55. Meng, X.: Biomass gasification: The understanding of sulfur, tar, and char reaction in fluidized bed gasifiers. In: TU Delft Repositories, Delft University of Technology (2012)
56. Balas, M., Lisy, M., Skala, Z., Jiri, P.: Wet scrubber for cleaning of syngas from bi-omass gasification. Adv. Environ. Sci. Dev. Chem. 195–201 (2014)
57. ECN: Biomass gasification—Tar and particles in product gases—Sampling and analysis. http://www.eeci.net/results/pdf/CEN-Tar-Standard-draft-version-2_1-new-template-version-05-11-04.pdf (2004). Accessed 10 Oct 2004
58. Gustafsson, E.: Characterization of particulate matter from atmospheric fluidized bed biomass gasifiers. Doctoral Dissertation, Linnaeus University (2011)
59. Akudo, C.O., Theegala, C.S.: Quantification of tars, particulates, and higher heating values in gases produced from a biomass gasifier. BioResources **9**(3), 5627–5635 (2014)
60. Wang, Robbins, C., Hoekman, S.K., Chow, J.C., Watson, J.G., Schuetzle, D.: Dilution sampling and analysis of particulate matter in biomass-derived syngas. Front. Environ. Sci. Eng. China **5**(3), 320–330 (2011)

Chapter 12
Condensable and Liquid Compounds from Biomass and Waste Thermal Degradation

Capucine Dupont, Andrés Anca-Couce, Jean-Michel Commandré, Alba Dieguez-Alonso, Thierry Ghislain, Maria Gonzalez Martinez and Jean-Michel Lavoie

Abstract This chapter deals with condensable/liquid compounds produced during biomass and waste thermal degradation processes. These compounds include a huge number of species, mainly organic, but also inorganic. These species may vary according to biomass or waste type, and even more according to process operating conditions, namely temperature, heating rate, reactive gas or pressure. For instance, liquids obtained during torrefaction and their related yields will be strongly different from those obtained during slow pyrolysis, fast pyrolysis or gasification. A general overview is therefore firstly given of the condensable/liquid compounds produced in each of the main thermal processes together with their typical mass yields. Then the emphasis is put on the challenging task of determination of composition in organic and inorganic species. The most common measurement techniques of these condensable/liquid compounds are detailed. For each technique, the principle and a typical experimental procedure are given, as well as the main advantages and drawbacks, notably regarding species detected, accuracy limits and thus suitability with the different thermal processes. Both online analysis methods, and offline

C. Dupont (✉) · M. Gonzalez Martinez
Department of Environmental Engineering and Water Technology, IHE Delft Institute for Water Education, Westvest 7, 2611 AX Delft, The Netherlands
e-mail: c.dupont@un-ihe.org

A. Anca-Couce
Institut für Wärmetechnik/Institute of Thermal Engineering, TU Graz, Inffeldgasse 25/B, 8010 Graz, Austria

J.-M. Commandré
Biomass Energy Unit, CIRAD, 73 Rue Jean-François Breton, 34398 Montpellier Cedex 5, France

A. Dieguez-Alonso
Institute of Energy Engineering, Technische Universität Berlin, Fasanenstrasse 89, 10623 Berlin, Germany

T. Ghislain · J.-M. Lavoie
Department of Chemical Engineering and Biotechnological Engineering, Université de Sherbrooke, 2500, Boulevard de L'Université, Sherbrooke J1K 2R1, Canada

A. Nzihou (ed.), *Handbook on Characterization of Biomass, Biowaste and Related By-products*, https://doi.org/10.1007/978-3-030-35020-8_12

methods are considered. In the case of offline analysis, particular attention is paid to the sample collection step, for instance through condensation in staged cooled vessels.

12.1 Introduction

Biomass/waste thermal degradation processes produce a broad range of products, which can be lumped into a solid phase called char, permanent gases and a complex mixture of volatile compounds which condense at ambient temperature and pressure conditions. The latter includes water and over 100 oxygenated hydrocarbons. The obtained yields of each product strongly depend on the employed thermal process [1–4] (see Table 12.1).

The liquid bio-oil can be employed for power or heat generation, upgraded to a liquid fuel for combustion engines [5], used as a biological pesticide [6], and different fractions can be used for binders, adhesives roofing and smoke flavours [7], or to be a promising source for chemicals in a bio-refinery concept [8, 9].

The obtained bio-oil, when condensed, is a mixture of water and hundreds of organic condensable molecules. A typical dry bio-oil composition, when produced at moderate temperatures (<600 °C) is $CH_{1.3}O_{0.5}$ [5], with a small amount of nitrogen and even smaller amounts of sulphur and chlorine [10]. The mixture includes from light condensable species, mainly carbonyls and alcohols, to heavy condensable species, like some phenolic compounds, sugars or heterocyclics part of furans [10, 11]. Concerning the organic molecules present in the bio-oil, higher molecular weights lead to a lower saturation pressure and a higher boiling point [12]. The content of nitrogen, sulphur and chlorine depends to a significant extent on the initial biomass composition.

12.1.1 *Fast Pyrolysis*

Fast pyrolysis at temperatures around 500 °C maximizes the production of liquid compounds. This process requires as well a fast heating rate, a low residence time of the volatile species, controlled (generally low) initial moisture content of biomass and commonly a small particle size [3]. Fluidized bed, rotative cone or ablative

Table 12.1 Typical yields of main products from dry biomass in different thermal processes

	Char (%)	Liquids (%)	Gases (%)
Fast pyrolysis	15	70	15
Slow pyrolysis	30	40	30
Torrefaction	70	20	10
Gasification	<5	<5 (without water)	>90

reactors are usually employed to achieve these conditions. A liquid yield of around 70% can be obtained for fuels with low ash content, while this yield decreases for a higher ash content (under 50% for an ash content over 2%) [13]. Bio-oil from fast pyrolysis is a very viscous (in the order of tens of cSt) and dense (1.2 kg L^{-1}) liquid, which has low pH and calorific value (<20 MJ kg^{-1}) [5]. A typical composition of that oil is shown in Table 12.2. The most relevant compounds besides moisture are [10, 13, 14]:

- Pyrolytic lignin: oligomers (di-, tri- and tetramers are the most abundant) of single ring phenolic units (aromatic compounds that contain a phenyl group), which are produced from pyrolysis of lignin.
- Sugars produced from the rearrangement of macromolecular units from cellulose and hemicelluloses.
- Carbonyls and alcohols, which are the main unit rings scission products from cellulose, hemicelluloses and lignin degradation. The most relevant carbonyls are acids (which account to around 5% of bio-oil from fast pyrolysis and lead to the low pH), aldehydes or ketones. Other relevant scission products are heterocyclic compounds such as furans (produced from cellulose and hemicelluloses) and one-ring phenolic units (produced from lignin).

The fast pyrolysis bio-oil from woody biomass has commonly solely one phase. The discontinuous and dispersed pyrolytic lignin, which is water insoluble, is stabilized by water and other organic molecules forming a micro-emulsion (mixture of immiscible liquids). Phase separation can take place when the initial content of ash in biomass is higher than 2%. At the bottom a heavy oil phase is formed, mainly composed of non polar compounds as pyrolytic lignin [13]. Small amounts of solid particles (such as char, ash or sand) can be also found in bio-oil.

12.1.2 *Slow Pyrolysis*

Slow pyrolysis at moderate temperatures (around 500 °C) increases the production of char. Char yield is further enhanced with higher residence times of the volatiles, larger particle size, higher pressure and higher initial ash content [15]. The typically employed reactor types are fixed and slow-moving beds, screw reactors or rotary kilns [16]. The production of liquids is however reduced in comparison to fast

Table 12.2 Typical composition in mass percent of fast pyrolysis bio-oil

	Concentration (%)
Water	25
Sugars	30
Pyrolytic lignin	20
Carbonyls and alcohols	20
Others	5

pyrolysis. Liquids composition is similar to fast-pyrolysis bio-oil (FPBO) one. However, there are significant differences that mainly arise from secondary charring reactions, which take place to a higher extent in this case. Water content is higher, due to a higher production of pyrolytic water, and sugar content is significantly lower, due to secondary reactions and catalytic effect of some inorganic elements [11, 17]. Moreover, due to the higher water content phase, separation usually takes place in slow pyrolysis bio-oils.

When pyrolysis is conducted at higher temperatures (>600 °C) there is a significant reduction of the liquids yield due to organic molecules cracking. The cracking of bio-oil oxygenated molecules produced during pyrolysis at lower temperatures, also called primary tars, produces mainly permanent gases (CO, H_2 and light hydrocarbons) [18] and secondary and tertiary tars. Secondary tars are mainly phenolics and tertiary tars are aromatic hydrocarbons without oxygen with one ring (benzene, toluene and xylene: BTX) and several rings (polyaromatic hydrocarbons: PAH) [15, 19].

12.1.3 Torrefaction

Torrefaction is a partial pyrolysis process conducted at low temperatures, usually up to 300 °C. In this process, some properties of solid product are improved in comparison to original biomass. Torrefied biomass has a higher energy density and it is easier to store, thanks to its hydrophobic aspect, it has a low biological degradation and a good grindability, which facilitates pelletization [20]. The employed reactor configurations are similar to those for slow pyrolysis [16, 21]. The main liquid compounds produced are water, acetic acid and other carbonyls alcohols. Heavier condensable species are produced in lower quantities [2, 21–23].

12.1.4 Gasification

Gasification processes lead to a gas production which can be used for efficient production of heat and power, different liquid and gas fuels (e.g. Fischer-Tropsch diesel, synthetic natural gas) or chemical organic compounds. Gasification reactions are carried out with an oxidizing agent (air, oxygen or steam) and at high temperatures, usually at least 800 °C [24, 25]. Reactor configuration usually depends on the power range. Fixed-bed reactors with updraft or downdraft configuration are employed for small scales (up to 5 MW_{th}), fluidized bed reactors for medium scales (up to 50 MW_{th}) and entrained flow reactors only for large scales. Under gasification conditions, high temperatures lead to a very low yield of liquids, excluding water. The obtained liquids are mainly tertiary tars (see slow pyrolysis) which, despite their low yield, are commonly a major concern since they can condense during further steps.

In this chapter, the focus is put on the challenging task of determination of composition in organic and inorganic species among the liquids produced during thermal processes presented above. For a complete characterization several techniques should be combined [14, 26]. Methods that will be described are divided into two groups:

- Online analysis;
- Offline analysis.

The most common measurement techniques of these condensable/liquid compounds are detailed below. For each technique, the principle, typical experimental procedure, are given, as well as the main advantages and drawbacks, notably in terms of species detected, accuracy limits and thus suitability with the different thermal processes.

12.2 Online Analysis of Composition

12.2.1 Fourier Transformed Infrared (FTIR) Spectroscopy

12.2.1.1 Principle

Fourier transform infrared (FTIR) spectroscopy is based on the absorption of infrared radiation. Molecules absorb this radiation at specific wavelengths, changing the dipole momentum and modifying the vibrational energy level. The wavelengths of absorbed radiation match the vibrational frequency. FTIR is commonly conducted in the mid infrared region (2.5–25 μm, 4000–400 cm^{-1}) and most of the molecules absorb radiation, being the main exceptions diatomic molecules as O_2, N_2 and H_2 [27].

An FTIR spectrometer consists on the infrared source, a Michelson interferometer, a detection and analysis system. In the Michelson interferometer a beam is split into two and each resulting beam is reflected in a mirror. There is one fixed and one movable mirror. The two beams recombine at the beamsplitter and are sent to the detector. The mirror displacement of the movable mirror leads to a difference in optical path length between the two beams, resulting in an interference due to the two signals superposition. The signal is measured by the detector for different values of path differences. The signal is then amplified, digitalized and afterwards the Fourier transformation is carried out in order to obtain a function of the frequency.

12.2.1.2 Experimental Procedure

The FTIR spectrometer delivers a spectrum such as the one shown in Fig. 12.1. The main vibration modes and their wavenumber ranges of the most relevant molecules

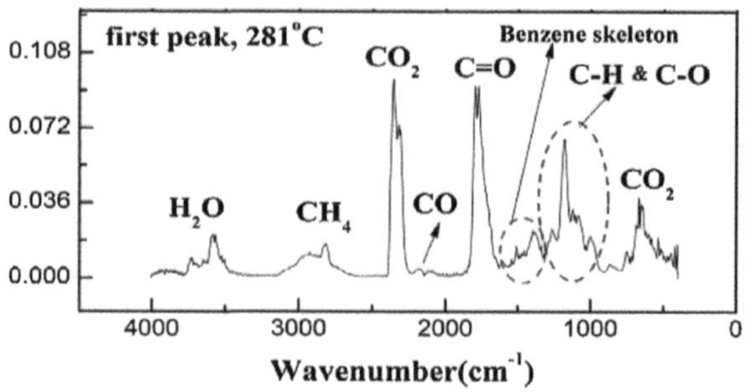

Fig. 12.1 FTIR spectrum palm kernel shell (PKS) pyrolysis volatiles at 281 °C and heating rate of 20 °C/min. Reprinted with permission from Elsevier [28]

Table 12.3 Main vibration modes and their wavenumber ranges of the most relevant molecules (only the main absorption area is indicated, absorption takes place as well in other areas) [28–30]

Species	Wavenumbers (cm^{-1})	Functional groups and vibrations modes
H_2O	4000 ~ 3400	O–H stretching
CH_4	3200 ~ 2700	C–H stretching
CO_2	2400 ~ 2250	C=O stretching
CO	2250 ~ 2000	C–O stretching
Carbonyls	1900 ~ 1650	C=O stretching
Aromatics	1690 ~ 1450	C=C and benzene skeleton stretching
Alcohols, phenolic compounds	1300 ~ 1000	C–O, C–C and carbon chain skeleton stretching

are shown in Table 12.3. The main permanent gases, CO, CO_2 and CH_4, have a very limited number of absorption peaks. Condensable species produced during biomass pyrolysis show absorption on a broader area. Analysis can become complex due to relevant overlaps appearing between spectra of some molecules. This is the case in particular among compounds having the same functional group in their structure, as the technique is actually based on the frequencies of vibrations between the bonds measurement.

For analysis, reference spectra for the considered species are required. At least one reference spectrum for each molecule is needed. If the concentration versus absorbance of a compound is not linear, it may be necessary to include several spectra at different concentrations. This is particularly the case for H_2O or CO. Nonlinearity correction factors may also be needed. To determine the product composition, different spectral ranges—from one to three areas—are commonly considered for each component. The representative regions (that means not too narrow and in presence of major peak absorption) of each species spectrum have to be selected. After analysis, the remaining areas of the spectrum should be checked in which absorbance has not been assigned to any compound, in order to verify that the characterization is conducted to a significant extent.

12.2.1.3 Advantages and Limits

The main advantage of FTIR measurements is that it is possible to detect quantitatively several volatile molecules released during pyrolysis, including permanent gases and some condensable species, with a non-intrusive method. The FTIR measurement can be conducted with commercial devices at temperatures that avoid condensation during the measurement, although this may induce to dilute the gas sample. Moreover, a full spectrum can be fast recorded (even in the range of milliseconds) allowing an online measurement of the pyrolysis process.

The main disadvantage of FTIR measurements is the fact that a complete characterization of pyrolysis volatiles is not possible. The most relevant species that are commonly measured are CO, CO_2, CH_4 and other light hydrocarbons, nitrogen species (HCN, NH_3 and HNCO), H_2O, and relevant carbonyls, alcohols and phenolic species. The analysis of organic condensable species is usually limited to light condensable species. Spectra are not available for every molecule released in pyrolysis and, besides, there are significant overlaps in the spectra of several molecules, especially the ones with the same functional group, which makes the analysis of a great number of species extremely complex, if not impossible.

FTIR is often applied in combination with thermo-gravimetric analysis (TGA) to measure mass loss and product composition during pyrolysis with samples of a few mg to achieve an intrinsic kinetic regime. TGA-FTIR has been applied for several biomass samples, including woody biomass [31, 32] and other kind of biomass [33, 34] as well as the main biomass macromolecular components, i.e. cellulose, hemicelluloses and lignin [30, 34, 35]. As previously mentioned, the analysis is limited to the main permanent gases, including in some cases nitrogenated compounds, H_2O and light condensable organic species such as acetic acid, formic acid, formaldehyde, acetone, methanol or phenol [31, 34], or a group of species determined at a defined wavelength range (e.g. carbonyls) [28].

FTIR has been as well applied to describe online the release of volatiles in single particle pyrolysis experiments, considering the main permanent gases, H_2O and four [36] or six [37] carbonyls and alcohol species; as well as in fixed-bed slow pyrolysis [38, 39] and torrefaction [2, 24].

12.2.2 *Laser-Induced Fluorescence (LIF)*

12.2.2.1 Principle

As defined by Lakowicz [40], fluorescence is a type of luminescence that occurs when a compound returns from an electronically excited state (the electron in the excited orbital is in an excited singlet state) to the ground state, emitting light. This process is very fast (spin allowed), being the fluorescence lifetimes in the order of 10^{-8} s [40, 41]. The excitation is produced by previous absorption of light with wavelengths in the absorption band of the species [40], having the emitted light typically less energy (due to the faster vibrational relaxation, with excited vibrational state lifetimes typically in the order of 10^{-12} s) and, therefore, longer wavelengths than the absorbed light, also known as Stokes shift [40, 41].

In order to apply this technique to characterize different volatile species, it is necessary to know which compounds are susceptible to emit fluorescence and in which conditions. LIF has been widely used in combustion to track the evolution of species, such as OH or NO [42] within flames. CO can also be detected by two-photon LIF, as shown by Brackmann et al. [43] in a flame and Prins et al. [44] in fast pyrolysis. LIF of oxygenated compounds as acetaldehyde, acetone or other aldehydes and ketones has been used for fuel imaging in combustion systems, such as internal combustion engines, as reviewed by Kohse-Höinghaus [45]. However, the interest of applying LIF spectroscopy to biomass/waste thermochemical conversion processes lies on the possibility of characterizing aromatic species, such as polycyclic aromatic hydrocarbons (PAHs), phenolics and BTXs (benzene, toluene, xylene), main components of the tar fraction. These aromatic molecules are typical species emitting fluorescence [40], having also strong disposition to absorb ultraviolet (UV) radiation [42]. Besides the excitation wavelength, important properties influencing the excitation and fluorescence emission of fluorophores include the molar attenuation coefficient (also called molar absorptivity) of each species at the applied excitation wavelength, the quantum yield, and the fluorescence lifetime [41]. The molar absorptivity is a measure of the probability that absorption (of light at the specific wavelength) takes place; the quantum yield is defined as the ratio of radiative to non-radiative decay rate or, in other words, the number of emitted photons in the form of fluorescence signal to the number of absorbed photons [40]; the fluorescence lifetime is a measure of the electronic relaxation time of a molecule and, therefore, of the available time for interaction or diffusion in the environment [40] and its measurement can give information about fluorescence deactivation processes, energy transfer or reactions in the excited state [41]. These properties are affected by excitation wavelength, temperature and environment composition due to the potential presence of quenching phenomena such as: (i) collisional quenching, due to the presence of quencher substances—oxygen, halogens, heavy atoms, (ii) static quenching, due to the formation of new compounds between the fluorophores and the quenchers or (iii) non-molecular quenching, due to decrease in the excitation light power by other absorbing species or the fluorophore itself [40, 41].

In Fig. 12.2, the absorption cross section (closely related to molar absorptivity) of different aromatic species (phenol, naphthalene, anthracene, pyrene) and formaldehyde-d2 is presented versus the wavelength for different temperatures. It is possible to see the variation of absorption cross section with the wavelength as well as the influence of temperature.

In Fig. 12.3 (left), the normalized fluorescence spectra for phenol, naphthalene, phenanthrene, anthracene, and pyrene are presented, obtained with an excitation

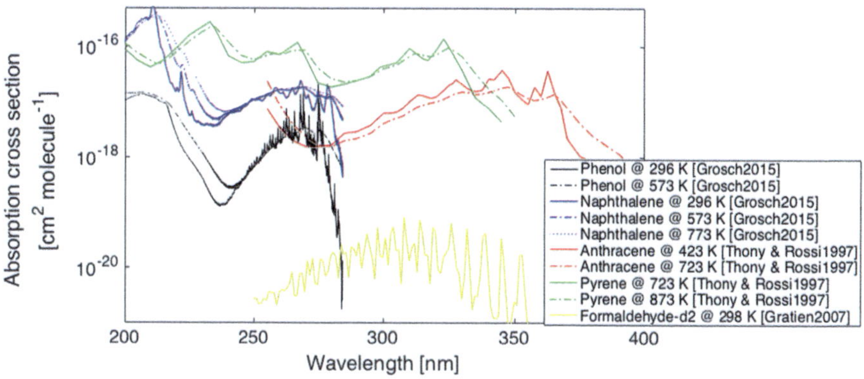

Fig. 12.2 Absorption cross section versus wavelength for different species. The data are taken from [46]. Phenol spectra were obtained by Grosch et al. [47] using phenol in nitrogen with a stable concentration, a high temperature gas flow cell, a deuterium lamp as light source and a UV grating spectrometer setup. Naphthalene spectra were obtained by Grosch et al. [47] using naphthalene in nitrogen with a stable concentration and the same setup as before. Anthracene and pyrene absorption spectra were obtained by Thöny and Rossi [48] in an air atmosphere with a deuterium lamp. The formaldehyde-d2 spectra was obtained by Gratien et al. [49]

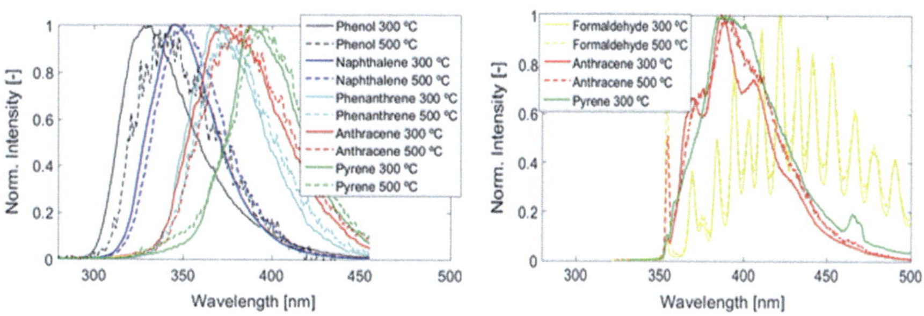

Fig. 12.3 Spectra of pure species in N_2 atmosphere. Left: Spectra obtained with an excitation wavelength of 266 nm. Right: Spectra obtained with an excitation wavelength of 355 nm

wavelength of 266 nm. In Fig. 12.3 (right), the normalized fluorescence spectra for formaldehyde, anthracene, and pyrene, with an excitation wavelength of 355 nm, are shown. In both cases, the fluorescence spectra were obtained at 300 and 500 °C in N_2 atmosphere. These spectra were obtained with the LIF spectroscopy setup and measurement cell presented in [50]. It is possible to see that as the number of rings in the aromatic species increases, the fluorescence signal is shifted to longer wavelengths (red-shift). This behaviour is very useful to identify the different PAHs groups (according to the number of rings) from a global fluorescence signal [42], especially when measuring these species in the gas phase and at high temperatures, due to the rather broad and featureless fluorescence spectra. Besides, it is possible to see that higher measurement temperatures generally lead to a small red-shift in the fluorescence signal (Fig. 12.3, left). It is also shown how the excitation wavelength plays a very important role. For example, with 355 nm excitation wavelength, phenol and naphthalene are not detected (see Fig. 12.3, right). This is due to the fact that the excitation wavelength is too long (energy too low) to favour the transition of the molecule to an electronically excited state. In other words, this excitation wavelength is outside the absorption band of the species. As it has been previously discussed, the emitted light (fluorescence) has typically lower energy (longer wavelengths) than the absorbed light due to vibrational relaxation and potential internal conversion. Furthermore, it is possible to see that formaldehyde can be detected with an excitation wavelength of 355 nm, but not with 266 nm. Both excitation wavelengths are within the absorption band of the species, but the absorption cross section is significantly higher for 355 nm than for 266 nm [46].

The distinctive absorption cross sections and emission wavelengths of the different aromatic species can be used therefore to increase the selectivity of the LIF spectroscopy tool. On top, fluorescence lifetime can also contribute to gain further selectivity when the signal of different species overlap and no discrimination can be achieved using different excitation wavelengths. Moreover, the fluorescence emission spectrum does not change substantially, especially the position, with different excitation wavelengths, due to the rapid vibrational relaxation in comparison to the electronic relaxation, although some exceptions exist [40].

In Fig. 12.4 (left), a typical fluorescence spectrum from a mixture of aromatic species is presented. The spectrum was obtained during a pyrolysis experiment at particle level (particle reactor), at a temperature of 350 °C, using the setup introduced in [50]. Integration of the fluorescence spectra for each time step according to the equation below gives the evolution of the total fluorescence intensity (TFI) over time.

$$TFI = \int_{\lambda o}^{\lambda f} I(\lambda) d\lambda \tag{12.1}$$

In Fig. 12.4, such evolution is presented qualitatively (normalized signal) for a pyrolysis process (right top), enabling to track the release of aromatic species, and a gasification process (right bottom [51, 52]) respectively. In the latter, a mobile LIF spectroscopy sensor for total PACs (polycyclic aromatic compounds) content monitoring, developed at the TU Berlin, was used connected to the gasifier at the Chalmers University of Technology (Sweden). The light source used was an array of light emitting diodes, allowing to reach higher sensitivity in species discrimination, in contrast to only one wavelength. This system is a further improvement of a mobile LIF sensor, also developed at TU Berlin [53], using a Nd:YAG laser with a wavelength of 266 nm.

A similar portable LIF system was developed at TU Munich [54], using a pulsed nitrogen laser with an excitation wavelength of 337.1 nm, a pulse duration of 3 ns and a repetition rate of 60 Hz, with a maximum average power of 10 mW.

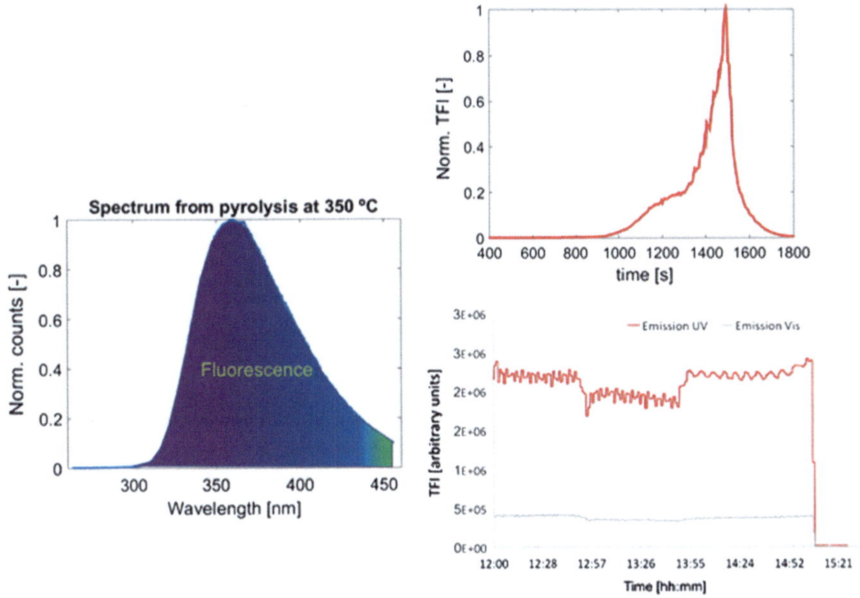

Fig. 12.4 Left: Example of a fluorescence spectrum of a mixture of aromatic species obtained with an excitation wavelength of 266 nm, under pyrolysis conditions at 350 °C. Right top: Temporal evolution of the TFI for a pyrolysis experiment. Right bottom: Temporal evolution of the TFI for a gasification experiment (reproduced with permission from [52])

The emission spectra were detected with a Charge Coupled Device (CCD) camera (FlameStar 2F) connected to a spectrograph (Chromex 250is). Noteworthy to mention in this work is the quantitative calibration of the LIF spectroscopy system with 14 different species (fluorescence intensity vs. species concentration), including phenolics and PAHs. The system was validated for synthetic tar mixtures, as in [52], and for real gasification gas, comparing tar species concentrations derived from the LIF spectroscopy system with the concentrations measured with a SPA (solid phase adsorption) system, obtaining good agreement except for toluene. The same system has been tested at different gasifiers [55], showing quite good correlation between LIF measurements and SPA measurements for different tar species concentration. The concentration of the different species in the gas mixture was obtained by fitting the global tar fluorescence spectra with the spectra of the calibrated pure species at each time step. Baumhakl and Karellas [56] also developed a system for online tar monitoring based on LIF spectroscopy but using deep-UV and near-UV LEDs, instead of UV lasers. As an interesting result, the power of deep-UV LED at 265 nm (0.3 mW) was not enough to get good fluorescence signals. Therefore, if these LEDs should be used, optimization of the optical path and of the measurement setup will be necessary [56]. With an excitation wavelength of 365 nm (optical output of 25 mW) a semi-quantitative correlation between tar load (determined with tar protocol) and LIF signal has been achieved [56].

Within the framework of biomass pyrolysis, LIF spectroscopy has been used to characterize the qualitative evolution of fluorescending species (mainly produced in heterogeneous secondary reactions) at reactor level [57, 58], with an excitation wavelength of 266 nm, and at particle level [50, 59], with an excitation wavelength of 355 nm, to gain a deeper understanding on the pyrolysis mechanism, especially on factors influencing heterogeneous secondary reactions and their thermochemistry.

12.2.2.2 Typical Experimental Procedure

In Fig. 12.5, a schematic presentation of a LIF spectroscopy setup is presented.

(1) Light source able to excite the targeted species with the required wavelengths. As the name indicates, this light source has been traditionally a laser. The use of lasers lies on their potential to emit spatially and temporally coherent light, their narrow bandwidths and the possibility to reach high intensities and very short pulses [41]. However, most lasers are expensive, with high maintenance requirements and not adequate for environments with a lot of dust, temperature gradients, etc., typically encountered close to biomass thermochemical conversion installations of relevant size. Industrial lasers can be used to overcome part of these limitations related to the measurement environment [60]. LEDs are an alternative, with possible wavelengths down to 240 nm approximately. This option is significantly cheaper, accompanied by the fact that LEDs present higher optical power stability with an adequate current source in comparison to lasers [56]. Moreover, combination of several LEDs with different wavelengths is possible, increasing the sensitivity of the measurement as introduced in the previous section. However, LEDs have lower power in the deep-UV region which may be not enough to have good fluorescence signal [56]. Therefore,

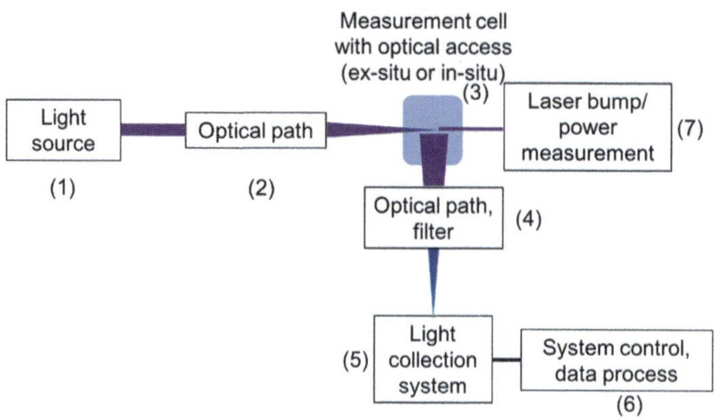

Fig. 12.5 Schematic representation of a LIF spectroscopy setup

optimization of the optical path and measurement cell design may be needed, together with high sensitivity detectors. LEDs also have significantly broader wavelength bands than lasers.

(2) Optical path from the light source to the measurement cell. The light beam needs to be guided and focussed to the measurement area where to excite the targeted species. Auxiliary elements to track the energy/power of the laser (or other light source) over time may be also necessary, above all if quantitative characterization is targeted. If imaging experiments have to be done, the use of a laser sheet gives more information (spatially) than a laser beam. Therefore, optics to convert a laser beam into a laser sheet are necessary. The lower the wavelengths used are, usually the more expensive these optical elements are.

(3) The measurement cell, which can be an in situ or an ex situ setup. The latter is used in the LIF spectroscopy portable sensors presented in the previous section. The gas stream is extracted from the reactor (e.g. gasifier), conditioned, and introduced in the measurement cell. This one is kept at constant temperature and pressure during the measurements, being these conditions the same as for the calibration of the system. Besides, a constant gas flow through the cell should be maintained. These are critical experimental conditions that can significantly influence the fluorescence intensity measurements, having a strong impact on the species quantification. For this reason, these values should be perfectly known and considered during the calibration step.
In an in situ setup, such as the pyrolysis measurements at particle level presented in the previous section, the temperature level may change over time, as well as the flow rate and flow field. If quantification is targeted, these values must be known, as well as their impact on the fluorescence signal. Such in situ measurements make quantification very challenging.
The measurement cells need to have optical access. For the UV region, this is typically the case in the form of UV fused silica windows.

(4) The optical path to guide the fluorescence signal to the detector. Optical filters in order to block the scattered light from the light source may be needed.

(5) Detector or light collection system. Generally, it includes a combination of CCD camera and spectrograph, a compact spectrometer or a photomultiplier. The coupling of a CCD camera with a spectrograph, as done in [50], constitutes a high sensitivity detector, although quite expensive. Compact spectrometers [52, 56] are cheaper but also have lower sensitivity and resolution. Photomultipliers are quite economic as well, with high sensitivity. However, discrimination between wavelengths is not possible. Therefore, other auxiliary devices to limit the detection wavelength range are required. A photomultiplier or a combination of intensified CCD camera and spectrograph are necessary when measuring fluorescence lifetimes.

The detection system is placed at 90° with respect to the laser beam path.

For species quantification, calibration of the system is required. Besides knowing the aforementioned experimental variables, it is of high importance to determine the linear range of fluorescence emission, i.e., the concentration range where the fluorescence intensity increases linearly with the species concentration (determined by other method).

12.2.2.3 Advantages and Limits

LIF spectroscopy offers the possibility to develop on-line, fast, and non-invasive techniques for characterization of aromatic species, main components of the condensable fraction in gasification processes. Therefore, no sampling and cooling down is required, avoiding further reactions and long sample preparation times. This technology shows high potential to develop sensors that can be used to monitor and optimize biomass thermal conversion plants in real time. It also presents high sensitivity, down to the order of ppm. The main disadvantage is that such setup can be very expensive and with rather high maintenance requirements. An alternative is the use of LEDs and compact detectors, as introduced in the previous section, optimal for sensor development where high accuracy is not necessary. Another drawback are the challenges related to species quantification, including the necessary calibration and operation of the system in the linear region of fluorescence-concentration. Influence of other factors such as temperature, pressure or fluorescence signal quenching has to be also characterized and taken into account during calibration and measurement.

12.3 Offline Analysis of Composition

12.3.1 Sample Collection

Offline analysis requires the collection of the sample. This step is crucial since it will determine the sample representativeness—in particular in terms of compounds collected—and therefore the measurement quality.

12.3.1.1 Recovery in Liquid Phase: Condensation

Principle

The condensation of a mixture of gas vapours is a unitary operation very common in industry, in chemical engineering and petrochemistry for example. Theoretical and experimental aspects of heat and mass transfer related to condensation are very complex and have been widely studied [61]. Theoretical condensation temperature of a gas mixture can be usually calculated by using thermodynamic models such as the UNIFAC model. These theoretical aspects will not be discussed here.

Condensation has the advantage of not requiring any solvent. This facilitates the subsequent analysis of the sample and the possible chemical recovery of the condensed product. This is also interesting from the environmental point of view, since it is not necessary to treat effluents or to recycle a solvent.

Condensation of mixtures has been used by many authors, for the recovery of fast pyrolysis oils but also of condensable species of torrefaction [62]. This condensation is usually performed in one step in fast pyrolysis industrial processes, such as Ensyn or Dynamotive processes [63]. To recover the pyrolysis oil, the condensable fraction has to be cooled and condensed rapidly to avoid secondary reactions which promote permanent gas production at the expense of bio-oil production. Some properties of the bio-oils and more generally, of the condensable species, can be improved by fractionation of liquid recovered. As in the case of conventional crude oil, distillation of condensed liquids is not attractive because some of the constituents of these liquids are highly reactive and, when heated, undergo thermal degradation that produces a solid residue [64]. For this reason, fractional condensation of liquids could be an interesting technique as an effective downstream process to separate the constituents of the condensed liquids into different fractions [64–66].

In order to separate the heavy and light fractions of the liquid, the cooling system can be divided into several different stages. This technique is called fractional condensation or staged condensation. It aims at recovering the oils in several separate stages, depending on their physicochemical properties. Fractional recovery in the different stages enables to characterize each fraction separately [65]. Staged condensation systems for pyrolysis oils can be composed of:

- one or more cyclones to trap particles in suspension;
- several condensing stages cooled to different temperatures;
- a washing tower using an oil or a bio-oil to condense the gases;
- an electrostatic filter used to trap aerosols by charging them electrically.

The process described by Chen et al. [65] is as follows. Condensable gases are produced continuously at reaction temperature in the reactor. Then solid and gases are separated in a cyclone maintained at a temperature level close to the reactor one to avoid gas condensation. Then, the gases are sent to several selective condensers in series and maintained at different temperatures until the final condensation temperature. An electrostatic precipitator (EP) is used to trap the fine particulate matter or the fine droplets of liquid using the force of an induced electrostatic charge. Usually, the condensers are made of stainless steel to promote heat transfer and to facilitate cleaning. The temperatures of each condenser are monitored by a thermocouple fixed inside the condenser, and condensable species samples are collected at the bottom of each condenser.

The technical difficulty of this kind of system is the temperature control of each recovery stage and the avoidance of cold spots between the stages.

The staged condensation of condensable species is currently implemented in the case of bio-oil production through biomass flash pyrolysis [67–69], and rarely used for gasification tars. To our knowledge, fractional recovery of condensable species has never been applied to gas produced during biomass torrefaction.

Typical Experimental Procedure

Whatever the number of condensers, the aim of the condensation device is to cool down the gases produced during biomass/waste thermal degradation. Several technical criteria must be fulfilled:

- the coolant has to be maintained at a temperature close to the target temperature to avoid cold spots on the exchange surfaces;
- the surface of heat exchange has to be sufficient to ensure cooling of gases and condensation of a fraction of condensable species;
- the material used for condenser has to be chosen to avoid fouling and to make cleaning easy.

12.3.1.2 Recovery in Liquid Phase: Bubbling

Principle

As for condensation, the aim of this process is to recover the condensable species. In this case, a representative part of a gas stream containing particles and organic compounds is sampled under isokinetic conditions.

The sampled mixture of gases undergoes bubbling in a cold solvent to improve the heat exchange with gases and to cool them down. The gaseous species having an affinity with the solvent are then trapped.

This bubbling system is often used for the analysis of gases resulting from thermochemical transformation of biomass, especially in the case of tars from gasification. The best known example is the standard [70] commonly referred to as "Tar protocol". It relies on a train of bubblers for gas sampling and analysis [71]. In the series of six impinger bottles, the three bottles on the left are immersed in a heated bath, while the three bottles on the right are immersed in a cold bath. The first impinger bottle allows to cool down the gases sampled in the process and to collect mainly water contained in gases. The first five bottles are filled with solvent; the last one is empty. Tars and water remaining are condensed by absorption in solvent in the sampling train. The heat released by gas cooling and condensation is removed in two external baths, usually maintained at temperature (given in the typical experimental procedure section) [72].

At the end of the sampling test, all the bottles are weighted to determine the mass of condensable species recovered during the experiment. The different samples collected may then be gathered and kept under appropriate conditions before analysis in laboratory devices.

Typical Experimental Procedure

The typical experimental procedure is well described in standard CEN/TS 15439 [70]. From a practical point of view, the following aspects should be highlighted:

- six impinger bottles are used to collect the condensable species. The first five bottles contain a solvent, and the last bottle is empty;
- the solvent used is usually isopropanol because it allows to collect the largest range of tars produced in thermal processes. Some tests have been performed with other solvents. For instance, Kern et al. [72] have used toluene as solvent, but benzene, toluene and xylene (BTX), which are major compound of tars, could not be detected;
- the bottles 1, 2 and 4 are immersed in a heated bath usually maintained between 35 and 40 °C. The cold bath containing the bottles 3, 5 and 6 is usually cooled to around −20 °C;
- the impinger bottles should be preferably equipped with frit glass allowing bubbles production to improve heat and mass exchanges in the solvent. These frit materials should be located either as a filter at the end of the inner tube or around the inner tube covering the rest of the diameter of the impinger bottle.

12.3.1.3 Recovery in Solid Phase Micro-Extraction (SPME)

Principle

Solid phase micro-extraction is an analytical technique that emerged in the 1990s. It is a method of solvent-free extraction of organic compounds by sorption on polymer-coated fused silica fiber. The phenomenon of absorption and/or adsorption on the polymer can take place, depending on the type of polymer chosen:

- absorbent fibers such as Carbowax, which are viscous polymeric gels;
- adsorbent fibers such as Carboxen, which are porous solid materials, generally fixed on the fiber using a polymer such as PDMS (polydimethylsiloxane).

The measurement can be performed:

- by immersion, in a gas or a liquid loaded with compounds to extract;
- or in headspace, i.e. in the gaseous atmosphere above the liquid.

In both cases, the fibers are in equilibrium between the coating(s) and the matrix or headspace. It is not therefore a total extraction.

A syringe type SPME fiber is shown in Fig. 12.6.

The fiber can then be thermosorbed directly into a GC injector. The SPME technique has the advantage of not requiring additional equipment or handling of the sample.

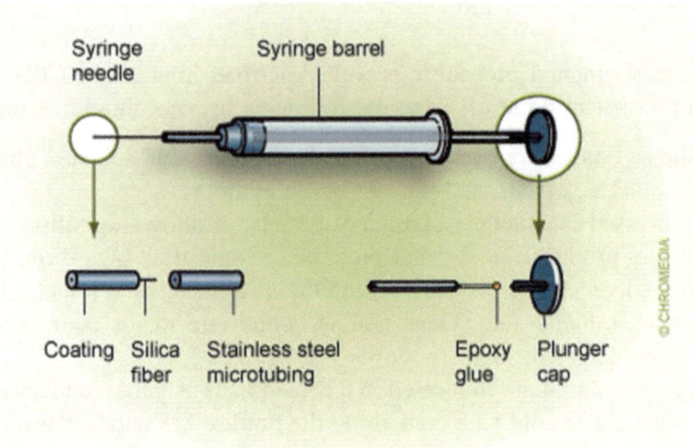

Fig. 12.6 Scheme of SPME syringe type system [73]

Examples of Applications

The SPME technique is frequently used for the analysis of volatile compounds in the fields of food, water and air. It is also found for trapping compounds derived from biomass processing.

Comte [74] studied pyrolysis vapors and gasification tars on two pilot-units. The objective was to develop a method of trapping and analysing these species to achieve lower quantification limits than the tar protocol. The first pilot is a slow pyrolysis rotary kiln coupled to a cracking and steam reforming reactor. The second is a fluidized bed gasification reactor. In both cases, a sampling line was equipped with a storage bulb and a pump allowed gas sampling. The storage bulb was equipped with a hole to introduce the SPME fiber for sorption. A PDMS type fiber was used. For pyrolysis, phenols, aliphatic compounds, furans and PAHs could be identified. For gasification, PAHs could be identified. The quantification step was then performed by determining the partition coefficients and calibration of some standard compounds.

Voinot [75] studied the drying of jack pine through a pilot dryer and red oak through an industrial dryer, in order to characterize the volatile compounds emitted. Recovery of these compounds was carried out by bubbling and by SPME fiber in headspace. The samples, whether gaseous or liquid, were then analyzed by GC MS. For jack pine drying, a 75 μm CAR/PDMS fiber was selected. This has identified many monoterpenes, sesquiterpenes and terpenoids. Some terpenoids were derived from the thermal degradation of terpenes originally present. For example, verbenone was derived from degradation of camphene, pinocarvone limonene [76]. The major compounds in terms of areas were caryophyllene, α-pinene, 3-carene, β-pinene, α-humulene and limonene. For red oak, the two fibers retained were PA

85 μm and PDMS/CAR/DVB 50 μm. A dozen of species was identified, including eugenol, isoeugenol and vanillin, which are derived from the degradation of lignin.

Finally, Tessini et al. [77] studied the low molecular weight aldehydes contained in pyrolysis oils. A O-(2,3,4,5,6-Pentafluorobenzyl)hydroxylamine (PFBHA) solution was first grafted by adsorption onto a SPME fiber. PFBHA is a derivatizing agent which, in contact with aldehydes, forms an oxime, which is analyzed later. The SPME fiber was then exposed in headspace of the oil sample. The fiber was finally thermodesorbed in a GCMS. Three types of fibers were compared: PA, PDMS/CAR, PDMS/DVB. The PDMS/DVB type fiber was chosen because it adsorbs less non-targeted aromatic compounds. The injection of standards made it possible to measure the contents of formaldehyde (between 1 and 4% w/w), acetaldehyde (between 0.01 and 0.1% w/w) and propionaldehyde (up to 0.04% w/w).

12.3.1.4 Adsorption on Solid Phase (SPA) and Thermodesorption

Principle

Solid phase adsorption (SPA) is a sampling technique that appeared in the 1980s. The principle is based on the trapping of gaseous species on a porous solid phase by physical adsorption. The adsorption is carried out without solvent and generally at room temperature. The interaction between the adsorbed molecules and the adsorbent is of low energy. The reaction is therefore reversible, which allows the analysis of the molecules after desorption.

Two types of adsorption sampling can be distinguished:

- dynamic sampling, on adsorbent tube. This type of sampling is well suited to channelized fixed sources;
- passive sampling, by diffusion. This type of sampling is more suitable for atmospheric analysis.

In both cases, the main families of adsorbents are the same. One can distinguish:

- activated carbons, which have large specific surfaces (greater than 1000 $m^2\ g^{-1}$). However, their thermal desorption is partial, which makes quantification difficult;
- carbon molecular sieves (examples: Carboxen, Carbosieve, Spherocarb), which are obtained by cracking or polymerization of species such as benzene on the surface. These are microporous materials having large specific surfaces, and rather hydrophobic;
- graphitized black carbons (Carbopack, Carbotrap, Carbograph), which are non-porous, very inert, and have a small specific surface area because of their structure in sheets;

- porous organic polymers (Tenax, Porapak, Chromosorb), which also have a low specific surface area. Tenax TA, for example, has a surface area of 35 m^2 g^{-1}. It is nevertheless the best known and most used adsorbent because it traps a wide range of molecules.

The choice of the type of adsorbent is mainly related to the targeted species and the adsorption conditions.

To analyze the adsorbed molecules, the adsorbent phase must then undergo a thermal desorption step, or thermodesorption. This step is carried out in a thermodesorber, which is usually coupled to a gas chromatography.

Thermodesorption is performed in two separated stages [78]:

- the adsorbent phase is heated in a flow of carrier gas (the carrier gas used in GC, generally helium) to desorb the collected species and refocus them on a low mass cold trap. This cold trap is usually cooled between −30 and −40 °C. Thanks to the low mass of this trap, inertia of desorption is weak, which allows to improve the chromatographic resolution. An inlet split may also be applied if the adsorbent phase is overloaded;
- the cold trap is then heated at high heating rate to desorb the species recovered. The cold trap consists of an adsorbent phase similar to the phase of the analyzed tube. The species collected are then conducted to the GC. An outlet split may also be applied if necessary.

Examples of Applications

Solid phase adsorption and thermodesorption have many applications in the field of the environment: analysis of air in work area, atmosphere, indoor air, odors. Simon et al. [79], for example, investigated the trapping of monoterpenes on Tenax GC and Tenax TA and their GC FID analysis. The reliability of the quantification of terpenes by adsorption and thermodesorption could thus be shown.

Dufour et al. [80] studied the adsorption sampling of volatile compounds, oxygenated or not, produced by pyrolysis of biomass at high temperature (700 °C). The adsorbent phase retained is of the Carbotrap 300 type. It is a multi-bed adsorbent phase composed of Carbotrap C, Carbotrap B and Carbosieve SIII, making it possible to trap a wide range of compounds, from C_{20} to C_2. The adsorbent phase was then analyzed using a thermodesorber coupled to a GC-MS. In addition, the volatile compounds were also recovered by bubbling and analyzed by GC MS. The results were compared with those of adsorption sampling. The qualitative analysis by SPA sampling showed the presence of a hundred of compounds, among which acetic acid, phenol, aromatic oxygen compounds or PAHs. Recovery by bubbling allows to collect the same compounds. However, quantitative analysis by SPA sampling is more accurate than bubbling, especially for the quantification of light PAHs.

Finally, Candelier et al. [81] studied the adsorption sampling of volatile compounds emitted during hardwood and softwood torrefaction between 180 and 230 °C. The wood sample was heated in the thermodesorber for 15 min. The cold trap is of the Tenax type. The thermodesorber was coupled to a GC-MS for the identification of compounds. An approximate quantitative analysis was performed from the peak areas obtained by the mass spectrometer. These tests showed the presence of acetic acid, furfural, methylfurfural, hydroxymethylfurfural, 2 methoxyphenol, 2,6 dimethoxyphenol, vanillin, syringaldehyde, acetovanillin, acetosyringone.

12.3.2 *Karl-Fischer Analysis*

The water content of the liquid compounds is generally determined using a Karl-Fischer volumetric titrator according to the ASTM E203 08 standard method. This method enables to measure the water content of a sample between 0.01 and 100% v/v. It uses the oxidation reaction of sulfur dioxide with iodine in the presence of water according to the following equation:

$$I_2 + SO_2 + 2H_2O \rightarrow 2HI + H_2SO_4$$

The end of the reaction results in a slight excess of iodine detected and potentiometrically assayed by a platinum electrode. The water content is calculated from the volume of iodine introduced.

12.3.3 *Gas Chromatography (GC) Followed by Flame Ionization Detector (FID)/Mass Spectrometry (MS)*

This off-line method is a combination of a gas chromatograph coupled in line with a flame ionization detector or mass spectrometer. The former has the function of separating the different compounds within the sample, while the latter is the detector for quantification of each compound.

12.3.3.1 Principle of GC

The main parts in the gas chromatograph include the injection port, the separation column and the oven.

The injection port is the part that allows the sample to be driven onto the separation column. There are different types of injectors [82]:

- The normal packed column injector, where the sample injected is vaporized in a heated glass liner and driven onto the column by a pressurized inert carrier gas. This system is used with packed columns.
- The split/splitless capillary injector, more appropriate for capillary columns, where the sample is vaporized in the heated liner, but part of it is discarded through a split vent in split mode. Therefore, the carrier flow gas only introduces a part of the sample onto the column. This disposition has two reasons: the gas flow entering the column is kept constant, and the split vent flow controls also the amount of sample entering the column. Thanks to this option and optimising the split ratio, performance of the column can be improved, avoiding overload peaks and enabling to decrease the solvent peak width. Indeed, in some cases, there can be coelution of solvent and compounds, i.e. the solvent retention time in the column can be the same as these compounds, and as the solvent concentration is high in the sample the presence of other compounds is masked. On the contrary, in splitless mode, all vaporized sample is introduced into the column. This method is advantageous for low concentration samples, for which a higher sensitivity is required.
- The on-column injector, where the sample is injected into a cool column without previous vaporisation step. This option is good when the sample contains compounds with significantly different boiling points to avoid discrimination in the vaporization step.
- The programmed temperature vaporizing injection (PTV), where a controlled injection rate is kept with a vaporization temperature below the boiling point of the solvent. It is used when introducing large sample volumes to improve sensitivity.

The column, which can be packed or capillary column, has the function of separating the different compounds present in the sample. Thanks to this separation, compounds can be qualitatively characterized according to their retention time within the column. Since each component should reach the end of the column, and therefore the detector, at different retention times (depending on the compound and on the column properties), the detector may be able to quantify each compound individually. Packed columns are nowadays mainly used for permanent gases and small hydrocarbons analysis, and for sample separation (preparation). Capillary columns are the ones mainly used in GC/FID applications. These columns are significantly thinner and longer than packed columns. They show significantly better sample separation, resolution and sensitivity, with lower analysis time and sample size [83].

The oven, in which the column is placed, provide the temperature program to help separating the components in the sample. Temperature program will depend on the sample and the column properties.

12.3.3.2 GC/FID

Principle

The flame ionization detector (FID) is the most popular detector used in gas chromatography. The working principle is based on the detection of ions formed during combustion of carbon containing organic compounds in a hydrogen-air flame. The response mechanism is generally accepted to be based on the formation and chemical ionization of CHO^+ ($CH^* + O^* \rightarrow CHO^+ + e^-$) which quickly loses a proton to water molecules forming CO and proton-bound water clusters [82–84]. The FID response for saturated aliphatic hydrocarbons is proportional to the carbon number [84, 85]. However, in increasing molecular structure complexity and in presence of heteroatoms, this response decreases [82], potentially due to other reaction pathways competing with the above proposed mechanism [84]. In order to quantify these variations in the FID response, the "effective carbon number" (ECN) concept was developed in the 60s [82, 84], which can be described as the number of equivalent aliphatic carbon atoms that would give the same response in the FID detector as the considered compound [83, 84]. Polycyclic aromatic hydrocarbons (PAH), despite being hydrocarbons, showed lower FID response in comparison to aliphatic hydrocarbons, being structural condensation one of the potential factors contributing to this lower response per carbon atom. Heteroatoms (O, S, N, halogens) substitution in hydrocarbons also reduces the FID response [85]. Functional groups have a significant impact on the FID response as well, to the point that it has been suggested that a carbon atom being part of a carbonyl or a carboxyl group do not contribute to the FID response [84, 86]. A great deal of research has been done, as briefly reviewed by Faiola et al. [84], in order to determine the response factors (or ECN contribution from different groups of atoms) according to the composition and structure of several organic species, due to the potential reaction pathways that can be followed in the flame. Despite the fact these mechanisms are not fully understood, ECN values have been given to groups of atoms (aliphatic, aromatic, olefinic, acetylenic, carbonyl, alcohol, ether, amine, etc.) with respect to a reference substance, typically an alkane, having the latter a ECN equal to the number of carbon atoms in the molecule. The theoretical ECN can be derived from the presence of the aforementioned groups in the considered molecule. Besides, many studies have reported experimental response factors (area/peak height of the FID signal divided by the mass of the injected solute) for plenty of organic species, as for example in [87], where the response factors of 164 are given, grouped in: normal paraffins, branched paraffins, cyclopentanes, cyclohexanes, aromatics, unsaturates, alcohols, aldehydes, acids, esters, nitrogen compounds, ketones and others (solvents). Despite the possibility of using experimentally determined response factors or ECNs from literature to quantify organic compounds based on the solely calibration of aliphatic hydrocarbons, careful quality control should be performed when measuring with FID, including always the calibration with representative standards of each type of species to be analysed.

Online operation of a GC/FID is also possible, analysing therefore gas samples, if instead of an injector port as previously described, a heated sampling/switching valve for chromatography is used. These measurements would be online from the point of view that the samples are not condensed and further prepared for the chromatographic analysis, but the measurement can still take tenths of minutes, depending on the type of the column. Therefore, such application can be considered as semi-continuous.

Another possibility is the utilization of an on-line semi-continuous FID tar analyser, developed at the University of Stuttgart [88]. This system does not have a chromatograph; therefore, detailed separation of the sample compounds is not possible. The system contains three sampling loops and two filters, besides hot temperature switching valves. The sample loaded in two of the three loops goes through these filters, which can be filled with different materials to retain different fractions of the organic volatiles compounds. Therefore, in the subsequent analysis of the sample from the loops, by driving the carrier gas through each loop and pushing the sample into the detector, the total hydrocarbons content (from the loop led unaltered to the detector, i.e. without filter), and two fractions of these hydrocarbons can be measured. The analysis time lasts at least one minute and the minimum tar concentration that can be detected is approximately 50 mg/Nm^3. In this case, it is assumed that the FID response is almost directly proportional to the amount of organically bound carbon, using an average response factor determined with propane.

Typical experimental procedure

A FID is a robust and sensitive detector with no high requirements in maintenance. During the measurement, it is important to be sure that the hydrogen-air flame stays in optimal conditions, and that the flow rate and temperature are kept constant, since this detector is quite sensitive to these parameters [88].

For calibration, FIDs have the nice property that the response is proportional to the number of carbons for aliphatic hydrocarbons, with a potential small reduction in the response factor for aromatic hydrocarbons. For other types of species, this reduction can be more significant, however there is plenty of information available in literature regarding response factors (or ECNs) for several species. This behaviour translates into a significantly lighter calibration procedure in comparison to other systems such as GC/MS. For example, as introduced in [87], only calibration with propane was performed, considering that the targeted species are hydrocarbons. Moreover, for quality control purposes, calibration with representative compounds of each class of species to be analysed should be performed, e.g. hydrocarbons, ketones, alcohols, heterocyclic hydrocarbons, etc.

The calibration of the GC is of high relevance in this system, since the qualitative identification of each component relies on the retention time in the column. It is important to achieve a high-quality separation within the column, since the detector is not able to distinguish between two different compounds. For this, the appropriate column must be selected, depending on the expected compounds to be analysed, as well as other process parameters such as temperature program, flow, split/splitless injection, etc.

When analysing complex mixtures of organic compounds, it is highly probable that the column is not able to fully separate all compounds in the sample or, in other words, that more than one compound reaches the end of the column, and therefore the detector, at the same time. This is a critical point since there is no other pathway in this system to recognize a compound, besides its retention time. Therefore, pretreatment of the sample to enhance this separation is very important, i.e., first step separation before performing the analysis. For this reason, different separation techniques such as distillation, solubility in different solvents, SPA (solid phase adsorption) filters could contribute to improve the measurement. The optimal situation would be measurements with a GC/MS to qualitatively characterize the sample followed by measurements with the GC/FID to quantify each compound.

Advantages and Limits

The FID is a robust and easy to use analytical device, easy to calibrate due to its response to different classes of organic compounds, as previously said, with low detection limits, high repeatability and long term stability. The main disadvantage of the FID is its impossibility to qualitatively characterize the different compounds, relying therefore on the performance of the GC.

12.3.3.3 GC/MS

Principle

Mass spectrometry is a technique enabling to identify and quantify a chemical compound. The first coupling with GC appeared in the 60s. It consists in 3 main steps:

- ionization, during which the solute is charged electrically;
- analysis, during which the ions formed are separated versus their ratio mass/charge;
- detection, during which the ions formed and separated are counted.

Each step can be performed in various ways. However, the most common ones are:

- ionization by electronic impact,
- analysis by quadrupole mass,
- detection by electron multiplicator.

Molecular ions are formed, which enables to determine the molecular mass of the analysed solutes, and through fragmentation, to access their molecular structure. The mass spectra obtained are very reproducible. They are compiled in spectra libraries (NIST, FFNSC, Adams). For example, the library NIST 2014 contains more than 240,000 compounds and spectra. The identification of unknown compounds is then made by comparison of their mass spectra to those referenced in the libraries. The mass spectrum of a solute is therefore an additional characteristic to the retention time and intensity of response.

Typical Experimental Procedure

An important feature of GC/MS is the possibility to use two modes that enable to derive chromatograms with different detection sensitivities. The SCAN mode enables to record complete spectra between two extreme masses. This mode makes it possible to view all compounds in the sample as an identification step, but is sensitivity is relatively low. The SIM (Select Ion Monitoring) mode enables to detect targeted substances, whose spectral characteristics are known, with a maximum of sensitivity. These substances are detected through 3 characteristic fragments, which leads to a high gain of sensitivity.

Advantages and Limits

The main advantage is to identify a very large range of compounds contained in a solute. Moreover, a quantitative analysis of selected compounds is also possible, requiring a calibration for each of them. However, analysis duration can be quite long (depending on several factors including number of followed compounds, required degree of precision …). In addition, the fact that several compounds can be coeluted may be a limit in some cases.

12.3.3.4 Bidimensional GC/MS

The bidimensional GC/MS has been developed in the 90s, as a solution to the coelution issue mentioned above in the case of GC/MS. More precisely, two techniques have emerged, namely GC/GC and GCxGC. The latter one has been applied in biomass thermal conversion studies since it is suitable with non-targeted analyses. It enables to increase chromatography sensitivity through refocalisation and fast reinjection in a second column.

Staš et al. [89] have for instance proposed a review of bio-oil pyrolysis analyses by GCxGC. They showed how this enables to get rid of coelutions observed in GC and the improvement of identification associated.

12.3.4 *High Performance Liquid Chromatography (HPLC) and High Resolution Liquid Chromatography Mass Spectrometry (HRLC/MS)*

12.3.4.1 Principle of HPLC

High performance liquid chromatography is a well-established method to characterize liquid samples for many decades. There are many types of HPLC which can be classified based upon the column type (that means on the separation mechanism involved) and the mobile phase. In the field of biomass, the main chromatography systems used are normal-phase chromatography, reversed-phase liquid chromatography, ion chromatography, size exclusion chromatography, etc. Among these techniques, several of them are particularly suitable for the characterization of biomass thermal conversion processes. This section will focus therefore on reversed-phase liquid chromatography (RP-HPLC), ion chromatography (IC) such as ion-exclusion chromatography, anion-exchange chromatography and size exclusion chromatography (SEC).

Reversed-Phase Liquid Chromatography (RP-HPLC)

Reversed-phase liquid chromatography (RP-HPLC) is one of the oldest types of chromatography. It is usually the easiest and fastest method for general characterization of both neutral and ionic species and is widely available in laboratories.

The principle for compound separation in RP-HPLC is based on the affinity between the compounds in solution and the stationary phase. The stationary phase is usually made of silica bounded with aliphatic chains such as C_8 or C_{18}. The mobile phase is polar and contains water or a mixture of water and an organic solvent. Detection is usually ensured by an UV detector [90]. Typical conditions include the use of a C_{18} column, 2 mL/min mobile phase, a gradient or in isocratic mode using water or an increasing concentration of an organic solvent in water. Typically, polar compounds will not be very well retained and will leave the column first, the less polar to non-polar compounds will be more or less retained [91].

In the field of biomass, RP-HPLC has been used to characterize and quantify anions [91] and neutral compounds [92, 93]. While useful, RP-HPLC is not the most used chromatography since, for instance, one might say that the more universal a column is, the less the resolution power for a specific compound class is. To that end, more specific types of chromatography techniques are used.

Ion-Exclusion Chromatography (IEC)

Ion exclusion chromatography (IEC), first introduced in 1953 by Wheaton and Bauman [94] is particularly suitable for the analysis of weak inorganic and organic acids, alcohols, aldehydes, amino acids and carbohydrates [93, 95–98]. Thus, IEC is very efficient to follow thermal degradation processes.

Three main mechanisms are involved in IEC: the Donnan exclusion, the steric exclusion and the adsorption effect [99]. When using a cation-exchange column, i.e. sulfonated groups are attached to a polyvinylbenzene support, the acid mobile phase, more often H_2SO_4, is repelled by the charge "barrier" called the Donnan effect (Fig. 12.7). On the contrary, neutral compounds have the ability to pass through the "barrier" and can be adsorbed on the polymer, resulting in a longer retention time. Partially charged compounds will thus be retained differently.

For biomass processes, the use of IEC will result in a separation between sugars, organic acids, solvent and neutral molecules such as 5-HMF and furfural (Fig. 12.8). Typical suitable columns include Phenomenex Rezex^{TM} ROA-Organic-H^+ or, Biorad Aminex 87H. Detection is usually ensured by refractive index detector (RID) and conductivity detector [99].

Anion-Exchange Chromatography (AEC) and Cation Exchange Chromatography (CEC)

Anion-exchange chromatography (AEC) is a versatile and a powerful technique that allows separation and quantification of various types of anions, both organic and inorganic. In AEC, the stationary phase usually consists in a polymeric support functionalized with quaternary ammonium. The separation process is based on the interactions between the anion in solution and the counter-ion on the stationary

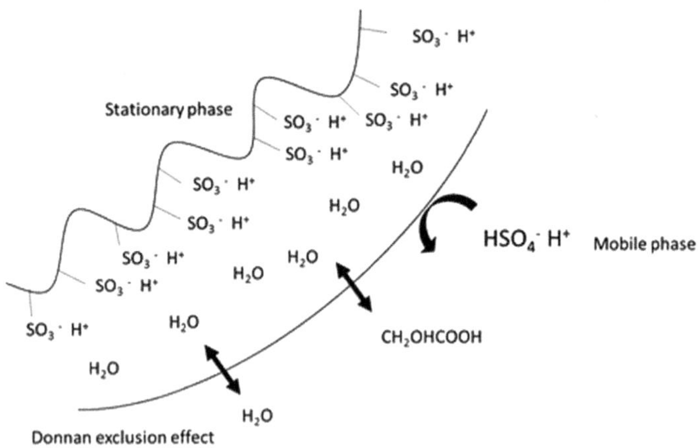

Fig. 12.7 Schematic of the Donnan exclusion effect

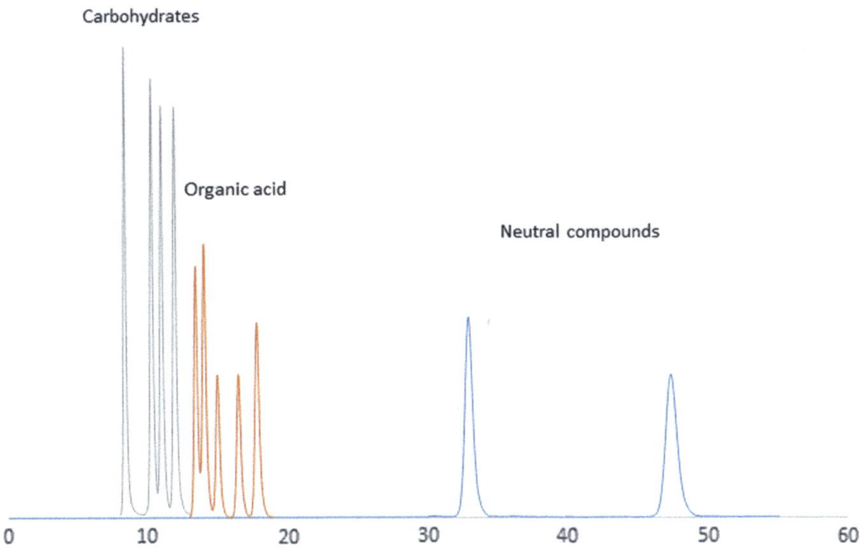

Fig. 12.8 Ion exclusion chromatogram for a mixture of carbohydrates (cellobiose, glucose, xylose and arabinose), organic acids (glycolic acid, lactic acid, formic acid, acetic acid, levulinic acid), fufural and 5-HMF using a Phenomenex Rezex^{TM} ROA-Organic-H^{+} column; flow rate: 0.6 mL/min, mobile phase; H_2SO_4 2.5 mM, detection: RID, concentrations: 50 ppm

phase. The difference in terms of affinity will result in the separation quality. Today, different types are commercially available such as polymer-based anion exchangers, latex-agglomerated exchangers, silica-based anion exchangers, etc. [95]. Thus, AEC is suitable to separate carbohydrates, polysaccharides, organic acids and inorganic ions for example. Typical experiment involves a basic mobile phase, NaOH or KOH, an ion exchange column, an ion suppressor and a detector (usually a conductivity detector but also an electrochemical detector for carbohydrates detection for instance).

Cation exchange chromatography (CEC) is used to separate and quantify cations especially alkali metals, alkaline-Earth metals, transition metals and amines. CEC involved similar types of columns (except carrying sulfonated groups instead) and instrumental configurations [95].

Size Exclusion Chromatography (SEC)

Size exclusion chromatography (SEC) is one of the most powerful techniques in terms of chemical conversion characterization for biomass. The early work on SEC is attributed to Porath and Flodin in 1959 [100]. SEC is a technique able to monitor major changes in terms of molecular sizes (or more precisely in terms of hydrodynamic volumes). The separation process is based on the entropy decrease of the mixture due to the stationary phase made of a porous material containing fixed pore sizes. During an analysis, a mixture of molecules is sorted by size.

As presented in a simplified scheme (Fig. 12.9), the mixture is injected at the beginning of the column, the largest molecular size compounds will not be able to penetrate inside the pores resulting in a faster elution. On the contrary, smaller molecules will penetrate inside pores and they will be temporally retained. However, in some cases, in addition to the entropy decrease effects during an analysis, enthalpy interactions can occur between molecules and the stationary resulting in estimating retention times. In addition, one of the main challenges in SEC resides in setting a proper calibration. Usually polystyrenes, polyethylene oxides or polyethylene glycols are used to generate a calibration curve allowing converting retention times in terms of molecular masses. It is rather obvious that when analyzing a complex mixtures, the hydrodynamic volumes of the molecules and those for the standards are different resulting in an incorrect calibration. The most convenient way to run SEC experiments is either to analyse known increasing size polymers corresponding to the molecules to analyse or to use the classic standards considering that the molecular masses will be an approximation.

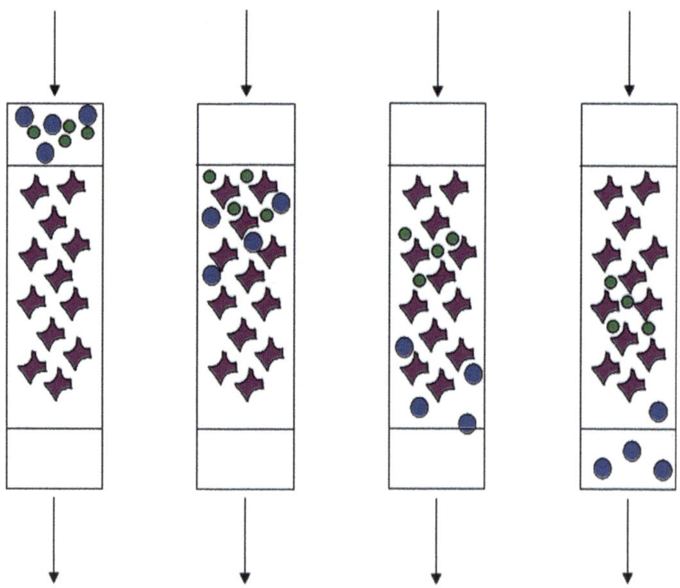

Fig. 12.9 Principle scheme of SEC chromatography for two different molecules. After sample injection, molecules will be separated according to their molecular sizes, the biggest ones will pass through the column faster than the smallest ones

To ensure the detection, the most classical detectors are evaporative light scattering detector (ELSD), refractive index detector (RID) and UV detectors. Additional types of detectors can be used such as low angle laser scattering detector (LALLS), multi-angle laser light scattering detection (MALLS), right-angle laser light scattering detection and triple detection, as well as on-line viscosimetric detection.

HPLC-MS

When working in biomass conversion processes, the resulting samples fall into the definition of a complex mixture. The previous sections dealt with the main chromatography techniques to resolve the complex mixtures. However, the main limitations resides in both separation and detection fields. HPLC often requires using standards to determine the retention times and to generate calibration curves.

Today, one of the most powerful techniques that can solve the previous limitations resides in coupling HPLC to mass spectrometry (MS). The fact that the technique is not common, even though HPLC/MS is very powerful, is due to the high complexity in coupling those very different operating conditions techniques and the interpretation of the results especially for complex mixtures and many aspects, especially the interface between HPLC and MS need to be wisely chosen [101]. In many cases, electrospray ionization (ESI), atmospheric pressure chemical ionization (APCI), atmospheric pressure photo-ionization sources are used even in combination, to interface both instrumentation, but careful attention should be ensured due to the presence of salts incompatible with mass spectrometry.

HPLC/MS is well developed in some fields (especially in biology) and has yet to be developed for biomass conversion processes.

12.4 Conclusion

This chapter focused on the main techniques of analysis of the condensable compounds produced during biomass and waste thermal degradation processes, with specific emphasis on the characterization of the composition of this liquid fraction. Each technique, either online or offline—these latter ones requiring a crucial step of collection—has its own advantages and limits. As this handbook aims to be a practical tool, these advantages and limits are synthesized in Table 12.4 to support technique selection.

Table 12.4 Advantages and limits of the main methods used for analysis of condensable products composition

Analysis principle	Advantages	Limits
Fourier transformed infrared (FTIR)	Online measurement Quantification of several volatile molecules, including permanent gases and condensable species, with a non-intrusive method, at temperatures that avoid condensation during the measurement	Sample dilution may be required, which decreases sensitivity level Limited types of volatile molecules measured (for lightorganic condensable species) Significant overlaps in the spectra of several molecules, especially the ones with the same functional group (making identification complex and unreliable)
Laser induced fluorescence	Online measurement Non-invasive technique Suitable for characterization of aromatic species	Very expensive method High maintenance requirements Calibration and utilisation of the system in the linear region fluorescence-concentration required for species quantification
Gas chromatography flame ionization detector (GC/FID)	Robust detector Sensitive Low need in maintenance Easy calibration Online—but non continuous—measurement possible	Long response time
Gas chromatography mass spectrometry (GC/MS)	Identification and quantification of a very large number of volatile molecules	Long response time Slow analysis Calibration of each species required for quantification
Bidimensional gas chromatography mass spectrometry	Separation of species coeluted using GCMS	Higher complexity than GCMS
Reversed phase high performance liquid chromatography (RP-HPLC)	Easy and fast method for characterization of both neutral and ionic species and is widely available in laboratories	Low resolution
Ionic exclusion chromatography (IEC)	Suitable for the analysis of weak inorganic and organic acids, alcohols, aldehydes, amino acids and carbohydrates	Limited types of species analyzed
Anion-exchange chromatography (AEC) and cation-exchange chromatography (CEC)	Powerful technique for separation and quantification of anions/cations, both organic and inorganic	Limited types of species analyzed

(continued)

Table 12.4 (continued)

Analysis principle	Advantages	Limits
Size exclusion chromatography (SEC)	Powerful technique to characterize species produced during biomass conversion	Challenging calibration
High performance liquid chromatography coupled with mass spectrometry (HPLC/MS)	Powerful technique combining the advantages of the two techniques HPLC and MS	High complexity Presence of salts problematic for the MS

References

1. Libra, J.A., Ro, K.S., Kammann, C., Funke, A., Berge, N.D., Neubauer, Y., Titirici, M.M., Führer, C., Bens, O., Kern, J., Emmerich, K.H.: Hydrothermal carbonization of biomass residuals: a comparative review of the chemistry, processes and applications of wet and dry pyrolysis. Biofuels **2**, 71–106 (2011)
2. Nocquet, T., Dupont, C., Commandre, J.M., Grateau, M., Thiery, S., Salvador, S.: Volatile species release during torrefaction of wood and its macromolecular constituents: part 1—experimental study. Energy **72**, 180–187 (2014)
3. Bridgwater, A.V.: Review of fast pyrolysis of biomass and product upgrading. Biomass Bioenergy **38**, 68–94 (2012)
4. Kaltschmitt, M., Thrän, D.: Chapter 7. Biomass-based green energy generation. In: Hofer, R. (ed.) Sustainable Solutions for Modern Economies, pp. 86–124. RSC Publishing, Cambridge (2009)
5. Mohan, D., Pittman, C.U., Steele, P.H.: Pyrolysis of wood/biomass for bio-oil: a critical review. Energy Fuels **20**, 848–889 (2006)
6. Stoytcheva, M.: Pesticides in the Modern World—Pesticides Use and Management. InTechOpen (2011)
7. Pyro Wiki: Welcome to PyroWiki. http://pyrowiki.pyroknown.eu/index.php?title=Welcome_to_PyroWiki (2018). 12 Jan 2018
8. Balat, M., Kırtay, E., Balat, H.: Main routes for the thermo-conversion of biomass into fuels and chemicals. Part 1: pyrolysis systems. Energy Convers. Manag. **50**, 3147–3157 (2009)
9. Vitasari, C.R., Meindersma, G.W., de Haan, A.B.: Conceptual process design of an integrated bio-based acetic acid, glycolaldehyde, and acetol production in a pyrolysis oil-based biorefinery. Chem. Eng. Res. Des. **95**, 133–143 (2015)
10. Oasmaa, A., van de Beld, B., Saari, P., Elliott, D.C., Solantausta, Y.: Norms, standards, and legislation for fast pyrolysis bio-oils from lignocellulosic biomass. Energy Fuels **29**, 2471–2484 (2015)
11. Anca-Couce, A., Mehrabian, R., Scharler, R., Obernberger, I.: Kinetic scheme of biomass pyrolysis considering secondary charring reactions. Energy Convers. Manag. **87**, 687–696 (2014)
12. Anca-Couce, A., Brunner, T., Kanzian, W., Obernberger, I., Trattner, K.: Characterization and condensation behaviour of gravimetric tars produced during spruce torrefaction. J. Anal. Appl. Pyrol. **119**, 173–179 (2016)
13. Oasmaa, A., Fonts, I., Pelaez-Samaniego, M.R., Garcia-Perez, M.E., Garcia-Perez, M.: Pyrolysis oil multiphase behavior and phase stability: a review. Energy Fuels **30**, 6179–6200 (2016)
14. Bayerbach, R., Meier, D.: Characterization of the water-insoluble fraction from fast pyrolysis liquids (pyrolytic lignin). Part IV: structure elucidation of oligomeric molecules. J. Anal. Appl. Pyrol. **85**, 98–107 (2009)

15. Anca-Couce, A.: Reaction mechanisms and multi-scale modelling of lignocellulosic biomass pyrolysis. Prog. Energy Combust. Sci. **53**, 41–79 (2016)
16. Nachenius, R.W., Ronsse, F., Venderbosch, R.H., Prins, W.: Biomass pyrolysis. In: Murzin, D.Y. (ed.) Chemical Engineering for Renewables Conversion, pp. 75–139. Academic Press, Burlington (2013)
17. Branca, C., Di Blasi, C.: Kinetics of the isothermal degradation of wood in the temperature range 528–708 K. J. Anal. Appl. Pyrol. **67**, 207–219 (2003)
18. Neves, D., Thunman, H., Matos, A., Tarelho, L., Gómez-Barea, L.: Characterization and prediction of biomass pyrolysis products. Prog. Energy Combust. Sci. **37**, 611–630 (2011)
19. Evans, R.J., Milne, T.A.: Molecular characterization of the pyrolysis of biomass. Energy Fuels **1**, 123–137 (1987)
20. Koppejan, J., Sokhansanj, S., Melin, S.: Status overview of torrefaction technologies (2012)
21. van der Stelt, M.J.C., Gerhauser, H., Kiel, J.H.A., Ptasinski, K.J.: Biomass upgrading by torrefaction for the production of biofuels: a review. Biomass Bioenergy (2011). https://doi.org/10.1016/j.biombioe.2011.06.023
22. Anca-Couce, A., Obernberger, I.: Application of a detailed biomass pyrolysis kinetic scheme to hardwood and softwood torrefaction. Fuel **167**, 158–167 (2016)
23. González Martínez, M., Dupont, C., Thiéry, S., Meyer, X.M., Gourdon, C.: Impact of biomass diversity on torrefaction: study of solid conversion and volatile species formation through an innovative TGA-GC/MS apparatus. Biomass Bioenergy **119**, 43–53 (2018)
24. Heidenreich, S., Foscolo, P.U.: New concepts in biomass gasification. Prog. Energy Combust. Sci. **46**, 72–95 (2015)
25. Basu, P.: Biomass Gasification and Pyrolysis. Academic Press, Burlington (2010)
26. Staš, M., Chudoba, J., Kubička, D., Blažek, J., Pospíšil, M.: Petroleomic characterization of pyrolysis bio-oils: a review. Energy Fuels **31**, 10283–10299 (2017)
27. ChemLibretexts: How an FTIR spectrometer operates. https://chem.libretexts.org/Core/Physical_and_Theoretical_Chemistry/Spectroscopy/Vibrational_Spectroscopy/Infrared_Spectroscopy/How_an_FTIR_Spectrometer_Operates (2018). Accessed on 1 Mar 2018
28. Ma, Z., Chen, D., Gu, J., Bao, B., Zhang, Q.: Determination of pyrolysis characteristics and kinetics of palm kernel shell using TGA–FTIR and model-free integral methods. Energy Convers. Manag. **89**, 251–259 (2015)
29. Singh, S., Wu, C., Williams, P.T.: Pyrolysis of waste materials using TGA-MS and TGA-FTIR as complementary characterisation techniques. J. Anal. Appl. Pyrol. **94**, 99–107 (2012)
30. Biagini, E., Barontini, F., Tognotti, L.: Devolatilization of biomass fuels and biomass components studied by TG/FTIR technique. Ind. Eng. Chem. Res. **45**, 4486–4493 (2006)
31. De Jong, W., Pirone, A., Wojtowicz, M.: Pyrolysis of *Miscanthus giganteus* and wood pellets: TG-FTIR analysis and reaction kinetics. Fuel **82**, 1139–1147 (2003)
32. Gao, N., Li, A., Quan, C., Du, L., Duan, Y.: TG–FTIR and Py–GC/MS analysis on pyrolysis and combustion of pine sawdust. J. Anal. Appl. Pyrol. **100**, 26–32 (2013)
33. Giuntoli, J., de Jong, W., Arvelakis, S., Spliethoff, H., Verkooijen, A.H.M.: Quantitative and kinetic TG-FTIR study of biomass residue pyrolysis: dry distiller's grains with solubles (DDGS) and chicken manure. J. Anal. Appl. Pyrol. **85**, 301–312 (2009)
34. Tsamba, A.J., Yang, W., Blasiak, W., Wójtowicz, M.A.: Cashew nut shells pyrolysis: individual gas evolution rates and yields. Energy Fuels **21**, 2357–2362 (2007)
35. Liu, Q., Zhong, Z., Wang, S., Luo, Z.: Interactions of biomass components during pyrolysis: a TG-FTIR study. J. Anal. Appl. Pyrol. **90**, 213–218 (2011)
36. Bennadji, H., Smith, K., Shabangu, S., Fisher, E.M.: Low-temperature pyrolysis of woody biomass in the thermally thick regime. Energy Fuels **27**, 1453–1459 (2013)
37. Anca-Couce, A., Sommersacher, P., Scharler, R.: Online experiments and modelling with a detailed reaction scheme of single particle biomass pyrolysis. J. Anal. Appl. Pyrol. **127**, 411–425 (2017)

38. Colantoni, A., Evic, N., Lord, R., Retschitzegger, S., Proto, A.R., Gallucci, F., Monarca, D.: Characterization of biochars produced from pyrolysis of pelletized agricultural residues. Renew. Sustain. Energy Rev. **64**, 187–194 (2016)
39. Chen, D., Liu, D., Zhang, H., Chen, Y., Li, Q.: Bamboo pyrolysis using TG–FTIR and a lab-scale reactor: analysis of pyrolysis behavior, product properties, and carbon and energy yields. Fuel **148**, 79–86 (2015)
40. Lakowicz, J.R.: Principles of Fluorescence Spectroscopy, 3rd edn. Springer, New York (2006)
41. Skoog, D.A., Holler, F.J., Crouch, S.R.: Principles of Instrumental Analysis, 6th edn. Brooks Cole, Belmont (2007)
42. Kohse-Höinghaus, K., Jefferies, J.B.: Applied Combustion Diagnostics. CRC Press, New York (2002)
43. Brackmann, C., Sjöholm, J., Rosell, J., Richter, M., Bood, J., Aldén, M.: Picosecond excitation for reduction of photolytic effects in two-photon laser-induced fluorescence of CO. Proc. Combust. Inst. **34**, 3541–3548 (2013)
44. Prins, M.J., Li, Z.S., Bastiaans, R.J.M., van Oijen, J.A., Aldén, M., de Goey, L.P.H.: Biomass pyrolysis in a heated-grid reactor: visualization of carbon monoxide and formaldehyde using laser-induced fluorescence. J. Anal. Appl. Pyrol. **92**, 280–286 (2011)
45. Kohse-Höinghaus, K.: Laser techniques for the quantitative detection of reactive intermediates in combustion systems. Prog. Energy Combust. Sci. **20**, 203–279 (1994)
46. Keller-Rudek, H., Moortgat, G.K., Sander, R., Sörensen, R.: The MPI-Mainz UV/VIS spectral atlas of gaseous molecules of atmospheric interest. Earth Syst. Sci. Data **5**, 365–373 (2013)
47. Grosch, H., Sárossy, Z., Egsgaard, H., Fateev, A.: UV absorption cross-sections of phenol and naphthalene at temperatures up to 500 °C. J. Quant. Spectrosc. Radiat. Transf. **156**, 17–23 (2015)
48. Thöny, A., Rossi, M.J.: Gas-phase UV spectroscopy of anthracene, xanthone, pyrene, 1-bromopyrene and 1,2,4-trichlorobenzene at elevated temperatures. J. Photochem. Photobiol. Chem. **104**, 25–33 (1997)
49. Gratien, A., Nilsson, E., Doussin, J.F., Johnson, M.S., Nielsen, C.J., Stenstrøm, Y., Picquet-Varrault, B.: UV and IR absorption cross-sections of HCHO, HCDO, and DCDO. J. Phys. Chem. A **111**, 11506–11513 (2007)
50. Lang, N., Rupp, C., Almuina-Villar, H., Dieguez-Alonso, A., Behrendt, F., Röpcke, J.: Pyrolysis behavior of thermally thick wood particles: time-resolved characterization with laser based in-situ diagnostics. Fuel **210**, 371–379 (2017)
51. Borgmeyer, Y., Neubauer, J.: Continuous on-line tar monitoring for process control. In: Proceedings of 24th European Biomass Conference and Exhibition (2016)
52. Borgmeyer, J., Continuous on-line tar monitoring in hot process gases from biomass gasification by means of fluorescence spectroscopy. PhD thesis, Technische Universität Berlin, Mar 2019, unpublished
53. Sun, R., Zobel, N., Neubauer, Y., Cardenas Chavez, C., Behrendt, F.: Analysis of gas-phase polycyclic aromatic hydrocarbon mixtures by laser-induced fluorescence. Opt. Lasers Eng. **48**, 1231–1237 (2010)
54. Mitsakis, M.: Online analysis of the tar content of biomass gasification producer gas. Technische Universität München (2011)
55. Meng, X., Mitsakis, P., Mayerhofer, M., de Jong, W., Gaderer, M., Verkooijen, A.H.M., Spliethoff, H.: Tar formation in a steam-O_2 blown CFB gasifier and a steam blown PBFB gasifier (BabyHPR): comparison between different on-line measurement techniques and the off-line SPA sampling and analysis method. Fuel Process. Technol. **100**, 16–29 (2012)
56. Baumhakl, C., Karellas, S.: Tar analysis from biomass gasification by means of online fluorescence spectroscopy. Opt. Lasers Eng. **49**, 885–891 (2011)
57. Dieguez-Alonso, A., Anca-Couce, A., Zobel, N.: On-line tar characterization from pyrolysis of wood particles in a technical-scale fixed-bed reactor by applying laser-induced fluorescence (LIF). J. Anal. Appl. Pyrol. **102**, 33–46 (2013)

58. Dieguez-Alonso, A., Anca-Couce, A., Zobel, N., Behrendt, F.: Understanding the primary and secondary slow pyrolysis mechanisms of holocellulose, lignin and wood with laser-induced fluorescence. Fuel **153**, 102–109 (2015)
59. Zobel, N., Anca-Couce, A.: Slow pyrolysis of wood particles: characterization of volatiles by laser-induced fluorescence. Proc. Combust. Inst. **34**, 2355–2362 (2013)
60. Karellas, S., Karl, J.: Analysis of the product gas from biomass gasification by means of laser spectroscopy. Opt. Lasers Eng. **45**, 935–946 (2007)
61. Bontemps, A.: Condensation de mélanges de vapeurs—Transfert de chaleur et de masse. Techniques de l'Ingénieur. BE9910 v1 (2001)
62. Tumuluru, J.S., Sokhansanj, S., Wright, C.T., Kremer, T.: GC analysis of volatiles and other products from biomass torrefaction process. Adv. Gas Chromatogr. Prog. Agr. Biomed. Ind. Appl. 211–234 (2012)
63. Robson, A.: Dynamotive 2000. Progress report. PyNe Newsletter (2000)
64. Bridgwater, A.: Renewable fuels and chemicals by thermal processing of biomass. Chem. Eng. J. **91**, 87–102 (2003)
65. Chen, T., Deng, C., Liu, R.: Effect of selective condensation on the characterization of bio-oil from pine sawdust fast pyrolysis using a fluidized-bed reactor. Energy Fuels **24**, 6616–6623 (2010)
66. Westerhof, R.J.M., Kuipers, N.J.M., Kersten, S.R.A., van Swaaij, W.P.M.: Controlling the water content of biomass fast pyrolysis oil. Ind. Eng. Chem. Res. **46**, 9238–9247 (2007)
67. Jendoubi, N., Broust, F., Commandre, J.M., Mauviel, G., Sardin, M., Lédé, J.: Inorganics distribution in bio oils and char produced by biomass fast pyrolysis: the key role of aerosols. J. Anal. Appl. Pyrol. **92**, 59–67 (2011)
68. Oasmaa, A., Sipilä, K., Solantausta, Y., Kuoppala, E.: Quality improvement of pyrolysis liquid: effect of light volatiles on the stability of pyrolysis liquids. Energy Fuels **19**, 2556–2561 (2005)
69. Pollard, A.: Comparison of bio-oil produced in a fractionated bio-oil collection system. Iowa State University (2009)
70. AFNOR: Biomass gasification. Tar and particles in product gases. Sampling and analysis. DD CEN/TS 15439:2006 (2007)
71. Good, V., Ventress, L., Knoef, H., Zielke, U., Lyckhansen, P., van de Kamp, W., de Wild, P., Coda, S., van Paasen, S., Kiel, J., Sjöström, K., Liliedahl, T., Unger, C., Neeft, J., Suomalainen, M., Simel, L.: Sampling and analysis of tar and particles in biomass producer gases. Technical report prepared under CEN BT/TF 143 organic contaminants ("tar") in biomass producer gases (2005)
72. Kern, S., Pfeifer, C., Hofbauer, H.: Gasification of wood in a dual fluidized bed gasifier: influence of fuel feeding on process performance. Chem. Eng. Sci. **90**, 284–298 (2013)
73. Chromedia Analytical Sciences: Principles of SPME. http://www.chromedia.org/chromedia?waxtrapp=npuhcHsHiemBpdmBlIEcCKJ&subNav=abffyDsHiemBpdmBlIEcCtBDF (2018). Accessed on 2 Feb 2018
74. Comte, J.: Mise au point d'une méthodologie d'échantillonnage utilisant la micro-extraction sur phase solide pour la quantification de goudrons sur des procédés de thermolyse de la biomasse. Université Joseph Fourier, Grenoble I (2008)
75. Voinot, D.: Caractérisation des composés organiques volatils issus du séchage du bois. Application au chêne rouge et au pin gris. Université Laval (2007)
76. McGraw, G.W., Hemingway, R.W., Ingram, L.L., Canady, C.S., McGraw, W.B.: Thermal degradation of terpenes: camphene, Δ3-carene, limonene, and α-terpinene. Environ. Sci. Technol. **33**, 4029–4033 (1999)
77. Tessini, C., Müller, N., Mardones, C., Meier, D., Berg, A., von Baer, D.: Chromatographic approaches for determination of low-molecular mass aldehydes in bio-oil. J. Chromatogr. A **1219**, 154–160 (2012)

78. TurboMatrix: Turbomatrix series thermodesorbers. https://www.perkinelmer.com/lab-solutions/resources/docs/GDE_TurboMatrixTDATDUsersGuide.pdf (2015). Accessed on 30 June 2015
79. Simon, V., Riba, M.L., Waldhart, A., Torres, L.: Breakthrough volume of monoterpenes on Tenax TA: influence of temperature and concentration for α-pinene. J. Chromatogr. A **704**, 465–471 (1995)
80. Dufour, A., Girods, P., Masson, E., Normand, S., Rogaume, Y., Zoulalian, A.: Comparison of two methods of measuring wood pyrolysis tar. J. Chromatogr. A **1164**, 240–247 (2007)
81. Candelier, K., Dumarcay, S., Petrissans, A., Petrissans, M., Kamdem, P., Gerardin, P.: Thermodesorption coupled to GC-MS to characterize volatiles formation kinetic during wood thermodegradation. J. Anal. Appl. Pyrol. **101**, 96–102 (2013)
82. Grob, R.L., Barry, E.F.: Modern Practice of Gas Chromatography. Wiley, Hoboken (2004)
83. Poole, C.F.: Ionization-based detectors for gas chromatography. J. Chromatogr. A **1421**, 137–153 (2015)
84. Faiola, C.L., Erickson, M.H., Fricaud, V.L., Jobson, B.T., VanReken, T.M.: Quantification of biogenic volatile organic compounds with a flame ionization detector using the effective carbon number concept. Atmospheric Meas. Tech. **5**, 1911–1923 (2012)
85. Wang, W.T., LeDonne, N.C., Ackerman, B., Sweeley, C.C.: Structural characterization of oligosaccharides by high-performance liquid chromatography, fast-atom bombardment-mass spectrometry, and exoglycosidase digestion. Anal. Biochem. **141**, 366–381 (1984)
86. Sternberg, J.: Placental transfers: modern methods of study. Am. J. Obstet. Gynecol. **84**, 1731–1748 (1962)
87. Dietz, A.J.: Biotransformation studies on 1-chloro-2, 3-propanediol dinitrate. J. Pharm. Sci. **56**, 1664–1665 (1967)
88. Herthan, T., Moersch, O., Spliethoff, H., Berger, R.: The tar analyser—a suitable tool for the development and control of gasifiers and gas cleaning system. In: 1st World Conference on Biomass for Energy and Industry, Sevilla, Spain (2000)
89. Staš, M., Kubička, D., Chudoba, J., Pospíšil, M.: Overview of analytical methods used for chemical characterization of pyrolysis bio-oil. Energy Fuels **28**(1), 385–402 (2014)
90. Snyder, L.R., Kirkland, J.J., Dolan, J.W.: Introduction to Modern Liquid Chromatography, 3rd edn. Wiley, Hoboken (2009)
91. Gennaro, M.C., Angelino, S.: Separation and determination of inorganic anions by reversed-phase high-performance liquid chromatography. J. Chromatogr. A **789**, 181–194 (1997)
92. Chen, S.F., Mowery, R.A., Castleberry, V.A., van Walsum, G.P., Chambliss, C.K.: High-performance liquid chromatography method for simultaneous determination of aliphatic acid, aromatic acid and neutral degradation products in biomass pretreatment hydrolysates. J. Chromatogr. A **1104**, 54–61 (2006)
93. Chheda, J.N., Román-Leshkov, Y., Dumesic, J.A.: Production of 5-hydroxymethylfurfural and furfural by dehydration of biomass-derived mono- and poly-saccharides. Green Chem. **9**, 342–350 (2007)
94. Wheaton, R.M., Bauman, W.C.: Ion exclusion—a unit operation utilizing ion exchange materials. Ind. Eng. Chem. **45**, 228–233 (1953)
95. Weiss, J., Schpigun, O.: Handbook of Ion Chromatography, 3 Volume Set, 4th edn. Wiley, Hoboken (2016)
96. Damay, J., Duret, X., Ghislain, T., Lalonde, O., Lavoie, J.M.: Steam explosion of sweet sorghum stems: optimisation of the production of sugars by response surface methodology combined with the severity factor. Ind. Crops Prod. **111**, 482–493 (2018)
97. Scarlata, C.J., Hyman, D.A.: Development and validation of a fast high pressure liquid chromatography method for the analysis of lignocellulosic biomass hydrolysis and fermentation products. J. Chromatogr. A **1217**, 2082–2087 (2010)

98. Tanaka, K., Haddad, P.R.: Ion exclusion chromatography: liquid chromatography. In: Wilson, I. (ed.) Encyclopedia of Separation Science, pp. 3193–3201. Academic Press, London (2000)
99. Tanaka, K., Ohta, K., Fritz, J.S., Lee, Y.S., Shim, S.B.: Ion-exclusion chromatography with conductimetric detection of aliphatic carboxylic acids on an H^+-form cation-exchange resin column by elution with polyols and sugars. J. Chromatogr. A **706**, 385–393 (1995)
100. Porath, J., Flodin, P.: Gel filtration: a method for desalting and group separation. Nature **183**, 1657–1659 (1959)
101. Consultant, S.K.: The HPLC-MS Handbook for Practitioners. Wiley, Hoboken (2017)

Chapter 13
Dioxins and Dioxin-like Compounds

Shengyong Lu, Alfons Buekens, Tong Chen, Xiaoqing Lin, Mingxiu Zhan and Mengmei Zhang

Abstract Dioxins are a group of unwanted contaminants almost exclusively produced from anthropogenic sources. These compounds are generally persistent in the environment, resistant towards metabolism, hydrophobic and lipophilic, and hence bioaccumulating in the fatty tissues of animals and humans. Traditionally, waste incineration and iron ore sintering were the largest sources of dioxins. Biomass burning, especially uncontrolled burning, is becoming increasingly more significant as a source, as regulatory interventions as well as technological advances lead to marked emission reductions in the industrial customary contributions. After an introduction overviewing chemical structures and properties, toxicity and health effects, and emission sources of dioxins and dioxin-like compounds, this chapter introduces some formation pathways of dioxins from biomass related thermal processes and discusses various influencing factors for formation and emission of dioxins. Different methods used for sampling and analysis of dioxins from different sources are then reviewed. Generally, determination of dioxins highly relies on high-resolution gas chromatography/high-resolution mass spectrometry (HRGC/HRMS). Several alternatives to the HRGC/HRMS-based methods are also introduced. Numerous techniques were developed and applied for remediation, reduction and prevention of dioxins emissions and these are surveyed and compared.

S. Lu (✉) · A. Buekens · T. Chen · X. Lin · M. Zhang
State Key Laboratory of Clean Energy Utilisation, Zhejiang University,
Hangzhou, China
e-mail: lushy@zju.edu.cn

M. Zhan
College of Metrology and Measurement Engineering, China Jiliang University,
Hangzhou, China

A. Buekens
Department of Chemical Engineering, Vrije Universiteit Brussel, Brussels, Belgium

A. Nzihou (ed.), *Handbook on Characterization of Biomass, Biowaste and Related By-products*, https://doi.org/10.1007/978-3-030-35020-8_13

13.1 Introduction

13.1.1 Chemical Structures and Properties

The term "dioxins" is often used to designate polychlorinated dibenzo-p-dioxins (PCDD) and polychlorinated dibenzofurans (PCDF), as well as certain polychlorinated biphenyls (PCB) with dioxin-like toxic properties. Their chemical structures are shown in Fig. 13.1.

PCDD and PCDF have a triple-ring structure that comprises two benzene rings connected by two ether bridges, shaping a third oxygenated ring. There are 75 PCDD congeners, differing in the number and location of their chlorine atoms. PCDF are connected by a single ether bridge and have 135 congeners.

PCDD and PCDF are undesirable substances and were never produced willfully [1, 2]. Conversely, PCB industrially was formed by the chlorination of biphenyl [3] and have 209 congeners.

Only seven PCDD congeners and ten PCDF congeners with chlorine substitution in 2,3,7,8 positions, as well as twelve PCB congeners, are considered to have dioxin-like toxicity, relating to their ability to assume a planar structure.

Dioxins are high melting, colourless and flavourless. They show very strong thermal and chemical stability. Due to their lipophilicity, limited vapour pressure and low water solubility, dioxins are primarily associated with particulate matter in air and organic matter in soil and water [4]. The physical and chemical properties of each congener vary according to the degree and position of chlorine substitution [5]. Generally, the water solubility and volatility of individual dioxins congeners decrease with higher chlorination [6].

As persistent environmental pollutants, PCDD/F and dioxin-like PCB (dl-PCB) are widespread in air, soil, water, sediment, and biota. They resist to abiotic and biotic transformation. Therefore, in the atmosphere dioxins can be transported over long distances before their wet or dry deposition [4]. Ingestion of contaminated vegetation by grazing animals seems to be a major pathway by which dioxins enter the terrestrial food chain [7]. Except for direct deposition into water, dioxins

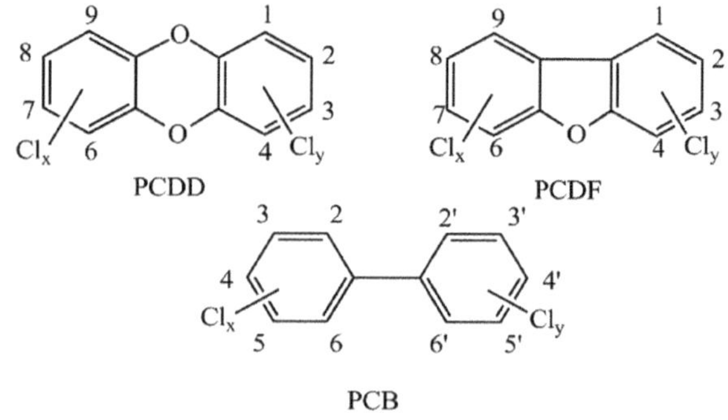

Fig. 13.1 Chemical structures of PCDD, PCDF, and PCB

deposited onto soil can also enter water bodies via surface run off. The bioaccumulation by aquatic biota is believed to be the primary mechanism by which dioxins enter the aquatic food chain [8].

13.1.2 Dioxins Toxicity and Health Effects

Dioxins are a worldwide concern, as these persistent environmental pollutants accumulate in the food chain, due to their lipophilic properties. The major pathways for dioxins enter the terrestrial and aquatic have been discussed in Sect. 13.1.1. By consumption of potentially contaminated foods, dioxins finally enter into the human body. Actually, more than 90% of human exposure to dioxins is due to dietary ingestion. In the human body, dioxins are absorbed and stored in fat tissues and persist for a long time: the half-life of dioxins in the human body has been assessed as 7–11 years [9]. Moreover, the half-life of dioxins is associated with dioxins concentrations and body composition. Higher concentration contributes to faster elimination, yet more body fat results in increased persistence [10, 11].

Toxic equivalency (TEQ) values have been applied to estimate the toxicity of dioxins. TEQ is a weighted value based on the toxicity of each individual congener of dioxins. As mentioned above, only seven PCDD congeners, ten PCDF congeners, and twelve PCB congeners show such dioxin-like toxicity. To calculate a TEQ, a toxic equivalent factor (TEF) is assigned to each congener. The TEF is the ratio of the toxicity of each congener to the toxicity of the most toxic congener, 2,3,7,8-tetrachlorodibenzo-p-dioxin (2,3,7,8-TCDD), of which the TEF is assigned to be unity [12]. A TEQ is calculated by multiplying the gram weight of each dioxin congener by its corresponding TEF and then summing the products (Eq. 13.1). The resulting value is expressed in gram TEQ

$$\mathrm{TEQ} = \sum_{i=1}^{29} C_i \times TEF_i \tag{13.1}$$

in which C_i is the concentration of i congener and TEF_i presents the TEF of i congener.

Dioxins may exert significant effects on human health by their high affinity for binding to the aryl hydrocarbon receptor (AhR), a specific cellular protein. The AhR is an intracellular ligand activated transcription factor related to the expression of a lot of genes. Previous studies on mice suggest that the AhR is a key regulatory protein, associated with normal development and homeostasis [13, 14], interacting with other regulatory proteins including specific cellular kinases, or cell cycle control proteins [15]. In the human body dioxins may cause overt or subclinical illness, including cancer, immune deficiency, reproductive and developmental abnormalities, endocrine disruption, decreased pulmonary functions and bronchitis, skin rashes (chloracne or acne) etc. Also, Kimbrough and Jensen reported transient acute health problems caused by dioxins, such as headache, pruritis, fatigue, irritability, and diarrhea, particularly after industrial exposure [16]. These health effects of exposure to dioxins are reviewed comprehensively by Schecter et al. [17].

13.1.3 Dioxins Emission Sources

Earlier studies show that the presence of dioxins in the environment primarily arises from anthropogenic sources [18–20]. They are emitted to the environment through various stacks (i.e. guided emissions) and in varying amounts associated with the source's potency and dispersion. Next, there are diffuse or non-point sources, such as forest fires, smouldering landfill fires, or human domestic activities.

PCDD/F were only produced for research studies, for example, in studies of physical-chemical properties, reaction mechanisms and toxicology, or for use as ^{13}C analytical standards. At a certain level of temperature trace amounts of PCDD/F are formed in any reacting system containing carbon, hydrogen, oxygen and chlorine. These reactions are catalyzed by inorganic carriers [21] and dramatically enhanced by metal catalysts such as copper and iron [22–25]. Much more PCDD/F result from incomplete combustion, occurring during open burning, in domestic fireplaces, accidental fires, wildfires, cremation, and poorly operated waste incinerators, in particular when starting up [26]. Municipal solid waste and medical waste incineration was once the major source of PCDD/F emission. Other sources include:

- metal smelting and refining, in particular those involving scrap,
- chlorine bleaching of pulp and paper,
- synthesis side products of several chemicals (especially PCB, chlorophenols, chlorophenoxy acid herbicides and hexachlorophene),
- disinfection of potable water loaded with humic and fulvic substances, and
- uncontrolled combustion (e.g. open burning, accidental fires, wildfires) [24, 27].

PCB always contain dl-PCB and small amounts of PCDF. These compounds were widely used as insulating oil, heat carrier, lubricant and plasticizer, as additive in a variety of industrial products (e.g., resins, rubber, coatings, sealant, auto-copying paper, ceramic glaze, fire retardant, dye dispersant) [28–31]. They enter the environment through various pathways: leaks from transformers or heat exchangers, fires or incomplete combustion of chlorine-containing organics, etc.

Benefitting from stringent regulations, the chief sources of dioxins have altered drastically [32, 33]. Industry has adopted appropriate measures to avoid dioxins generation and emission. The use of pesticides leading to dioxins contamination has been banned. Free chlorine pulp bleaching also caused dioxins formation [34]. However, the application of alternative bleaching agents, such as peroxides, reduced their formation. Although municipal solid waste and medical waste incineration once was the foremost source, their dioxin emissions have been significantly decreased due to improved incineration conditions, i.e. increasing and steadier combustion temperature, better furnace control (e.g., turbulence and oxygen), and sufficient residence time [6, 27]. Flue gas cleaning techniques, e.g. wet or semi-dry scrubbing followed by injection of activated carbon (AC) adsorbent, and selective catalytic reduction (SCR), etc., considerably contribute to the falling emission of dioxins. Various techniques (Hagenmaier drum, fly ash vitrification) have been applied for treating the dioxins-containing fly ash and solid residues from waste incineration.

13.2 Formation of Dioxins

13.2.1 Formation Pathways of Dioxins During Thermal Processes

Ever since the discovery of dioxins in fly-ash and flue gas from municipal solid waste incineration (MSWI) (in 1977) [35], enormous efforts have been made to identify their major pathways of formation. Any thermal process proceeding in the presence of carbon, chlorine, hydrogen, oxygen and metal catalyst has the potential to generate dioxins. There are at least two temperature windows in which dioxins and dioxin-like compounds may form actively: the "homogeneous" gas phase route taking place between 500 and 800 °C, and the "heterogeneous" low-temperature catalytic route proceeding mainly between 200 and 400 °C [23, 24, 36].

High-temperature (>500 °C) pyrogenic processes progressing in anoxic clouds surviving flaming combustion preserve precursors shielded from oxygen. These are formed from a variety of reactive gases through fast thermal reactions. These 'trace chemistries of fire' routes are 'homogeneous', since they do not rely on any specific catalytic processes of fly-ash or soot particles [37]. The PCDD/F and precursors formed from high-temperature homogeneous reactions result from pyrolytic formation, rearrangement and condensation of chlorinated precursors, e.g. chlorophenols (CP) and chlorobenzenes (CBz), or else from even smaller reactive entities (ethyne, ethene, propene, and butadiene) in the gas phase. The principal steps include (i) self-condensation of precursors, as a result of the coupling of molecule/molecule, molecule/radical or radical/radical species, (ii) cyclisation of the initial intermediates to produce PCDD/F and (iii) chlorination/dechlorination reactions [38]. It is generally accepted that the amount of PCDD/F formed following these high-temperature homogeneous routes is far lower than those deriving from low-temperature heterogeneous or catalytic pathways during thermal processes, e.g. incineration and combustion [24, 39]. The erratic nature of most combustion processes and the infinite variety in composition and activity of catalysts, spontaneously formed during the desublimation of metal halogenides, sublimated in the fire, are responsible for a wide variety in amount and fingerprint of the dioxins formed.

The low-temperature heterogeneous routes or reactions used to describe catalytic processes for forming dioxins can be subdivided into (a) de novo synthesis and (b) the precursor pathways. Both pathways proceed at the surface of solid carbon (e.g., soot, charred materials) or of fly ash particles, catalysed by certain metal ions, with copper clearly identified as strongest catalyst [22, 40]. The second seems to proceed much faster [41, 42], yet failed reproducing the typical thermal fingerprints with isomer distribution patterns that contain almost every isomer [43, 44]. The relative importance of the low-temperature de novo or precursor routes is debated and apparently varies case by case [23, 45, 46].

The de novo synthesis of dioxins refers to the formation of dioxins from macro-molecular carbon, soot, biomass char, medium or large Polycyclic Aromatic Hydrocarbons (PAH) structures, or any other organic structure. The de novo reaction route proceeds, firstly, through the transfer of chloride or bromide (halide) over transition metal halides to the macromolecular structure of residual carbon, followed by oxidative degradation of the chlorinated carbon matrix [39, 47]. In the first step, viz. the formation of organic chlorine, the C–H bond is turned into C–Cl bonds mainly from the chlorinating effect of Cl_2 rather than HCl [22]. The Deacon Reaction, through which HCl is oxidized to Cl_2 in the presence of oxygen, plays an important role in chlorination of carbon [48, 49], with $CuCl_2$ as most effective catalyst. Metal catalyst and oxygen are also essential for the second step—degradation of carbon matrix: the former reduces the temperature required for oxidation of carbon [47, 50]; the latter destroys the partly chlorinated carbon matrix and reactivates reduced transition metals, and thus boosts the emission of dioxins [51].

Precursor pathways for dioxins formation start from organic molecules that already are structurally similar to dioxins, e.g. PCB [3], CP and CBz. These form dioxins by relatively straightforward reactions involving, e.g., the condensation of CP or CBz- and CP-entities or radicals, followed by ring closure [52]. From a basic chemistry viewpoint, the precursor pathway is faster and easier explained than de novo synthesis. Still, metal catalyst is required in condensation and ring closure [36].

Precursor formation is certainly possible and active, yet the PCDD/F generated from typical precursors usually shows very low PCDF to PCDD ratios [50, 53, 54]. This ratio was proposed as a means to discriminate between de novo (PCDF) and precursor CP-based (PCDD) formation [23]. A decreasing order of CBz > CP > PCDF > PCDD > dl-PCB is found from de novo synthesis [23, 39].

13.2.2 Biomass Burning and Dioxins

Combustion is 'the rapid oxidation of fuel in which heat and usually flames are produced' [55]. Biomass combustion comprises several steps: (1) drying, (2) heating-up, (3) devolatilization, yielding volatiles (tars and gases) and char, (4) flaming combustion of volatiles, and (5) glowing or smouldering combustion of char [56]. The wet biomass is converted into an ignitable, dry fuel during drying and heating-up. Pyrolytic decomposition and devolatilization of biomass starts at about 160–250 °C [57, 58] and plays an important role in the early stages of the combustion of biomass, producing large amounts of thermal decomposition products and supplying highly reactive intermediates that polymerise to cyclic compounds, i.e. aromatics, PAH, and soot. Incomplete combustion occurs whenever one or more of the three T's (temperature, time and turbulence) fail to completely convert all evolving Volatile Matter (VM) into the principal products of combustion, i.e., CO_2 and H_2O. The pathway starting from reactive aliphatic compounds (e.g., ethene, ethyne) and leading to soot under oxygen-lean conditions can be described as follows [59]:

Reactive compounds → simple BTEX-aromatics (e.g., benzene, toluene, ethylbenzene, xylenes) → PAH → soot.

With the presence of organic/inorganic chlorine and metal catalyst, the generated PAH and soot can subsequently be converted into PCDD/F, through both precursor and de novo pathways [60–62]. Dibenzofuran (DF) and dibenzo-p-dioxin (DD) is also produced as volatile organics during the pyrolytic and oxidative decomposition of solid biomass [63, 64]. They easily form PCDD/F via direct chlorination [65, 66]. As stated before, chlorine gas (Cl_2) is much more potent than hydrogen chloride (HCl). Conversely, HCl may be activated by catalytic particles or replaced by metal chlorides.

The combustion process can be split into two types of phenomena: flaming versus smouldering combustion, for burning of the VM and the charred residue, respectively. Flaming combustion involves volatiles as fuel, mixing and reacting these with the surrounding air oxygen, releasing heat and light, while producing a variety of combustion products (e.g., CO_2, CO, H_2O, etc.) as well as smoke [67]. Conversely, smouldering is a low-temperature, slow, flameless form of combustion, continuing long after flaming combustion ceased [68]. Typically, it leads to incomplete combustion, with residual combustible often remaining occluded in ash. Flaming combustion occurs in the gas phase, while smouldering combustion starts at the solid surface, where oxygen meets char and then progressively moves inside [69, 70], e.g. following a shrinking core model. Formation of dioxins may happen in both combustion modes yet activated through distinct processes and in different physical contexts and time-scales.

The situation is different for incineration of biomass in a modern incinerator designed for energy generation or/and waste disposal. Generally, the temperature and the residence time in the incineration chamber (>850 °C, 2 s), combined with well-organised flow and turbulence patterns, are high enough to destroy most dioxins contained already in biofuel and to largely avoid their formation or that of their precursors. However, regeneration of dioxins is unavoidable along the flue gas cooling at the post-combustion zone, via precursor routes from phenols and lignin and/or via de novo reactions from unburned carbon, in the presence of chlorine and catalytic metals [60, 71, 72]. Waste heat boiler, with operating temperature (900–200 °C) overlapping the optimal temperature for synthesis of dioxins and dioxin-like compounds, provides potential sites for their formation and eventually emission. Already around the nineties, Düwel et al. discovered the major role of the waste heat boiler in the generation of PCDD/F [73].

Incineration under controlled conditions combined with adequate air pollution control device (APCD) is considered a favourable destruction procedure for many biomass wastes [26], emitting lower pollutants (per unit mass) than uncontrolled combustion and open fires. Conversely, the incomplete combustion in uncontrolled biomass combustion and open burning result in numerous additional Products of Incomplete Combustion (PICs) being emitted, dioxins included [59].

13.2.3 Factors of Influence

The effect of various factors of influence on the formation of dioxins, related to de novo synthesis or precursor routes, has been studied in numerous earlier studies. Generally considered were temperature, time, reaction atmosphere, the composition of fuel and fly ash (carbon, chloride and catalysts), and the presence of suppressants and precursors, etc. These factors are firstly examined on the basis of extrapolating earlier studies focusing on MSWI, and then discussed for their effects on formation of dioxins from biomass combustion.

13.2.3.1 Temperature

Process temperature is the most influential parameter in dioxins formation [6, 74]. A sufficient temperature level is needed to ensure that chemical reactions proceed at adequate rates. Only negligible formation of dioxins is to be expected below ca. 200 °C; whereas, temperatures above 400–500 °C often witness destruction at a faster pace than formation. Optimal temperatures varied in different studies [52, 75]; typically, a maximum output is registered between 300 and 380 °C.

In the open burning of biomass, e.g. forest, savannah and grassland fires, burning of agricultural residues and garden waste, etc., the absence of an enclosure around the fire fails to halt radiant and convective cooling and protect the fire from being quenched by surrounding cold air, thus leading to lower combustion temperatures than usual in MSWI and more important temperature gradients ranging from the cold boundaries to the hot center of the plume [76]. Incomplete combustion and concomitant formation of PICs are stimulated by a combination of low combustion temperatures and cold surroundings, combined with inadequate turbulence and weak aeration. Very little is known about the temperature window and time frame, leading to the maximal formation of dioxins during open burning [77]. An important factor seems to be the survival of pyrolytic puffs subsequent to the ignition of a large amount of highly calorific waste [78].

For formation of dioxins during biomass incineration under controlled conditions, MSWI seems the principal model to follow: most dioxins being formed in the post-combustion regions during heterogeneous processes involving boiler and flues deposits, as well as on fly ash particles during their flight with a residence time of few seconds only in the flues [60], similar to that in MSWI. Rapid quenching of flue gases to below 260 °C leads to a reduction in the PCDD/F concentrations [60]. The capacity of biomass incinerators, rarely exceeding 10 MW, is lower than that of municipal solid waste incinerators (MSWIs). Equipped with advanced APCD, well-organized and continuously operated biomass incinerators can emit rather low concentration of dioxins in the stack gas and meet the emission limit value of 0.1 ng TEQ/Nm^3. However, older, batch-operated furnaces, with very little air pollution and combustion control, inevitably lead to high dioxins emission factors [59]. Emission of PCDD/F largely correlates with combustion quality of biomass incinerators: poor combustion conditions, characterized by increasing CO concentration and decreasing flue gas temperature, lead to higher emission concentrations of PCDD/F [79]. The size of facilities also show influence due to the corresponding combustion conditions [71]. One example is the penetration of secondary air into the flue gas leaving the combustion chamber; Keppel Seghers proposed Prism injectors as a means to halve the width to be served by secondary air jets.

13.2.3.2 Oxygen

Oxygen is one of the most important ingredients in PCDD/F formation through the de novo route. Besides in the oxidative breakdown of carbon, oxygen is also required for: (i) the metal ligand mechanism, requiring oxygen for reactivating reduced transition metals [51] and (ii) the production of Cl_2 through the Deacon reaction [22] and direct chlorination of aromatic rings through substitution reactions. A rising oxygen content in the carrier gas (0–10%) can significantly promote the total PCDD/F yield from de novo tests on MSWI fly ash and model fly ash (MFA) samples [80–82]. The amount of PCDD on average rises linearly with the oxygen concentration, while that of PCDF follows a reaction order of about ½ [82]. Precursor formation by condensation of two CP molecules may proceed without any oxygen requirements [36]. Conversion of CBz, PCB, or PAH into PCDD/F requires oxygen, albeit in moderate amounts [83, 84]. Contradictory results exist on the effect of oxygen on condensation reactions: both enlarged and suppressed formation of PCDD were reported from CP as precursor [41, 52, 85, 86].

Oxygen supply is one of the operating parameters which should be optimized for minimizing PCDD/F emission from biomass incinerators. The influence of oxygen on distribution of PCDD/F in a full-scale MSWI plant was investigated by Zhang et al. [87]. Increasing oxygen supply (from 6.0 to 10.5 vol. %) essentially led to a higher yield of PCDD/F. It is suggested that under adequate burning conditions ($\geq$6.0 vol. % in this study) the lower oxygen level is favourable for reducing PCDD/F formation.

The presence of ample oxygen is essential for realizing complete combustion. In the open burning of biomass, however, aeration occurs only by natural air circulation, possibly stimulated by blowing wind and geometrical factors. The formation of PCDD/F is largely determined by the combustion conditions. Thus, the influence of oxygen is contrary to that in incineration. Combustion of loose biofuel, providing easier access to the oxygen in air, yields PCDD/F emission factors up to two orders of magnitude lower than for the same feedstock in compressed, compacted, or impacted conditions [88, 89].

13.2.3.3 Time and Turbulence

The reaction time required for formation of dioxins from different pathways varies widely. The de novo route requires more time (typically from several minutes to hours) [90] than the self-condensation of chlorophenols/chlorophenoxy-radicals into PCDD, which at 300 °C completes in few minutes [91]. Temperature and time together are linked: at low temperature (250 °C) de novo formation may even start after a latency period and formation is still observed after heating for longer than 4 h [90]; at high temperature (500 °C), maxima are reached after rather brief time periods, even within 10 s [92].

Time scales required for complete combustion and decomposition of dioxins should be related directly to temperatures and also to the quality of mixing resulting from turbulent flow, eddy mixing, and strong velocity gradients. At 1200 °C, a temperature typical for hazardous waste incinerators, a fairly complete combustion is realized in air in a matter of few milliseconds, whereas several seconds are required at 800–900 °C, the higher temperature range of fluidized bed incineration. Thus, the high velocity (up to 100 m/s) injection of secondary air at the outlet of the MSWIs combustion chamber is necessary and a residence time of 2 s at 850 °C is still mandatory for burning out of the PICs present (E.U. Directive on incineration of waste). These measures are also helpful in biomass incinerators for improving their combustion conditions and reducing dioxins emissions.

During open fires, there has been not much systematic monitoring of the effects of time and temperature on dioxins formation. Yet this time factor—upon burning biomass fuel—was invoked to explain the occurrence of higher emission factors from laboratory tests than from field fires [93]. For the open burning of biomass fuel, the laboratory tests specify a longer residence time in the temperature window of dioxins formation than in the field tests. In these, the bulk density of feedstock is lower, and the airing conditions are better than in laboratory tests.

13.2.3.4 Combustion Modes

The potential of dioxins formation varies throughout the combustion process. A distinction is made between a first period of flaming combustion and the later stages (after evolution of most VM) of smouldering combustion. Sampling of dioxins at certain time intervals allows their monitoring with respect to time and burning advancement. Higher emission factors and a higher CO/CO_2 ratio occur in these later stages, when the combustion pile is still hot but less mass loss is occurring, suggesting that smouldering combustion is more proficient in generating dioxin emissions than flaming combustion [88, 94]. The PCDD/F formed in flaming combustion, smouldering, and oxidative pyrolysis of wood impregnated with preservatives were experimentally estimated by Tame et al. [95]. The product gases arising from smouldering combustion following oxidative pyrolysis, in a dramatic contrast, contained concentrations 2–3 orders of magnitude higher than those from flaming combustion [95]. It suggests that low temperature thermal treatment of biomass should be avoided during common combustion processes.

13.2.3.5 Components and Contaminants

Dioxins from combustion processes are mainly formed by heterogeneous routes, involving sources of carbon, chlorine, as well as an active catalyst. Carbonaceous compounds supply preformed aromatic rings required in dioxins synthesis. These derive partly through direct de novo release from existing structures and are partly provided via various precursors, or else supplied by high-temperature pyrogenic synthesis, starting from small reactive molecules formed from pyrolytic processes [77].

Chlorine is a ubiquitous element, extant in all living matter. The Cl content in biomass (dry basis) varies in the interval of 0.01–0.9 wt% and normally decreases in the order: animal biomass (mean value: 0.69 wt%) > herbaceous and agricultural biomass (mean value: 0.20 wt%) > wood and woody biomass (mean value: 0.02 wt%) [96]. In combustion systems chlorine exerts two contradictory effects on the carbon matrix: enhancing its degradation [97] and, after chlorination, rendering it more resistant to burn-off [98]. The effect of chlorine concentration on formation of dioxins from MSWI and uncontrolled combustion has been vastly investigated [88, 99–101]. Catalyst, maybe not required in homogeneous high temperature reactions, is always essential in the heterogeneous formation of dioxins in both precursor routes and de novo pathways. The former catalyzes the ring condensation of monoaromatic structures; the latter even provides a dual role for this catalyst: (a) chlorinating carbonaceous structures and (b) their subsequent degradation. Various metals, especially transition metal group metals, e.g. copper, iron, nickel, zinc, chromium, lead, etc., show catalytic activity for these reactions, with $CuCl_2$ as by far the strongest [82, 102, 103].

In most biogenic fuels carbon is mainly available as carbohydrates (cellulose and hemicellulose) and lignin [104]. Such different components show distinct characteristics during PCDD/F formation: under the same experimental condition and using the same catalyst, lignin yields considerably more PCDD/F than cellulose and hemicellulose [64, 105]. Generally, there are only traces amounts of chlorine (ca. 0–0.5 wt%) [56] and less copper [96] in native biomass. These contents of chlorine and copper could decisively affect their formation ability for dioxins [106]. However, external contaminants, such as pesticides, wood preservatives, paint, or general waste, are often present and even mixed into biomass fuels. These contaminants are most likely to bring in more chlorine, metal catalysts and even organic additives which thermally decompose to provide precursors for dioxins synthesis, largely boosting the emission of dioxins [59].

Most of the commonly used pesticides are either chlorine-containing (e.g., atrazine, 2,4-dichlorophenoxyacetic acid, pentachlorophenol dichlorprop, etc.) and/or embedded with dioxins precursor structures (e.g., dichlorprop, decamethrin, niclosamide, etc.), or even charged with dioxins impurities. Their involvement in biomass burning experimentally has been proved to stimulate the formation of PCDD/F and dl-PCB [107–109]. Metal-based solutions, e.g. chromated copper arsenate (CCA), copper boron azole (CBA), alkaline copper quaternary (ACQ), are popular preservatives for protecting timber. They have been proved to promote dioxins formation during the combustion of impregnated wood, resulting probably from their copper-based and chlorinated constituents, and from prolonged smouldering of wood char following the cessation of flaming combustion [63, 95]. Anthropogenic waste is often involved into biomass burning, potentially bringing in chlorine and metal catalysts, and even flame-retardants materials complicating deliberate efforts to burn them adequately.

Polyvinylchloride (PVC) is an extensively used, highly chlorine-containing (57%, before compounding with additives) plastic, frequently branded as a major chlorine donor; hence it leads to substantial formation of dioxins during combustion [110]. Co-combustion of PVC and biomass can lead to emission factors of dioxins that are 2–4 orders of magnitude higher than those from pure biogenic fuel [111]. Brominated flame retardants (BFR) are widely used for reducing the ignition of combustible materials and preventing accidental fires. Also, it can stimulate the emission of PCDD/F and mixed chlorinated/brominated PXDD/F from co-combustion with biomass [25].

13.2.4 Discussion

There are at least two distinct major routes, responsible for forming dioxins. The precursor route is chemically often quite straightforward, e.g. the condensation of two CP molecules, or the introduction of an oxygen into a PCB molecule. Both routes were investigated a number of times, yet their exploration should cover a sufficiently wide range of experimental conditions, such as temperature, time and oxygen, in order to establish their influence and ensure that they englobe the domain of active dioxins generation. Indeed, the response obtained, a PCDD/F output, is only some casual observation, the result of the difference between two amounts [$PCDD/F_{formed}$] and [$PCDD/F_{destroyed}$]. Their relative importance remains often totally unknown, unless the effect of temperature and time are explored systematically.

Other obstacles to convincing and decisive experimental data involve effects of hidden experimental factors, such as the matrix [21], the precise nature of the catalyst at work [46, 82] and the way it was prepared for reacting to form dioxins [112]. With so many variables active, it is a real feat to generate reproducible experimental data sets.

Surprisingly, most investigators throw aside a rich source of important mechanistic information, i.e. the dioxins fingerprint. The latter seems capable of revealing much more mechanistic detail, after an analysis of the relative amount and level of correlation between individual congeners. Finally, too little effort is paid to analyze the low chlorinated (Cl = 0–4) and discover the prevailing patterns of chlorination.

13.3 Sampling and Analysis

The determination of PCDD/F in the flue gas emitted from stationary sources such as MSWI presents considerable difficulties in sampling because of their low concentrations. Many countries and organizations have set up sampling methods for PCDD/F, for example, Method 23 of US EPA and EN 1948, as published by the European Committee for Standardization [113, 114]. Isokinetic sampling of PCDD/F means that the velocity of the gas entering into the sampling nozzle is the same as that of the surrounding gas in the duct at this sampling point. All samples from the gas stream are withdrawn isokinetically and collected in the probe, on a glass fibre filter (for particles) and a packed column of adsorbent material (for gas phase), respectively [113]. After sampling, the samples are transferred to the lab for further analysis. Since 1994 high resolution gas chromatography (HRGC) coupled with high resolution mass spectrometry (HRMS), is most often applied for instrumental determination of PCDD/F [115]. The official methods for PCDD/F analysis in USA, European Union, Japan and China rely on the individual measurement of the seventeen 2,3,7,8-substituted PCDD/F congeners by HRGC/HRMS. HRGC/HRMS is considered as an international gold standard method.

13.3.1 Determination of PCDD/F by Isotope Dilution HRGC/HRMS

Isotope dilution HRGC/HRMS has been used for determination of PCDD/F in aqueous, solid, tissue and a variety of environmental matrices [116, 117]. A specified amount of sample is spiked with solution containing specified amounts of isotopically ($^{13}C_{12}$) labelled PCDD/F. The sample is then extracted according to specific extraction procedures. After extraction, ^{37}Cl-labelled 2,3,7,8-TCDD (for flue gas or air sample, different isotopically labelled PCDD/F is used) is added to each extract, to evaluate the efficiency of the clean-up process. Such sample clean-up may include back-extraction with acid and/or base, and gel permeation, alumina, silica gel, florisil and activated carbon chromatography [116]. After clean-up, the extract is concentrated to near dryness. Immediately prior to injection, internal standards are added to each extract, and an aliquot of the extract is injected into the gas chromatograph (GC).

The analytes are separated by gas chromatography (GC) and detected by high-resolution ($\geq$ 10,000) mass spectrometry (MS). Two exact mass/load (m/z) values are monitored for each analyte [116]. An individual PCDD/F is identified by comparing the GC retention time and the ion-abundance ratio of two exact m/z values with the corresponding retention time of an authentic standard and the theoretical or acquired ion-abundance ratio of the two exact m/z values. Quantitative analysis is based on selected ion current profile (SICP) areas [116].

Quality Assurance/Quality Control is very important for PCDD/F analysis, because of their trace level concentration and tedious preparation procedures. The minimum requirements of a recognized dioxin laboratory consist of an initial demonstration of laboratory capability, analysis of samples spiked with labelled compounds to evaluate and document data quality, and analysis of standards and blanks as tests of continued performance [116]. In addition, contamination and interferences should be avoided or eliminated during the entire PCDD/F analytical procedure. All materials (e.g. glassware) used during the analysis should be properly cleaned, ensuring minimal contamination and interferences. Establishing blank values, including solvents, operation and field test, and duplicate analyses are recommended for quality assurance. Besides, detection limits, such as instrument-, method- and sample-detection-limit and recovery rate of surrogate-, internal- and clean-up-standards should be within the required range.

13.3.1.1 Stationary Sources

Method 23, EN 1948 and JIS K0311 are applied for determining the emissions of PCDD/F from stationary sources in the USA, the European Union and Japan, respectively [113, 114, 118]. In China, HJ77.2-2008 is used for the determination of PCDD/F sampled from ambient air and flue gas from stationary sources [119]. Figure 13.2 shows the schematic diagram for sampling PCDD/F from flue gas. The quartz fibre cartridge or filter is used to collect the particles and the gas absorption trap (usually containing amberlite XAD-2 resin) is used to absorb the PCDD/F in the gas phase.

All reagents used should be cleaned sufficiently to reduce interferences and a leak check must be done before sampling. The entire sampling and analytical procedure of PCDD/F is shown in Fig. 13.3.

The final determination of PCDD/F is performed by HRGC/HRMS operated in Electron Impact (EI) mode. A DB-5MS (60 m $\times$ 0.25 mm i.d. $\times$ 0.25 μm film thickness) fused silica gas chromatographic column is used for separating the isomers of PCDD/F; the carrier gas is helium at a constant flow of 1.2 ml/min. The use of splitless injection is recommended and the GC/MS interface components should withstand a temperature of 350 °C [117]. It is suggested that the GC column should be fitted directly into the MS ion source without being exposed to the

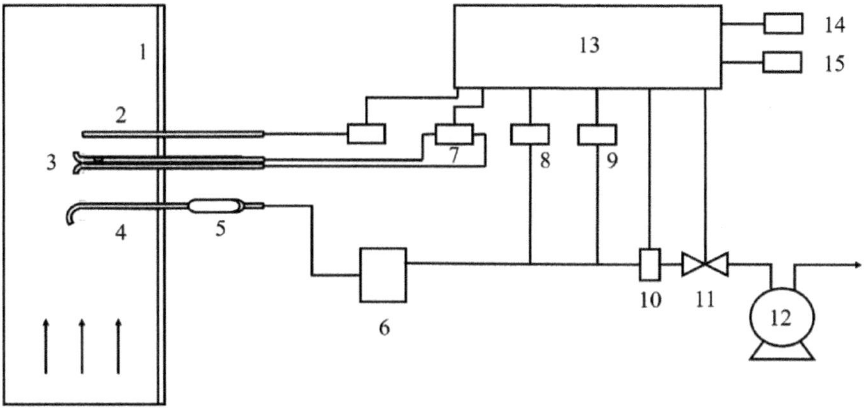

Fig. 13.2 Schematic diagram for PCDD/F sampling from flue gas of stationary sources/1: stack, 2: thermometer, 3: pitot tube, 4: sampling probe, 5: cartridge or filter, 6: gas adsorption trap with a condensation device, 7: micro pressure sensor, 8: pressure sensor, 9: temperature sensor, 10: flow sensor, 11: flow regulator, 12: sampling pump, 13: microprocessing system, 14: miniature printer or a connector, 15: monitor

ionizing electron beam [117]. The resolution of MS must be maintained at a minimum of 10,000 (10% valley definition) by using perfluorokerosene (PFK) before any analysis. The recommended MS tuning conditions are based on the groups of monitored ions shown in Table 13.1.

The concentration of the seventeen 2,3,7,8-substituted PCDD/F is detected by HRGC/HRMS, and calculated by the formula (Eq. 13.2):

$$C_n = \frac{(A1_n + A2_n) \times C_l}{(A1_l + A2_l) \times RRF} \tag{13.2}$$

where,

$A1_n$ and $A2_n$	the area of the primary and secondary m/z's for the PCDD/F;
$A1_l$ and $A2_l$	the area of the primary and secondary m/z's for the labelled PCDD/F;
C_l	the amount of the labelled PCDD/F added;
RRF	relative response factor of calibration standard (CS1–CS5).

In the above formula, RRF is calculated using the following formula (Eq. 13.3):

$$RRF = \frac{(A1_n + A2_n) \times C_l}{(A1_l + A2_l) \times C_n} \tag{13.3}$$

where,

$A1_n$ and $A2_n$	the area of the primary and secondary m/z's for the PCDD/F in CS1 to CS5;

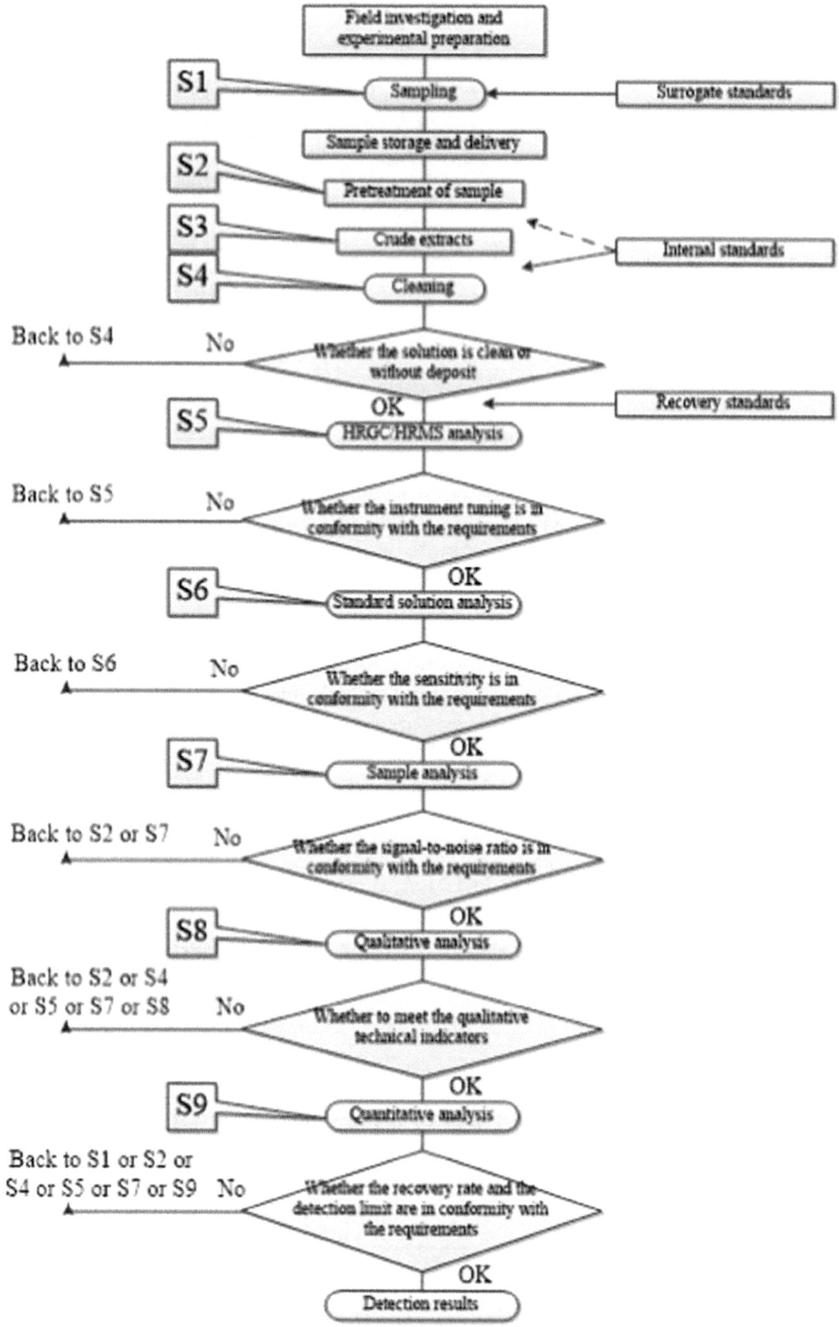

Fig. 13.3 The flow chart for analysis of PCDD/F emitted from stationary sources

Table 13.1 Ions monitored for HRGC/HRMS analysis of PCDD/F

Homologues	M^+	$(M + 2)^+$	$(M + 4)^+$
T_4CDDs	319.8965	321.8936	
P_5CDDs		355.8546	357.8517[a]
H_6CDDs		389.8157	391.8127[a]
H_7CDDs		423.7767	425.7737
O_8CDD		457.7377	459.7348
T_4CDFs	303.9016	305.8987	
P_5CDFs		339.8597	341.8568
H_6CDFs		373.8207	375.8178
H_7CDFs		407.7818	409.7788
O_8CDF		441.7428	443.7398
$^{13}C_{12}$-T_4CDDs	331.9368	333.9339	
$^{37}Cl_4$-T_4CDD	327.8847		
$^{13}C_{12}$-P_5CDDs		367.8949	369.8919
$^{13}C_{12}$-H_6CDDs		401.8559	403.8530
$^{13}C_{12}$-H_7CDDs		435.8169	437.8140
$^{13}C_{12}$-O_8CDD		469.7780	471.7750
$^{13}C_{12}$-T_4CDFs	315.9419	317.9389	
$^{13}C_{12}$-P_5CDFs		351.9000	353.8970
$^{13}C_{12}$-H_6CDFs	383.8369	385.8610	
$^{13}C_{12}$-H_7CDFs	417.8253	419.8220	
$^{13}C_{12}$-O_8CDF	451.7860	453.7830	
PFK (Lock mass)	292.9829 (for quantitative analysis of T_4CDD/Fs)		
	354.9792 (for quantitative analysis of P_5CDD/Fs)		
	392.9760 (for quantitative analysis of H_6CDD/Fs)		
	430.9729 (for quantitative analysis of H_7CDD/Fs)		
	442.9723 (for quantitative analysis of O_8CDD/Fs)		

[a]Means PCB interference

$A1_l$ and $A2_l$ the area of the primary and secondary m/z's for the labelled PCDD/F in CS1 to CS5;
C_l the concentration of the labelled PCDD/F in the CS1 to CS5;
C_n the concentration of the native PCDD/F in the CS1 to CS5.

The relative concentration of PCDD/F emitted from stationary sources could be calculated using the formula (Eq. 13.4).

$$C = \frac{C_n}{V_{sd}} \tag{13.4}$$

where,

ρ the relative concentration of PCDD/F (ng/m^3);
C_n the concentration of PCDD/F in the sample (ng);
V_{sd} the sampling volume of flue gas (m^3).

13.3.1.2 Solid Waste

EPA Method 8290A is used to determine PCDD/F in a variety of environmental matrices including fly ash, soil, sediment, water, sludge etc. [117]. In China, Method HJ 77.3-2008 is applied for the determination of PCDD/F in solid waste by the isotope dilution HRGC/HRMS [120]. According to HJ/T 20-1998, the solid waste should be pretreated before extraction [121]. For example, residues or fly ash produced from waste incineration need to be grinded and sieved. The particle size of 95% of the sample should be smaller than 2 mm in diameter. Figure 13.4 shows the analytical procedure for PCDD/F in solid waste samples. If the carbon content of the sample is high (such as for fly ash), the sample should be pretreated by 2 mol/L HCl before extraction. The amount of HCl is no less than 10 ml for 1 g of sample [120]. The HRGC/HRMS operating condition is the same as before.

13.3.1.3 Soil and Sediment

Besides EPA Method 8290A, EPA Method 1613, Revision B is also used to determine PCDD/F in soil, sediment, water and other sample matrices by HRGC/HRMS [116]. In China, Method HJ77.4-2008 is used for the determination of PCDD/F in soil and sediment [122]. For soil samples, sample collection, air-drying and sieving should be done according to HJ/T 166-2004 [123]; for sediment samples, sediment collection, air-drying and sieving are based on GB 17378.3-2007 and GB 17378.5-2007 [124, 125]. The moisture content of soil and sediment is determined according to the following procedure: (i) weighing 5 g of the soil or sediment sample and drying it for 4 h at 105–110 °C in an oven, (ii) cooling the sample in a desiccator, and (iii) weighing the dried sample again.

Soil and sediment samples should also be treated with 2 mol/L HCl and the amount of HCl is the same as that described in Sect. 13.3.1.2. The PCDD/F analytical procedure for soil and sediment sample is identical with that for solid waste sample. The HRGC/HRMS operating condition is the same as that for stationary sources.

13.3.1.4 Water

Both EPA Method 8290A and Method 1613, Revision B can be used to determine PCDD/F in water. In Japan, Method JIS K0312 is used for analysis of PCDD/F and

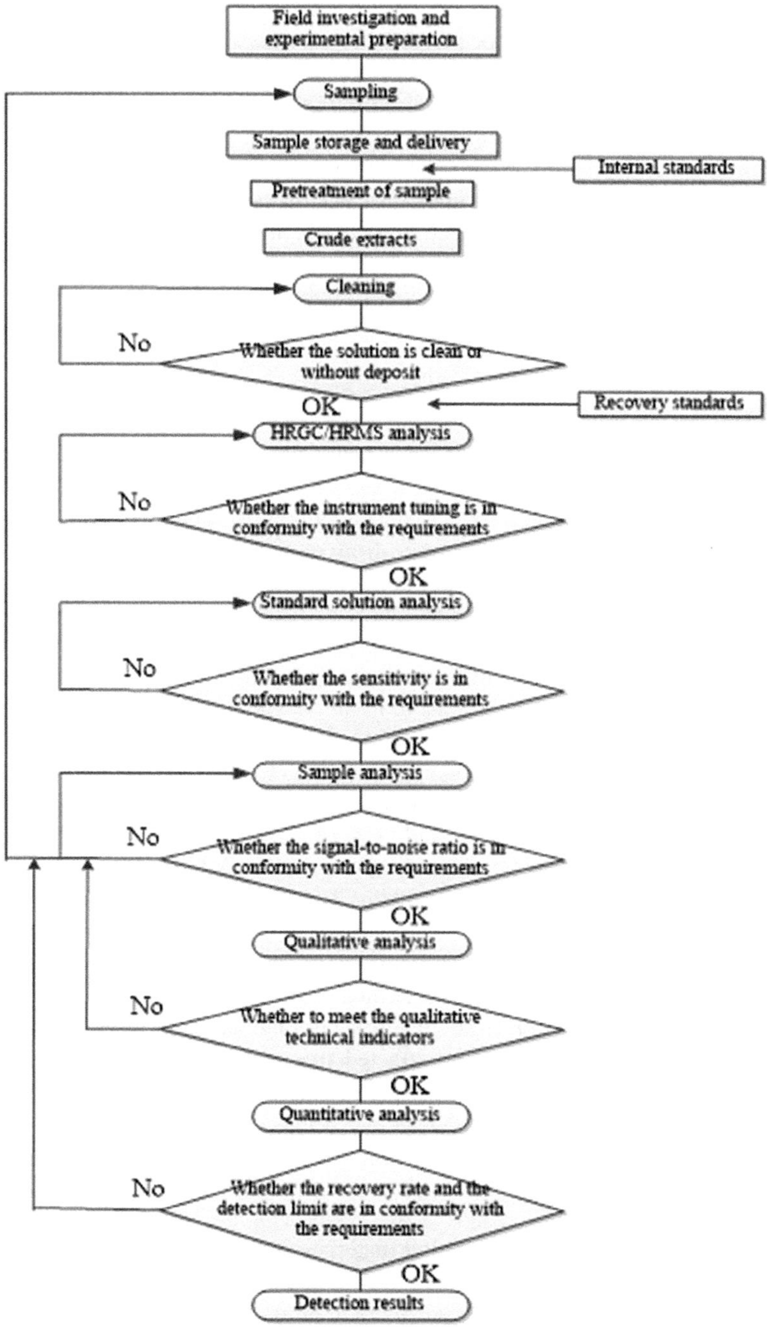

Fig. 13.4 The flow chart for analysis of PCDD/F in solid waste samples

coplanar PCB in industrial water and wastewater [126]. In China, Method HJ 77.1 2008 is used for determination of PCDD/F in water by the isotope dilution method applied to HRGC/HRMS [127]. Different sampling methods are applied for different types of water, for example, Method HJ/T 164-2004 is adopted for sampling groundwater [128]. After sampling, internal standards are added to the samples. Then, the PCDD/F in water are extracted by means of a glass fibre filter and a solid phase extraction (SPE) technique filter. The detailed analytical procedure is shown in Fig. 13.5. The HRGC/HRMS operating condition for water samples is the same as that for stationary sources.

13.3.1.5 Open Fires

The open fires mentioned here refer to the uncontrolled combustion and open burning of biomass and biowaste, intentionally fired for agricultural and forestry purposes (e.g., prescribed burning, slash-and-burn agriculture, fire–fallow cultivation, etc.) or unintentionally burned during wildfires. The emission factors of dioxins (based on unit weight of biofuel) from open fires can be up to three orders of magnitude higher than those from well-organized and APCD-equipped incinerators [59]. These sources of dioxins sources are, by their nature, non-point sources and spread out over large areas. The commonly used sampling method for stationary sources (introduced in Sect. 13.3.1.1) is not suitable for these sources. For measuring dioxins from open fires of biomass, various experimental equipment and sampling methods have been designed and applied.

The US EPA's Open Burning Test Facility (OBTF) has been used for measuring dioxins and some other pollutants from a wide variety of fires, e.g. combustion of household waste [88, 101, 129] and agricultural residue burning [130], etc. Figure 13.6 shows a schematic diagram of the OBTF. A high-volume air handler meters dilution air to the burn hut, simulating ambient dilution. Additional fans and flow deflectors are set up inside this burn hut, to enhance recirculation and maintain high levels of gas phase mixing. Ventilation holes are drilled all around the base of the barrel for supplying primary combustion air. The sample weight is continuously measured by a scale platform, so that the burning rate can be estimated. Gas samples for measuring dioxins are extracted from the transfer duct, using EPA's ambient TO-9 [131] method (specifying an open-faced filter holder, followed by polyurethane foam surrounding an XAD-2 sorbent). After sampling, the subsequent pretreatment and clean-up process of sample are the same as that from stationary sources (Sect. 13.3.1.1), as well as the analytical procedure of PCDD/F using HRGC/HRMS.

Field measurements involve igniting materials in the open air and detecting the emission levels, directly from sampling the erratically moving plume from the open fire. This method is suitable for evaluating emission of dioxins from several biomass-burning sources, such as forest fires, savannah and grassland fires, agricultural residue burning, and domestic waste burning [89, 93, 94, 132] During such field tests, sampling is much more challenging than during laboratory tests, because

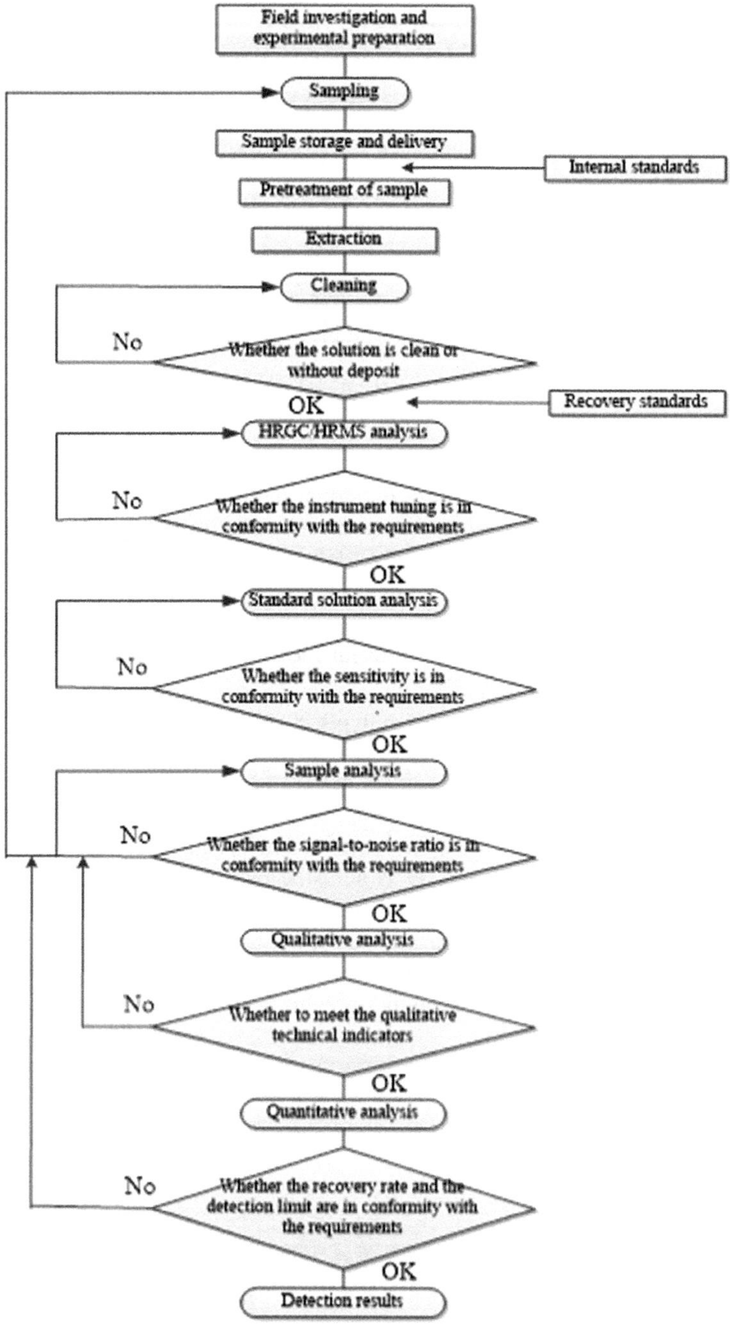

Fig. 13.5 The flow chart for analysis of PCDD/F in water samples

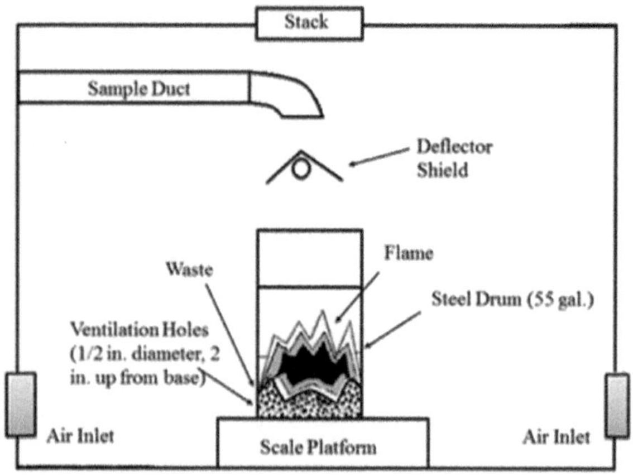

Fig. 13.6 US EPA's Open Burning Test Facility. Reprinted with permission from Elsevier [76]

of the unpredictable changes in wind direction and shifting position of the flame front. Thus, a mobile sampler should be preferred for adjusting to changes in wind direction and moving fire. The US EPA developed a movable hand-held boom (Nomad) sampler, enabling the sampling crews to insert and adjust the suction end of the probe directly into the plume, without need to get too close to smoke or fire (Fig. 13.7) [76]. The sample inlet is several meters above the fire, collecting the smoke at the source.

To keep the orifice of sampler always in the heart of the plume and to sample the high plumes from wildfires, an aerostat-borne instrument was developed and tested by Aurell et al. [133, 134] (Fig. 13.8). A helium-filled aerostat carries the "flyer" containing instruments adapted for sampling different pollutants for different purposes. The aerostat is tethered by two spectra lines to a pair of winches equipped on vehicles. The location of the flyer is adjusted by controlling the length of the lines and the movement of vehicles. This sampling method can be employed in sampling emissions from in situ burns.

In the cases of both Nomad sampler and aerostat-borne sample, PCDD/F are sampled by drawing the plume through a polyurethane foam/XAD-2/polyurethane

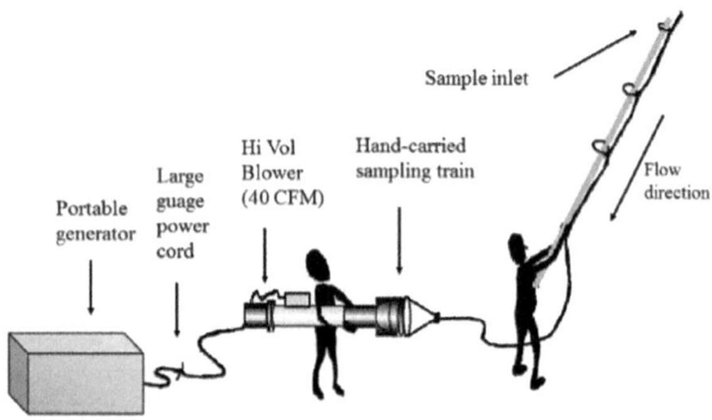

Fig. 13.7 Hand-held boom (Nomad) sampler. Adapted with permission from Elsevier [76]

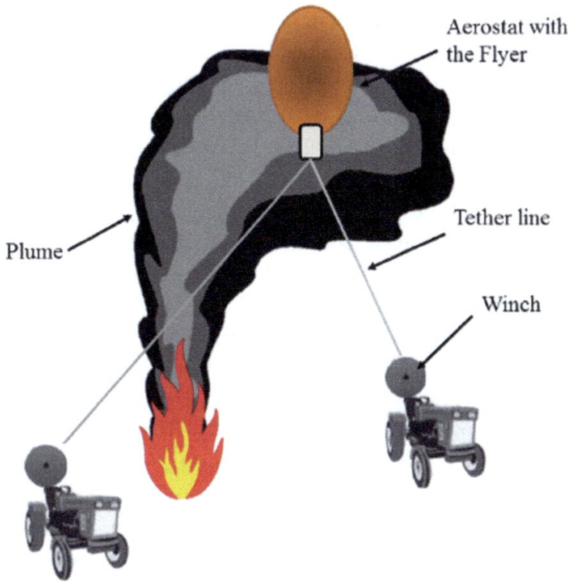

Fig. 13.8 Schematic of the plume sampling aerostat-borne flyer. Adapted with permission from Elsevier [133]

foam sorbent cartridge and then analyzed by HRGC/HRMS after extraction and clean-up, similar to that from stationary sources.

Sampling soot, soil, ash, and settled debris at fire sites or at sites of eventual downwind deposition is another assessing method, used under real life conditions for sources presenting measuring difficulties or for evaluating eventual environmental impacts after fire extinguishment. After sampling, the pretreatment of samples and analytical procedure of PCDD/F are identical with that described in Sect. 13.3.1.3.

13.3.2 On-line Measurement of Dioxins

Isotope dilution HRGC/HRMS has been used successfully for many years as an off-line method for trace analysis of dioxins. However, this method is also labour intensive and time-consuming and requires expensive equipment and highly trained analysts, resulting in high expense and a typical delay of weeks, passing between sampling and receipt of final results [135, 136]. Numerous studies have demonstrated that PCDD/F emissions increase significantly during transient operation (i.e. start-up and shut-down) of MSWIs [137–139]. The transient operation of MSWIs contributes to up to two thirds of annual PCDD/F emissions from MSWI [137, 138]. The rise of PCDD/F due to transients and operating variations suggests that fast on-line monitoring is essential for reflecting and coping with the changing operating conditions [136]. Direct trace analysis of PCDD/F is technically difficult and expensive due to the large molecular weights and the large group of more than 200 analytes. Therefore, real-time, on-line analysis of surrogate compounds instead of dioxins seems to be convenient [140, 141]. In 1987 already, Öberg and Bergström found a close covariation between CBz, CP and PCDD/F from waste incineration [142]. Amongst the potential surrogates, CBz show the best possibilities as indicators for TEQ of PCDD/F [141, 143].

The method of resonance-enhanced multiphoton ionization (REMPI) coupled to time-of-flight mass spectrometry (REMPI-TOFMS) has been confirmed by many research groups [136, 144–146] as a reasonable method for fast detection of surrogates. REMPI-TOFMS is a laser-based, soft ionization process which combines enhanced sensitivity and selectivity with real time response [136]. The REMPI technique consists of the laser-induced ionization of a molecule of interest using a two-step process.

First an ultraviolet (UV) photon is absorbed, and subsequent absorption of a second UV photon raises the internal energy above the ionization energy of the molecule so that a molecular ion is formed. This process is significantly enhanced, when the first step is in resonance with an electronic excited state, preferably the lowest electronic state (S1) (see the energy levels of an electron of molecule A in Fig. 13.9) [147]. Each molecule has its own characteristic UV absorption spectrum. Using an appropriate laser wavelength, selective absorption and selective ionization of a particular molecule can be achieved [148]. For REMPI of molecules, mostly pulsed lasers are used. This permits ion production within a short time and in small volumes, resulting in ideal preconditions for applying time of-flight mass spectrometry (TOFMS) [148].

REMPI-TOFMS has been successfully used for on-line detection of surrogate compounds of PCDD/F from waste incinerators [136, 149]. A correlation of R = 0.85 and 0.89 was established between 40 pairs of simultaneous 5-min GC-REMPI-TOFMS measurements of 1,2,4-trichlorobenzene, and 5 min conventional sampling and analysis for the TEQ and total amounts of PCDD/F, respectively [136]. Using the mathematical model between CBz and the TEQ value of

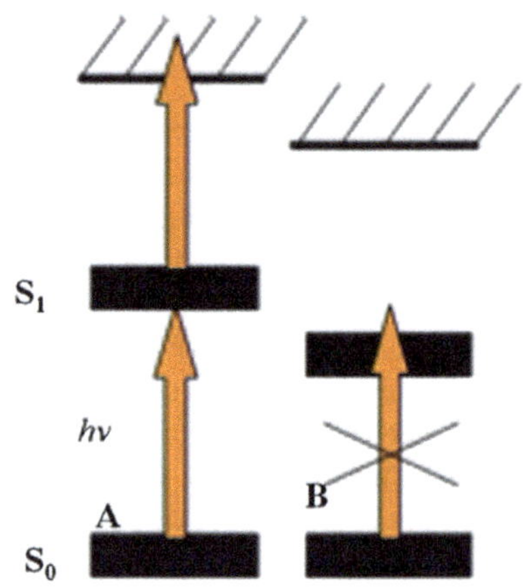

Fig. 13.9 The principle of resonance-enhanced multiphoton ionization (REMPI)

PCDD/F, the TEQ value of PCDD/F could be calculated by on-line measurement of CBz through REMPI-TOFMS.

The GC-REMPI-TOFMS system can be applied for providing frequent measures of correlative PCDD/F concentrations. Thus, it allows a testing and an understanding of measures devised to minimize the formation of PCDD/F and developing operational feedback for reducing their emissions.

13.3.3 CALUX Bioassay

Besides on-line measurement of surrogate compounds of PCDD/F, there is also another rapid, cheap and reliable complementary technique—the bioassay. This is a bioanalytical tool, based on the ability of key biological molecules (e.g., antibodies, receptors, enzymes) to recognize a unique structural property of dioxin-like compounds, or on the ability of cells or organisms to show a specific response to dioxin-like compounds. There are many bioanalytical methods in actual use, e.g. Chemical-Activated Luciferase Gene Expression (CALUX), Ethoxy Resorufin-O-Deethylase (EROD), Enzyme-Linked Immuno Sorbent Assay (ELISA), etc. CALUX and EROD are based on toxicity of dioxins signaling routes and ELISA is based on the enzyme immunoassays. The CALUX bioassay is the most sensitive and reliable method among all of these screening methods.

CALUX is a reporter-gene-based cell bioassay that uses genetically modified cells that respond to chemicals that activate the AhR [150]. Figure 13.10 shows the mechanism of CALUX bioassay. Halogenated aromatic hydrocarbons (HAH), including dioxins, bind to an intracellular receptor called the aryl hydrocarbon receptor (Ah Receptor) and activate this receptor. The HAH-Ah Receptor complex then travels to the nucleus of the cell. An activated HAH-Ah Receptor then binds to specific sequences in the DNA called dioxin responsive elements (DRE) or xenobiotic responsive elements (XRE). The binding of the HAH-Ah Receptor complex to the XRE induces the expression of XRE-associated genes. Luciferase gene expression is increased. The induced messenger RNA (mRNA) is released into the cytoplasm. The mRNAs are translated into new proteins. New proteins produce the biological and toxicological effects that are observed following exposure to dioxins and related HAHs. Finally, a dioxin TEQ is determined from a dose response curve of the luminescence produce by the enzymatic activity of the induced luciferase reporter protein.

In 2002, the CALUX bioassay was adopted as the screening method for the EU food and feed regulation (2002/69/EC) and in 2005 it was also approved as an official method by the Ministry of the Environment in Japan. In 2007, US EPA adopted the CALUX bioassay for biosolids analysis and designated it as US EPA Method 4435. In China, screening method of PCDD/F in solid waste by CALUX was established by National Research Centre for Environmental Analysis and Measurement in 2015.

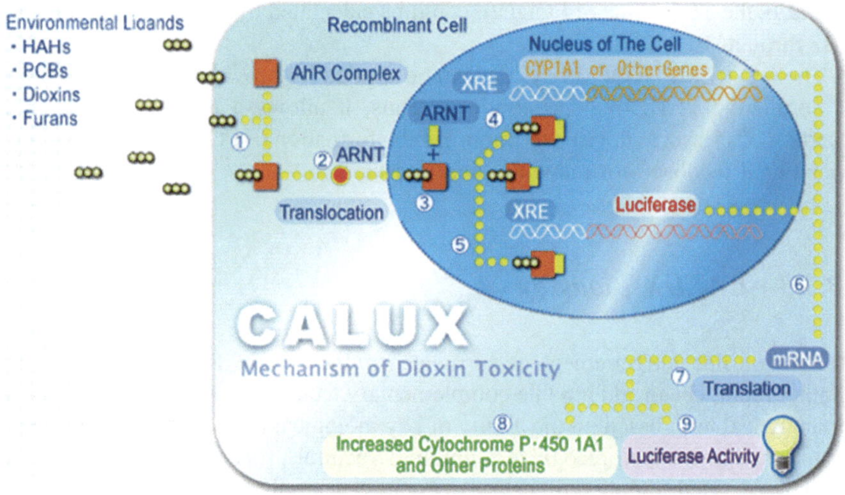

Fig. 13.10 The mechanism of CALUX bioassay

The CALUX bioassay has been successfully applied to determine the dioxin TEQ in various environmental and biological matrices, e.g. fish oil and feed [151], human serum [152], fly ash, soil and sewage sludge [153, 154], human breast milk [155], etc. Comparative analysis of PCDD/F in soil around waste incineration plants in China using CALUX bioassay and HRGC/HRMS was conducted by Du et al. [156]. The results show that the CALUX-TEQ and the WHO-TEQ values are significantly correlated (R = 0.87), while the costs of CALUX analysis are 40–70% lower than those of HRGC/HRMS [156].

ELISA is an enzyme-linked immunosorbent assay and also adopted as Method 4025 by US EPA [157], which uses ELISA kits manufactured by Cape Technologies to screen for PCDD/F at 500 pg toxic equivalents per gram of soil substrate (pg TEQ/g). However, ELISA has a limited dynamic range of dose-response. Thus, it is unsuitable to screen for wide ranges of PCDD/F concentrations [158].

Hence, the CALUX bioassay for PCDD/F determination has been regarded as an important tool for screening samples.

13.4 Dioxins Emission Standards

Table 13.2 lists the standards for dioxins emission from waste incineration source in different countries. The most stringent limit for dioxins emission from waste incineration is set as 0.1 ng TEQ/m^3 (dry gas at 11 vol. % of oxygen) by the European Union countries since 2000 [159], yet individual MSWI plant may have

Table 13.2 Standards for dioxins emission from waste incineration in different countries

Standard	EU2000/76/EC [159]	The United Stated standard [160]	Japanese standard/2002[a] [161]	Chinese standards		
				GB 18485-2001 for MSWI [163]	GB 18485-2014 for MSWI[b] [162]	GB 18484-2001 for HWI [164]
Limit of dioxins (ng TEQ/m^3)	0.1	0.2	0.1	1.0	0.1	0.5

Note All the limits are normalized to dry gas, 11% O_2 content

[a]When the treatment capacity is no less than 4 t/h, standards is 0.1 ng TEQ/m^3; no less than 2 t/h and less than 4 t/h, is 1 ng TEQ/m^3; less than 2 t/h, is 5 ng TEQ/m^3

[b]When the treatment capacity is more than 100 t/d, standards is 0.1 ng TEQ/m^3; between 50 and 100 t/d, is 0.5 ng TEQ/m^3; less than 50 t/d, is 1 ng TEQ/m^3

received more ambitious limit values, together with their operating permit. The United Stated standard of dioxins emission is set as 0.2 ng TEQ/m^3 by the US EPA [160]. In Japan, the government imposed numerous measures to reduce dioxin emission, leading to the emission decreasing by more than 95% from year 1999 to 2003. Japan is also implementing the stringent limit of 0.1 ng TEQ/m^3 since 2002 [161] as well as a requirement of <5 μg/Mg of MSW.

In China, the guidance on strengthening the control of dioxins emissions issued in 2010 required that dioxins should be gradually incorporated into the environmental impact assessment of related industries. The latest standard for dioxins emission from MSWI (GB 18485-2014) is limited to below 0.1 ng TEQ/m^3 [162], much lower than the older version (GB 18485-2001, of 1.0 ng TEQ/m^3) [163]. However, the dioxins emission standard (GB 18484-2001) from hazardous waste incinerator is still set as 0.5 ng TEQ/m^3 [164]. In the near future, the dioxin emission standard will become more and more stringent, as a result of the technical development of dioxins remediation, reduction and prevention.

In Colombia, the PCDD/F emission limits depended on the type of facility, capacity and operating conditions before August 2012. After then, limit values were set as 0.1 and 0.2 ng I-TEQ/Nm^3 for incinerators capacities between 100 and 500 kg/h and below 100 kg/h, respectively [165]. Table 13.3 lists PCDD/F limit values for four different time periods as well as recommended operating temperatures [165, 166]. However, in other Asian countries and some African countries, such as in Vietnam, India and some Sub-Saharan countries, there are still a lack of standards.

Not only the value of the standard is important, but also the way it is enforced, respected and monitored. In Belgium, emissions of all MSWI plants are continuously monitored, using AMESA sampling systems.

Table 13.3 Limit values for PCDD/F from incinerator stack gas emissions in Colombia (Comb: combustion; Post-comb: Post-combustion)

Capacity (kg/h)	Limit value for PCDD/F (ng I-TEQ/Nm^3)					Temperature (°C)	
	Before August 2005	August 2005 to July 2006	August 2006 to July 2009	August 2009 to July 2012	August 2012	Comb	Post-comb
<100	20	5	2	1	0.2	750	1000
100–500	10	2	1	0.5	0.1	800	1100

Reprinted with permission from Elsevier [166]

13.5 Dioxins Remediation, Reduction and Prevention

13.5.1 Treatment of Flue Gas

Incineration processes release large volumes of flue gases (typically, 4000–6000 m^3/Mg of MSW) figuring as one of the main sources of dioxin emissions in the environment. The amount of dioxins in the raw flue gases varies widely, from 1 to 500 ng TEQ/m^3 [27]. Memory effects may increase this load: when burning clean biomass, dioxins load may be considerably inflated if other waste was burned before [167]. Intermittent operation is to be discouraged, since starting up and shutting down will lead to disproportionally large emissions [168–170]. The treatment of flue gas comprises mainly the following duties: dust collection, neutralization of acid gases (HCl, SO_2), and adsorption of dioxins on activated carbon. These operations are usually conducted in a temperature range of 140–220 °C. Lower temperatures allow a deeper removal, but the salts formed by neutralizing acid gases show excessive tackiness below 140 °C ($CaCl_2$). In this range 20–80% of dioxins in the flue gas still are in the gas phase. More dioxins report to the solid phase, if the fly ash has a higher carbon content. Therefore, it is important to treat the flue gas to reduce its dioxins concentration to an acceptable limit value (0.1 ng TEQ/Nm^3) or below, before it enters the environment. The following techniques have been adopted for the remediation, reduction and prevention of dioxins in the flue gas.

13.5.1.1 Prevention

The waste to be fired should be homogenized, before being charged into the incinerator unit. The presence of heavy metals and chlorine encourages the formation of dioxins. Biomass is often lean in these elements, so that its firing as a clean single stream is advantageous.

Feeding into the incinerator should be steady, spread over time. The ignition of a large amount of waste may lead to the formation of PICs. Their formation can considerably be reduced in case combustion is conducted in the following conditions:

- Temperature: 850 °C.
- Time of residence at the above Temperature: minimum 2 s.
- Turbulence: extra turbulence is generated at the exit of the furnace by the injection of high velocity (60–100 m/s) secondary air, with the purpose of finely mixing oxygen-lean with oxygen-rich flue gas strands.
- Oxygen: minimum level of 6 vol. %.

Following these rules of prevention will reduce the emission of carbon monoxide and PICs emitted. Both pollutants peak at those moments that the furnace is temporarily in overload, following the ignition of too large amounts of high calorific waste [78].

Dioxins form mainly in the waste heat boiler, while the flue gas cools down from 450 to 200 °C. It is advantageous to have faster cooling, or even abrupt quenching of flue gas. Boiler deposits are believed to play a major role in dioxins formation. A clean boiler shows lower exit temperatures than a strongly fouled boiler and is usually leaner in dioxins.

13.5.1.2 Particulate Matter Control

In principle, fly ash can be reduced from a level of 3000–5000 mg/m^3 in the raw gas to <10 mg/m^3 in clean gas, in European practice even <2 mg/m^3. This corresponds with a collection efficiency of ca. 99.8%. In China, more effort is still needed re Particulate Matter (PM) control, to lower these emissions. With a typical fly ash load of 0.4–4 ng TEQ/g fly ash, a deep PM removal is required to attain the mandatory limit values for dioxins.

Baghouse filters are normally used as particulate matter (PM) control devices for MSWIs. Typically, they remove >98% of all dioxins adsorbed onto ash particles. The fabric sustains a bed of formerly collected materials, allowing the bags to act as deep bed filter media and retaining a wide range of particles sizes (i.e. down to less than 1 μm in diameter) [6].

Although electrostatic precipitators (EP) also collects fly ashes with high efficiency (>98%), the electric field in an EP can facilitate the formation of PCDD/F [171]. Consequently, baghouse filters are more effective than EP in controlling PCDD/F [172]. The efficiency for removal of PCDD/F of the bag filter system largely depends on the operating temperature of the bag filter system. When the bag filter is operating at the temperature range between 150 and 170 °C, the reduction efficiency is about 94%; the efficiency can reach around 98% when the operating temperature decreases to 130 °C [173]. The deep bed filter formed by the fly ash layer operates better at low temperature and with a high activated carbon content in the bed. PM should be eliminated almost completely.

In recent years, the dual bag filter (DBF) and catalytic filter technologies have been developed for removing dioxins emitted from MSWIs. In the DBF system, solid particles such as fly ash and particulate phase PCDD/F are removed by the first bag filter, and then the PCDD/F in the gaseous fraction are removed by the second bag filter using AC adsorption. The PCDD/F concentration in the stack gas dropped enormously, from 3.38–7.73 ng TEQ/m^3 to 0.03 ng TEQ/Nm^3, after the plant was equipped with the dual bag filter system [174]. A 2000 Nm^3/h pilot-scale test was conducted to test and develop a dual bag filter system. The removal efficiency of PCDD/F declines as pressure drop rises in the first bag filter of the DBF system; while, the removal efficiency of PCDD/F grows with a rise in pressure drop over the second bag filter [175].

According to [176] a memory effect is inevitable when a bag filter is applied to the MSWI flue gas, as a result of the gaseous phase PCDD/F released from the contaminated, aged bag filter. After the bag filters have been operated for over four years, the PCDD/F concentrations in the stack flue gases can be even higher than those prior to the aged bag filter, rendering a negative PCDD/F removal [176]. This is surprising and alternative hypotheses should be considered: aged bag filters are rather likely to exhibit pinholes or cracks. Moreover, fly ash and activated carbon could gradually migrate from the dirty gas to the clean gas compartment. There is hardly any evidence for dioxins desorption from fly ash at the typical temperatures (140–170 °C) of bag filtration.

13.5.1.3 Scrubbers

Scrubbers have been in use for many years in waste incinerators for reducing dioxins emissions. Historically, there have been wet and dry scrubbers and also semi-wet or semi-dry scrubbers.

Wet scrubbers wash out the acid gases (HCl, SO_2) with water, attaining high efficiency for HCl, even at pH-values <2 of the scrubbing liquor. After a first, acidic scrubbing step SO_2 is washed out in a second scrubbing step at pH values between 6 and 8. At the highest pH values also a small part of CO_2 is removed, resulting in a higher requirement of neutralizing chemicals, mostly hydrated lime milk.

Wet scrubbers realize the most complete removal of acid gas and require little more neutralizing chemicals than calculated from stoichiometry. Conversely, they are subject to acid corrosion, clogging by barely soluble Ca-compounds and they generate an almost permanent wet plume, psychologically an inconvenient feature. Moreover, wet scrubbing requires by far the highest investment of all methods: since they transfer contaminants (fly ash, heavy metals) to the scrubbing liquor, any resulting wastewater must be cleaned before being discharged. This results in a much larger complexity of flue gas cleaning.

In the USA, venturi scrubbers sometimes combined dust removal with wet scrubbing. The removal efficiencies of the venturi scrubber on the total PCDD/F emission and the total PCDD/F TEQ emission are 46.0 and 44.5%, respectively [177].

However, wet scrubbers are prone to memory effects: dioxins levels measured after wet scrubbing have been found at times to be higher than the levels measured before the scrubber. Scrubbers are manufactured from glass reinforced thermoset polyester resins. It is believed that there is an absorption of PCDD/F on the plastic materials in the scrubber [178, 179]. The absorbed PCDD/F would be released into the flue gas during the operation of scrubber [180]. Hunsinger et al. developed a process, in which dioxins are absorbed during wet scrubbing into polypropylene fill materials. Absorption is conducted at ca. 65 °C; desorption may proceed at 130 °C.

As a next step, dry scrubbers were launched. These show an admirable simplicity, since they only require adequate contact between the flue gas to be cleaned and hydrated lime powder. The plant involves a silo, lime extraction and dispersion into the flue gas. Initially, results were quite convincing, operating at typically 3–5 times the stoichiometric amount required, yet as the acid gas limit values were lowered to 10 mg HCl/m^3 reaching so low a limit is no longer possible. Another dry possibility is offered, however, by using $NaHCO_3$ as neutralizing agent. This reactant is more expensive, yet much more efficient.

Dry scrubbing can also be improved using a wet preconditioning of the flue gas: atomizing water into the ducts, prior to dry scrubbing, enhances the reactivity of hydrated lime.

The Best Available Technology seems to be the semi-dry scrubber. The absorbent (lime slurry) is sprayed and atomized in the spray tower, serving as contactor of flue gas and lime slurry. The acid gas is first absorbed by the liquid phase and then, after evaporating the liquid, by the solid phase. The lime slurry mixes with the combustion gases within the reactor, ideally without forming wall deposits. The neutralizing capacity of the lime can reduce the acid gas constituents (e.g. HCl and SO_2 gas) to the emission thresholds required.

13.5.1.4 Chemical Inhibitors

Addition of some chemical compounds, such as alkaline compounds, N- or S-containing compounds and special functional groups containing compounds, can achieve strong inhibition on dioxins formation [49, 181].

A large number of studies have been conducted regarding the potential suppressing effects of $CaCO_3$ and CaO on the formation of PCDD/F due to the ability of calcium-containing compounds to bind to hydrochloric acid, thus avoiding a fraction of later chlorination and the consequent formation of PCDD/F [182–184]. The suppression efficiency of PCDD/F by CaO surpassed 90% in a range of 280–450 °C and a reaction time of 2 h using pentachlorophenol as a model precursor [185], increasing with the amount of CaO rising. The presence of alkaline compounds is also the main reason for the low emission of dioxins from the cement kiln [186].

As early as 30 years ago, SO_2 has been observed to strongly suppress PCDD/F formation, possibly via reducing elemental chlorine at the source [187]. Sulphur containing compounds, such as elemental sulphur (S), sodium sulphide (Na_2S), sodium thiosulphate ($Na_2S_2O_3$), carbon sulphide (CS_2), pyrite (FeS_2), sulphur trioxide (SO_3), sulphur dioxide (SO_2) and high-sulphur coal can all be successfully used to reduce the formation of PCDD/F [188–190]. Besides S-containing compounds, nitrogen containing compounds such as urea, ammonia, ethanolamine, ethylenediamine tetraacetic acid (EDTA), monoethanolamine (MEA), dimethylamine (DMA) and triethanolamine (TEA) are also confirmed inhibitors [191–193]. Poisoning the catalyst metal has been proven to be the main suppression mechanism for S- and N-compounds [193, 194]. Moreover, SO_2 can also react with Cl_2 to form HCl, and the chlorine available for dioxins formation will reduce [187, 188]. The S- or N-compounds also show dioxins suppression effects in real-scale experiments, with a reduction efficiency of higher than 50% [195, 196].

By comparing the suppression effect of 21 groups of substances, Pandelova et al. indicated that adding compounds containing either S or N led to relatively small inhibitory effects; adding both N- and S-compounds together, however, strongly suppressed PCDD/F [197]. For instance, when using 1% w/w of amidosulphonic acid (ASA) as an additive high PCDD/F inhibition efficiency, about 96%, was obtained [198]. Lot of work has been conducted using the sludge drying gases and thiourea to suppress PCDD/F formation [199, 200]. With a (S + N)/Cl molar ratio of only 0.47, thiourea could inhibit 97.3% of PCDD/F and even 99.8% of TEQ at lab-scale experiments [201]. The suppression effect of thiourea on the formation of dioxins was also systematically tested in a pilot-scale system, situated at the bypass of a hazardous waste incinerator. A dioxin reduction of more than 80% had already been achieved when the (S + N)/Cl molar ratio was increased to ca. 2.20 [199]. The full-scale suppression experiment showed that thiourea could reduce the total TEQ value in flue gas by 73.4% with a (S + N)/Cl molar ratio of 9.4 [199]. Thiourea was also tested as a dioxins inhibitor in a full-scale MSWI plant with high capacity (34 t/h). The suppressant, featuring a high S- and N-content, was converted into liquor and then injected (35 kg/h) into the furnace (850 °C) through the inlets already used for selective non-catalytic reduction (SNCR) of flue gas NO_x. The concentration of PCDD/F decreased to 0.08 ng TEQ/Nm^3, below the national standard in the E.U. and in China of 0.1 ng TEQ/Nm^3 [202].

13.5.1.5 Sorbent or Flow Injection Process

The flow injection process is generally based on the injection of AC or bituminous coal mixed with limestone, lime or inert material into the flue gas flow at temperature of approx. 120 °C. The material is suspended in the flow homogeneously and subsequently settles in a layer on the surface of the cloth filter.

The most effective end-of-pipe treatment is the combination of a semi-dry scrubber and a bag filter coupled with AC injection. Carbon consumption in this entrained flow process is around 50 mg/Nm^3. The bag filters should be operated below 200 °C, and preferably between 120 and 150 °C [6]. AC adsorbs PCDD/F from gas streams but can simultaneously generate PCDD/F via de novo synthesis if the temperature for injection is higher than 200 °C. If the temperature for injection is higher than 150 °C, the adsorption efficiency of AC would largely decrease with the rising temperature [203]. Moreover, there is a rising ignition hazard, prompted by the collection of heavy metals on AC. Although the lower temperature is beneficial for dioxins adsorption, the deliquescence of $CaCl_2$ resulting from low temperature should also be seriously considered. The injection amount of sorbent is an important factor for the adsorption efficiency. The spraying AC is suggested to be 50–80 mg/Nm^3 when the initial concentration of dioxins in the flue gas is 0.2–15.8 mg TEQ/Nm^3 [204]. A linear relationship was found between the injection amount of AC and removal efficiency of dioxins when the injection amount was lower than 65 mg/Nm^3. The removal efficiency would not increase considerably with more AC injection when the amount has reached 150 mg/Nm^3 [205]. Removal efficiency has been modelled by Lu et al. [203] and Ma et al. [206].

To increase AC's PCDD/F-sorption capacity and lifetime, it is beneficial to develop carbon-based sorbents that simultaneously destroy PCDD/F while adsorbing the toxic chemicals from gas streams. Hydrogen-treated and surface-functional AC can adsorb PCDD/F efficiently, with TEQ removal efficiencies exceeding 99% and mass removal efficiencies exceeding 98% [207]. Carbon nanotubes, which are also effective dioxins sorbent, can achieve higher adsorption efficiency than AC. Carbon nanotubes reaches the total removal efficiency over 86.8%, much higher than the removal efficiencies of only 54.2% for the traditional AC [208].

13.5.1.6 Fluidized-Bed Process with Adsorbent Recycling

From the process engineering point of view the fluidized-bed process is situated between the fixed bed as well as moving-bed adsorbent process and the flow injection process described before. This process operates at a temperature of 100–120 °C and uses a fluidized bed adsorber downstream of the conventional flue gas cleaning equipment. The pre-cleaned flue gas is intimately mixed with fine-grained HOC and an additive, which aims to adsorb the dioxins. The adsorbent is separated from the flue gas in a dust collector and re-circulated to the fluidized bed [6]. The dioxins loaded sorbent particles are removed in a downstream fabric filter and re-circulated to the fluidized bed adsorber. This process can also be used for simultaneous absorption of acid gases such as HCl, HF and SO_2 by suitable additives. The decisive advantage of the circulating fluidized bed over the entrained flow adsorber is the better utilization of the sorbent due to the more favourable mass transfer conditions and longer solids retention times in the system [209, 210].

13.5.1.7 Catalytic Decomposition of Dioxins

Catalytic filter technique for removing dioxins offers the ability to destroy toxic air pollutants and reduce their overall release to and impact on the environment. This technology can eliminate the need for sorbents, which move dioxins from the gas phase to the solid phase, effectively reducing the potential future liabilities associated with dioxin-contaminated solid waste. Catalytic filters are installed in existing baghouses and require no new equipment or operating procedures. It is cost-effective. The non-poisoned catalytic filter shows destruction and removal efficiency for PCDD/F and their TEQ of higher than 99%. The catalytic decomposition of PCDD/F strongly depends on their volatility and oxidative behaviour, both related to the degree of chlorination [211]. It appears that the PCDD/F removal efficiency achieved with a catalytic filter is higher than that of an AC injection + bag filter system.

Catalytic decomposition is capable of complete destruction of PCDD/F, and considered to be the most promising technique for eliminating dioxins from biomass incineration [212]. Generally, the catalysts for the PCDD/F decomposition comprise transition metal oxides as active elements and their carriers (e.g. TiO_2, Al_2O_3, or activated carbon). A V_2O_5–WO_3/TiO_2 (VWTi) catalyst has been most widely applied in SCR units and has been proven to be an efficient catalyst for dioxins decomposition. The decomposition efficiency of PCDD/F with the addition of VWTi can reach 96% at 280 °C, or 93% for V_2O_5–MoO_3/TiO_2 at 300 °C [213, 214].

In order to attain its industrial application, the VO_x–CeO_x/TiO_2 catalyst has been developed to enable efficient destruction of PCDD/F at relatively low temperatures ($\leq$200 °C) [215, 216]. A decomposition efficiency of 93% was observed over the VO_x–CeO_x/TiO_2 catalyst at 200 °C [215].

The decomposition efficiency of PCDD/F with the addition of the VWTi catalyst was also tested in a commercial full-scale MSW incinerator, and the decomposition efficiency of PCDD/F reached 85.4% [217]. The composition of flue gas can largely affect the decomposition efficiency. The effects of oxygen, water vapour and sulphur dioxide on PCDD/F destruction were studied in terms of destruction efficiency [215]. The results revealed that oxygen played a positive role on PCDD/F destruction by accelerating the conversion of $V^{4+}O_x$ and $V^{5+}O_x$, $Ce^{3+}O_x$ and $Ce^{4+}O_x$. On the contrary, water vapour and SO_2 negatively affected the destruction of PCDD/F by competitive adsorption or poisons active sites of catalyst [215].

The catalytic activity can be enhanced with the addition of a strong oxidizer, such as hydrogen peroxide, proposed by Degussa, or ozone (O_3). The decomposition efficiency of dioxins largely increased when the VO_x–CeO_x/TiO_2 catalyst was coupled with O_3 [218]. Furthermore, the decomposition efficiency of dioxins precursors can also be enhanced. For instance, a higher decomposition efficiency of benzene was achieved at lower temperature when MnO_2 was applied with O_3, and finally the benzene was decomposed into CO_2, CO and H_2O at room temperature [219]. The decomposition efficiency of CBz with the addition of iron-based catalysts is less than 10%, while it increased to 91.7% when added with 1200 ppm O_3 [220].

The catalytic conversion of PCDD/F on adapted nitrogen oxide reduction catalysts was studied and developed simultaneously in Germany and in Japan, respectively by Prof. Hagenmaier (Tubingen University) and Prof. Hiraoka and collaborators (Kyoto University). Today, the catalytic reduction of NO_x is state of the art technology, featuring ceramic blocks with honeycomb cross-section. The most usual reducing agents are ammonia or urea. The same units are capable of oxidizing PCDD/F. A delicate point could be the simultaneous presence of ammonia and remaining acid gases, if the reactor is operating at low temperature (200–220 °C), since ammonia salts could deposit and foul the catalyst. The catalytic activity can be recovered when the catalyst is reheated to ca 300 °C.

13.5.1.8 Electron Beam

High-current electron accelerators are used in diverse industries to enhance the physical and chemical properties of materials and to reduce undesirable contaminants such as pathogens, toxic by-products, or other emissions [221]. This method is based on gas treatment with fast electrons and active components that are formed in a gas, exposed to ionizing radiation and responsible for the degradation of PCDD/F. The International Atomic Energy Agency (IAEA) has shown a strong interest in the treatment of organic pollutants by an electron beam due to the low energy consumption [222]. The electron-beam technique was applied to reduce the emission of PCDD/F in a flue gas of 1000 Nm^3/h from the MSWI at a temperature of 200 °C [223]. More than 90% decomposition of PCDD/F was obtained using an electron accelerator at a dose of 14 kGy [223]. The decomposition was initiated through reactions with OH radicals produced by the irradiation of flue gases, followed by oxidation such as the ring cleavage of the aromatic ring, the dissociation of ether bond, and dechlorination. The efficiency of the electron beam treatment of industrial flue gases for removing PCDD/F was also investigated using methods of mathematical modelling [224]. The model can be used to predict the main features of the process and evaluate the process efficiency. In sum, electron beam technique has a number of advantages over traditional ones, namely, the simultaneous removal of harmful admixtures, the simplicity and reliability in operation, low capital and operating costs, etc. [225].

13.5.1.9 Non-thermal Plasma Treatment

Non-thermal plasma technique, producing various active species, radicals and ions, is an attractive method for decomposing organic pollutants in the flue gas of biomass incineration [226]. In the last two decades, the application of non-thermal plasma on the treatment of flue gas has been extensively explored. For instances, dielectric barrier discharge non-thermal plasma was used to evaluate the treatment efficiency of several common volatile organic compounds (VOCs) at constant experimental conditions. The highest removal efficiencies were observed for compounds with the highest percentage of hydrogen in their molecular structures [227]. Compared with non-thermal plasma alone, a combination of non-thermal plasma and catalysis could result in higher energy efficiency and mineralization rates and lower by-product yields during the decomposition of VOCs [228]. Non-thermal plasma combined with catalysts has shown highly removal efficiencies of trichloroethylene, benzene and toluene [229]. The decomposing performance of organic compounds can be largely affected by the original concentration, temperature, and residence time [230]. Therefore, the operation parameters should be determined when the non-thermal plasma techniques are used for flue gas treatment.

13.5.1.10 Summary

Prevention is a preferred method to control the emission of carbon monoxide and PICs. It involves homogenizing the waste materials to be fired, feeding these little by little, and respecting suitable combustion conditions, i.e. a residence time of >2 s at 850 °C and >6 vol. % of oxygen and adequate turbulence to mix the flue gas at the exit of the combustion chamber. Limiting the heavy metal (especially copper) content and the amount of halogen in the feed is also useful. The catalytic activity required for forming dioxins may be partly and sometimes largely suppressed by adding sulphur and/or nitrogen compounds, or else basic compounds such as $Ca(OH)_2$ or NaOH.

Two methods of removing dioxins from flue gas are at present state of the art:

- Adsorption of dioxins and other semi-volatiles on AC, and
- Oxidation of dioxins and other semi-volatiles on a suitable catalyst, generally based on vanadium and tungsten oxides on a titania carrier.

Much research has been conducted on the destruction of chloroaromatics on industrial and home-made catalysts [229]. Another research avenue is the photochemical catalytic destruction of PCDD/F on a suitable TiO_2 catalyst [215, 216, 231]. The catalytic destruction may also be enhanced by the addition of oxidants, such as ozone [218–220].

13.5.2 Treatment of Fly Ash

Waste incineration generates fly ash, i.e. particulate matter carried over from the combustion chamber. Part of this fly ash settles along the flue gas path, forming boiler ash and filter dust. The latter particles are usually mixed with (i) neutralization salts, derived from reacting acid flue gas compounds with basic substances, such as hydrated lime, and with (ii) dioxins sorbent materials, collected together in filter devices. The collected ashes are addressed as air pollution control (APC) residues and in many countries are classified as hazardous waste because of their high inventory of dioxins, leachable heavy metals, and chlorides [232]. Fly ash contains dioxins, typically in a concentration between 0.3 and 30 ng TEQ/g, and it is required to be properly treated to meet the elution standards for heavy metals before being disposed of in landfills.

For several decades, numerous methods have been developed to destroy dioxins in fly ash, and these can roughly be classified into thermal and non-thermal methods.

Fly ash is usually landfilled, after appropriate treatment. The latter should ensure that the leaching of heavy metals satisfies some conditions that are formulated differently in each country. Dioxins are not too problematic, given their low mobility and strong adsorption. They will not be leached out by groundwater, yet they may be mobilized in the presence of solvents or oil.

Conventional thermal methods including sintering, vitrification and melting. These treatment methods are effective, but they show serious disadvantages such as high cost, formation of some secondary fly ash, and generation of a flow of contaminated flue gas. Therefore, alternative thermal techniques with low energy consumption such as low-temperature thermal treatment, hydrothermal treatment, and supercritical water oxidation are proposed. In addition, non-thermal methods including non-thermal plasma treatment and mechanochemical destruction are also being developed.

13.5.2.1 Thermal Treatment

Thermal treatment is a process by which heat is applied to fly ash in order to obtain a material that is environmentally stable. These processes usually result in a significant reduction in fly ash volume, requiring less space in landfills [233, 234]. Moreover, the dioxins present in fly ash are effectively destroyed due to the high temperatures used [235]. It is reported that more than 95% destruction of dioxins can be obtained during various lab-scale and industrial fly-ash treatment processes [236].

In general, thermal methods can be subdivided into three groups [232]: (a) sintering, at temperatures typically in a range of 900–1000 °C, (b) vitrification, with temperatures of 1300–1500 °C and the material melted together with fluxing additives, and (c) straight melting, similar process to vitrification but without any additives used. The equipment used in thermal treatment includes electric resistance, coke-bed melting furnace, rotary kiln with electric heater, sintering in LPG burning furnace, plasma melting furnace, etc. [236].

The work of Vogg and Stieglitz [237] revealed that in an oxidative atmosphere, 95% of dioxins in fly ash can be decomposed after a thermal treatment at 600 °C for 2 h. However, it is important to note that new dioxins could be formed while passing to lower temperatures zone (300–500 °C). Therefore, the produced flue gas and secondary fly ash during thermal treatment contain high concentration of dioxins and require further treatment. First Hagenmaier et al. [238] and then Vogg and Stieglitz [237] found that, thermal treatment of fly ash at 300 °C for 2 h in an inert atmosphere resulted in >90% decomposition of dioxins. This finding has been widely proven by numerous studies [52, 234, 239]. Compared to conventional thermal treatment, this low-temperature thermal treatment is both energy saving and dioxins reformation preventing.

13.5.2.2 Hydrothermal Treatment

Hydrothermal treatment is a physico-chemical process based on the T/RH/t (temperature, relative humidity, time) relation. Fly ash is put into water or a solution and subject to subcritical hydrothermal treatment at high pressure and temperature. Supercritical water has properties similar to organic solvents and can act as a benign medium for various reactions. Therefore, the dioxins in fly ash could dissolve into solution and be decomposed effectively. Yamaguchi et al. [240] firstly used hydrothermal treatment to destroy dioxins in MSWI fly ash. The total concentration of dioxins in fly ashes decreased from 1100 to 0.45 ng/g, when the fly ashes were subjected to hydrothermal treatment for 20 min in a solution of NaOH containing methanol at 300 °C. Considering the toxicity of methanol, Hu et al. [241] used Fe-sulphate as an alternative to assist the decomposition of dioxins, finding that temperature was the most important factor influencing the destruction of dioxins and the usage of ferric and ferrous sulphate also enhanced dioxins destruction. In addition, water-washing pretreatment of fly ash could only improve the decomposition of dioxins slightly. Jin et al. [242] suggested that the injection of O_2 can significantly accelerate the degradation reaction of dioxins, reducing the reaction temperature and time.

Hydrothermal treatment is more energy saving compared to sintering or vitrification since steam produced in incineration heat recovery boiler can be used as heating source. Moreover, the treated fly ash could be used in the cement industries [243]. However, the corrosion of reactor by alkali or chloride under high pressure and temperature, and the disposal of effluent produced during treatment are tough issues for further developing the application of hydrothermal treatment.

13.5.2.3 Supercritical Water Oxidation

Water is supercritical at a temperature of over 374 °C and a pressure over 22.1 MPa. Supercritical water is an atypical reaction medium for organic pollutants because of the combination of the solvation and physical properties. When organic compounds and oxygen are dissolved in water above the critical point, they are immediately brought into intimate molecular contact in a single homogeneous phase, with no interface transport limitations. Hence, at sufficiently high temperatures, the kinetics is fast and the oxidation reaction proceeds rapidly to completion. Supercritical water oxidation (SCWO) has been proven to be an effective method for dioxins degradation. After 30 min treatment with supercritical water (400 °C, 30 MPa) and hydrogen peroxide Sako et al. [244] obtained a 99.7% decomposition of dioxins in fly ash. Afterwards, they also developed another process to remove dioxins from fly ashes. Dioxins in fly ashes were first extracted using supercritical fluid (CO_2) and then were adsorbed and concentrated by AC. Dioxins contained in fly ash can be removed effectively through the extraction-adsorption process, and the dioxins contained in AC were then completely destructed by SCWO [245].

13.5.2.4 Non-thermal Plasma Treatment

Plasma technology has several advantages over other, more conventional control devices. It performs effectively and economically at very low concentrations under ambient temperature condition and only requires low maintenance. Plasma technology is a new method for decomposition of dioxins, and it can be divided into thermal and non-thermal plasma. Thermal plasma is used to melt fly ash into slag, decomposing the dioxins in it. Thermal plasma is effective yet requires high temperature and consumes much energy. An alternative method is non-thermal plasma, which shows lower energy consumption and simpler structure.

Non-thermal plasma, including dielectric barrier discharge, pulsed corona plasma, gliding arc plasma, is also an effective method for dioxins destruction. Zhou et al. [246] treated fly ash with non-thermal nanosecond plasma, finding that the higher the toxicity of the compounds is, the higher is the destruction efficiency. The congener 2,3,7,8-TCDD which has the highest toxicity shows the highest destruction efficiency of 81%. Yan et al. [247] studied the effect of gliding arc plasma on the destruction of dioxins in fly ash, resulting in a destruction efficiency of 79%. Ren et al. [248] destructed dioxins in fly ash with vortex gliding arc plasma in different atmosphere conditions, obtaining a better result in oxidative conditions than in nitrogen.

13.5.2.5 Mechanochemical Destruction

Mechanochemical destruction (MCD) draws on mechanical energy to induce degradation by the reaction between persistent organic pollutants (POPs) and additives. The reactors are usually planetary ball mills for laboratory studies, radical planetary ball mills or vibration mill for full-scale applications. Additives play the most important role in MCD, and they can be generally subdivided as metals (Ca, Na, Mg, Al, Fe), metal oxides (La_2O_3, CaO, Al_2O_3, Bi_2O_3), metal hydroxides (KOH, NaOH, $Ca(OH)_2$), metal hydrides (CaH_2), oxidants (MnO_2), neutral compounds (SiO_2, $CaCO_3$), and other calcium-based natural waste materials (eggshells) [249].

MCD has been proven to have an extraordinary degradation effect on various chlorinated [250], brominated [251], and fluorinated [252] POPs. Also, MCD has been successfully applied to destroy dioxins present in fly ash. Dioxins in fly ash from medical waste incineration can be effectively degraded with the assistant of CaO, and the degradation efficiency increases with the rising ratio of CaO [253]. Some intrinsic compounds in fly ash such as $CaCO_3$, CaO, SiO_2, and Al_2O_3 have also been proven to positively affect dioxins degradation during MCD [253, 254]. Mitoma et al. [255] achieved a nearly complete degradation (99.9%) of dioxins in MSWI fly ash with a mixed additive CaO–Ca. Chen et al. [256] suggested that the chlorides in MSWI fly ash have negative influence on dioxins degradation during MCD. Removing these chlorides with water washing pretreatment prior to ball milling can significantly increase the degradation efficiency. Though MCD owns superiorities such as a high degradation efficiency, simple operation, and environment friendliness [249], many of its studies are limited to the laboratory stage, and only a few full-scale applications can be found in Japan [257] and New Zealand [258]—featuring a planetary mill with capacity of 500 L per vessel and vibration mill for treating 100 m^3 of soil per week.

13.5.2.6 Solidification and Stabilization with Cementitious Materials

Solidification and stabilization (S/S) with cementitious materials is the most commonly used method for the treatment of MSWI fly ash. The process reduces the migration of contaminants (e.g., heavy metals and dioxins) in fly ash through encapsulation-solidification, and also converts the contaminants (mainly heavy metals) into less soluble forms [259]. This method is featured by low-cost and easy-operation: mixing the cementitious materials with fly ash and water, and sometimes with chemical additives (e.g., inorganic or organic chelating agents) [260]. However, there are some drawbacks of this method:

- It almost has no effect for the immobilization of soluble salts, which could result in a poor performance of the S/S matrix after the leaching of salts. A washing step is recommended to remove the soluble salts before S/S [261, 262];
- The treatment is carried out to meet the landfill criteria in most countries, and the quality of S/S matrix is usually far away from being accepted for utilization in constructions;
- The treatment increases the volume of fly ash significantly, putting much more pressure for the subsequent landfill;
- S/S treatment barely has no effect on the destruction of persistent organic pollutants (e.g., dioxins);
- Basically, S/S treatment can greatly lower the leaching concentration of most heavy metals (except for Cr), and it is a mature technology at present, while more optimization of the process is required regarding to the pretreatment of fly ash and the improvement of this method.

13.5.2.7 Discussion

The standard treatment of fly ash is its preparation for landfill and involves mechanical, physical and chemical treatment to reduce the leachability of heavy metals. Dioxins are strongly adsorbed on the fly ash and are regarded only rarely as a problem. Nevertheless, suitable preventive methods may ensure its carbon burn-out and thus also their load of dioxins.

Japan is exceptional in this respect, since it prescribes that the net output from MSWI should not exceed 5 µg TEQ/ton of MSW fired. This statement makes mandatory a treatment of the fly ash. The main possibilities are the use of the Hagenmaier drum, involving a thermal treatment at ca. 400 °C in a reducing atmosphere, or else a high temperature thermal treatment, involving a wide range of furnaces and concomitant heating methods, such as coke fire, thermal plasma, electric resistance heating, or post-combustion while firing generator gas. Some other methods are less usual and technologically demanding, involving oxidation in sub- or supercritical water. Mechanochemical treatment has been proven to be a simple and successful technique yet depending on milling fly ash with the right reactant additives.

13.6 Conclusions

Dioxins are a class of structurally and chemically related polychlorinated aromatic hydrocarbons, including PCDD, PCDF and certain dioxin-like PCB as well. The physical/chemical properties of these compounds vary according to their degree and position of chlorine substitution. They are persistent and stable, showing strong bioaccumulation along the food chain. Most often concerned are the seventeen 2,3,7,8-substituted PCDD/F congeners and the twelve dl-PCB congeners. A TEQ value, calculated on the basis of the individual TEF value assigned to each congener, has been developed to facilitate risk assessment of exposure to dioxins. Dioxins may cause overt or subclinical illness and transient acute health problems to the human body.

Thermal processes, e.g. incineration and combustion, are major sources of dioxins. Dioxins during thermal processes are mainly formed through two pathways: a high-temperature homogeneous route (>500 °C) and a low-temperature heterogeneous route (200–400 °C). Generally, the latter contributes to much more dioxins than the former. The heterogeneous route is further subdivided into de novo synthesis and precursor pathways, both proceeding at the surface of solid carbon and/or fly ash and requiring certain metals as catalyst. Numerous endeavours have been launched to study their routes and reaction mechanisms, mainly on formation from MSWI, reaching some consensus but still much in debate.

The relation between dioxins formation and biomass burning were discussed. Factors of influence include temperature, oxygen, time and turbulence, combustion modes of burning, and major chemical components and contaminants in biomass fuels, etc. The main structural organic components in biomass, i.e. carbohydrates (cellulose and hemicellulose) and lignin, go through different ways of decomposition during combustion and show different activities in dioxins formation. The combustion of contaminated biofuels, by pesticides, wood preservatives, paint, or general waste, etc., can produce emission factors of dioxins several orders of magnitude higher than those from pure biomass. Equipped with advanced APCD, well-organised and continuously operated biomass incinerators emit rather low concentration of dioxins in the stack gas and meet the emission limit value of 0.1 ng TEQ/Nm^3. However, combustion in older, batch-operated furnaces, domestic stove and fireplace and even open burning of biomass should be largely avoided because of the rather high emission factors of dioxins.

Standard reference methods for dioxins analyses are based on HRGC/HRMS due to its high sensitivity and selectivity. The standard methods for measuring PCDD/F (and PCB) from stationary sources and in solid waste, soil and sediment and water samples were discussed. However, these methods are costly and time consuming and require highly qualified infrastructure and operators. There has been a search for an alternative to HRGC/HRMS. On-line measurement of dioxin surrogate compounds using REMPI-TOFM is one of the methods. The concentration and/or TEQ of PCDD/F can be estimated based on the built mathematical model between PCDD/F and the measured surrogate. Another analytical method is the CALUX bioassay, using biological molecules to recognize the unique structural property of dioxin-like compounds, or using cells or organisms to present a specific response to dioxin-like compounds. It has been used as an important tool for fast screening samples.

Due to the numerous regulations, guidelines and stringent standards issued by many countries, a considerable reduction of global dioxins emissions has been achieved during the last decades, especially for the dioxins emitted from waste incineration sources. Properly manipulated incineration parameters, e.g. temperature, oxygen, turbulence and residence time, etc., have led to significant reduction of dioxins emissions. Numbers of techniques, for treating both flue gas and fly ash and other solid residues from waste incineration, have also been developed and widely applied. Commonly used techniques for removing or destroying dioxins from flue gas include particulate matter control, scrubbers, injection of AC, addition of suppressants, catalytic decomposition, etc. Conventional treatment methods for fly ash and other solid residues are mainly thermal methods, which are limited with disadvantages such as high cost, reformation of dioxins during cooling process and generation of contaminated flue gas, etc. Alternatives are thermal techniques with low energy consumption such as low-temperature thermal treatment in an inert or reducing atmosphere, hydrothermal treatment and supercritical water oxidation, and non-thermal methods including plasma treatment and mechanochemical destruction.

References

1. Hites, R.A.: Dioxins: an overview and history. Environ. Sci. Technol. **45**, 16–20 (2010)
2. Shields, W.J., Tondeur, Y., Benton, L., Edwards, M.R.: 14—Dioxins and furans. In: Morrison, R.D., Murphy, B.L. (eds.) Environmental Forensics. Academic Press, Burlington (1964)
3. Qi, Z., Buekens, A., Liu, J., Chen, T., Lu, S., Li, X., Cen, K.: Some technical issues in managing PCBs. Environ. Sci. Pollut. Res. **21**, 6448–6462 (2014)
4. Euro, C.: Dioxins and Furans in the Environment. Science Dossier (2003)
5. Govers, H.A., Krop, H.B.: Partition constants of chlorinated dibenzofurans and dibenzo-p-dioxins. Chemosphere **37**, 2139–2152 (1998)
6. McKay, G.: Dioxin characterisation, formation and minimisation during municipal solid waste (MSW) incineration: review. Chem. Eng. J. **86**, 343–368 (2002)

7. Harrad, S.J., Smith, D.J.T.: Evaluation of a terrestrial food chain model for estimating foodstuff concentrations of PCDD/Fs. Chemosphere **34**, 1723–1737 (1997)
8. Fletcher, C.L., McKay, W.A.: Polychlorinated dibenzo-p-dioxins (PCDDs) and dibenzofurans (PCDFs) in the aquatic environment—a literature review. Chemosphere **26**, 1041–1069 (1993)
9. Pirkle, J.L., Wolfe, W.H., Patterson, D.G., Needham, L.L., Michalek, J.E., Miner, J.C., Peterson, M.R., Phillips, D.L.: Estimates of the half-life of 2,3,7,8-tetrachlorodibenzo-p-dioxin in Vietnam veterans of Operation Ranch Hand. J. Toxicol. Environ. Health Part A Curr. Issues **27**, 165–171 (1989)
10. Emond, C., Michalek, J.E., Birnbaum, L.S., DeVito, M.J.: Comparison of the use of a physiologically based pharmacokinetic model and a classical pharmacokinetic model for dioxin exposure assessments. Environ. Health Perspect. **113**, 1666 (2005)
11. Aylward, L.L., Brunet, R.C., Carrier, G., Hays, S.M., Cushing, C.A., Needham, L.L., Patterson Jr., D.G., Gerthoux, P.M., Brambilla, P., Mocarelli, P.: Concentration-dependent TCDD elimination kinetics in humans: toxicokinetic modeling for moderately to highly exposed adults from Seveso, Italy, and Vienna, Austria, and impact on dose estimates for the NIOSH cohort. J. Exposure Sci. Environ. Epidemiol. **15**, 51 (2005)
12. Van den Berg, M., Birnbaum, L., Bosveld, A.T., Brunström, B., Cook, P., Feeley, M., Giesy, J.P., Hanberg, A., Hasegawa, R., Kennedy, S.W., Kubiak, T., Larsen, J.C., van Leeuwen, F.X., Liem, A.K., Nolt, C., Peterson, R.E., Poellinger, L., Safe, S., Schrenk, D., Tillitt, D., Tysklind, M., Younes, M., Waern, F., Zacharewski, T.: Toxic equivalency factors (TEFs) for PCBs, PCDDs, PCDFs for humans and wildlife. Environ. Health Perspect. **106**, 775–792 (1998)
13. Andersson, P., McGuire, J., Rubio, C., Gradin, K., Whitelaw, M.L., Pettersson, S., Hanberg, A., Poellinger, L.: A constitutively active dioxin/aryl hydrocarbon receptor induces stomach tumors. Proc. Natl. Acad. Sci. **99**, 9990–9995 (2002)
14. Andersson, P., Ridderstad, A., McGuire, J., Pettersson, S., Poellinger, L., Hanberg, A.: A constitutively active aryl hydrocarbon receptor causes loss of peritoneal B1 cells. Biochem. Biophys. Res. Commun. **302**, 336–341 (2003)
15. Puga, A., Tomlinson, C.R., Xia, Y.: Ah receptor signals cross-talk with multiple developmental pathways. Biochem. Pharmacol. **69**, 199–207 (2005)
16. Kimbrough, R.D., Jensen, A.A.: Halogenated Biphenyls, Terphenyls, Naphthalenes, Dibenzodioxins and Related Products. Elsevier (2012)
17. Schecter, A., Birnbaum, L., Ryan, J.J., Constable, J.D.: Dioxins: an overview. Environ. Res. **101**, 419–428 (2006)
18. Ligon, W., Dorn, S.B., May, R.J., Allison, M.J.: Chlorodibenzofuran and chlorodibenzo-p-dioxin levels in Chilean mummies dated to about 2800 years before the present. Environ. Sci. Technol. **23**, 1286–1290 (1989)
19. Alcock, R.E., Jones, K.C.: Dioxins in the environment: a review of trend data. Environ. Sci. Technol. **30**, 3133–3143 (1996)
20. Czuczwa, J.M., Hites, R.A.: Environmental fate of combustion-generated polychlorinated dioxins and furans. Environ. Sci. Technol. **18**, 444–450 (1984)
21. Pekárek, V., Grabic, R., Marklund, S., Punčochář, M., Ullrich, J.: Effects of oxygen on formation of PCB and PCDD/F on extracted fly ash in the presence of carbon and cupric salt. Chemosphere **43**, 777–782 (2001)
22. Olie, K., Addink, R., Schoonenboom, M.: Metals as catalysts during the formation and decomposition of chlorinated dioxins and furans in incineration processes. J. Air Waste Manag. Assoc. **48**, 101–105 (1998)
23. Huang, H., Buekens, A.: On the mechanisms of dioxin formation in combustion processes. Chemosphere **31**, 4099–4117 (1995)
24. Stanmore, B.R.: The formation of dioxins in combustion systems. Combust. Flame **136**, 398–427 (2004)
25. Zhang, M., Buekens, A., Li, X.: Brominated flame retardants and the formation of dioxins and furans in fires and combustion. J. Hazard. Mater. **304**, 26–39 (2016)

26. Tuomisto, J., Vartiainen, T., Tuomisto, J.T.: Synopsis on dioxins and PCBs. Report/National Institute for Health and Welfare (THL) (Raportti/Terveyden ja hyvinvoinnin laitos): 14/2011 (2011)
27. Kulkarni, P.S., Crespo, J.G., Afonso, C.A.M.: Dioxins sources and current remediation technologies—a review. Environ. Int. **34**, 139–153 (2008)
28. Godish, T.: Indoor Environmental Quality. CRC Press (2016)
29. Pivnenko, K., Olsson, M.E., Götze, R., Eriksson, E., Astrup, T.F.: Quantification of chemical contaminants in the paper and board fractions of municipal solid waste. Waste Manage. **51**, 43–54 (2016)
30. Rudel, R.A., Seryak, L.M., Brody, J.G.: PCB-containing wood floor finish is a likely source of elevated PCBs in residents' blood, household air and dust: a case study of exposure. Environ. Health **7**, 2 (2008)
31. UNEP Chemicals: Guidelines for the Identification of PCBs and Materials Containing PCBs. First Issue, UNEP, pp. 1–19 (1999)
32. U.S.E.P.A.: Update to an Inventory of Sources and Environmental Releases of Dioxin-like Compounds in the United States for the Years 1987, 1995, and 2000. National Center for Environmental Assessment, Washington DC (2013)
33. Fiedler, H., Cao, Z., Huang, J., Wang, B., Deng, S., Yu, G.: PCDD/PCDF inventories 1990 vs. 2012. Organohalogen Compd. **74**, 1521–1524 (2012)
34. Axegård, P., Renberg, L.: The influence of bleaching chemicals and lignin content on the formation of polychlorinated dioxins and dibenzofurans. Chemosphere **19**, 661–668 (1989)
35. Olie, K., Vermeulen, P.L., Hutzinger, O.: Chlorodibenzo-p-dioxins and chlorodibenzofurans are trace components of fly ash and flue gas of some municipal incinerators in The Netherlands. Chemosphere **6**, 455–459 (1977)
36. Addink, R., Cnubben, P.A.J.P., Olie, K.: Formation of polychlorinated dibenzo-p-dioxins/dibenzofurans on fly ash from precursors and carbon model compounds. Carbon **33**, 1463–1471 (1995)
37. Peng, Y., Chen, J., Lu, S., Huang, J., Zhang, M., Buekens, A., Li, X., Yan, J.: Chlorophenols in municipal solid waste incineration: a review. Chem. Eng. J. **292**, 398–414 (2016)
38. Altarawneh, M., Dlugogorski, B.Z., Kennedy, E.M., Mackie, J.C.: Mechanisms for formation, chlorination, dechlorination and destruction of polychlorinated dibenzo-p-dioxins and dibenzofurans (PCDD/Fs). Prog. Energy Combust. Sci. **35**, 245–274 (2009)
39. Ooi, T.C., Lu, L.: Formation and mitigation of PCDD/Fs in iron ore sintering. Chemosphere **85**, 291–299 (2011)
40. Takaoka, M., Yamamoto, T., Shiono, A., Takeda, N., Oshita, K., Matsumoto, T., Tanaka, T.: The effect of copper speciation on the formation of chlorinated aromatics on real municipal solid waste incinerator fly ash. Chemosphere **59**, 1497–1505 (2005)
41. Dickson, L.C., Lenoir, D., Hutzinger, O.: Quantitative comparison of de novo and precursor formation of polychlorinated dibenzo-p-dioxins under simulated municipal solid waste incinerator postcombustion conditions. Environ. Sci. Technol. **26**, 1822–1828 (1992)
42. Altwicker, E.R.: Relative rates of formation of polychlorinated dioxins and furans from precursor and de novo reactions. Chemosphere **33**, 1897–1904 (1996)
43. Ryu, J., Mulholland, J.A., Kim, D.H., Takeuchi, M.: Homologue and isomer patterns of polychlorinated dibenzo-p-dioxins and dibenzofurans from phenol precursors: comparison with municipal waste incinerator data. Environ. Sci. Technol. **39**, 4398–4406 (2005)
44. Rappe, C.: Sources of PCDDs and PCDFs. Introduction. Reactions, levels, patterns, profiles and trends. Chemosphere **25**, 41–44 (1992)
45. Tuppurainen, K., Asikainen, A., Ruokojärvi, P., Ruuskanen, J.: Perspectives on the formation of polychlorinated dibenzo-p-dioxins and dibenzofurans during municipal solid waste (MSW) incineration and other combustion processes. Acc. Chem. Res. **36**, 652–658 (2003)
46. Zhang, M., Buekens, A., Olie, K., Li, X.: PCDD/F-isomers signature—effect of metal chlorides and oxides. Chemosphere **184**, 559–568 (2017)

47. Stieglitz, L., Eichberger, M., Schleihauf, J., Beck, J., Zwick, G., Will, R.: The oxidative degradation of carbon and its role in the de-novo-synthesis of organohalogen compounds in fly ash. Chemosphere **27**, 343–350 (1993)
48. Stieglitz, L., Zwick, G., Beck, J., Bautz, H., Roth, W.: Carbonaceous particles in fly ash—a source for the de-novo-synthesis of organochlorocompounds. Chemosphere **19**, 283–290 (1989)
49. Addink, R., Paulus, R.H.W.L., Olie, K.: Prevention of polychlorinated dibenzo-p-dioxins/dibenzofurans formation on municipal waste incinerator fly ash using nitrogen and sulfur compounds. Environ. Sci. Technol. **30**, 2350–2354 (1996)
50. Altwicker, E.R., Milligan, M.S.: Proceedings of the twelfth international symposium formation of dioxins: competing rates between chemically similar precursors and de novo reactions. Chemosphere **27**, 301–307 (1993)
51. Stieglitz, L.: Selected topics on the de novo synthesis of PCDD/PCDF on fly ash. Environ. Eng. Sci. **15**, 5–18 (1998)
52. Addink, R., Olie, K.: Mechanisms of formation and destruction of polychlorinated dibenzo-p-dioxins and dibenzofurans in heterogeneous systems. Environ. Sci. Technol. **29**, 1425–1435 (1995)
53. Hell, K., Altwicker, E.R., Stieglitz, L., Addink, R.: Comparison of 2,4,6-trichlorophenol conversion to PCDD/PCDF on a MSWI-fly ash and a model fly ash. Chemosphere **40**, 995–1001 (2000)
54. Sakurai, T., Kobayashi, T., Watanabe, T., Kondo, T.: Formation of PCDD/Fs from chlorophenols (CPs) on fly ash produced by municipal solid waste incinerators. Organohalogen Compd. **27**, 183–186 (1996)
55. National, W.C.G.: Glossary of Wildland Fire Terminology. U.S. (2008)
56. Williams, A., Jones, J.M., Ma, L., Pourkashanian, M.: Pollutants from the combustion of solid biomass fuels. Prog. Energy Combust. Sci. **38**, 113–137 (2012)
57. Sheng, C., Azevedo, J.: Modeling biomass devolatilization using the chemical percolation devolatilization model for the main components. Proc. Combust. Inst. **29**, 407–414 (2002)
58. Saddawi, A., Jones, J.M., Williams, A., Wojtowicz, M.A.: Kinetics of the thermal decomposition of biomass. Energy Fuels **24**, 1274–1282 (2009)
59. Zhang, M., Buekens, A., Li, X.: Dioxins from biomass combustion: an overview. Waste Biomass Valori. **8**, 1–20 (2017)
60. Chagger, H.K., Kendall, A., McDonald, A., Pourkashanian, M., Williams, A.: Formation of dioxins and other semi-volatile organic compounds in biomass combustion. Appl. Energy **60**, 101–114 (1998)
61. Iino, F., Imagawa, T., Takeuchi, M., Sadakata, M., Weber, R.: Formation rates of polychlorinated dibenzofurans and dibenzo-p-dioxins from polycyclic aromatic hydrocarbons, activated carbon and phenol. Chemosphere **39**, 2749–2756 (1999)
62. Iino, F., Imagawa, T., Takeuchi, M., Sadakata, M.: De novo synthesis mechanism of polychlorinated dibenzofurans from polycyclic aromatic hydrocarbons and the characteristic isomers of polychlorinated naphthalenes. Environ. Sci. Technol. **33**, 1038–1043 (1999)
63. Tame, N.W., Dlugogorski, B.Z., Kennedy, E.M.: Formation of dioxins and furans during combustion of treated wood. Prog. Energy Combust. Sci. **33**, 384–408 (2007)
64. Tame, N.W., Dlugogorski, B.Z., Kennedy, E.M.: Conversion of wood pyrolysates to PCDD/F. Proc. Combust. Inst. **32**, 665–671 (2009)
65. Ryu, J., Mulholland, J.A., Chu, B.: Chlorination of dibenzofuran and dibenzo-p-dioxin vapor by copper (II) chloride. Chemosphere **51**, 1031–1039 (2003)
66. Ryu, J., Mulholland, J.A., Dunn, J.E., Iino, F., Gullett, B.K.: Potential role of chlorination pathways in PCDD/F formation in a municipal waste incinerator. Environ. Sci. Technol. **38**, 5112–5119 (2004)
67. Lasagni, M., Collina, E., Piccinelli, E., Anzano, M.N., Piazzalunga, A., Pitea, D.: Kinetic modeling of the formation and destruction of polychlorinated dibenzo-p-dioxin and dibenzofuran from fly ash native carbon at 300 °C. Environ. Sci. Technol. **47**, 4349–4356 (2013)

68. Ohlemiller, T.J.: Smoldering Combustion. Center for Fire Research, U.S. (1986)
69. Ogle, R.A., Schumacher, J.L.: Fire patterns on upholstered furniture: smoldering versus flaming combustion. Fire Technol. **34**, 247–265 (1998)
70. Rein, G.: Smouldering combustion phenomena in science and technology. Int. Rev. Chem. Eng. **1**, 3–18 (2009)
71. Lavric, E.D., Konnov, A.A., Ruyck, J.D.: Dioxin levels in wood combustion—a review. Biomass Bioenergy **26**, 115–145 (2004)
72. California Air Resources Board: Evaluation on a Woodwaste Fired Incinerator at Koppers Company, Oroville, California. Test Report No. C-88-065. Engineering Evaluation Branch Monitoring and Laboratory Division (1990)
73. Düwel, U., Nottrodt, A., Ballschmiter, K.: Simultaneous sampling of PCDD/PCDF inside the combustion chamber and on four boiler levels of a waste incineration plant. Chemosphere **20**, 1839–1846 (1990)
74. Zhang, M., Buekens, A.: De novo synthesis of dioxins: a review. Int. J. Environ. Pollut. **60**, 63–110 (2016)
75. Stieglitz, L., Vogg, H.: Formation and Decomposition of Polychlorodibenzodioxins and -Furans in Municipal Waste Incineration. Laboratorium für Isotopentechnik, Institut für Heiße Chemie, KfK 4379 (1992)
76. Lemieux, P.M., Lutes, C.C., Santoianni, D.A.: Emissions of organic air toxics from open burning: a comprehensive review. Prog. Energy Combust. Sci. **30**, 1–32 (2004)
77. Zhang, M., Buekens, A., Li, X.: Open burning as a source of dioxins. Crit. Rev. Environ. Sci. Technol. **47**, 543–620 (2017)
78. Buekens, A.: Solving emission problems in a fluid bed MSWI. In: 5th International Conference on: Combustion, Incineration/Pyrolysis, Emission and Climate change (i-CIPEC), Chiang-Mai, Thailand (2008)
79. Kolenda, J., Gass, H., Wilken, M., Jager, J., Zeschmar-Lahl, B.: Determination and reduction of PCDD/F emissions from wood burning facilities. Chemosphere **29**, 1927–1938 (1994)
80. Addink, R., Olie, K.: Role of oxygen in formation of polychlorinated dibenzo-p-dioxins/dibenzofurans from carbon on fly ash. Environ. Sci. Technol. **29**, 1586–1590 (1995)
81. Vogg, H., Metzger, M., Stieglitz, L.: Recent findings on the formation and decomposition of PCDD/PCDF in municipal solid waste incineration. In: Incineration of Municipal Waste, Specialized Seminars on Incinerator Emissions of Heavy Metals and Particulates, Copenhagen, 18–19 Sept 1985 and Emission of Trace Organics from Municipal Solid Waste Incinerators, Copenhagen, 20–22 Jan 1987, pp. 207–216 (1988)
82. Zhang, M., Yang, J., Buekens, A., Olie, K., Li, X.: PCDD/F catalysis by metal chlorides and oxides. Chemosphere **159**, 536–544 (2016)
83. Ghorishi, S.B., Altwicker, E.R.: Rapid formation of polychlorinated dioxins/furans during the heterogeneous combustion of 1,2-dichlorobenzene and 2,4-dichlorophenol. Chemosphere **32**, 133–144 (1996)
84. Schoonenboom, M.H., Olie, K.: Formation of PCDDs and PCDFs from anthracene and chloroanthracene in a model fly ash system. Environ. Sci. Technol. **29**, 2005–2009 (1995)
85. Ross, B.J., Naikwadi, K.P., Karasek, F.W.: Proceedings of the eight international symposium effect of temperature, carrier gas and precursor structure on PCDD and PCDF formed from precursors by catalytic activity of MSW incinerator fly ash. Chemosphere **19**, 291–298 (1989)
86. Tuppurainen, K.A., Ruokojärvi, P.H., Asikainen, A.H., Aatamila, M., Ruuskanen, J.: Chlorophenols as precursors of PCDD/Fs in incineration processes: correlations, PLS modeling, and reaction mechanisms. Environ. Sci. Technol. **34**, 4958–4962 (2000)
87. Zhang, H., Ni, Y., Chen, J., Zhang, Q.: Influence of variation in the operating conditions on PCDD/F distribution in a full-scale MSW incinerator. Chemosphere **70**, 721–730 (2008)
88. Lemieux, P.M., Gullett, B.K., Lutes, C.C., Winterrowd, C.K., Winters, D.L.: Variables affecting emissions of PCDD/Fs from uncontrolled combustion of household waste in barrels. J. Air Waste Manag. Assoc. **53**, 523–531 (2003)

89. Solorzano-Ochoa, G., de la Rosa, D.A., Maiz-Larralde, P., Gullett, B.K., Tabor, D.G., Touati, A., Wyrzykowska-Ceradini, B., Fiedler, H., Abel, T., Carroll Jr., W.F.: Open burning of household waste: effect of experimental condition on combustion quality and emission of PCDD, PCDF and PCB. Chemosphere **87**, 1003–1008 (2012)
90. Zhang, M., Buekens, A.: De novo synthesis in iron ore sintering. Int. J. Environ. Pollut. **60**, 111–135 (2016)
91. Ross, B., Naikwadi, F., Karasek, F.: Kinetic study of PCDD formation from a model chlorinated precursor by catalytic activity of MSW incinerator fly ash. Organohalogen Compd. **3**, 147–150 (1990)
92. Weber, R., Sakurai, T.: Short term de novo formation of PCDD/PCDF and correlation to polychlorinated benzenes as surrogates. Organohalogen Compd. **59**, 49–52 (2002)
93. Meyer, C.P., Beer, T., Muller, J.F.: Dioxins Emissions from Bushfires in Australia. Technical Report No. 1. Australian Government, Department of the Environment, Water, Heritage and the Arts (2004)
94. Gullett, B.K., Wyrzykowska, B., Grandesso, E., Touati, A., Tabor, D.G., Ochoa, G.S.: PCDD/F, PBDD/F, and PBDE emissions from open burning of a residential waste dump. Environ. Sci. Technol. **44**, 394–399 (2010)
95. Tame, N.W., Dlugogorski, B.Z., Kennedy, E.M.: PCDD/F formation in flaming combustion, smoldering, and oxidative pyrolysis of 'eco-friendly' treated wood. Proc. Combust. Inst. **30**, 1237–1243 (2005)
96. Vassilev, S.V., Baxter, D., Andersen, L.K., Vassileva, C.G.: An overview of the chemical composition of biomass. Fuel **89**, 913–933 (2010)
97. Wikström, E., Ryan, S., Touati, A., Gullett, B.K.: Key parameters for de novo formation of polychlorinated dibenzo-p-dioxins and dibenzofurans. Environ. Sci. Technol. **37**, 1962–1970 (2003)
98. Richard Hall, C., Holmes, R.J.: The preparation and properties of some activated carbons modified by treatment with phosgene or chlorine. Carbon **30**, 173–176 (1992)
99. Rigo, H.G., Chandler, A.J., Lanier, W.S.: The Relationship Between Chlorine in Waste Streams and Dioxin Emissions from Waste Combustor Stacks. ASME Research Report CRTD 36 (1995)
100. Wang, L., Lee, W., Lee, W., Chang-Chien, G., Tsai, P.: Effect of chlorine content in feeding wastes of incineration on the emission of polychlorinated dibenzo-p-dioxins/dibenzofurans. Sci. Total Environ. **302**, 185–198 (2003)
101. Gullett, B.K., Lemieux, P.M., Lutes, C.C., Winterrowd, C.K., Winters, D.L.: Emissions of PCDD/F from uncontrolled, domestic waste burning. Chemosphere **43**, 721–725 (2001)
102. Zhang, M., Buekens, A., Ma, S., Li, X.: Iron chloride catalysed PCDD/F-formation: experiments and PCDD/F-signatures. Chemosphere **191**, 72–80 (2018)
103. Fujimori, T., Takaoka, M., Takeda, N.: Influence of Cu, Fe, Pb, and Zn chlorides and oxides on formation of chlorinated aromatic compounds in MSWI fly ash. Environ. Sci. Technol. **43**, 8053–8059 (2009)
104. Vassilev, S.V., Baxter, D., Andersen, L.K., Vassileva, C.G., Morgan, T.J.: An overview of the organic and inorganic phase composition of biomass. Fuel **94**, 1–33 (2012)
105. Tame, N., Dlugogorski, B.Z., Kennedy, E.M.: PCDD/F formation from heterogeneous oxidation of wood pyrolysates. Fire Saf. Sci. **9**, 1189–1200 (2008)
106. Launhardt, T., Thoma, H.: Investigation on organic pollutants from a domestic heating system using various solid biofuels. Chemosphere **40**, 1149–1157 (2000)
107. Zhang, T., Huang, J., Deng, S., Yu, G.: Influence of pesticides contamination on the emission of PCDD/PCDF to the land from open burning of corn straws. Environ. Pollut. **159**, 1744–1748 (2011)
108. Chen, K., Mackie, J.C., Kennedy, E.M., Dlugogorski, B.Z.: Determination of toxic products released in combustion of pesticides. Prog. Energy Combust. Sci. **38**, 400–418 (2012)
109. Muñoz, M., Gullett, B.K., Touati, A., Font, R.: Effect of 2,4-dichlorophenoxyacetic acid (2,4-D) on PCDD/F emissions from open burning of biomass. Environ. Sci. Technol. **46**, 9308–9314 (2012)

110. Zhang, M., Buekens, A., Jiang, X., Li, X.: Dioxins and PVC in combustion and fires: a review. Waste Manage. Res. **33**, 630–643 (2015)
111. Hedman, B., Näslund, M., Nilsson, C., Marklund, S.: Emissions of polychlorinated dibenzodioxins and dibenzofurans and polychlorinated biphenyls from uncontrolled burning of garden and domestic waste (backyard burning). Environ. Sci. Technol. **39**, 8790–8796 (2005)
112. Kuzuhara, S., Sato, H., Kasai, E., Nakamura, T.: Influence of metallic chlorides on the formation of PCDD/Fs during low-temperature oxidation of carbon. Environ. Sci. Technol. **37**, 2431–2435 (2003)
113. US EPA: Method 23 Determination of Polychlorinated Dibenzo-p-Dioxins and Polychlorinated Dibenzofurans from Stationary Sources. U.S. Environmental Protection Agency, Washington, DC (2017)
114. CEN: European Standard EN 1948: Stationary Source Emissions, Determination of the Mass Concentration of PCDDs/PCDFs. CEN, Brussels, Belgium (1997)
115. García-Bermejo, Á., Ábalos, M., Sauló, J., Abad, E., González, M.J., Gómara, B.: Triple quadrupole tandem mass spectrometry: a real alternative to high resolution magnetic sector instrument for the analysis of polychlorinated dibenzo-p-dioxins, furans and dioxin-like polychlorinated biphenyls. Anal. Chim. Acta **889**, 156–165 (2015)
116. US EPA: Method 1613, Tetra-through Octa-Chlorinated Dioxins and Furans by Isotope Dilution HRGC/HRMS. U.S. Environmental Protection Agency, Washington, DC (1994)
117. US EPA: EPA Method 8290A Polychlorinated Dibenzodioxins (PCDDs) and Polychlorinated Dibenzofurans (PCDFs) by High-Resolution Gas Chromatography/High-Resolution Mass Spectrometry (HRGC/HRMS). U.S. Environmental Protection Agency, Washington, DC (1998)
118. JSA: JIS K0311: Method for Determination of Tetra-through Octa-Chlorodibenzo-p-Doxins, Tetra-through Octachlorinated Furans and Coplanar Polychlorinated Biphenyls in Stationary Source Emissions. Japanese Standards Association, Tokyo, Japan (1999)
119. Ministry of Environmental Protection of the People's Republic of China: HJ 77.2-2008, Ambient Air and Flue Gas Determination of Polychlorinated Dibenzo-p-Dioxins (PCDDs) and Polychlorinated Dibenzofurans (PCDFs) Isotope Dilution HRGC-HRMS. The State Environmental Protection Standards of the People's Republic of China (2008)
120. Ministry of Environmental Protection of the People's Republic of China: HJ 77.3-2008, Solid Waste Determination of Polychlorinated Dibenzo-p-Dioxins (PCDDs) and Polychlorinated Dibenzofurans (PCDFs) Isotope Dilution HRGC-HRMS. The State Environmental Protection Standards of the People's Republic of China (2008)
121. Ministry of Environmental Protection of the People's Republic of China: HJ/T20-1998, Technical Specifications on Sampling and Sample Preparation from Industrial Solid Waste. The State Environmental Protection Standards of the People's Republic of China (1998)
122. Ministry of Environmental Protection of the People's Republic of China: HJ 77.4-2008, Soil and Sediment Determination of Polychlorinated Dibenzo-p-Dioxins (PCDDs) and Polychlorinated Dibenzofurans (PCDFs) Isotope Dilution HRGC-HRMS. The State Environmental Protection Standards of the People's Republic of China (2008)
123. Ministry of Environmental Protection of the People's Republic of China: HJ/T 166-2004, The Technical Specification for Soil Environmental Monitoring. The State Environmental Protection Standards of the People's Republic of China (2004)
124. The National Standard of the People's Republic of China: GB 17378.3-2007, The Specification for Marine Monitoring, Part 3: Sample Collection, Storage and Transportation (2007)
125. The National Standard of the People's Republic of China: GB 17378.5-2007, The Specification for Marine Monitoring, Part 5: Sediment Analysis (2007)
126. JSA: JIS K0312: Method for Determination of Tetra-through Octa-Chlorodibenzo-p-Doxins, Tetra-through Octa-Chlorinated Furans and Coplanar Polychlorinated Biphenyls in Industrial Water and Wastewater. Japanese Standards Association, Tokyo, Japan (1999)

127. Ministry of Environmental Protection of the People's Republic of China: HJ 77.1-2008, Water Determination of Polychlorinated Dibenzo-p-Dioxins (PCDDs) and Polychlorinated Dibenzofurans (PCDFs) Isotope Dilution HRGC-HRMS. The State Environmental Protection Standards of the People's Republic of China (2008)
128. Ministry of Environmental Protection of the People's Republic of China: HJ/T 164-2004, Technical Specification for Environmental Monitoring of Groundwater. The State Environmental Protection Standards of the People's Republic of China (2004)
129. Lemieux, P.M.: Evaluation of Emissions from the Open Burning of Household Waste in Barrels, vol. 1. Technical Report, EPA/600/R-97-134a. US Environmental Protection Agency, National Risk Management Research Laboratory (1997)
130. Gullett, B., Touati, A.: PCDD/F emissions from burning wheat and rice field residue. Atmos. Environ. **37**, 4893–4899 (2003)
131. Winberry, W.T., Murphy, N.T., Riggan, R.M.: Compendium method TO-9: "Method for the determination of polychlorinated dibenzo-p-dioxins (PCDDs) in ambient air using high-resolution gas chromatography/high resolution mass spectrometry (HRGC/HRMS)". In: Compendium of Methods for the Determination of Toxic Organic Compounds in Ambient Air. U.S. Environmental Protection Agency, Atmospheric Research and Exposure Assessment Laboratory, Research Triangle Park, NC (1988)
132. Black, R.R., Meyer, C.P., Touati, A., Gullett, B.K., Fiedler, H., Mueller, J.F.: Emissions of PCDD and PCDF from combustion of forest fuels and sugarcane: a comparison between field measurements and simulations in a laboratory burn facility. Chemosphere **83**, 1331–1338 (2011)
133. Aurell, J., Gullett, B.K.: Aerostat sampling of PCDD/PCDF emissions from the gulf oil spill in situ burns. Environ. Sci. Technol. **44**, 9431–9437 (2010)
134. Aurell, J., Gullett, B.K., Pressley, C., Tabor, D.G., Gribble, R.D.: Aerostat-lofted instrument and sampling method for determination of emissions from open area sources. Chemosphere **85**, 806–811 (2011)
135. Anezaki, K., Yamaguchi, K., Takeuchi, S., Iida, M., Jin, K., Kojima, H.: Application of a bioassay using DR-EcoScreen cells to the determination of dioxins in ambient air: a comparative study with HRGC-HRMS analysis. Environ. Sci. Technol. **43**, 7478–7483 (2009)
136. Gullett, B.K., Oudejans, L., Tabor, D., Touati, A., Ryan, S.: Near-real-time combustion monitoring for PCDD/PCDF indicators by GC-REMPI-TOFMS. Environ. Sci. Technol. **46**, 923–928 (2011)
137. Grosso, M., Cernuschi, S., Giugliano, M., Lonati, G., Rigamonti, L.: Environmental release and mass flux partitioning of PCDD/Fs during normal and transient operation of full scale waste to energy plants. Chemosphere **67**, S118–S124 (2007)
138. Wang, L., Hsi, H., Chang, J., Yang, X., Chang-Chien, G., Lee, W.: Influence of start-up on PCDD/F emission of incinerators. Chemosphere **67**, 1346–1353 (2007)
139. Šyc, M., Fišerová, E., Karban, J., Punčochář, M., Pekárek, V.: The effect of transient operations on the levels and congener profiles of PCBz, PCPh and PCDD/F in raw flue gases of MSWI plant. Chemosphere **118**, 261–267 (2015)
140. Ben-Kang, L., Liang, C., Yan-Qiu, W., Zhe, S., Li, W.: Vacuum ultraviolet photon ionization mass spectrometer for detection of dioxin precursors. Chin. J. Anal. Chem. **42**, 173–178 (2014)
141. Lavric, E.D., Konnov, A.A., Ruyck, J.D.: Surrogate compounds for dioxins in incineration. A review. Waste Manage. **25**, 755–765 (2005)
142. Öberg, T., Bergström, J.G.: Emission and chlorination pattern of PCDD/PCDF predicted from indicator parameters. Chemosphere **16**, 1221–1230 (1987)
143. Wang, T., Chen, T., Lin, B., Lin, X., Zhan, M., Li, X.: Emission characteristics and relationships among PCDD/Fs, chlorobenzenes, chlorophenols and PAHs in the stack gas from two municipal solid waste incinerators in China. RSC Adv. **7**, 44309–44318 (2017)

144. Oser, H., Crosley, D.R., Coggiola, M.J., Faris, G.W.: Development of a real-time continuous emissions monitor for dioxins and other chlorinated aromatics. Organohalogen Compd. **40**, 551–554 (1999)
145. Zimmermann, R., Boesl, U., Lenoir, D., Kettrup, A., Grebner, T.L., Neusser, H.J.: The ionization energies of polychlorinated dibenzo-p-dioxins: new experimental results and theoretical studies. Int. J. Mass Spectrom. Ion Processes **145**, 97–108 (1995)
146. Blumenstock, M., Zimmermann, R., Schramm, K., Kettrup, A.: Identification of surrogate compounds for the emission of PCDD/F (I-TEQ value) and evaluation of their on-line real-time detectability in flue gases of waste incineration plants by REMPI–TOFMS mass spectrometry. Chemosphere **42**, 507–518 (2001)
147. Oudejans, L., Touati, A., Gullett, B.K.: Real-time, on-line characterization of diesel generator air toxic emissions by resonance-enhanced multiphoton ionization time-of-flight mass spectrometry. Anal. Chem. **76**, 2517–2524 (2004)
148. Boesl, U., Heger, H.J., Zimmermann, R., Nagel, H., Püffel, P.: Laser mass spectrometry in trace analysis. In: Encyclopedia of Analytical Chemistry (2000)
149. Gullett, B., Touati, A., Oudejans, L., Tabor, D.: Real time monitoring of PCDD/PCDF for transient characterization and process control. In: AWMA Hazardous Waste Combustor Conference, Charleston, SC (2007)
150. Windal, I., Denison, M.S., Birnbaum, L.S., Van Wouwe, N., Baeyens, W., Goeyens, L.: Chemically activated luciferase gene expression (CALUX) cell bioassay analysis for the estimation of dioxin-like activity: critical parameters of the CALUX procedure that impact assay results. Environ. Sci. Technol. **39**, 7357–7364 (2005)
151. Hoogenboom, L., Traag, W., Bovee, T., Goeyens, L., Carbonnelle, S., Van Loco, J., Beernaert, H., Jacobs, G., Schoeters, G., Baeyens, W.: The CALUX bioassay: current status of its application to screening food and feed. TrAC Trends Anal. Chem. **25**, 410–420 (2006)
152. Koppen, G., Covaci, A., Van Cleuvenbergen, R., Schepens, P., Winneke, G., Nelen, V., Schoeters, G.: Comparison of CALUX-TEQ values with PCB and PCDD/F measurements in human serum of the Flanders Environmental and Health Study (FLEHS). Toxicol. Lett. **123**, 59–67 (2001)
153. Brown, D., Kishimoto, Y., Ikeno, O., Chu, M., Nomura, J., Murakami, T., Murata, H.: Validation study for the use of the dioxin responsive CALUX™ assay for analysis of Japanese ash and soil samples. Organohalogen Compd. **45**, 200–203 (2000)
154. Zhang, T., Yu, G., Wang, B., Fiedler, H., Huang, J., Deng, S.: Bioanalytical characterization of dioxin-like activity in sewage sludge from Beijing, China. Chemosphere **75**, 649–653 (2009)
155. Hui, L.L., Hedley, A.J., Nelson, E., Malisch, R., Wong, T.W., Cowling, B.J.: Agreement between breast milk dioxin levels by CALUX bioassay and chemical analysis in a population survey in Hong Kong. Chemosphere **69**, 1287–1294 (2007)
156. Du, Y., Chen, T., Lu, S., Yan, J., Li, X., Cen, K., Nakamura, M., Handa, H.: Comparative analysis of PCDD/Fs in soil around waste incineration plants in China using CALUX bioassay and HRGC/HRMS. J. Hazard. Mater. **192**, 1729–1738 (2011)
157. US EPA: Method 4025—Screening for Polychlorinated Dibenzodioxins and Polychlorinated Dibenzofurans (PCDD/Fs) by Immunoassay. U.S. Environmental Protection Agency, Washington, DC (2002)
158. Lo, C., Buan, E., Li, J., Zhang, W.: Optimized efficiency of mapping a site contaminated with dioxins by immunoassay compared to gas chromatography-high resolution mass spectrometry. Anal. Bioanal. Chem. **408**, 1095–1105 (2016)
159. Tian, H., Gao, J., Hao, J., Lu, L., Zhu, C., Qiu, P.: Atmospheric pollution problems and control proposals associated with solid waste management in China: a review. J. Hazard. Mater. **252**, 142–154 (2013)
160. US EPA: Health Assessment Document for 2,3,7,8-Tetrachlorodibenzo-p-dioxin (TCDD) and Related Compounds, EPA/600/Bp-92/001c Estimating Exposure to Dioxin-Like Compounds, EPA/600/6-88/005Cb. Office of Research and Development, Washington, DC (1994)

161. Kishimoto, A., Oka, T., Yoshida, K., Nakanishi, J.: Cost effectiveness of reducing dioxin emissions from municipal solid waste incinerators in Japan. Environ. Sci. Technol. **35**, 2861–2866 (2001)
162. The National Standard of the People's Republic of China: GB 18485-2014, Standard for Pollution Control on Municipal Solid Waste Incineration (2014)
163. The National Standard of the People's Republic of China: GB 18485-2001, Standard for Pollution Control on Municipal Solid Waste Incineration (2001)
164. The National Standard of the People's Republic of China: GB 18484-2001, Pollution Control Standard for Hazardous Waste Incineration (2001)
165. Ministerio de Ambiente, Vivienda y Desarrollo territorial: Resolución 0886, Colombia (2004)
166. Aristizábal, B., Cobo, M., Hoyos, A., de Correa, C.M., Abalos, M., Martínez, K., Abad, E., Rivera, J.: Baseline levels of dioxin and furan emissions from waste thermal treatment in Colombia. Chemosphere **73**, S171–S175 (2008)
167. Nussbaumer, T.: Dioxin- und PAK-Emissionen der privaten Abfallverbrennung (in German). Bundesamt für Umwelt, Wald und Landschaft (BUWAL), Bern (2004)
168. Zimmermann, R., Blumenstock, M., Heger, H.J., Schramm, K., Kettrup, A.: Emission of nonchlorinated and chlorinated aromatics in the flue gas of incineration plants during and after transient disturbances of combustion conditions: delayed emission effects. Environ. Sci. Technol. **35**, 1019–1030 (2001)
169. Wang, H.C., Hwang, J.F., Chi, K.H., Chang, M.B.: Formation and removal of PCDD/Fs in a municipal waste incinerator during different operating periods. Chemosphere **67**, S177–S184 (2007)
170. Tejima, H., Nishigaki, M., Fujita, Y., Matsumoto, A., Takeda, N., Takaoka, M.: Characteristics of dioxin emissions at startup and shutdown of MSW incinerators. Chemosphere **66**, 1123–1130 (2007)
171. Altwicker, E.R.: Formation of PCDDF in municipal solid waste incinerators: laboratory and modeling studies. J. Hazard. Mater. **47**, 137–161 (1996)
172. Kilgroe, J.D.: Control of dioxin, furan, and mercury emissions from municipal waste combustors. J. Hazard. Mater. **47**, 163–194 (1996)
173. Fujii, T., Murakawa, T., Maeda, N., Kondo, M., Nagai, K., Hama, T., Ota, K.: Removal technology of PCDDs/PCDFs in flue gas from MSW incinerators by fabric bag filter and SCR system. Chemosphere **29**, 2067–2070 (1994)
174. Lin, W., Wang, L., Wang, Y., Li, H., Chang-Chien, G.: Removal characteristics of PCDD/Fs by the dual bag filter system of a fly ash treatment plant. J. Hazard. Mater. **153**, 1015–1022 (2008)
175. Kim, B., Lee, S., Maken, S., Song, H., Park, J., Min, B.: Removal characteristics of PCDDs/Fs from municipal solid waste incinerator by dual bag filter (DBF) system. Fuel **86**, 813–819 (2007)
176. Li, H., Wang, L., Chen, C., Yang, X., Chang-Chien, G., Wu, E.M.: Influence of memory effect caused by aged bag filters on the stack PCDD/F emissions. J. Hazard. Mater. **185**, 1148–1155 (2011)
177. Lee, W., Chang-Chien, G., Chen, S., Wang, L., Lee, W., Wang, Y.: Removal of polychlorinated dibenzo-p-dioxins and dibenzofurans in flue gases by Venturi scrubber and bag filter. Aerosol Air Qual. Res. **4**, 27–37 (2004)
178. Kreisz, S., Hunsinger, H., Vogg, H.: Wet scrubbers—a potential PCDD/F source? Chemosphere **32**, 73–78 (1996)
179. Löthgren, C., van Bavel, B.: Dioxin emissions after installation of a polishing wet scrubber in a hazardous waste incineration facility. Chemosphere **61**, 405–412 (2005)
180. Takaoka, M., Liao, P., Takeda, N., Fujiwara, T., Oshita, K.: The behavior of PCDD/Fs, PCBs, chlorobenzenes and chlorophenols in wet scrubbing system of municipal solid waste incinerator. Chemosphere **53**, 153–161 (2003)
181. Ruokojärvi, P.H., Asikainen, A.H., Tuppurainen, K.A., Ruuskanen, J.: Chemical inhibition of PCDD/F formation in incineration processes. Sci. Total Environ. **325**, 83–94 (2004)

182. Jozewicz, W., Gullett, B.K.: Reaction mechanisms of dry Ca-based sorbents with gaseous HCl. Ind. Eng. Chem. Res. **34**, 607–612 (1995)
183. Shemwell, B., Levendis, Y.A., Simons, G.A.: Laboratory study on the high-temperature capture of HCl gas by dry-injection of calcium-based sorbents. Chemosphere **42**, 785–796 (2001)
184. Lu, S.Y., Chen, T., Yan, J.H., Li, X.D., Ni, Y., Cen, K.F.: Effects of calcium-based sorbents on PCDD/F formation from pentachlorophenol combustion process. J. Hazard. Mater. **147**, 663–671 (2007)
185. Liu, W., Zheng, M., Zhang, B., Qian, Y., Ma, X., Liu, W.: Inhibition of PCDD/Fs formation from dioxin precursors by calcium oxide. Chemosphere **60**, 785–790 (2005)
186. Li, Y., Zhan, M., Chen, T., Zhang, J., Li, X., Yan, J., Buekens, A.: Formation, reduction and emission behaviors of CBzs and PCDD/Fs from cement plants. Aerosol Air Qual. Res. **16**, 1942–1953 (2016)
187. Griffin, R.D.: A new theory of dioxin formation in municipal solid waste combustion. Chemosphere **15**, 1987–1990 (1986)
188. Gullett, B.K., Bruce, K.R., Beach, L.O.: Effect of sulfur dioxide on the formation mechanism of polychlorinated dibenzodioxin and dibenzofuran in municipal waste combustors. Environ. Sci. Technol. **26**, 1938–1943 (1992)
189. Gullett, B.K., Dunn, J.E., Raghunathan, K.: Effect of cofiring coal on formation of polychlorinated dibenzo-p-dioxins and dibenzofurans during waste combustion. Environ. Sci. Technol. **34**, 282–290 (2000)
190. Gullett, B.K., Raghunathan, K., Dunn, J.E.: The effect of cofiring high-sulfur coal with municipal waste on formation of polychlorinated dibenzodioxin and polychlorinated dibenzofuran. Environ. Eng. Sci. **15**, 59–70 (1998)
191. Tuppurainen, K., Aatamila, M., Ruokojärvi, P., Halonen, I., Ruuskanen, J.: Effect of liquid inhibitors on PCDD/F formation. Prediction of particle-phase PCDD/F concentrations using PLS modelling with gas-phase chlorophenol concentrations as independent variables. Chemosphere **38**, 2205–2217 (1999)
192. Tuppurainen, K., Halonen, I., Ruokojärvi, P., Tarhanen, J., Ruuskanen, J.: Formation of PCDDs and PCDFs in municipal waste incineration and its inhibition mechanisms: a review. Chemosphere **36**, 1493–1511 (1998)
193. Ruokojârvi, P., Asikainen, A., Ruuskanen, J., Tuppurainen, K., Kilpinen, C.M.A.P., Yli-Keturi, N.: Urea as a PCDD/F inhibitor in municipal waste incineration. J. Air Waste Manag. Assoc. **51**, 422–431 (2001)
194. Ryan, S.P., Li, X., Gullett, B.K., Lee, C.W., Clayton, M., Touati, A.: Experimental study on the effect of SO_2 on PCDD/F emissions: determination of the importance of gas-phase versus solid-phase reactions in PCDD/F formation. Environ. Sci. Technol. **40**, 7040–7047 (2006)
195. Wu, H., Lu, S., Li, X., Jiang, X., Yan, J., Zhou, M., Wang, H.: Inhibition of PCDD/F by adding sulphur compounds to the feed of a hazardous waste incinerator. Chemosphere **86**, 361–367 (2012)
196. Chang, M.B., Cheng, Y.C., Chi, K.H.: Reducing PCDD/F formation by adding sulfur as inhibitor in waste incineration processes. Sci. Total Environ. **366**, 456–465 (2006)
197. Pandelova, M., Lenoir, D., Schramm, K.: Inhibition of PCDD/F and PCB formation in co-combustion. J. Hazard. Mater. **149**, 615–618 (2007)
198. Samaras, P., Blumenstock, M., Lenoir, D., Schramm, K., Kettrup, A.: PCDD/F prevention by novel inhibitors: addition of inorganic S- and N-compounds in the fuel before combustion. Environ. Sci. Technol. **34**, 5092–5096 (2000)
199. Zhan, M., Fu, J., Chen, T., Lin, X., Li, X., Yan, J., Buekens, A.: Suppression of dioxins by SN inhibitors in pilot-scale experiments. Environ. Sci. Pollut. Res. **23**, 16463–16477 (2016)
200. Chen, T., Zhan, M., Lin, X., Li, X., Lu, S., Yan, J., Buekens, A., Cen, K.: Inhibition of the de novo synthesis of PCDD/Fs on model fly ash by sludge drying gases. Chemosphere **114**, 226–232 (2014)
201. Fu, J.Y., Li, X.D., Chen, T., Lin, X.Q., Buekens, A., Lu, S.Y., Yan, J.H., Cen, K.F.: PCDD/Fs' suppression by sulfur–amine/ammonium compounds. Chemosphere **123**, 9–16 (2015)

202. Lin, X., Yan, M., Dai, A., Zhan, M., Fu, J., Li, X., Chen, T., Lu, S., Buekens, A., Yan, J.: Simultaneous suppression of PCDD/F and NO_x during municipal solid waste incineration. Chemosphere **126**, 60–66 (2015)
203. Lu, S., Ji, Y., Buekens, A., Ma, Z., Jin, Y., Li, X., Yan, J.: Activated carbon treatment of municipal solid waste incineration flue gas. Waste Manage. Res. **31**, 169–177 (2013)
204. Everaert, K., Baeyens, J.: The formation and emission of dioxins in large scale thermal processes. Chemosphere **46**, 439–448 (2002)
205. Chang, Y., Hung, C., Chen, J., Chang, C., Chen, C.: Minimum feeding rate of activated carbon to control dioxin emissions from a large-scale municipal solid waste incinerator. J. Hazard. Mater. **161**, 1436–1443 (2009)
206. Ma, X.C., Zeng, X.P., Liu, J.T., You, K.K., Ren, J.X.: Technology to control PCDD/Fs from MSW incineration processes. Adv. Mater. Res. **610–613**, 2621–2626 (2013)
207. Atkinson, J.D., Hung, P.C., Zhang, Z., Chang, M.B., Yan, Z., Rood, M.J.: Adsorption and destruction of PCDD/Fs using surface-functionalized activated carbons. Chemosphere **118**, 136–142 (2015)
208. Zhou, X., Li, X., Xu, S., Zhao, X., Ni, M., Cen, K.: Comparison of adsorption behavior of PCDD/Fs on carbon nanotubes and activated carbons in a bench-scale dioxin generating system. Environ. Sci. Pollut. Res. **22**, 10463–10470 (2015)
209. Liljelind, P., Unsworth, J., Maaskant, O., Marklund, S.: Removal of dioxins and related aromatic hydrocarbons from flue gas streams by adsorption and catalytic destruction. Chemosphere **42**, 615–623 (2001)
210. Shiomitsu, T., Hirayama, A., Iwasaki, T., Akashi, T., Fujisawa, Y.: Volatilization and decomposition of dioxin from fly ash with agitating fluidized bed heating chamber. Group **500**, 5 (2002)
211. Weber, R., Plinke, M., Xu, Z., Wilken, M.: Destruction efficiency of catalytic filters for polychlorinated dibenzo-p-dioxin and dibenzofurans in laboratory test and field operation—insight into destruction and adsorption behavior of semivolatile compounds. Appl. Catal. B **31**, 195–207 (2001)
212. Hiraoka, M., Takeda, N., Okajima, S., Kasakura, T., Imoto, Y.: Catalytic destruction of PCDDs in flue gas. Chemosphere **19**, 361–366 (1989)
213. Boos, R., Budin, R., Hartl, H., Stock, M., Wurst, F.: PCDD- and PCDF-destruction by a SCR-unit in a municipal waste incinerator. Chemosphere **25**, 375–382 (1992)
214. Cho, C., Ihm, S.: Development of new vanadium-based oxide catalysts for decomposition of chlorinated aromatic pollutants. Environ. Sci. Technol. **36**, 1600–1606 (2002)
215. Yu, M., Lin, X., Li, X., Chen, T., Yan, J.: Catalytic decomposition of PCDD/Fs over nano-TiO_2 based V_2O_5/CeO_2 catalyst at low temperature. Aerosol Air Qual. Res. **16**, 2011–2022 (2016)
216. Yu, M., Li, W., Li, X., Lin, X., Chen, T., Yan, J.: Development of new transition metal oxide catalysts for the destruction of PCDD/Fs. Chemosphere **156**, 383–391 (2016)
217. Zhan, M., Fu, J., Ji, L., Deviatkin, I., Lu, S.: Comparative analyses of catalytic degradation of PCDD/Fs in the laboratory vs. industrial conditions. Chemosphere **191**, 895–902 (2018)
218. Yu, M., Lin, X., Yan, M., Li, X., Chen, T., Yan, J.: Low temperature destruction of PCDD/Fs over V_2O_5-CeO_2/TiO_2 catalyst with ozone. Environ. Sci. Pollut. Res. **23**, 17563–17570 (2016)
219. Einaga, H., Futamura, S.: Catalytic oxidation of benzene with ozone over alumina-supported manganese oxides. J. Catal. **227**, 304–312 (2004)
220. Wang, H.C., Liang, H.S., Chang, M.B.: Chlorobenzene oxidation using ozone over iron oxide and manganese oxide catalysts. J. Hazard. Mater. **186**, 1781–1787 (2011)
221. Chmielewski, A.G., Han, B.: Electron beam technology for environmental pollution control. Top. Curr. Chem. **374**, 68 (2016)
222. Belderbos, J.S., Kepka, L., Martel, M.K., Videtic, G.M., Jeremic, B.: Report from the International Atomic Energy Agency (IAEA) consultants' meeting on elective nodal irradiation in lung cancer: non–small-cell lung cancer (NSCLC). Int. J. Radiat. Oncol. Biol. Phys. **72**, 335–342 (2008)

223. Hirota, K., Hakoda, T., Taguchi, M., Takigami, M., Kim, H., Kojima, T.: Application of electron beam for the reduction of PCDD/F emission from municipal solid waste incinerators. Environ. Sci. Technol. **37**, 3164–3170 (2003)
224. Gerasimov, G.Y.: Simulation of the behavior of dioxins under the conditions of electron-beam gas cleaning of sulfur and nitrogen oxides. High Energy Chem. **37**, 141–145 (2003)
225. Frank, N.W.: Economics of the electron beam process. Radiat. Phys. Chem. **1993**(45), 1017–1019 (1995)
226. Yamamoto, T.: VOC decomposition by nonthermal plasma processing—a new approach. J. Electrostat. **42**, 227–238 (1997)
227. Karatum, O., Deshusses, M.A.: A comparative study of dilute VOCs treatment in a non-thermal plasma reactor. Chem. Eng. J. **294**, 308–315 (2016)
228. Van Durme, J., Dewulf, J., Leys, C., Van Langenhove, H.: Combining non-thermal plasma with heterogeneous catalysis in waste gas treatment: a review. Appl. Catal. B **78**, 324–333 (2008)
229. Du, C., Lu, S., Wang, Q., Buekens, A.G., Ni, M., Debecker, D.P.: A review on catalytic oxidation of chloroaromatics from flue gas. Chem. Eng. J. **334**, 519–544 (2018)
230. Saleem, F., Zhang, K., Harvey, A.: Temperature dependence of non-thermal plasma assisted hydrocracking of toluene to lower hydrocarbons in a dielectric barrier discharge reactor. Chem. Eng. J. **356**, 1062–1069 (2019)
231. Ji, S., Yu, M., Ren, Y., Buekens, A., Lu, S., Cen, K., Li, X.: Combined catalytic oxidation and activated carbon adsorption of dioxins. Environ. Prog. Sustain. **34**, 346–350 (2015)
232. Chandler, A.J., Eighmy, T.T., Hjelmar, O., Kosson, D.S., Sawell, S.E., Vehlow, J., Van der Sloot, H.A., Hartlén, J.: Municipal Solid Waste Incinerator Residues. IAWG (International Ash Working Group), Elsevier, Amsterdam, the Netherlands (1997)
233. Cheung, W.H., Lee, V.K., McKay, G.: Minimizing dioxin emissions from integrated MSW thermal treatment. Environ. Sci. Technol. **41**, 2001–2007 (2007)
234. Lundin, L., Marklund, S.: Thermal degradation of PCDD/F, PCB and HCB in municipal solid waste ash. Chemosphere **67**, 474–481 (2007)
235. Sakai, S., Hiraoka, M.: Municipal solid waste incinerator residue recycling by thermal processes. Waste Manage. **20**, 249–258 (2000)
236. Buekens, A., Huang, H.: Comparative evaluation of techniques for controlling the formation and emission of chlorinated dioxins/furans in municipal waste incineration. J. Hazard. Mater. **62**, 1–33 (1998)
237. Vogg, H., Stieglitz, L.: Thermal behavior of PCDD/PCDF in fly ash from municipal incinerators. Chemosphere **15**, 1373–1378 (1986)
238. Hagenmaier, H., Kraft, M., Brunner, H., Haag, R.: Catalytic effects of fly ash from waste incineration facilities on the formation and decomposition of polychlorinated dibenzo-p-dioxins and polychlorinated dibenzofurans. Environ. Sci. Technol. **21**, 1080–1084 (1987)
239. Wu, H., Lu, S., Yan, J., Li, X., Chen, T.: Thermal removal of PCDD/Fs from medical waste incineration fly ash–effect of temperature and nitrogen flow rate. Chemosphere **84**, 361–367 (2011)
240. Yamaguchi, H., Shibuya, E., Kanamaru, Y., Uyama, K., Nishioka, M., Yamasaki, N.: Hydrothermal decomposition of PCDDs/PCDFs in MSWI fly ash. Chemosphere **32**, 203–208 (1996)
241. Hu, Y., Zhang, P., Chen, D., Zhou, B., Li, J., Li, X.: Hydrothermal treatment of municipal solid waste incineration fly ash for dioxin decomposition. J. Hazard. Mater. **207**, 79–85 (2012)
242. Jin, Y., Ma, X., Jiang, X., Liu, H., Li, X., Yan, J.: Hydrothermal degradation of polychlorinated dibenzo-p-dioxins and polychlorinated dibenzofurans in fly ash from municipal solid waste incineration under non-oxidative and oxidative conditions. Energy Fuels **27**, 414–420 (2012)

243. Ma, W., Brown, P.W.: Hydrothermal reactions of fly ash with $Ca(OH)_2$ and $CaSO_4{\cdot}2H_2O$. Cement Concrete Res. **27**, 1237–1248 (1997)
244. Sako, T., Sugeta, T., Otake, K., Sato, M., Tsugumi, M., Hiaki, T., Hongo, M.: Decomposition of dioxins in fly ash with supercritical water oxidation. J. Chem. Eng. Jpn. **30**, 744–747 (1997)
245. Sako, T.: Destruction of dioxins and PCBs in solid wastes by supercritical fluid treatment. Organohalogen Compd. **66**, 1171–1177 (2004)
246. Zhou, Y.X., Yan, P., Cheng, Z.X., Nifuku, M., Liang, X.D., Guan, Z.C.: Application of non-thermal plasmas on toxic removal of dioxin-contained fly ash. Powder Technol. **135**, 345–353 (2003)
247. Yan, J., Peng, Z., Lu, S., Du, C., Li, X., Tong, C., Ni, M., Cen, K.: Destruction of PCDD/Fs by gliding arc discharges. J. Environ. Sci. China **19**, 1404–1408 (2007)
248. Ren, Y., Li, X., Yu, L., Cheng, K., Yan, J., Du, C.: Degradation of PCDD/Fs in fly ash by vortex-shaped gliding arc plasma. Plasma Chem. Plasma P **33**, 293–305 (2013)
249. Cagnetta, G., Robertson, J., Huang, J., Zhang, K., Yu, G.: Mechanochemical destruction of halogenated organic pollutants: a critical review. J. Hazard. Mater. **313**, 85–102 (2016)
250. Lu, S., Huang, J., Peng, Z., Li, X., Yan, J.: Ball milling 2, 4, 6-trichlorophenol with calcium oxide: dechlorination experiment and mechanism considerations. Chem. Eng. J. **195**, 62–68 (2012)
251. Zhang, K., Huang, J., Zhang, W., Yu, Y., Deng, S., Yu, G.: Mechanochemical degradation of tetrabromobisphenol A: performance, products and pathway. J. Hazard. Mater. **243**, 278–285 (2012)
252. Zhang, K., Huang, J., Yu, G., Zhang, Q., Deng, S., Wang, B.: Destruction of perfluorooctane sulfonate (PFOS) and perfluorooctanoic acid (PFOA) by ball milling. Environ. Sci. Technol. **47**, 6471–6477 (2013)
253. Yan, J.H., Peng, Z., Lu, S.Y., Li, X.D., Ni, M.J., Cen, K.F., Dai, H.F.: Degradation of PCDD/Fs by mechanochemical treatment of fly ash from medical waste incineration. J. Hazard. Mater. **147**, 652–657 (2007)
254. Mao, Q., Peng, Z., Lu, S., Li, X., Yan, J.: Mechanochemical degradation of OCDD/OCDF in fly ash from medical waste incinerators. Acta Chim. Sinica **70**, 659–666 (2012)
255. Mitoma, Y., Miyata, H., Egashira, N., Simion, A.M., Kakeda, M., Simion, C.: Mechanochemical degradation of chlorinated contaminants in fly ash with a calcium-based degradation reagent. Chemosphere **83**, 1326–1330 (2011)
256. Chen, Z., Mao, Q., Lu, S., Buekens, A., Xu, S., Wang, X., Yan, J.: Dioxins degradation and reformation during mechanochemical treatment. Chemosphere **180**, 130–140 (2017)
257. Radicalplanet Technology Mechanochemical Principle Provisional Version, POPs Technology Specification and Data Sheet, pp. 1–11. http://www.radicalplanet.co.jp/en
258. Bulley, M., Black, B.: Mechano-Chemical Remediation (MCDTM) of POP's Contaminated Soil, pp. 1–11. EDL, New Zealand (2013)
259. Shi, H., Kan, L.: Leaching behavior of heavy metals from municipal solid wastes incineration (MSWI) fly ash used in concrete. J. Hazard. Mater. **164**, 750–754 (2009)
260. Youcai, Z., Lijie, S., Guojian, L.: Chemical stabilization of MSW incinerator fly ashes. J. Hazard. Mater. **95**, 47–63 (2002)
261. Mulder, E.: Pre-treatment of MSWI fly ash for useful application. Waste Manage. **16**, 181–184 (1996)
262. Mangialardi, T.: Disposal of MSWI fly ash through a combined washing-immobilisation process. J. Hazard. Mater. **98**, 225–240 (2003)

Chapter 14
Particulate Matter

Jun Dong, Yong Chi, Augustina Ephraim, Ange Nzihou and Lina María Romero Millán

Abstract Particulate matter (PM), defined as the sum of solid and liquid particles suspended in the air, is often divided into two main groups: the coarse fraction with a size ranging from 2.5 to 10 μm (PM_{10}–$PM_{2.5}$), and the fine fraction with a size smaller than 2.5 μm ($PM_{2.5}$). PM can be generated from various sources such as unpaved roads, vehicles, agricultural processes, uncovered soil, mining operations, as well as burning of fuels. Statistical data show that in Europe, 1378 thousand tonnes of $PM_{2.5}$ are generated in year 2016, accounting for 5% of the total main air pollutants including sulphur oxides (SO_x), nitrogen oxides (NO_x), ammonia (NH_3), and non-methane volatile organic compounds (NMVOCs) [1]. This chapter will focus on the thermochemical process of solid fuels such as biomass and biowaste, because these types of fuels today are increasingly being used for energy purposes and have contributed to PM emission in many regions of the world. The presence of PM is significantly relevant to health problems, smog formations, acid rain issues, and climate changes. Characterization of PM is challenging, because PM is not only made up of a complex group of components (mainly black carbon, organics, sulfate, nitrate, alkalis, trace metals, crustal material and salt), but also comes in a wide range of sizes from nanoparticles (diameter less than 0.05 μm) up to millimeter-sized particles. In addition to this, PM characteristics are also proven to depend on various factors such as fuel type, thermal technology and temperature. Detailed knowledge of PM released during biomass and biowaste thermal conversion is essential for controlling air emissions. Over the last decades, attempts have been made to better understand the mechanisms and pathways of PM formation. Also, techniques are developed for PM sampling, collection and analysis, as well as the understanding of its physical and chemical characteristics. This chapter will thus be structured into four parts dedicated to PM fundamentals, collection techniques, physical and chemical characterizations. Available sampling

---

J. Dong (✉) · A. Ephraim · A. Nzihou · L. M. Romero Millán
RAPSODEE CNRS, UMR-5302, Université de Toulouse, IMT Mines Albi, Campus Jarlard, 81013 Albi Cedex, 09, France
e-mail: jun.dong@mines-albi.fr

Y. Chi
Department of Energy Engineering, Zhejiang University, Hangzhou 310027, China

A. Nzihou (ed.), *Handbook on Characterization of Biomass, Biowaste and Related By-products*, https://doi.org/10.1007/978-3-030-35020-8_14

devices, standards, analytical methods and techniques will be presented. Advantages, disadvantages, and major applications regarding biomass and waste thermal conversion will also be presented.

14.1 Fundamentals of Particulate Matter

14.1.1 Introduction: Definition, Origin, and Health Impacts

Particulate matter (PM) refers to the sum of solid and liquid particles dispersed in ambient air, which includes both organic- and inorganic-derived particles such as road dust, soot, smoke, and liquid droplets [2]. PM varies greatly in origin, size, and components (Table 14.1). It can be generated from both natural and man-made sources. Wind-blown dust, sea salt, volcanic ash, pollen, fungal spores, soil particles, the products of forest fires and the oxidation of biogenic reactive gases are examples of naturally origin PM. Anthropogenic PM sources constitute fossil fuel

Table 14.1 Classification, source origin, and main components of PM

PM fraction	Source origin	Main components
Coarse fraction ($PM_{2.5}$–PM_{10})	Agricultural activities	Agricultural soil, OC
	Traffic resuspension	Road dust
	Wind-blown dust, construction and mining activities, industrial resuspension	Si, Al, Ti, Fe
	Tyre and brake wear	Cu, Zn
	Combustion in energy and manufacturing industries (coal, coke, heavy oil)	EC
	Wind-land fires and volcanoes	Volcanoes' ashes, burned OC
	Biological sources	Plant debris and fungal spores
	Ocean spray	Na, Cl, Mg
Fine ($PM_{2.5}$) and ultrafine fraction ($PM_{0.1}$)	Diesel-fueled vehicle engines	BC
	Biomass combustion	OC, PAHs
	Maritime traffic	BC, OC, SO_4-2
	Combustion in energy and manufacturing industries	Pb, Cd, As, Cr, V, Ni, Se, SO_4-2
	Processes in non-metallic industries	Si, Al, Fe
	Metal processing activities	Pb, Cd, Cr, Zn

Reproduced with permission from Royal Society of Chemistry [4]
OC Organic carbon; *EC* elemental carbon; *BC* black carbon; *PAHs* polycyclic aromatic hydrocarbons. In general, combustion process has the potential to emit BC, which includes both OC and EC fractions

combustion (especially in vehicles and power plants), industrial processes (producing metals, cement, lime and chemicals), construction work, quarrying and mining activities, cigarette smoking and domestic fuel burning [3]. Particles in air can be either:

Primary particles, which are released directly from their source, for instance fuel burning, traffic, and industry.

Secondary particles, which are indirectly formed within the atmosphere as a result of chemical reactions. Examples include sulphates and nitrates formed from the oxidation of atmospheric SO_2 and NO_2.

In terms of size, PM is often classified into two main groups:

The coarse fraction, with a diameter larger than 2.5 μm;
The fine fraction, with a size up to 2.5 μm ($PM_{2.5}$). The particles which are smaller than 0.1 μm ($PM_{0.1}$) are called ultrafine particles, which includes the nanoparticles.

The size of particles determines primarily how long they will reside in air and how they are transported. Coarse particles tend to settle to the ground within hours, while fine particles can remain in the atmosphere for several weeks [5]. The limits of air quality regulations are typically based on the mass of PM_{10} and/or $PM_{2.5}$, for these fractions are small enough to be inhaled and respired to cause human health impacts [6]. For instance, PM_{10} can reach the upper part of the airways and lung, while $PM_{2.5}$ can perhaps penetrate the lung and reach the alveoli and $PM_{0.1}$ can directly pass into the bloodstream. The inhalation of PM is often associated with increased morbidity and mortality, the collective incidence and severity of which are strongly influenced by the duration and amount of PM exposure [7]. According to the estimation of the World Health Organization (WHO), exposure to PM has caused about 16% of lung cancer deaths, 11% of chronic obstructive pulmonary disease deaths, and more than 20% of ischemic heart disease and stroke [8]. Recent studies report that more than 5 million people in the world suffer from premature death due to indoor and outdoor air pollution [9, 10]. Realizing this fact, PM is currently considered to be one of the important indicators of ambient air pollution and proper characterization of PM has received increasing attention [11, 12].

14.1.2 Source Categories and Apportionment

PM is derived from a wide range of both natural and anthropogenic sources, among which the most significant contributors can be summarized into five categories: traffic, industry, domestic fuel burning, natural sources including soil dust (re-suspended) and sea salt, and unspecified sources of pollution of human origin [11].

Traffic related emissions mainly contain vehicle exhaust emissions and non-exhaust vehicular particle emission. PM from vehicles may be generated directly in the engine during combustion of fuels and lubricants, or, formed in the air by nucleation and condensation during dilution and cooling of the hot exhaust gaseous emissions from the tailpipe. Non-exhaust particles typically arise from road-tyre interaction and brake wear by mechanical process, followed by re-suspending with organic and inorganic gaseous PM precursors and crustal dust particles.

Industry is one of the major contributors of air pollutants in urban areas. Its sources include mainly combustion emissions in power plants or factories (e.g. coal, oil, municipal waste incineration), and industrial process emissions of various types of industries (petrochemical, metallurgic, ceramic, pharmaceutical, IT hardware, etc.).

Domestic fuel burning includes wood, coal, agricultural straw and gas fuel for cooking or heating, especially in developing countries. PM from cooking and heating emissions is a major source of indoor environment pollution.

Natural sources include soil dust and sea salt. **Soil dust** can be carried by the wind and re-suspended by vehicular traffic. **Sea salt** particles primarily generated from the mechanical disruption of the ocean surface, and secondarily formed by non-sea sulphate and organic species due to gas-to-particle conversion process [13].

The "**unspecified sources of human origin**" category mainly includes secondary particles formed from unspecified pollution sources of human origin. Those particles are secondarily formed within the atmosphere as a result of chemical reactions, which can either be inorganic or organic. For instance, SO_2 emitted from shipping, industry and power generation can combine agricultural NH_3 to form inorganic sulphate PM. A combination of NH_3 with NO_x from traffic, power generation, industrial and residential sources may form inorganic nitrate PM. VOCs are examples of secondary organic PM, which may be formed from industrial activities, combustion processes, and transport emissions.

The global source appointment of urban ambient $PM_{2.5}$ and PM_{10} has currently been reported by Karagulian et al. [11]. The results (Fig. 14.1) of different source contributions suggest that 25% of urban ambient $PM_{2.5}$ was contributed by traffic, 22% from unspecified sources of human origin, 20% from domestic fuel burning, 18% from natural dust and sea salt, and 15% by industrial activities including power generation.

Thermal treatment process is the main focus of this chapter. Regarding fuel combustion, the statistical data in the US show that this sector contributed to about 25% of the total anthropogenic $PM_{2.5}$ emissions in 2011, significantly decreased from about 40% in 1990 (Fig. 14.2) [14]. In China, coal burning contributes to about 3–26% of urban atmospheric $PM_{2.5}$, especially in the northern cities (Fig. 14.3). Biomass burning is likely to be another important source category for controlling China's $PM_{2.5}$, as it has contributed 5–7% of the total $PM_{2.5}$ in the urban atmosphere [15].

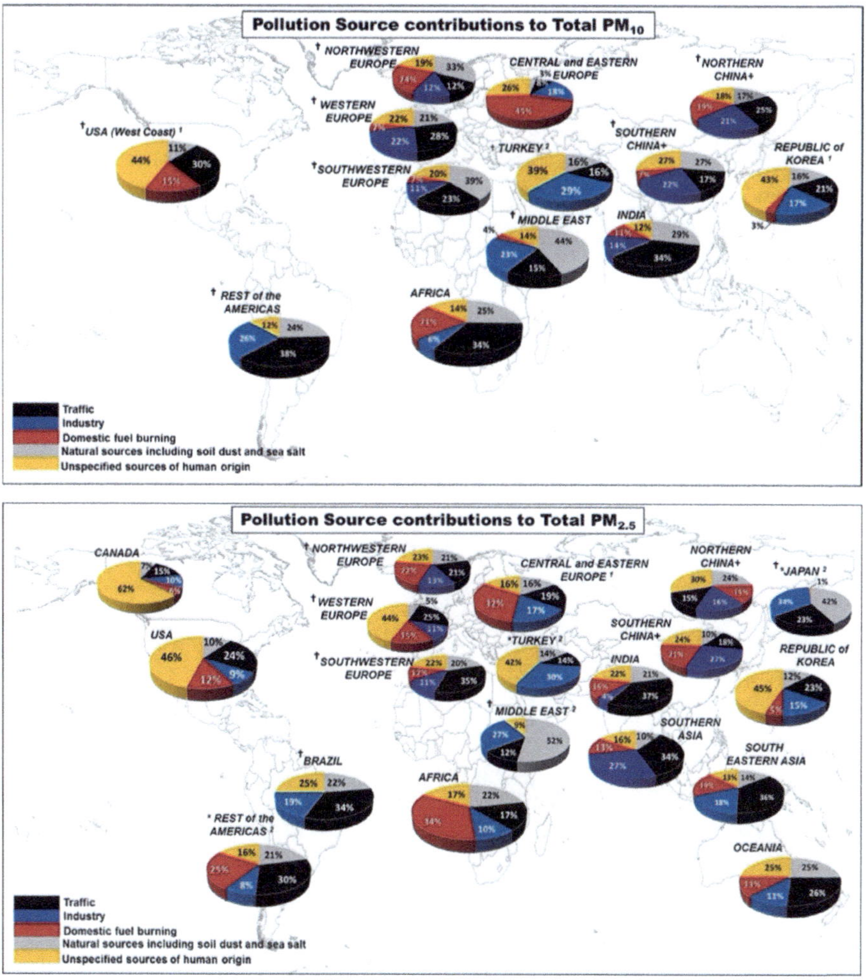

Fig. 14.1 Population-weighted averages for relative source contributions to total PM_{10} and $PM_{2.5}$ in urban sites. Some data (e.g. PM_{10} in USA, Korea and Turkey; $PM_{2.5}$ in central and eastern Europe, Japan, Middle East, and rest of the Americas) only provide indicative results. More details as well as the meaning of each symbol (e.g. *, $^+$) can be found in the sourced literature. Reprinted with permission from Elsevier [11]

14.1.3 Components of PM

PM comprises a number of complex components including crustal materials (soil, dust), organic matters, metals, acid salts, sulfate, nitrate and ammonium [16]. The composition of PM can vary dramatically, depending on the emission source categories or process conditions.

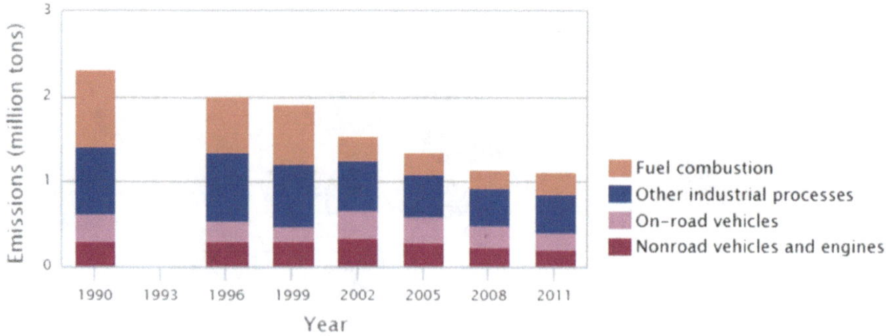

Fig. 14.2 Anthropogenic $PM_{2.5}$ emissions in the US by source category (1990–2011). *Source* US EPA [14]. Data for year 1993 have not been updated by the US EPA, so that they are absent from the figure

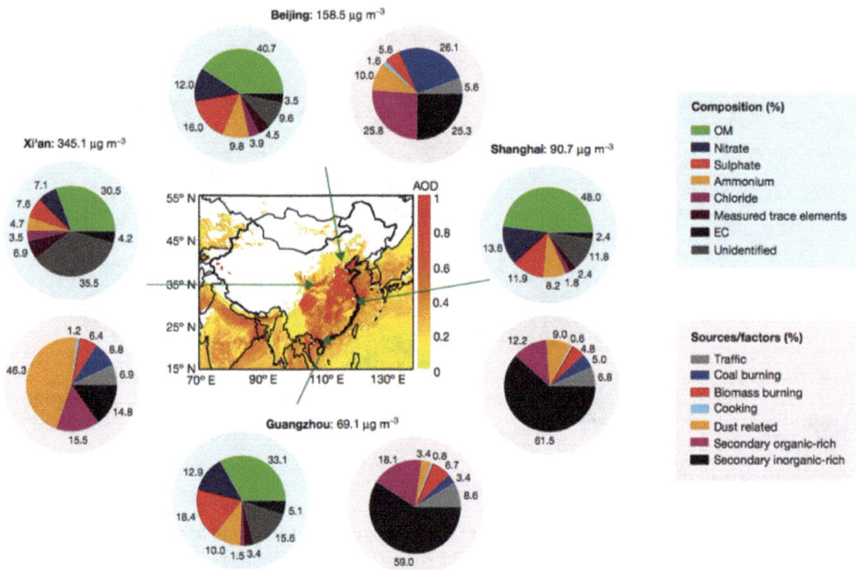

Fig. 14.3 Source apportionment and composition of $PM_{2.5}$ collected during the high pollution events of 5–25 January 2013 at the urban sites of Beijing, Shanghai, Guangzhou and Xi'an in China. Reprinted with permission from Springer Nature [15]

Figure 14.4 shows the composition of an example airborne PM_{10} and $PM_{2.5}$, based on average monitoring data in the UK [17]. The chemical makeup of PM has shown a division between the coarse and fine modes. The coarse fraction (PM_{10}) consists mainly of crustal material, i.e., iron rich dust. The dominant components of fine fraction ($PM_{2.5}$) are organic matters, sulfate and nitrate.

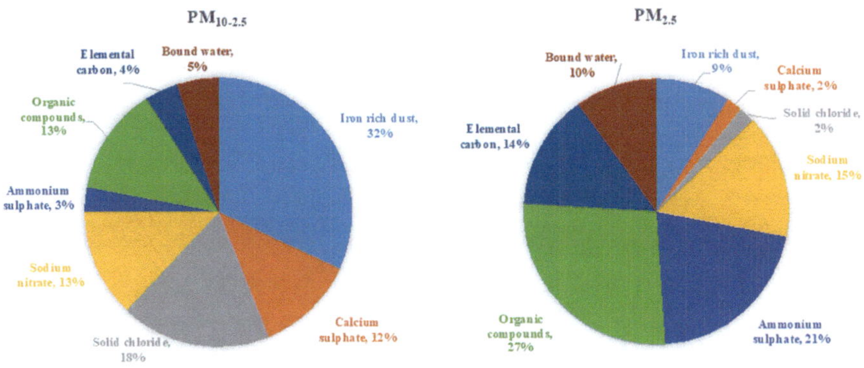

Fig. 14.4 Composition of an example airborne PM_{10} and $PM_{2.5}$ at UK sites

In addition to the size distinction, PM from a gasoline vehicle, diesel vehicle, woodstove, or a coal-fired boiler may have a large difference [18]. PM from solid fuel combustion such as coal, oil, biomass and waste can produce coarse particles from the release of non-combustible materials, i.e. fly ash; fine particles from the condensation of materials vaporized during combustion; as well as secondary formed particles through the atmospheric reactions of sulphur oxides and nitrogen oxides initially released as gases [2]. Organic carbon (OC) includes products of incomplete combustion, such as polycyclic aromatic hydrocarbon (PAHs) and unburned fuel. Elemental carbon (EC) is nonvolatile carbon in graphite like form, found in mainly nanometer size fractions (<56 nm; 56–100 nm; and 100–180 nm) [19]; however, EC is never found in pure form in the atmosphere, because various OC and other compounds are adsorbed onto, and/or mixed with the EC core [20]. EC containing a coating of mixed OC and inorganics, with adsorbed carbonaceous material, is referred to as back carbon (BC), or soot in fuel combustion.

14.1.4 PM Formation Mechanism During Thermal Conversion of Biomass and Biowaste

This chapter will focus mainly on the characteristics of PM formed and emitted during thermochemical process of solid fuels such as biomass and biowaste. Thermal conversion of solid fuels, such as pyrolysis, gasification, incineration or their combination process, mainly produces PM containing soot, spherical organic carbon particles and inorganic ash particles. Figure 14.5 illustrates the possible formation pathways of PM during combustion of typical fuels (coal, biomass, and oil) [21]. The formation of combustion generated PM can be summarized by three mechanisms:

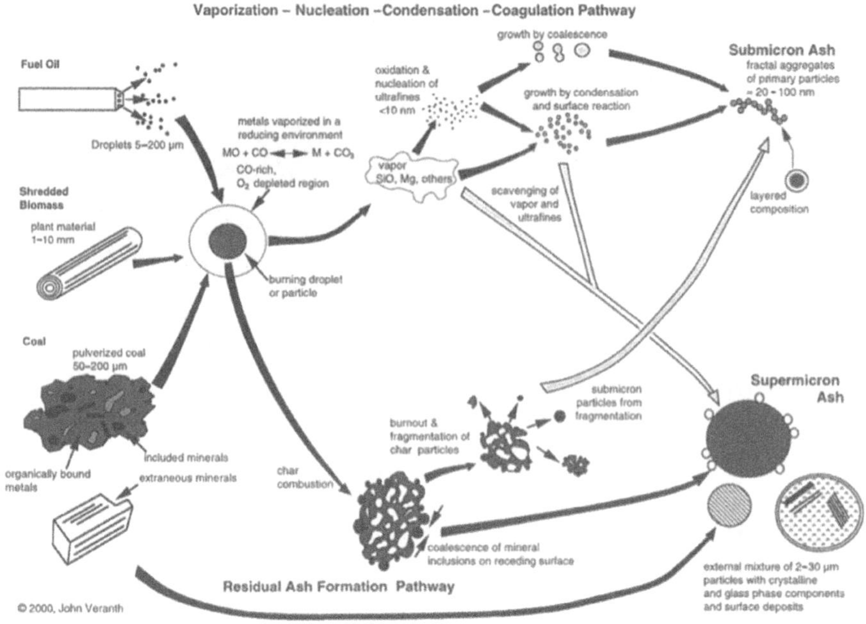

Fig. 14.5 Formation pathway of PM during combustion of solid and liquid fuels. Reprinted with permission from Taylor & Francis [21]

1. Coarse particles formed from residual fly ash particles;
2. Fine fly ash particles produced from the vaporization, condensation, nucleation of easily volatilized ash components;
3. Fine organic particles and soot produced from incomplete combustion.

Solid fuels contain considerable amounts of ash forming inorganic elements. During the thermal conversion process, these inorganic species produce an oxidation product, i.e., ash. The majority of the ashes could be retained inside the furnace, named, bottom ash; however, a small portion could be entrained from the furnace with the flue gas and forms the **coarse fly ashes** (>1 μm). These coarse fly ash emissions can achieve a particle size range from a few μm to about 200 μm [22].

The **fine ash particles** (<1 μm) are mainly formed from the easily volatilized inorganic elements, such as alkali and heavy metals, released from the fuel to the gas phase during combustion. These fine ash particles are formed by nucleation, condensation, surface reactions, coagulation and agglomerations.

Fine organic particles and soot are formed due to incomplete combustion, originating from combustible materials of the fuels. For instance, soot is a complex mixture of amorphous EC and organics, formed mainly inside the flame in the carbon rich area. During devolatilization and combustion of volatiles, the hydrocarbon compounds of fuel particles are likely to form aromatic rings and PAHs. Soot is thereafter formed by further growth of PAHs, which then becomes larger by surface reactions, coagulation and agglomeration. Organic particles are also one source of PM, which are mainly incomplete combustion products that remain in the flue gas. Organic particles can be present in liquid or gaseous form depending on environmental conditions, as the main part of the organic material of fine particles could usually be condensed at far below 500 °C [23–25]. The incomplete combustion conditions may include poor combustion, or sub-stoichiometric reactions of pyrolysis and gasification. In flue gas conditions, the organic vapors condense mainly on existing fine particles through condensation and/or adsorption [26].

The PM formation depends various factors including the fuel, combustion condition and flame type, etc. Compared with fossil fuels (e.g. coal), biomass fuels, such as wood, straw or biowaste, have several special characteristics: a higher volatile content, a lower ash content, and a higher alkali metals content. Thus, in industrial scale biomass combustion units, the amounts of soot in the emissions are typically negligible owing to the high volatilized components and a high combustion efficiency. However, the incomplete combustion condition in traditional small scale combustion units, especially in household stoves, would cause an important fraction of soot and OC in PM emissions.

The ash forming inorganic elements in biomass fuels can be divided into three categories [22]: (i) non-volatilized elements such as Ca, Si, Mg, Fe, Al, etc.; (ii) easily volatilized elements such as K, Na, S, Cl, etc.; (iii) easily volatilized heavy metals such as Zn, Cd, Ob, etc. Biomass has a lower ash content than coal, typically 0.5–1% for wood and 4–8% for agricultural by-products. However, it is rich in alkali elements. For example, the K content in a range of straws can be as high as 2% of the dry weight according to the study of Christensen et al. [26]. Their S content is found to be fairly constant at 0.1%. For MSW (municipal solid waste), the Cl contained could be even richer than biomass or coal, which is mainly determined by the PVC (polyvinyl chloride) in plastic and the salts (NaCl) in food waste. During thermal conversion, these species undergo gas phase reactions resulting in the formation of alkaline (K, Na) metal chlorides, sulphates, carbonates, and heavy metals oxides [22]. These elements in biomass or biowaste can greatly influence the size distribution and composition of inorganic PMs.

As a result of the aforementioned particle formation mechanisms, PM during thermal conversion of biomass and biowaste typically exhibits a bimodal particle size distribution: a peak of submicron particles (PM_1), and the other peak of coarse mode dominated by PM_{10}. The main constituents of biomass PM are K, Cl, S, Ca, Na, Si, P, Fe and Al [27]. Becidan et al. [28] report that K, S, Cl and Na dominated the PM composition for a number of biomass types, such as wood, demolition wood and coffee wastes. This is confirmed by the study of Jöller et al. [29, 30] that the PM are predominantly K_2SO_4 and other alkali compounds (Na_2SO_4, KCl, NaCl) for clean wood chips. Other studies also verify the abundant existence of Fe, Ca, Al, Zn, etc., in fine particles during waste and biomass combustion [31, 32]. Vaporization and subsequent condensation from the gas phase is the main pathway of PM_1 formation; while entrained non-volatilized ash residuals and fly ash particles are the main compounds of PM_{10}. It is also shown by Gao et al. [33] that the ultrafine PM is produced mostly from the volatiles, rather than during char combustion.

14.1.5 Emission Limits and Regulations in Selected Countries

To keep a clean and healthy air environment, air quality standards concerning PM emission have been gradually built up. However, the development of associated regulations, especially for PM_{10} and $PM_{2.5}$, just started within the latest decade. Those standards regulate the PM emission either in the ambient air, or from specific facilities, such as coal-fired power plants, biomass and waste-to-energy plants and other sources. Generally, the emission standards are established by the governments based on the best available technologies and monitoring emission data.

The World Health Organization (WHO) has produced air quality guidelines with the aim to reduce the health impacts of air pollution worldwide. Its updated version 2005 has regulated thresholds and limits for PM_{10} and $PM_{2.5}$, which should be measured on both a 24-h (short-term) and annual mean (long-term) basis to estimate health effects. Many different countries have regulated a separate emission limit for $PM_{2.5}$ in the ambient air. Table 14.2 summarizes the associated air quality standards of $PM_{2.5}$ in several selected countries [34].

However, the emission limit of $PM_{2.5}$ from coal, biomass or waste combustion is still under development: $PM_{2.5}$ from coal combustion is still regulated as part of PM—no country has a separate emission limit for $PM_{2.5}$ from coal-fired power plants. Additionally, waste-to-energy and biomass-to-energy plants are still not widely applied in most countries, since, the emission limits of PM for biomass or waste thermal plants are far from well-established or regulated worldwide.

The emission limits of PM from coal-fired power plants and waste-to-energy plants in selected countries are listed in Table 14.3 [34]. Different countries base their standards on different time periods, ranging from half-hourly average to 24-h average, or annual averages. The differences also attain at the varying reference conditions and units. For instance, most countries use 0 °C and 101.3 kPa in dry flue gas as the reference condition; but Indonesia and Thailand set their reference temperature at 25 °C. The USA set limits based emissions per unit of electricity produced; but Japan and Thailand use parts per million by volume in flue gas. The EU, Australia, China, Germany, India, Indonesia and South Africa regulate pollutant concentration per cubic meter of flue gas. Currently, China has the strictest limits for coal-fired power plants, although its air quality standards are less strict than those for the WHO and EU. Regarding waste-to-energy plants, EU has the toughest limits.

Table 14.2 International and national air quality standards of $PM_{2.5}$ in selected countries (Unit: $\mu g/m^3$)

Country	WHO	EU	Australia	China	India	Indonesia	Japan	South Africa	Thailand	USA
24 h	25		25	75	60	230	35	40	50	35
Annual	10	25	10	35		90	15	20	25	12

Table 14.3 Emission standards of PM from coal-fired power plants and waste-to-energy plants in selected countries (Unit: mg/m^3)

Country[a]	Coal-fired power plants (24-h average)		Waste-to-energy plants	
	Existing	New	Half-hourly average	24-h average
Australia*		80		n.a.[e]
China*	30/20[b]	10	30	20
Germany*	20	10	30	10
India*	100/50[c]	30		50
Indonesia**	150	100		350
Japan*	100	50		17/19/50[f]
South Africa*	100	50		10
Thailand**	180	80		320
USA*	18.5[d]	12.3[d]		25.2
EU IED*	20	10	30	10

[a]Countries with a superscript "*" use 0 °C, 101.3 kPa on a dry flue gas basis as the reference conditions; while countries with a superscript "**" use 25 °C, 101.3 kPa on a dry flue gas basis as the reference conditions

[b]The first value applies to all plants; the second value applies to plants in key regions

[c]The first value applies to plants installed before 2003; the second value applies to plants installed between 2003 and 2016

[d]The USA set limits based emissions per unit of electricity produced. For conversion into a mg/m^3 basis, data from Zhu et al. [35] is used

[e]Currently Australia does not have national air emissions standards applicable to industrial facilities such as WtE plants

[f]For existing municipal waste combustion plant, limit value of 19 mg/Nm^3 is applied to plants with capacity >250 tpd, limit value of 50 mg/Nm^3 is applied to plants with capacity 35–250 tpd; for new or substantially modified municipal waste combustion plants (after September 2008), limit value of 17 is applied to plants with all sizes; for commercial and industrial waste incineration plants, limit value of 50 mg/Nm^3 is applied to plants with all sizes. Emission standard concentrations are corrected to 7% oxygen

The emission standards for each country and region have been progressively strengthened over the years. The regulated emission limits will become stricter with the developing best available technologies.

14.1.6 PM Characterization

The growing utilization of biomass and biowaste as energy resources highlights the importance of the need for their environmentally sustainable management. PM is one of the major constituents of air pollutants. Gaining better understanding of the characteristics of PM is helpful to design an efficient process for its capture and removal from the flue gas, as well as to develop more effective controlling methods. For example, PM generated from biomass gasification process can be removed using high temperature filters; however, the filtration efficiency and the operation and dimensioning of the filter could greatly depend on the PM mass concentration and size distribution. The composition of PM could also be crucial, since biomass materials are rich in alkali, which may form silicates with low melting temperatures that may negatively affect the filter operation [36].

Determining the characteristics of PM commonly requires a sampling and collection process beforehand. Collection of PM by filters or sets of impactor substrates are typically used techniques, which can provide detailed information about PM composition and size. The disadvantages of using these techniques are that they can be expensive and time consuming, since the analysis methods require a minimum amount of material, which may take time for collecting sufficient samples. Over the last decades, the development of different PM collection techniques, including both off-line and on-line approaches, have received considerable attention.

Environmental and health effects of PM depend on both the physical and chemical characteristics of PM. Various analytical techniques have thus been developing for PM characterization. Typical properties that are characterized include PM size, shape, morphology, chemical composition, crystalline structure, organic compounds, etc., using instruments such as scanning electron microscope (SEM), x-ray fluorescence (XRF), inductively coupled plasma (ICP), x-ray diffraction (XRD), thermal/optical reflectance and transmittance (TOR/TOT), gas chromatography-mass spectrometry (GC-MS), etc.

Different PM collection techniques are presented in Sect. 14.2. The major sampling devices and standards are also introduced. Sections 14.3 and 14.4 discuss a variety of PM physical and chemical characterization techniques, respectively. Each specific set of PM characteristic is linked to its major applications regarding biomass and biowaste thermal conversion.

14.2 Particulate Matter Collection Techniques

14.2.1 Sampling and Measurement Types

The collection of PM produced from the thermal conversion of biomass and biowaste can be done in two ways: by collecting source samples or ambient samples. For source sampling, the PM is collected directly from the gas conduit such as the flue-gas stack of a combustion process or the product-gas pipeline of a pyrolysis or gasification process. For ambient samples, the PM is collected from the atmosphere away from the exhaust or pipeline of any gas source. In this chapter, we will focus only on source sampling.

How a source is sampled depends on the type of particle measurement of interest. Particle mass or concentration is currently the regulated measurement, although there is growing interest in particle size distribution [22, 37]. For most

Table 14.4 Technical standards for particulate emissions monitoring

Standard	Content
ISO 23210	Stationary source emissions—determination of $PM_{10}/PM_{2,5}$ mass concentration in flue gas—measurement at low concentrations by use of impactors
ISO 9096	Stationary source emissions—manual determination of mass concentration of particulate matter
EN 13284-1	Stationary source emissions—determination of low range mass concentration of dust—part 1: manual gravimetric method
EN 13284-2	Stationary source emissions—determination of low range mass concentration of dust—part 2: automated measuring systems
EN 14181	Stationary source emissions—quality assurance of automated measuring systems

particle analyzers, dilution of the gas sample is required in order to operate within their required ranges. Dilution conditions have been shown to affect measurement of particle size distribution, although they tend to have little effect on overall particulate mass [37, 38]. Hence, special attention must be given to the choice of dilution systems if the particle sizes are to be determined. Examples of dilution systems will be discussed in Sect. 14.2.3.

The relevant standards for sampling and analysis of particulate emissions are shown in Table 14.4 [39]. These standards are qualified by either International Organization for Standardization (ISO 23210, ISO 9096) or European Committee for Standardization (EN 13284-1, EN 13284-2, EN 14181), thus are widely suitable to be used globally.

14.2.2 *Examples of Off-Line Collection Techniques*

14.2.2.1 Bag Sampling

A bag sampling system is often used to collect a specified volume of PM sample in a bag to be later analyzed at a different location. Thus, the use of this system is ideal when the PM analyzer is far from the source or when the source produces irregular emissions. Nonetheless, a bag sampling system can lead to a loss of PM through adsorption of the particles on the bag wall as well as changes to the particle size distribution due to coagulation [18]. Thus, the samples must be analyzed as quickly as possible to reduce errors in measurements.

Figure 14.6 illustrates a bag sampling system, which is available commercially [18]. This system consists of a stainless steel, airtight container, which is fitted with a sampling bag by means of a clamp-down lid. The Tedlar bag has a twist valve at its inlet and an actuated ball valve is placed in the sampling line connected to this inlet. Furthermore, a vacuum pump removes the gas between the stainless steel container and the sampling bag. When the actuated ball and twist valves are opened, the sample is drawn into the bag until the bag pressure is the same as that

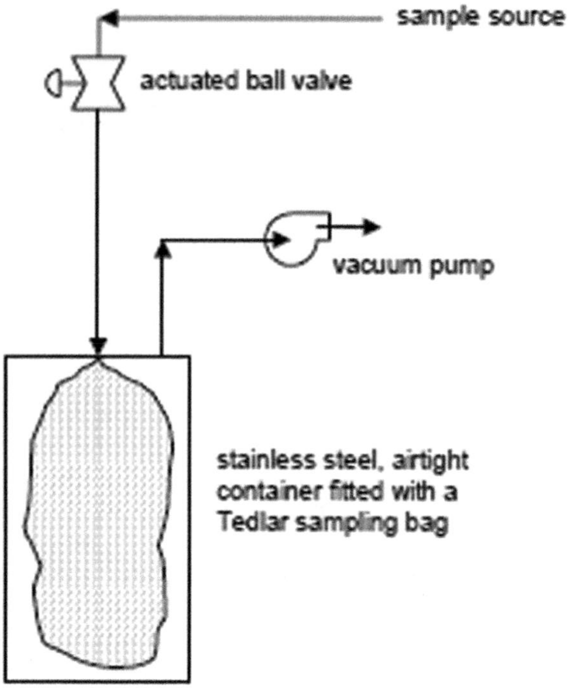

Fig. 14.6 Bag sampling system. Reprinted with permission from Institute for Clean and Secure Energy [18]

between the stainless steel container and the sampling bag. Once the bag is filled, the ball and twist valves are closed, and the sample bag is removed from the stainless steel container. Then the twist valve inlet can be connected to a dilution system or directly to an analyzer for further analysis.

14.2.2.2 Filter Sampling

PM samples can also be collected on a filter, and afterwards extracted with a solvent, or directly sent into the analyzing equipment, depending on the characterization techniques selected (see detailed physical and chemical characterization techniques in Sects. 14.3 and 14.4). If the particle size distribution is to be measured in addition to particle concentration, the filter samples can be collected on cascade impactors, such as a micro-orifice uniform deposit impactor (MOUDI) [40] or a K-JIST cascade impactor [41]. The MOUDI contains 8–10 stages plus an after filter, with each stage containing a filter. For a K-JIST cascade impactor, 5 stages are present, as shown in Fig. 14.7, which enable the sampling of PM of major concern including PM_{10}, $PM_{2.5}$ and PM_1 [41]. After particle collection, the particles on each filter can be extracted using a solvent, such as hexane, and then analyzed by an advanced analytical method such as gas chromatography/mass spectrometry, in a laboratory.

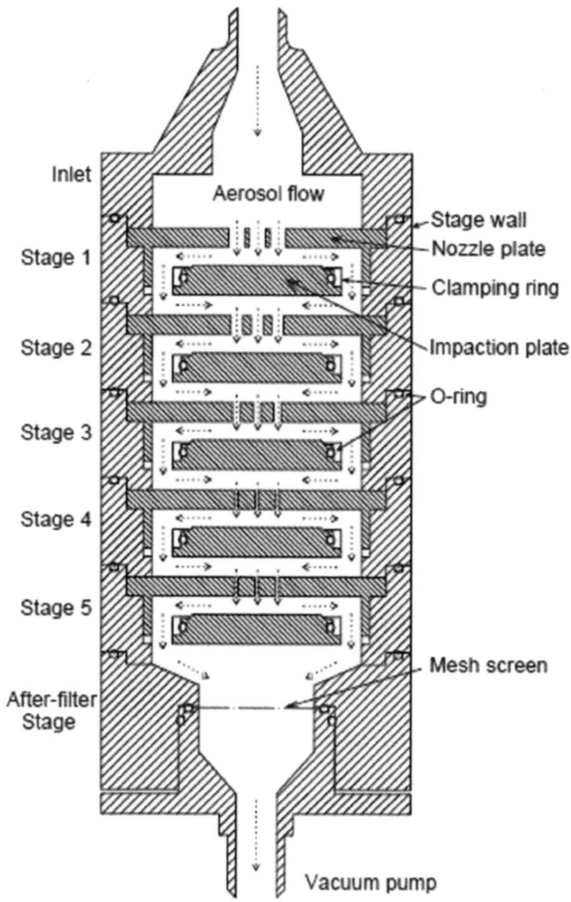

Fig. 14.7 A schematic of 5-stage cascade impactor (K-JIST). Reprinted with permission from Elsevier [41]

14.2.3 Examples of On-Line Collection Techniques

14.2.3.1 Diluted Sampling

On-line analyzers of PM usually require upstream dilution of the source sample. This is mainly to prevent supersaturation of volatile matter in sample gas (e.g. water, SO_x and VOCs), which may result in condensation, nucleation and coagulation of nanoparticles, and thereby compromise the PM number measurements. Dilution conditions, including humidity, temperature, gas residence time and system size/configuration, could affect the formation of nanoparticles and thus must be controlled. These influences are discussed in detail by Kelly et al. [18]. Another reason for lowering the PM concentration of the source sample is to match the ranges of the analytical instruments used.

Typically, dilution is performed using dry compressed air. Furthermore, in order to avoid adherence of PM to the sampling components or chemical reactions in the sampling train, it is recommended to use Teflon or stainless steel tubing [42]. Three common types of dilution systems are: Eductor system, Hildemann system and constant-volume dilution sampler, as shown in Fig. 14.8. These types are discussed extensively by Kelly et al. [18].

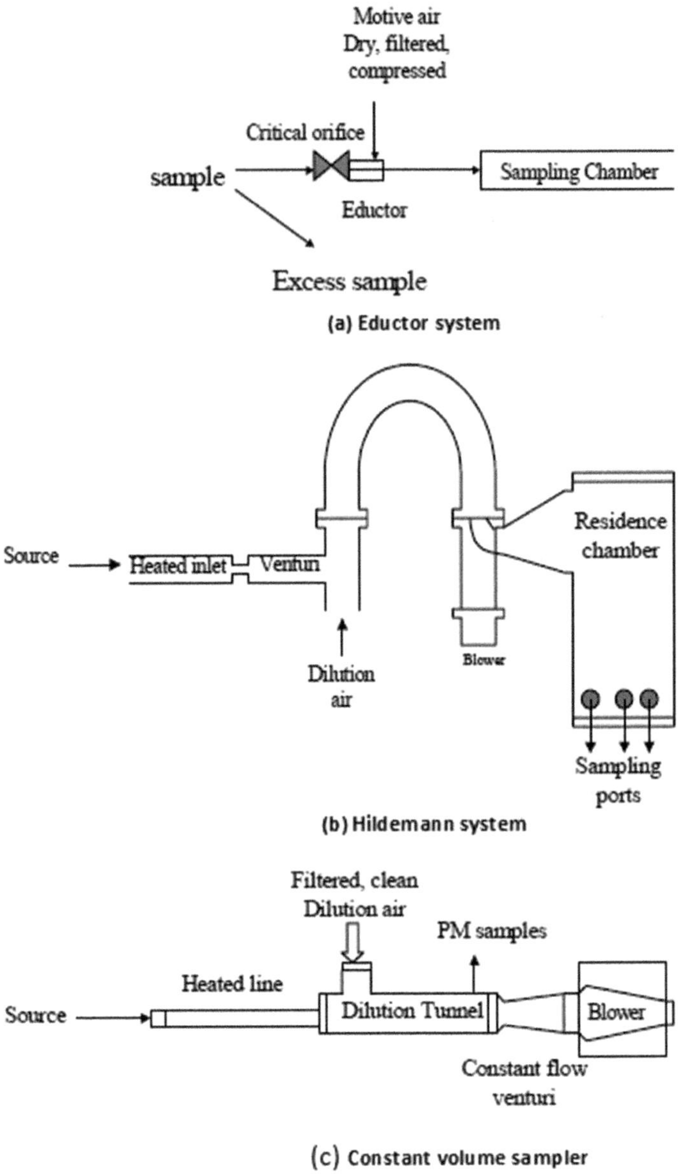

Fig. 14.8 Common types of dilution systems. Reprinted with permission from Institute for Clean and Secure Energy [18]

14.2.3.2 Raw Gas Sampling

Despite what has been discussed above, undiluted or raw gas sampling is also used because of the relative simplicity of the sampling system. However, because of the high concentration of moisture, particulates, and temperature, it is often difficult to obtain reproducible and reliable results. Although permitted for some regulatory applications, it is mainly used for non-regulatory and/or field testing [42].

An example of a commonly used system for raw gas sampling is the GRAVIMAT SHC502 by SICK AG, which is illustrated in Fig. 14.9 [43]. Using a probe, particle matter (or dust) is sampled by extracting a partial gas volume isokinetically, i.e. at the same velocity as the main volume flow. Isokinetic extraction prevents the occurrence of sedimentation in the sample line and ensures that the gas contains the precise representative amount of PM at the source. All dust particles in the extracted partial volume flow are retained in the dust collector filter (GS5 filter head probe).

In addition to the dust or PM collector, the probe head also contains the measuring orifice for the gas velocity, and partial volume flow as well as the measuring sensor for the gas temperature in the duct.

14.2.4 Filter Selection

The above-mentioned PM collection techniques highlight the importance of the use of filters, which are widely applied in most PM collection cases. Currently, different

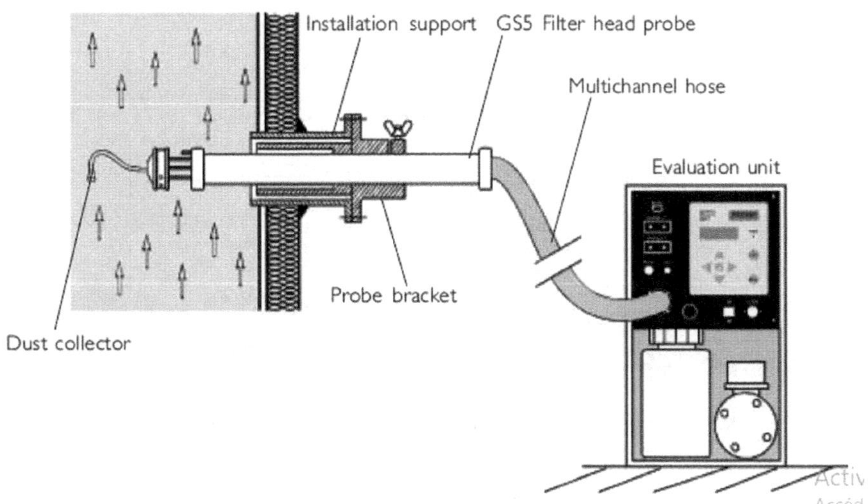

Fig. 14.9 GRAVIMAT SHC 502 System. Reprinted with permission from SICK AG [43]

types of filters are available for PM collection. They can be generally classified into three categories [44]:

- **Fiber filters**, for example glass fiber filter, quartz fiber filter and cellulose fiber filter. They are made of a deep mesh of fibers with random orientations.
- **Membrane filters**, for example polycarbonate membrane filter, Teflon membrane filter and polytetrafluoroethylene (PTFE) filter. They have a tortuous complex structure through the whole filter material.
- **Capillary pore filters**. They are normally composed of a thin, smooth polycarbonate or polyethylene terephthalate film, with some circular pores within the structure.

A scanning electron micrographs (SEM) diagram of these filters is shown in Fig. 14.10.

The appropriate selection of a filter for the collection of PM from biomass and biowaste thermal processes will depend on many key factors, such as sampling systems, filter media, and analysis techniques.

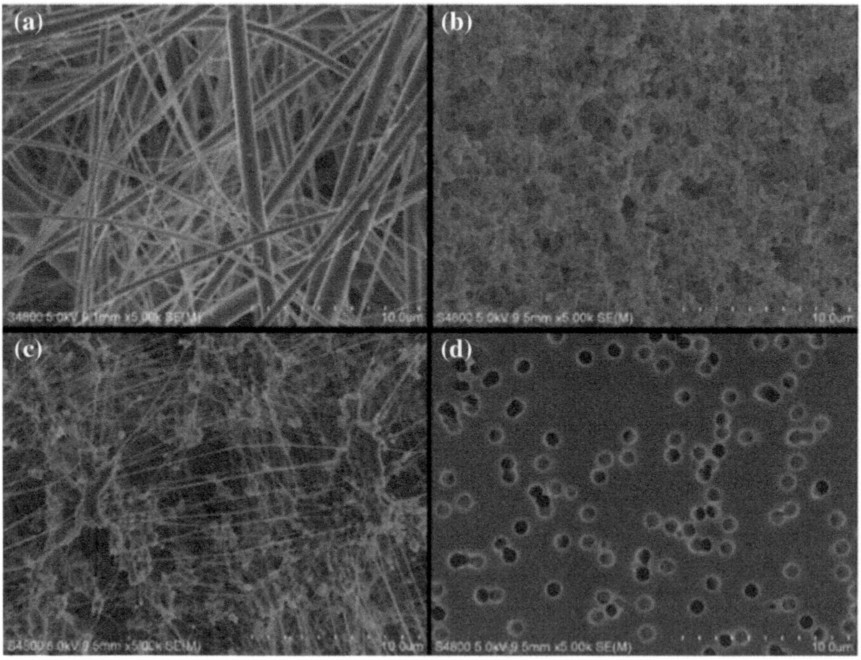

Fig. 14.10 Scanning electron micrographs (SEM) diagram of different types of filters. *Source* Adapted from Lindsley et al. [44]. **a** glass fiber filter with a 1-μm equivalent pore diameter; **b** mixed-cellulose esters filter with 0.8-μm equivalent pore diameter; **c** polytetrafluoroethylene filter with 3-μm equivalent pore diameter; **d** polycarbonate capillary pore filter with 1-μm pore size

The filter pore size and filter efficiency are important characteristics of the filter, which will directly affect the PM collection efficiency. In addition, the intended analyzing techniques are also a determining factor. For example, quartz fiber filter can endure high temperatures without decomposing [45], which makes it an optimal material for thermal analysis techniques, such as thermal/optical reflectance and transmittance (TOR/TOT). A polycarbonate membrane filter is often used when PM samples are to be examined by scanning electron microscopy (SEM), since the particles could be easily distinguished from the flat filter surface. However, no single filter media will be appropriate for all analyses. Different types of filters should be paired with different analysis techniques, including both physical and chemical PM characterization procedures.

14.3 Physical Characterization of Particulate Matter

Understanding the physical characteristics of PM is very important to estimate its effect on human health and the environment. Generally, the physical characteristics of PM could include the mass, size, shape, density, surface properties, microstructure, volatility, hygroscopicity, optical properties, etc. Here the most important parameters associated with thermal conversion of biomass and waste—PM mass, size distribution, and morphology, are discussed.

14.3.1 Weighing Gravimetry for Mass

The mass of PM is determined gravimetrically based on the post-weight and pre-weight difference of a sampling process.

Filters are used in most cases for PM collection. Its mass is determined by an analytical electron-microbalance. The sampling period and gas flow should be recorded, as mentioned in Sect. 14.2, so the PM quantity could be calculated as a function of the air/gas flow (mg/m^3) or the combusted fuel mass (mg/g fuel).

Both the unexposed and exposed filters should be equilibrated in a conditioning environment beforehand to reduce variability to an acceptable level, for example, placed at 15–30 °C for at least 24 h free of acidic or basic gases [46]. Specific conditions should be determined follow the type of filters and the guideline from the manufacturer. The filters should be then taken directly from the conditioning chamber to the balance to minimize the risk of contamination.

To guarantee the accuracy of results, quality control procedures are also regulated. For example, NIST standards [47] specify the recommended frequency of quality control filters for the PM program, as shown in Table 14.5.

Table 14.5 Recommended frequency of quality control filters for the PM program

Type of filter	Description	Acceptance criteria
Lot stability test filters	12 filters are repeatedly weighed to determine the minimum necessary equilibration time for filters from the same manufacturing lot	Weight trend approaches zero
Batch stability test filters	3 filters from a batch are repeatedly weighed during equilibration to verify the stability of the filter shipment batch	Weight loss <5 μg
Laboratory blank filters	1 laboratory blank filter is weighed for every weighing session	Weight loss <15 μg
Field Blank Filters	Unexposed filters from each shipment batch are designated as field blanks by the client	Weight difference <30 μg
Replicate filter weighings	Every filter (pre-weighing) or every third filter (post-weighing) is reweighed	Weight difference <15 μg

14.3.2 Size Distribution

The size of the particle is a key physical parameter that influences the characteristics of PM, for example, determines the deposition in the human respiratory system when inhaled. As has been mentioned in Sect. 14.1, $PM_{2.5}$ and PM_1 are more toxic than PM_{10}, as their smaller diameters allow them to penetrate deep into the lungs. For biomass and biowaste thermal plants, the size of PM determines primarily at which place it could be trapped in flue gas cleaning systems. For instance, the efficient operation of a cyclone requires a minimum particle size of 2.5 μm, and an electrostatic precipitator could remove very fine particles less than 1 μm with 95–99% efficiency [48].

Currently, there are a wide range of techniques available for particle size analysis. Some of the more common methods, and their strengths and limitations, are shown in Table 14.6 [49, 50].

The result of a particle size distribution can be presented in different ways: number size distribution, mass size distribution, volume size distribution, or intensity size distribution. The result will depend on the mechanism of measuring techniques and equipment used. The data could be converted from one type of distribution to another, however several assumptions should be made.

For biomass and biowaste thermal processes, laser diffraction and impactors are the dominant techniques used in PM size distribution determination.

The principle of laser diffraction particle sizing technique has been introduced in Chap. 2. This technique could report the particle size as a volume or number equivalent sphere diameter. The PM samples can be either dry powder or wet dispersed in a liquid. Generally, a larger quantity (a minimum of 20 ml) of the PM sample is required for dry analysis. Wet dispersion is applicable for PM collected on a filter, since PM usually has a very small quantity. For example, Min et al. [48]

Table 14.6 Particle size distribution techniques

Technique	Strengths and limitations [49, 50]	
Sieve analysis	✓	The sample could include a range of different densities, different refractive indices or water-soluble materials, suitable for bulk materials
	✓	The results are a separated size fraction for further study
	✗	Not suitable for fine particles <100 μm
	✗	Not suitable for elongated needle-shape samples
Sedimentation	✓ ✓ ✗	Determines particle size as a function of settling velocity Suitable for particles <10 μm Requires careful selection of the dispersion media
	✗	A bulk sample is required, e.g., 2–5 g
Elutriation analysis	✓ ✓ ✗	Determines particle size as a function of settling velocity The technique is non-destructive A bulk sample (about 10 g) is necessary
Laser diffraction method	✓ ✓ ✓	Suitable for measuring sizes between 0.1 and 3000 μm Fast, can be performed on very small samples Can conduct a continuous measurement for analyzing process streams
Electrical impedance	✓	Suitable for a wide range of sizes 0.4–1200 μm
	✗	Requires dispersing of the sample in a liquid medium
Microscopic sizing and image analysis	✓	Samples can be analyzed without being handled
	✗	Analyzing can be time-consuming
	✗	The results will be tedious and error prone if manually analyzed

used the wet dispersion method to detect the size distribution of PM collected from the gasification syngas and cleaning systems. The results were given in volume size distribution as shown in Fig. 14.11.

Impactors are also widely used to measure PM in the flue gas, which present particle size distribution in mass. The working principle of impactors is based on gravimetrically segregation the particles into different sizes from a particle laden gas stream. Some of the most used impactors include: Low Pressure Cascade Impactors (LPI), Electrical Low Pressure Impactor (ELPI), and Micro-Orifice Uniform Deposit Impactor (MOUDI). The former two could collect PM in the range of 30 nm–10 μm; and the latter is able to trap ultrafine PM smaller than 0.4 μm [51].

As an example, Fig. 14.12 shows the size distribution of PM from wood pellets gasification using a LPI [52]. Three types of gasifier were used: a bubbling fluidized bed (BFB), a circulating fluidized bed (CFB), and an indirect BFB. Results clearly reveal a bimodal particle mass size distribution: a fine mode at a size of around 0.1 μm, and a coarse mode at approximately 3 μm. Note that slightly different particle size distributions are observed for each gasifier, which may be attributed to tar condensation to form new particles.

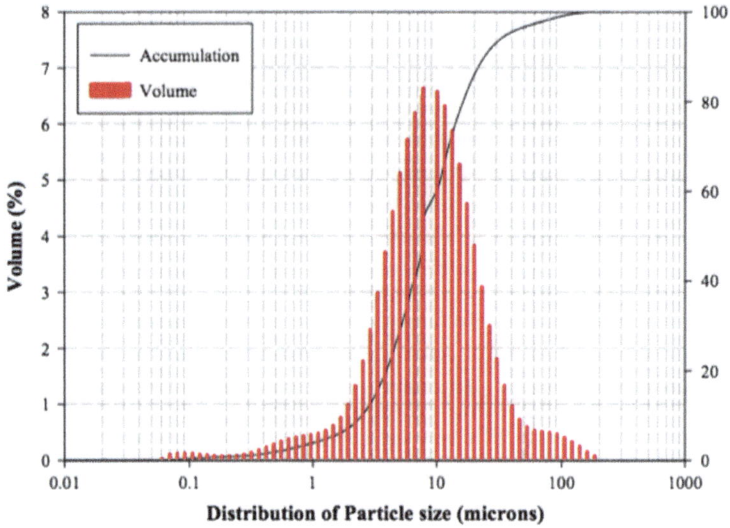

Fig. 14.11 An example of particle size distribution (volume) of PM from gasifier outlet, using laser diffraction technique. Reprinted with permission from Springer Nature [48]

Fig. 14.12 An example of particle size distribution (mass) of PM from wood pellets gasification at three types of furnace, using a low pressure impactor. Reprinted with permission from Elsevier [52]

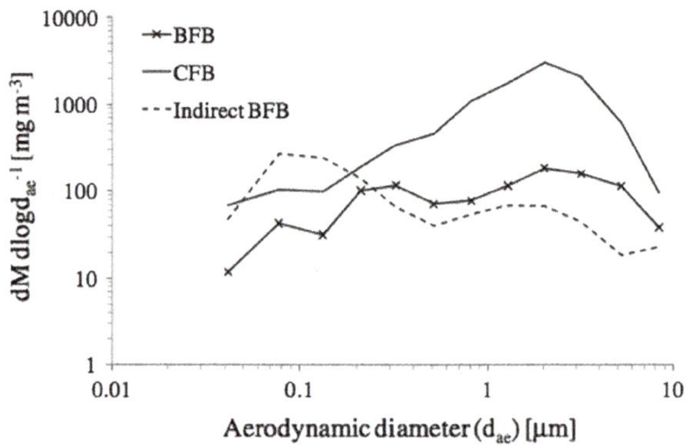

14.3.3 Morphology

Morphology, for example, shape, surface structure and microstructure, is another major physical property of PM. The morphology of PM has proven to significantly affect the chemical properties. Scanning electron microscopy (SEM) is ideally suited and widely used for characterization of PM from biomass and biowaste thermal conversion. The principle of SEM technique is to use electrons to form magnified images, which could provide better feature resolution, a wider range of magnification, and a greater depth-of-field than conventional optical microscopes [45]. The basic information and mechanism of the SEM technique can be found in

Chap. 2. One of the major advantages of SEM applied for biomass and biowaste thermal processes is that the SEM technique can achieve spatial resolution at submicron level, which makes it well-suited for $PM_{2.5}$ and PM_{10} analysis.

SEM technique is able to identify the different particle shapes of both fine and coarse PM fractions. For example, Morgalla et al. [53] characterized the PM collected from different stages of an impactor during wood pellet gasification in an indirect bubbling fluidized bed. As shown in Fig. 14.13, it is found that the fine particles (especially for PM collected in stage 1) seem to be agglomerated; while the coarse particles (stage 9) appear more separated and irregular.

SEM technique is also convenient to characterize the PM samples trapped by fiber filters. An example is given in Fig. 14.14 [54], which identified a high degree of the heterogeneity of PM both in terms of particle size and morphology ranging from nanoscale (ca. 0.04 μm) to micron scale (ca. 22 μm).

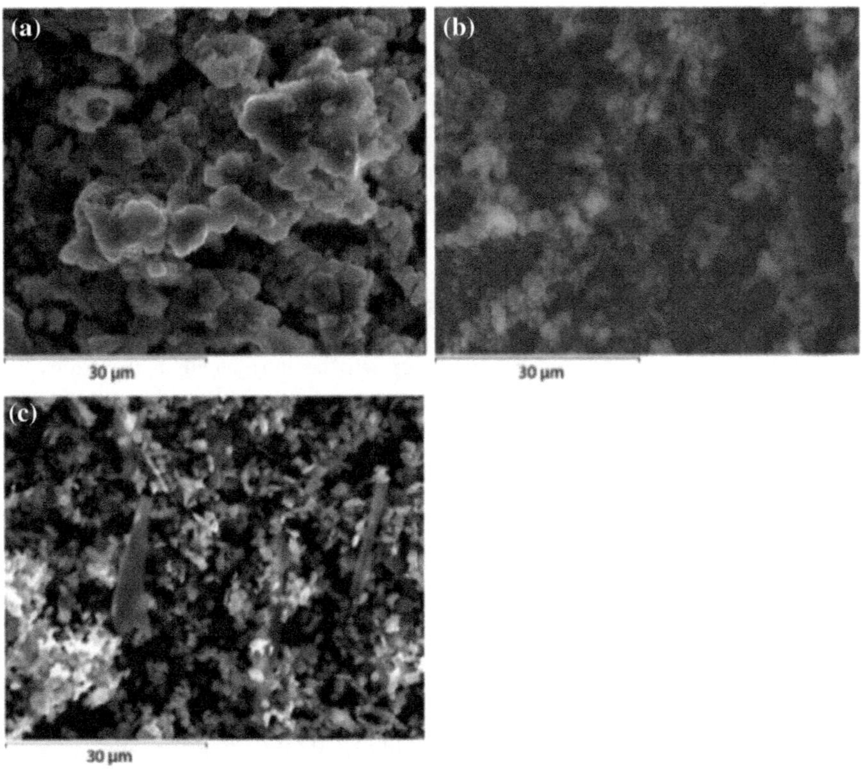

Fig. 14.13 SEM images of PM collected on different low pressure impactor stages corresponding to different geometric mean diameter (dae). **a** Stage 1 (0.04 μm); **b** stage 5 (0.32 μm); **c** stage 9 (2.02 μm). Reprinted with permission from Elsevier [53]

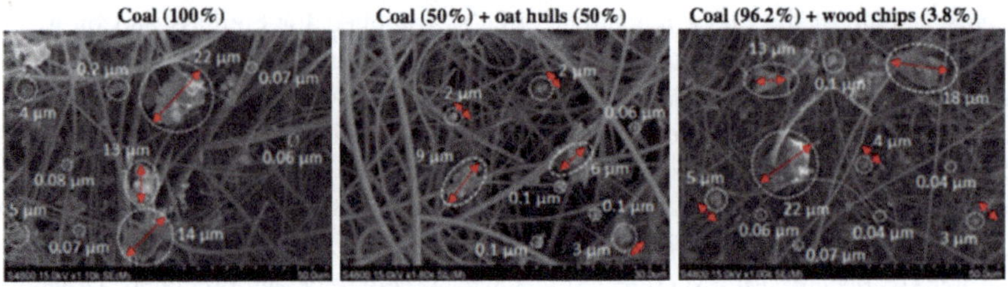

Fig. 14.14 SEM images of PM collected on quartz fiber substrates during biomass and coal co-firing. Reprinted with permission from Elsevier [54]

14.4 Chemical Characterization of Particulate Matter

The chemical characterization of PM is a very important issue due to the harmful effects that it represents for human health [55]. PM chemical characterization is complex, as fine particles are a mixture of different organic and inorganic compounds, and then, the use of several simultaneous analytical techniques is required.

At the present time, the chemical composition of PM produced from the thermochemical transformation of alternative materials like biomass and biowaste is not completely identified. Accordingly, the appropriate characterization of PM can help to understand the impact of raw materials characteristics and transformation process conditions on the chemical patterns of emission, and their associated environmental impacts and risks on human health.

Figure 14.15 briefly illustrates the various techniques which are commonly available for the characterization of both inorganic and organic composition of PM. Several of the most widely used techniques will be introduced in the following parts. Note that the fundamentals, measurement principles, and equipment used for these techniques have been described in detail in Chap. 2. Here the focus will be mainly associated with their applications in PM characterization.

14.4.1 Inorganic Composition Characterization Techniques

14.4.1.1 X-Ray Fluorescence (XRF)

XRF technique is usually used for PM chemical analysis, as it presents major advantages over other analysis techniques like ICP-OES, ICP-MS and AAS. In particular, XRF is a non-destructive technique. The PM samples, usually captured in filters, can be directly analyzed in the XRF apparatus without particular pretreatments, reducing the contamination risks of the samples.

The sampling strategy, e.g. the filter material, flow rate and sampling time should be adapted to the analytical technique and must be selected to obtain

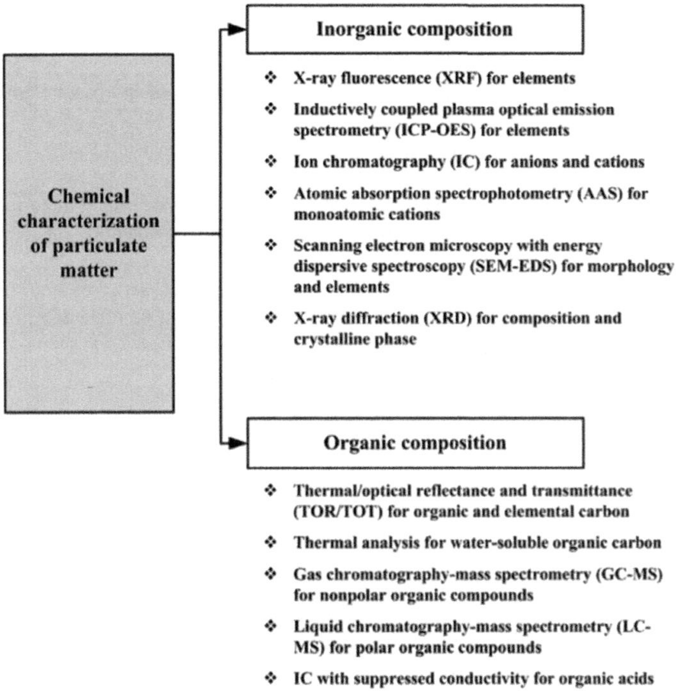

Fig. 14.15 Some of the commonly used techniques for chemical characterization of particulate matter

sufficient material for analysis. According to the PM loading in the air/gas stream, sampling times can go to hours to days [56].

Since both the sampling filter and the collected particles need to be analyzed, the appropriate choice of the filter material is an important issue to avoid interference between them [57]. Also, it is generally recommended to use thin membrane filters instead of thick fiber filters to avoid measurement errors due to the particle penetration in the collecting media [58]. The suitability of different filter materials for inorganic composition analysis using X-ray fluorescence is presented in Table 14.7.

In all cases, the blank analysis of the filter background is necessary to identify its major and minor constituents and accurately determine the collected PM composition. In this regard, it is recommended to analyze 2–5 filters to determine the background variability. It must be pointed out that the method detection limit is determined from the standard deviation of the filter blank analysis [62–64], and then, lower blank values are related to lower detection limits.

After PM sampling, filters must be carefully protected from contamination, since the sample size is small and the levels to be measured are low. Filter contamination can be minimized using samplers that do not expose the filter directly to the atmosphere and reducing the air exposure time before and after sampling [62]. The filters and membranes can be directly analyzed with the XRF instrument without

Table 14.7 Suitable sampling filter materials for XRF analysis of particulate matter

Filter material	Suitability for X-ray fluorescence [59–61]	
Glass fiber filter	✗	High and often variable concentration of inorganic element. High blank values
Polycarbonate membranes	✓	Low blank values
	✓	Thin membranes
	✓	Collect particles in the membrane surface only. No corrections for x-ray adsorption have to be taken into account
Cellulose fiber filters	✓	Very low blank values
	✗	Thick filters
	✗	Corrections for x-ray adsorption should be done due to the filter thickness
PTFE filters	✓	Very low blank values
	✓	Thin membrane filters
	✓	The presence of F does not interfere with the analysis

any particular pretreatment or sample preparation, according to the protocols presented in Chap. 2. For accurate quantification, the XRF apparatus should be adequately calibrated using standards i.e. vacuum-deposited thin-film elements and compounds, polymer films and thin glass films [62].

The obtained spectrum from XRF analysis should be corrected by the subtraction of the filter background and adjusted for x-ray adsorption if necessary [62, 65]. If a quantitative analysis is performed, the PM composition can be obtained as a function of the analyzed sample surface area (mg/cm^2). From this value and considering the gas flow and sampling time, the composition can also be expressed as a function of the air/gas flow (mg/m^3) or the combusted fuel mass (mg/g fuel).

14.4.1.2 Inductively Coupled Plasma Optical Emission Spectrometry (ICP-OES)

PM analysis can also be performed using ICP-OES, as this technique has a very wide analytical range with detection limits in the ppb range for almost all elements. ICP-OES allows the identification of elements from lithium to uranium excepting halogens. However, carbon, nitrogen, hydrogen and oxygen cannot be identified using this technique and are usually determined according to CHNS analysis.

As presented in Chap. 2, the main challenge of this analytical technique is the adequate digestion of the sample prior to the analysis. Since the PM cannot be easily recovered from the sampling filters and membranes, both filters and collected particles may be digested together completely before ICP-OES analysis. In this regard, the correct selection of the filter materials is of great importance.

The suitability of different filter materials for inorganic composition analysis using ICP-OES technique is presented in Table 14.8. As stated with the XRF technique, it must be pointed out that the method detection limit is determined from

Table 14.8 Suitable sampling filter materials for ICP-OES analysis of particulate matter

Filter material	Suitability for ICP-OES analysis [62, 63, 66]	
Glass fiber filter	✗	High and often variable blank values
Quartz fiber filter	✓	Low and relatively low blank values
	✗	The presence of soluble silicates can interfere with Si quantification
Cellulose filters	✓	Very low blank values
	✓	Easily digested due to their low resistance to acids
PTFE filters	✓	Very low blank values
	✗	Digestion performed with strong acids and high temperatures

the standard deviation of the filter blank analysis, and then, lower blank values are related to lower detection limits.

The filter choice will also depend on many other factors like sampling efficiency, mechanical resistance, chemical stability, cost and availability, and compatibility with the analytical method, as presented in Sect. 14.2.

Regarding the filter and collected particle digestion prior to ICP-OES analysis, highly concentrated acids like HCl, HNO_3, HF, and H_2O_2 are usually used. Reagent quantities are determined according to the sample size and the elements to be analyzed. For unknown samples, it is recommended to develop a digestion method according to the filter and sample size and the elements to be quantified, as presented in 'Digestion method development' section in Chap. 2.

In this regard, the standard EN 14902:2005 related to the measurement of Pb, Cd, As and Ni in the PM_{10} fraction of suspended PM, can be used as a reference for the digestion method development. Even when this standard is focused on the measurement of some particular constituents, it can give valuable insights to the sample preparation procedure for the analysis of a large range of elements.

Most digestion methods use HNO_3 and H_2O_2 to perform the complete digestion of samples and filters; however, if Si needs to be quantified, the use of HF and H_3BO_3 acid is recommended [67]. It is important to note that during the digestion step, it is necessary to ensure the complete submersion of the sampling filter in the digestion vessel, in order to attain the total decomposition of the sample.

When the total digestion of the sample is completed, the resulting solution is diluted volumetrically with demineralized water, according to the requirements of the analysis apparatus. Then, the solution can be analyzed using the ICP-OES protocol, as has been described in detail in Chap. 2.

As presented for XRF analysis, the PM composition can be obtained as a function of the analyzed sample surface area (mg/cm^2) or collected particle mass (mg/g PM). From this value and considering the gas flow and sampling time, the composition can also be expressed as a function of the air/gas flow (mg/m^3) or the combusted fuel mass (mg/g fuel).

14.4.1.3 Scanning Electron Microscopy with Energy Dispersive Spectroscopy (SEM-EDS)

The EDS is a variation of the X-ray fluorescence, which is generally coupled to SEM. The combination of these techniques allows the simultaneous analysis of the morphology and surface structure of a sample, and its elemental composition. The SEM technique of PM has been described previously in Sect. 14.3; hence, the use of EDS will be introduced here. Even if the quantitative analysis is possible for this technique with certain corrections and restraints, SEM-EDS is mostly used to analyze the content and dispersion of metals in the sample.

As described in Chap. 2, energy-dispersive X-ray spectroscopy uses the X-ray spectrum emitted by a sample bombarded by a focused high-energy beam of electrons, to determine a localized chemical composition. Point analysis or element mapping are possible with this technique.

As for the other presented techniques, attention must be paid to the composition of the sampling filter, as it can impact the obtained results. Filters with low concentration of inorganic elements are then recommended, as in the case of the XRF analysis.

The SEM-EDS pictures of PM produced from biomass and biowaste combustion are presented here as a reference.

In Fig. 14.16, the SEM-EDS analysis of a single particle from the co-firing of 50% coal—50% oat hull is presented [54]. It is possible to observe the size, morphology and inorganic composition of the particle, as well as the quartz filter used for the sample collection. It must be noted that the Si signal in the ED spectrum comes from the filter, showing the importance of the appropriate choice of the sampling filter or membrane material.

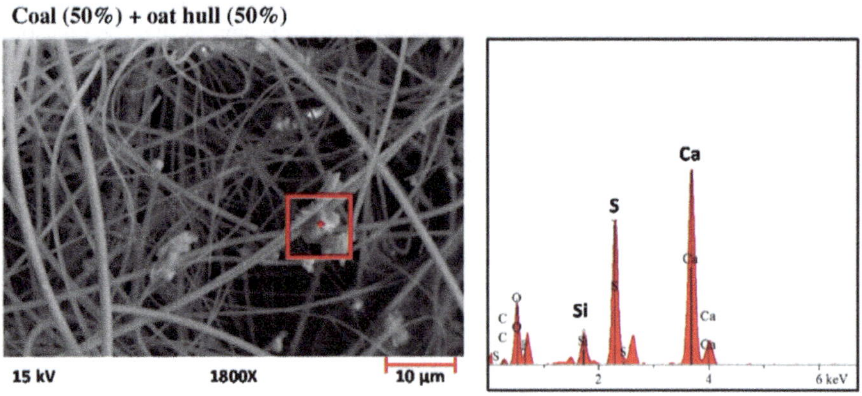

Fig. 14.16 SEM-EDS analysis of an individual particle from the co-firing of coal and oat hull. Reprinted with permission from Elsevier [54]

14.4.1.4 X-Ray Diffraction (XRD)

The already presented techniques are all suitable for the analysis of the inorganic content of biowaste combustion PM; however, they are not sufficient to distinguish different compounds and crystalline structures of a given element. In this regard, XRD is an excellent tool for the analysis of inorganic compounds in PM and the identification of their crystalline phases. This information can give insight into the origin and formation mechanisms of the components present in the PM.

One of the major advantages of this technique is the fact that this is a non-destructive analysis, giving the possibility to perform further characterization and analysis.

As for the already presented techniques, the filter material selection is important for the accurate analysis of samples. For XRD, the blank corresponds to the diffraction pattern of the filter material, and then, the filtering media should be chosen according to the crystallographic phases to be analyzed. In practice, the x-ray diffraction peaks of the filter spectra should not overlap with the diffraction peaks of the crystalline species present in the analyzed samples.

In this regard, several authors have studied the XRD patters of different filter materials used for atmospheric particulate matter analysis [68–70]. Accordingly, Teflon, glass fiber, quartz fiber and cellulose acetate are considered as the most suitable materials for particulate matter analysis from biomass and biowaste combustion.

Figure 14.17 shows an example of the impact of the filter material in the sample XRD pattern. It can be observed that the amorphous matrix of the quartz filter used for particle collection is easily distinguished by a broad peak, while the present crystalline phases in the analyzed particles produce sharp peaks. If there is an

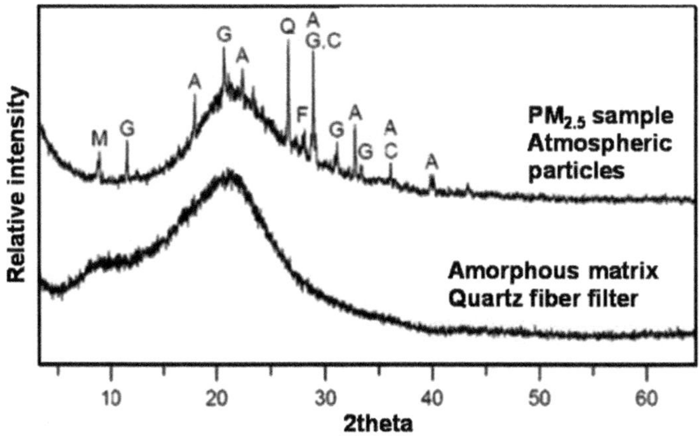

Fig. 14.17 Impact of filter material in sample x-ray diffraction pattern. A: ammonium nitrate, C: calcite, F: feldspar group, G: gypsum, M: mica group, Q: quartz. Reprinted with permission from Elsevier [71]

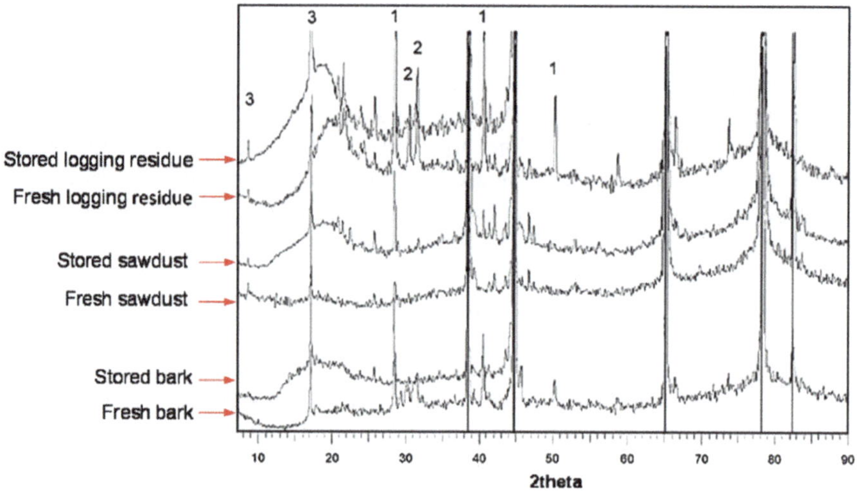

Fig. 14.18 XRD diffractogram of PM from the combustion of different pelletized biofuels. Reprinted with permission from American Chemical Society [72]

overlap of the diffraction pattern of the filter and the sample, it is not possible to precisely identify the crystalline species in the sample.

For its part, Fig. 14.18 shows the XRD diffractogram of the particles collected form the combustion of different pelletized biowaste [72]. The vertical lines indicate the spectra from the collector material (aluminum foil impactor), and the observed peaks are related with the crystalline phases present in the analyzed particles. The authors report the identified phases as KCl (1), $K_3Na(SO_4)_2$ (2), and a possible unidentified Zn phase (3).

As each different phase produces a different combination of peaks, the crystalline compounds in the sample can be identified from the analysis of the x-ray diffraction spectra. It is important to note that the previous knowledge of the inorganic elements present in the sample is required to the correctly identify the phase.

Generally, a qualitative analysis of the crystalline phases present in the sample is adequate for the basic PM chemical characterization. However, in some applications, it would be necessary to perform a quantitative analysis. In this respect, different authors have proposed several methods for quantitative analysis of PM samples using standards [70, 73–75].

14.4.1.5 Examples of Inorganic Characterization of Biomass and Biowaste Combustion PM

The particle formation and chemical composition is influenced by the fuel properties, i.e. biomass or biowaste. From the available characterization techniques, the main inorganic constituents of PM from biomass and biowaste thermochemical conversion processes are presented in Table 14.9.

Table 14.9 Examples for the main inorganic composition of PM generated from biomass and biowaste thermochemical conversion processes

Biomass/ biowaste	Particle size	Sampling method/ material	Main reported inorganic constituents	Analytical technique	References
Wheat straw combustion	PM_1	Low pressure impactor	K, Cl, Si	SEM-EDS	[76]
Wheat straw combustion	PM_{1-10}	Low pressure impactor	K, Mg, Ca, Si, Cl, S	SEM-EDS	[76]
Pine wood bioslurry combustion	PM_{10}	Low pressure impactor	Mg, Ca, Al, Si, Fe, Na, K, S	ICP-OES	[77]
Beech wood combustion	$PM_{2.5}$	Quartz fiber filters	Al, Ca, Cu, Fe, K, Na, Pb, Zn	ICP-OES	[78]
Eucalyptus wood combustion	PM_{10}	Cellulose acetate filters	Ca, Mg, Cr, Fe, Co, Si	ICP-OES XRF	[79]
MSW combustion	$PM_{2.5}$	High temperature and low pressure impactor	Ca, Si, Mg, Al, K, Na, Zn, Cl, Pb, Ti, P, S	ICP-OES SEM-EDS	[80]
Rice straw combustion	$PM_{2.5}$	Cellulose and PTFE filter	Al, Ca, Cr, Cu, K, Si, Mg, S	ICP-OES	[81]
Rice straw combustion	PM_{10}	Quartz filter	Fe, Si, Al, Ca,, Cr, Cu, S, Pb	ICP-OES	[81]
Light diesel oil	$PM_{2.5}$	Teflon filter	As, Fe, Al, Mn, Pb, Cr, Zn, Cu, Ni	ICP-OPS	[82]

Some authors presented the PM composition as a function of the air/gas flow (mg/m^3), while some others, as a function of the fuel mass input (mg/g fuel), or the collected PM mass (mg/g PM). In some cases, only a qualitative analysis is performed. Accordingly, for comparison purposes, it is important to verify the sampling and analysis conditions, as well as the reported units for PM composition. The list of the inorganic composition present in PM is not exhaustive and is based on the reported results of the referred authors. The composition of PM from light diesel oil combustion is presented only for comparison purposes.

14.4.2 Organic Composition Characterization Techniques

14.4.2.1 Thermal/Optical Reflectance and Transmittance (TOR/TOT)

TOR/TOT technique is used to quantify organic and elemental carbon (OC and EC) in PM. OC and EC are typical and important components from thermal conversion of biomass and waste. As has been reported, OC could account for about two-thirds by mass of aerosol produced from biomass burning [83]. The principle of TOR/TOT technique is based on heating the carbon-containing sample to different temperature stages under both inert and oxidizing conditions, and quantifying the converted CO_2 with an infrared absorption detector, or, converting CO_2 to CH_4 by a methanator oven before being measured with a more sensitive flame ionization detector (FID).

Currently, more than 20 methods are available for OC and EC quantification [45]. Generally they provide comparable results of the value of total carbon (TC = OC + EC); the main difference among the methods are specific definition of OC and EC.

For PM analysis, the Interagency Monitoring Protection Visual Environment (IMPROVE) thermal protocol is the most widely applied method. Its recommended procedures are illustrated in Table 14.10 and Fig. 14.19. In the first 1–4 steps, the four OC fractions are thermally desorbed under a flow of non-oxidizing helium carrier gas with controlled temperature ramps. In the second 5–7 steps, the oven is then switched to an oxidizing 2% oxygen/98% helium with another series of temperature ramps to determine the three EC fractions. Special attention should be paid to a pyrolyzed carbon fraction (OP), which is formed during the first non-oxidizing part of the analysis (between step 1–4). The OP fraction is quantified by carbon evolved between the time oxygen is added and the reflected signal returns to its original value. IMPROVE regulates the final OC as the sum of OC1 + OC2 + OC3 + OC4 + OP; and EC as the sum of EC1 + EC2 + EC3 − OP.

Table 14.10 IMPROVE method for TOR/TOT technique

Step	Carbon fraction	Carrier gas	Temperature (°C)
1	OC1	100% He	25 (ambient) −140
2	OC2	100% He	140–280
3	OC3	100% He	280–480
4	OC4	100% He	480–580
5	EC1	98% He/2% O_2	580
6	EC2	98% He/2% O_2	580–740
7	EC3	98% He/2% O_2	740–840

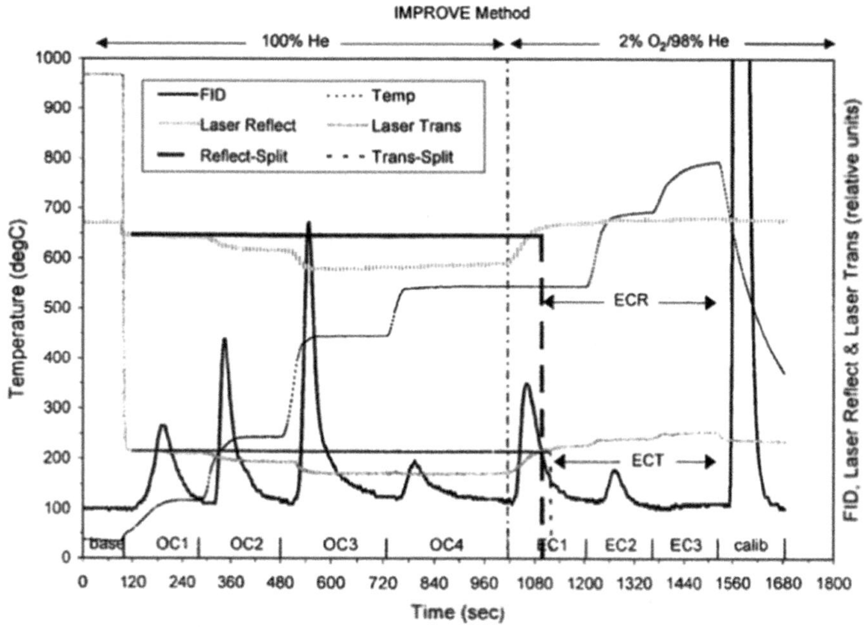

Fig. 14.19 Typical IMPROVE protocol thermogram. Reprinted with permission from Taylor & Francis [84]

Table 14.11 TOR/TOT characteristics of $PM_{2.5}$ samples from three types of biomass burning.

	Unit	Biomass type		
		Rice straw	Pine needles	Sesame stems
$PM_{2.5}$	mg/m^3	41.39 ± 15.01	27.32 ± 13.15	29.47 ± 11.28
OC	mg C/m^3	24.15 ± 9.51	14.22 ± 7.37	13.70 ± 6.13
EC	mg C/m^3	1.42 ± 0.46	2.04 ± 1.01	1.30 ± 0.49
OC/EC	–	16.80 ± 2.81	7.01 ± 0.96	10.15 ± 1.58

Reproduced with permission from Elsevier [85]

To conduct TOR/TOT analysis, PM samples should be collected on quartz-fiber filters that need to be pre-heated at 900 °C for 3 h by the carrier gas before sampling. During analysis, a 0.5 cm^2 punch from each exposed filter is taken and placed in a quartz boat that is inserted into the oven for appropriate analysis.

As an example, Table 14.11 shows the TOR/TOT analysis results from $PM_{2.5}$ samples of three types of biomass burning (rice straw, pine needles, and sesame stems). The results show a high ratio of OC/EC for all types of the biomasses.

14.4.2.2 Gas Chromatography-Mass Spectrometry (GC-MS)

GC-MS is used for qualitative and quantitative determination of nonpolar organic compounds in the PM. In the past years, techniques have been established for appropriate preparation of samples ready for GC-MS analysis, of which a pre-treatment procedure involving solvent extraction and concentration is necessary. Realizing that this pre-treatment is often time-consuming and increases the risk for error, currently, thermal desorption (TD) technique has been coupled with GC-MS as an alternative to the solvent extraction technique [86, 87]. The TD-GC/MS technique does not require sample extraction. Thus, the PM sample preparation procedures are similar as TOR/TOT technique: a small piece of quartz-fiber filter, around 1–2 cm^2 punch, is taken out of a whole filter; 1 μL of internal standard solutions is added to the sample to normalize the MS response before the sample is attached to the analyzing injection tube. However, the use of TD technique should raise attention to procedure conditions, thus affecting the accuracy of final results [88].

The GC-MS technique is typically used to quantify polycyclic aromatic hydrocarbons (PAHs) in PM, since they are reported to contribute the carcinogenicity adverse effects. An example of the various PAHs contained in the PM during the combustion of five agricultural residue-derived biomass fuels (sunflower stalk pellets, straw pellets, buckwheat shells, corn stalk pellets, wheat grain screenings) and two reference fuels (wood, sewage sludge pellets), is illustrated in Fig. 14.20.

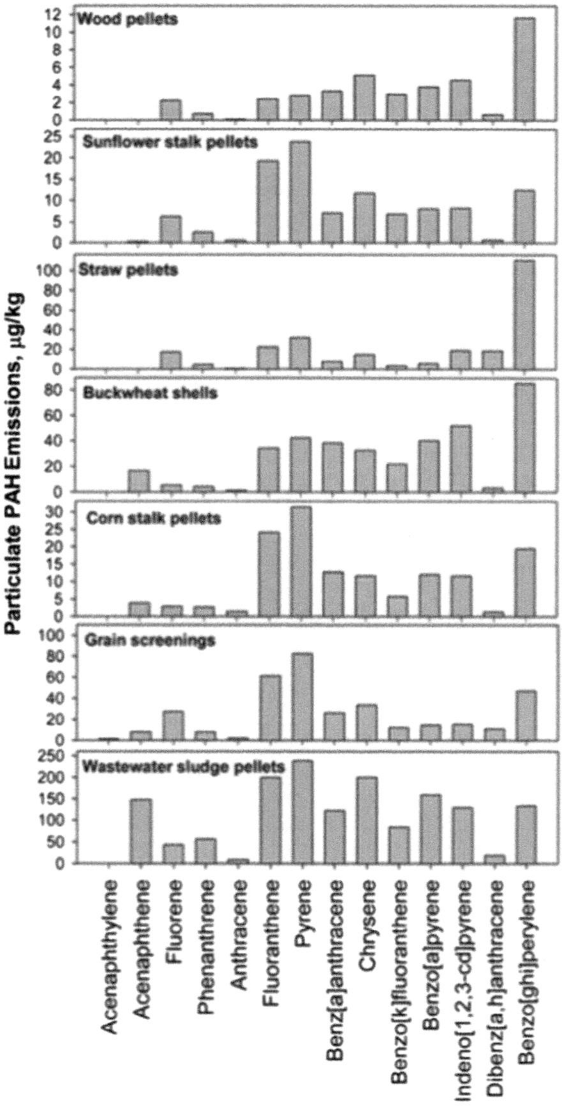

Fig. 14.20 PAHs contained in PM during combustion of several biomass fuels (µg/kg). Reprinted with permission from American Chemical Society [89]

References

1. European Environment Agency (EEA).: National Emissions Reported to the Convention on Long-Range Transboundary Air Pollution (LRTAP Convention). https://www.eea.europa.eu/data-and-maps/data/national-emissions-reported-to-the-convention-on-long-range-transboundary-air-pollution-lrtap-convention-12. Accessed 30 Oct 2018
2. WHO Regional Office for Europe: Health Aspects of Air Pollution with Particulate Matter, Ozone and Nitrogen Dioxide. Bonn, Germany (2003)

3. Kelly, F.J., Fussell, J.C.: Size, source and chemical composition as determinants of toxicity attributable to ambient particulate matter. Atmos. Environ. **60**, 504–526 (2012)
4. Harrison, R.M., Hester, R.E., Querol, X.: Airborne Particulate Matter: Sources, Atmospheric Processes and Health. Royal Society of Chemistry, Cambridge (2016)
5. Nussbaumer, T.: Overview on technologies for biomass combustion and emission levels of particulate matter. Swiss Federal Office for the Environment (FOEN), Zürich, Switzerland (2010)
6. Zereini, F., Wiseman, C.L.S.: Urban Airborne Particulate Matter—Origin, Chemistry, Fate and Health Impacts. Springer, Berlin (2010)
7. Morakinyo, O.M., Mokgobu, M.I., Mukhola, M.S., Hunter, R.P.: Health outcomes of exposure to biological and chemical components of inhalable and respirable particulate matter. Int. J. Environ. Res. Public Health **13**, 592 (2016)
8. Traboulsi, H., Guerrina, N., Iu, M., Maysinger, D., Ariya, P., Baglole, C.J.: Inhaled pollutants: the molecular scene behind respiratory and systemic diseases associated with ultrafine particulate matter. Int. J. Mol. Sci. **18**(2), 243 (2017)
9. Zhang, Y., Cai, J., Wang, S., He, K., Zheng, M.: Review of receptor-based source apportionment research of fine particulate matter and its challenges in China. Sci. Total Environ. **586**, 917–929 (2017)
10. Brauer, M., Freedman, G., Frostad, J., et al.: Ambient air pollution exposure estimation for the global burden of disease 2013. Environ. Sci. Technol. **50**(1), 79–88 (2015)
11. Karagulian, F., Belis, C.A., Dora, C.F.C., Prüss-Ustün, A.M., Bonjour, S., Adair-Rohani, H., Amann, M.: Contributions to cities' ambient particulate matter (PM): a systematic review of local source contributions at global level. Atmos. Environ. **120**, 475–483 (2015)
12. Burnett, R.T., Pope III, C.A., Ezzati, M., et al.: An integrated risk function for estimating the global burden of disease attributable to ambient fine particulate matter exposure. Environ. Health Perspect. **122**(4), 397–403 (2014)
13. Chen, J., Li, C., Ristovski, Z., et al.: A review of biomass burning: Emissions and impacts on air quality, health and climate in China. Sci. Total Environ. **579**, 1000–1034 (2017)
14. United State Environmental Protection Agency (USEPA).: 2011 National Emissions Inventory (NEI) Data. https://www.epa.gov/air-emissions-inventories/2011-national-emissions-inventory-nei-data. Accessed 30 Oct 2018
15. Huang, R.J., Zhang, Y., Bozzetti, C., et al.: High secondary aerosol contribution to particulate pollution during haze events in China. Nature **514**, 218–222 (2014)
16. Sielicki, P., Janik, H., Guzman, A., Namieśnik, J.: The progress in electron microscopy studies of particulate matters to be used as a standard monitoring method for air dust pollution. Crit. Rev. Anal. Chem. **41**, 314–334 (2011)
17. AQEG (Air Quality Expert Group): Particulate Matter in the United Kingdom. Department of Environment, Food and Rural Affairs, UK (2005)
18. Kelly, K.E., Sarofim, A.F., Lighty, J.S., Arnott, W.P., Rogers, C.F., Zielinska, B., Prather, K. A.: User Guide for Characterizing Particulate Matter: Evaluation of Several Real-Time Methods. University of Utah, Institute for Clean and Secure Energy, USA (2003)
19. Mauderly, J.L., Chow, J.C.: Health effects of organic aerosols. Inhal. Toxicol. **20**, 257–288 (2008)
20. Li, N., Nel, A.E.: The cellular impacts of diesel exhaust particles: beyond inflammation and death. Eur. Respir. J. **27**, 667–668 (2006)
21. Lighty, J.S., Veranth, J.M., Sarofim, A.F.: Combustion aerosols: factors governing their size and composition and implications to human health. J. Air Waste Manage. Assoc. **50**, 1565–1618 (2000)
22. Obaidullah, M., Bram, S., Verma, V.K., De Ruyck, J.: A review on particle emissions from small scale biomass combustion. Int. J. Rene. Energ. Res. **2**, 147–159 (2012)
23. Simoneit, B.R.T.: Biomass burning—a review of organic tracers for smoke from incomplete combustion. Appl. Geochem. **17**, 129–162 (2002)
24. Rogge, W.F., Hildemann, L.M., Mazurek, M.A., Cass, G.R.: Sources of fine organic aerosol. 8. Boilers burning No. 2 distillate fuel oil. Environ. Sci. Technol. **31**(10), 2731–2737 (1997)

25. McElroy, M.W., Carr, R.C., Ensor, D.S., Markowski, G.R.: Size distribution of fine particles from coal combustion. Science **215**, 13–19 (1982)
26. Christensen, K.A., Stenholm, M., Livbjerg, H.: The formation of submicron aerosol particles, HCl and SO_2 in straw-fired boilers. J. Aerosol Sci. **29**(4), 421–444 (1998)
27. Nussbaumer, T.: Combustion and co-combustion of biomass: fundamentals, technologies, and primary measures for emission reduction. Energy Fuels **17**(6), 1510–1521 (2003)
28. Becidan, M., Todorovic, D., Skreiberg, Ø., Khalil, R.A., Backman, R., Goile, F., Skreiberg, A., Jovovic, A., Sørum, L.: Ash related behaviour in staged and non-staged combustion of biomass fuels and fuel mixtures. Biomass Bioenerg. **41**, 86–93 (2012)
29. Jöller, M., Brunner, T., Obernberger, I.: Modeling of aerosol formation during biomass combustion in grate furnaces and comparison with measurements. Energy Fuels **19**(1), 311–323 (2005)
30. Jöller, M., Brunner, T., Obernberger, I.: Modeling of aerosol formation during biomass combustion for various furnace and boiler types. Fuel Process. Technol. **88**, 1136–1147 (2007)
31. Wang, G., Deng, J., Ma, Z., Hao, J., Jiang, J.: Characteristics of filterable and condensable particulate matter emitted from two waste incineration power plants in China. Sci. Total Environ. **639**, 695–704 (2018)
32. Uski, O., Jalava, P.I., Happo, M.S., et al.: Effect of fuel zinc content on toxicological responses of particulate matter from pellet combustion in vitro. Sci. Total Environ. **511**, 331–340 (2015)
33. Gao, X., Wu, H.: Combustion of volatiles produced in situ from the fast pyrolysis of woody biomass: direct evidence on its substantial contribution to submicrometer particle (PM_1) emission. Energy Fuels **25**(9), 4172–4181 (2011)
34. Zhang, X.: Emission Standards and Control of $PM_{2.5}$ from Coal-Fired Power Plant. IEA Clean Coal Centre, London, UK (2016)
35. Zhu, F., Wang, L.: Analysis on technology-economy and environment benefit of ultra-low emission from coal-fired power units. Environ. Prot. **42**(21), 28–33 (2014)
36. Gustafsson, E., Strand, M., Sanati, M.: Physical and chemical characterization of aerosol particles formed during the thermochemical conversion of wood pellets using a bubbling fluidized bed gasifier. Energy Fuels **21**(6), 3660–3667 (2007)
37. Mylläri, F., Karjalainen, P., Taipale, R., Aalto, P., Häyrinen, A., Rautiainen, J., Pirjola, L., Hillamo, R., Keskinen, J., Rönkkö, T.: Physical and chemical characteristics of flue-gas particles in a large pulverized fuel-fired power plant boiler during co-combustion of coal and wood pellets. Combust. Flame **176**, 554–566 (2017)
38. Nyström, R.: Particle emissions from residential wood and biodiesel combustion. Doctoral thesis of Umeå University. Umeå, Sweden (2016)
39. Castellani, B., Morini, E., Filipponi, M., Nicolini, A., Palombo, M., Cotana, F., Rossi, F.: Comparative analysis of monitoring devices for particulate content in exhaust gases. Sustainability **6**, 4287–4307 (2014)
40. Marple, V.A., Rubow, K.L., Behm, S.M.: A microorifice uniform deposit impactor (MOUDI): description, calibration, and use. Aerosol Sci. Technol. **14**, 434–446 (1991)
41. Kwon, S.B., Lim, K.S., Jung, J.S., Bae, G.N., Lee, K.W.: Design and calibration of a 5-stage cascade impactor (K-JIST cascade impactor). J. Aerosol Sci. **34**, 289–300 (2003)
42. Hinds, W.C.: Aerosol Technology: Properties, Behavior, and Measurement of Airborne Particles. Wiley, New York (1999)
43. SICK AG.: Operating Instructions GRAVIMAT SHC502: Gravimetric Dust Concentration Measuring System. SICK Engineering GmbH, Ottendorf-Okrilla, Germany (2014)
44. Lindsley, W.G.: Filter Pore Size and Aerosol Sample Collection. NIOSH Manual of Analytical Methods, 5th edn. (2015)
45. Ruzer, L.S., Harley, N.H.: Aerosols Handbook: Measurement, Dosimetry, and Health Effects. CRC Press, Boca Raton (2012)
46. United State Environmental Protection Agency (USEPA): APTI Course 435—Atmospheric Sampling: Instructor's Guide. Air Pollution Training Institute, NC, USA (1980)

47. RTI International: Standard Operating Procedure for Particulate Matter (PM) Gravimetric Analysis. North Carolina, USA (2008)
48. Min, J., Nam, S.B., Kim, N.R., Kim, D.J., Yoon, Y.S., Park, S.N., Gu, J.H.: A study on the characteristics of particulate matter in the syngas produced from the waste gasification with cleaning systems for energy utilization. J. Mater. Cycles Waste Manag. **19**, 1155–1165 (2017)
49. Malvern Instruments. A Basic Guide to Particle Characterization. Malvern Instruments Limited, Worcestershire, UK (2012)
50. Epstein, N., Grace, J.R., Otten, L., Fayed, M.: Handbook of Powder Science & Technology. Chapman & Hall, New York (1997)
51. Amaral, S.S., De Carvalho, J.A., Costa, M.A.M., Pinheiro, C.: An overview of particulate matter measurement instruments. Atmosphere **6**(9), 1327–1345 (2015)
52. Gustafsson, E., Lin, L., Strand, M.: Characterization of particulate matter in the hot product gas from atmospheric fluidized bed biomass gasifiers. Biomass Bioenerg. **35**, S71–S78 (2011)
53. Morgalla, M., Lin, L., Seemann, M., Strand, M.: Characterization of particulate matter formed during wood pellet gasification in an indirect bubbling fluidized bed gasifier using aerosol measurement techniques. Fuel Process. Technol. **138**, 578–587 (2015)
54. Al-Naiema, I., Estillore, A.D., Mudunkotuwa, I.A., Grassian, V.H., Stone, E.A.: Impacts of co-firing biomass on emissions of particulate matter to the atmosphere. Fuel **162**, 111–120 (2015)
55. Vasconcellos, P.C., Souza, D.Z., Sanchez-Ccoyllo, O., et al.: Determination of anthropogenic and biogenic compounds on atmospheric aerosol collected in urban, biomass burning and forest areas in São Paulo, Brazil. Sci. Total Environ. **408**, 5836–5844 (2010)
56. Fine, P.M., Cass, G.R., Simoneit, B.R.T.: Chemical characterization of fine particle emissions from fireplace combustion of woods grown in the northeastern United States. Environ. Sci. Technol. **35**(13), 2665–2675 (2001)
57. Austria International Atomic Energy Agency: Sampling, storage and sample preparation procedures for X ray fluorescence analysis of environmental materials—IAEA-TECDOC-9SO. Vienna, Austria (1997)
58. Lodge Jr., J.P.: Methods of Air Sampling and Analysis. CRC Press, Florida (1988)
59. Salazar, A.: Glass-fibre filter for analysis of aerosol particles using energy-dispersive x-ray fluorescence. X-Ray Spectrom. **21**(3), 143–147 (1992)
60. Beckhoff, B., Kanngießer, B., Langhoff, N., Wedell, R., Wolff, H.: Handbook of Practical X-ray Fluorescence Analysis. Springer, Berlin (2006)
61. Adams, F.C., Van Grieken, R.E.: Absorption correction for X-ray fluorescence analysis of aerosol loaded filters. Anal. Chem. **47**(11), 1767–1773 (1975)
62. Landsberger, S., Creatchman, M.: Elemental Analysis of Airborne Particles. Gordon and Breach Science Publishers, Amsterdam (1999)
63. EN 14902:2005: Ambient air quality—Standard method for the measurement of Pb, Cd, As and Ni in the PM_{10} fraction of suspended particulate matter. European Standard (2005)
64. California Environmental Protection Agency: Procedure for the determination of trace elements in particulate matter emitted from motor vehicle exhaust using inductively coupled plasma mass spectrometry (ICP-MS). California, USA (2010)
65. Criss, J.W.: Particle size and composition effects in X-ray fluorescence analysis of pollution samples. Anal. Chem. **48**(1), 179–186 (1976)
66. Buck, C.S., Paytan, A.: Evaluation of commonly used filter substrates for the measurement of aerosol trace element solubility. Limnol. Oceanogr. Methods **10**, 790–806 (2012)
67. Coedo, A.G., Dorado, M.T., Padilla, I., Alguacil, F.J.: Use of boric acid to improve the microwave-assisted dissolution process to determine fluoride forming elements in steels by flow injection inductively coupled plasma mass spectrometry. J. Anal. At. Spectrom. **13**, 1193–1197 (1998)
68. Santos, M.D., Matos, J.R., Carvalho, L.R.F., Sant'Agostino, L.M., Korn, L.M.: An X-ray diffraction study of filters used for atmospheric aerosol sampling. Powder Diffr. **11**(3), 230–234 (1996)

69. Davis, B.L., Johnson, L.R.: On the use of various filter substrates for quantitative particulate analysis by X-ray diffraction. Atmos. Environ. **16**(2), 273–282 (1982)
70. Sturges, W.T., Harrison, R.M.: Semi-quantitative X-ray diffraction analysis of size fractionated atmospheric particles. Atmos. Environ. **23**(5), 1083–1098 (1989)
71. Choung, S., Oh, J., Han, W.S., Chon, C.M., Kwon, Y., Kim, D.Y., Shin, W.: Comparison of physicochemical properties between fine ($PM_{2.5}$) and coarse airborne particles at cold season in Korea. Sci. Total Environ. **541**, 1132–1138 (2016)
72. Boman, C., Nordin, A., Boström, D., Öhman, M.: Characterization of inorganic particulate matter from residential combustion of pelletized biomass fuels. Energy Fuels **18**(2), 338–348 (2004)
73. Esteve, V., Rius, J., Ochando, L.E., Amigó, J.M.: Quantitative X-ray diffraction phase analysis of coarse airborne particulate collected by cascade impactor sampling. Atmos. Environ. **31**(23), 3963–3967 (1997)
74. Esteve, V., Carda, J., Reventos, M.M., Amigó, J.M.: Quantitative X-ray diffraction phase analysis of airborne particulate collected by a cascade impactor sampler using the Rietveld full-pattern fitting method. Powder Diffr. **12**(3), 151–154 (1997)
75. Jordán, M., Alvarez, C., Sanfeliu, T.: A powder diffraction quantification method (ALJOR) for atmospheric particulate matter. Environ. Geol. **42**(7), 810–816 (2002)
76. Wang, X., Hu, Z., Adeosun, A., Liu, B., Ruan, R., Li, S., Tan, H.: Particulate matter emission and K/S/Cl transformation during biomass combustion in an entrained flow reactor. J. Energy Inst. (2017). https://doi.org/10.1016/j.joei.2017.10.005
77. Feng, C., Wu, H.: Synergy on particulate matter emission during the combustion of bio-oil/biochar slurry (bioslurry). Fuel **214**, 546–553 (2018)
78. Czech, H., Miersch, T., Orasche, J., et al.: Chemical composition and speciation of particulate organic matter from modern residential small-scale wood combustion appliances. Sci. Total Environ. **612**, 636–648 (2018)
79. Gonçalves, C., Alves, C., Evtyugina, M., Mirante, F., Pio, C., Caseiro, A., Schmidl, C., Bauer, H., Carvalho, F.: Characterisation of PM_{10} emissions from woodstove combustion of common woods grown in Portugal. Atmos. Environ. **44**(35), 4474–4480 (2010)
80. Brunner, T., Fluch, J., Obernberger, I., Warnecke, R.: Investigations of aerosol formation pathways during MSW combustion based on high-temperature impactor measurements. Fuel Process. Technol. **105**, 154–160 (2013)
81. Oanh, N.T.K., Ly, B.T., Tipayarom, D., Manandhar, B.R., Prapat, P., Simpson, C.D., Liu, L. J.S.: Characterization of particulate matter emission from open burning of rice straw. Atmos. Environ. **45**(2), 493–502 (2011)
82. Wu, D., Zhang, F., Lou, W., Li, D., Chen, J.: Chemical characterization and toxicity assessment of fine particulate matters emitted from the combustion of petrol and diesel fuels. Sci. Total Environ. **605**, 172–179 (2017)
83. Duan, F., Liu, X., Yu, T., Cachier, H.: Identification and estimate of biomass burning contribution to the urban aerosol organic carbon concentrations in Beijing. Atmos. Environ. **38**(9), 1275–1282 (2004)
84. Chow, J.C., Watson, J.G., Crow, D., Lowenthal, D.H., Merrifield, T.: Comparison of IMPROVE and NIOSH carbon measurements. Aerosol Sci. Technol. **34**(1), 23–34 (2001)
85. Park, S.S., Yu, J.: Chemical and light absorption properties of humic-like substances from biomass burning emissions under controlled combustion experiment. Atmos. Environ. **136**, 114–122 (2016)
86. Ono-Ogasawara, M., Myojo, T., Smith, T.J.: A simple direct injection method for GC/MS analysis of PAHs in particulate matter. Ind. Health **46**(6), 582–593 (2008)
87. Chow, J.C., Yu, J.Z., Watson, J.G., Ho, S.S.H., Bohannan, T.L., Hays, M.D., Fung, K.K.: The application of thermal methods for determining chemical composition of carbonaceous aerosols: a review. J. Environ. Sci. Heal. A **42**(11), 1521–1541 (2007)
88. Elorduy, I., Durana, N., García, J.A., Gómez, M.C., Alonso, L.: Optimization and validation of thermal desorption gas chromatography-mass spectrometry for the determination of

polycyclic aromatic hydrocarbons in ambient air. J. Anal. Methods Chem. (2018). https://doi.org/10.1155/2018/8734013

89. Krugly, E., Martuzevicius, D., Puida, E., Buinevicius, K., Stasiulaitiene, I., Radziuniene, I., Minikauskas, A., Kliucininkas, L.: Characterization of gaseous-and particle-phase emissions from the combustion of biomass-residue-derived fuels in a small residential boiler. Energy Fuels **28**(8), 5057–5066 (2014)

Chapter 15
Solid Residues (Biochar, Bottom Ash, Fly Ash, …)

Elsa Weiss-Hortala, Anthony Chesnaud, Laurène Haurie, Nathalie Lyczko, Rajesh Munirathinam, Ange Nzihou, Séverine Patry, Doan Pham Minh and Claire E. White

Abstract Solid co-products from biochemical, chemical and thermochemical processes of biomass and biowaste have gained momentum in utilization as secondary raw materials. These solids are carbon-based or mineral-based materials, and for their suitable use in a number of fields, various properties should be determined. This chapter addresses advanced techniques used to determine physical and chemical properties of these solid residues. For each technique, the basics and protocols are described. Post-treatment procedures and interpretation of the results obtained are also provided for some residues.

15.1 Introduction

The solid residues are mainly composed of a certain fraction of ash, fixed carbon and volatile matter from the feedstock (see Chap. 2 for definitions). These residues should adequately be characterized to be considered as a secondary raw material for further uses. In this chapter the focus is put on the residues from thermochemical conversion, performed either with dry or wet routes.

Physico-chemical properties of solid residues can be determined for the bulk material or at the surface. Depending on the application, the bulk properties are not sufficient to explain the activity of the solid residue. Thus, generic characterization

E. Weiss-Hortala (✉) · L. Haurie · N. Lyczko · R. Munirathinam · A. Nzihou · S. Patry · D. Pham Minh
RAPSODEE CNRS, UMR-5302, Université de Toulouse, IMT Mines Albi, Campus Jarlard, 81013 Albi Cedex, 09, France
e-mail: elsa.weiss@mines-albi.fr

A. Chesnaud
MINES ParisTech, PSL Research University, MAT—Centre des matériaux, CNRS UMR 7633, BP 87, 91003 Evry, France

C. E. White
Department of Civil and Environmental Engineering and the Andlinger Center for Energy and the Environment, Princeton University, Princeton, NJ 08544, USA

A. Nzihou (ed.), *Handbook on Characterization of Biomass, Biowaste and Related By-products*, https://doi.org/10.1007/978-3-030-35020-8_15

techniques have been explained in Chap. 2 and the present chapter suggests a step forward in the characterization of the physico-chemical properties of solids. It should be noticed that the examples considered in this chapter will be related to biogenic residues, pyrocarbons (biochars) for instance rather than on ash.

15.1.1 Definition of Solid Residues from Thermal Processes

Solid residues from thermal processes are mainly described as a mixture of ashes (mineral content) and pyrocarbons in various proportions as seen in Fig. 15.1. Basically, ash proportion increases in the solid residue as the temperature increases, while the organic matter is gasified. When a thermochemical process is performed, the mineral species are mainly recovered in the solid residue. However, a small amount of these minerals might be volatilized and transferred in form of fine particles (particulate matter, see Chap. 14) in the flue gas as "fly ashes" [1]. The term "bottom ash" refers to coarse or granular non-combustible residue from the combustion.

Pyrocarbon is namely a family of carbon-based materials obtained from thermal processes and containing carbons of different natures: biochar, soot and black carbons. A clear definition of each carbon material is quite difficult to express, however it could be summarized as follows:

- Biochar (charcoal): charcoal is a solid biofuel containing graphitized carbon with different yield from low graphitic content (lignite) to high graphitic content (anthracite). This material is mainly characterized by its high heating value and the values of O/C and H/C molar ratios [2].
- Biochar (also hydrochar): this is a solid residue containing amorphous, turbostratic and graphitic carbons in various proportions, as well as minerals. This is the main term used to qualify the solid residues from thermochemical conversion of biomass and biowaste [3, 4]. Hydrochar is sometimes used and refers to solid residue from hydrothermal conversion (wet thermal process).

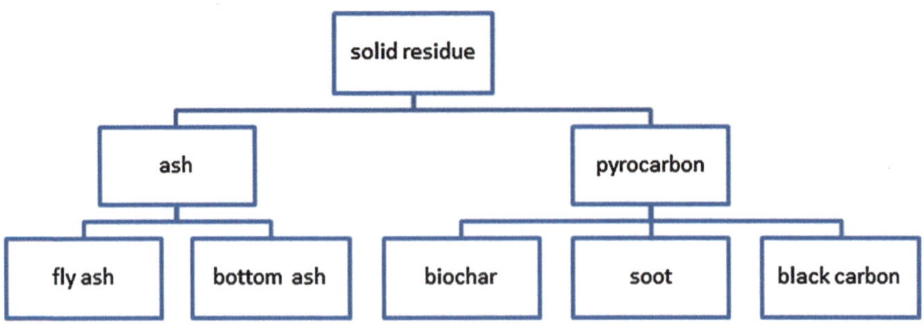

Fig. 15.1 Categories of solid residues from thermal conversion

- Soot: this is basically small particles containing carbon and some minerals that are recovered in the flue gas (see Chap. 14). Depending on the operating conditions the proportion of graphitized carbon could be high [5–8].
- Black carbon: this is a material with high carbon content (90 wt%) which could be described as aggregates of primary particles, such as carbon black [9]. This kind of material is composed of graphene layers, resulting in a high thermal resistance. The black carbon is therefore a part of the soots [9–11]. Black carbon is different from carbon black which is a manufactured product [6].

15.1.2 Targeted Valorization of These Solid Residues

Among the solid residues, biochars may currently be used as solid biofuel [12], soil amendment [3], sorbent [13–15], activated carbons [16], catalysts [17–21], support for catalyst [21, 22], storage for energy [21], electrode/supercapacitors [23] or additive in building sector [24]. It should be noted that the biochar market is currently developed for soil amendment (estimated at 283.2 kt in 2015, and expected to reach USD 3.14 billion by 2025) [25].

Soots are produced in a low amount during thermal processes but their presence results in severe environmental and health issues. Literature is mostly focused on the characterization of soots formation and chemical structure [6–11, 26] for their removal. However, the authors showed that the particles are graphenic carbon materials with a certain amount of minerals, and containing oxygenated functions especially in soots from biomass [27]. Finally soots are expected to be used in a same way as for biochar, including also as reinforcement material [28].

Regarding the use of fly ashes, the ASTM standard [1] figures out the applications versus the properties of the solid. The main current applications are wastewater treatment, gas cleaning, soil amendment and stabilization, pozzolanic liners, slurry wall, or an additive for cement concrete or gypsum products [1].

Based on the specifications requested for these further uses, the reader will find generic (Chap. 2) and advanced techniques (this chapter) to properly characterize these solid residues from the above mentioned processes. However to support the discussion of some techniques, the next sections are providing some basics about the two main categories of solid residues from thermoconversion: biochar and ash.

15.1.3 Basic Description of Biochar

Briefly, biochar from thermoconversion processes is a complex carbon-based material containing various carbon phases such as pyrolyzed and non-pyrolyzed carbons. This solid material may contain porosity (from macro to nano scale) that could be improved and controlled using post treatments (i.e. activation) [13, 15, 22,

29–31]. The surface of the biochar particles is characterized by various surface chemical functions related to the carbonaceous matrix (aromatic structures, elemental carbon) [19, 29, 32, 33], O-containing groups (phenol, lactone, acids…) [21, 34–36], and minerals [19, 37–40]. Figure 15.2 shows schematic properties of a carbonaceous solid residue [41].

Solid residues from thermochemical conversion are usually described as turbostratic carbons (3D sp^2-bonded carbon) [42] with highly disordered layering [43, 44] while graphite is a well-ordered 3D structure with a perfect stacking of graphene layers (see definitions in [42]). Structure of carbon-based materials relates to crystallinity and coherence of the stacking [44], such as graphite (long-range order crystalline structures) or turbostratic, as shown in Fig. 15.3. Turbostratic term refers to different criteria: a short-range order of the graphene layers (a stack of less than 5

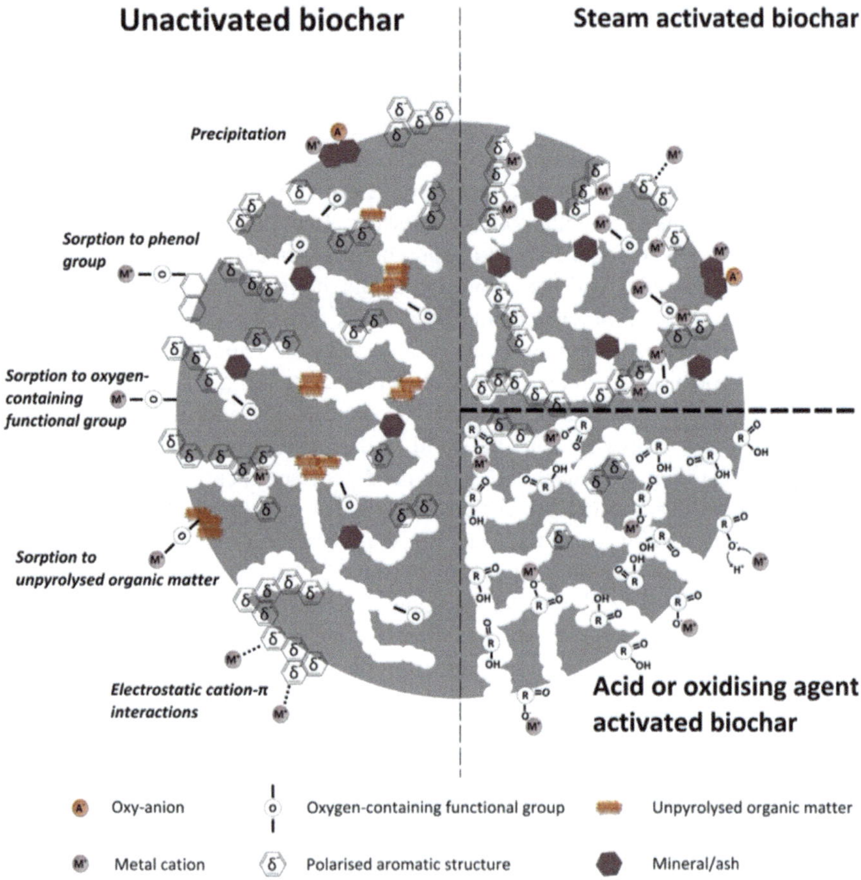

Fig. 15.2 Biochar and activated biochars highlighting physico-chemical properties: porosity, oxygenated groups, minerals and carbonaceous matrix. Adapted with permission from Elsevier [41]

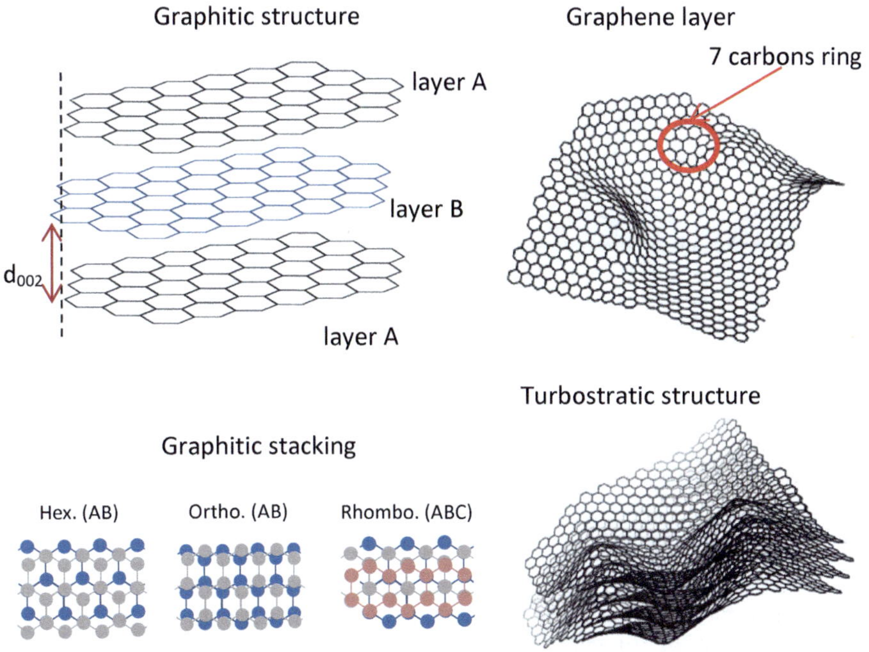

Fig. 15.3 Graphitic structure showing the graphene layers, graphitic stacking, graphene layers with defect (7 carbons ring, from [45]) and turbostratic structure (from [45]). Reprinted with permission from Elsevier [45]

layers) with no preferential orientation forming edge defects, a disparity in the sizes of the aromatics (more or less than 6 carbons) and aromatic ring systems (less than 6 fused benzene rings), a presence of aliphatic chains, and an insertion of heteroatoms or vacancies [44, 46–49]. In turbostratic carbons, a coherent domain containing graphenic nanocrystallites (with different graphitic stacking [50]) could be observed at a very short range, characterized with the L_a value (basal plane) and L_c value (stacking) as seen in Fig. 15.4.

15.1.4 Basic Description of Ash

The bottom ash is a coarse material, recovered at the bottom of the reactor in thermochemical processes and combustion in particular. The grain size typically ranges from fine sand to gravel in size. Bottom ash consists of noncombustible inorganics, unburned carbon, and combustion ash, depending on the composition of the biowaste and the operating conditions. It contains greater quantities of carbon, from plastics, cardboards, or paper incompletely burnt. Depending on the original feedstock, it may also contain fine ash and residues such as iron wire, glass, or cement block.

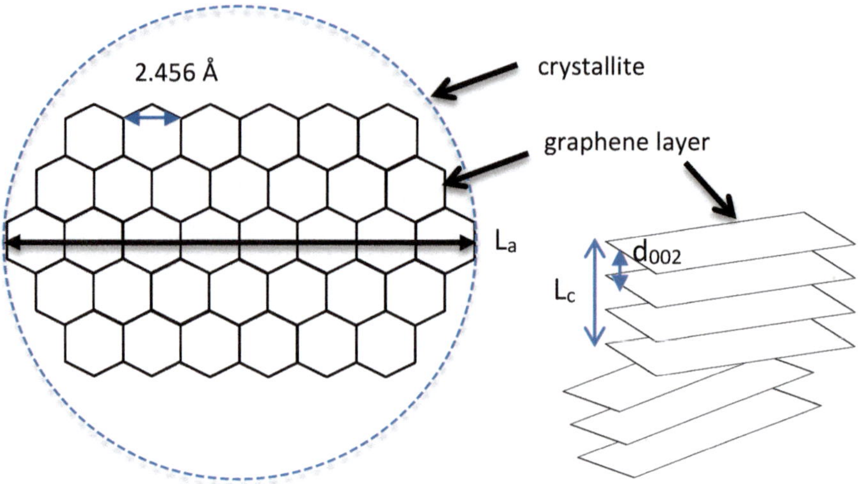

Fig. 15.4 Graphenic nanocrystallite description: L_a and L_c

The melt glass phases serve as a matrix for bottom ash particles, and are typically composed of silicon, aluminum, calcium, and also some alkaline and alkaline-earth elements [51]. The melt glass phases could be described as a:

- isotrope material, transparent or translucent locally shifting to brown
- mixture composed of metallic inclusions
- highly vesicular structure due to the entrapment of air and gas bubbles
- hybrid material with coexistence of various well-formed mineral species, quench crystallites, and refractory components.

On the contrary, fly ash is fine, complex, and a residual material entrained with the flue gas. Fly ash is a powder composed of fine particles, mostly spherical and varying in diameter. Under a microscope they look like tiny solidified bubbles or spheres of various sizes. The average particle size is about 10–20 μm but can vary from <1 μm to about 100 μm. As explained previously, ASTM [1] classifies fly ash into two categories—Class F (low calcium content) and Class C fly ash (high calcium content). Particularly, the total weight percentage of $SiO_2 + Al_2O_3 + Fe_2O_3$ should be ≥ 50 wt% for Class C or ≥ 70 wt% for Class F, and will direct the further uses of the fly ash [52]. The structure of the fly ash is the same as described for bottom ash, meaning that the main matrix is composed of a melt glass phase [51].

Table 15.1 shows some examples of bottom and fly ash compositions from coal and biomass combustion [52–54]. Oxides of Si, Al, and Fe, with a higher boiling point, are supposed to be present mainly in bottom ash, in contrast with oxides of Na, K, which are mainly present in fly ash, due to their lower boiling point. However this distribution is related to the operating conditions, the original composition of the feedstock and the reactor technology as well.

Table 15.1 Chemical compositions (wt% in dry basis) of fly and bottom ashes from various feedstocks combustion: lignite coal (from [52]), fir chips (from [53]), and biomass mixture (from [54])

Oxides	Lignite coal		Fir chips		Biomass mixture[a]	
	Bottom ash	Fly ash	Bottom ash	Fly ash	Bottom ash	Fly ash
SiO_2	10.8–48.3	14.8–50.0	14.8	16.4	89.0	57.5
Al_2O_3	2.5–24.9	3.4–25.7	3.2	5.9	1.5	4.8
Fe_2O_3	0.5–8.2	0.9–11.8	3.3	3.6	1.0	2.9
CaO	8.6–45.1	13.0–54.1	48.4	33.7	3.9	17.3
MgO	0.4–4.6	0.5–9.1	9.9	11.7	0.6	2.3
Na_2O	0.2–1.2	0.2–1.2	1.0	1.4	0.2	0.4
K_2O	0.0–3.6	0.2–4.9	11.1	13.0	2.1	3.9
P_2O_5	–	–	1.0	3.3	0.4	2.0

[a]Biomass mixture = 80% tree waste + 20% palm kernel shells

The structure of ash, and especially spherical fly ash particles, could be summarized as shown in Fig. 15.5. The Al–Si glassy phase (aluminosilicate sphere) is composing in the most of cases the core structure; however, a spongy Ca sphere (calcisphere) or a hollow sphere are also encountered [55].

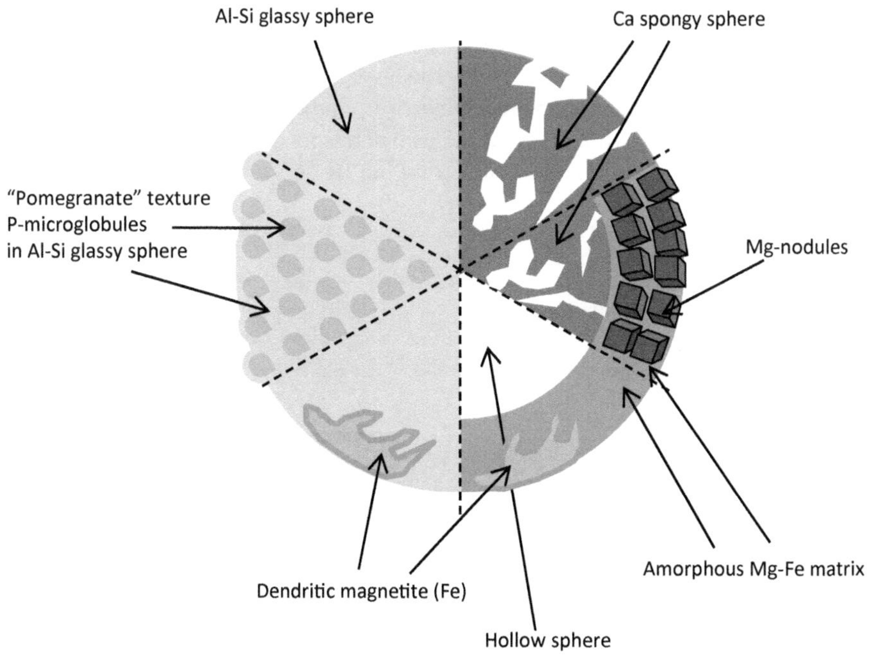

Fig. 15.5 Schematic structure of fly ash spherical particles

15.2 Sorption Mechanism, Surface Energy and Heat of Sorption Using Dynamic Vapor Sorption (DVS)

Dynamic Vapor Sorption (DVS) is useful to investigate sorption mechanisms, sorption kinetics, surface energy, heat of sorption, diffusion coefficients… (see Chap. 2) at the surface of the solid using solvent sorption. In this section some examples will be provided to determine sorption mechanism and active surface area, kinetics, surface energy and heat of sorption.

15.2.1 Sorption Mechanism Modeling

It is well-known that solid residue from thermochemical conversion contained a certain amount of condensable organics that may block the pores and thus modify the available surface [56]. The aim of the outgassing prior to N_2 or Ar adsorption is to clean the surface to get the maximal specific surface area. Using DVS (as explained in Chap. 2), the sample does not require preparation such as outgassing technique compared to gas adsorption. In addition, this dynamic technique allows the use of various solvents (water and organic solvents) to be adsorbed at ambient temperature, indicating the active surface available.

Sorption of chemicals onto a surface could involve weak energy (physisorption) or strong energy bonds (chemisorption). Chemisorption is easily recognized as a monolayer mechanism (type I isotherm as explained in Chap. 2). Since the surface of a solid residue is particularly heterogeneous, active sites have various energies resulting in various physisorption mechanisms (monolayer, multilayer or cluster, as seen in Fig. 15.6) [57]. Multilayer mechanism corresponds to isotherm type II or IV while cluster mechanism refers to isotherm type III or V.

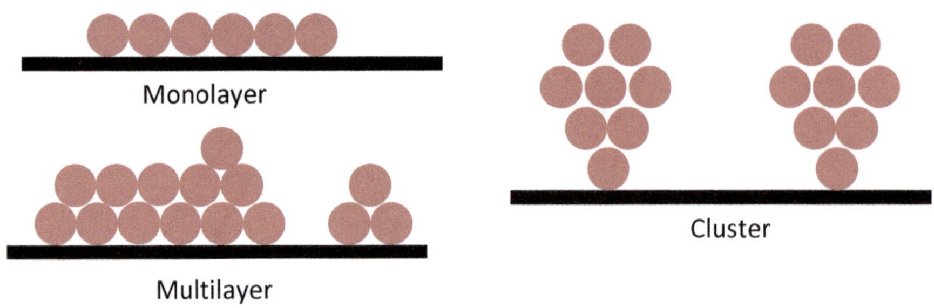

Fig. 15.6 Possible mechanisms for physisorption

For the sorption isotherm obtained as detailed in Chap. 2, different models could be applied to determine the sorption mechanism, as seen in Fig. 15.7. Note that the list of models is not exhaustive. In the present chapter the following models are described: Langmuir, Brunauer-Emmett-Teller (BET), Freundlich, Guggenheim–Anderson–de Boer (GAB) and Young & Nelson. It could be noticed that water is mainly following a type III isotherm when adsorbed on low energy surfaces such as solid residues (biochar for example) and food [58–61]. In this case, other models from the literature could be used: Dubinin-Radushkevich type models [58, 62–64], well applied to carbon-based material such as biochar and activated carbons to define the micropore filling; Dubinin-Serpinsky type models [58, 65], to take into account the water cluster formation; and Do & Do type models [58, 62, 66–68], specifically designed to improve isotherm description of water onto microporous carbons. For alcohols, a mixed isotherm, close to type II, is generally observed due to their polar contribution and non-polar alkane chain. Type III isotherm is a typical behavior for molecules with low polarity such as alkanes.

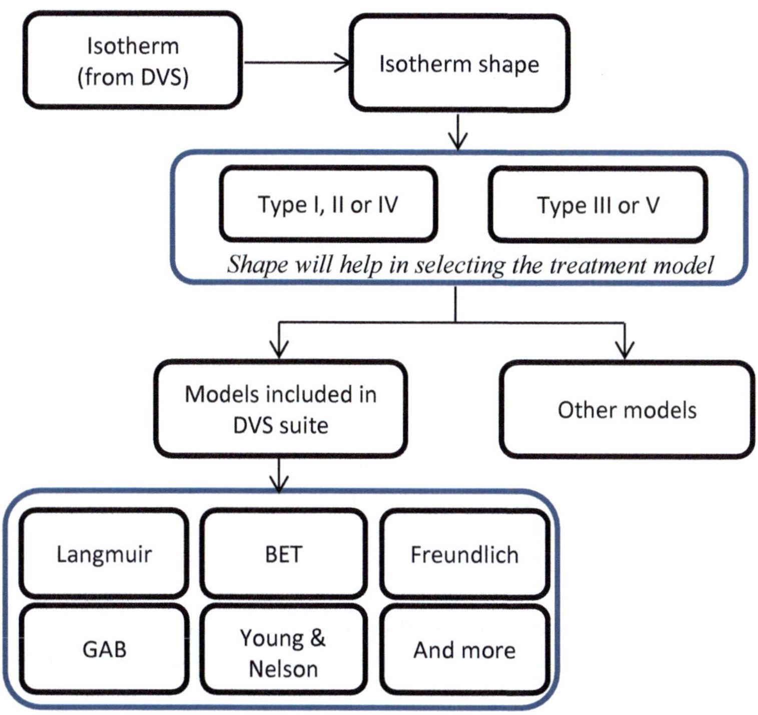

Fig. 15.7 Procedure to model isotherm from DVS data

15.2.1.1 Langmuir Model

This model is only applied to type I isotherm, which basically describes chemisorption. However, at room temperature, chemisorption could appear but will probably not occur alone. Therefore this isotherm type will rarely be observed for solid residues.

15.2.1.2 Brunauer-Emmett-Teller Model (BET)

For type II and IV isotherms, basically BET equation is used (Eq. 15.1). As detailed in Chap. 2 for gas sorption, the linear part of the curve in a range of p/p° equal to 0.05–0.35 is used to calculate the mole number of solvent in the monolayer. This value is then introduced in Eq. (15.2) for calculating the specific surface area (namely active surface area using DVS). However, using solvents, this range could be modified and it is highly recommended to check directly with the DVS software the quality of the straight line (R^2 value).

$$\frac{p_r}{n \times (1 - p_r)} = \frac{C_{BET} - 1}{C_{BET} \times n_m} p_r + \frac{1}{C_{BET} \times n_m} \tag{15.1}$$

where p_r is the reduced pressure (this is equal to relative pressure, p/p°), n_m is the mole number of solvent in the monolayer, n is the mole number of solvent at each relative pressure, and C_{BET} is a constant related to the adsorbate-adsorbent interaction strength (no dimension). In other words, C_{BET} is related to the net heat of sorption (difference between the energy of the first layer and the other remaining layers).

From the quantity adsorbed in the monolayer, the active surface area (ASA) could be calculated with the following equation:

$$ASA = \frac{S}{m_{sample}} = \frac{n_{m\times} \sigma \times N_A}{m_{sample}} \tag{15.2}$$

where m_{sample} is in g, S, the surface in $m^2\ g^{-1}$, is calculated using n_m the moles of sorbent in the monolayer, N_A Avogadro number and σ the average surface of a sorbent molecule (m^2). Some values of σ are presented in Table 15.2 [66].

Figure 15.8a presents an example of isotherm obtained on a wood biochar using hexane as sorbent at 25 °C. BET model has been applied to calculate the active surface area. In this case, the domain of relative pressure has been reduced to get a straight line (Fig. 15.8b). The reduction of the linear domain might be explained by

Table 15.2 Average molecular surface of some solvents (σ)

Solvent	Water[a]	Hexane	Isopropanol[b]	Acetone
Surface (Å^2)	10.4–12.5	51.5	44	42.5

[a]Depends on strength of interaction
[b]Calculated from bulk properties

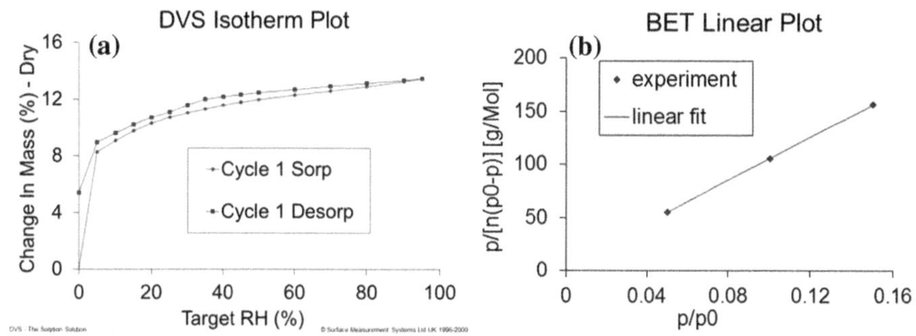

Fig. 15.8 Isotherm plot (**a**) and BET linear plot (**b**) for hexane sorption at 25 °C on a wood biochar (gasification of pine wood at 750 °C)

the combination of mechanisms over different kinds of active sites since the surface is highly heterogeneous. A modified BET equation, taking into account two kinds of sites, could be found in the literature [69].

15.2.1.3 Freundlich Model

Freundlich is an empiric model available in the DVS suite (Eq. 15.3) that represents type II/IV and III/V isotherms.

$$n_{ads} = K \times p_r^{1/n} \tag{15.3}$$

where n_{ads} is the amount adsorbed (mol), p_r is the relative pressure (dimensionless), K is a constant that indicates the sorption capacity while n measures the intensity of sorption.

As seen in Fig. 15.9, a multilayer mechanism $\left(\frac{1}{n} < 1\right)$ as well as a cluster mechanism $\left(\frac{1}{n} > 1\right)$ could be modeled with Freundlich model. It should be noted

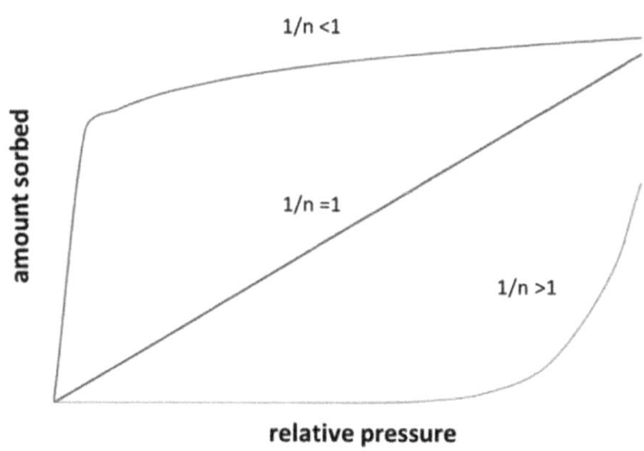

Fig. 15.9 Freundlich isotherm shapes depending on the n value

that a Langmuir-Freundlich model is also available in the DVS Suite. This model combines features of Langmuir hypotheses (monolayer sorption, non-porous materials, chemisorption) and empirical Freundlich equation. This model is particularly suitable for surfaces having active sites of significant different chemical nature or for binary mixture sorption [70].

15.2.1.4 Guggenheim–Anderson–de Boer Model (GAB)

The GAB model has been widely applied in food science, and became the most recommended model to describe water sorption by food materials [59, 71]. This model is a refinement of Langmuir and BET theories. Indeed, the model is based on a similar state of the sorbed molecules from the second layer, while BET model postulates that all layers are equivalent. Therefore the first layer is separately described (as for the Langmuir model). The GAB equation (Eq. 15.4) contains a new k constant compared to BET equation (Eq. 15.1), which includes the specific interaction between the first and the second layer [72].

$$\frac{p_r}{n \times (1 - k \times p_r)} = \frac{C_{GAB} - 1}{C_{GAB} \times n_m} p_r + \frac{1}{C_{GAB} \times n_m \times k} \tag{15.4}$$

where p_r is the reduced pressure (this is equal to relative pressure, p/p°), n_m is the mol number of solvent in the monolayer, n is the mol number of solvent at each relative pressure, and C_{GAB} is a constant related to the adsorbate-adsorbent interaction strength while k represents the adsorbate-adsorbate interaction (1st and 2nd layer).

Equation (15.4) is rewritten as Eq. (15.5) and fitted up to p/p° = 0.9 in the DVS Suite. From n_m value, this is still possible to determine an active surface area using Eq. (15.2).

$$\frac{p_r}{n} = \frac{k \times (C_{GAB} - 1)}{C_{GAB} \times n_m} p_r^2 + \frac{C_{GAB} - 2}{C_{GAB} \times n_m} p_r + \frac{1}{C_{GAB} \times n_m \times k} \tag{15.5}$$

A GAB modified equation has also been proposed in the literature for microporous solids to better accounts for the heterogeneity of the surface by combining Langmuir-Freundlich and GAB descriptions [70]. Finally, the new equation (Eq. 15.6) is structured as GAB equation (Eq. 15.4), but the relative pressure contains the symbol of heterogeneity α:

$$\frac{p_r^{\alpha}}{n \times (1 - k \times p_r^{\alpha})} = \frac{C_{GAB} - 1}{C_{GAB} \times n_m} p_r^{\alpha} + \frac{1}{C_{GAB} \times n_m \times k} \tag{15.6}$$

15.2.1.5 Young & Nelson Model

Young & Nelson model is particularly suitable to describe sorption isotherm of water and other highly polar adsorbates that have very strong adsorbate-adsorbate interactions. This model considers 3 contributions: monolayer adsorption, multi-layer adsorption and absorption into the bulk for a sorption-desorption cycle (Eqs. 15.7, 15.8 [71, 73]).

$$n_{ads} = A \times (\beta + \theta) + B \times \theta \times p_r \tag{15.7}$$

$$n_{des} = A \times (\beta + \theta) + B \times \theta \times p_{r,max} \tag{15.8}$$

In these equations, n is the moles sorbed and p_r is the partial pressure. The different parameters are describing the monolayer adsorption (θ), the multilayer adsorption ($\theta \times p_r$), and the bulk absorption (β). The A constant is related to the couple sorbent-monolayer of adsorbate (g solvent adsorbed/g dry sample). The B constant is describing the absorption (g solvent absorbed/g dry sample). The E constant is directly describing the ability of the adsorbate to be bounded to the surface through the monolayer at a given temperature.

Figure 15.10 presents the isotherms recorded for water sorption on two cellulose samples (standard and silicified microcrystalline cellulose) as well as the Young & Nelson partial components plot.

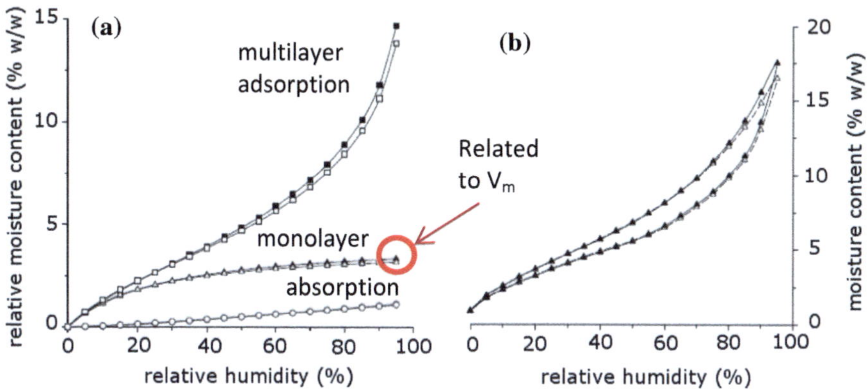

Fig. 15.10 Example of Young & Nelson components plot (**a**) and experimental isotherms (**b**) of water sorption onto standard (empty symbols) and silicified microcrystalline cellulose (full symbols) [73]. Figure **a** shows the 3 components: monolayer, multilayer and absorption. Reprinted with permission from Elsevier [73]

15.2.2 Kinetics of Sorption

Basically, sorption process is decomposed in several steps, especially considering the presence of pores:

- Diffusion of adsorbate from bulk to surface
- Diffusion of molecules in the pore structure
- Diffusion along the surface of the pore
- Diffusion through constrictions in the porosity
- Adsorption on an active site, depending on physisorption or chemisorption.

The rates of adsorption and desorption in porous media are rather controlled by transport within the pore network than by the intrinsic kinetics of sorption at the surface. The diffusion in the porous media follows different mechanisms (with different kinetics parameters) depending on the pore size, and classified using Knudsen number:

- Free gas diffusion
- Knudsen diffusion
- Surface diffusion (site-to-site hopping process)
- Activated diffusion (size of adsorbate $\approx$ pore diameter).

The dynamic measurement of the sample mass with the time in a DVS experiment results in kinetics parameters determination. As for the isotherm analysis, different kinetics models are available to explain the sorption mechanism and to determine the extent of the sorption kinetics. However, the models are not included in the DVS Suite (see Fig. 15.11).

Basically the diffusion is a mass transport process generated by chemical potential gradients. Fick diffusion is a concentration gradient dependent process, and is rarely observed. However the diffusion in porous network also includes the diffusion through the barrier of pores. When the latter phenomenon is a rate-determining step, the Linear Driving Force (LDF) model is used. When both phenomena are significant, two other models have been used in the literature:

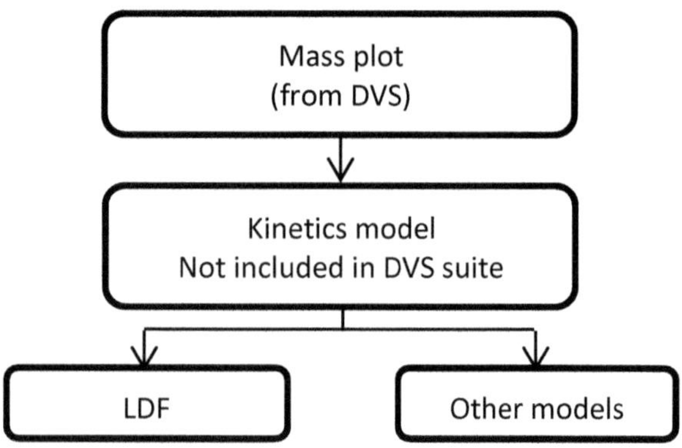

Fig. 15.11 Procedure for kinetics analysis from DVS data

combined barrier resistance/diffusion (CRBD) [74] or parallel exponential kinetics (PEK) [75, 76].

Kinetics would be described by two different processes, associated with a double stretched model (DSE, Eq. 15.9) [77]. Linear Driving Force (LDF) derives from DSE model where $k_1 = k_2$, and $\beta_1 = \beta_2 = 1$, and is the simplest model describing the sorption kinetics (Eq. 15.10). Thus this model assumes an average concentration of adsorbates at the surface, and a multilayer mechanism on equivalent sorption active sites. This equation is easily solved with analytical methods, and has been widely used for water sorption on activated carbons [78–82].

$$\frac{M_t}{M_e} = A_1\left(1 - \mathrm{e}^{-(k_1 t)^{\beta_1}}\right) + (1 - A_1)\left(1 - \mathrm{e}^{-(k_2 t)^{\beta_2}}\right) \tag{15.9}$$

$$\frac{M_t}{M_e} = 1 - \mathrm{e}^{-kt} \tag{15.10}$$

where M_t and M_e are the amounts (g) of solvent adsorbed at t and at the equilibrium time, t is the time (min or s), and k is the rate constant (time^{-1}).

Figure 15.12 shows an example of LDF calculation from DVS data. The kinetics parameters should be calculated for each step of partial pressure (or relative humidity). The kinetics constant (k) is then calculated for each step and it is useful to follow its value over the entire sorption cycle. As shown in Fig. 15.13, this variation results in 3 different domains for water sorption on a commercial activated carbon.

15.2.3 Surface Energy

Adsorption (physical or chemical phenomena) is always related to an interaction between at least the adsorbate and energetic actives sites of the sorbent surface. To understand or predict the behavior of a probe on a specific material, or to better explain sorption isotherms, the determination of surface energy is very helpful. This would be done using Temperature programmed desorption (TPD) [83, 84], microcalorimetry [85, 86], Inverse Gas Chromatography at infinite dilution (IGC) [85, 87, 88], wettability [89] or DVS [88, 90, 91].

Polar and non-polar solvents are supposed to have various affinities towards the surface depending on the surface tension of the solid residues. The surface tension is calculated as two contributions related to polar and non-polar surface energies, which can be calculated using isotherms. The determination of surface energy of solid residues is performed using a non-polar (for example hexane) and a polar (for example isopropanol) solvent at the same temperature. The DVS suite allows a post-treatment calculation to determine the surface tension in term of the dispersive and the polar contributions. The procedure is presented in Fig. 15.14.

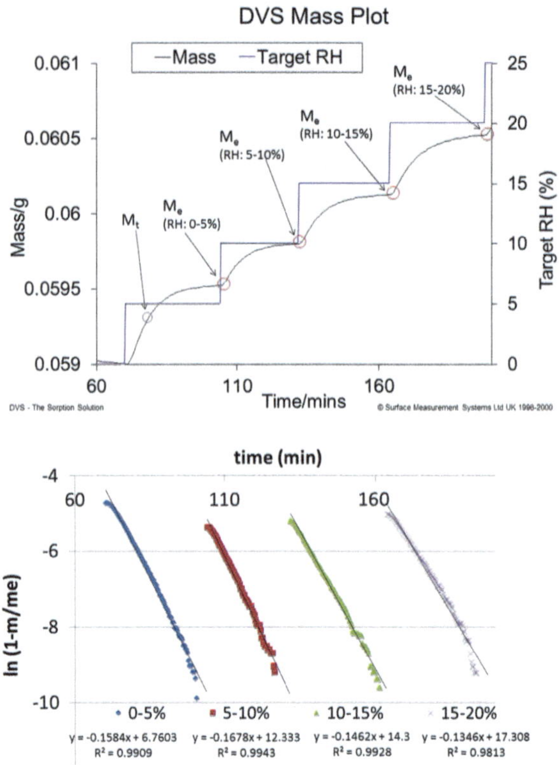

Fig. 15.12 Example of DVS mass plot and kinetics LDF model plot (water sorption on a wood biochar)

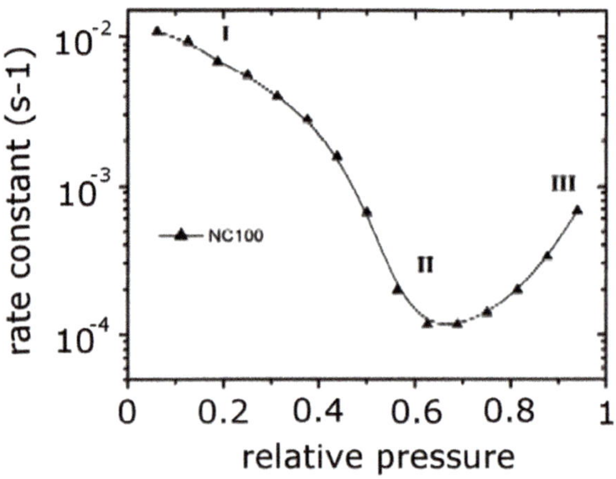

Fig. 15.13 Rate constant from LDF model (water sorption on commercial activated carbon from [80]). Reprinted with permission from Elsevier [80]

The interaction between a liquid and a solid surface (work of adhesion) is described by Young's equation (Eq. 15.11). However for high wettability of the liquid on the solid surface (contact angle <10°), modified Young's equation can be applied (Eq. 15.12).

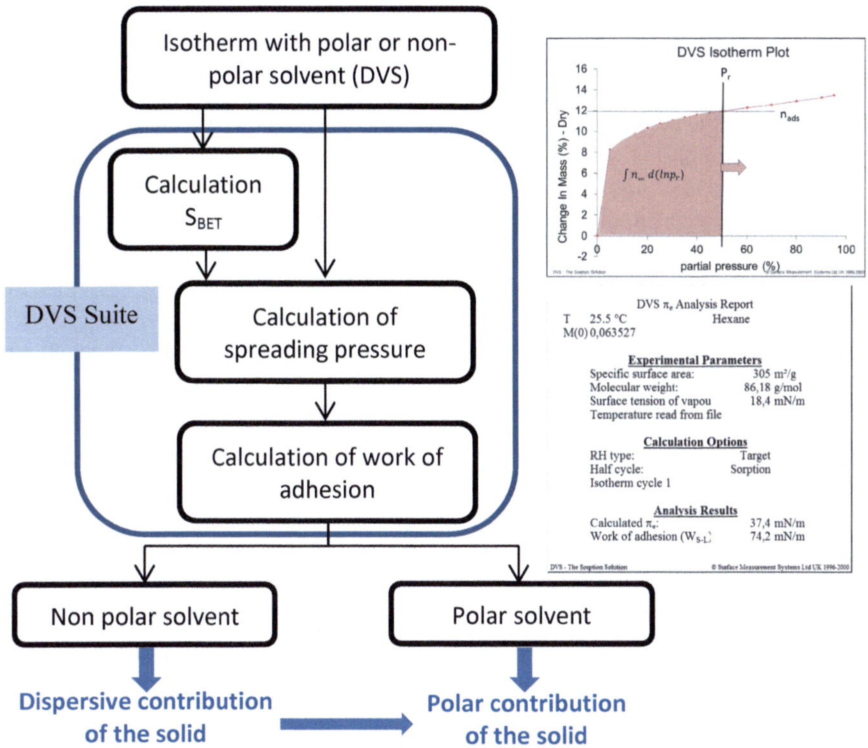

Fig. 15.14 Procedure for surface energy calculation from DVS data

$$W_{S-L} = \gamma_{L-V}(1 + \cos\theta) + \pi_e \tag{15.11}$$

$$W_{S-L} = 2\gamma_{L-V} + \pi_e \tag{15.12}$$

where $W_{S\text{-}L}$ is the work of adhesion (mN m^{-1}) at the liquid/solid interface, $\gamma_{L\text{-}V}$ is the surface tension of liquid with air (see Table 15.3), θ is the contact angle (°) and π_e is the equilibrium spreading pressure.

For the solvent used, $\gamma_{L\text{-}V}$ is known and the equilibrium spreading pressure can be calculated from the isotherm with Eq. (15.13).

Table 15.3 Molecular size (for surface calculation) and dispersive and polar contributions to the surface tension of some solvents used for surface energy determination [66]

Solvent	Hexane	Isopropanol	Octane
Molecular size (Å^2)	51.5	43.2	62.8
γ_{L-V} (mN m^{-1}), 25 °C	18.4	21.2	21.3
γ_L^d (mN m^{-1}), 25 °C	18.4	19.1	21.3
γ_L^p (mN m^{-1}), 25 °C	0	2.1	0

$$\pi_e = \frac{RT}{S_{BET}} \int n_{ads} d(\ln p_r) \qquad (15.13)$$

where S_{BET} is the active surface area of the solid towards the solvent obtained with BET equation, n_{ads} is the amount of solvent adsorbed at p_r, the partial pressure of solvent (e.g. relative humidity).

The work of adhesion is also described using the extension of Fowkes' equation:

$$W_{S-L} = 2\left(\gamma_S^d \gamma_L^d\right)^{1/2} + 2\left(\gamma_S^p \gamma_L^p\right)^{1/2} \qquad (15.14)$$

where dispersive (non-polar) and polar contributions to the overall surface tension are separated. By using a non-polar solvent such as hexane, the polar contribution to surface tension is negligible and the work of adhesion is calculated as follows:

$$W_{S-L} = 2\left(\gamma_S^d \gamma_L^d\right)^{1/2} \qquad (15.15)$$

The adhesive work (W_{S-L}) is calculated from isotherm and taking into account the surface tension of hexane (see Table 15.3) the value of dispersive contribution to surface tension of the solid (γ_S^d) is obtained. Fowkes' equation (Eq. 15.14) is then applied to isopropanol media to determine the polar contribution of the solid (γ_S^p). The overall surface tension is consequently the sum of both contributions.

As an example, Fig. 15.15 presents the adsorption isotherms of hexane and isopropanol on the same wood biochar. Tables 15.4 and 15.5 present the results obtained from isopropanol and hexane adsorption at 25 °C, for various biochars obtained from Pinus Sylvestris (PS) or CCA treated wood pyro-gasification at 750 °C in various atmospheres (air: A or nitrogen: N).

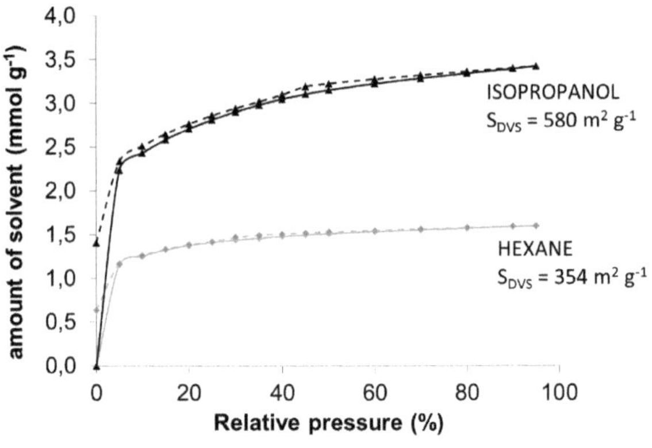

Fig. 15.15 Example of DVS isotherms on a wood biochar (PS750A), at 25 °C using isopropanol and hexane

Table 15.4 Values of work of adhesion between solvents (hexane and isopropanol) and 4 wood biochars surfaces

Biochar	W_{S-L} (mN m^{-1}) (with hexane)	W_{S-L} (mN m^{-1}) (with isopropanol)
PS750A	73.7	87.4
PS750N	74.4	85.2
CCA750A	74.2	88.3
CCA750N	74.0	85.4

Table 15.5 Values of work of adhesion, and dispersive and polar contributions to the surface tension of 4 wood biochars

Biochar	γ_S^d (mN m^{-1}) (with hexane)	γ_S^p (mN m^{-1}) (with isopropanol)	γ_S (mN m^{-1})
PS750A	73.8	18.0	91.8
PS750N	75.3	10.4	85.7
CCA750A	74.7	19.4	94.1
CCA750N	74.4	11.9	86.3

15.2.4 *Heat of Sorption*

The heat of sorption is a thermodynamic variable which describes the heat effects during the sorption process. The isosteric heat of sorption (Q_{st}) measures the amount of energy required to adsorb molecules on the material, at constant amount of adsorbate adsorbed. From the point of view of the vapour, the isosteric heat of sorption is an exothermic phenomenon ($Q_{st} < 0$). DVS equipment is suitable to determine the total isosteric heat of sorption of water or organic solvents. For a defined gas, the isosteric heat of sorption is obviously related to the type of active sites and the mechanism involved (physisorption or chemisorption). Basically liquefaction of adsorbent is only involved in the physisorption phenomena. Therefore, pure physisorption will exhibit a similar isosteric heat of sorption to the liquefaction enthalpy [92]. When the interaction between the adsorbent and the probe is stronger, a difference will be observed. A higher absolute value is a consequence of a strong interaction between the adsorbates and surface functional groups [93]. An interesting paper related to the impact of the functional groups concentration on the isosteric heat of sorption was published by Klomkliang et al. [93]. The isosteric heat of adsorption is a specific property of an adsorbent–adsorbate combination; therefore the determination of the isosteric heat of sorption is a proper parameter to understand sorption phenomena, to identify energetic heterogeneity of the surface and to apply appropriate surface functionalization.

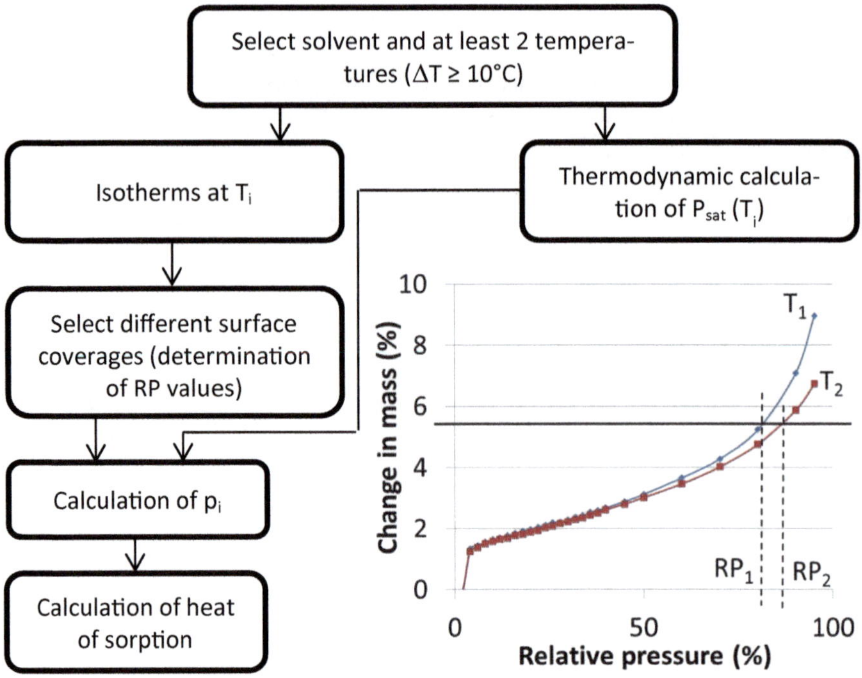

Fig. 15.16 Procedure for surface energy calculation from DVS data and using the DVS software

The determination of the isosteric heat of sorption is basically performed from the isotherms recorded at least at 2 temperatures. The procedure, included in the DVS software, is presented in Fig. 15.16. It is advised to use at least a difference of 10 °C between the temperatures selected for the calculation of the isosteric heat of sorption.

When isotherms are measured at different temperatures, using the same solvent, the isosteric heat of sorption is calculated using a Clausius-Clapeyron type equation (Eq. 15.16). For isotherms measured at different temperatures, and assuming that the isosteric heat of sorption is exclusively exothermic and independent of temperature in the considered range, Eq. (15.16) is modified to Eq. (15.17).

$$\left(\frac{\partial \ln p}{\partial T}\right)_\theta = -\frac{Q_{st}}{RT^2} \tag{15.16}$$

$$\left(\frac{\partial \ln p}{\partial (1/T)}\right)_\theta = \frac{Q_{st}}{R} \tag{15.17}$$

where p is the partial pressure of the solvent, T is the temperature (K), and Q_{st} is the isosteric heat of sorption (J mol^{-1}). The derivation is calculated at a certain coverage rate (θ), i.e. amount adsorbed at the surface.

The DVS software will automatically calculate the partial pressure (p) following Eq. (15.18). The relative pressure is extracted from the isotherm (RP), while the saturation pressure of the solvent at the considered temperatures (P_{sat} (T_i) in Fig. 15.16) is calculated by thermodynamic correlation (Antoine law).

$$p_i = RP \times P_{sat}(T_i) \tag{15.18}$$

Then, using two temperatures, the isosteric heat of sorption is calculated from isotherms with Eq. (15.19).

$$Q_{st} = R\frac{\ln(p_2/p_1)}{1/T_2 - 1/T_1} \tag{15.19}$$

As an example, Fig. 15.17 shows isotherms of n-hexane sorption on carbon black (N330) recorded at 20 and 30 °C. These type II isotherms indicate a multi-layer mechanism of sorption. Table 15.6 shows the calculated isosteric heat of sorption obtained for hexane sorption at two temperatures (20 and 30 °C). The values, in the range of −19 to −23.5 kJ mol^{-1}, are close to the liquefaction enthalpy of n-hexane at 25 °C (−21.7 kJ mol^{-1}). This means that the n-hexane molecules are mostly condensed at the surface, and physisorption is the main mechanism involved.

In the literature, the isosteric heat of water sorption is frequently determined on biomass feedstock, to estimate the state of sorbed water, which is a measure of the physical, chemical and microbial stability during its storage. The data are often displayed as the absolute values of isosteric heat of water sorption versus the moisture content (i.e. the change in mass or the coverage). Figure 15.18a presents the mean values obtained on leaves and stems of lemon balm in the range 25–45 °C [94]. Figure 15.18b shows the values of adsorption and desorption recorded on bamboo

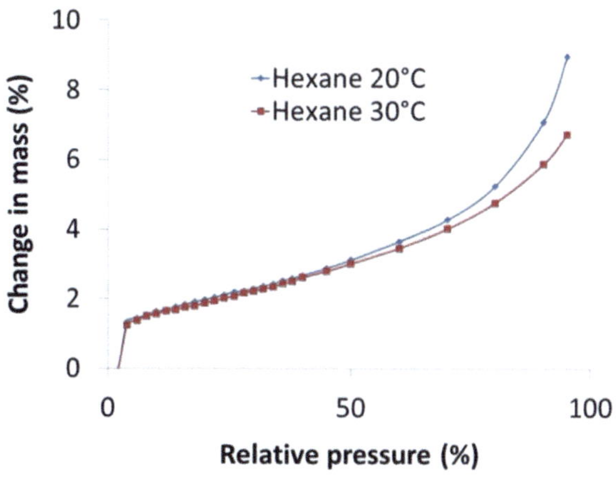

Fig. 15.17 Isotherms of n-hexane sorption recorded at 20 and 30 °C on carbon black (N330)

Table 15.6 Values of total isosteric heat of sorption calculated for hexane sorption on carbon black (N330) in the range 20–30 °C

Coverage (change in mass in %)	Isosteric heat of sorption (Q_{st} in kJ mol^{-1})
1.34	−22.37
2.69	−19.72
4.03	−19.25
5.38	−19.06
6.72	−19.94

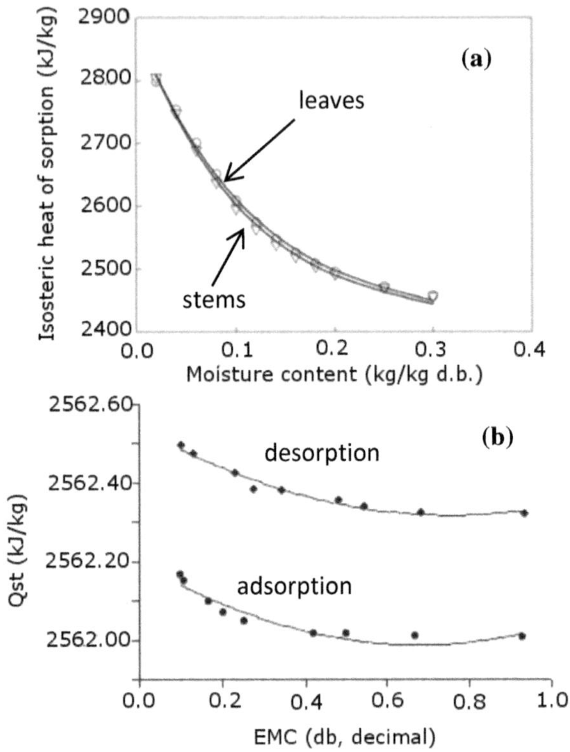

Fig. 15.18 Absolute values of isosteric heat of water sorption versus the moisture content. **a** Values determined from water adsorption on lemon stem and leaves between 25 and 45 °C, adapted from [94]. **b** values determined from water adsorption and desorption on bamboo shoots between 20 and 35 °C, adapted from [95]. Reprinted with permission from Elsevier [94, 95]

shoots in the range 20–35 °C [95]. The liquefaction enthalpy of water is equal to −43.99 kJ mol^{-1} at 25° and −43.53 kJ mol^{-1} at 35 °C. These values correspond to −2441.85 and −2416.32 kJ kg^{-1} respectively. Figure 15.18 shows that at low moisture content (or coverage) the isosteric heat of water sorption is higher than the liquefaction enthalpy, due to more active polar sites on the surface. When the coverage increases, the isosteric heat of sorption approaches the value of latent heat for condensation of pure water vapor. Therefore, these values revealed the existence of significant amounts of bounded water.

The procedure presented is the simplest way to determine the isosteric heat of sorption. However, this calculation is only convenient for qualitative considerations of thermodynamic properties [96]. Since the heat of sorption is assumed to be invariant with temperature, the resulting net isosteric heat of sorption is an average value over the temperature range. This calculation also suffers from errors in determining the p values from graphics as well as experimental errors in the isotherm data.

To determine heat of sorption values with high accuracy, some authors have used an expression based on the fluctuation theory in the grand canonical ensemble [93, 97]. In other cases, the net isosteric heat of sorption is extracted from the constants of the GAB equation (see Sect. 15.2.1.4 for Guggenheim, Anderson and de Boer equation) [96]. The net isosteric heat of sorption in this case is defined as the difference in absolute values of total isosteric heat of sorption minus liquefaction enthalpy.

15.3 Surface Chemical Functions Using Temperature Programmed Desorption (TPD)

As explained previously, oxygenated group is one of the surface physico-chemical properties which accounts for the activity to the solid residues. TPD (Temperature Programmed Desorption) analysis is a relevant technique to characterize solid residues, especially to determine the nature of the oxygenated groups at the surface [36, 98, 99].

When a solid is heated under inert atmosphere, oxygenated groups are desorbed from the surface. Depending on the nature of the oxygenated function, desorption will occur at different temperatures and releasing different gaseous products (CO or CO_2). A coupled system TPD/gas detector (μ-Gas Chromatography or Mass Spectroscopy) is then useful to identify and quantify the oxygenated functions [98]. This method implies the heating up of the solid residue at a constant rate under an inert atmosphere (N_2 or He). During the heating, the oxygen functional groups will desorb at a specific temperature as CO, CO_2 or both (see Fig. 15.19).

As explained in Chap. 2, TPD device is equipped with a TCD detector for continuous acquisition. However, as seen in Fig. 15.19, evolved gas from different functional groups may overlap. The use of gas chromatography (μ-GC, with short analysis time, i.e. 30 s per analysis) or mass spectroscopy (MS) is required to post-treat the TCD signal. For this purpose, the TPD equipment should be connected to the gas analyzer. In the procedure described here, a μ-GC has been used to analyze the gas emitted from TPD experiments. The procedure is summarized in Fig. 15.20.

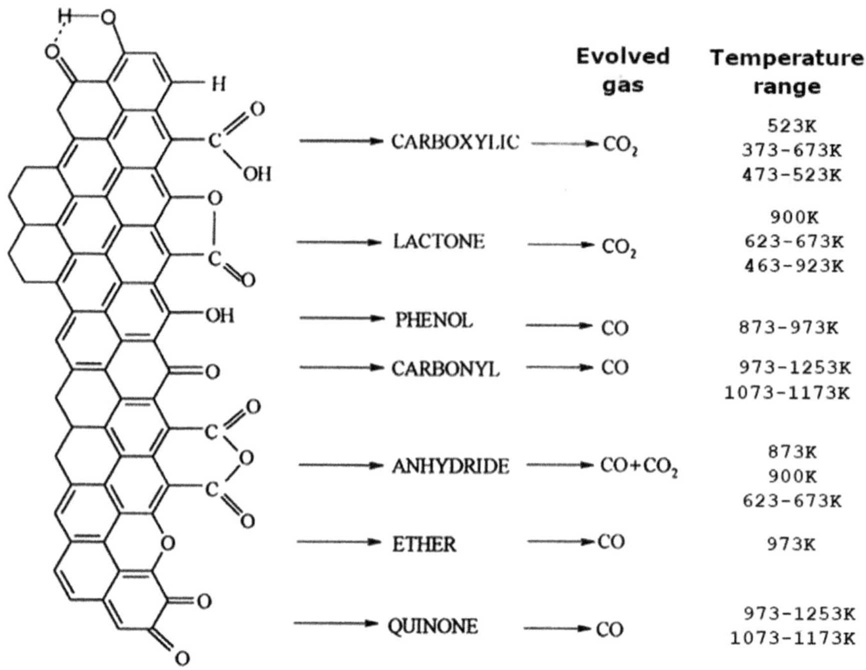

Fig. 15.19 Surface groups on carbon-based materials and their decomposition by TPD (from [99]). Adapted with permission from Elsevier [99]

15.3.1 Sample Preparation and Analysis

The amount of sample depends on its density and particle size, the height of the sample bed should be within the isothermal zone, and also care should be taken to ensure that there are no limitations to mass transfer properties such as both inter-particle and intra-particle diffusion. An example of TPD analysis method for a biochar from poplar wood is given below [98].

Procedure

- Predetermined sample weight (for e.g. 100 mg) is measured and loaded into the sample tube and installed on the TPD instrument.
- Pretreatment: the sample is degassed by flowing inert gas (He) over the sample while ramping the temperature to ensure all the foreign physisorbed molecules (which could be water and other impurities) are removed. After this step, the temperature is brought down to a starting temperature (generally ambient temperature) for TPD analysis.

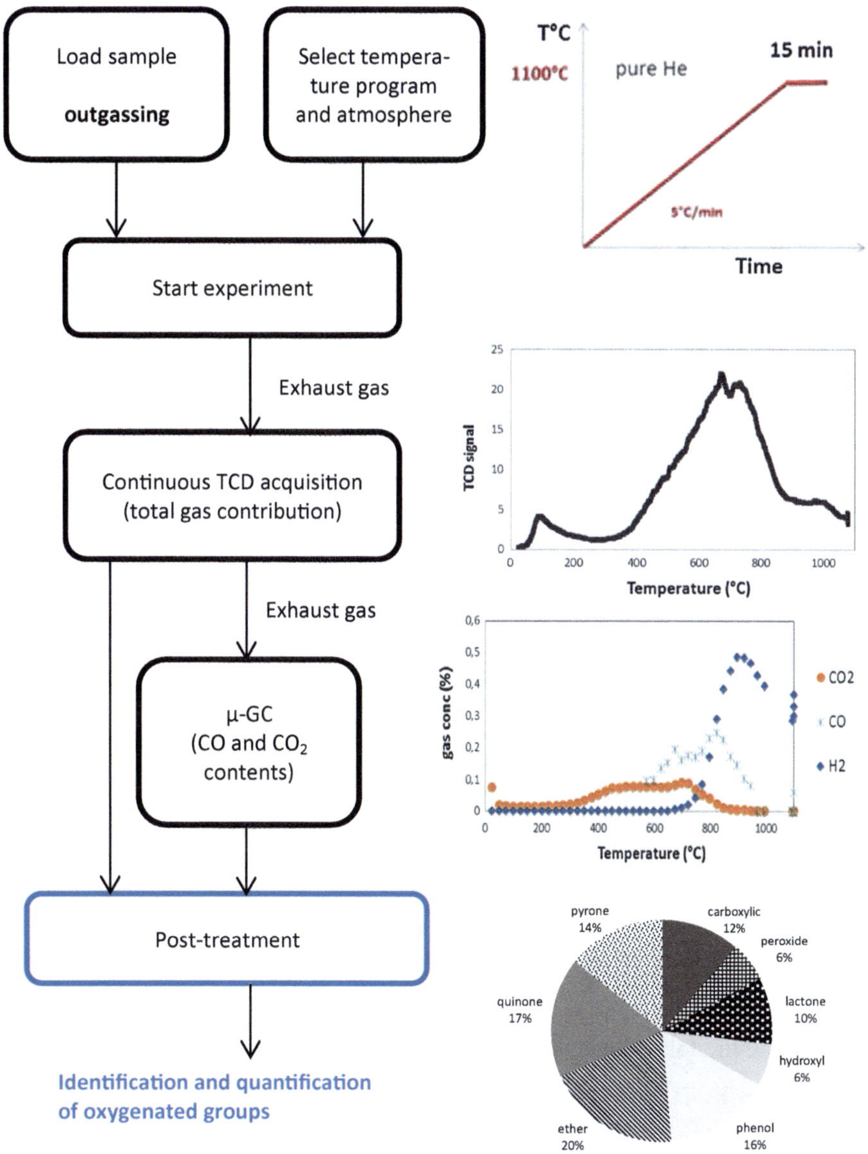

Fig. 15.20 Procedure used for oxygenated group quantification on wood biochar using TPD/µGC coupling

- Analysis: The predetermined gas flow rate (for e.g. 25 mL min^{-1}) over the sample is set along with the temperature program (for e.g. 25–1000 °C at 5 °C min^{-1}). Gas produced by the decomposition of oxygenated functional groups is monitored by both the TCD detector from the Micromeritics AutoChem II 2920 instrument and by μ-GC from Agilent (MyGC model) procured via SRA Instrument Company. This μ-GC is equipped with four different columns and TCD detectors which allow the separation and the detection of He, Ar/N_2, O_2, H_2, CH_4, CO, CO_2 and even lighter hydrocarbons (C_2–C_4 range).

15.3.2 Examples of Post-treatment

To get the total amount of each oxygenated group, the TCD signal is used. However, this spectrum is a continuous acquisition of a unique value over time (i.e. over the temperature). For this reason, the heating rate should be maintained at a constant value along the temperature program. The value of the TCD signal is thus the sum of each gas contribution at each instantaneous time. However, the desorption temperature is subjected to a controversy; literature papers may give various values, as seen in Fig. 15.19. The chromatogram obtained approximately every 3 min gives the CO and CO_2 content in the gaseous phase at a specified time (or temperature). From this total value, a curve-fitting is directly performed on the TPD software using more precise desorption temperatures and peak shapes.

In the case of an oxygenated biochar, and from the literature and the μ-GC results, the signal has been curve-fitted using Gaussians functions [99–101]. The peaks related to O-containing groups desorption are presented in Fig. 15.21. Peak temperatures were bound by ±50 °C and the Half Width at Half Maximum (HWHM) was kept constant.

Typical TPD spectra of the raw biochar (steam gasification poplar wood biochar) and oxygenated biochars (oxygenation of the raw biochar at 280 °C for 2, 4, 8 and 16 h) are depicted in Fig. 15.22, and the composition of CO and CO_2 derived from these biochars are given in Table 15.7 [98]. The results in Table 15.7 were obtained by applying the following steps:

- Calibrate the TCD detector with the well-known gas standard containing CO or CO_2 (or another gas to analyze).
- Calculate the concentration of CO or CO_2 (volume concentration) from the TCD signals obtained from the analysis of the biochar sample, using the corresponding calibrated curves.
- The gas flow rate (mL min^{-1}), the analysis time (min) and the volume concentration of CO or CO_2 being available, calculate the quantity of CO or CO_2 using ideal gas low.

Functions	T_{des} (°C)	Gas
Carboxylic acid	390	CO_2
Anhydride	450	CO_2
	590	CO
Peroxide	530	CO_2
Lactone	660	CO_2
	720	CO_2
Hydroxyl	620	CO
Phenol	750	CO
Ether	844	CO
Quinone	880	CO
Pyrone	973	CO

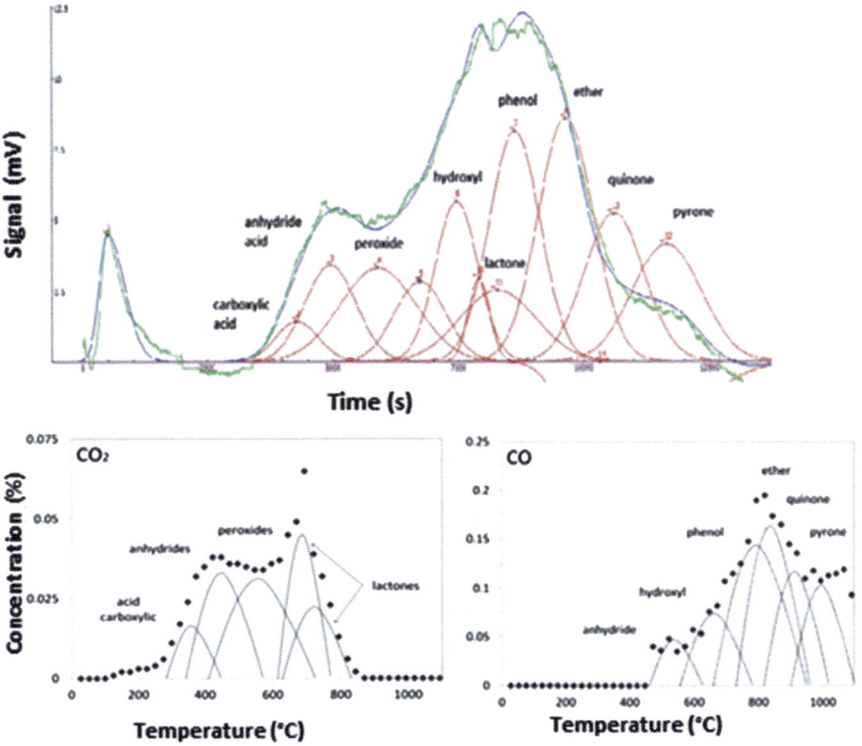

Fig. 15.21 Example for curve fitting of the TCD signal from the TPD analysis for a poplar wood biochar obtained from steam gasification at 750 °C (from [36, 98]). Reprinted with permission from Elsevier [36] & Ducousso Ph.D. thesis [98]

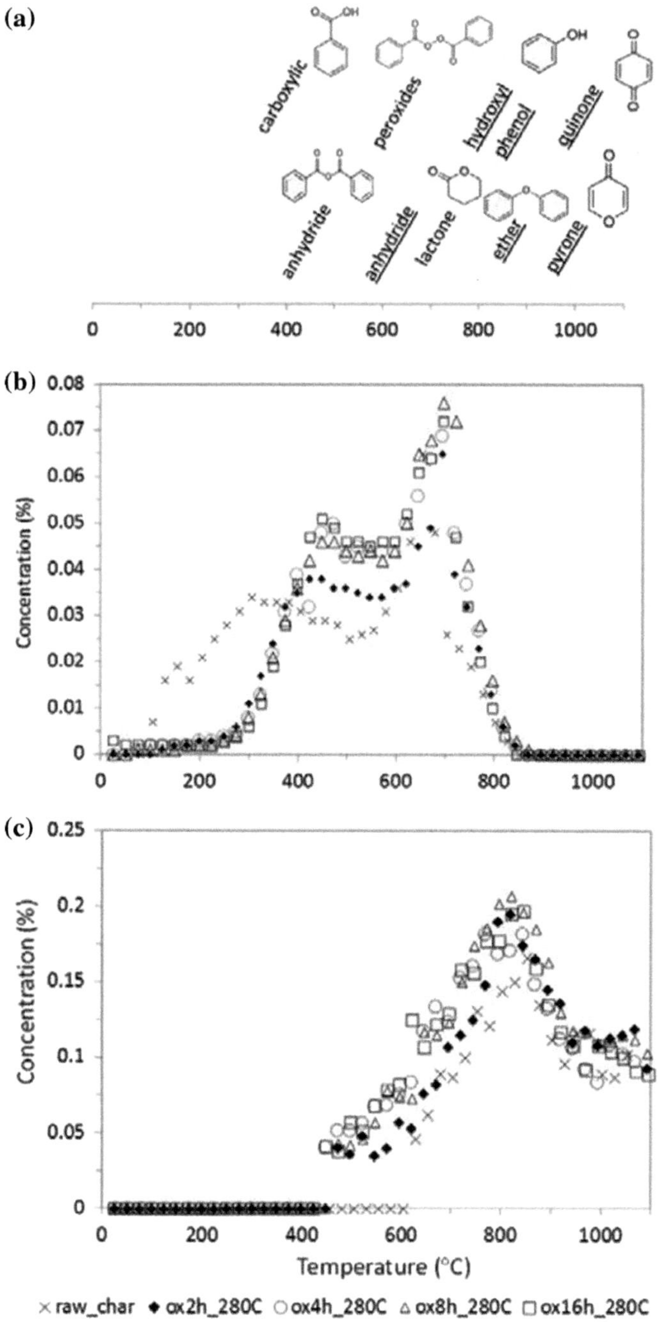

Fig. 15.22 TPD profiles of the raw and oxygenated biochars at 280 °C for 2, 4, 8 and 16 h: **a** peak assignment (underlined text: functions desorbing into CO; not underlined text: functions desorbing into CO_2); **b** CO_2 desorption; **c** CO desorption (from [36]). Reproduced with permission from Elsevier [36]

Table 15.7 Comparison of CO_2, CO and total production for raw and functionalized biochars oxygenated at 280 °C for different duration times (from the integration of the area under the curve of Fig. 15.22)

	(mmol $g_{biochar}^{-1}$)			
	CO_2	CO	Total	CO_2/CO
Raw biochar	0.9	2.1	3.0	0.42
2 h oxygenation	0.8	3.0	3.8	0.28
4 h oxygenation	1.0	3.3	4.3	0.30
8 h oxygenation	1.0	3.5	4.5	0.29
16 h oxygenation	1.0	3.4	4.4	0.29

Reprinted with permission from Elsevier [36]

- The quantity of the biochar used for TPD analysis being well-known, calculate the quantity of CO or CO_2 as presented in Table 15.7.

As partial conclusion, TPD is an adequate technique to analyze oxygenated functional groups on the surface of carbonaceous materials. Due to the overlapping of different functional groups under thermal effect, the quantification of these groups generally needs a coupling of TPD with other analytical methods like µ-GC, MS, FTIR, XPS etc.

15.4 Qualitative and Quantitative Distribution of Carbon Phases Using Raman Spectroscopy

Raman spectroscopy is a useful non-destructive technique able to distinguish different crystallographic carbon structures based on their vibrational modes (see Chap. 2). A refined spectrum (cosmic ray removal and baseline correction) is obtained from the raw spectrum, as explained in Chap. 2. For homogeneous samples, the average refined spectrum is highly representative of their content, and a curve-fitting could be applied to identify and quantify the proportion of each carbon phase.

The surface reactivity of the carbon-based materials (sorption, catalysis, adhesion…) depends on their crystallographic structure (amorphous, turbostratic, graphenic or graphitic). Literature showed that Raman spectra of graphite and graphene materials at the first order are composed of the G band at 1560 cm^{-1} referring to pure sp^2 carbons [102]. Raman spectra of pyrocarbons are typically characterized by the G band and the D band at 1585 cm^{-1} and 1350 cm^{-1} respectively, namely Graphite and Defect bands [19, 32, 102, 103]. The latter refers to the turbostratic carbons (sp^2-bonded carbons) that include large or small aromatic ring systems (2–8 fused benzene rings), structural defects (surface energy at the edge planes, orientation of layers), heteroatoms insertion, fullerene-like fringes and aliphatic systems [42, 104–106]. Simplified spectra of some carbon-based materials are presented in Fig. 15.23 [107].

To define the carbon phases, the refined spectrum is curve-fitted in the Raman shift region of 800–2000 cm^{-1}. In the literature, carbon-based materials spectra are

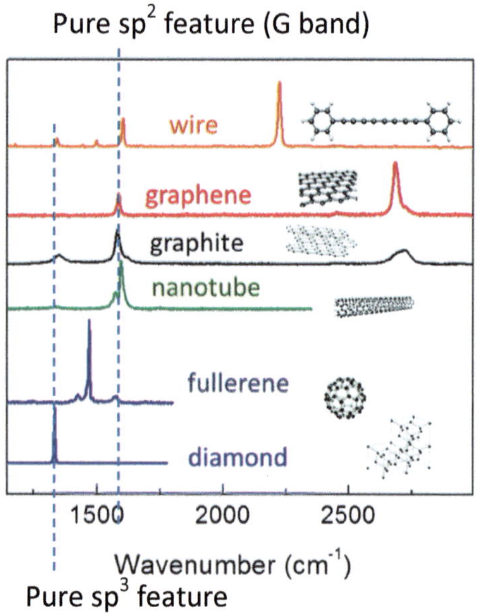

Fig. 15.23 Simplified Raman spectra of various carbon-based materials (from [107]). Republished with permission of RSC Pub from Carbon-atom wires: 1-D systems with tunable properties, Casari et al., Nanoscale 8, 2016 [107]; permission conveyed through Copyright Clearance Center, Inc.

defined with three, five, or ten peaks [19, 32, 103, 106, 108, 109, 110–112] and curve-fitting is performed using either Gaussian or Lorentzian peaks [108, 111].

For curve-fitting, personal utilities developed on MatLab ® software or commercial software such as Origin ® could be used. Depending on the software and the programme, different methodologies would be applied. For instance, the methodology performed with MatLab ® is presented in Fig. 15.24 and developed in the next sections for the user's guidance.

15.4.1 Qualitative Carbon Phases Analysis

For this first analysis, the shape of the refined spectrum as well as the intensity and position of the bands are discussed. At the end of this part, the explanations are given for an average Raman spectrum acquired with a wavelength of 532 nm on a surface of 5 μm × 5 μm on a wood biochar from steam gasification.

15.4.1.1 Shape of the Spectrum

The shape of the spectrum is dependent on the wavelength of the laser used. As seen in Fig. 15.25, the same sample (pyrocarbon) has been analyzed at varying laser wavelengths. Therefore, Raman spectra should be carefully compared with the literature at similar excitation wavelength.

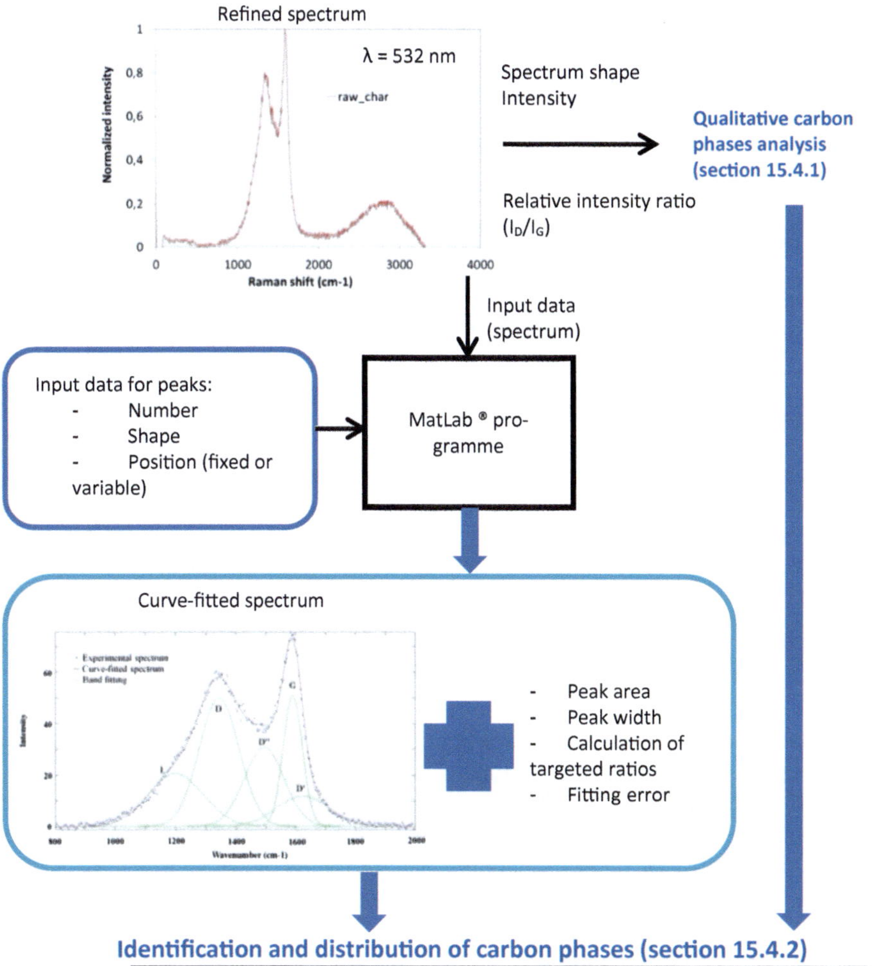

λ =532 nm	Description	Ideal graphitic lattice	Turbostratic carbon			Amorphous carbon
			A_{1g}-symmetry	E_{2g}-symmetry	A_{1g} symmetry + polyenes	
	Band	G	D	D’	I	D’’
	Raw char	0.167	0.334	0.104	0.188	0.207

Fig. 15.24 Procedure for carbon phases identification and distribution

The shape gives information on the relative contribution of the crystallographic structures such as ideal graphitic lattice and graphenic carbons (G band), disordered graphitic lattice (A_{1g}-symmetry) due to turbostratic carbons (D band) and amorphous carbons (height of the valley). With this qualitative approach, the objective is to determine the presence of these carbon phases as well as their proportion in the carbon-based material. The G band width is also an indication of the ordering since fine peaks are attributed to long-order structures (graphite crystalline structure).

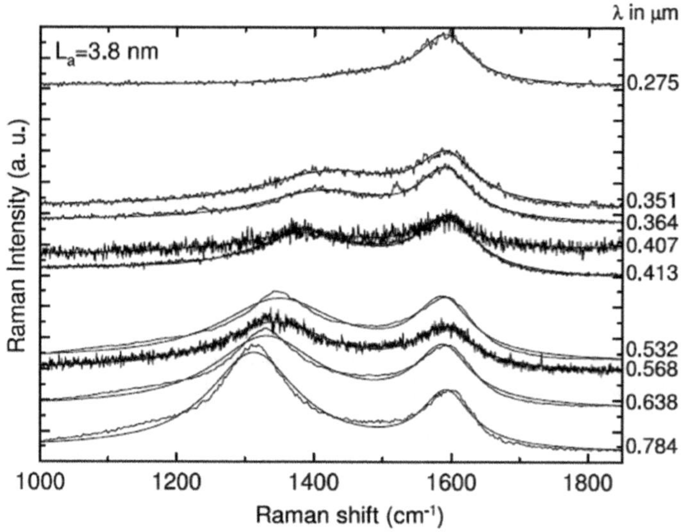

Fig. 15.25 Raman spectra of a pyrocarbon at different laser wavelengths (from [113]). Reprinted with permission from Springer [113]

15.4.1.2 Intensity and Relative Intensity Ratio of Raman Bands

A refined spectrum contains the intensity of each band which is proportional to the amount of carbon structures. For graphene layers, the G band could present lower intensity compared to a graphitic structure (due to a small number of excited bonds). This absolute intensity is directly related to the power of the incident laser and the probability of bond excitation. Therefore, the comparison of the absolute value between samples requires to use the same power of the incident laser. In case this value is not accessible with the Raman instrument, it is advised to normalize the intensity for comparison purposes. The intensity is the sum of various carbon bonds that may vibrate at a given wavenumber, and the absolute intensity would be significantly different depending on the spatial arrangement of the bonds. Figure 15.26 compares the Raman spectra (acquired at the same wavelength, and same power of incident laser) of graphene layer edges. The intensity of the D band is much lower for zig-zag edges compared to armchair edges, meaning that a large proportion of armchair edges will significantly increase the intensity of the D band.

The use of relative intensity ratio (I_D/I_G), corresponding to the height of each band, should be avoided or carefully used in case of significant baseline deviation and noisy spectrum. For significant baseline deviation, the boundary limits of the baseline should carefully be selected [115]. For very noisy spectra, the maximal value might be difficult to be determined.

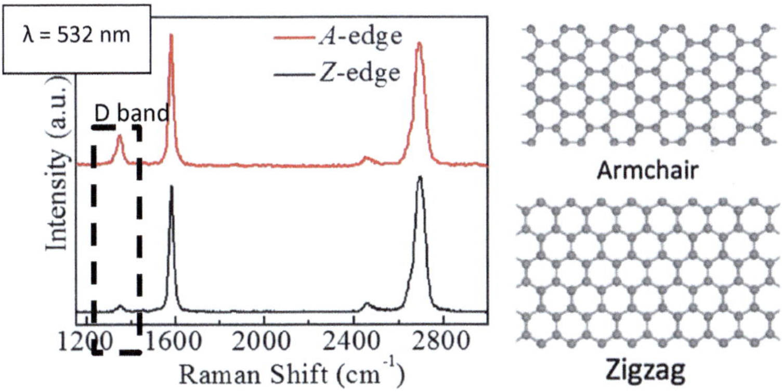

Fig. 15.26 Raman spectra showing the difference in D band intensity of bi-layers graphene containing zig-zag or armchair edges (from [114]). Raman spectra acquired at 532 nm. Reprinted with permission from Springer [114]

15.4.1.3 Band Position

The peak position varies with various factors such as the incident laser wavelength (Fig. 15.25), nanocrystallite size (Fig. 15.27), intercalation of oxygen atoms (Fig. 15.28) or carbon-based materials nature (Table 15.8).

As explained by McDonald-Wharry et al. the G band position is greatly influenced by the size of the cycles and the presence of O-containing functional groups (see Fig. 15.28) since the G band position is largely strain related [105]. Then a blue-shift of the G band would be correlated to the modification of the carbon texture/nanotexture [116–118]. As seen in Table 15.8, the G band of the graphite spectrum is positioned at 1580 cm^{-1} and is blue-shifted to 1590 cm^{-1} when moving to glassy carbon, carbon fibers, carbon nanotubes or graphene oxides [105].

Fig. 15.27 Raman spectra of pyrocarbon with various crystalline sizes (L_a) (from [113]). Reprinted with permission from Springer [113]

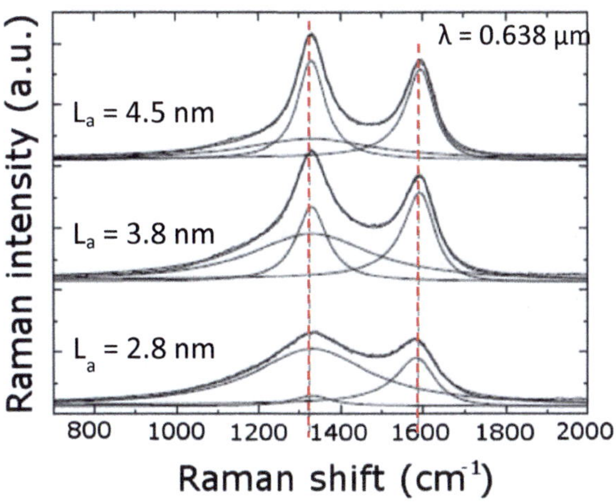

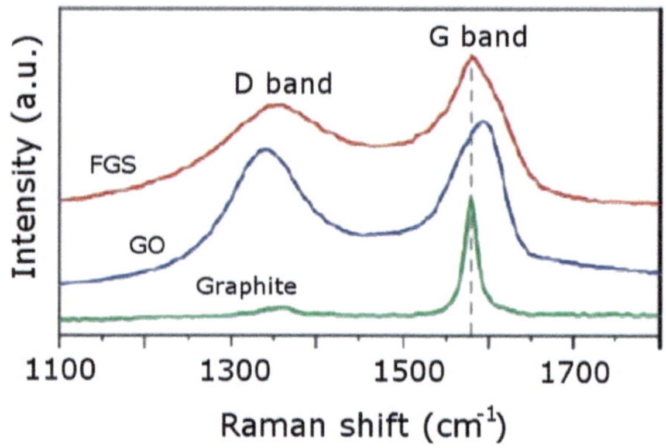

Fig. 15.28 Raman spectra of graphite, graphene oxide (GO) and functionalized graphene oxide (FGO) acquired at 514.5 nm (from [116]). Reprinted with permission from American Chemical Society [116]

Table 15.8 D and G band positions of some carbonaceous materials

Carbon material	G band position (cm^{-1})	D band position (cm^{-1})
C_{60} fullerenes	1571.8 ± 1.0	1340 ± 95
Dried colloidal graphite	1583.2 ± 0.5	1316.4 ± 3.2
Synthetic graphite	1578.4 ± 1.7	1316.2 ± 6.8
Synthetic graphite (air oxidized)	1580.0 ± 2.0	1319.4 ± 3.0
Single-walled carbon nanotubes	1590.6 ± 1.0	1289.6 ± 0.6
Glassy carbon	1598.8 ± 2.5	1308.8 ± 2.6
Graphene oxide (after 200 s of laser exposure)	1601.6 ± 7.7	1345 ± 3.8
Sucrose char 410 °C	1512 ± 10	1350 ± 22
Sucrose char 500 °C	1593.4 ± 1.7	1315.0 ± 5.7
Sucrose char 1000 °C	1599.0 ± 1.6	1309.0 ± 2.9

Raman spectra obtained with a wavelength (λ) of 785 nm (adapted from [105]). Adapted with permission from Elsevier [105]

This shift is related to the growth of aromatic clusters, and to the modification of the nanotexture due to internal stress [105]. A red-shift in the G band position may be due to curvature induced tensile strain in fullerene-like structures as well as a partial relaxation of the compressive strain in graphitic-like network [105].

A red-shift of the D band could be caused by the nature of graphene defects (armchair or zig-zag edges) [114] as well as fullerene-like defects which are involved in the curvature of the graphene layers [105]. This formation of curved graphene fringes is assumed to be due to a release of the compressive strain in the graphitic network [105].

Overall this quantitative analysis gives some information about the carbon phases containing the sample. However, this analysis is rather devoted to a comparative study (effect of thermal treatment, nature of samples, …). Figure 15.29 presents the spectra of two biochars: raw biochar (poplar wood steam gasification at

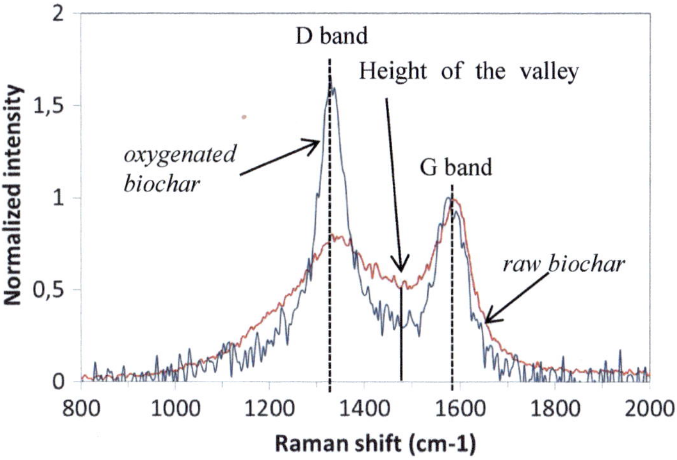

Fig. 15.29 Raman spectra of the raw wood biochar and oxygenated biochar (oxygenation at 400 °C for 2 h) obtained with a wavelength of 532 nm. The intensity normalization was applied to the G band; a difference is highlighted for the D band intensity and the height of the valley

750 °C) and oxygenated biochar (oxygenation of raw biochar at 400 °C for 2 h). It could clearly be observed that the oxygenated biochar contains more structural features that comprise the D band (graphenic nanocrystallites, oxygen intercalation, curved graphene layers…) than the raw biochar and a lower amorphous content (lower valley height).

15.4.2 Quantitative Distribution of Carbon Phases

As explained in Fig. 15.24, the quantitative analysis refers to the distribution of the carbon phases (to 100%) into the sample. High precision quantitative analyses require: internal standard (for suitable baseline determination); laser power (this influences the absolute intensity of the bands); exposure time (laser power may induce chemical modifications and/or heating); the use of heat dispersive media such as KBr; spectral acquisition at various orientation positions (the orientation of the bonds may not fit with the detector position). An example of quantitative analysis is detailed for the raw biochar from Fig. 15.29. Curve-fitting was performed using parameters from Tables 15.9 and 15.10. Figure 15.30 presents the graph from which the data gathered in the tables were extracted.

Five ratios were then considered to describe the carbon phases. These parameters were calculated using peak area over total peak area. Table 15.11 shows that the carbon phase contained in this sample is mainly turbostratic since the disordered part is significantly higher than the "graphitic" content (G band). The amorphous content is about 20%. A good agreement is obtained between the qualitative description and the calculation.

Table 15.9 Parameters used for the curve-fitting programme

Number of peaks	5				
Peak shape	Gaussian				
Band assignment	Chemical significance (might be purely mathematical) see Table 15.10				
Peak position	Set position of bands (Raman shift in cm^{-1}) for λ = 532 nm				
	D′	G	D″	D	I
	1625	1591	1495	1345	1200
Baseline fitting	Linear from 800 to 2000 cm^{-1}				
Width limit	No constraint				
Fitting error	<2% (on normalized spectra)				

Table 15.10 Band assignment and chemical description

Description	Band name
Disordered graphitic lattice (surface graphene layers edges, E_{2g}-symmetry) = turbostratic carbons	D′
Graphite E_{2g}-symmetry; ideal lattice, also graphenic carbons	G
Amorphous carbon structures	D″
Disordered graphitic lattice (graphene layers edges, A_{1g} symmetry) = turbostratic carbons	D
Disordered graphitic lattice (A_{1g} symmetry), polyenes, ionic impurities = turbostratic carbons	I

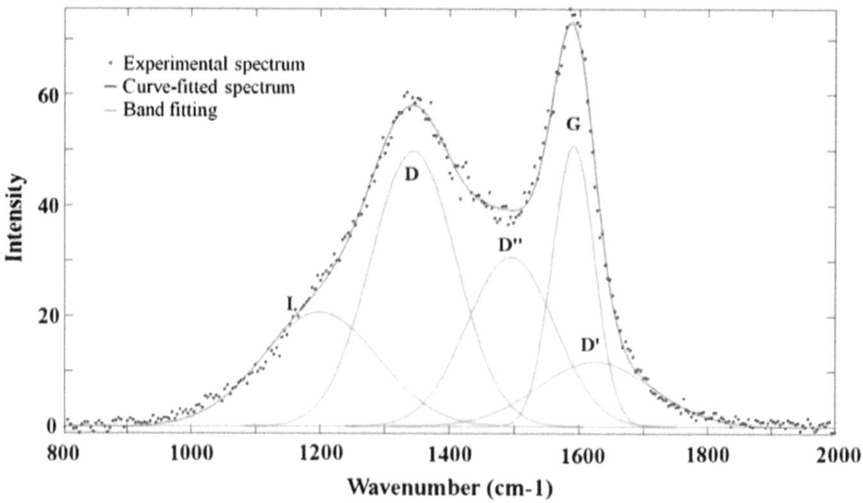

Fig. 15.30 Results of curve-fitting (5 Gaussian bands) of the refined spectrum

Table 15.11 Ratio calculated from the peak area after curve-fitting of the refined spectrum

Description	Graphenic and graphitic carbons	Disordered graphitic lattice, turbostratic carbons			Amorphous carbon
Ratio (surface)	I_G/I_{tot}	I_D/I_{tot}	$I_{D'}/I_{tot}$	I_I/I_{tot}	$I_{D''}/I_{tot}$
Raw char	0.167	0.334	0.104	0.188	0.207
	Total graphitic, graphenic, turbostratic = 0.793				

Table 15.12 Some recommendations for curve-fitting

Examples	Comments
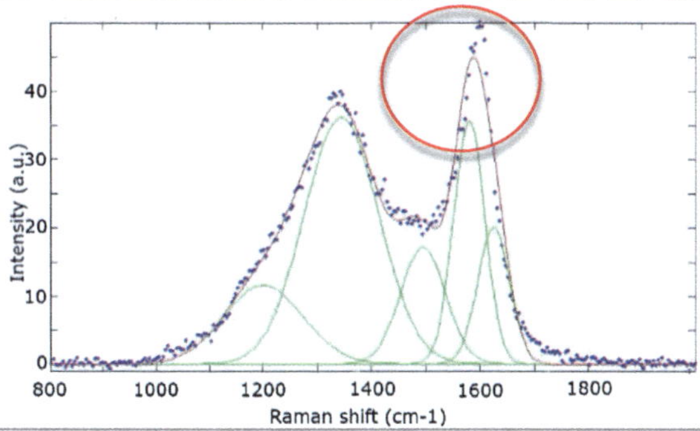	No good agreement between the experimental maximum and curve-fitting spectrum → Change the peak position
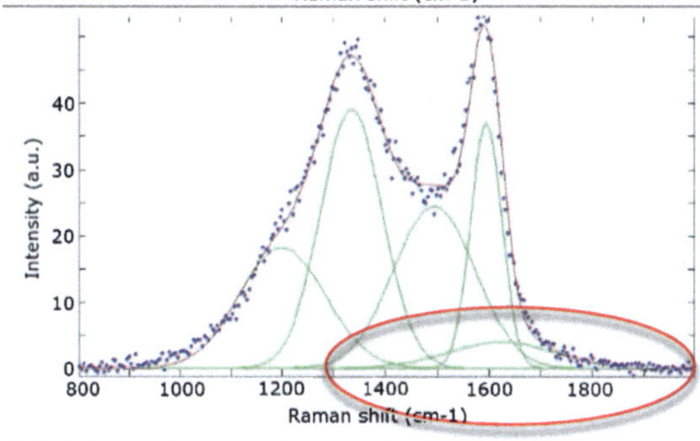	The band width could be too large → For spectra comparison, pay attention to the width evolution, maybe use a constraint
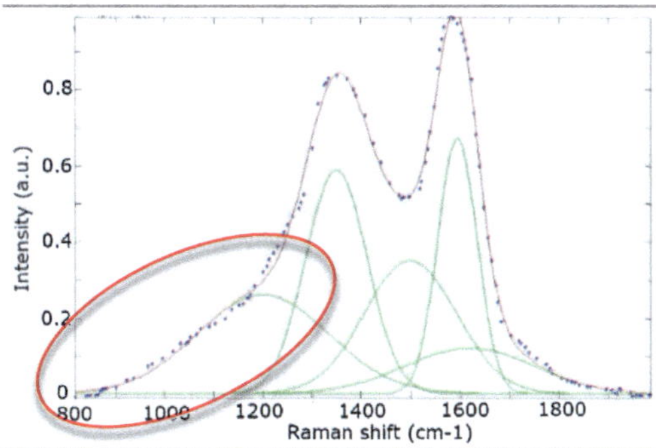	Pay attention to this shoulder that would be significant for biochar treated at low temperature (due to condensed hydrocarbons) → Curve-fitting with 10 peaks will improve this description

(continued)

Table 15.12 (continued)

Examples	Comments
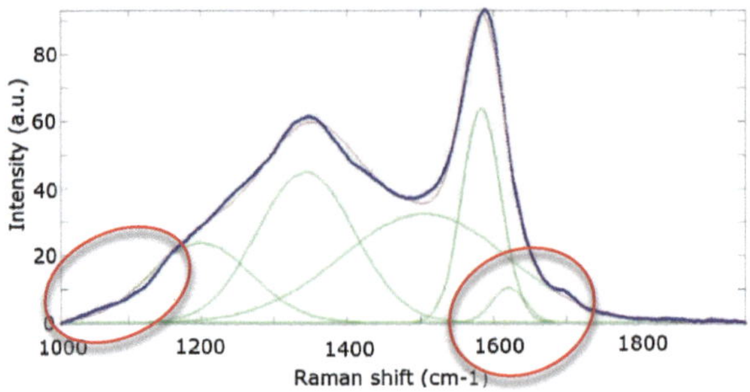	The shoulders are not well defined. → Change the peak position or use 10-peaks curve-fitting, or other peak shape (Lorentzian, Fano or pseudo-Voigt)

Here are also presented some elements for guidance in performing the curve-fitting (Table 15.12). This list is not exhaustive.

This quantification is an overall picture of the carbonaceous sample. However it is strongly recommended to couple Raman analysis with other techniques such as High Resolution Transmission Electron Microscopy (HRTEM), X-ray Photoelectronic Spectroscopy (XPS), Nuclear Magnetic Resonance (NMR), Pair Distribution Function (PDF) to draw accurate conclusions about the investigated samples.

15.5 Characterization of Texture/Nanotexture and Chemistry of Carbon-Based Materials Using Transmission Electron Microscopy (TEM)

Various carbon-based materials are produced during thermal treatment and the understanding of their characteristics at various scale is helpful for further application, depending on their morphology (particle, film, fibre, nanotube…), texture (isotropic, anisotropic, porous, …), nanotexture (which relates to graphene perfection and defect occurrence) and crystallographic structure (amorphous, turbostratic or graphitic) [42, 44]. The application envisaged for these carbonaceous materials will depend on their abovementioned characteristics as well as to the heterophases (minerals, metals, …) they contain. As mentioned previously thermal treatment results in solid carbonaceous residues containing mineral and/or metallic nanoparticles, along with the carbon phase, which can carry the main reaction sites. For the exploration and qualitative analyses of carbon phases and textures in carbonaceous residues, Transmission Electronic Microscope (TEM) is a dedicated characterization tool for investigating thin specimens down to the atomic scale [32].

Electron microscopy is based on the excitation of the atoms through an accelerated incident electron beam under vacuum. Using TEM, the accelerated electrons from the incident beam passes through the thin sample to be characterized. The resulting transmitted electrons (those which have crossed the thin specimen without having interacted with the specimen atoms, and those having undergone deviations or emitted by the sample) are then collected through various detectors (after having passed multiple projecting lenses) to be "sorted", counted and analyzed. Since the electrons are highly accelerated, the resolution of TEM is enhanced compared to that of SEM (see Chap. 2). For the exploration and qualitative analyses of carbon phases in solid residues using TEM, bright and dark-field imaging and selected area electron diffraction (SAED) modes are often used, the latter is generally used to select the transmitted electrons that will participate in the formation of the image. High-resolution transmission electron microscopy (HRTEM) combines the transmitted and the scattered beams which interact with each other to form an interference image. This high-magnification phase contrast image allows to observe the atomic scale. The outgoing electron wave at very low angles interferes with itself during propagation through the objective lens resulting in a point in the image plane. HRTEM has been extensively and successfully used to analyze crystal structures and lattice imperfections in various kinds of advanced materials at atomic resolution scale. It can be used for the characterization of point defects, stacking faults, dislocations, precipitates grain boundaries, and surface structures. In this case, HRTEM is used to characterize the anisotropy, the distribution and, if any, the local periodicity (correlation length) of the elements of carbon-based materials and phases on a very fine scale.

15.5.1 Experimental Recommendations

Table 15.13 indicates specific experimental procedures that should be followed, after introduction of the sample in the sample holder and TEM alignment (see Chap. 2), to obtain proper SAED patterns and HRTEM images as well [119]. Both procedures are compared to a conventional TEM imaging procedure.

Since the HRTEM mode is used to observe the matter at the atomic scale, any vibration should be avoided to guaranty the high resolution and the focus along the selected zone axis. For this mode, an objective lens and a wide aperture diaphragm should be selected. The image contrast is directly correlated to the projected potential of the sample, and the contrast drastically changes according to the defocusing (negative or positive) and the sample thickness.

15.5.2 Procedure for Indexing SAED Patterns

For SAED pattern, the resulting series of spots or ring should be indexed (see Fig. 15.31) in order to extract the interplanar distances (d_{hkl}) that could be compared to a selected number of reference patterns in the database. This technique

Table 15.13 Specific steps of the procedure for each of the 3 TEM modes

	Conventional TEM imaging (CTEM)	SAED	HRTEM
1	–	The beam should be width and parallel	Isolate the specimen holder (compustage) from any vibration
2	Place the sample under the electron beam and find a region of interest, the finest possible	Place the sample under the electron beam and find a region of interest, the finest possible	Select a thin zone of interest, and record its coordinates. Adjust the astigmatism of the objective lens (rings on the FFT perfectly circular) using an amorphous zone on the edge of the sample
3	Introduce an objective diaphragm to improve the image contrast[a]. Do lift the display screen	Spread the beam over all the surface of the display screen. Introduce the selected area diaphragm, center its aperture on the display screen and switch to diffraction mode[b]	Go back on the region of interest and center the selected area diaphragm on this first
4	Automatic search of the brightest region and adjust brightness. Do not forget to identify zones of interest and save their position. Possibly tilt the sample	Tilt the sample to select the zone axis of interest. When the electron diffraction pattern is well oriented, do insert the beam stop and place its point on the spot relative to the transmitted beam	In diffraction mode, tilt the sample to place the desired zone axis along the optical axe. After each tilt increment, check the eucentricity of the sample
5	Switch to the diffraction mode and center the diaphragm on the transmitted spot[c] or a diffracted beam depending on the imaging mode (BF or DF); then go back to the image mode	Focus the diffraction pattern	Center the diffraction diaphragm and then insert and center the objective diaphragm on the central spot. Go back to TEM mode. Increase the magnification little by little until lattice fringes appear
6	Record image with a CCD camera[d]	Capture pattern with CCD detector	Record HRTEM images with the CCD camera. Adjust to Scherzer defocus

[a]If necessary, do reduce the beam intensity in case the sample is sensitive to electron
[b]Remove the objective diaphragm if present
[c]Transmitted spot corresponds to the electrons that pass through the specimen without having interacting
[d]The camera is generally placed behind the imaging filter. Before acquiring an image, an alignment of the latter is required. For that purpose, do remove the objective diaphragm, place the beam in a zone without sample, and spread the incident beam over all the surface of the display screen

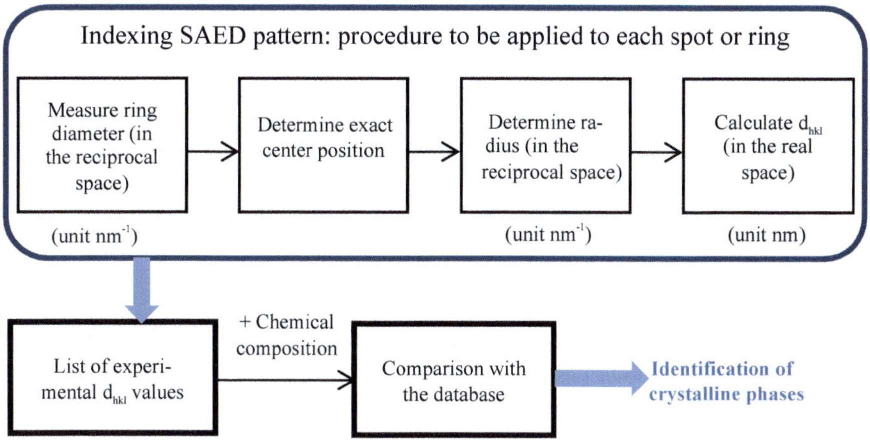

Fig. 15.31 Procedure to get interplanar distance from SAED patterns

allows chemical identification of crystals or polycrystals contained in the region selected with the selected area aperture [119]. For amorphous structures, the SAED pattern will be diffuse and there is no indexation possible.

15.5.3 Example of Texture, Crystalline Structure and Nanotexture Analysis of Carbon-Based Material

Transmission electron microscopy (TEM) samples were prepared by ultramicrotomy using a LEICA Ultracut R Ultramicrotome. Ultra-thin slices (thickness <100 nm) of the sample were cut with a diamond knife (DiaTome) provided by LFG distribution. The resulting ultra-thin specimen was transferred onto a 300 square mesh gold grid from Agar Scientific. TEM observation and analyses were performed using a FEI Tecnai F 20 ST TEM, operated at 200 kV, with a field electron gun (FEG) as the electron source. TEM bright field (BF) imaging enabled to observe the carbonaceous specimen and selected area electron diffraction (SAED) patterns were recorded on a zone randomly chosen.

The 002 lattice fringe mode—referred later as high resolution (HR)—was used to image the graphene layers (extent and shape) on the thinnest parts of the samples. The HRTEM images were acquired with drastic care at a magnification of 700,000 times, and at about the Scherzer optimum underfocus (−55.88 nm) with a spatial resolution of 0.28 nm to obtain reliable data.

15.5.3.1 Texture Analysis Using Conventional Bright Field (BF) Images and Indexing Sizes

From the conventional bright field images, a post-treatment could be applied to measure sizes of the nanopores. This procedure is not detailed here and is dependent on the software used. However the indexation of sizes such as distance or radius is often available in dedicated software. Figure 15.32 presents a BF image recorded on an ultrafine slice of oxygenated wood biochar. Because of the thickness of the sample (<100 nm), the change in contrast can be linked to only a lower density of the carbonaceous matrix. Therefore, the brightest zones can be attributed to nanoporosities and the order of magnitude of their size can be indexed by image treatment under the assumption of a spherical shape.

15.5.3.2 Qualitative Analysis of Crystalline Structure and Nanotexture Using Conventional Bright Field Images and SAED Pattern Indexation

Figure 15.33a shows a conventional bright field TEM image of a wood biochar (pyrolysis at 700 °C) and the corresponding SAED pattern. The SAED pattern is characterized by the presence of a series of three diffuse rings only. This limited number of rings is typical of graphene crystals with turbostratic structure, i.e., in

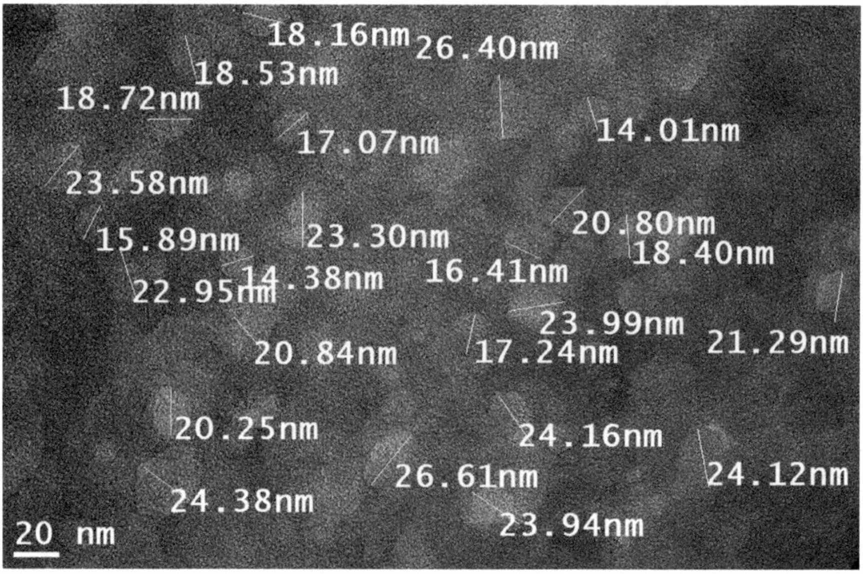

Fig. 15.32 Bright field TEM image of an oxygenated poplar wood biochar (ultra-thin slice) and identified nanopores (bright regions)

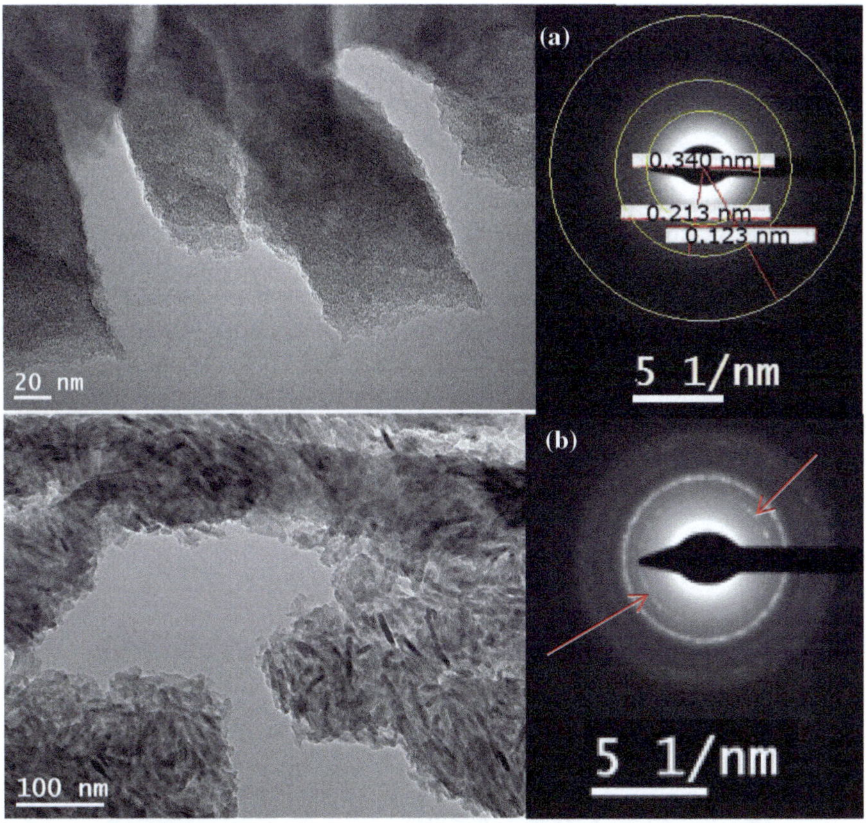

Fig. 15.33 **a** Bright field TEM image of a wood biochar and corresponding SAED patterns acquired in the entire region. SAED pattern shows diffuse rings and their indexation for graphenic phases. **b** Bright field TEM image of a biochar from the pyrolysis of food waste and coagulation-flocculation sludge and corresponding SAED patterns acquired in the entire region (from [19]). SAED pattern shows both diffuse rings and well-defined points attributed to a well-defined crystalline structure. Image **b** reprinted with permission from Elsevier [19]

which graphenes are piled-up randomly with no other coherence but that of their stacking [42]. The ring radii measured correspond to inter-reticular distances of ~0.340 nm, 0.213 nm, and 0.123 nm, may be respectively assigned to the 002, 10, and 11 reflections of the single graphene layer, as there is no 3D stacking periodicity (as marked by the two-indice hk reflections only). The fact that the intensity of each ring is homogeneous over 360° indicates that the orientation of the graphene stacks is statistically isotropic (texture), at least in the direction analyzed. The fact the intensity of each ring is faint and large indicates that the graphene stacks are small (nanotexture). Thus, one can deduce that this wood biochar consists of a carbonaceous matrix with a short-range nanotexture (small graphene dimensions), isotropic texture, and turbostratic structure.

Figure 15.33b shows a conventional bright field TEM image of a biochar obtained from the pyrolysis of Food Waste and Coagulation-Flocculation Sludge and the corresponding SAED pattern. The SAED pattern is characterized by the presence of a series of diffuse rings as for the wood biochar. However, the arrows show that other rings are present corresponding to periodicities different from that of the graphenic phase and which contain well-defined points at specific locations. This means that a crystalline structure phase different from the carbon phase has a preferred orientation in the material, indicating a longer distance order for this phase. The indexation of the rings is not shown in the figure, but the analysis reveals the presence of crystallized calcium phosphate (hydroxyapatite).

The graphene fringes of the wood biochar were further observed in High Resolution Transmission Electron Microscopy (HRTEM) mode. Figure 15.34 reveals the presence of graphene fringes that are randomly oriented (arrows show some graphene fringes). In terms of nanotexture, the curvature of graphene fringes indicates a poor quality of the stacking. This sample has a short-order isotropic organization of the graphene layers, an observation that is in agreement with the SAED pattern.

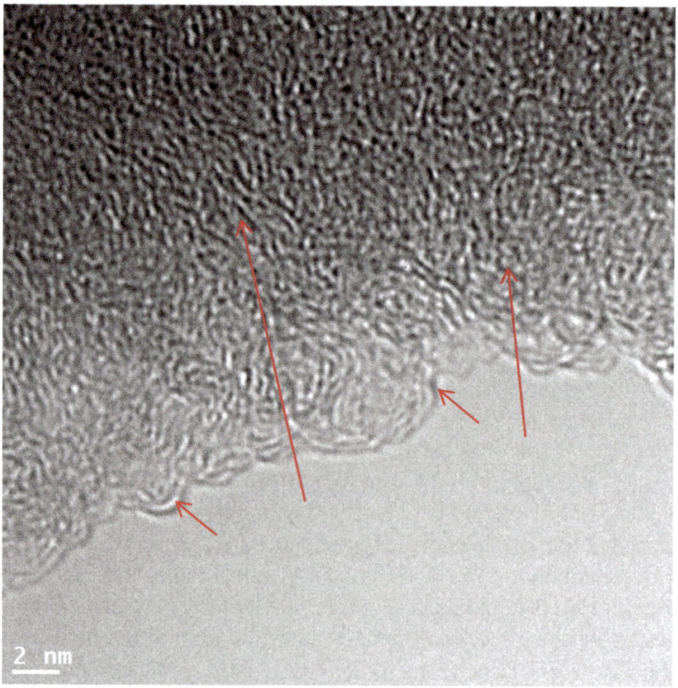

Fig. 15.34 HRTEM image acquired with the 002 lattice fringe mode of a wood biochar showing graphene fringes (arrows). Graphene fringes are not oriented with a preferred direction, giving a short-ordered graphitic structure (as shown in Fig. 15.33a)

15.6 Structure Analysis Using Pair Distribution Function (PDF)

Pair distribution function (PDF) analysis is an elastic X-ray/neutron scattering technique used to determine the atomic structure of disordered materials. The experimental details of the technique have been discussed in detail in Chap. 2. The PDF data contain a wealth of information on the atomic structural arrangements present in the material of interest. However, obtaining this information is not obvious since the PDF is a histogram containing all the atom-atom distances, and therefore deciphering which peak in the data corresponds to a particular atom-atom correlation is dependent on prior knowledge on the material of interest.

15.6.1 Data Analysis Overview

The simplest approach for analyzing PDF data is a qualitative approach, as indicated in Fig. 15.35. More detailed approaches include quantitative analysis and structure refinement. Structural models can help in the analysis of PDF data, since fitting of experimental data to a model phase (or multiple phases) enables for peak assignments to be carried out. This fitting process is analogous to the Rietveld refinement, for which a structural model is refined against reciprocal space scattering data. For the case of PDF data, this fitting is carried out in real space using software such as PDFgui [120]. A preliminary analysis of PDF data for an oxygenated wood biochar is presented here using the PDFgui software.

Qualitative analysis

- Peak identification
- Qualitative features (e.g., amorphous versus nano-crystalline)
- …

Quantitative analysis

- Direct comparison of multiple experimental datasets
- Reaction kinetics from in situ measurements
- …

Structure refinement

- Small-box modeling (e.g., PDFgui)
- Large-box modeling (e.g., reverse Monte Carlo)
- Molecular modeling and subsequent refinement (e.g., ab initio and force-field modeling)
- …

Fig. 15.35 Scheme of data analysis options for PDF experimental results. Note that the list is not exhaustive

15.6.2 Data Analysis Example

An example of the X-ray PDF data analysis approach for an oxygenated poplar wood biochar is provided. The analysis indicates that biochar tends to conform to a graphitic-like structure, but also may contain wood-derived atomic structures such as cellulose and lignin [121]. Hence, these three structures (graphite, cellulose and lignin) have been refined against the experimental data, with the most promising results given below. The PDFgui software requires the existing structural models as input files that can be manually entered in the software or directly imported as a CIF or PDB file (available from literature or online databases, some of which require a subscription). Additional input parameters requested by the software include the Q_{max} value used to generate the experimental dataset, and the instrument parameters, Q_{broad} and Q_{damp} (as discussed previously in Chap. 2). Parameters that can be refined in the software include scale factor, lattice parameters, atomic displacement parameters, spherical particle diameter (if nanosized particles/regions are present), nearest-neighbor correlation factor (accounts for in-phase vibrations of nearest-neighbor atoms) and site occupancy. It is important to note that there are other analysis approaches for PDF data in addition to PDFgui. The latter is often referred as "small-box" modeling [122], since the structural model size is restricted to tens to hundreds of atoms. Other approaches include reverse Monte Carlo, whereby a structural model of thousands to tens of thousands of atoms is generated and refined against the experimental data (Fig. 15.35). A more in depth discussion of the types of modeling available, and the limitations of each, is given in [122].

Figure 15.36 shows that neither the local atomic structure of graphite nor that of the cellulose accounts for the structural arrangements present in oxygenated biochar. Although the amorphous nature of the sample was accounted for in the structural models via the use of the spherical particle diameter refinement option, there are still large discrepancies for both structural fits, especially beyond 3 Å. However, by performing a two phase refinement using both graphite and cellulose, it turns out that a relatively good fit can be achieved with the experimental dataset. The quality of the fit can be evaluated from the R_w value (i.e., the weighted R-value, see PDFFIT manual for more details [123]), for which a smaller value indicates a better fit. Hence, with an R_w value of 0.32, the "graphite + cellulose" refinement presented in Fig. 15.36 shows that these two phases provide a reasonable representation of the atomic structure of the oxygenated biochar. More advanced modeling approaches would involve the generation of a combined structural model incorporating graphitic-like regions together with the local structure found in cellulose, potentially via molecular simulation approaches such as density functional theory or force-field molecular mechanics. These advanced approaches would enable for the bonding interactions between the graphitic regions and cellulose to be accurately accounted for, and ideally lead to a better fit with the experimental PDF data.

Once a suitable fit has been obtained, the contributions from individual phases in PDFgui can be assessed together with the element-specific atom-atom correlations.

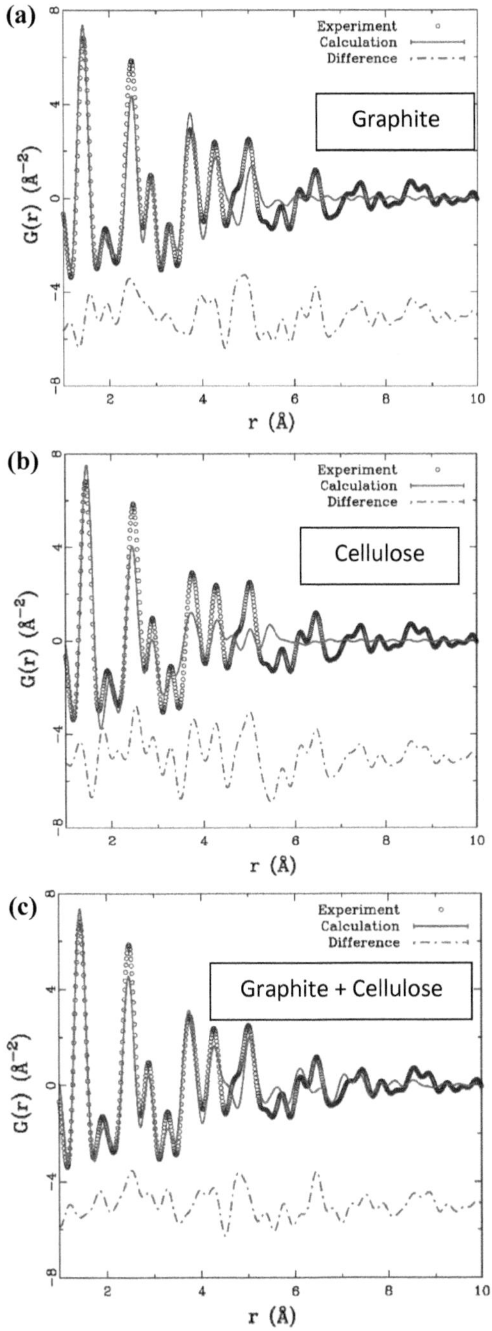

Fig. 15.36 Experimental X-ray PDF of an oxygenated poplar wood biochar (markers) together with the PDFgui fits in grey for **a** graphite, **b** cellulose [124], and **c** "graphite + cellulose". The weighted R-value (R_w) for the PDFgui refinements are 0.394, 0.477 and 0.323, respectively, for (**a**), (**b**), and (**c**), calculated over the r range of $1 \leq r \leq 10$ Å. Parameters that were refined include the scale factor, lattice parameters, atomic displacement parameters, spherical particle diameter, and the nearest-neighbor correlation factor

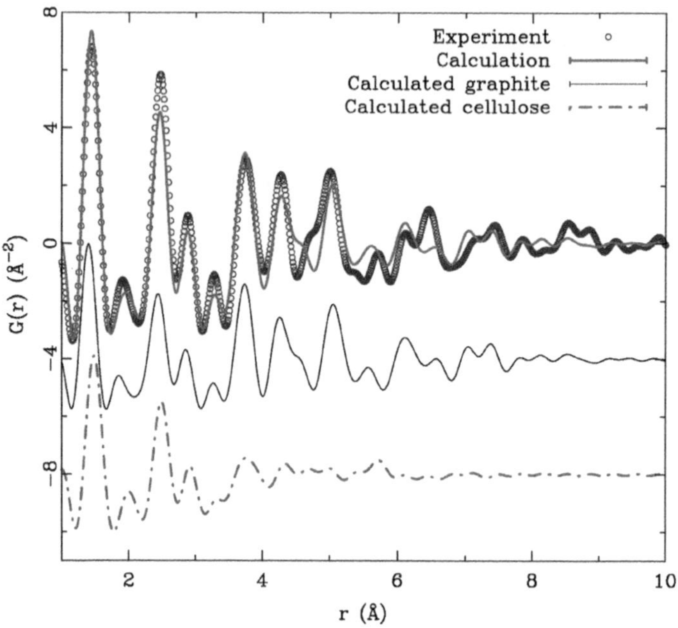

Fig. 15.37 Experimental X-ray PDF of an oxygenated poplar wood biochar (markers) together with the PDFgui fit ("graphite + cellulose") in grey previously given in Fig. 15.36 and the individual contributions from simulated graphite (in black) and cellulose (in dotted grey)

As shown in Fig. 15.37, the atomic structure of the oxygenated biochar is dominated by the local ordering found in graphite; however, there are additional short range atom-atom correlation contributions from a cellulose-like structure. The element-specific atom-atom correlations for the two phases in Fig. 15.37 are shown in Fig. 15.38. This latter figure clearly displays the oxygen-containing correlations that contribute to the local atomic structure of the oxygenated biochar, specifically via the nearest-neighbor atom-atom distance at ~1.45 Å that is assigned to the C–O bond length and the C–O correlation at ~2.47 Å. From Fig. 15.37, it can be observed that the C–C bond length of graphite is slightly too short to account for the nearest-neighbor peak of the oxygenated biochar, whereas the C–O and C–C bond lengths in cellulose in Fig. 15.37 are slightly too long. The difference in the C–C bond length between graphite and cellulose is attributed to the different carbon bonding arrangements in these phases. In graphite, the C atoms exist in planar sheets with each atom covalently bonded to three neighbors (C–C distance of ~1.40 Å in Fig. 15.38, while the C atoms in cellulose are saturated leading to longer C–C bond lengths (~1.55 Å in Fig. 15.38). Finally, Fig. 15.37 shows that the graphitic structure existing in the oxygenated biochar has a structural coherence length of ~9 Å since the atom-atom correlations associated with this phase disappear at around this *r* value.

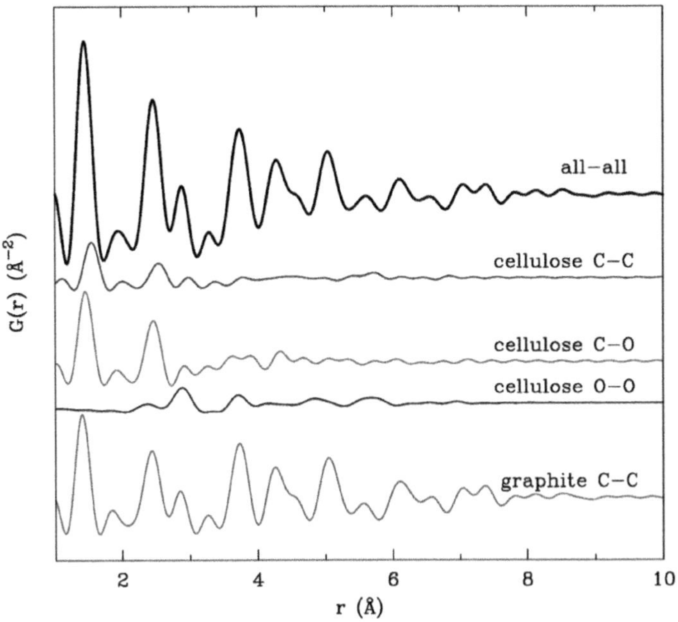

Fig. 15.38 Element-specific atom-atom correlations (known as partial correlations) found in the graphite and cellulose phases together with the simulated X-ray PDF for the two-phase refinement ("graphite + cellulose") seen in Figs. 15.36 and 15.37

15.7 Minerals Distribution in a Carbonaceous Matrix Using Raman Spectroscopy

Carbon-based materials such as biochars may contain significant amounts of minerals. Some of the minerals might be highly reactive which could enhance the biochars physico-chemical properties for further applications [37]. Raman mapping can be applied in particular on the solid residues to investigate their chemical heterogeneity surface and in-depth. From the refined spectra, different post-analysis techniques could be applied to determine chemical heterogeneity.

From the raw spectra, the first step consist in suppressing the noise (cosmic ray removal and background noise to apply a baseline correction), as explained in Chap. 2. Figure 15.39 shows the data which can be extracted from the refined spectrum. For this purpose, a biochar containing high ash content (about 50 wt%) was analyzed to illustrate all the steps [19].

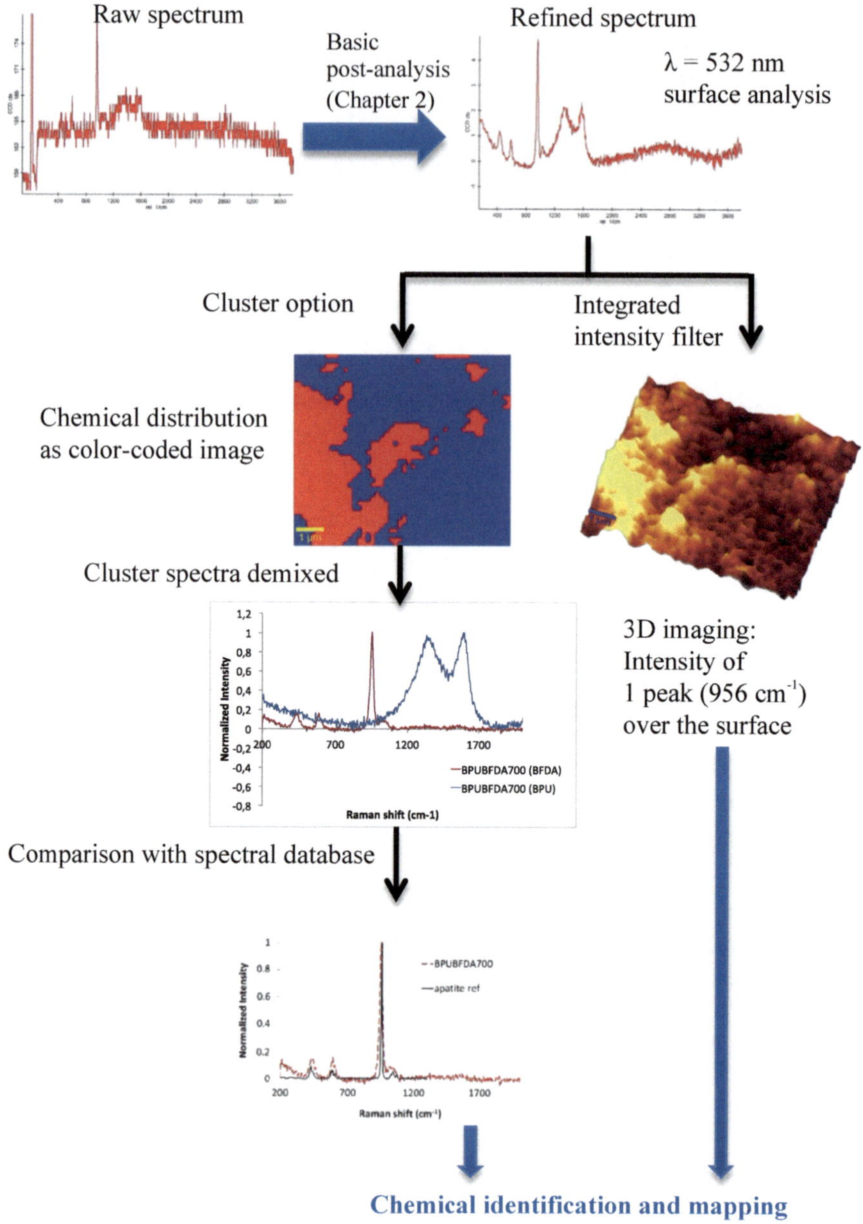

Fig. 15.39 Procedure for post-analysis of chemical identification and mapping

15.7.1 Chemical Distribution Using Cluster Option

The refined spectrum clearly shows extra bands that do not belong to carbon phases (Fig. 15.39, intense peak at 956 cm^{-1}), meaning that the sample may contain other chemical compounds. To obtain a spatial distribution of the components in the sample, the cluster analysis option can be used to localize similar regions in spectral data sets. Then, similar spectra are classified into a defined and chosen number of clusters and a color-coded image is formed based on the selected clusters.

This current sample is supposed to contain at least two chemical compounds. This means that two different spectra are expected to describe each spectrum acquired on the surface of the sample. Therefore two clusters have been chosen (one for the carbon spectra and one for the carbon-minerals mixture). The mathematical treatment applied by the software is as follows:

- The two first spectra (top-left image) define the two clusters (step 1 in Fig. 15.40).
- The next one spectrum is compared to both initial clusters. If no difference is detected, this spectrum is consolidated with its corresponding cluster (step 2 in Fig. 15.40). Then the comparison is gradually spread.
- When a spectrum does not match with the current clusters, a new calculation over the all analyzed spectra is run (step 3 in Fig. 15.40). Then 2 new clusters are defined and the distribution of spectra over both clusters is modified.
- Reaching the last spectrum comparison, the final spatial distribution of spectrum as 2 clusters is obtained (see step 4 in Fig. 15.40).

With the cluster option, a colored image (Fig. 15.41) is generated and the number of spectra associated as well. For this sample, two average spectra were generated (Fig. 15.42a). This indicates that a region of the surface analyzed only contains carbon matrix (only carbon G and D band signature) while another regions are chemically described as a carbon-mineral mixture. This red spectrum is still a mixture of at least two chemical compounds.

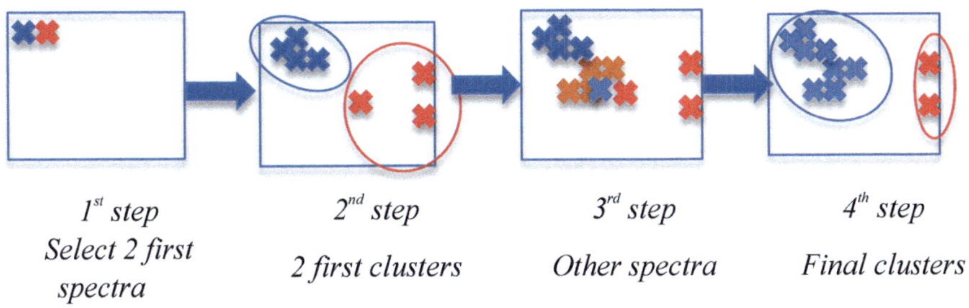

Fig. 15.40 Steps of the cluster option process

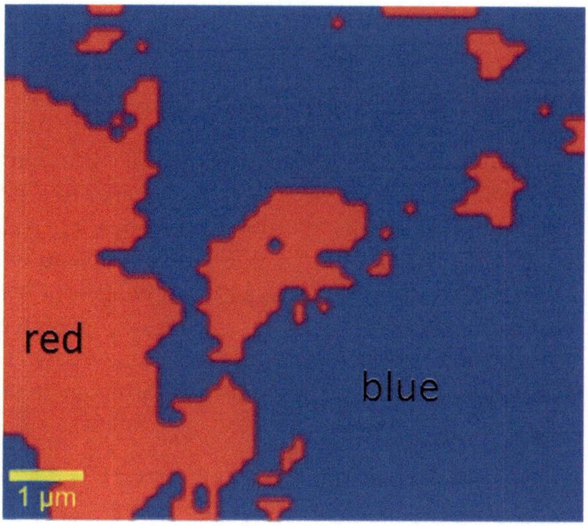

Fig. 15.41 Cluster distribution presented as a colored coded image over the surface. Biochar obtained from the pyrolysis of a mixture (wood, food waste and coagulation-flocculation sludge) at 700 °C during 30 min

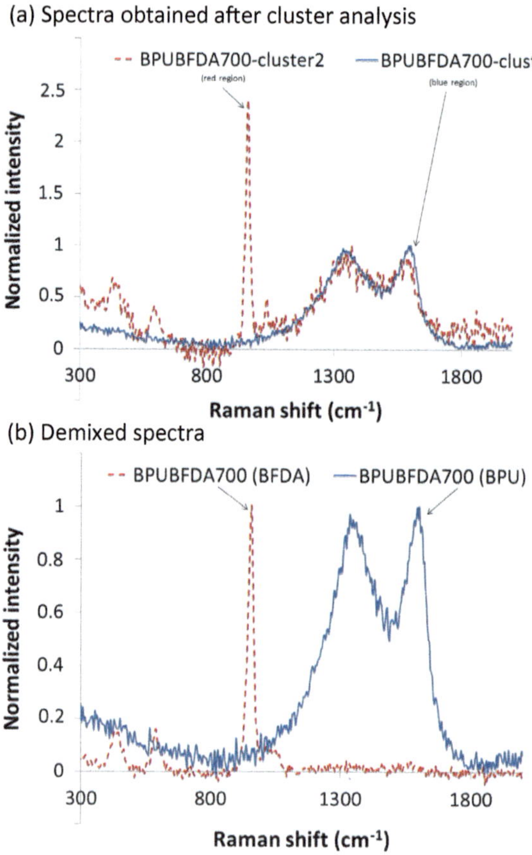

Fig. 15.42 Spectra obtained with advanced post-analysis on biochar from the pyrolysis of a mixture (wood, food waste and coagulation-flocculation sludge) at 700 °C during 30 min. Spectrum of cluster 1 refers to the blue region (more carbonaceous). Spectrum from cluster 2 refers to the red region (more mineral features)

Another post-analysis option (mathematical treatment) can be used to "demixed" these spectra. The graph demixer option corresponds to a subtraction or addition of spectral information from/to other spectra with a free adjustable weighting factor (weight per spectrum and immediate preview function). This option was applied to the red spectrum and the carbon spectrum (blue) has been subtracted. The two "pure" spectra are displayed in Fig. 15.42b.

15.7.2 Compounds Identification

The cluster spectra are compared with reference spectra to identify the chemical compound. This could be performed with the Raman software (now available for Witec and Horiba equipment). When the software does not provide database, the spectrum of interest can be compared with some spectra available in free database. The research in database obviously requires the knowledge of the mineral atomic composition for an efficient research. A special attention should be paid on the laser wavelength for proper comparison with the reference spectrum. Here is a list of free databases:

- Société Française de Minéralogie et Cristallographie (SFMC)
- Handbook of minerals Raman spectra, ENS-Lyon
- Integrated database of Raman spectra (RRUFF Project)
- Mineral Raman database, University of Parma
- Raman spectra database of minerals and inorganic materials (RASMIN)
- Raman spectroscopic library of natural and synthetic pigments, University College London.

The red spectrum (mineral, from Fig. 15.42b) was compared to different reference spectra of hydroxyapatite obtained at 532 nm (from RRUFF Database [125]), using normalized intensity. Figure 15.43 shows good agreement between the

Fig. 15.43 Comparison of experimental demixed spectrum with reference spectra

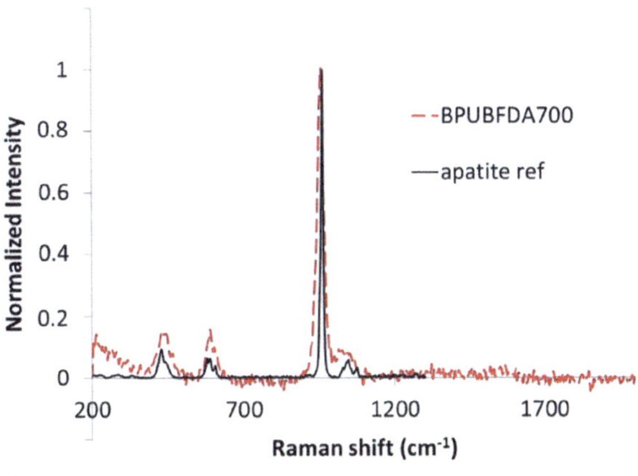

cluster/demixed spectrum and the Raman spectrum of apatitic material. This mineral is thus present at the surface, and is distributed at the surface (see Fig. 15.41). This process could also be repeated from the surface up to a certain depth. For this biochar, it was observed that the mineral particle was embedded in the carbon matrix and not only deposited on the surface, as in Hervy et al. [19].

15.7.3 3D Imaging with Integrated Intensity Filter

From the refined average spectrum obtained after the basic post-analysis, a 3D colored image could be generated with specific dark or bright areas (Fig. 15.44). For this kind of image, an integrated intensity filter should be used on the targeted peak. For instance, the filter was applied on the most intense mineral peak (see Fig. 15.44a). The peak intensity is then coded from dark (low absolute intensity) to bright color (highest absolute intensity). A special attention should be paid on the choice of the selected peak which should not overlap with any other for the filtering to avoid misinterpretation. The use of the integrated intensity filter would rather be used after cluster analysis to overcome this issue. A two dimensional image is firstly generated, and is secondly transposed by the software into a three dimensional image (see Fig. 15.44b). The areas in the 3D-image then represent the intensity of

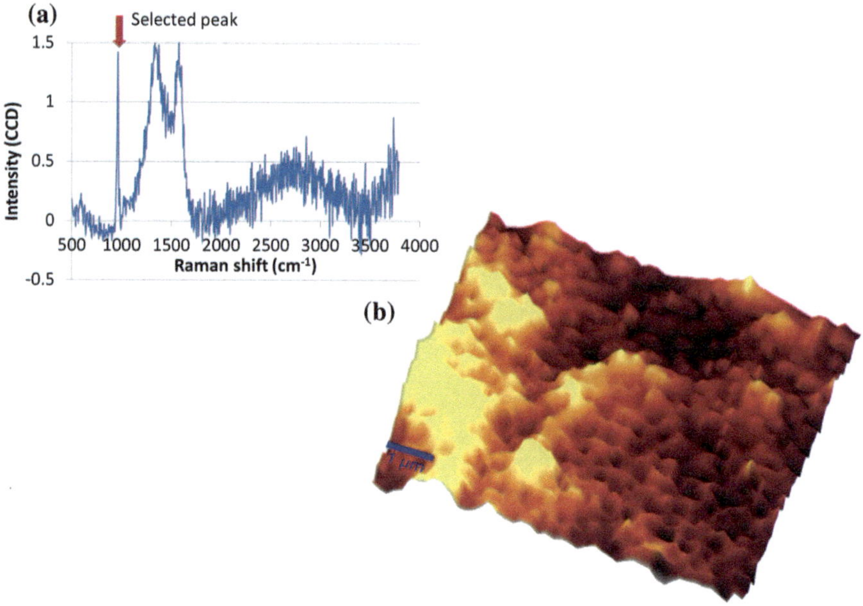

Fig. 15.44 **a** Average refined spectrum of a biochar sample obtained after basic post-analysis. **b** 3D image generated with an integrated intensity filtering on a biochar sample (intensity of the peak obtained at a wavenumber of 956 cm^{-1})

this specific peak (956 cm^{-1}) for the entire analyzed surface. This 3D representation is very useful if combined with the cluster distribution (Fig. 15.41) since this first will overcome the mathematical selection between both colored clusters at boundary limits. Another interest of this 3D representation would be to display the topography of the analyzed area of the sample; this operation can be carried out by selecting other peaks to compare the profile of recorded intensity (for example a dark region located at the same place for all the peaks will indicate a low point).

15.8 Minerals Characterization at the Nanoscale Using Transmission Electron Microscope (TEM) Techniques

Carbon-based materials such as biochars may contain significant amounts of minerals deposited at the surface of the carbonaceous matrix. Raman spectroscopy has been demonstrated to be a useful technique for the analysis of mineral inclusions having a certain size (see Sect. 15.7). However the resolution of the technique does not cover the nanoscale. Conventional bulk characterization tools such as X-ray Diffraction (XRD) do not provide access to the chemistry of these fine particles.

To complete the analyses of solid residues at the nanostructural scale, and especially to detect and quantify the chemical composition of minerals contained in the carbonaceous matrix, Transmission Electron Microscopy (TEM) is used. As seen in Chap. 2, TEM is able to operate with different modes. The specific following modes have been selected to make possible the characterization of minerals (chemistry, dispersion, structure, texture) dispersed in solid residues:

- Scanning Transmission Electronic Microscopy (STEM) coupled with Energy Dispersive Spectroscopy (EDS), as well as High-Angle Annular Dark Field (HAADF): for chemical information at atomic resolution
- High resolution transmission electron microscopy (HRTEM): for nanostructure and crystallite organization
- Selected area electron diffraction (SAED): for chemical information about the pseudo-crystalline structures.

15.8.1 Experimental Recommendations for Various TEM Modes

Among the different modes, only STEM will be detailed in this section. Specific procedures dedicated to SAED and HRTEM are provided in Sect. 15.5. The main difference between TEM and STEM mode refers to the focalization of the incident

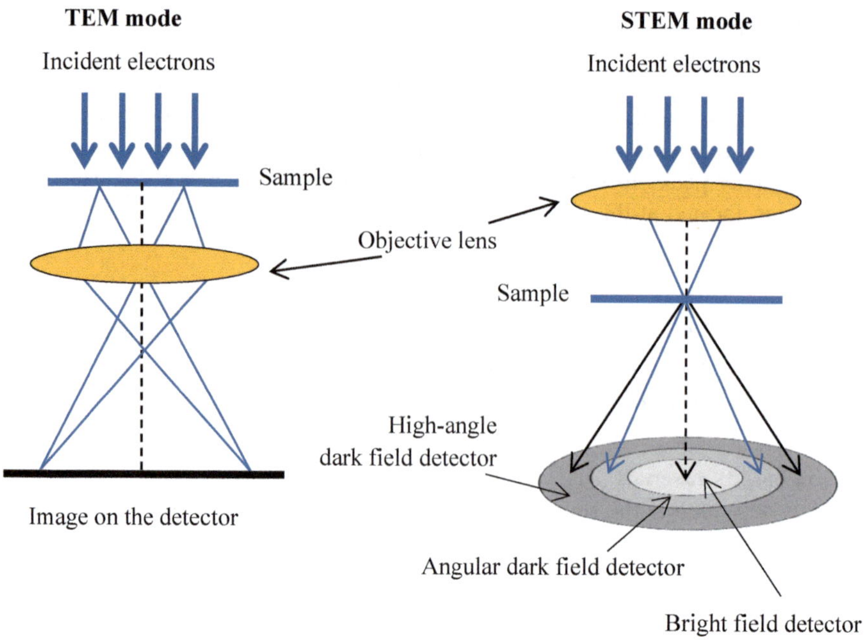

Fig. 15.45 Difference between conventional TEM imaging and STEM imaging

beam (see Fig. 15.45). Indeed, in TEM mode the incident beam interacts with the sample and the transmitted electrons pass through a series of lenses to form an image on the CCD detector. Using STEM mode, the incident beam is preliminary focused to form a spot that hits the sample while scanning it by means of deflector lenses. Due to the convergent incident beam, three different emerging signals can be detected: the direct beam (collected with a bright field detector), the scattered electrons (collected with an annular dark field placed around the bright field detector) and the high-angle scattered electrons (collected with a high-angle annular dark field). The high-angle annular dark field (HAADF) detector is a disk with a large hole. Thus, it detects electrons that are scattered to higher angles and almost only incoherent Rutherford scattering contributes to the image. Therefore, Z contrast is achieved on the CCD image. An HAADF image will thus display dark pixels for zone without sample, while the brightness will increase with the atomic number.

Thus, operating in STEM mode requires some adjustments before inserting the sample in the specimen holder. The settings are modified from the software:

- the FEG conditions (incident beam) are changed by loading the STEM FEG,
- the column optics are switched into STEM mode,
- the condenser aperture is adjusted,
- the detector should be selected.

Table 15.14 Specific steps of the procedure for STEM imaging and EDS spectrum acquisition

STEM/HAADF	STEM/EDS
STEM alignment	STEM alignment
Insert HAADF detector	Tilt sample ($\alpha \approx 15°$)
Focus image	Focus image
Adjust contrast	Adjust contrast
Record STEM image with HAADF detector	Record STEM image with a detector
	Select dispersion
	Select type of scan
	Collect the EDS spectrum

It should be noticed that turning the FEG from TEM mode to STEM mode will require almost 30 min to be stabilized.

In the STEM mode, images are formed by scanning a focused beam through the specimen and then collecting the electrons that emerge from the latter, in such a condition the STEM operates similarly to a SEM. The STEM alignment should be done and differs a little from the TEM alignment. However, as for TEM alignment, the electron gun should be adjusted using rotation center and pivot points, and the condenser aperture and astigmatism should be checked to have the image in-focus. The astigmatism is checked by selecting an amorphous region of the sample and by adjusting the Ronchigram [119]. Table 15.14 indicates specific experimental manipulations that should be carried out after STEM alignment to get targeted images [119, 126]. For the EDS spectrum acquisition: a point, scan line or mapping profile could be selected.

15.8.2 *Example of Chemical Characterization of Minerals in a Carbonaceous Matrix Using STEM/HAADF and STEM/EDS*

Scanning Transmission Electronic Microscopy (STEM) coupled with Energy Dispersive Spectroscopy (EDS), as well as High-Angle Annular Dark Field (HAADF), would significantly improve the understanding of the material chemistry at the nanoscale. While the EDS detector will collect the energy dispersion of the emitted radiation (see Chap. 2), the HAADF detector provides mainly atomic number (Z) contrast. In addition, the contrast is stronger for thin samples. The image interpretation is thus greatly simplified compared to the strong and confusing diffraction contrast observed in conventional STEM and TEM images. For this example, the samples were prepared according to the procedure described in Sect. 15.5.3. STEM characterization was performed using a FEI Tecnai F 20 ST TEM, operated at 200 keV, with a field electron gun (FEG) as the electron source.

Figure 15.46 shows the STEM-HAADF image of a biochar that contains a certain amount of minerals the chemical composition of which should be characterized. On the STEM image, the brightness contrast is due to the variation of atomic number. As an example, the intensity collected (proportional to the number of electrons) through the HAADF detector could be evaluated along the straight line. The significant decrease of the intensity in the range 30–45 nm indicates the presence of a hole (nanocavity) which is observable on the HAADF image (dark ellipsoidal zone). Along the same line, a EDS spectra was acquired and shows the presence of some minerals with varying concentrations, depending on the position on the profile, which indicates their heterogeneous distribution.

15.8.3 Example of Fly Ash Analysis Using TEM/SAED and HRTEM

Ash, and especially fly ash from thermoconversion of biomass and biowaste, is mainly composed of fine particles (PM). Conventional bulk chemical analyses are not selective characterization tools in the sense that they give access to an average composition which could arise from the presence of various types of particles (see Sect. 15.1.4). Indeed, XRD will allow the identification of only crystalline phases (typically for content above 1 wt%). In addition, the presence of broad peaks caused by nanocrystallites will strongly limit identification of compounds from XRD patterns [127]. On the contrary, Transmission Electron Microscopy (TEM) is able to explore and analyze individual particles and to understand the mechanism of their formation.

For this example (from [127]), only the smaller size (PM2.5) fraction of fly ash was analyzed. TEM samples were prepared using the two following methods:

- Suspension: several drops of suspensions prepared by ultrasonicating a small amount of fly ash in high-purity acetone were deposited on the TEM grid coated with a lacey carbon film
- Thin films: cross-section samples were elaborated by embedding fly ash in EMBed-812 resin and then cutting ultrathin sections with a diamond knife in a Reichert Ultracut E ultramicrotome.

The samples were analyzed using a 200 keV JEOL-2010F field emission high resolution transmission electron microscope. Electron diffraction patterns were recorded in SAED mode, and the d spacings were compared to the International Center for Diffraction Data (ICDD) inorganic compound powder diffraction file (PDF) database to identify the crystalline phases. The procedures for acquisition and indexing of SAED patterns are described in Sect. 15.5.2.

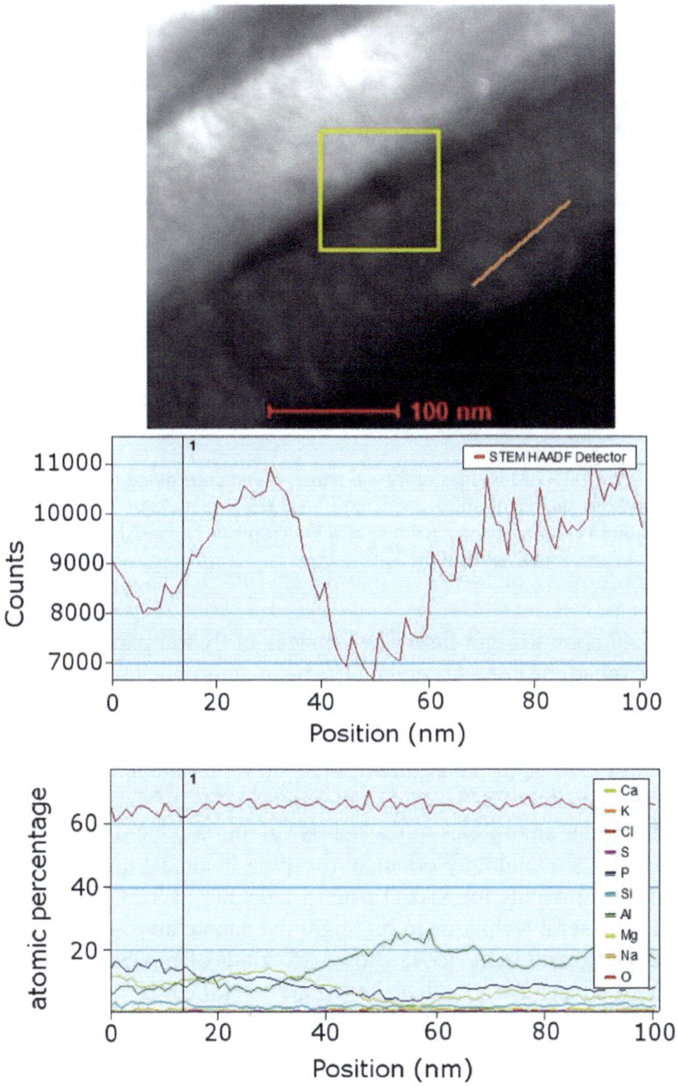

Fig. 15.46 STEM/HAADF image of a wood biochar containing minerals (pyrolysis of food waste and coagulation-flocculation sludge at 700 °C during 30 min) and the corresponding EDS spectra acquired on the orange line. HAADF detector shows a decrease in the counts (maybe due to a porosity). EDS detector shows the mineral distribution which is almost homogeneous

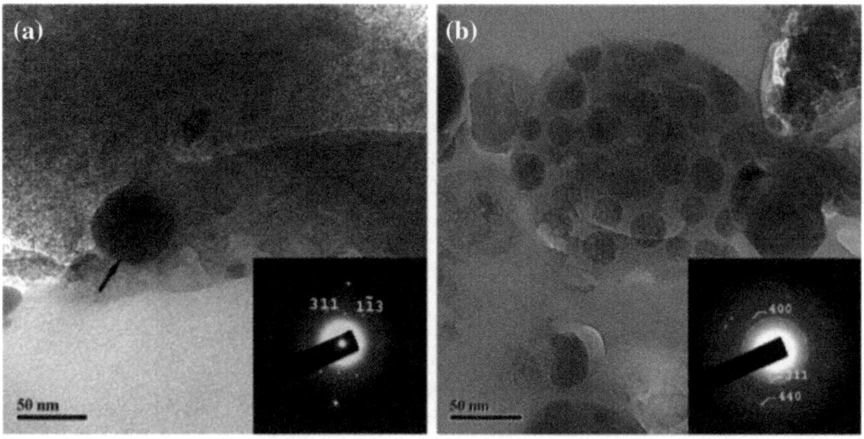

Fig. 15.47 Bright field TEM images of fly ash particles and corresponding SAED patterns (from [127]). **a** Amorphous matrix of silica containing a spherical particle of $FeAl_2O_4$. **b** Amorphous matrix of silica and $FeAl_2O_4$ spheres forming a submicrometer Fe–Si–Al particle. Reprinted with permission from American Chemical Society [127]

Figure 15.47 shows bright field TEM images of fly ash particles. Figure 15.47a shows a dark sphere of $FeAl_2O_4$ embedded in an amorphous silica matrix (brighter zone). The diffuse part of the SAED pattern results from the amorphous structure of the silica matrix. The diffraction points have been attributed to the main reflection lattice plane of $FeAl_2O_4$ (311). Figure 15.47b shows a submicrometer composite of about 150 nm, namely a Fe–Si–Al particle. Indeed, lots of dark spheres ($FeAl_2O_4$) are embedded in an amorphous silica matrix, giving a composite material. In this case, the particles are randomly oriented, resulting in almost diffraction rings in the SAED pattern (see details for SAED pattern indexing in Sect. 15.5.2).

HRTEM is a useful technique to highlight the nanotexture (orientation and size) of individual particles. Figure 15.48 shows two kinds of crystalline zones for which nanocrystals of silver (a) and anhydrite (b) are identified.

In both cases, the lattice fringes are well identified on the image meaning that the nanocrystals are well organized. The distances measured on the image (b) are presently compared to inter reticular distances of the corresponding minerals for identification. Another interesting example showing the nanoparticles of magnetite lattice {220} at 0.296 nm could be found in Chen et al. [128].

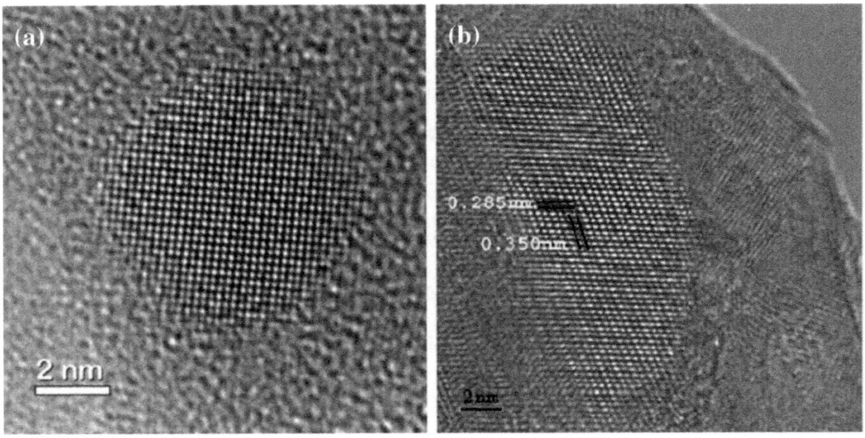

Fig. 15.48 HRTEM images of fly ash nanocrystallites: **a** identification of silver nanoparticles in silica, **b** identification of $CaSO_4$ lattices (0.350 nm {020} and 0.285 nm {012}), image from [127]. Photo credit image **a** P. Benzo (CEMES-CNRS, Toulouse). Image **b** reprinted with permission from American Chemical Society [127]

15.9 Bulk Crystalline Analysis Using X-Ray Powder Diffraction (XRPD)

The X-ray powder diffraction is a non-destructive technique used to identify and quantify the composition of a mixture which can be composed of more than one crystalline compounds and amorphous phase. The XRPD theory and device as well as the details of the acquisition procedure are described in Chap. 2. From diffractograms obtained, the background is subtracted and a profile refinement method is used as described in the following paragraphs. Different data can be extracted from a diffractogram: the crystalline phase identification, the semi-quantitative analysis (RIR, for reference intensity ratio), the quantitative analysis by the Rietveld method and the determination of the crystallite size and residual stress. The software used as part of this work is the PANalytical HighScore Plus [129]. Other available software such as DQUANT, EVA, PolySNAO or TOPAS [130] could also be used.

To identify crystalline phase, different databases can be used, however the number of reference samples included are different. The ICDD (International Centre for Diffraction Data) database includes more than 400,000 references of organic and inorganic compounds [131] and replaced the old JCPDS (Joint Committee on Powder Diffraction Standards, until 1978) database. The databases used in this section are the PDF-2 (Powder Diffraction Pattern) and the COD (Crystallographic Open Database) which contain 147,575 and 223,493 reference patterns, respectively. So, the identification of crystalline phases is carried out with a database containing 371,068 reference patterns. Each reference pattern contains the reference

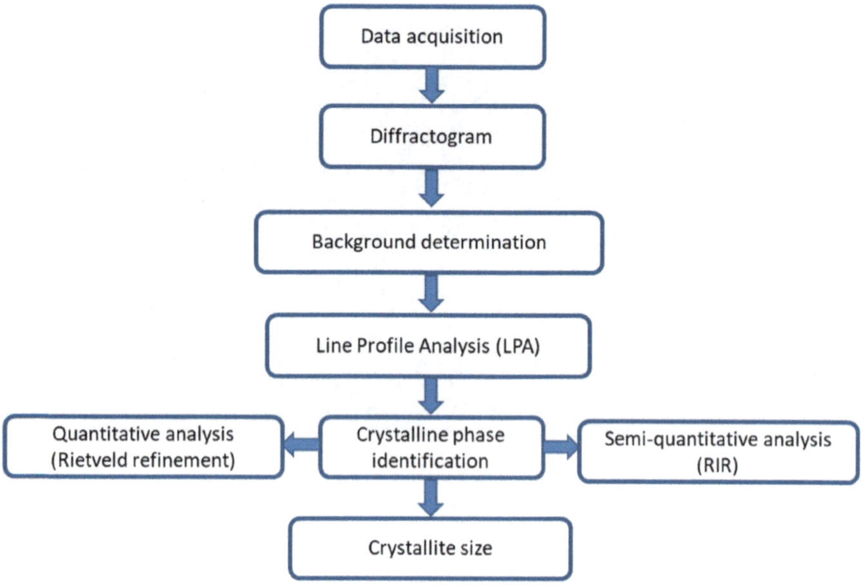

Fig. 15.49 Scheme of the procedure for the determination of properties from XRD

code, the name and chemical formula of the compound, the crystallographic data (crystal system, space group, lattice parameters), the reference of the article in which the resolved crystal structure is published, the inter-reticular spacing, the intensities of the diffraction peaks, the Miller's indices and the stick pattern of the compound. A very good tutorial explaining every step to use HighScore Plus software created by the MIT Center for Materials Science and Engineering is available on internet [132].

The methodology developed in this section is summarized in the following scheme (Fig. 15.49).

15.9.1 Crystalline Phase Identification: Profile Refinement Method

For the crystalline phase identification, the first step is to compare XRD experimental data with reference patterns after having subtracted the background. For this, a profile refinement method is used. The method is based on the least squares method (Eq. 15.20) to minimize the difference (so called the residue) between the experimental and calculated data from the model in the full diffractogram.

$$R = \sum w_i[I_i(obs) - I_i(cal)]^2 \tag{15.20}$$

where R is the function to be minimized, w_i is the weighting assigned to $I_i(obs)$ (counts number in the step i), $I_i(obs)$ is the intensity observed and $I_i(cal)$ is the corresponding intensity calculated.

The profile refinement method corresponds to a profile fitting of each diffraction peak. For this, the Line Profile Analysis (LPA) is used. The profile functions used have Gaussian (Eq. 15.21) and Lorentzian (Eq. 15.22) shapes. Figure 15.50 presents the peak profile of the different functions. A Pseudo-Voigt function (Eq. 15.23) is used to fit each diffraction peak. The Pseudo-Voigt function is a combination of the Gaussian and Lorentzian shapes as shown in Eq. (15.23).

For the profile fitting of diffraction patterns using the Pseudo-Voigt function, η is adjusted as a function of 2θ with the following relation: $\eta = \eta_0 + \eta_1\,\theta + \eta_2\,\theta^2$ (θ is the diffraction angle).

At the same time, the deconvolution of Gaussian and Lorentzian contributions as a function of 2θ is carried out by means of the empirical formula of Cagliotti (Eq. 15.24). This equation allows to fit the full width at half maximum (FWHM) of each diffraction peak. The refinement is performed with the HighScore Plus software by modifying the profile parameters. The parameters U, V and W correspond to the peak height extension and η_0, η_1 and η_2 to the peak shape. So, each diffraction peak is fitted using the profile function.

$$G = f(x) = \frac{\sqrt{4\ln 2}}{FWHM\sqrt{\pi}}\exp\left[\frac{-4\ln 2x^2}{FWHM^2}\right] \tag{15.21}$$

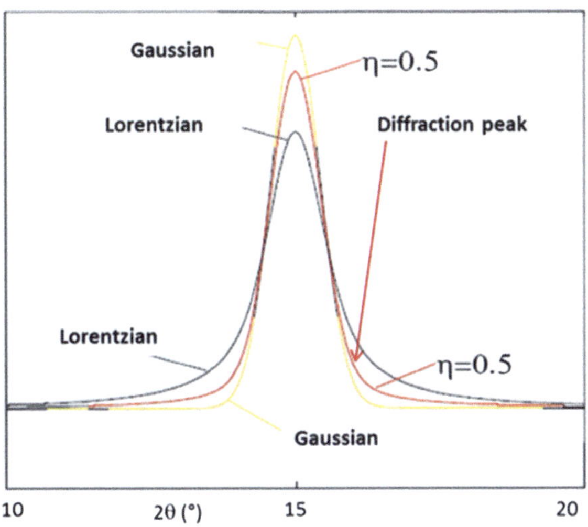

Fig. 15.50 Profile of diffraction peaks

$$F = f(x) = \frac{2}{\pi FWHM} / \left[1 + 4\frac{x^2}{FWHM^2}\right] \quad (15.22)$$

$$F(x) = \eta L + (1 - \eta)G \quad (15.23)$$

$$FWHM^2 = U\tan^2\theta + V\tan\theta + W \quad (15.24)$$

where θ is in ° and FWHM in °.

Profile fitting provides precise information on the position, width, height, and areas of diffraction peaks with parameter estimates, the latter being then compared with that of obtained on reference patterns from the database. Given the large number of reference patterns, the research for the one(s) that coincide(s) most with the diffractogram being analyzed requires the knowledge of the different chemical elements constituting the phase(s) (and not the sample because the latter can be a mixture of phases). The XRF and ICP techniques presented in Chap. 2 are used, among other techniques, for determining elemental composition of the sample. So, elements identified and quantified by these analytical techniques are selected in the software before the search for the crystalline phase. In a second step, the reference pattern can be chosen. It is important to verify that the positions and relative intensities of major diffraction peaks of the reference pattern coincide as much as possible with the corresponding peaks of the experimental data recorded. From this point, the crystalline phases contained in the powder have been identified.

As an example, the Fig. 15.51 shows a diffractogram recorded of a sample of refused-derived fuel (RDF) treated at 800 °C under nitrogen atmosphere. The X-ray powder diffraction (XRPD) data were collected using a Phillips Panalytical X'pert Pro MPD diffractometer. The X-ray radiation source used is the copper (Cu) whose the Kα wavelength is 1.543 Å. The Bragg-Brentano geometry and θ–θ

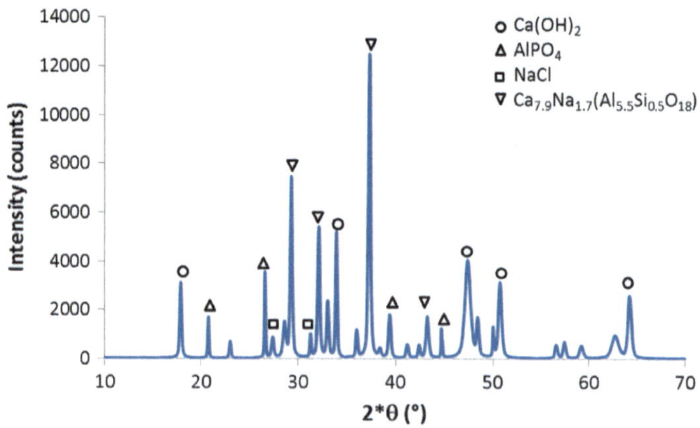

Fig. 15.51 Diffractogram of a refused-derived fuel (RDF) sample after thermal treatment (λ(Cu) = 1.543 Å)

Table 15.15 List of the reference patterns

Crystalline phase	Reference pattern number	Crystal structure
Halite, NaCl	96-900-3312	Cubic
Calcium Hydroxide, $Ca(OH)_2$	00-044-1481	Hexagonal
Aluminum phosphate, $AlPO_4$	01-076-0228	Hexagonal
Calcium sodium aluminum silicon oxide $Ca_{7.9}Na_{1.7}(Al_{5.5}Si_{0.5}O_{18})$	01-083-1361	Monoclinic

configuration were used. A nickel filter was placed in forward mounting in order to remove the Cu K_β ray. The apparatus operates at a voltage of 45 kV and with an intensity of 40 mA. The diffractogram was recorded from 10 to 75° in 2θ with a scan speed of 0.037°/s. The peaks identification in the solid is presented in Fig. 15.51 and the list of the corresponding reference patterns are presented in Table 15.15.

15.9.2 Semi-quantitative Analysis

The semi-quantitative analysis is performed with the Relative Intensity Ratio (RIR) method which consists in comparing ratio of the strongest intensities of the sample and a reference as defined in Eq. (15.25). In some PDF and COD reference patterns, the RIR of the compound is given. It is a very old method and has to be used with special care because for the same compound, the RIR value can be different in two reference patterns [133]. For example, for the cristobalite (SiO_2), reference patterns 01-076-0935 and 01-076-0941 corresponding to this compound, everything else remained the same (crystal structure and parameters), give a RIR of 5.01 and 5.13, respectively. So, the result of the semi-quantitative analysis can be different and vary up to a factor of 1.5 in the quantification phase result depending on the reference pattern selected.

Another possibility is to determine the RIR by experimental measurements. For this purpose, the sample to be analyzed has to be composed by only one crystalline phase. This sample is mixed with alpha alumina (corundum) in a weight ratio of 50/50. The RIR corresponds to the intensity ratio between the strongest diffraction peak of the compound and the strongest reflection of Corundum in the 50/50 mixture. From the diffractogram obtained, the RIR can be calculated from the following equation:

$$RIR(sample) = I_{max(sample)}/I_{max(corundum)} \tag{15.25}$$

When a solid contains different compounds (crystalline phases), the RIR has to be measured and calculate for each compound. It should be noticed that the method is reliable even when the strongest diffraction peaks of compounds are very close to

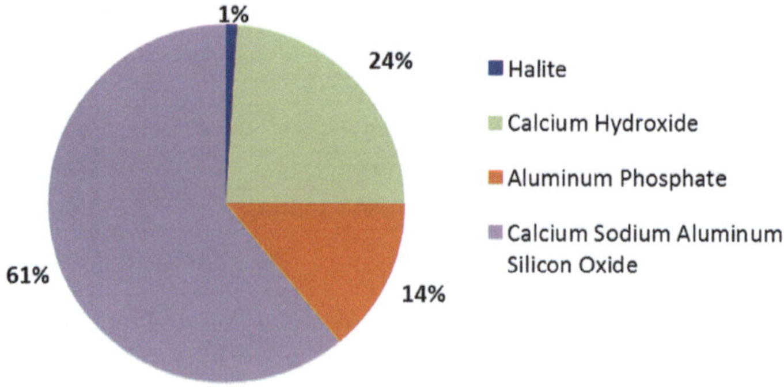

Fig. 15.52 Semi quantification (wt% of chemical compounds) from RIR calculation for a refused-derived fuel after thermal treatment

each other. The sample is then analyzed and the concentration of each compound in this mixture is calculated with Eq. (15.26).

$$C_i = \frac{1/RIR_i \times I_i}{\sum_{j=1}^{n} 1/RIR_j \times I_j} * 100 \tag{15.26}$$

where C_i is the concentration of the crystalline phase i (wt%), RIR_i is the RIR value of the crystalline phase i and I_i is the net intensity corresponding to the crystalline phase i (in counts).

The results obtained after having applied the RIR method from the diffractogram displayed in Fig. 15.51 are presented in Fig. 15.52. This result will be compared later to those obtained with the Rietveld refinement.

The difficulty of applying this method is that the crystallinity of each compound has to be the same in the mixture to be analyzed and their RIR value need to be available. This parameter is very difficult to determine, it is the reason why this method is called as semi-quantification method.

15.9.3 Quantitative Analysis: Rietveld Method

To quantify crystalline and amorphous phases in a solid, Rietveld refinement can be used. Results obtained give the quantitative phase analysis, along with the lattice parameters and the crystal structure model of a material.

The Rietveld method is applied to the whole angular range of the diffractogram. This refinement method consists in adjusting the calculated model with the observed diffractogram using the LPA (Line Profile Analysis) method [134, 135]. The difference between both diffractograms (calculated and experimental) is calculated by the least squares method, whenever the value of a parameter is changed, and with the objective of minimizing it (Eq. 15.20).

Table 15.16 Rietveld refinement of a refused-derived fuel after thermal treatment

R_p = 5.2%	R_{exp} = 1.7%			R_{wp} = 6.0%			GoF = 0.9	
Phase	a (Å)	b (Å)	c (Å)	α (°)	β (°)	γ (°)	R_{Bragg} (%)	Weight (%)
$Ca_{7.9}Na_{1.7}(Al_{5.5}Si_{0.5}O_{18})$	3.589	3.589	4.900	90	90	120	3.5	59.9
$AlPO_4$	4.927	4.9270	10.775	90	90	120	4.6	9.9
$Ca(OH)_2$	10.850	15.044	11.103	90	90.39	90	3.1	28.5
NaCl	5.713	5.713	5.713	90	90	90	2.4	1.8

In a first step, crystalline phases identification is carried out by the LPA method. From the reference pattern selected, the ".CIF" files corresponding to all crystalline phases identified will be downloaded in the Crystallographic Open Database [136]. These files contain the crystal structure, lattice parameter and all crystal parameters useful for the refinement such as the nature and positions of all atoms in the crystal structure. With the HighScore Plus software, the ".CIF" files downloaded previously are inserted in the experimental data file (.xrdml). From a script (program made by the user) the High Score Plus software can give results obtained which correspond to the quantity of each crystalline compounds and their lattice parameters. The results obtained are gathered in an excel file as presented in Table 15.16.

For the Rietveld refinement, some indicators allow to validate the refinement by numerical statistical indices [137–139]. Three agreement indices are reliability factors strongly correlated to the refinement quality of the complete diffractogram:

- R_p ($R_{pattern}$): is the residual difference between observed and calculated plots (Eq. 15.27). If $R_p < 10\%$, the refinement is considered as good.
- R_{wp} ($R_{Weighted\ Pattern}$) [140] uses a mathematical weighting function to place more emphasis on good fit between data points of high intensity (such as intense peaks) and less emphasis on data points having low intensity (such as background), see Eq. 15.28. If $R_{wp} < 10\%$, the refinement is considered as good.
- R_{exp} ($R_{expected}$) is a statistical evaluation of the noise of the data (Eq. 15.29). A small $R_{expected}$ value indicates good quality of data. If $R_{exp} < 5\%$, the refinement is considered as good.
- GoF (Goodness of Fit) is defined in Eq. (15.30). To consider the fit as good, the GOF should be less than 4.

One agreement indice is related to the refinement of the crystalline structure phase, the R_{Bragg}.

- R_{Bragg}: If $R_{Bragg} < 10\%$, the refinement is considered as good (Eq. 15.31).

$$R_P = \sum_{set\ of\ measuring\ points} [I_i(obs) - I_i(cal)] / \sum_{set\ of\ measuring\ points} I_i(cal) \quad (15.27)$$

$$R_{WP}^2 = \sum_{set\ of\ measuring\ points} [I_i(obs) - I_i(cal)]^2 / \sum_{set\ of\ measuring\ points} I(cal)_i^2 \quad (15.28)$$

$$R_{\exp} = \sqrt{N / \sum_i I(cal)_i^2} \quad (15.29)$$

$$\text{Goodness of Fit} = R_{wp}^2 / R_{\exp}^2 \quad (15.30)$$

where N is the number of data points

$$\mathrm{R_{Bragg}} = \sum_{\mathrm{i}} [\mathrm{Ii(obs)} - \mathrm{Ii(cal)}] / \sum_i \mathrm{Ii(cal)} \quad (15.31)$$

In the Table 15.16, results of the quantification achieved by Rietveld refinement of the diffraction pattern of a refused-derived fuel sample after thermal treatment (Table 15.15) is presented. The values of the reliability factors attest to the good quality of the Rietveld refinement. Mass proportions of compounds resulting from phase quantification carried out by the Rietveld method are different from those obtained by applying the RIR semi-quantification (results are displayed in Fig. 15.52). For example, the mass proportion of the calcium hydroxide phase determined by semi-quantification is 4 points higher than that obtained by the Rietveld method. This sample contains only crystalline compounds.

An amorphous phase can be also quantified by addition of an internal standard; this is a standard method in quantitative analysis. Experimental measurements have to be made with a mixture of the sample and a standard (alpha alumina) added in a known concentration (around 30 wt%) [141, 142]. An example of diffractograms recorded on a sample with and without the addition of a standard is presented in Fig. 15.53. The black dashed curve corresponds to diffraction peaks of the biochar and standard mixture. This diffractogram is only used to quantify the amorphous phase but has no other added value. The quantity of α-alumina used is also introduced in the software and the LPA is carried out. From a script, the HighScore Plus software gives the quantity of each compound (crystalline and amorphous phases) contained in the sample.

An example of diffractograms of a poplar wood biochar, obtained from the steam gasification in a fluidized bed reactor at the temperatures of 600 and 700 °C, is presented in Fig. 15.54. The spread peak between 20 and 30° is characteristic of the amorphous carbon. The main peaks are assigned to crystalline phases on the diffractograms. This amorphous phase was quantified by the Rietveld refinement method. Results obtained for the two temperatures are presented in Table 15.17. Three crystalline phases are detected: calcium hydroxide, calcium oxide and calcium carbonate.

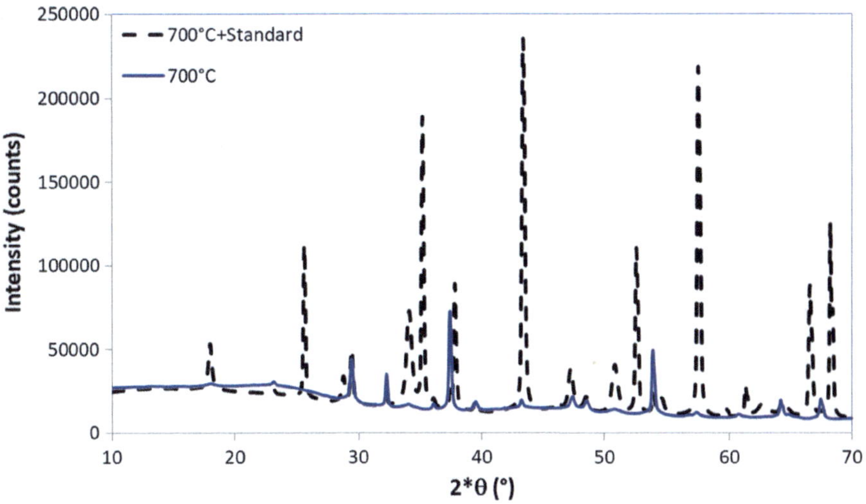

Fig. 15.53 Diffractograms of biochar with and without the addition of standard

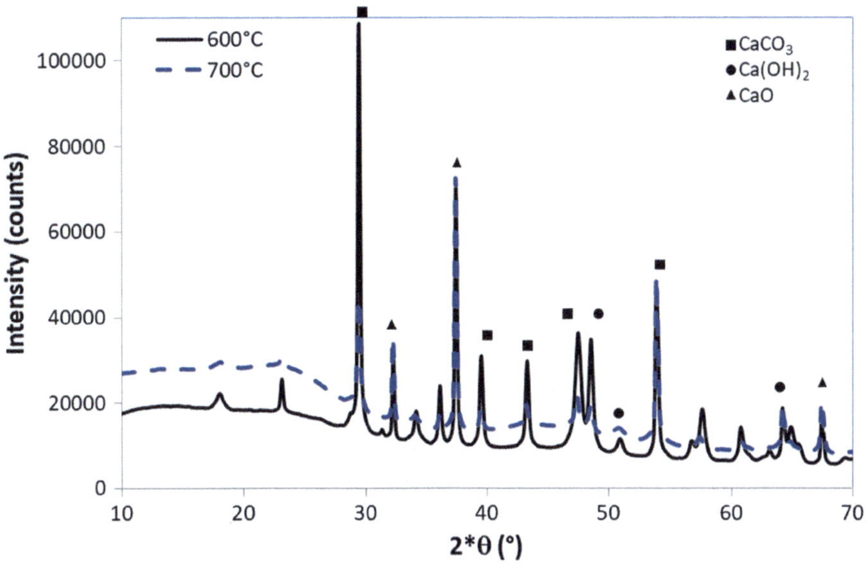

Fig. 15.54 Diffractograms of biochar after gasification at 600 °C (black line) and 700 °C (blue dashed line)

Table 15.17 Results of a Rietveld refinement carried out from a diffraction pattern of a biochar after gasification

600 °C								
R_p = 2.9%		R_{wp} = 4.9%			R_{exp} = 0.7%	GoF = 1.4		
Phase	a (Å)	b (Å)	c (Å)	α (°)	β (°)	γ (°)	R_{Bragg} (%)	Weight (%)
$Ca(OH)_2$	3.593	3.593	4.918	90.000	90.000	120.000	4.1	29.1
CaO	4.813	4.813	4.813	90.000	90.000	90.000	9.1	3.6
$CaCO_3$	4.991	4.991	17.090	90.000	90.000	120.000	7.7	32.2
Amorphous								35.0
700 °C								
R_p = 3.45%		R_{wp} = 5.95%				R_{exp} = 0.66%		
Phase	a (Å)	b (Å)	c (Å)	α (°)	β (°)	γ (°)	R_{Bragg} (%)	Weight (%)
$Ca(OH)_2$	3.593	3.593	4.918	90	90	120	4.2	22.0
CaO	4.811	4.811	4.811	90	90	90	9.3	0.7
$CaCO_3$	4.991	4.991	17.079	90	90	120	9.4	4.8
Amorphous								72.5

15.9.4 Crystallite Size

The crystallite size can be estimated from peak profile analysis. The width of the diffraction peaks give information about the crystallite size as illustrated in Fig. 15.55. The wider is the peak, the smaller the crystallite size, and vice versa for the micro-deformation evolution. The full width at half maximum (FWHM) and the peak profile are related to the crystallite size and micro-deformation. Figure 15.56 represents a scheme of a crystallite. A grain can be composed of few crystallites. The analysis of the width of XRD peaks allows the determination of the crystallite size. Thereby, results obtained from XRD are generally complementary to those of obtained with other characterization tools, except the high resolution transmission electron microscopy (HRTEM; refer to Sects. 15.5 and 15.8) which gives access, by observation, to the dimension of crystallite. So, it is important to compare the crystallite size determined by XRD with the one deduced from observation with another technique such as HRTEM. Since the microscope displays a projection of

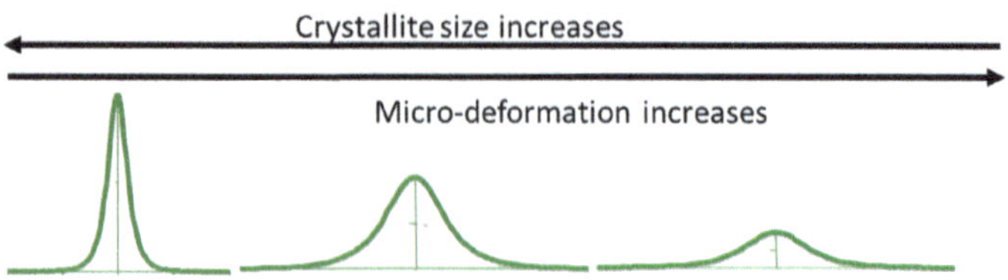

Fig. 15.55 Evolution of the peak width as a function of the crystallite size and micro-deformation

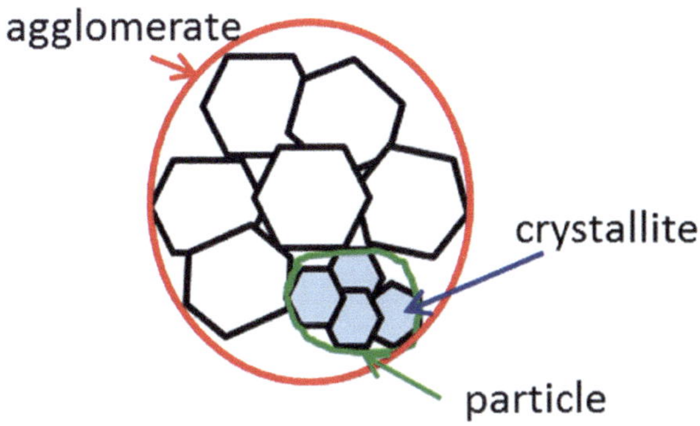

Fig. 15.56 Representation of different elements constituting a powder

the object observed, the size measured can be identical or different to that determined by XRD. The elements constituting the microstructure of a material are classified in size ascending order: crystallite < grain < particle (agglomerate or aggregate).

To estimate the crystallite size, the Scherrer formula is basically used (Eq. 15.32) [143]:

$$\beta = \frac{K \times \lambda}{L \times \cos\theta} \tag{15.32}$$

where

- K is the Scherrer constant which depend on crystal shape. K is ranged between 0.32 and 2.08. For a spherical crystal, K is equal to 0.94. The paper written by J. L. Langford deals with the values that the constant K can take [144];
- λ is the wavelength of the X-ray source used. For example, for the copper, $\lambda_{K\alpha} = 1.543$ Å;
- L is the crystallite size (in Å);
- θ is the Bragg diffraction angle (in rad) considered;
- β is the Full Width at Half Maximum (FWHM) of the same peak (Fig. 15.57). β is expressed in rad and can be calculated using the following expression: $\beta(\text{rad}) = 2{*}\pi\beta(°)/360$.

In Fig. 15.57 is presented an example of an analysis of a diffraction peak recorded on a biochar treated at 600 °C. The calculation is carried out with: $\lambda = 1.543$ Å, $K = 0.32$, $\theta = 29.4579° = 0.5138$ rad, $\beta = 0.1948° = 0.0034$ rad. So, the crystallite size is $L = (0.32 \times 1.543)/(0.0034 \times \cos 0.5138) = 167$ Å.

The Scherrer formula can be applied whatever the diffraction peak considered. However, the better and more representative results are obtained for diffraction peaks localized between 30 and 50° in 2θ. The application of the Scherrer formula is validated for crystallite sizes ranging from 1 nm to 1 μm. Another method used to estimate crystallite size is the Williamson-Hall plot [145]. This method requires

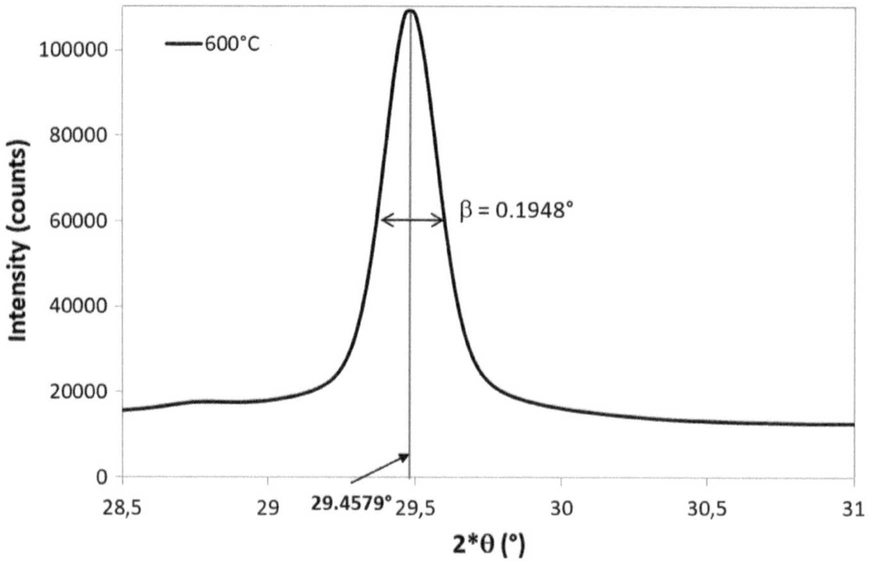

Fig. 15.57 Full Width at Half maximum of a diffraction peak recorded on a biochar treated at 600 °C

analysis of all diffraction peaks of the diffractogram. After Rietveld refinement, the HighScore Plus software calculates and draws the Williamson-Hall plot by means of Eq. (15.33).

$$\beta \times \cos \theta/\lambda = 1/L + \varepsilon \sin \theta/\lambda \tag{15.33}$$

The plot displays ($\beta \times \cos \theta/\lambda$) as a function of ($\sin \theta/\lambda$) and a linear regression is obtained if the size distribution is uniform. (1/L) is the intercept of the line with the y-axis which gives the mean crystallite size L. ε (the slope of the linear regression) corresponds to the micro-deformation. An example extracted from the HighScore Plus software is presented in the Fig. 15.58. In this case, the mean crystallite size is of 436 Å. The Williamson-Hall plot takes into account all diffraction peaks without distinction of crystalline phases, so the crystallite size determined is the average of the sizes of all crystallites of the sample.

Some materials can contain crystalline (turbostatic structure) and amorphous (non-aromatic structure) carbon as example for coal materials [43]. In this case, it can be difficult to calculate the crystallite size. As described by these authors [43] a procedure for the calculation is as follows, first the amorphous carbon quantity is determined from the diffractogram considering its intensity is constant in all diffractogram. So, the average crystalline size is calculated for the crystalline carbon structure. From assumption which is that the inter-layer spacing is the same for all size groups and from the amount of amorphous carbon determined, the average crystallite size is calculated. In the Table 15.18, results obtained for coal by this

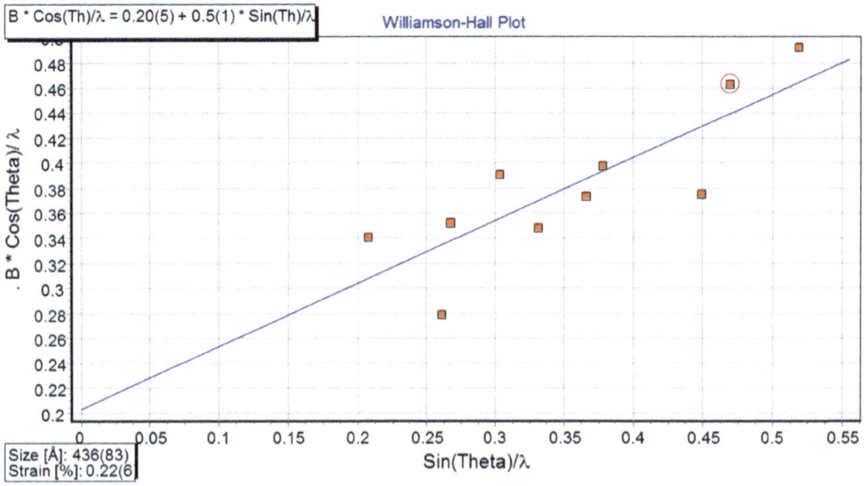

Fig. 15.58 Williamson-Hall plot

Table 15.18 Average crystallite size diameter (Å)

Coal type	Crystallite diameter (Å) [55]	Crystallite diameter (Å) (Sherrer's equation)
Coal1-DD	7.50	7.20
Coal3-JD	10.55	12.9
Coal4-ED	9.20	10.3

calculation are compared with the calculation from the Sherrer's equation. A good agreement is found as the deviation varies from 4 to 13% only.

Some solid residues contain a certain amount of graphitic or non-graphitic carbon. The determination of the crystallite size L in these products is complex and the literature has shown a lot of contradictions between the methods used [43, 111, 146, 147]. Zickler et al. demonstrated in their study that the crystallite size obtained is strongly correlated to the method used (the Scherrer's law or deconvolution) [111]. As an example, Fig. 15.59 displays values of crystallite sizes as a function of the pyrolysis temperature for wood determined by using three methods. This graph shows that the estimation of crystallite size using various methods is within the same range with small deviation. Manoj et al. present a study which describes how to determine the crystallite size of an amorphous carbon [148]. When a diffraction peak is overlapped on its base by a wide and intense background, the presence of a disordered material in the form of amorphous carbon can be suspected in the sample. If a peak asymmetry appears, it can be interpreted by the presence of the γ band of carbon (aliphatic chains) superimposed with a band characteristic of an aromatic layer. The curve fitting is a very important step in the determination of crystallite size and, in some case, may lead to overestimated values because of the

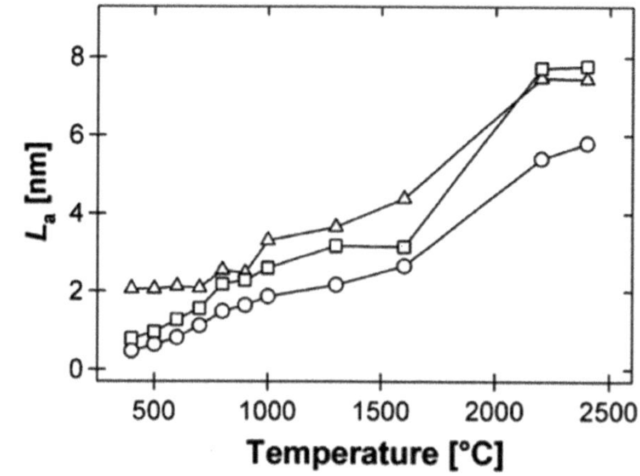

Fig. 15.59 Crystallite size estimates by different methods [111]. Calculated with Scherrer equation: (□) 11 peaks and (Δ) 10 peaks [111] and (O) calculated on the basis of the Ruland and Smarsly model [147]. Reprinted with permission from Elsevier [111]

difficulty to separate peak profile and to obtain a good fitting. Some studies describe the procedure to be followed to determine the crystallite size in carbon materials [149, 150]. So, when an asymmetry appears on diffraction peaks, characterization techniques complementary to XRD, such as TEM and Raman spectroscopy, should be used.

15.10 Conclusions

This chapter developed the characterization tools of physico-chemical properties of solid residues at the surface or into the bulk. It has been shown that the combination of these advanced characterization techniques is beneficial to better characterize the carbonaceous matrix and/or minerals composing the solid residues from biowaste and biomass treatment.

Acknowledgements The author thanks Pascal Puech and Marc Monthioux (CEMES-CNRS, Toulouse, France) for their relevant advices.

References

1. ASTM D5759-12, Standard guide for characterization of coal fly ash and clean coal combustion fly ash for potential uses. ASTM International, West Conshohocken (2012). www.astm.org
2. Antal Jr., M.J., Grønli, M.: The art, science, and technology of charcoal production. Ind. Eng. Chem. Res. **42**, 1619–1640 (2003)
3. Lehmann, J., Joseph, S.: Biochar for Environmental Management: Science, Technology and Implementation. Earthscan from Routledge, Taylor and Francis Group, London (2015)

4. Singh, B., Camps-Aberstain, M., Lehmann, J.: Biochar, a Guide to Analytical Methods. CRC Press, Taylor and Francis Group, Boca Raton (2017)
5. Koziński, J.A., Saade, R.: Effect of biomass burning on the formation of soot particles and heavy hydrocarbons. An experimental study. Fuel **77**, 225–237 (1998)
6. Watson, A.Y., Valberg, P.A.: Carbon black and soot: two different substances. Am. Ind. Hyg. Assoc. J. **62**, 218–228 (2001)
7. Reza Kholghy, M., Veshkini, A., Thomson, M.J.: The core-shell internal nanostructure of soot—a criterion to model soot maturity. Carbon **100**, 508–536 (2016)
8. Trubetskaya, A., Jensen, P.A., Jensen, A.D., Garcia Llamas, A.D., Umeki, K., Gardini, D., Kling, J., Bates, R.B., Glarborg, P.: Effects of several types of biomass fuels on the yield, nanostructure and reactivity of soot from fast pyrolysis at high temperatures. Appl. Energy **171**, 468–482 (2016)
9. Corbin, J.C., Lohmann, U., Sierau, B., Keller, A., Burtscher, H., Mensah, A.A.: Black carbon surface oxidation and organic composition of beech-wood soot aerosols. Atmos. Chem. Phys. **15**, 11885–11907 (2015)
10. Andreae, M.O., Gelencsér, A.: Black carbon or brown carbon? The nature of light-absorbing carbonaceous aerosols. Atmos. Chem. Phys. **6**, 3131–3148 (2006)
11. Long, C.M., Nascarella, M.A., Valberg, P.A.: Carbon black vs. black carbon and other airborne materials containing elemental carbon: Physical and chemical distinctions. Environ. Pollut. **181**, 271–286 (2013)
12. Basso, D., Patuzzi, F., Castello, D., Baratieri, M., Rada, E.C., Weiss-Hortala, E., Fiori, L.: Agro-industrial waste to solid biofuel through hydrothermal carbonization. Waste Manage. **47** Part A, 114–121 (2016)
13. Liu, W.-J., Zeng, F.-X., Jiang, H., Zhang, X.-S.: Preparation of high adsorption capacity bio-chars from waste biomass. Bioresour. Technol. **102**, 8247–8252 (2011)
14. Ahmad, M., Upamali Rajapaksha, A., Lim, J.E., Zhang, M., Bolan, N., Mohan, D., Vithanage, M., Lee, S.S., Ok, Y.S.: Biochar as a sorbent for contaminant management in soil and water: a review. Chemosphere **99**, 19–33 (2014)
15. Mohan, D., Sarswat, A., Ok, Y.S., Pittman Jr., C.U.: Organic and inorganic contaminants removal from water with biochar, a renewable, low cost and sustainable adsorbent—a critical review. Bioresour. Technol. **160**, 191–202 (2014)
16. Azargohar, R., Dalai, A.K.: Biochar as a precursor of activated carbon. Appl. Biochem. Biotechnol. **131**, 762–773 (2006)
17. Rodríguez-Reinoso, F.: The role of carbon materials in heterogeneous catalysis. Carbon **36**, 159–175 (1998)
18. Serp, P., Figueiredo, J.L.: Carbon Materials for Catalysis. Wiley, Hoboken (2009)
19. Hervy, M., Berhanu, S., Weiss-Hortala, E., Chesnaud, A., Gérente, C., Villot, A., Pham Minh, D., Thorel, A., Le Coq, L., Nzihou, A.: Multi-scale characterisation of chars mineral species for tar cracking. Fuel **189**, 88–97 (2017)
20. Lee, J., Kim, K.-H., Kwon, E.E.: Biochar as a catalyst. Renew. Sust. Energ. Rev. **77**, 70–79 (2017)
21. Liu, W.-J., Jiang, H., Yu, H.-Q.: Development of biochar-based functional materials: toward a sustainable platform carbon material. Chem. Rev. **115**, 12251–12285 (2015)
22. Barroso-Bogeat, A., Alexandre-Franco, M., Fernández-González, C., Gómez-Serrano, V.: Preparation of activated carbon–metal oxide hybrid catalysts: textural characterization. Fuel Process. Technol. **126**, 95–103 (2014)
23. Qian, K., Kumar, A., Zhang, H., Bellmer, D., Huhnke, R.: Recent advances in utilization of biochar. Renew. Sust. Energ. Rev. **42**, 1055–1064 (2015)
24. Westra, L.: The Common Good and Environmental Governance for the Support of Life. Cambridge Scholars Publishing, Newcastle (2016)
25. World Energy Resources Report 2016. World Energy Council, London (2016)
26. Parent, P., Laffon, C., Marhaba, I., Ferry, D., Regier, T.Z., Ortega, I.K., Chazallon, B., Carpentier, Y., Focsa, C.: Nanoscale characterization of aircraft soot: a high-resolution

transmission electron microscopy, Raman spectroscopy, X-ray photoelectron and near-edge X-ray absorption spectroscopy study. Carbon **101**, 86–100 (2016)
27. Trubetskaya, A., Brown, A., Tompsett, G.A., Timko, M.T., Kling, J., Broström, M., Larsen Andersen, M., Umeki, K.: Characterization and reactivity of soot from fast pyrolysis of lignocellulosic compounds and monolignols. Appl. Energy **212**, 1489–1500 (2018)
28. Fals, A.E., Hadjiev, V.G., Robles Hernández, F.C.: Porous media reinforced with carbon soots. Mater. Chem. Phys. **140**, 651–658 (2013)
29. Fidalgo, B., Menéndez, J.Á.: Carbon materials as catalysts for decomposition and CO_2 reforming of methane: a review. Chin. J. Catal. **32**, 207–216 (2011)
30. Serrano, D.P., Botas, J.A., Fierro, J.L.G., Guil-López, R., Pizzaro, P., Gómez, G.: Hydrogen production by methane decomposition: origin of the catalytic activity of carbon materials. Fuel **89**, 1241–1248 (2010)
31. Liu, S., Wang, Y., Wu, R., Zeng, X., Gao, S., Xu, G.: Fundamentals of catalytic tar removal over in situ and ex situ chars in two-stage gasification of coal. Energy Fuels **28**, 58–66 (2014)
32. Berhanu, S., Hervy, M., Weiss-Hortala, E., Proudhon, H., Berger, M.-H., Chesnaud, A., Faessel, M., King, A., Pham Minh, D., Villot, A., Gérente, C., Thorel, A., Le Coq, L., Nzihou, A.: Advanced characterization unravels the structure and reactivity of wood-based chars. J. Anal. Appl. Pyrol. **130**, 79–89 (2018)
33. Muradov, N.: Catalysis of methane decomposition over elemental carbon. Catal. Commun. **2**, 89–94 (2001)
34. Muradov, N., Smith, F., T-Raissi, A.: Catalytic activity of carbons for methane decomposition reaction. Catal. Today **102**, 225–233 (2005)
35. Suelves, I., Pinilla, J.L., Lázaro, M.J., Moliner, R.: Carbonaceous materials as catalysts for decomposition of methane. Chem. Eng. J. **140**, 432–438 (2008)
36. Ducousso, M., Weiss-Hortala, E., Nzihou, A., Castaldi, M.J.: Reactivity enhancement of gasification biochars for catalytic applications. Fuel **159**, 491–499 (2015)
37. Nzihou, A., Stanmore, B., Sharrock, P.: A review of catalysts for the gasification of biomass char, with some reference to coal. Energy **58**, 305–317 (2013)
38. Wang, F.-J., Zhang, S., Chen, Z.-D., Liu, C., Wang, Y.-G.: Tar reforming using char as catalyst during pyrolysis and gasification of Shengli brown coal. J. Anal. Appl. Pyrol. **105**, 269–275 (2014)
39. Klinghoffer, N.B., Castaldi, M.J., Nzihou, A.: Influence of char composition and inorganics on catalytic activity of char from biomass gasification. Fuel **157**, 37–47 (2015)
40. Hervy, M., Pham Minh, D., Gérente, C., Weiss-Hortala, E., Nzihou, A., Villot, A., Le Coq, L.: H_2S removal from syngas using wastes pyrolysis chars. Chem. Eng. J. **334**, 2179–2189 (2018)
41. Sizmur, T., Fresno, T., Akgül, G., Frost, H., Moreno-Jiménez, E.: Biochar modification to enhance sorption of inorganics from water. Bioresour. Technol. **246**, 34–47 (2017)
42. Bianco, A., Cheng, H.-M., Enoki, T., Gogotsi, Y., Hurt, R.H., Koratkar, N., Kyotani, T., Monthioux, M., Park, C.R., Tascon, J.M.D., Zhang, J.: All in the graphene family—a recommended nomenclature for two-dimensional carbon materials. Carbon **65**, 1–6 (2013)
43. Lu, L., Sahajwalla, V., Kong, C., Harris, D.: Quantative X-ray diffraction analysis and its application to various coals. Carbon **39**, 1821–1833 (2001)
44. Monthioux, M., Noé, L., Kobylko, M., Wang, Y., Cazares-Huerta, T.C., Pénicaud, A.: Determining the structure of graphene-based flakes from their morphotype. Carbon **115**, 128–133 (2017)
45. Smith, M.A., Foley, H.C., Lobo, R.F.: A simple model describes the PDF of a non-graphitizing carbon. Carbon **42**, 2041–2048 (2004)
46. Gracia-Espino, E., López-Urías, F., Kim, Y.A., Hayashi, T., Muramatsu, H., Endo, M., Terrones, H., Terrones, M., Dresselhaus, M.S.: Novel carbon-based nanomaterials: graphene and graphitic nanoribbons. In: Somiya, S., Kaneno, M. (eds.) Handbook of Advanced Ceramics: Materials, Applications, Processing and Properties, pp. 61–87. Elsevier, Waltham (2013)

47. Terrones, M., Botello-Méndez, A.R., Campos-Delgado, J., López-Urías, F., Vega-Cantú, Y. I., Rodríguez-Macías, F.J., Elías, A.L., Muñoz-Sandoval, E., Cano-Márquez, A.G., Charlier, J.C., Terrones, H.: Graphene and graphite nanoribbons: Morphology, properties, synthesis, defects and applications. Nano Today **5**, 351–372 (2010)
48. Araujo, P.T., Terrones, M., Dresselhaus, M.S.: Defects and impurities in graphene-like materials. Mater. Today **15**, 98–109 (2012)
49. Pré, P., Huchet, G., Jeulin, D., Rouzaud, J.N., Sennour, M., Thorel, A.: A new approach to characterize the nanostructure of activated carbons from mathematical morphology applied to high resolution transmission electron microscopy images. Carbon **52**, 239–258 (2013)
50. Anees, P., Valsakumar, M.C., Chandra, S., Panigrahi, B.K.: *Ab initio* study on stacking sequences, free energy, dynamical stability and potential energy surfaces of graphite structures. Model. Simul. Mater. Sci. (2014). https://doi.org/10.1088/0965-0393/22/3/035016
51. Youcai, Z.: Pollution Control and Resource Recovery: Municipal Solid Wastes Incineration Bottom Ash and Fly Ash. Butterworth-Heinemann (2017)
52. Liyanage, M., Jayaranjan, D., van Hullebusch, E.D., Annachhatre, A.P.: Reuse options for coal fired power plant bottom ash and fly ash. Rev. Environ. Sci. Biotechnol. **13**, 467–486 (2014)
53. Maschio, S., Tonello, G., Piani, L., Furlani, E.: Fly and bottom ashes from biomass combustion as cement replacing components in mortars production: rheological behaviour of the pastes and materials compression strength. Chemosphere **85**, 666–671 (2011)
54. Ulewicz, M., Jura, J.: Effect of fly and bottom ash mixture from combustion of biomass on strength of cement mortar. In: E3S Web of Conferences, vol. 18 (2017). https://doi.org/10.1051/e3sconf/201712301029
55. Valentim, B., Białecka, B., Gonçalves, P.A., Guedes, A., Guimarães, R., Cruceru, M., Całus-Moszko, J., Popescu, L.G., Predeanu, G., Santos, A.C.: Undifferentiated inorganics in coal fly ash and bottom ash: calcispheres, magnesiacalcispheres, and magnesiaspheres. Minerals (2018). https://doi.org/10.3390/min8040140
56. Chen, J., Fang, D., Duan, F.: Pore characteristics and fractal properties of biochar obtained from the pyrolysis of coarse wood in a fluidized-bed reactor. Appl. Energy **218**, 54–65 (2018)
57. Thielmann, F., Burnett, D.: Isotherm types and adsorption mechanisms of solvents on pharmaceutical excipients. Application Note 26, Surface Measurement Systems Ltd.
58. Sullivan, P.D., Stone, B.R., Hashisho, Z., Rood, M.J.: Water adsorption with hysteresis effect onto microporous activated carbon fabrics. Adsorption **13**, 173–189 (2007)
59. García-Pérez, J.V., Cárcel, J.A., Clemente, G., Mulet, A.: Water sorption isotherms for lemon peel at different temperatures and isosteric heats. LWT Food Sci. Technol. **41**, 18–25 (2008)
60. Furmaniak, S., Gauden, P.A., Terzyk, A.P., Rychlicki, G.: Water adsorption on carbons—critical review of the most popular analytical approaches. Adv. Colloid Interface Sci. **137**, 82–143 (2008)
61. Hill, C.A.S., Norton, A.J., Newman, G.: The water vapour sorption properties of Sitka spruce determined using a dynamic vapour sorption apparatus. Wood Sci. Technol. **44**, 497–514 (2010)
62. Nguyen, C., Do, D.D.: The Dubinin-Radushkevich equation and the underlying microscopic adsorption description. Carbon **39**, 1327–1336 (2001)
63. Kapoor, A., Ritter, J.A., Yang, R.T.: On the Dubinin-Radushkevich equation for adsorption in microporous solids in the Henry's law region. Langmuir **5**, 1118–1121 (1989)
64. Stoeckli, F.: Dubinin's theory and its contribution to adsorption science. Russ. Chem. Bull. **50**, 2265–2272 (2001)
65. Gauden, P.A.: Does the Dubinin-Serpinsky theory adequately describe water adsorption on adsorbents with high-energy centers? J. Colloid Interface Sci. **282**, 249–260 (2005)
66. Williams, D. R., Levoguer, C.L.: Measuring BET surface areas using organic probe molecules. Application Note 18, Surface Measurement Systems

67. Do, D.D., Do, H.D.: A model for water adsorption in activated carbon. Carbon **38**, 767–773 (2000)
68. Rutherford, S.W.: Modeling water adsorption in carbon micropores: study of water in carbon molecular sieves. Langmuir **22**, 702–708 (2006)
69. Charrière, D., Behra, P.: Water sorption on coals. J. Colloid Interface Sci. **344**, 460–467 (2010)
70. Zou, L., Gong, L., Xu, P., Feng, G., Liu, H.: Modified GAB model for correlating multilayer adsorption equilibrium data. Sep. Purif. Technol. **161**, 38–43 (2016)
71. Bravo-Osuna, I., Ferrero, C., Jiménez-Castellanos, M.R.: Water sorption–desorption behaviour of methyl methacrylate–starch copolymers: effect of hydrophobic graft and drying method. Eur. J. Pharm. Biopharm. **59**, 537–548 (2005)
72. Timmermann, E.O.: Multilayer sorption parameters: BET or GAB values? Colloids Surf. A **220**, 235–260 (2003)
73. Kachrimanis, K., Noisternig, M.F., Griesser, U.J., Malamataris, S.: Dynamic moisture sorption and desorption of standard and silicified microcrystalline cellulose. Eur. J. Pharm. Biopharm. **64**, 307–315 (2006)
74. Loughlin, K.F., Hassan, M.M., Fatehi, A.I., Zahur, M.: Rate and equilibrium sorption parameters for nitrogen and methane on carbon molecular sieve. Gas Sep. Purif. **7**, 264–273 (1993)
75. Xie, Y., Hill, C.A.S., Jalaludin, Z., Curling, S.F., Anandjiwala, R.D., Norton, A.J., Newman, G.: The dynamic water vapour sorption behavior of natural fibres and kinetic analysis using the parallel exponential kinetics model. J. Mater. Sci. **46**, 479–489 (2011)
76. Popescu, C.-M., Hill, C.A.S., Anthony, R., Ormondroyd, G., Curling, S.: Equilibrium and dynamic vapour water sorption properties of biochar derived from apple wood. Polym. Degrad. Stab. **111**, 263–268 (2015)
77. Fletcher, A.J., Uygur, Y., Mark Thomas, K.: Role of surface functional groups in the adsorption kinetics of water vapor on microporous activated carbons. J. Phys. Chem. C **111**, 8349–8359 (2007)
78. Foley, N.J., Thomas, K.M., Forshaw, P.L., Stanton, D., Norman, P.R.: Kinetics of water vapor adsorption on activated carbon. Langmuir **13**, 2083–2089 (1997)
79. Harding, A.W., Foley, N.J., Norman, P.R., Francis, D.C., Thomas, K.M.: Diffusion barriers in the kinetics of water vapor adsorption/desorption on activated carbons. Langmuir **14**, 3858–3864 (1998)
80. Cossarutto, L., Zimny, T., Kaczmarczyk, J., Siemieniewska, T., Bimer, J., Webera, J.V.: Transport and sorption of water vapour in activated carbons. Carbon **39**, 2339–2346 (2001)
81. Fletcher, A.J., Benham, M.J., Thomas, K.M.: Multicomponent vapor sorption on active carbon by combined microgravimetry and dynamic sampling mass spectrometry. J. Phys. Chem. B **106**, 7474–7482 (2002)
82. Fletcher, A.J., Yüzak, Y., Thomas, K.M.: Adsorption and desorption kinetics for hydrophilic and hydrophobic vapors on activated carbon. Carbon **44**, 989–1004 (2006)
83. Laine, N.R., Vastola, F.J., Walker Jr., P.N.: The importance of active surface area in the carbon-oxygen reaction. J. Phys. Chem. **67**, 2030–2034 (1963)
84. Ehrburger, P., Louys, F., Lahaye, J.: The concept of active sites applied to the study of carbon reactivity. Carbon **27**, 389–393 (1989)
85. Burg, P., Abraham, M.H., Cagniant, D.: Methods of determining polar and non-polar sites on carbonaceous adsorbents. The contribution of the linear solvation energy relationship approach. Carbon **41**, 867–879 (2003)
86. Groszek, A.J.: Graphitic and polar surface sites in carbonaceous solids. Carbon **25**, 717–722 (1987)
87. López-Garzón, F.J., Pydan, M., Domingo-Garcia, M.: Studies of the surface properties of active carbons by inverse gas chromatography at infinite dilution. Langmuir **9**, 531–536 (1993)

88. Tisserand, C., Calvet, R., Patry, S., Galet, L., Doods, J.A.: Comparison of two techniques for the surface analysis of alumina (Al_2O_3): inverse gas chromatography at finite concentration (IGC-FC) and dynamic vapor sorption (DVS). Powder Technol. **190**, 53–58 (2009)
89. Nowak, E., Combes, G., Stitt, E.H., Pacek, A.W.: A comparison of contact angle measurement techniques applied to highly porous catalyst supports. Powder Technol. **233**, 52–64 (2013)
90. Burnett, D., Thielmann, F.: Organic solvent sorption using a dynamic vapour sorption instrument-an overview. Application Note 103, Surface Measurement Systems Ltd., (last update 24/11/06)
91. Levoguer, C.L., Williams, D.R.: Measurement of the surface energies of pharmaceutical powders using a novel vapour adsorption method. Application Note 17, Surface Measurement Systems Ltd.
92. Burnett, D., Thielmann, F., Booth, J.: Determining the heat of sorption on organic and inorganic powders using DVS. Application Note 19, Surface Measurement Systems Ltd.
93. Klomkliang, N., Kaewmanee, R., Saimoey, S., Intarayothy, S., Do, D.D.: Adsorption of water and methanol on highly graphitized thermal carbon black: the effects of functional group and temperature on the isosteric heat at low loadings. Carbon **99**, 361–369 (2016)
94. Argyropoulos, D., Alex, R., Kohler, R., Müller, J.: Moisture sorption isotherms and isosteric heat of sorption of leaves and stems of lemon balm (*Melissa officinalis L.*) established by dynamic vapor sorption. LWT Food Sci. Technol. **47**, 324–331 (2012)
95. Choudhury, D., Sahu, J.K., Sharma, G.D.: Moisture sorption isotherms, heat of sorption and properties of sorbed water of raw bamboo (*Dendrocalamus longispathus*) shoots. Ind. Crops Prod. **33**, 211–216 (2011)
96. Quirijns, E.J., van Boxtel, A.J.B., Loon, W.K.P., van Straten, G.: Sorption isotherms, GAB parameters and isosteric heat of sorption. J. Sci. Food Agric. **85**, 1805–1814 (2005)
97. Sarkisov, L., Centineo, A., Brandani, S.: Molecular simulation and experiments of water adsorption in a high surface area activated carbon: hysteresis, scanning curves and spatial organization of water clusters. Carbon **118**, 127–138 (2017)
98. Ducousso, M.: Gasification biochar reactivity toward methane cracking. Ph.D. thesis, Université de Toulouse (2015). https://tel.archives-ouvertes.fr/tel-01411838
99. Figueiredo, J., Pereira, M.F., Freitas, M.M., Órfão, J.J.: Modification of the surface chemistry of activated carbons. Carbon **37**, 1379–1389 (1999)
100. Zhou, J.H., Sui, Z.J., Zhu, J., Li, P., Chen, D., Dai, Y.C., Yuan, W.-K.: Characterization of surface oxygen complexes on carbon nanofibers by TPD. XPS FT-IR. Carbon **45**, 785–796 (2007)
101. Szymański, G.S., Karpiński, Z., Biniak, S., Świątkowski, A.: The effect of the gradual thermal decomposition of surface oxygen species on the chemical and catalytic properties of oxidized activated carbon. Carbon **40**, 2627–2639 (2002)
102. Ferrari, A.C.: Raman spectroscopy of graphene and graphite: disorder, electron-phonon coupling, doping and nonadiabatic effects. Solid State Commun. **143**, 47–57 (2007)
103. Asadullah, M., Zhang, S., Min, Z., Yimsiri, P., Li, C.Z.: Effects of biomass char structure on its gasification reactivity. Bioresour. Technol. **101**, 7935–7943 (2010)
104. Dresselhaus, M.S., Jorio, A., Souza Filho, A.G., Saito, R.: Defect characterization in graphene and carbon nanotubes using Raman spectroscopy. Philos. Trans. R. Soc. A **368**, 5355–5377 (2010)
105. McDonald-Wharry, J., Manley-Harris, M., Pickering, K.: Carbonisation of biomass-derived chars and the thermal reduction of a graphene oxide sample studied using Raman spectroscopy. Carbon **59**, 383–405 (2013)
106. Vautard, F., Dentzer, J., Nardin, M., Schultz, J., Defoort, B.: Influence of surface defects on the tensile strength of carbon fibers. Appl. Surf. Sci. **322**, 185–193 (2014)
107. Casari, C.S., Tommasini, M., Tykwinski, R.R., Milani, A.: Carbon-atom wires: 1-D systems with tunable properties. Nanoscale **8**, 4414–4435 (2016)

108. Sadezky, A., Muckenhuber, H., Grothe, H., Niessner, R., Pöschl, U.: Raman microspectroscopy of soot and related carbonaceous materials: spectral analysis and structural information. Carbon **43**, 1731–1742 (2005)
109. Li, X., Hayashi, J., Li, C.Z.: FT-Raman spectroscopic study of the evolution of char structure during the pyrolysis of a Victorian brown coal. Fuel **85**, 1700–1707 (2006)
110. Li, X., Hayashi, J., Li, C.Z.: Volatilisation and catalytic effects of alkali and alkaline earth metallic species during the pyrolysis and gasification of Victorian brown coal. Part VII. Raman spectroscopic study on the changes in char structure during the catalytic gasification in air. Fuel **85**, 1509–1517 (2006)
111. Zickler, G.A., Smarsly, B., Gierlinger, N., Peterlik, H., Paris, O.: A reconsideration of the relationship between the crystallite size L_a of carbons determined by X-ray diffraction and Raman spectroscopy. Carbon **44**, 3239–3246 (2006)
112. Keown, D.M., Li, X., Hayashi, J.I., Li, C.Z.: Evolution of biomass char structure during oxidation in O_2 as revealed with FT-Raman spectroscopy. Fuel Process. Technol. **89**, 1429–1435 (2008)
113. Mallet-Ladeira, P., Puech, P., Weisbecker, P., Vignoles, G.L., Monthioux, M.: Behavior of Raman D band for pyrocarbons with crystallite size in the 2–5 nm range. Appl. Phys. A **114**, 759–763 (2014)
114. Cong, C., Li, K., Zhang, X.X., Yu, T.: Visualization of arrangements of carbon atoms in graphene layers by Raman mapping and atomic-resolution TEM. Sci. Rep. **3** (2013). https://dx.doi.org/10.1038/srep01195
115. Shen, A.-L., Weng, Y.-C., Chou, T.-C.: Effect of the analytic regions on the quality trend of diamond-like/graphitic carbon ratios in Raman Spectra. Z. Naturforsch., B: Chem. Sci **65**, 67–71 (2010)
116. Kudin, K.N., Ozbas, B., Schniepp, H.C., Prud'homme, R.K., Aksay, I.A., Car, R.: Raman spectra of graphite oxide and functionalized graphene sheets. Nano Lett. **8**, 36–41 (2008)
117. Chu, P.K., Li, L.: Characterization of amorphous and nanocrystalline carbon films. Mater. Chem. Phys. **96**, 253–277 (2006)
118. Campos-Delgado, J., Botello-Méndez, A.R., Algara-Siller, G., Hackens, B., Pardoen, T., Kaiser, U., Dresselhaus, M.S., Charlier, J.C., Raskin, J.P.: CVD synthesis of mono- and few-layer graphene using alcohols at low hydrogen concentration and atmospheric pressure. Chem. Phys. Lett. **584**, 142–146 (2013)
119. Besser, A. I., Chen, Z.: FEI Tecnai F-20 operations manual. Center for Electron Microscopy and Nanofabrication. Portland State University (last update 2010). https://www.pdx.edu/sites/www.pdx.edu.cemn/files/TEM_manual.pdf
120. Farrow, C.L., Juhas, P., Liu, J.W., Bryndin, D., Bozin, E.S., Bloch, J., Proffen, T., Billinge, S.J.L.: PDFfit2 and PDFgui: computer programs for studying nanostructure in crystals. J. Phys. Condens. Matter (2007). https://doi.org/10.1088/0953-8984/19/33/335219
121. Keiluweit, M., Nico, P.S., Johnson, M.G., Kleber, M.: Dynamic molecular structure of plant biomass-derived black carbon (biochar). Environ. Sci. Technol. **44**, 1247–1253 (2010)
122. Proffen, T., Kim, H.: Advances in total scattering analysis. J. Mater. Chem. **19**, 5078–5088 (2009)
123. Proffen, T., Billinge, S.: PDFFIT 1.3 users guide: http://www.diffpy.org/doc/pdffit/pdf_man.pdf (created in 2003). Accessed Nov. 2017
124. Nishiyama, Y., Sugiyama, J., Chanzy, H., Langan, P.: Crystal structure and hydrogen bonding system in cellulose Iα from synchrotron X-ray and neutron fiber diffraction. J. Am. Chem. Soc. **125**, 14300–14306 (2003)
125. Downs, R.T.: The RRUFF Project: an integrated study of the chemistry, crystallography, Raman and infrared spectroscopy of minerals. In: Program and Abstracts of the 19th General Meeting of the International Mineralogical Association in Kobe, Japan, O03-13 (2006). http://rruff.info/
126. Li, Y.: Operating procedure for Tecnai G^2 STEM, 4D Labs Research Institute, Simon Fraser University (version 1.7). https://users.4dlabs.ca/uploads/documents/SOP-STEM1.pdf

127. Chen, Y., Shah, N., Huggins, F.E., Huffman, G.P.: Transmission electron microscopy investigation of ultrafine coal fly ash particles. Environ. Sci. Technol. **39**, 1144–1151 (2005)
128. Chen, Y., Shah, N., Huggins, F.E., Huffman, G.P.: Characterization of fine and ultrafine fly ash by electron microscopy techniques. World of Coal Ash. https://www.researchgate.net/publication/266338677_Characterization_of_Fine_and_Ultrafine_Fly_Ash_by_Electron_Microscopy_Techniques (2005). Accessed June 2018
129. Degen, T., Sadki, M., Bron, E., König, U., Nénert, G.: The HighScore suite. Powder Diffr. **29**(S2), S13–S18 (2014)
130. Bruker: X-ray Diffraction Software. https://www.bruker.com/fr/products/x-ray-diffraction-and-elemental-analysis/x-ray-diffraction/xrd-software.html. Accessed Feb. 2018
131. ICDD: The International Centre for Diffraction Data. www.icdd.com. Accessed Feb. 2018
132. MIT Center for Materials Science and Engineering. http://prism.mit.edu/xray/oldsite/tutorials.htm. Accessed Feb. 2018
133. Chung, F.H.: Quantitative interpretation of X-ray diffraction patterns of mixtures, III-simultaneous determination of a set of reference intensities. J. Appl. Crystallogr. **8**, 17–19 (1974)
134. Bish, D.L., Howard, S.A.: Quantitative phase analysis using the Rietveld method. J. Appl. Crystallogr. **21**, 86–91 (2001)
135. Will, G.: The Rietveld method and the two stage method. In: Will, G. (eds.) Powder Diffraction, pp. 42–194. Springer, Berlin (2006)
136. Crystallography Open Database. http://www.crystallography.net/cod/. Accessed Feb. 2018
137. Hill, R.J., Fischer, R.X.: Profile agreement indices in Rietveld and pattern- fitting analysis. J. Appl. Crystallogr. **23**, 462–468 (1990)
138. Jansen, E., Schäfer, W., Will, G.: R values in analysis of powder diffraction data using Rietveld refinement. J. Appl. Crystallogr. **27**, 492–496 (1994)
139. Brian, H.T.: R factors in Rietveld analysis: how good is good enough? Powder Diffr. **21**, 67–70 (2006)
140. Young, R.A.: Introduction to the Rietveld method. In: Young, R.A. (ed.) The Rietveld Method, pp. 1–38. Oxford University Press, Oxford (1993)
141. De La Torre, A.G., Brusque, S., Aranda, M.A.G.: Rietveld quantitative amorphous content analysis. J. Appl. Crystallogr. **34**, 196–202 (2001)
142. Orlhac, X., Fillet, C., Deniard, P., Dulac, A.M., Brec, R.: Determination of the crystallized fraction of a largely amorphous multiphase material by Rietveld method. J. Appl. Crystallogr. **34**, 114–118 (2001)
143. Scherrer, P.: Bestimmung der Grösse und der inneren Struktur von Kolloidteilchen mittels Röntgenstrahlen. Nachr. Ges. Wiss. Göttinger. **26**, 98–100 (1918)
144. Langford, J.L., Wilson, A.J.C.: Scherrer after sixty years: a survey and some new results in the determination of crystallite size. J. Appl. Crystallogr. **11**, 102–113 (1978)
145. Williamson, G.K., Hall, W.H.: X-ray line broadening from filed aluminium and wolfram. Acta Mettall. **1**, 22–31 (1953)
146. Marsh, H., Rodriguez-Reinoso, F.: Activated Carbon. Elsevier, Oxford (2006)
147. Ruland, W., Smarsly, B.: X-ray scattering of non-graphitic carbon: an improved method of evaluation. J. Appl. Crystallogr. **35**, 624–633 (2002)
148. Manoj, B., Kunjomana, A.G.: Study of stacking structure of amorphous carbon by X-Ray diffraction technique. Int. J. Electrochem. Sci. **7**, 3127–31324 (2012)
149. Iwashita, N., Park, C.R., Fujimoto, H., Shiraishi, M., Inagakie, M.: Specification for a standard procedure of X-ray diffraction measurements on carbon materials. Carbon **42**, 710–714 (2004)
150. Yasuda, E-I., Inagaki, M.: X-ray diffraction method to study crystallite size and lattice constants of carbon materials. In: E-I. Yasuda, Inagaki, M., Kaneko, K., Endo, M., Oya, A., Tanabe, Y. (eds.) Carbons Allows: Novel Concepts to develop Carbon Science and Technology, pp. 161–173. Elsevier, Oxford (2003)

Printed by Printforce, the Netherlands